# Advances in Intelligent Systems and Computing

Volume 1183

The series "Advances in Intelligent Systems and Computing" contains publications on theory, applications, and design methods of Intelligent Systems and Intelligent Computing. Virtually all disciplines such as engineering, natural sciences, computer and information science, ICT, economics, business, e-commerce, environment, healthcare, life science are covered. The list of topics spans all the areas of modern intelligent systems and computing such as: computational intelligence, soft computing including neural networks, fuzzy systems, evolutionary computing and the fusion of these paradigms, social intelligence, ambient intelligence, computational neuroscience, artificial life, virtual worlds and society, cognitive science and systems, Perception and Vision, DNA and immune based systems, self-organizing and adaptive systems, e-Learning and teaching, human-centered and human-centric computing, recommender systems, intelligent control, robotics and mechatronics including human-machine teaming, knowledge-based paradigms, learning paradigms, machine ethics, intelligent data analysis, knowledge management, intelligent agents, intelligent decision making and support, intelligent network security, trust management, interactive entertainment, Web intelligence and multimedia.

The publications within "Advances in Intelligent Systems and Computing" are primarily proceedings of important conferences, symposia and congresses. They cover significant recent developments in the field, both of a foundational and applicable character. An important characteristic feature of the series is the short publication time and world-wide distribution. This permits a rapid and broad dissemination of research results.

**** Indexing: The books of this series are submitted to ISI Proceedings, EI-Compendex, DBLP, SCOPUS, Google Scholar and Springerlink ****

More information about this series at http://www.springer.com/series/11156

Xin-She Yang · R Simon Sherratt ·
Nilanjan Dey · Amit Joshi
Editors

# Proceedings of Fifth International Congress on Information and Communication Technology

ICICT 2020, London, Volume 1

Springer

*Editors*
Xin-She Yang
School of Science and Technology
Middlesex University
London, UK

R Simon Sherratt
Department of Biomedical Engineering
The University of Reading
Reading, UK

Nilanjan Dey
Department of Information Technology
Techno India Institute of Technology
Kolkata, West Bengal, India

Amit Joshi
Global Knowledge Research Foundation
Ahmedabad, India

ISSN 2194-5357 ISSN 2194-5365 (electronic)
Advances in Intelligent Systems and Computing
ISBN 978-981-15-5855-9 ISBN 978-981-15-5856-6 (eBook)
https://doi.org/10.1007/978-981-15-5856-6



This Springer imprint is published by the registered company Springer Nature Singapore Pte Ltd.
The registered company address is: 152 Beach Road, #21-01/04 Gateway East, Singapore 189721, Singapore

# Preface

This AISC volume contains the papers presented at ICICT 2020: Fifth International Congress on Information and Communication Technology in concurrent with ICT Excellence Awards. The conference was held during February 20–21, 2020, London, UK, and collaborated by the Global Knowledge Research Foundation, City of Oxford College. The associated partners were Springer, InterYIT IFIP, Activate Learning. The conference was held at Brunel University, London. This conference was focused on e-business fields such as e-agriculture, e-education, e-mining. The objective of this conference was to provide a common platform for researchers, academicians, industry persons and students to create a conversational environment wherein topics related to future innovation, obstacles are to be resolved for new upcoming projects, exchange of views and ideas. The conference attracted immense experts from more than 45 countries, the conference was involved in deep discussion, and issues were intended to solve at the international level. New technologies were proposed, experiences were shared, and future solutions for design infrastructure for ICT were also discussed. Research Submissions in various advanced technology areas were received and then were reviewed by the committee members; 120 papers were accepted. The conference was overwhelmed by the presence of various members. Amit joshi, Organizing Secretary, ICICT 2020, gave the welcome speech on behalf of conference committee and editors. Our special invitee guest—Sean Holmes, Vice Dean, International College of Business, Arts and Social Sciences, Brunel University, London, UK, also addressed the conference by a speech. The conference was also addressed by our Inaugural Guest and Speakers—Mike Hinchey, President, International Federation for Information Processing (IFIP); Xin-She Yang, Professor, Middlesex University, UK; Jyoti Choudrie, Professor, University of Hertfordshire, UK; and Milan Tuba, Vice Rector for International Relations, Singidunum University, Serbia. There were 14 technical sessions in total, and talks on academic and industrial sector were focused on both the days. We are obliged to the Global Knowledge Research Foundation for their immense support to make this conference a successful one. A total of 105 papers were presented in technical sessions, and 120 were accepted with strategizing on ICT and intelligent systems.

At the closing ceremony, 10 Best Paper Awards by Springer were announced among the best selected and presented paper. On behalf of editors, we thank all the sponsors, press, print and electronic media for their excellent coverage of this conference.

London, UK — Xin-She Yang
Reading, UK — R Simon Sherratt
Kolkata, India — Nilanjan Dey
Ahmedabad, India — Amit Joshi

# Contents

# About the Editors

**Xin-She Yang** obtained his D.Phil. in Applied Mathematics from the University of Oxford, and subsequently worked at the Cambridge University and the National Physical Laboratory (UK) as a Senior Research Scientist. He is currently a Reader in Modelling and Optimization at Middlesex University London and an Adjunct Professor at Reykjavik University (Iceland). He is also an elected Bye-Fellow at the Cambridge University and the IEEE CIS Chair for the Task Force on Business Intelligence and Knowledge Management. He was included in the "2016 Thomson Reuters Highly Cited Researchers" list.

**R Simon Sherratt** was born near Liverpool, England, in 1969. He is currently a Professor of Biosensors at the Department of Biomedical Engineering, University of Reading, UK. His main research area is signal processing and personal communications in consumer devices, focusing on wearable devices and health care. Professor Sherratt received the 1st place IEEE Chester Sall Memorial Award in 2006, the 2nd place in 2016 and the 3rd place in 2017.

**Nilanjan Dey** is an Assistant Professor at the Department of Information Technology, Techno India College of Technology, India. He has authored/edited more than 75 books with Springer, Elsevier, Wiley, CRC Press and published more than 300 peer-reviewed research papers. Dr. Dey is the Editor-in-Chief of the International Journal of Ambient Computing and Intelligence; a Series Co-Editor of Springer Tracts in Nature-Inspired Computing (STNIC); and a Series Co-Editor of Advances in Ubiquitous Sensing Applications for Healthcare, Elsevier.

**Amit Joshi** is Director of the Global Knowledge Research Foundation and the International Chair of InterYIT at the International Federation of Information Processing (IFIP, Austria). He has edited more than 40 books for Springer, ACM and other reputed publishers. He has also organized more than 50 national and international conferences and workshops in association with the ACM, Springer and IEEE in e.g. India, Thailand and the UK.

# Adaptive Cognitive Modeling of Electroconvulsive Treatment (ECT)

S. Sahand Mohammadi Ziabari(✉) and Charlotte Gerritsen

Social AI Group, Vrije Universiteit Amsterdam, De Boelelaan 1105, Amsterdam, The Netherlands
sahandmohammadiziabari@gmail.com, cs.gerritsen@vu.nl

**Abstract.** This paper presents a cognitive model on electroconvulsive treatment to reduce the stress level in body. The stress reduction is triggered by a cognitive electroconvulsive treatment that uses persistent manipulation of this treatment. The goal of this treatment is to decrease the strength between certain parts of the brain which are in charge of the stress. The proposed adaptive cognitive model aims to illustrate the effect of the therapy on different components of the brain. The model begins with a state of tough, powerful, and consistent stress within a post-traumatic disorder patient, and after following electroconvulsive treatment, the stress level starts to decrease from time to time according to each treatment session. The results show that, in the end, the disorder person will have a declined stress in contrast to not performing electroconvulsive treatment.

**Keywords:** Cognitive modeling · Extreme emotion · Electroconvulsive treatment · PTSD

## 1 Introduction

The need to understand human mental processes is an increasing trend in the designing of intelligent systems for facilitating behavioral and lifestyle changes. The reason for requiring this understanding is that this enables to more adequately help people improve their general well-being and support them in their daily life and activities.

Furthermore, developing intelligent systems capable of effecting supportive actions according to anticipated environmental and situational factors, as well as a person's mental states, must be robust and justifiable from a scientific perspective. In the design of dynamical models related to internal mental processes for intelligent human-supporting systems, the computational approach provides a good basis as it provides insight into behavioral changes and the related possibilities.

The crucial question in this instance is how such an intelligent support system can be designed, taking mental and environmental states into account, in a way that enables it to effectively support people in their daily lives.


X.-S. Yang et al. (eds.), *Proceedings of Fifth International Congress on Information and Communication Technology*, Advances in Intelligent Systems and Computing 1183, https://doi.org/10.1007/978-981-15-5856-6_1

Usually, the heuristic is followed that a model should not be made at a too low level of abstraction for the accuracy required, as that would entail unnecessary complexity. However, what deviation is introduced by doing so is a serious question that needs analysis. Therefore, in this research area, models at different abstraction levels and their (interlevel) relations are being investigated.

The first part of this work involves inquiry about theories and knowledge from the fields of cognitive neuroscience. Based on this information, computational models were designed. The evaluation of these models was conducted by simulation. These computational models have been scientifically justified by various fundamental theories from the fields of neuroscience and cognitive science. This approach helped to facilitate a better understanding of the link between different human behaviors, mental states, and conditions, thereby providing a framework of knowledge for the design of intelligent support systems.

As it has been declared in [1], PTSD is a disorder that might play a role after exposure to exceptionally threatening or horrifying events. PTSD may be the outcome of a wide variety of causes ranging from a fight to car. Because of the fragile interaction between personality and circumstances in which the trauma has been developed, it is impossible to predict whether PTSD will develop. There are some treatments that have been used for years and have proven to be successful, such as the use of antidepressant drugs like fluoxetine, deep brain stimulation of different brain components like the amygdala and the hippocampus or much older therapies such as yoga or music therapy [2–12]. The mobile application of these works is presented in [13]. This work presents a cognitive modeling for electroconvulsive therapy. By providing a conceptual model based on the neuroscience literature, the effectiveness and results of aforementioned therapy for PTSD patients are demonstrated.

In this paper, the electroconvulsive therapy (ECT) which is used for decreasing extreme emotions in a patient with PTSD is presented. This is based on the neuroscientific and psychological literature review. This therapy can be applied when the limitation of other therapies, such as antidepressant drugs, are not allowed for example during pregnancy or when immediate treatment is needed, e.g., when patients experience high depression or have suicidal thoughts.

In Sect. 2, the findings in neuroscience, psychology, and biological research concerning the components of the brain are involved. In Sect. 3, network-oriented modeling is explained. In Sect. 4, we present the graphical results, and in the final section, the conclusion is presented.

## 2 Neurological Findings

Recent studies [14–16] show the effect of ECT in the hippocampus and amygdala and show that it enforces massive structural plasticity in humans. The process of the effect of the ECT on the brain components is mentioned in different research, and in the following part, the main outcomes are presented.

Also, it has been mentioned that the neurogenesis of the amygdala may be enhanced by performing ECT, and the feedforward cortical-subcortical connection from FFA to amygdala will be improved. However, by using ECT, the general global connectivity in the left dorsolateral prefrontal cortex is decreased. In [17], the correlation between sad faces and the amygdala has been investigated. There is a remarkable negative interconnectivity between alters in the activity of amygdala to safe faces and symptoms of using ECT. Moreover, analysis depicted that the reactivity degree of the amygdala to negative emotional stimuli was accompanied by clinical results.

Also, in [18], the effect of ECT on dentate gyrus of the hippocampus has been shown. It has been stated that the volume of the bilateral hippocampus and amygdala has been increased by performing ECT, and this change is more visible in young adults.

The reason for increased volume in the hippocampus is based on the increase of neurons in the dentate gyrus, and this increase is maintained for three months. Furthermore, it has been shown that electroconvulsive shock is faster and stronger than antidepressant drugs [15]. It increases cell proliferation 2.5-4-fold compared with 1.5-fold using antidepressant drugs. ECT enforces an increase in proliferation of glial progenitor (NG2-positive cells) and enhances the number of glial cells and also alters the activation of glial cells and elevates the maturation of dendritic spines. These alterations also remodeled the neural circuits which intervened with the therapeutic impact of ECT. Deep into the molecular layer, it has been noted that ECT increased endothelial cell numbers by up to 30% and the length of the vessel by 16% in the dentate gyrus. The metabolism and perfusion are increased in the medial temporal lobe which contains para-hippocampal gyrus and hippocampus after doing ECT [19].

In the major depressive disorder, the density of total glia and oligodendrocytes in the amygdala and subgenual cingulate cortex is declined. However, using ECT increases the glial cells in the amygdala [20].

Also, in [21], changes in BDNF and its receptor, trkB, are discussed. Acute ECT increases BDNF mRNA almost twofold; however, the impacts of chronic ECT on the hippocampus varies. Stress-induced atrophy and in the worst case, cell death, cause to the loss of hippocampal control of the hypothalamic-adrenal axis (HPA) and exhibition in hypercortisolism in a depressive period.

In [22], different indications of ECS therapy has been mentioned. The result of a proton magnetic resonance spectroscopy shows that by having ECT treatment, the occipital cortex GABA concentrations are increased in depressed patients. In addition to the effect of ECT in increasing the level of glutamate plasma and normalizing the reduced level of glutamate/glutamine in the left cingulum in PTSD, it also plays a role in glutamate and an excitatory neurotransmitter.

However, this dysregulation normalizes based on the therapeutic effect of ECT.

As it has been discussed in [22], ECT is involved in increasing the neuron formation in the hippocampus, and this depends on doses of using ECT on neurogenesis. It has been also found that CBF in frontal areas of the brain after depression declines.

Also, in [22, p. 107], different brain parts involved during ECT have been mentioned. After using ECT, in many other areas in the brain such as frontal and parietal cortex, and in anterior and posterior cingulate gyrus of PTSD, a decreased regional glucose metabolism happened. This might also consider as a therapeutic impact of ECT.

The usage of ECT can be considered when the high risk of depression and severe psychological issue and suicide noticed. Moreover, in the case of mania and depression during pregnancy and other therapeutic options such as antidepressant drugs are useless,

ECT can play an important role [23]. The success rate (remission) for ECT therapy is close to 90%, and the resting time for relief is within 10–14 days.

In [24], the influence of electroconvulsive on hippocampus has been described more precisely. Another result of using ECT on the hippocampus is discussed in [24] which it states that cell proliferation of ECT on the hippocampus is increased twice in three days.

In [25], the privilege of using of electroconvulsive therapy comparing to other therapies and antidepressant drugs has been explained. Although there are many improvements in variant pharmacological agents which are patient-friendly and used in many aspects of disorders such as catatonia, treatment-resistant schizophrenia, and acute manic episode, ECT still is more useful in mood-stabilizing property than pharmacotherapy [25]. Another reason for using ECT is quicker response comparing to other existing psychotropic medications.

## 3 The Network-Oriented Modeling

In this part, we discuss the principle and description of network-oriented modeling approach [24, Chap. 2, 25, 1]. This approach is basically adopted from the foundations and principles of neuroscience, psychology and social sciences.

This kind of modeling can be described on two different levels called conceptual representation and numerical representation. A conceptual representation consists of all states (nodes) and connections between them as can be interpreted as the causal impacts of the nodes on each other. We have considered 26 states to represent the brain components in neuroscience, biological components such as hormones and cognitive components such as the goal of using ECT and preparation of doing that, and stress stimuli.

In the next phase, a mathematical representation of the model needs to be considered. This representation consists of three important parts such as *strength of a connection, aggregate multiple causal impacts, and speed of change.* The former represents the strength of the connection among states. The middle one noted the aggregation of multiple connection weights when there are more than one causal relation impacts other states. Finally, *speed of change* shows the speed of the changes among nodes and how fast they change.

- $\omega_{X,Y}$: illustrates the strength of the connection from a state $X$ to state $Y$, and it is between $-1$ (negative effect) and 1(positive effect).
- **state** $\mathbf{c}_Y(\ldots)$: represents an aggregation function for combining the causal effects on state $Y$.
- $\eta_Y$: represents a speed of change, and it varies between 0 and 1.

There are multiple combination functions which are used in different causal relations, but the most used ones are identity function, one entering connection, scale sum function, and the advanced logistic function for more than one. The advanced logistic function is mostly used when there are some changes during the time.

Figure 1 shows the proposed cognitive model. $ps_{ECS}$ is the preparation state of doing electroconvulsive therapy. $es_{ECS}$ is the state of using electroconvulsive therapy. The other states link to the components in the brain (anterior insula, bilateral inferior frontal gyrus, putamen, thalamus, bilateral parietal, occipital cortex, ACTH, cortisol, adrenal cortex, HPA, amygdala, hippocampus, bilateral cingulate gyrus, the left dorsal cortex, LC, and NA) which are involved in stress regulation.

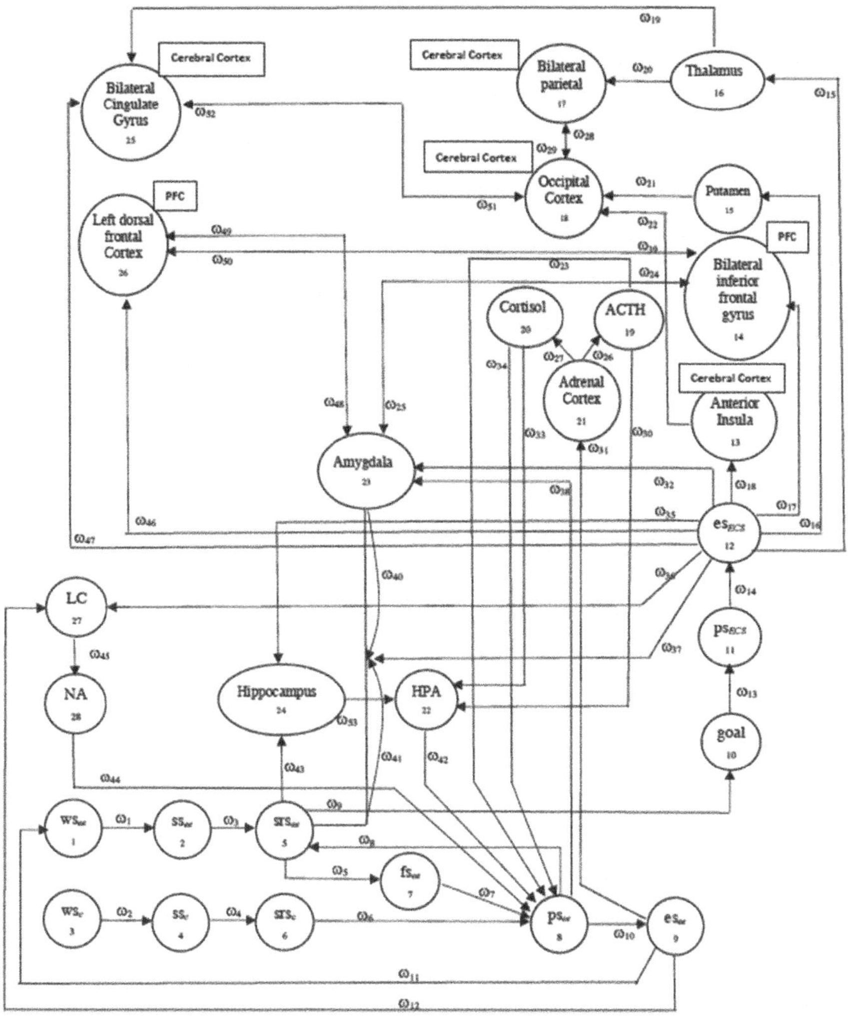


**Fig. 1.** Proposed cognitive model of ECT

The connection weights $\omega_i$ in Fig. 1 are based on the findings presented in Sect. 2. First, the cognitive part of the model is described: the beginning part of the explanation is based on Damasio theory about cognitive theory of mind [26]. Firstly, the cognitive part of the model and secondly the biological processes of the brain are described in this part (Table 1).

**Table 1.** Cognitive model explanation

| | | |
|---|---|---|
| $X_1$ | $ws_{ee}$ | World state |
| $X_2$ | $ss_{ee}$ | Sensor state of extreme emotion *ee* |
| $X_3$ | $ws_c$ | World state |
| $X_4$ | $ss_c$ | Sensor state for context *c* |
| $X_5$ | $srs_{ee}$ | Sensory representation |
| $X_6$ | $srs_c$ | Sensory representation |
| $X_7$ | $fs_{ee}$ | Feeling state |
| $X_8$ | $ps_{ee}$ | Preparation state |
| $X_9$ | $es_{ee}$ | Execution state |
| $X_{10}$ | goal | Goal of using therapy |
| $X_{11}$ | $ps_{ECS}$ | Preparation state of using electroconvulsive therapy |
| $X_{12}$ | $es_{ECS}$ | Execution of using electroconvulsive therapy |
| $X_{13}$ | Anterior insula (part of cerebral cortex) | Part of the brain |
| $X_{14}$ | Bilateral inferior frontal gyrus (part of PFC) | Part of the brain |
| $X_{15}$ | Putamen | Part of the brain |
| $X_{16}$ | Thalamus | Part of the brain |
| $X_{17}$ | Bilateral parietal (part of cerebral cortex) | Part of the brain |
| $X_{18}$ | Occipital cortex (part of cerebral cortex) | Part of the brain |
| $X_{19}$ | ACTH | Hormone |
| $X_{20}$ | Cortisol | Hormone |
| $X_{21}$ | Adrenal cortex | Brain part |
| $X_{22}$ | HPA | Hypothalamic pituitary adrenal axis |
| $X_{23}$ | Amygdala | Part of the brain |
| $X_{24}$ | Hippocampus | Part of the brain |
| $X_{25}$ | Bilateral cingulate gyrus (part of cerebral cortex) | Part of the brain |
| $X_{26}$ | Left frontal cortex (part of PFC) | Part of the brain |

The model contains a biological part as well. The state anterior insula has weight $\omega_{18}$ from $es_{ECS}$, and weight $\omega_{22}$ toward the cccipital cortex as it has been mentioned in Sect. 2 [27]. The bilateral inferior frontal gyrus part of the brain has three arriving connection weights $\omega_{17}$, $\omega_{39}$, and $\omega_{24}$ from $es_{ECS}$, left dorsal frontal cortex [22]. Putamen part of the brain has an incoming connection, $\omega_{16}$, from $es_{ECS}$. It is worth mentioning that the connection between the amygdala and $srs_{ee}$ is changing, and by using Hebbian learning, this will change ($\omega_{40}$, $\omega_{41}$). Thalamus has an arriving weight from $es_{ECS}$ ($\omega_{15}$). Bilateral parietal state has an arriving weight from thalamus with $\omega_{16}$ and an outgoing connection to occipital cortex, $\omega_{28}$. The state occipital cortex has four incoming weights from bilateral parietal, bilateral cingulate gyrus, anterior insula, putamen with $\omega_{29}$, $\omega_{51}$, $\omega_{22}$, $\omega_{21}$, respectively. The state ACTH has two leaving connection weights to HPA, $ps_{ee}$ named $\omega_{30}$, $\omega_{23}$, respectively. Furthermore, cortisol has leaving weights to HPA, $ps_{ee}$ named $\omega_{33}$, $\omega_{34}$, respectively. The state HPA has three connection weights from cortisol, ACTH, and hippocampus ($\omega_{53}$). amygdala has five incoming connection weights from $srs_{ee}$, $ps_{ee}$, $es_{ECS}$, bilateral inferior frontal gyrus, left frontal cortex named $\omega_{41}$, $\omega_{38}$, $\omega_{32}$, $\omega_{25}$, and $\omega_{48}$, respectively. Hippocampus has two incoming connection weights from $es_{ECS}$, $srs_{ee}$ named $\omega_{43}$, $\omega_{35}$, respectively. The brain part bilateral cingulate gyrus has three incoming connections from thalamus, occipital cortex, and $es_{ECS}$ named $\omega_{19}$, $\omega_{51}$, $\omega_{47}$, respectively. The state, left dorsal frontal cortex, has three arriving connections from bilateral inferior frontal gyrus, amygdala, $es_{ECS}$ named $\omega_{50}$, $\omega_{49}$, $\omega_{46}$.

This representation of the proposed cognitive model can be represented as a mathematical depiction [24, Chap. 2, 25, 1]:

- The *aggregated impact:*

$$\begin{aligned}\mathbf{aggimpact}_Y(t) &= c_Y\big(\mathbf{impact}_{X_1,Y}(t),\ \ldots,\ \mathbf{impact}_{Xk,Y}(t)\big)\\ &= c_Y\big(\omega_{X_1,Y}X_1(t),\ \ldots,\ \omega_{Xk,Y}X_k(t)\big)\end{aligned}$$

- Difference/differential equations for the cognitive mmodel:

$$Y(t+\Delta t) = Y(t) + \eta_Y\big[\mathbf{aggimpact}_Y(t) - Y(t)\big]\Delta t$$
$$Y(t+\Delta t) = Y(t) + \eta_Y\big[\mathbf{c}_Y\big((\omega_{X_1,Y}X_1(t),\ \ldots,\ \omega_{Xk,Y}X_k(t)\big) - Y(t)\big]$$

For getting more insight of these combination functions, see [24].

$$\mathbf{ssum}_\lambda(V_1,\ \ldots,\ V_k) = (V_1 + \cdots + V_k)/\lambda$$
$$\mathbf{alogistic}_{\sigma,\tau}(V_1,\ \ldots,\ V_k) = \Big[(1/(1+e^{-\sigma(V1+\ldots+Vk-t)})) - 1/(1+e^{\sigma t})\Big](1+e^{-\sigma t})$$

Hebbian learning is described as follows [21]:

$$d\omega(t)/dt = \eta[X_1(t)X_2(t)(1-\omega(t) - (1-\mu)\omega(t)]$$
$$\omega(t+\Delta t) = \omega(t) + \eta[X_1(t)X_2(t)(1-\omega(t)) - (1-\mu)\omega(t)]\Delta t$$

## 4 Graphical Result

The graphical results of the model are depicted in Fig. 2. The figure bears qualitative information. [27].

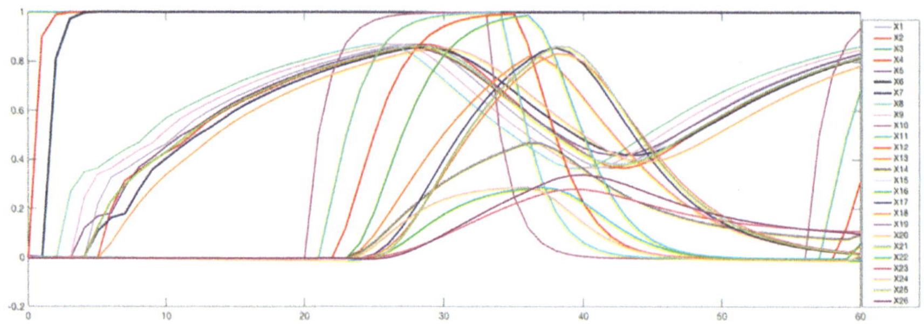


**Fig. 2.** Results of a simulation for the proposed cognitive model of the electroconvulsive therapy

We used MATLAB program in [2]. This program allowed us to use all differential equations that were needed for simulation, and also it contained adaptivity feature which we had in our proposed algorithm. With using appropriate connections, we make the model mathematical and well-fitted to the information gathered from the neuroscientific research findings. The time step for the graphical representation was equal to 1. Table 2 shows the $\lambda_i$ for the nodes with more than one entering weight. Firstly, the context stress becomes active, and consequently, it makes the stress level of the body to be increased gradually. Consequently, the stressed individual senses the stress, and therefore, as a next phase, the goal comes to play a role to reduce this by using ECT at about 25. The weights between most of the states are 1 except the following weights with their following weight's representation. $\omega_{18}$: −0.7, $\omega_{20}$: 0.7, $\omega_{23}$: 0.4, $\omega_{24}$: 0.4, $\omega_{35}$: −0.9, $\omega_{47}$: −0.9.

**Table 2.** Scaling factors for the proposed cognitive model

| State | $X_5$ | $X_8$ | $X_{14}$ | $X_{15}$ | $X_{17}$ | $X_{18}$ | $X_{19}$ | $X_{20}$ | $X_{21}$ |
|---|---|---|---|---|---|---|---|---|---|
| $\lambda_i$ | 2 | 3 | 3.4 | 3 | 1.4 | 2 | 2 | 2 | 2 |

Based on the simulation results, it is possible to show that the best solution for decreasing the stress is taken, performing of electroconvulsive treatment. The objective and the performing electroconvulsive treatment influence other brain components to decrease the activation at time 35. However, this influence is not permanent, and after some time, the stress up rises, which guides to activation of the objective and performing of another need of performing electroconvulsive therapy, and repeatedly until the patient resolves to stop executing electroconvulsive therapy. Thus, the results illustrate that the proposed cognitive model for the electroconvulsive treatment works well.

In Fig. 3, the stable condition is illustrated.

The suppression of connection and adaptivity among above-mentioned brain components; $srs_{ee}$ (cognitive) and amygdala is shown in Fig. 4. As it has been illustrated, grasp to tolerate with high level of stress, and declining begins at time around 30 and the following time until time 60 to remain stable.

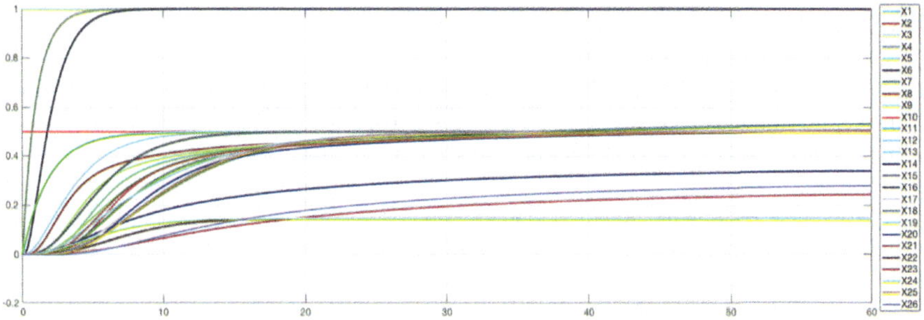


**Fig. 3.** Graphical representation for stable state for non-electroconvulsive treatment

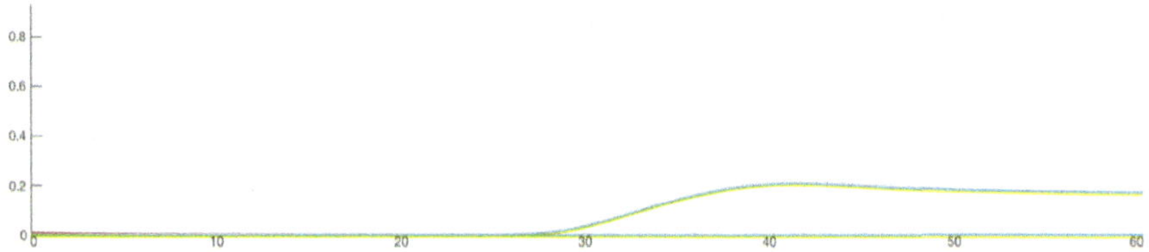

**Fig. 4.** Graphical representation for checking the adaptivity between amygdala and $srs_{ee}$, suppressed by electroconvulsive treatment $es_{ECS}$

## 5 Discussion

This work introduced a cognitive model of electroconvulsive treatment for post-traumatic disorder. This model can be used in order to evaluate various neurological patterns about the impact on the brain, and the influences that several brain components have on the stress reduction.

One important limitation of this work might concern the assumptions made while designing the computational models. However, by involving the right literature, those assumptions should be in line with underlying background in neuroscience according to real-world phenomenon. For instance, in this research, some assumptions regarding the extreme emotion responses and controlling of that were made. Connecting research from neuroscience to cognition can be tricky. We have considered mind processes which are based on the findings from neuroscience. For instance, sensory representation state that we have considered in our computational models is related to the perception happening in the human's mind, and we connect that to the real-world phenomenon which was stress in our case.

Various graphical representations have been performed, and one of which was illustrated in detail in this work. The presented cognitive model might be considered as the foundation of a greater model of the brain for having more insight in the processes inside the brain and to propose a possible treatment of different persons with disorders to do the treatments of extreme emotions. Future work of this model can be exploring the effect of different emotions on the brain and finding the physical effect of that in different brain components and biological processes.

## References


1. Treur, J.: The Ins and Outs of Network-Oriented Modeling: from biological networks and mental networks to social networks and beyond. Transactions on Computational Collective Intelligence. Paper for keynote lecture at ICCCI'18 (2018)
2. Mohammadi Ziabari, S.S., Treur, J.: A modeling environment for dynamic and adaptive network models implemented in Matlab. In: Yang, X.S., Sherratt, S., Dey, N., Joshi, A. (eds.) ICICT2019, AISC, vol.2, pp. 231–242. Springer Nature Singapore Pte Ltd (2019). https://doi.org/10.1007/978-981-32-9343-4
3. Treur, J., Mohammadi Ziabari, S.S.: An adaptive temporal-causal network model for decision making under acute stress. In: Nguyen, N.T., Pimenidis, E., Khan, Z., Trawiński, B. (eds.) ICCCI 2018. LNCS (LNAI), vol. 11056, pp. 13–25. Springer, Cham (2018). https://doi.org/10.1007/978-3-319-98446-9_2
4. Mohammadi Ziabari, S.S., Treur, J.: Computational analysis of gender differences in coping with extreme stressful emotions. In: Proceedings of the 9th International Conference on Biologically Inspired Cognitive Architecture (BICA 2018). Elsevier, Czech Republic (2018). https://doi.org/10.1016/j.procs.2018.11.088
5. Mohammadi Ziabari, S.S., Treur, J.: Cognitive modeling of mindfulness therapy by autogenic training. In: Satapathy, S.C., Bhateja, V., Somanah, R., Yang, X.-S., Senkerik, R. (eds.) Information Systems Design and Intelligent Applications. AISC, vol. 863, pp. 53–66. Springer, Singapore (2019). https://doi.org/10.1007/978-981-13-3338-5_6
6. Mohammadi Ziabari, S.S., Treur, J.: An adaptive cognitive temporal-causal network model of a mindfulness therapy based on music. In: Tiwary, U.S. (ed.) IHCI 2018. LNCS, vol. 11278, pp. 180–193. Springer, Cham (2018). https://doi.org/10.1007/978-3-030-04021-5_17
7. Mohammadi Ziabari, S.S., Treur, J.: An adaptive cognitive temporal-causal network model of a mindfulness therapy based on humor. In: Davis, F., Riedl, R., vom Brocke, J., Léger, P.-M., Randolph, A., Fischer, Th. (Eds.). Information Systems and Neuroscience, vol. NeuroIS Retreat 2019, Switzerland AG (2019). https:// doi.org/https://doi.org/10.1007/978-3-030-281 44-1
8. Mohammadi Ziabari, S.S., Treur, J.: Integrative biological, cognitive and affective modeling of a drug-therapy for a post-traumatic stress disorder. In: Fagan, D., Martín-Vide, C., O'Neill, M., Vega-Rodríguez, M.A. (eds.) TPNC 2018. LNCS, vol. 11324, pp. 292–304. Springer, Cham (2018). https://doi.org/10.1007/978-3-030-04070-3_23
9. Mohammadi Ziabari, S.S., Treur, J.: An Adaptive Cognitive Temporal-Causal Model for Extreme Emotion Extinction Using Psilocybin, In: Silhavy, R, Silhavy, P., Prokopova, Z. (eds.) CoMeSySo 2019. AISC, vol. 2. Springer, Springer Nature Switzerland AG (2019). https://doi.org/10.1007/978-3-030-31362-3
10. Mohammadi Ziabari S.S. (2019) A Cognitive Temporal-Causal Network Model of Hormone Therapy. In: Nguyen N., Chbeir R., Exposito E., Aniorté P., Trawiński B. (eds.) Computational Collective Intelligence. ICCCI 2019. Lecture Notes in Computer Science, vol. 11683. Springer, Cham. https://doi.org/10.1007/978-3-030-28377-3_54
11. Mohammadi Ziabari, S.S.: An adaptive temporal-causal network model for stress extinction using fluoxetine. In: MacIntyre, J., Maglogiannis, I., Iliadis, L., Pimenidis, E. (eds.) AIAI 2019. IAICT, vol. 559, pp. 107–119. Springer, Cham (2019). https://doi.org/10.1007/978-3-030-19823-7_8
12. Mohammadi Ziabari, S.S.: Integrative cognitive and affective modeling of deep brain stimulation. In: Wotawa, F., Friedrich, G., Pill, I., Koitz-Hristov, R., Ali, M. (eds.) Advances and Trends in Artificial Intelligence. From Theory to Practice. LNCS, vol. 11606, pp. 608–615. Springer, Cham (2019). https://doi.org/10.1007/978-3-030-22999-3_52

13. Andrianov, A., Mohammadi Ziabari, S.S., Treur, J.: Application of an adaptive temporal-causal network model for extreme emotion handling. In: The 9th International conference on computational science, Amsterdam, The Netherlands (2020)
14. Redlich, R., et al.: Effects of electroconvulsive therapy on amygdala function in major depression—a longitudinal functional magnetic resonance imaging study. Psychol. Med. **47**, 2166–2176 (2017)
15. Takamiya, A., et al.: Effect of electroconvulsive therapy on hippocampal and amygdala volumes: systematic review and meta-analysis. British J. Psychiatry **212**, 19–26 (2018). https://doi.org/10.1192/bjp.2017.11
16. Nibuya, M., Morinobu, S., Duman, R.: Regulation of BDNF and trkB mRNA in rat brain by chronic electroconvulsive seizure and antidepressant drug treatments. J. Neurosci. **15**(11), 7539–7547 (1995)
17. Baghai, T.C., Moller, H.-J.: Electroconvulsive therapy and its different indications. Dial. Clin. Neurosci. **10**(1), 105–117 (2008)
18. Nakamura, K., et al.: Effects of single and repeated electroconvulsive stimulation on hippocampal cell proliferation and spontaneous behaviors in the rat. Brain Res. **23**(1491), 88–97 (2013)
19. Wang, J. et al.: Electroconvulsive therapy selectively enhanced feedforward connectivity from fusiform face area to amygdala in major depressive disorder. In: Social Cognitive and Affective Neuroscience, 1983–1992 (2017). https://doi.org/10.1093/scan/nsx100
20. Tendolkar, I., et al.: Electroconvulsive therapy increases hippocampal and amygdala volume in therapy refractory depression: a longitudinal pilot study. Psychiatry Res. **214**, 197–203 (2013)
21. Singh, A., Kumar Kar, S.: How electroconvulsive therapy works? Understanding the neurobiological mechanisms. Clin. Psychopharma. Neurosci. **15**(3), 210–221 (2017)
22. Bisson, J.I., Cosgrove, S., Lewis, C., Lewis, C., Robert, N.P.: Post-traumatic stress disorder. BMJ (Clin. Res. ed) **351**, h6161 (2015). https://doi.org/10.1136/bmj.h6161
23. Abott, C.C., et al.: Hippocampal structural and functional changes associated with electroconvulsive therapy response. Trans. Psychiatry **4**, e483 (2014)
24. Treur, J.: Network-Oriented Modeling: Addressing Complexity of Cognitive. Springer Publishers, Affective and Social Interactions (2016)
25. Treur, J.: Verification of temporal-causal network models by mathematical analysis. Vietnam J. Comput. Sci. **3**, 207–221 (2016)
26. Damasio, A.: Self comes to mind. Constructing the conscious brain. Panthon Books, New York (2010)
27. Dukart, J., et al.: Electroconvulsive therapy-induced brain plasticity determines therapeutic outcome in mood disorders. PNAS **11**, 1156–1161 (2014)

# Digital Transformation in Swiss Hospitals: A Reference Modeling Approach

Mike Krey(✉)

Zurich University of Applied Sciences, Winterthur, Switzerland
Mike.Krey@zhaw.ch

**Abstract.** Through various approaches such as the eHealth Switzerland 2.0 strategy, the Swiss healthcare system aims to digitally catch up with other industries and drive the industry into the digital future. To enable hospitals to transform their business model and prepare for the future, this paper presents an approach for the implementation of the digital transformation in Swiss hospitals. Thus, a metamodel consisting of nine elements was created as a base. The focus of the metamodel and the later reference model lay on the central activity elements, which are each embedded in a phase and are directly or indirectly connected to all the other elements in the metamodel. For the reference modeling, the metamodel serves as a structural template, while an existing roadmap from the literature on the digital transformation was used as a content-based starting point. The final reference model consists of 30 activities within six different phases.

**Keywords:** Healthcare · Transformation · Model

## 1 Introduction

The digital transformation in the healthcare industry is gaining momentum, and new digital transformation trends are continuously emerging and are slowly establishing themselves. These trends include progresses in several areas of health services and innovations such as an increase in patient engagement through technology, artificial intelligence in health, health apps, big data and 3D printing [15]. To unleash the enormous potential behind these technologies and allow them to prosper, a solid digitized base is required [6, 26]. Compared to other business sectors, healthcare organizations remain at a significantly lower degree of digitization and record far lower investments in their digital future [2, 10, 24]. A positive indicator of the digitization in healthcare is a proportionate increase in investments compared to previous years [24].

Despite that, Switzerland, among other north European countries, scored highest in the digital evaluation index in 2017 [8]. Harvard Business Review created this index to analyze the digital evolution of 60 countries. Among other things, the index analyses the countries' pattern of digital evolution, evaluates the digital competitiveness and assesses the changes since the last publication. Even though Switzerland, Denmark, Sweden, and


X.-S. Yang et al. (eds.), *Proceedings of Fifth International Congress on Information and Communication Technology*, Advances in Intelligent Systems and Computing 1183, https://doi.org/10.1007/978-981-15-5856-6_2

Norway lead the digital evolution index, they score below two of a maximum of four points regarding their rate of change in digital evolution between 2008 and 2015 [8]. This result reflects the challenges of sustaining growth and offers chances to facilitate existing maturity, scale and network effects to advance the digitization. A study published by PricewaterhouseCoopers, and Google Switzerland analyzed the degree of digitization in various industries [12]. The degree of digitization is based on a self-assessment of participating companies. The survey evaluates the digital maturity in four areas: process and infrastructure, digital sales, customer involvement and people and culture. The questions were answered on a four-point scale. The industries energy and utilities (1.78) and healthcare (1.84) lag the furthest behind. However, [12] mentions that the implementation of the eHealth Switzerland 2.0 strategy could advance the digital maturity in Swiss healthcare. This strategy was introduced in 2018 by the Swiss Confederacy and cantons to increase the digitization within the healthcare sector [11]. One of the suggested reasons for the lagging digitization in the healthcare industry is the focus for human interaction that often stands in the way of digital advancements [12]. Additionally, [12] states the strict regulations as another reason for the slow-progressing digitization in healthcare.

Several technology and consulting companies published reports with various views and explanations about the current state of digitization in the healthcare industry in Switzerland [16]. Barriers obstruct the digital transformation from the inside (e.g. the absence of relevant knowledge) and outside (e.g. strict regulations) of an organization. These impediments have led to the current digital deficit compared to other industries. Therefore, special attention must be given to the barriers when approaching future digitization projects.

As a result of the above-described slow digital transformation and the identified factors supporting it, this paper presents a reference model for the digital transformation of Swiss hospitals that address the mentioned issues and allow a simplified and guided transformation process. In the next chapter, the metamodel will be elaborated, and the deduction of the reference model described. Chapter 4 presents the outcome of the reference modeling approach and provides examples showing how the model is applied in practice. The fifth chapter reflects the results, discusses the findings, and describes the limitations. The final chapter concludes the paper, provides an outlook, and proposes approaches to reduce the impact of the previously discussed limitations.

## 2 Problem Statement

Medical institutions, specifically hospitals, face various challenges in connection with the digital transformation. Continuous pressure to decrease costs and increase efficiency is forcing hospitals to promptly address the digital transformation. However, the hospitals are diverse, and tailor-made solutions do not yet exist or are too specific. Therefore, they do not fit the specific organizational needs [16]. Hence, a reference model is defined to exploit these industry-specific drivers and barriers. The goal of the model is to create a generic approach where organizational characteristics are not considered, and the drivers are used to support the specific strengths of a company while at the same time removing and overcoming barriers. The result is a reference model, which is instanced based on an organizations unique characteristics and requirements. Therefore, the model uses only generic elements which are adopted by the management or project lead to specific entities within the target organization. This approach addresses the different organizational settings of the various players within the hospital system. This paper will answer the following primary research question:

- What does an ideal approach for implementing the digital transformation in the Swiss healthcare system look like?

In order to answer this main research question, the following subordinate research questions are approached first:

- Which framework serves as a suitable foundation for the reference model?
- Which elements, components, and the corresponding relationships between them have to be taken into account in a reference model for digital transformation in Swiss hospitals?

The answers given in this paper only consider the hospital organizations within the Swiss healthcare industry. In case, the same sector in a different geographic region has similar drivers and barriers, and the reference model may be suitable as well. The reference model is a generic guideline and needs to be instantiated upon utilization according to an organization's specific requirements.

## 3 Metamodeling

In order to answer the previously defined research questions, a reference modeling approach was selected. This approach was applied on the base of a previously created metamodel. Below, the details of this research design are explained and where necessary, more profound statements are delivered in order to present a conclusive procedure.

### 3.1 Metamodel

The first step of creating a reference model for the implementation of the digital transformation in Swiss hospitals is establishing and defining the applicability of the reference model in the targeted domain using a metamodel as a blueprint. In the metamodeling process the overall depth, scale as well as the syntax and structure of the reference model are determined by charting its elements, components, and the corresponding relationships between them [13, 23, 27]. Hence, the metamodel facilitates the conceptual modeling and allows a more intuitive and practical view on the model based on it [13]. This helps users and implementors to better understand its complexity and extent when evaluating or making use of it.

The previous description shows that various modeling levels and abstractions exist. The guidelines followed in this research regarding multi-level modeling are presented in the paper by [14] and further described in [23]. As shown in Fig. 1, the mentioned guidelines are differentiated into four hierarchies. Apart from M0, each layer conforms to or is implemented according to the adjoining layer above it and additionally (except M3) defines or abstracts the layer underneath it [14, 23]. Therefore, the metamodel describes the notation of the metamodel, while the metamodel describes the structure of the model. This research only includes the M1 and M2 layers. The first layer (M0) is not formally carried out and thereby is only briefly described in the following paragraph. In the final layer, the reference model is adapted to a Swiss hospital (layer M2) and consequently takes place in practice.

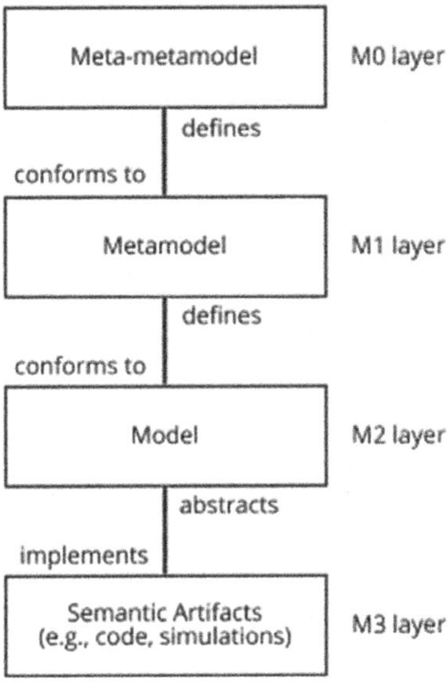


**Fig. 1.** Four-layer metamodeling architecture as described by [1]

The metamodel in this project was created using a simplified adoption of the UML class diagram notation. This notation, along with its elements, was chosen because it represents and aligns well with the main purpose of the metamodel, which is the listing of the allowed and necessary constructs within the reference model [13]. The mentioned simplification contributes to a better understanding by non-experts and increases the clarity of the model. Moreover, since the metamodel is not a model to a software artifact, using the notation to the full extent is not feasible and would not be appropriate. The following elements were used in the metamodeling process: the class element, the navigability, the multiplicity, and the generalization. The class element is used to model the single components that make up the reference model. Displayed with a rectangular outline, these components stand in an associative navigability to one another [19, 22]. The navigability gives additional information to the association and can either be unspecified, navigable or not navigable in both directions of the associated components [19, 22]. Additionally, the multiplicity or cardinality of the associations specifies the allowable number of instances of the described component in nonnegative integers [19, 22]. Lastly, associations can also occur in the form of a generalization. This form of association structures classes into hierarchies of inheritance, where the subclasses specify the parent in a more detailed manner but are essentially already covered in the parent class. To allow the creation of a complete metamodel, the metamodeling process was not undertaken completely uncoupled from the reference modeling process. Rather, the first version of the original metamodel was used as a base for designing the reference model. Elements of one or the other model where then added or removed if necessary, in order to perfectly match the models to the targeted domain. This iteration between the meta- and reference modeling process resulted in complete and well-aligned models.

### 3.2 Reference Model

After completing the metamodel, the reference model was created according to the prior defined elements and relations discussed above. A definition that is universally accepted for "reference model" cannot be found. Consequently, to use a broadly accepted definition of the term in the context of this paper, common denominators in different definitions by [5, 7, 18, 21, 25] were combined. For example, [25] describes reference models as a universal tool using "recommendation character" to construct and derivate other (enterprise-specific) models while [5, 7] call it a "normalized description of key concepts of a given domain" and [21] describes a reference model as a construction with recommended universal elements and relationships that create a point of reference. Resulting from the combination of the mentioned definitions, a reference model in this paper is defined as a universally applicable and reusable best practice framework for a certain domain, which in this case are Swiss hospitals.

As displayed in the metamodel (Fig. 2), the reference model (and its phases) was derived from an existing roadmap. This serves as a solid base to ensure the completeness of the approach presented in this paper. Thereby, it offers a fully accompanying guideline using established and proven methods to better address the hospitals lagging regarding digitization, when compared to organizations of other industries [4, 9]. However, the single activities and phases as well as other elements of the underlying roadmap were modified, adjusted, removed, replaced or extended to fit the specific application domain of this research when necessary.

For this purpose, the model by [20] was selected. Schallmo and colleagues present in their book a high-level, comprehensive roadmap with five phases starting with the digital reality phase and ending with the implementation phase. The roadmap by [14] met several key factors to serve as a template, which is why it was selected. Firstly, their model not only shows a certain procedure, but is also a well-documented roadmap for the digital transformation and business model innovation. The second key factor is the combination of the objectives "digital transformation" and "business model innovation." This allows to keep a customer-oriented view during the major digital changes. In addition, the roadmap by [20] is based on existing "digital transformation" and "business model innovation" approaches as well as on best practices and therefore suggests an established procedure. Lastly, the roadmap is divided into five phases, where the implementation does not take place until the last step. Thus, this ensures a solid base with a thoroughly developed framework, inclusion of all stakeholders, and organizational characteristics plus a carefully designed organizational fit.

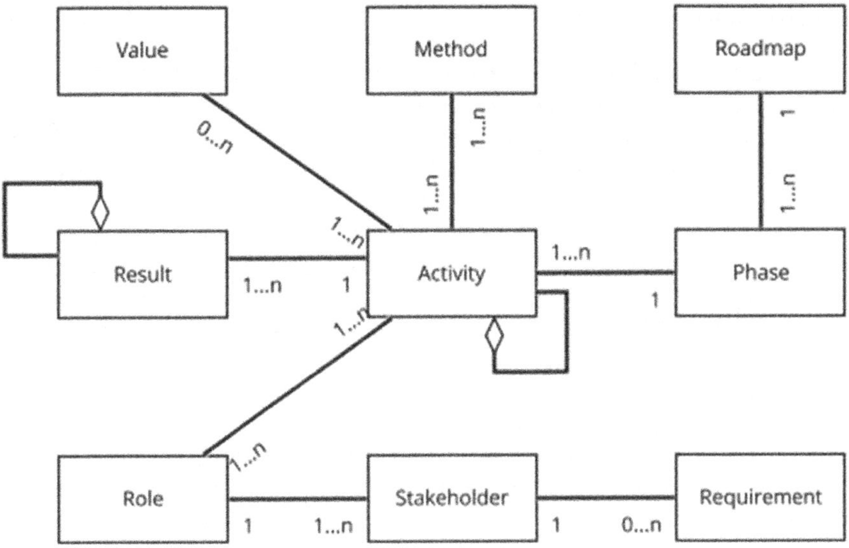


**Fig. 2.** Metamodel

In addition to the above-mentioned adjustments of the single parts of the selected approach by [20] to the specific domain, modifications of the superordinate phases were also discussed. Every phase of the original roadmap was critically analyzed concerning its value, contribution, and necessity to the digital transformation of a hospital. A phase was only used if it was required or added value to the purpose of this project. As a result, none of the original phases were removed; however, a final review phase as suggested by [1, 17] was added to the reference model. This reflection of the implementation of the digital transformation and the resulting business model innovation takes place iteratively with the previous implementation phase and assures an appropriate and proper transformation process. The review phase insures that the digital transformation takes place as it was planned and envisioned during the digital ambition phase by monitoring the progress and if necessary, rectifying the individual steps to reach the intended outcome. By reflecting the strategic and organizational implementation, potential conflicts and sources of error can be encountered early. As a result, the potential impact of discrepancies on the outcome and success of the digital transformation in an organization can be reduced or avoided altogether.

## 4 Results

The results of the metamodeling and reference modeling process are shown and explained in more detail in the following chapter. Two examples of activities from within the reference model are also presented in depth to give a better understanding.

### 4.1 Metamodel

The term metamodel is used when a model describes another model [3]. In this case, the metamodel is used to describe and define the reference model. The metamodel was developed in iterating workshops among the authors involving in-depth literature. As a result, nine different and interdependent elements were defined for the metamodel (cf. Fig. 2).

The activity is specified as the central element of the metamodel and can contain several sub-activities. At least one outcome results from each activity, whereby a result itself as a whole can also arise from the combination of multiple results. A role describes a function of a person within an organization and specifies which activities are performed by the employee. A stakeholder represents a party or person of interest, whereby it is possible that a stakeholder is also a role. Further, a stakeholder's need is recorded as a stakeholder specific requirement. A method for working out the objectives of an activity is also proposed, although these are not final. Since the reference model for the digital transformation contains an extensive number of activities and stretches over a considerable time period, the impact or contribution of every activity is also mentioned in order to keep an overview and creating additional transparency. Furthermore, an activity is assigned to a certain phase, which has arisen from an established roadmap or process model that can be found in the literature.

### 4.2 Reference Model

The reference model is created using the metamodel as a base, where the different elements are drawn up. This means that the different constructs are described in relation to the activity at the center.

As described in Chap. 3, the roadmap by [20] was selected as the initial basis for deriving the specially adapted reference model. This process model combines the best elements of the existing approaches by Bouée and Schaible, Esser and Pricewaterhouse-Coopers, among others, as well as existing business cases and best practices, consolidating them into an advanced process model. This process model is comprised of the five phases of [20], with an addition of a sixth phase, the review phase, which can be seen more clearly in Fig. 3.

In the first phase, the digital reality phase, the existing business model of a company is sketched out, the analysis of the value chain with associated actors is carried out, and the customer requirements are ascertained. Hence, the activities of this phase examine and document the actual state. This provides an understanding of the digital reality in different areas. Noticeably, in this phase the element "stakeholder" is represented by the management. This is because the current state of the processes and procedures is already known to the employees; therefore, there are no immediate changes for the employee. That is why the outline of the existing business model is mainly interesting for the management.

Based on the previous phase, the digital ambition defines the objectives with regard to the digital transformation. These relate to time, finance, space, and quality. The digital ambition states which objectives apply to the business model, its elements and how they

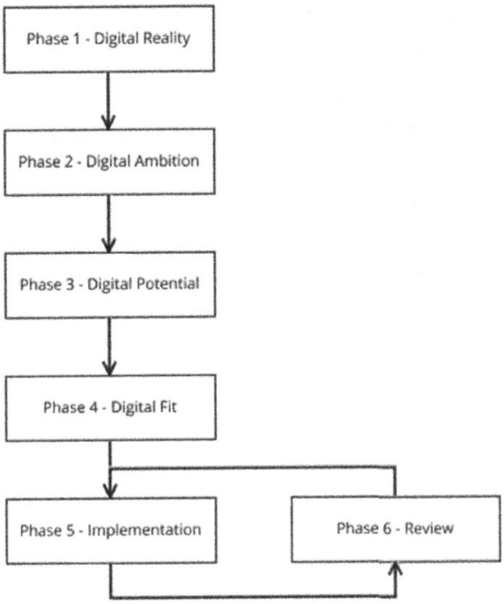


**Fig. 3.** Phases of the model

are prioritized. From the ascertainment of corporate objectives for the digital transformation, a digital vision is created as a result. By defining the priorities for the business model dimensions, a priority overview is created for the management.

In the third phase, best practices and enablers for the digital transformation, which serves as a baseline for the prospective digital business model, are identified. For this purpose, different digital options are derived for each business model element and logically combined with each other. To ensure that those responsible know why an activity needs to be carried out, the output and contribution of each activity in regard to future activities are listed. Thus, the main value of the activities to collect best practices and enablers for digital transformation can be adapted from activity "3.3.1 Development of the digital business model and associated options, taking into account enablers, disablers and best practices." Furthermore, this activity itself has the benefit of creating combinations of options on how the digital transformation could be implemented.

In the digital fit phase, the options for designing the digital business model are evaluated. Assuring the fit with the existing business model, the fulfillment of customer requirements and the achievement of goals are crucial. The evaluated combinations can be prioritized according to these factors. By suggesting methodical approaches for each activity, an important indication on how the activities can be implemented to achieve the desired result is given to the organizations. For example, information consolidation can be used to create combinations of options. It is proposed to use a decision-making method for the evaluation and selection of the combinations, because the responsibility for the right selection lays in the hands of the respective project management team.

The fifth phase contains the finalization and implementation of the digital business model, i.e., the combination of the compiled options from the previous phases that are pursued further. This means that created artifacts, processes, and workflows are transferred to the operative business. The digital implementation also includes designing the digital customer experience and the digital value network with partner integration. Resources and capabilities required for digital implementation are also considered. A large number of roles are involved in this phase, as it is a very comprehensive and cross-departmental phase with influence on different areas of the company. For this reason, many different stakeholders and roles are involved in the activities.

In the sixth phase, the implemented actions are examined, and the project plan and action plan are further adapted to the situation. In order to successfully implement the developed options, it is necessary to obtain comprehensive information. Project management methods are suggested to be used to adjust the project and action plan. The management as a stakeholder is primarily interested in a meticulous review phase, but the employees are as well, because this could have considerable impact on the future work activities and processes.

## 4.3 Activities

For all six phases mentioned, corresponding activities were elaborated, in order to comprehensively describe the digital transformation, make the process more tangible, and deliver an accompaniment to the application of the reference model. For each activity, the dependent elements according to the metamodel were defined. These include roles, stakeholders, results, benefits, methods, and the linked phase.

Depending on the phase, a different number of activities were assigned in order to be able to fulfill these activities as best as possible. Eleven activities were assigned to phase 1, the digital reality phase. Only two activities were assigned to the second phase and seven to determine the digital potential phase. Two activities are needed for the digital fit phase and five for the digital implementation phase. Finally, there are two activities in the review phase. An overview over the number of activities in each phase is shown in Table 1.

**Table 1.** Number of associated activities per phase

| Phase | Associated activities |
|---|---|
| Phase 1—Digital Reality | 11 |
| Phase 2—Digital Ambition | 2 |
| Phase 3—Digital Potential | 7 |
| Phase 4—Digital Fit | 2 |
| Phase 5—Implementation | 6 |
| Phase 6—Review | 2 |

In order to better demonstrate the understanding of the individual activities, two fully elaborated activities will be selected and explained in more detail in the following. Firstly, activity "2.1.1 Survey of company objectives for digital transformation" from the digital ambition phase will be exemplified in depth. This activity comprises the collection of business objectives on the basis of four categories: time, finance, space, and quality.

The result of this survey is a digital vision for the organization. In order for employees to know what the activity is being carried out for, the impact on future activities is shown. Thus, the result of this activity will be useful for activities "3.3.1 Development of the digital business model and related options taking into account enablers, disablers and best practices" and "4.1.2 Evaluation and selection of combinations." The project manager, the CEO, a business model expert and a digitization expert are the required roles to carry out the activity. These roles were defined using the Responsible, Expert, Work, Approver (REWA) matrix and the assignments can be more closely examined in Table 2. REWA is a variant of the RACI model and has a similar significance. Since many people and groups are involved in the digital transformation, a final list of the parties to be informed is difficult to compile and ultimately not very informative. Thus, in this case the letter "I" (informed) from the RACI model is of little use. Therefore, the REWA variant is more suitable due to the pragmatic and meaningful naming. After all, with "E" for expert and "W" for work, it is evident who carries out the task and who provides professional assistance.

**Table 2.** REWA matrix for activity 2.1.1 Survey of business objectives for digital transformation

| | R | E | W | A |
|---|---|---|---|---|
| Project manager | × | | × | |
| CEO | | × | | x |
| Business model expert | | × | × | |
| Digitization expert | | × | × | |
| Steering committee | | | | x |

Furthermore, stakeholders of activity "2.1.1 Survey of business objectives for digital transformation" are the management, the board of directors or executive board, a representative member of the parent company, the trade union, political regulators and interest groups as well as investors. The methodology of a strategic analysis and objectives is recommended for the processing of the activity.

Activity "3.3.1 Development of the digital business model and associated options, taking into account enablers, disablers and best practices" is to be carried out on the results, among other things, of the activity described above. Based on the collected best practices, enablers, and disablers, all options for the future design of the individual business model elements are now derived. Apart from their listing, they are not yet evaluated. The business model elements are evaluated to determine if digitization is needed. If applicable, the form of digitization will then be defined. It is also examined how the enablers can be used to improve the business model elements. The result is a list of options for the intended digital business model. In regard to the entire transformation process, this result lays the foundation for activity "4.1.1 Creation of combinations of options." As shown in the REWA matrix in Table 3, the roles project manager, business model expert, and digitization expert are necessary to complete activity 3.3.1.

**Table 3.** REWA matrix for activity 3.3.1 Development of the digital business model and related options taking into account enablers, disablers, and best practices

| | R | E | W | A |
|---|---|---|---|---|
| Project manager | × | | | x |
| Business model expert | | × | × | |
| Digitization expert | | × | × | |

As stakeholders, the management, the client, the partners, the trade unions, political regulators, interest groups and investors are listed. In addition, goal setting, strategic analysis and information consolidation are the suggested methods to cope with the activity.

## 5 Discussion

The following paragraphs will discuss and further explain the results described in the previous chapter. The difference between the roadmap by [20] and the approach described in this paper are presented and the additional value created explicated. Despite not having applied and validated the reference model in practice, the representability and validity of the model in the targeted domain are justified. Also, the possible impact of the created reference model on other domains is mentioned and the importance of this contribution is highlighted. Lastly, the research questions listed at the beginning of this paper are addressed.

As described in the prior chapter, the reference model uses six different phases to guide the digital transformation process, starting at the current state and finishing with an iterative implementation and reflection of the developed digital business model. Noticeably, many of the 30 total activities take place in the first phase, while other phases such as digital fit phase consist of a smaller number of activities. Importantly, a large number of activities do not necessarily reflect the time needed to finish a phase or its complexity. For instance, the examination and documentation of the digital reality phase which consist of eleven activities can be considered a relatively short phase compared to other phases, such as the implementation phase, where individual activities can take up to several months or even years. Furthermore, to offer a complete process model that offers the highest possible degree of support, important factors and influences on the digital transformation process, such as stakeholders and their requirements as well as suitable methods for individual steps, were listed. However, this additional information is neither conclusive, nor does it represent the only correct way. It is intended primarily as a guidance and food for thought. The effective application of the model and design of the digital transformation process depends on the respective project manager. The reference model supports the business model innovation and should be seen as a guideline, yet the design of the practically implemented procedure depends on the preferences of an organization and how they adopt it.

The reference model was derived from the roadmap of [20], and therefore, similarities between the two are implied. The common themes do not result from a simple replication of the underlying roadmap, but symbolize the uniformity of the superordinate purpose, which is the digital transformation. Therefore, certain activities, elements, instances, or parts of the digital transformation can be found in any approach. However, due to the very specific application domain of this reference model, none of the activities in [20] were identically replicated. The more general phases on the other hand were transferred after being carefully checked regarding their suitability. With the addition of the last iterative review phase, the base structure of the reference model now deviates from the one in [20] too. In addition, the degree of detail of the reference model created here is more profound and could only be realized by choosing a specific application domain. Despite not going into detail of each step of every activity, by defining the methodology, possible stakeholders, roles with the help of the REWA matrix, results, and the contribution of a single activity within the transformation process, considerably more content, transparency, and higher comprehensibility is established. As a result, the digital transformation in a Swiss hospital becomes more tangible and realistic, especially where such objectives pose blackboxes. Representing a seemingly small difference, the additional content creating an instructional guiding concept that not only talks about, but also shows and allows a successful implementation of the digital transformation, rather than a shallow roadmap.

In order to create a relevant and rigorous digital transformation approach, the reference modeling process was guided by well-established best practices and existing models. However, to increase representativeness and confirm the universal applicability of the reference model in practice of the Swiss hospital environment, additional input and adjustments in exchange with health and digital professionals in the targeted domain would be appropriate. Despite the positive effects in the context of this work, it was refrained from having the result validated by individual experts. Validation within the limited timeframe of this project risks that the general model is negatively manipulated based on personal subjective assessments and thus no longer corresponds to the original purpose. Due to the development based on practice-proven approaches, the presented reference model can be classified as representative and valid despite the lack of exemplary validation. Nevertheless, to create a full and well-aligned digital transformation approach to be used in a large number of digital transformation projects in Swiss hospitals, an extensive practical validation is crucial. Hence, as suggested in the following conclusion and outlook, objectively validating the reference model should be considered as one of the most important steps.

While this reference model was specifically created for the digital transformation in Swiss hospitals and took certain characteristics of the targeted domain into account, it does not rule out the (partial) applicability on organizations from other areas of the healthcare system or even outside the industry. This means that the presented approach's benefits are not limited to the very narrow domain but could also expand and trigger or encourage the digital transformation in other business sectors. Undeniably, pushing digital advancements in one sort of organizations within one industry (healthcare) promotes new digital solutions, a digital thinking, interorganizational digital processes and hence a much more extensive advancement of digital mature enterprises. Being the first specifically targeted contribution in a digitally restrained industry can therefore also serve as a basis for further research in this or related fields. Moreover, by taking this reference model and further evolve and improve it, one or more approaches for multiple domains can result. Therefore, this paper does not only contribute to the digital transformation of Swiss hospitals but also to the possible digital advancements of an entire industry and lays a foundation for more advanced and sophisticated specific and cross-industrial approaches.

The defined main research question at the beginning of this paper focuses on an ideal approach for the implementation of the digital transformation in Swiss hospitals. By developing a reference model based on best practices and established approaches, the research question was attempted to be answered. Considering the previously mentioned omitted validation and possible further developments in an iterative setting with healthcare professionals, the presented approach would require additional input as well as practical experience to be described as ideal. Nevertheless, the created output can be considered a valid reference model and starting point for the implementation of the digital transformation in Swiss hospital. Hence, it is a step in the right direction and as described previously, having a foundation can also trigger a digital evolution on a larger scale. In the context of the first subordinate research question, an analysis was conducted to determine which existing model can be used as a template for a reference model in the chosen application domain. As explained in the previous chapters, [20] model is highly suitable for adaptation in this specific context. In combination with [1], who suggests a reflection phase, a solid and proper foundation was established. As mentioned before, to create a reference model a previously designed metamodel was required in order to set and define the key elements and components as well as their relationship. Consequently, the answer to the second subordinate research questions was given by modeling the metamodel.

## 6 Conclusion and Outlook

Although the digitization of the healthcare sector has gained momentum, investment remains significantly lower compared to other sectors. One approach to make up for this deficit is the eHealth Switzerland 2.0 strategy, which was introduced by the federal government and the cantons in the fall of 2018. Medical institutions, in particular hospitals, are currently facing various challenges in the digital transformation. Therefore, the aim of this paper is the development of an ideal approach for implementing the digital transformation in the Swiss healthcare system. In this generic approach, drivers and best practices were used to push the digital transformation and overcome the barriers.

The first step was to create a metamodel which served as the basis for the reference model. As part of this project, the metamodel was created with a simplified UML class diagram notation to provide a simple understanding of the necessary constructs. Finally, this metamodel consists of nine different elements, which have several dependencies on each other. The activity serves as the central element. In addition to the activity, the further elements are "result," "value," "method," "roadmap," "phase," "role," "stakeholder," and "requirement."

Following the completion of the underlying metamodel, the reference model was created according to the previously defined elements and relationships. Within the selected application domain, the reference model can be regarded as a universally applicable and reusable best practice framework for Swiss hospitals. A high-quality and comprehensive template from [20] serves as the basic roadmap. This roadmap was reviewed in the course of a critical analysis and extended with an additional review phase. Thus, the roadmap consists of six phases: Digital Reality, Digital Ambition, Digital Potential, Digital Fit, Implementation, and Review. The implementation and review are carried out as iterative phases to ensure an appropriate and correct transformation process.

The individual phases consist of a varying number of activities with an overall of 30 activities listed. The various elements according to the metamodel were defined and assigned to the individual activities. In addition to the roles, a REWA model was created to show the responsibilities for each activity.

This paper and the presented approach for the digital transformation were created using extensive insights from different literature sources. So far, no practical feedback has been integrated into the work, which could be done in a future step. For example, experts from hospitals could examine the approach and suggest improvements through constructive feedback. However, additional inputs by selected experts are not solely contributing to more representative results but could also have a subjective impact on the otherwise generalized and objective reference model.

As mentioned above, the fact that no validation of the metamodel or reference model has taken place by professionals from the Swiss hospital system can be seen as a weakness as well as a strength of the work. It is possible that required elements have been forgotten due to the lack of involvement of experts from this industrial sector or that too little attention has been paid to some of the key factors. On the other hand, objectivity is guaranteed. This means that when validated by a hospital, subjective influences from this particular hospital automatically result from the feedback. Since not all hospitals are organized and set up the same way, this could lead to distortions of the work and individual opinions can be misinterpreted as generally applicable. Therefore, such a validation has to be carried out extensively and the gained feedback has to be critically analyzed, in order to take into account only suitable and value-adding objections.

Further, this model could now be used in cooperation with a hospital to create an individually broken-down approach specific to the selected hospital. Thus, this approach would be broken down from the theoretical level to the practical level, and a concrete use case would be developed. Additionally, the reference model offers not only the possibility for further specification but also generalization. As a result, it can serve as a guideline not only for hospitals, but also for other areas of the healthcare industry or even companies outside this sector.

## References

1. Appelfeller, W., Feldmann, C.: Die digitale Transformation des Unternehmens. Springer, Berlin Heidelberg, Berlin, Heidelberg (2018)
2. Bachofner, T.: "Schweizer Gesundheitswesen: Bereit für den Take-off, https://www.swisscom.ch/de/business/enterprise/themen/health/digitalisierung-gesundheitswesen.html (2018)
3. Benker, T., Jürck, C., Wolf, M. (eds.): Geschäftsprozessorientierte Systementwicklung: von der Unternehmensarchitektur zum IT-System. Springer Vieweg, Wiesbaden (2016)

4. Berghaus, S., Back, A., Kaltenrieder, B.: Digital Maturity & Transformation Report 2017. Universität St. Gallen & Crosswalk (2017)
5. Bernus, P., Nemes, L., Schmidt, G. (eds.): GERAM: The Generalised Enterprise Reference Architecture and Methodology. In Handbook on Enterprise Architecture, pp. 21–63. Springer, Berlin Heidelberg, Berlin, Heidelberg (2003)
6. Bughin, J., van Zeebroeck, N.: Artificial Intelligence: Why a Digital Base is Critical, p. 4. McKinsey Quarterly (2018)
7. Castano, S., De Antonellis, V., Fugini, M.G., Pernici, B.: Conceptual schema analysis: techniques and applications. ACM Trans. Database Syst. **23**(3), 286–333 (1998)
8. Chakravorti, B., Bhalla, A., Chaturvedi, B.S.: 60 Countries' Digital Competitiveness, Indexed (2017)
9. Chilukuri, S., Kuiken, S.V.: Four Keys to Successful Digital Transformations in Healthcare, p. 7 (2017)
10. Dümmler, P.: Digital. Swiss: Health, https://digital.swiss/en/topics/health
11. eHealth Suisse, Strategie eHealth Schweiz, eHealth Suisse (2018)
12. Greif, H., Kühnis, N., Warnking, P.: Digital Transformation: How Mature Are Swiss SMEs? PwC Switzerland (2016)
13. Jeusfeld, M.A.: Metamodel. In: Liu, L., Özsu, M.T. (eds.) Encyclopedia of Database Systems, pp. 1727–1730. Springer, US, Boston, MA (2009)
14. Karsai, G., Nordstrom, G., Ledeczi, A., Sztipanovits, j.: Specifying graphical modeling systems using constraint-based meta models. In: CACSD. Conference Proceedings. IEEE International Symposium on Computer-Aided Control System Design (Cat. No.00TH8537), pp. 89–94 (2000)
15. Kelly, C.: Digital Transformation in Healthcare: 6 Key Examples & Trends (2017)
16. Krey, M., Seiler, R.: Usage and acceptance of drone technology in healthcare: exploring patients and physicians perspective. In: Proceedings of the 52nd Hawaii International Conference on System Sciences. The 52nd Hawaii International Conference on System Science, Grand Wailea, January 8–11 2019
17. PwC, Digitale Transformation : der größte Wandel seit der industriellen Revolution. Frankfurt, PricewaterhouseCoopers (2013)
18. Rosemann, M.: Application Reference Models and Building Blocks for Management and Control. In: Bernus, P., Nemes, L., Schmidt, G. (eds.) Handbook on Enterprise Architecture, pp. 595–615. Springer, Berlin Heidelberg, Berlin, Heidelberg (2003)
19. Rumpe, B.: Modeling with UML: Language, Concepts, Methods. Springer International Publishing, USA (2016)
20. Schallmo, D., Rusnjak, A., Anzengruber, J., Werani, T., Jünger, M. (eds.): Digitale Transformation von Geschäftsmodellen: Grundlagen. Gabler Verlag, Instrumente und Best Practices (2017)
21. Schütte, R., Rotthowe, T.: The Guidelines of Modeling—An Approach to Enhance the Quality in Information Models, Conceptual Modeling—ER '98, pp. 240–254. Springer, Berlin Heidelberg (1998)
22. Seidl, M., Scholz, M., Huemer, C., Kappel, G.: UML @ Classroom: An Introduction to Object-Oriented Modeling. Springer International Publishing (2015)
23. Sprinkle, J., Rumpe, B., Vangheluwe, H., Karsai, G.: Metamodelling—State of the Art and Research Challenges. In: Giese, H., Karsai, G., Lee, E., Rumpe, B., Schätz, B. (eds.) Model-Based Engineering of Embedded Real-Time Systems: International Dagstuhl Workshop, Dagstuhl Castle, Germany, November 4–9, 2007. Revised Selected Papers., pp. 57–76. Springer, Berlin Heidelberg, Berlin, Heidelberg (2010)
24. Statista, Weltweite Anzahl von Finanzierungsabschlüssen für.: Digital Health-Unternehmen in den Jahren von 2009 bis 2017. Statista (2017)

25. Thomas, O.: Version Management for Reference Models: Design and Implementation, In Reference Modeling: Efficient Information Systems Design Through Reuse of Information Models. Physica-Verlag, Münster (2007)
26. Topol, E.: The Creative Destruction of Medicine: How the Digital Revolution Will Create Better Health Care. Basic Books, New York (2011)
27. Vangheluwe, H., de Lara, J.: XML-based modeling and simulation: meta-models are models too. In: Proceedings of the 34th conference on Winter simulation: exploring new frontiers, Winter Simulation Conference, pp. 597-605 (2002)

# Estimating Exceedance Probability in Air Pollution Time Series

Giuseppina Albano[1(✉)], Michele La Rocca[2], and Cira Perna[2]

[1] Dipartimento di Studi Politici e Sociali, Università degli Studi di Salerno, via Giovanni Paolo II, 84084 Fisciano, Salerno, Italy
pialbano@unisa.it

[2] Dipartimento di Scienze Economiche e Statistiche, Università degli Studi di Salerno, via Giovanni Paolo II, 84084 Fisciano, Salerno, Italy
larocca@unisa.it, perna@unisa.it

**Abstract.** In the last years, increasing attention has been paid to air pollution, due to its impact on human health and on the environment. Current EU legislation establishes fixed limits for some air components that have been shown to have adverse effects on human health. It is therefore important to identify regions where the probability of exceeding those limits is high. In this paper, we propose a bootstrap scheme to obtain the distribution of the considered air pollutant at a given time point. In particular, the proposed resampling scheme is based on the residuals of a semiparametric model which is able to incorporate some stylized facts usually observed in such kind of data, such as missing data, trends and conditional heteroscedasticity. The estimated bootstrap distribution is then used to estimate the probability that the air pollutant exceeds the fixed legal limits. An application to ($PM_{10}$) in Torino area in the North-Italian region Piemonte is shown.

**Keywords:** ARMA–GARCH model · Bootstrap · Exceedance probability

## 1 Introduction

Recent studies show the link between human health and exposition to pollutants; thus, it becomes important to analyse air quality, so to adopt appropriate political strategies to improve it if health conditions are not satisfied. In order to improve air quality and to defend human health, the EU directive 1999/30/EC has defined some regulations on limit values of air pollutants, including $PM_{10}$ concentration levels (pollution particles measuring 10 micron or less in diameter and measured in $g/m^3$ units). The binding daily limit value for $PM_{10}$ is


X.-S. Yang et al. (eds.), *Proceedings of Fifth International Congress on Information and Communication Technology*, Advances in Intelligent Systems and Computing 1183, https://doi.org/10.1007/978-981-15-5856-6_3

50 μg/m$^3$ which has not to be exceeded on more than 35 days within a calendar year. From a statistical perspective, the estimation of the probability that the $PM_{10}$ level exceeds a specific threshold becomes a fundamental issue. To this aim, it is important to accurately model typical behaviours generally observed in air pollutant time series such as often huge presence of missing values, a trend-cycle component, the presence of significant autocorrelations among near observations, heteroscedastic effects (see, e.g., [3,10]).

In [2], it has been proposed a semiparametric model in which the trend-cycle is estimated by using a local polynomial approach; such estimate is then used to impute missing values if present in the data at hand. The detrended component is assumed to follow an ARMA–GARCH-type model which is able to take into account the different stylized patterns of the $PM_{10}$ time series. The choice of the "best" model, among some possible specifications of the GARCH models, has been made by using the model confidence set. The analysis shows that the simplest ARMA–GARCH(1,1) model with standard normal errors in most cases outperforms more complex models. In this paper, the proposed model is used to estimate the exceedance probability, i.e. the probability for a certain air pollutant to exceed a certain legal limit $c$. In particular, we propose a model-based residual bootstrap scheme to estimate the distribution of the considered air pollutant from which an estimator of the exceedance probability is naturally raised. Moreover, to evaluate the estimated exceedance probability as a classifier for values under the limit, the area under the receiver operating characteristic curves is also used.

The paper is organized as follows. Section 2 presents the proposed modelling strategy. In Sect. 3, a bootstrap scheme is proposed and discussed in order to estimate the distribution of the considered air pollutant and the estimation of the exceedance probability is provided. In Sect. 4, the results of an application to real data are discussed. A sensitivity analysis is also performed in order to show the stability in the results, varying the tuning parameters. Some remarks close the paper.

## 2 The Model

Let $Y_t$ be the one-dimensional random process representing the daily average of an air pollutant at day $t$. The process can be modelled as (see [2]):

$$Y_t = m(t) + \zeta_t, \qquad t \in \mathbb{N}, \tag{1}$$

where $m(t)$ is a deterministic trend-cycle component, which is able to model the underlying level of the series, and $\zeta_t$ is a stochastic component. We assume that $\zeta_t$ follows a stationary and invertible autoregressive moving average (ARMA)–generalized autoregressive conditional heteroscedastic (GARCH) specification originally proposed in [4], described by:

$$\zeta_t = \Phi_1\zeta_{t-1} + \Phi_2\zeta_{t-2} + \cdots + \Phi_p\zeta_{t-p} + \varepsilon_t + \theta_1\varepsilon_{t-1} + \cdots + \theta_q\varepsilon_{t-q} \tag{2}$$

$$\varepsilon_t = \sigma_t\, z_t, \tag{3}$$

$$\sigma_t^2 = \omega + \sum_{i=1}^{p'} \alpha_i\varepsilon_{t-i}^2 + \sum_{j=1}^{q'} \beta_j\sigma_{t-j}^2. \tag{4}$$

where $\{z_t\}$ is a sequence of independent and identically distributed random variables such that $\mathbb{E}(z_t) = 0$ and $\mathbb{E}(z_t^2) = 1$, $\sigma_t$ is the conditional standard deviation of $\varepsilon_t$. The ARMA component (2) describes the stochastic behaviour of the process $Y_t$, while the conditional standard deviation of $\varepsilon_t$ is modelled by using a the GARCH model (3) and (4) which allows to consider the heteroscedastic effects. Although different GARCH-type specifications for $\sigma_t^2$ could be considered, existing studies, almost unanimously, agree on the good performance of standard GARCH model [7] which seems to be able to accurately capture the volatility clustering present in most data set in different fields. Recently, this result has been confirmed in [2] for environmental time series in which some alternative specifications of the GARCH-type models and different error distributions have been considered. By means of the model confidence set [8], the models have been compared in their ability to predict future values of the series. It has been showed that the standard normal GARCH$(1,1)$ seems to be able to provide performance comparable to more complicated models (see [2]). We assume that the following assumptions hold:

A1. The function $m(\cdot) \in C^1[0,+\infty) = \{f : [0,+\infty) \to \mathbb{R} | f$ is differentiable and its derivative is continuous$\}$;
A2. All zeros of $A_\Phi(x) := 1 - \sum_{i=1}^{p} \Phi_i x^i$ lie outside the unit circle;
A3. All zeros of $B_\theta(x) := 1 - \sum_{j=1}^{q} \theta_j x^j$ lie outside the unit circle;
A4. $\omega > 0$, $0 < \alpha_i < 1$ $\forall i = 1,2,\ldots,p'$ and $0 < \beta_j < 1$ $\forall j = 1,2,\ldots,q'$ with $P \equiv \sum_{i=1}^{p'} \alpha_i + \sum_{j=1}^{q'} < 1$.

Assumption A1 impones that the trend-cycle component in (1) is a smooth function of time; hence, $m(\cdot)$ can be locally approximated by a polynomial function. Assumptions A2 and A3 guarantee, respectively, the stationarity and the invertibility of the ARMA process (2). Finally, assumption A4 insures the positivity and the weak ergodic stationarity of the conditional variance $\sigma_t^2$. Let

$$\mathbf{y} = (y_1, \ldots, y_n)'$$

be the daily time series of $PM_{10}$ emissions observed at times $t_1 < t_2 < \cdots < t_n = T$, where missing values are eventually present. Under the assumption A1, the estimation of $m(t)$ can be made by means of a two-step procedure in which a local polynomial smoother is used to estimate the missing observations and then the same estimator is applied to obtain an estimate of the function $m(\cdot)$ (see [1]).

The asymptotic properties of the obtained estimator, along with expressions for the bias and the variance, have been derived and discussed in [11]. Moreover, it seems to work better that other standard methods when there is a huge presence of missing values [1]. Finally, since the trend-cycle estimator is essentially based on a local estimator approach and so it is nonparametric, it is flexible to handle any structure of the data.

The detrended time series $\widehat{\zeta}_i = y_i - \widehat{m}(t_i)$ is then used to identify and estimate the ARMA–GARCH component (2)–(4). In particular, the identification of the order $p$ and $q$ in the ARMA model (2) and $p'$ and $q'$ in the GARCH model (4) can be made by using automatic information criteria such as AIC or BIC. Moreover, the parameters in (2)–(4) can be estimated by QMLE method. Under the assumptions [A2]–[A4], the QMLE estimator converges almost surely to the true parameter $(\omega, \alpha_1, \ldots, \alpha_{p'}, \beta_1, \ldots, \beta_{q'})$ as $n$ goes to $\infty$ (see [6]).

## 3 The Estimation of the Exceedance Probability by a Bootstrap Resampling Scheme

In order to estimate the exceedance probability of a certain limit $c$, that is the probability that the process $Y_t$ exceeds the legal limit:

$$EP(t) := P[Y_t \geq c]$$

it is crucial to obtain the distribution function of $Y_t$ in (1). To this aim, we suggest the use of a residual bootstrap scheme in which the residuals of the fitted model are resampled.

The procedure can be outlined as follows.

Step 1. Estimate the function $m(t)$ in (1) as described in [2], obtaining $\widehat{m}(t)$;
Step 2. Estimate the detrended series $\widehat{\zeta}_i$ at time $t_i$, $i = 1, \ldots, n$, by

$$\widehat{\zeta}_i = y_i - \widehat{m}(t_i);$$

Step 3. Obtain the QMLE $(\widehat{\Phi}_1, \ldots, \widehat{\Phi}_p, \widehat{\theta}_1, \ldots, \widehat{\theta}_q, \widehat{\omega}, \widehat{\alpha}_1, \ldots, \widehat{\alpha}_{p'}, \widehat{\beta}_1, \ldots, \widehat{\beta}_{q'})$ for the model (2);
Step 4. Calculate the residuals $\widehat{\varepsilon}_i$ :

$$\widehat{\varepsilon}_i = \widehat{\zeta}_i - \sum_{l=1}^{p} \widehat{\Phi}_l \widehat{\zeta}_{i-l} - \sum_{j=1}^{q} \widehat{\theta}_j \widehat{\varepsilon}_{i-j}, \quad i = \max\{p, q\} + 1, \ldots, n;$$

Step 5. Compute the estimated heteroscedasticity $\widehat{\sigma}_t^2$ from (4):

$$\widehat{\sigma}_t^2 = \widehat{\omega} + \sum_{i=1}^{p'} \widehat{\alpha}_i \widehat{\varepsilon}_{t-i}^2 + \sum_{j=1}^{q'} \widehat{\beta}_j \widehat{\sigma}_{t-j\cdot}^2, \quad i = \max\{p, q, p', q'\} + 1, \ldots, n.$$

Step 6. Compute, for $i = \max\{p, q, p', q'\} + 1, \ldots, n$,

$$\widehat{\eta}_i = \frac{\widehat{\varepsilon}_i}{\widehat{\sigma}_i},$$

and its standardized version

$$\tilde{\eta}_i = \frac{\widehat{\eta}_i - \widehat{\mu}}{\widehat{h}}$$

where

$$\widehat{\mu} = \frac{1}{n - \max\{p, q, p', q'\}} \sum_{i=1}^{n} \widehat{\eta}_i$$

and

$$\widehat{h}^2 = \frac{1}{n - \max\{p, q, p', q'\} - 1} \sum_{i=1}^{n} (\widehat{\eta}_i - \widehat{\mu})^2;$$

Step 7. Obtain the empirical distribution function $\mathcal{F}_T(x)$ based on $\tilde{\eta}_t$ defined by

$$\mathcal{F}_T(x) := \frac{1}{T} \sum_{i=1}^{n} \mathbf{1}(\tilde{\eta}_i \leq x);$$

Step 8. Generate the bootstrap process $\varepsilon_i^*$ by computing

$$\varepsilon_i^* = \widehat{\sigma}_i \, \eta_i^*, \qquad \eta_i^* \overset{iid}{\sim} \mathcal{F}_T(x),$$

Step 9. Generate the bootstrap process $\zeta_i^*$ by computing

$$\zeta_i^* = \sum_{l=1}^{p} \widehat{\Phi}_l \widehat{\zeta}_{i-l} + \sum_{j=1}^{q} \widehat{\theta}_j \widehat{\varepsilon}_{i-j} + \varepsilon_i^*;$$

Step 10. Generate the bootstrap process $Y_i^*$ as

$$Y_i^* = \widehat{m}(t_i) + \zeta_i^*. \tag{5}$$

We point out that, under mild conditions on the existence of the moments of $z_t$ in (3), the proposed approach leads to a consistent estimator $Y_t^*$ of the process $Y_t$ (see [12]).

The generated bootstrap process can be used for statistical testing and for constructing confidence intervals. Here we employ it for the estimation of the exceedance probability. In particular, let

$$EP_*(t_i) := P_*(Y_i^* > c).$$

be the bootstrap analogous of $EP_{t_i}$, where $P_*$ is the probability distribution induced by the resampling scheme.

To estimate $EP(t)$ a Monte Carlo approach is implemented. $B$ replicates of $Y_i^*$, namely $Y_{1i}^*$, $Y_{2i}^*$, ..., $Y_{Bi}^*$, $i = 1, \ldots, n$ can be generated and $EP(t_i)$ is estimated by

$$\widehat{EP}(t_i) = \frac{1}{B} \sum_{j=1}^{B} I(Y_{ji}^* > c).$$

## 4 Application to $PM_{10}$ time series from Torino area

In order to evaluate the performance of the proposed approach for estimating the exceedance probability, a real data set has been considered. It consists of $PM_{10}$ concentrations (in $\mu g/m^3$) measured from the five monitoring stations from 1 January 2015 to 19 October 2016 in Torino. The data set is available at: http://www.arpa.piemonte.gov.it. Figure 1 shows the distribution of the $PM_{10}$ concentration for each of the considered stations: Consolata, Grassi, Lingotto, Rebaudengo, Rubino. We can see that all the stations present median concentration levels below the threshold level of 50 $\mu g/m^3$. Moreover, all the distributions are positively skewed, with many values exceeding the admitted concentration (Table 1).

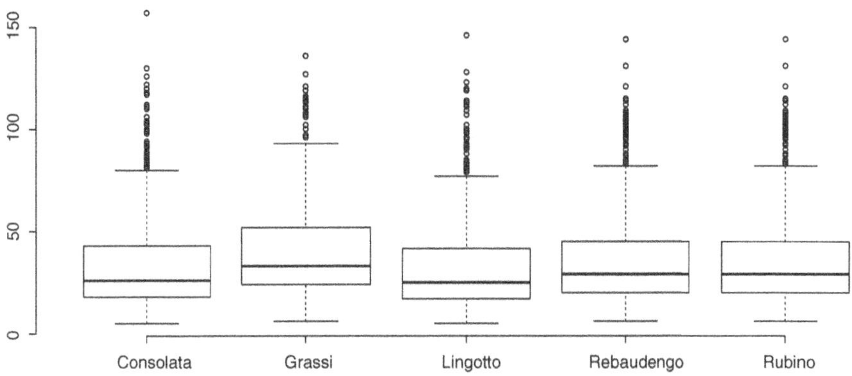


**Fig. 1.** $PM_{10}$ concentration distribution by station

**Table 1.** Descriptive statistics on $PM_{10}$ time series data

| Station | Min | $Q_1$ | Median | Mean | $Q_3$ | Max | NA's |
|---|---|---|---|---|---|---|---|
| Consolata | 5.00 | 18.00 | 26.00 | 34.81 | 43.00 | 157.00 | 32 |
| Grassi | 6.00 | 24.00 | 33.00 | 41.28 | 52.00 | 136.00 | 224 |
| Lingotto | 5.00 | 17.00 | 25.00 | 33.47 | 41.50 | 146.00 | 59 |
| Rebaudengo | 6.00 | 20.00 | 29.00 | 36.84 | 45.00 | 144.00 | 53 |
| Rubino | 5.00 | 15.00 | 24.00 | 31.45 | 38.00 | 142.00 | 25 |

This data set has been used in [2], where, after a Box–Cox transformation in order to stabilize the unconditional variance and to achieve approximate normality, the model (1)–(4) has been identified and estimated. The trend-cycle component estimator has been obtained by using the two-step procedure outlined in the previous section. In both the steps, local polynomial estimators of degree one have been used, with Epanechnikov kernel and both the bandwidth chosen by means of a least-squares cross-validation using the method suggested in [9]. The results are reported in Fig. 2. On the detrended series, an ARMA–GARCH model has been identified and estimated. The orders $p$ and $q$ have been determined by looking at the minimum value of the Bayesian information criterion (BIC), which among the numerous alternative selection criteria proposed in the literature, more heavily penalizes larger model orders. The orders $p'$ and $q'$ of the GARCH component have been fixed equal to 1, since models with higher orders generally do not seem to have better performance, particularly in a forecasting context (see [7]). The estimation results of the ARMA–GARCH model on the detrended time series are reported in Table 2.

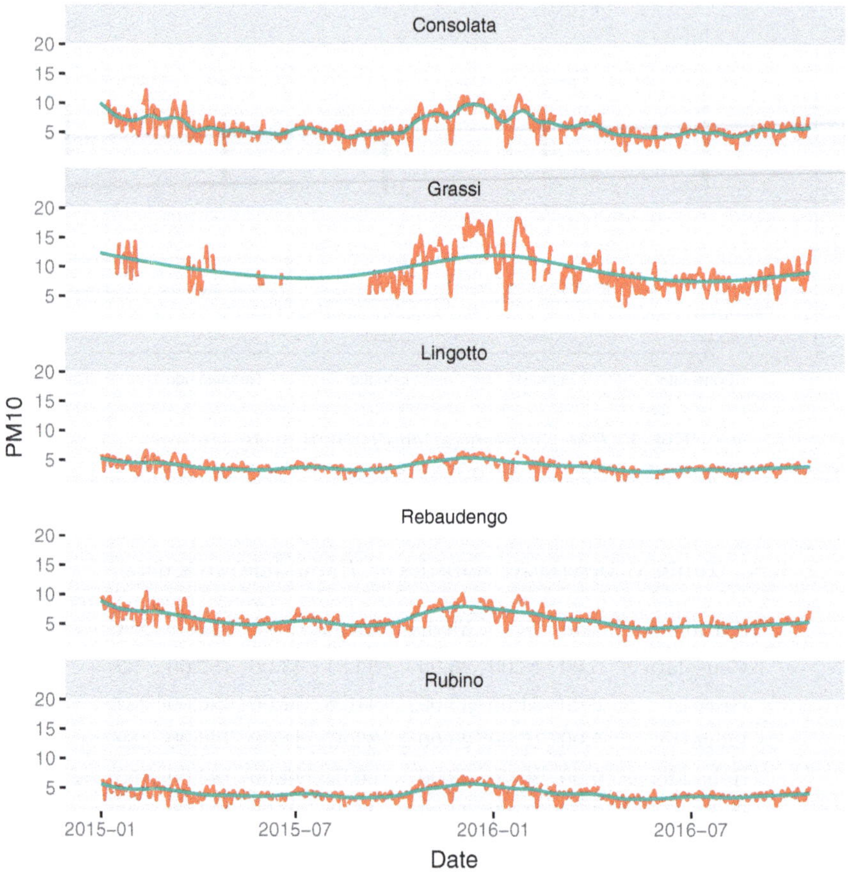


**Fig. 2.** Local polynomial estimates of the trend-cycle component $\widehat{m}(t)$ for the transformed $PM_{10}$ concentration in the 5 considered stations

**Table 2.** Estimated coefficients of the ARMA–GARCH model (2)–(4)

| Coefficient | Station | | | | |
|---|---|---|---|---|---|
| | Consolata | Grassi | Lingotto | Rebaudengo | Rubino |
| $\Phi_1$ | 0.9941*** | 0.9670*** | 0.9798*** | 0.5404*** | 0.6420*** |
| $\Phi_2$ | | | | 0.3531*** | 0.2742*** |
| $\theta_1$ | −0.8175*** | | −0.5787*** | | |
| $\omega$ | 0.3978*** | 3.8186*** | 0.5457*** | 0.5562*** | 1.3995*** |
| $\alpha_1$ | 0.1268* | 0.1114*** | 0.0732*** | 0.0840*** | 0.2401*** |
| $\beta_1$ | 0.8723*** | 0.8393*** | 0.9236*** | 0.9150*** | 0.7589*** |

Signif. codes: '***' 0.001 '**' 0.01 '*' 0.05 '.' 0.1

As suggested in Sect. 3, the exceedance probability for the considered stations can be estimated by:

$$\widehat{EP}^{(s)}(t_i) = \frac{1}{B}\sum_{j=1}^{B} I(y_{ji}^{*(s)} > c^{(s)}).$$

where $c^{(s)} = \frac{50^{\nu_s}-1}{\nu_s}$ being $\nu_s$ the power of the Box–Cox transformation.

In the bootstrap procedure, we fix $B = 999$.

The observed threshold exceedances, set as 0 if $y_i^{(s)}$ is under the 50 level, 1 otherwise, and their estimates are shown in Fig. 3. It is evident that for all the stations the bootstrap procedure is able to appropriately estimate the exceeding probability. For almost all the stations, the shape of the estimated exceedance probability is quite similar; only for Consolata station, more peaks are present in the central time interval due to the many sequences of $0 - 1$ present in the observed probability.

The estimates $\widehat{EP}^{(s)}(t_i)$ can be used as a classifier of the values of $PM_{10}$ exceeding the admitted limit of $50\,\mu g/m^3$. The idea is to obtain a binary sequence, that is equal to 1 if the value of $PM_{10}$ exceeds $50\,\mu g/m^3$ and 0 otherwise.

In order to evaluate the goodness of $\widehat{EP}^{(s)}(t_i)$ as a classifier of values of $PM_{10}$ under the limit, we use the approach based on the receiver operating characteristic (ROC) curves, a common way to visualize the performance of a binary classifier. In particular, to quantify the accuracy of the classifier we use the area under the ROC curve (AUC), usually employed for rating diagnostic tests [5] and defined as the area under the ROC curve. It represents the probability that a classifier will rank a randomly chosen positive instance higher than a randomly chosen negative instance. In Table 3, the AUC of $\widehat{EP}(\cdot)$ as a classifier of the exceedance of the legal limit is shown for each station $s$ along with the AUC obtained using only the trend-cycle component to fit the data ($AUC^{(m)}$). The aim is to verify if the introduction of the ARMA–GARCH component which

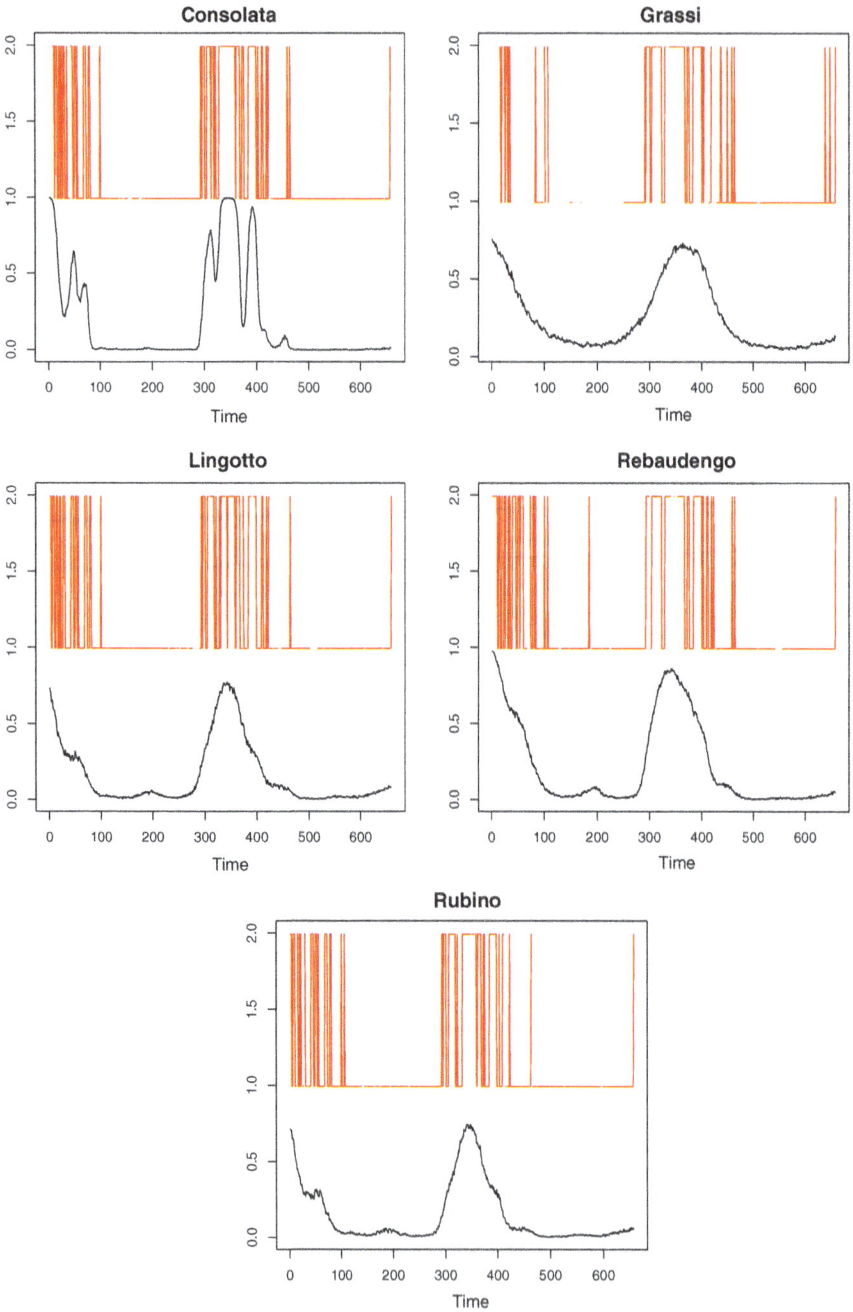


**Fig. 3.** Observed threshold exceedance (0 and 1) (on the top) and estimated exceedance probability via bootstrap (on the bottom)

**Table 3.** Values of AUC by using $EP^*$ (second column) and by using only the trend-cycle component $m(t)$ (third column)

| Station | $AUC$ | $AUC^{(m)}$ |
|---|---|---|
| Consolata | 0.96 | 0.80 |
| Grassi | 0.92 | 0.78 |
| Lingotto | 0.93 | 0.70 |
| Rebaudengo | 0.94 | 0.84 |
| Rubino | 0.93 | 0.66 |

takes into account the non-deterministic dependence and the clustering volatility in the detrended series leads to a significant improvement in the AUC.

It is evident that, for all the stations, $AUC$ is greater than 0.9, confirming that $\widehat{EP}$ is an excellent classifier for the exceedance of the legal limit. It is also clear that the introduction of the ARMA–GARCH component in the models leads to a significant improvement in the AUC.

**Concluding remarks**

The paper provides a bootstrap scheme to obtain the sample distribution of the air pollutant $PM_{10}$. It is based on the residuals of a semiparametric model which incorporates all the stylized facts generally observed on the series, such as missing data, trends and conditional heteroscedasticity. The bootstrap scheme is then applied to estimate the exceedance probability of the legal limit, i.e. the probability that the $PM_{10}$ time series exceeds the legal limit 50 $\mu g/m^3$.

To evaluate the estimated exceedance probability as a classifier for values under the limit, we use the area under the receiver operating characteristic curves, a common way to visualize the performance of a binary classifier.

The results, on a real data set consisting of $PM_{10}$ concentrations (in $\mu g/m^3$) measured from the five monitoring stations in Torino area, show values of $AUC$ greater than 0.90 in all the stations, highlighting that our estimator of the exceedance probability is an excellent classifier of $PM_{10}$ concentrations.

## References


1. Albano, G., La Rocca, M., Perna, C.: On the imputation of missing values in univariate $PM_{10}$ time series. In: Moreno-Díaz, R., Pichler, F., Quesada-Arencibia, A. (eds.) Computer Aided Systems Theory - EUROCAST 2017. LNCS, vol. 10672, pp. 12–19. Springer, Cham (2008). https://doi.org/10.1007/978-3-319-74727-9_2
2. Albano, G., La Rocca, M., Perna, C.: Volatility modelling for air pollution time series. In: Moreno-Díaz, R., Pichler, F., Quesada-Arencibia, A. (eds.) Computer Aided Systems Theory - EUROCAST 2019. LNCS (submitted)
3. Barthwal, A., Acharya, D., Lohani, D.: IoT system based forecasting and modeling exceedance probability and return period of air quality using extreme value distribution. In: 2019 IEEE Sensors Applications Symposium (SAS). https://doi.org/10.1109/SAS.2019.8706035

4. Engle, R.F.: Autoregressive conditional heteroskedasticity with estimates of the variance of United Kingdom inflation. Econometrica **50**, 987–1007 (1982). https://doi.org/10.2307/1912773
5. Fawcett, T.: An introduction to ROC analysis. Pattern Recogn. Lett. **27**, 861–874 (2006). https://doi.org/10.1016/j.patrec.2005.10.010
6. Francq, C., Zakoïan, J.M.: Maximum likelihood estimation of pure GARCH and ARMA-GARCH processes. Bernoulli **10**(4), 605–637 (2004)
7. Hansen, P.R., Lunde, A.: A forecast comparison of volatility models: does anything beat a GARCH(1,1)? J. Appl. Economet. **20**(7), 873–889 (2005). https://doi.org/10.1002/jae.800
8. Hansen, P.R., Lunde, A., Nason, J.M.: The model confidence Set. Econometrica **79**(2), 453–497 (2011). https://doi.org/10.3982/ECTA5771
9. Li, Q., Racine, J.: Cross-validated local linear nonparametric regression. Statistica Sinica **14**, 485–512 (2004)
10. Oguntunde, P.E., Odetunmibi, O.A., Adejumo, A.O.: A study of probability models in monitoring environmental pollution in Nigeria. J. Prob. Stat. 2014, ID 864965 (2014). https://doi.org/10.1155/2014/864965
11. Pèrez-Gonzàlez, A., Vilar-Fernàndez, J.M., González-Manteiga, W.: Asymptotic properties of local polynomial regression with missing data and correlated errors. Ann. Institute Stat. Math. **61**, 85–109 (2009). https://doi.org/10.1007/s10463-007-0136-2
12. Shimizu, K.: Bootstrapping stationary ARMA-GARCH models. Springer, Wiesbaden (2010). https://doi.org/10.1007/978-3-8348-9778-7_5

# Gemstone Merchandise Software

Mohammed Nazik Zayan(✉) and Gayashini Shyanka Ratnayake

Informatics Institute of Technology, 57, Ramakrishna Road, Colombo-06, Colombo, Sri Lanka
I.nazikzayan64@gmail.com, I.gayashini.r@iit.ac
https://www.iit.ac.lk

**Abstract.** Today, new and existing small-scaled gem merchants are finding it difficult to lead the business in a successful way due to poor decision-making, unpredictable market conditions, inexperience, and maintenance of client base. Therefore, it has urged the need of a solution focusing on helping to maintain and improve these areas leading to the profitability and success of small-scaled merchandise. This research focuses on analyzing the difficulties faced by small-scaled gem merchants relating to purchasing gemstones, adding and manipulating records, marketing and sales and to design and develop a software solution giving importance in helping to make corrective decisions leading to the profitability of the company such as purchasing a gemstone for the right price and adequate ways of reaching out to potential clients leading to marketing and sale of colored stones.

**Keywords:** Gem merchandise software · Colored stone software · Gem dealer software · Gem and jewelry · Gemstone dealer

## 1 Introduction

Many small-scaled gem merchants have the potentiality to succeed but mainly due to the competitiveness [1], small-scaled gem merchants are suffering from "Less sales and profit." The root cause to all these issues is poor decision-making when it comes to purchase and failure of putting innovation in place to reach out to customer's needs [2, 3]. These two issues mainly need to be addressed to compete in the competitive environment [4, 5].

## 2 Below Are the Subproblems Identified

### 2.1 Difficulty in Estimating Client Demand and Inability to Understand Market Conditions

Even though a gem merchant purchases a gemstone for the correct pricing, there can be difficult situations to sell the gemstone on a faster pace, and this is mainly because of factors such as not understanding the potential pricing that could be offered to a


X.-S. Yang et al. (eds.), *Proceedings of Fifth International Congress on Information and Communication Technology*, Advances in Intelligent Systems and Computing 1183, https://doi.org/10.1007/978-981-15-5856-6_4

gemstone in the target market. For example, A dealer's estimation on a 1-carat top quality blue sapphire at Facets, Sri Lankan gem exhibition was 15,000 while another dealer estimated the same quality stone for 30,000 ($1 was 110 rupees) [1]. The ultimate reason for such pricing estimation difference is due to each dealer's selling source, and this means, it is vital to keep the selling source in mind when purchasing gemstones as the acceptable pricing is not always the same in all the markets. Besides, if a gemstone has been purchased for a higher price, the stone would remain in stock for some time [6, 7].

Increase in the production of a similar colored stone could reduce the price of the alternative gemstone stone of the same color, and this could be mainly due to more brilliance, high production, and constant supply [8–12]. Certain markets give importance to birthstones and if a particular stone is moving faster on a seasonal basis, purchasing them on other times won't help to sell on a faster and profitable pace [3]. Public confidence in the gemstone is also an acceptable factor [13].

As colored stones do not have a standard grading scheme unlike diamonds, their value is based on the valuer and expressed in numbers based on the market condition and price during valuation [1]. Therefore, keeping the selling source in mind during purchasing is primary.

If a gem merchant purchases gemstones without analyzing the selling source, it could lead to various problems. For example, if a gemstone has been purchased for a higher price compared to the pricing at the selling source, the merchant will have to face a loss instead of making profit, and vice versa [2, 14]. Selling source exchange rate fluctuations [13] and tax percentage in the target market [7] are also factors influencing the price of a gemstone. Sale of gemstones is highly connected to fashion trends; due to this, they are more volatile [13, 15].

### 2.2 Difficulty in Predicting the Acceptability of a Gemstone

At some point, gem merchants may find a particular gemstone pleasing and would increase the tendency to purchase [16] but minor defects could affect the sales part of it so when purchasing gemstones, it is highly necessary to understand that beauty and acceptance are the keys to the success of any gemstone. So, when purchasing gemstones, no matter what, it should be of the expected quality [2, 17].

### 2.3 Difficult to Obtain the Highest Price for a Gemstone

Today, most of the gem merchants are finding it difficult to sell gemstones for the highest price that can be offered in the market regardless of its rarity. The reason behind this is that the gemstones are not being reached to all potential buyers to be offered the highest of price possible [1, 18].

### 2.4 Time Delay in Responding to Client Queries

Most of the time clients get in touch requesting for the availability of required gemstones [18]. In such situations, gem merchants are finding it difficult to get back in a prompt manner with the required details due to the following a decentralized process [2, 3, 19].

### 2.5 Difficulty in Keeping Potential Clients Informed

Different customers have different requirements; due to this, it is difficult to keep all the customers informed on their requirements on a live basis [2, 3, 5, 18].

### 2.6 Unable to Perform Follow Up's on Time with Clients

With a decentralized approach, gem merchants are finding it difficult to understand payment statuses, and to get in touch with customers on customer requirements on time [2, 3, 5, 16, 19, 20].

### 2.7 Visiting Clients to Sell Gemstones

Visiting local and international clients [18] and spending time in showing the stocks when it comes to selling gemstones is a tiring and costly process as not all the clients would purchase gemstones upon visiting [7, 16, 21].

### 2.8 Difficulty in Assessing the Price of a Gemstone

Not all gem merchants are equally talented in assessing the price of a gemstone, assessment skills differ from person to person. Price of gemstones is predominantly based on its color, clarity, cut, weight and mined location. For an assessment to be very accurate, the merchants must be updated on all these factors which may be possible but not in all the situations [1, 2, 9, 11, 13, 17, 22–24]. When a gem merchant purchases a gemstone, he/she relies on past transactions and not everyone in the industry is equally talented as things vary from person to person depending on genetics characteristics and age. In addition, the ability of any individual to grade a gemstone consistently is also influenced by psychological factors, state of the color adaption of human eye, and health state. Due to this, inconsistencies inevitably happen [25]. The lack of a standard grading scheme or codified knowledge to assess gemstones prevents the achievement of high value in the market [1]. Appraising and grading gemstones have been difficult compared to diamond due to the lack of a standard grading scheme [13, 15–17].

### 2.9 Lack of Appropriate Software

Usage of unspecialized software's, MS Excel and other traditional methods are inefficient as they aren't addressing the entire problem such as data handling, reporting, and several others as addressed above [14]. Due to this, it is hard to run a small-scaled gem merchandise without a dedicated software in place [19, 21, 26].

### 2.10 Spending Too Much Time on Data Entry

GIA [27] and GRS [28] are two leading international gemological bodies where both have outlined sample reports which display the information's related to a gemstone. When a gem merchant purchases a gemstone, all the details specific to it which are in the associated certificate will be entered into the stock section of a merchant's system.

The stock section can be a book, column of an excel sheet or any other applications a merchant uses to record and retrieve data [16, 21]. This proves that time is being spent on data entry, and there can be possibilities of human errors. Therefore, it is vital to have a method to capture and enter data in order to increase the effectiveness and reduce the time spent here.

## 3 Solution

The ultimate intention is to introduce a solution that would be of assistance to gem merchants in managing stocks and customers and to help make corrective decisions leading to the profitability of the company in terms of purchasing, marketing, and sales.

## References

1. Ekanayake, S., Abeysinghe, D.: Entrepreneurial strategic innovation model for attaining premium value for the Sri Lankan gem and jewellery industry. Core.ac.uk. Available at: https://core.ac.uk/download/pdf/89468740.pdf. Last Accessed 2 Sep 2019
2. Altingoz, M., Ali, S., Sebnem Duzgun, H., Smith, N., Syvrud, P.: Color and local heritage in gemstone branding: a comparative study of Blue Zoisite (Tanzanite) and color change Diaspore (Zultanite/Csarite). sciencedirect.com. Available at: https://www.sciencedirect.com/science/article/pii/S2214790X18303356. Last Accessed 29 Aug 2019
3. Schroeder, R.: Tanzanite as conflict gem: Certifying a secure commodity chain in Tanzania. sciencedirect.com. Available at: https://www.sciencedirect.com/science/article/pii/S0016718509000268. Last Accessed 1 Sep 2019
4. Collet, L., Curtze, L., Reed, K.: Responsible Sourcing of Colored Gemstones. [online] Responsiblejewellery.com. Available at: https://www.responsiblejewellery.com/files/Responsible-Sourcing-of-Colored-Gemstones_ARS-Final-Report_-Collet-Curtze-Reed.pdf. Last Accessed 2 Sep 2019
5. Anute, N., Deshmukh, A.: Customer relationship management practices in gems & jewellery retail sector. S3.amazonaws.com. Available at: https://s3.amazonaws.com/academia.edu.documents/58208405/Customer_Relationship_Management_Practices_in_Gems_and_Jewellery_Sector.pdf?response-content-disposition%E2%80%89=%E2%80%89inline%3B%20filename%3DCustomer_Relationship_Management_Practic.pdf&X-Amz-Algorithm%E2%80%89=%E2%80%89AWS4-HMAC-SHA256&X-Amz-Credential%E2%80%89=%E2%80%89AKIAIWOWYYGZ2Y53UL3A%2F20190902%2Fus-east-1%2Fs3%2Faws4_request&X-Amz-Date%E2%80%89=%E2%80%8920190902T155228Z&X-Amz-Expires%E2%80%89=%E2%80%893600&X-Amz-SignedHeaders%E2%80%89=%E2%80%89host&X-Amz-Signature%E2%80%89=%E2%80%891128935af006d7bfacd3b514500cbb7a4058bb99bdcca9f7227996923c489f5b. Last Accessed 2 Sep 2019
6. David, F.: Consumer Guide To Colored Gemstones, 1st edn. Modern Jeweler Magazine, USA (1990)
7. Hsu, T., Lucas, A., Qiu, Z., Li, M., Yu, Q.: Exploring the chinese gem and jewelry industry. researchgate.net. Available at: https://www.researchgate.net/profile/Tao_Hsu/publication/276840942_Exploring_the_Chinese_Gem_and_Jewelry_Industry/links/5703d87f08ae44d70ee05880.pdf. Last Accessed 6 Sep 2019
8. Antoinette, M.: Colored Gemstones, 3rd edn. Gemstone Press, USA (2010)

9. Pardieu, V., Hughes, R.W.: Tsavorite—The untamed beauty. Available at: https://www.researchgate.net/publication/264973314_Tsavorite_-_The_untamed_beauty. Last Accessed 28 Aug 2019
10. Vigne, E., Martin, L.: Discovering Tsavorite, Kenya's own gemstone. Journals.co.za. Available at: https://journals.co.za/docserver/fulltext/kenya/37/1/92.pdf?expires=1567066143&id=id&accname=guest&checksum=9BE5FCF6D7752A776C495EBA9AB86B48. Last Accessed 29 Aug 2019
11. Groat, L., Giuliani, G., Marshall, D., Turner, D.: Emerald deposits and occurrences: A review. sciencedirect.com. Available at: https://www.sciencedirect.com/science/article/pii/S0169136808000218. Last Accessed 29 Aug 2019
12. Phillips, R.S., Talantsev, A.: Russian Demantoid, Czar of the Garnet Family. Gia.edu. Available at: https://www.gia.edu/doc/Russian-Demantoid-Czar-of-the-Garnet-Family.pdf. Last Accessed 2 Sep 2019
13. Shor, R., Weldon, R.: Ruby and sapphire production and distribution: A quarter century of change. Gia.edu. Available at: https://www.gia.edu/doc/Ruby-and-Sapphire-Production-and-Distribution.pdf. Last Accessed 5 Sep 2019
14. Pandya, M.: A Critical review of literature: Gems and jewellery retailing. S3.amazonaws.com. Available at: https://s3.amazonaws.com/academia.edu.documents/55858800/IJMRA-10621.pdf?response-content-disposition%E2%80%89=%E2%80%89inline%3B%20filename%3DA_Critical_Review_of_Literature_Gems_and.pdf&X-Amz-Algorithm%E2%80%89=%E2%80%89AWS4-HMAC-SHA256&X-Amz-Credential%E2%80%89=%E2%80%89AKIAIWOWYYGZ2Y53UL3A%2F20190902%2Fus-east-1%2Fs3%2Faws4_request&X-Amz-Date%E2%80%89=%E2%80%8920190902T165527Z&X-Amz-Expires%E2%80%89=%E2%80%893600&X-Amz-SignedHeaders%E2%80%89=%E2%80%89host&X-Amz-Signature%E2%80%89=%E2%80%89d2451c3f13902aca9c6f5223eccea146b97924ab77084ef4385289662e527f88. Last Accessed 2 Sep 2019
15. Anna, M.: Gems and jewelry appraising, 3rd edn. Gemstone Press, New York (2008)
16. Gamage, K.: Relational norms, opportunism & business performance: an empirical evidence of gem dealers in SriLanka. researchgate.net. Available from https://www.researchgate.net/profile/H_M_S_Priyanath/publication/333930893_RELATIONAL_NORMS_OPPORTUNISM_BUSINESS_PERFORMANCE_AN_EMPIRICAL_EVIDENCE_OF_GEM_DEALERS_IN_SRI-LANKA/links/5d0d0226458515c11cec10fc/RELATIONAL-NORMS-OPPORTUNISM-BUSINESS-PERFORMANCE-AN-EMPIRICAL-EVIDENCE-OF-GEM-DEALERS-IN-SRI-LANKA.pdf. Last Accessed 19 September 2019
17. Jahns, R.: The purchase of a gem stone. Calteches.library.caltech.edu. Available at: http://calteches.library.caltech.edu/1512/1/jahns.pdf. Last Accessed 6 Sep 2019
18. Tiangsoongnern, L.: Purchasing behavior and attitudes of organizational gemstone buyers. researchgate.net. Available at: https://www.researchgate.net/profile/Leela_Tiangsoongnern/publication/265521816_Purchasing_Behavior_and_Attitudes_of_Organizational_Gemstone_Buyers/links/55e7d3f208ae3e1218421424.pdf. Last Accessed 2 Sep 2019
19. Wong, T., Cheung, C., Lau, H.: Decision support system for a jewellery manufacturer. sciencedirect.com. Available at: https://www.sciencedirect.com/science/article/abs/pii/S092552739800142X. Last Accessed 2 Sep 2019
20. Richman, B.: An autopsy of cooperation: diamond dealers and the limits of trust-based exchange. academic.oup.com. Available from https://academic.oup.com/jla/article/9/2/247/4430792. Last Accessed 22 Sep 2019
21. Lucas, A., Sammoon, A., Jayarajah, A., Hsu, T., Padua, P.: Sri Lanka: Expedition to the Island of the Jewels. researchgate.net. Available at: https://www.researchgate.net/publication/273904238_Sri_Lanka_Expedition_to_the_Island_of_the_Jewels. Last Accessed 6 Sep 2019

22. Feneyrol, J., Giuliani, G., Ohnenstetter, D., Fallick, A., Martelat, J., Monié, P., Dubessy, J., Rollion-Bard, C., Le Goff, E., Malisa, E., Rakotondrazafy, A., Pardieu, V., Kahn, T., Ichang'i, D., Venance, E., Voarintsoa, N., Ranatsenho, M., Simonet, C., Omito, E., Nyamai, C., Saul, M.: New aspects and perspectives on tsavorite deposits. www.sciencedirect.com. Available at: https://www.sciencedirect.com/science/article/pii/S0169136813000280. Last Accessed 29 Aug 2019
23. Pardieu, V.: Hunting for "Jedi" Spinels in Mogok. researchgate.net. Available from https://www.researchgate.net/profile/Vincent_Pardieu/publication/276840705_Hunting_for_Jedi_Spinels_in_Mogok/links/585d591408aebf17d38a3295/Hunting-for-Jedi-Spinels-in-Mogok.pdf. Last Accessed 19 Sep 2019
24. Ogden, M.J.: Rethinking laboratory reports for the geographical origin of gems—ProQuest. Search.proquest.com. Available from https://search.proquest.com/openview/40a049ed21effb4bc3f0fa785e822696/1?pq-origsite=gscholar&cbl=2041200. Last Accessed 19 Sep 2019
25. Beers, D.: GB2036360A—The assessment of colour in diamonds and other gems—Google Patents. Patents.google.com. Available at: https://patents.google.com/patent/GB2036360A/en?scholar. Last Accessed 22 Aug 2019
26. Agarwal, P., Devgun, R., Bhatnagar, J.: A study on problems faced by exporters of gems and jewellery industry. [online] Pdfs.semanticscholar.org. Available at: https://pdfs.semanticscholar.org/956a/ad336281dbaa4d2c47d2f0d340fc5739fb5a.pdf. Last Accessed 2 Sep 2019
27. GRS.: Gemstone Reports| GRS GemResearch Swisslab AG. [online] Gemresearch.ch. Available at: http://gemresearch.ch/gemstone-reports/. Last Accessed 6 Sep 2019
28. GIA.: Samples of GIA Identification Report and GIA Identification & Origin Report - GIA 4Cs. [online] GIA 4Cs. Available at: https://4cs.gia.edu/en-us/colored-gemstone-pearl-reports/sample_gemological_1920x1080-01/. Last Accessed 6 Sep 2019

# Factors Causing Stunting Among Under-Five Children in Bangladesh

Dm. Mehedi Hasan Abid, Aminul Haque, and Md. Kamrul Hossain(✉)

Daffodil International University, Dhaka, Bangladesh
{mehedi15-226,3kamrul.ged}@diu.edu.bd,
aminul.cse@daffodilvarsity.edu.bd

**Abstract.** Malnutrition is one of the major problems in developing countries including Bangladesh. Stunting is a chronic malnutrition, which indicates low height for age and interrupt the growth. The purpose of this research is to find out the factors associated with the malnutrition status and test the accuracy of the algorithms used to identify the factors. Data from Bangladesh Demographic Health Survey (BDHS), 2014, is used. Factors like demographic, socioeconomic, and environmental have differential influence on stunting. Based on analysis, about 36% of under-five children were suffering from stunting. Decision tree algorithm was applied to find the associated factors with stunting. It is found that mothers' education, birth order number, and economic status were associated with stunting. Support vector machine (SVM) and artificial neural network (ANN) are also applied with the stunting dataset to test the accuracy. The accuracy of decision tree is 74%, SVM is 76%, and ANN is 73%.

**Keywords:** Stunting · Z-score · Chi-Square · DT · ANN · SVM

## 1 Introduction

Nutrition is the consumption of food, perceived in relationship with the body's dietary demands. [1]. Stunting is a significant common health concern that grows the global fitness and decrease unanticipated fatality and malady throughout childhood [2, 3]. More than a couple of million children 0–59 months die every year because of malnutrition surrounding the environment [4, 5]. Bangladesh has improved significantly in several indicators of human development. Progress has also been made in health and nutrition. However, the nutrition status of children and women is still not good. The percentage of stunting in Bangladesh is alarming. A recent survey says that to overcome stunting of children, Bangladesh has to travel a lot more [6]. The concern is that 15% of the child's birth weight is less than normal and about one-third (1/3) of child deaths due to fatal nutrition. Malnutrition is one kind of nutritional deficiency or anemia. When there is malnutrition or anemia in a mother, it has a terrible impact on her child. Alongside,


X.-S. Yang et al. (eds.), *Proceedings of Fifth International Congress on Information and Communication Technology*, Advances in Intelligent Systems and Computing 1183, https://doi.org/10.1007/978-981-15-5856-6_5

the main cause of severe malnutrition in the country is poverty and food poisoning. Bangladesh has over population compared to its area; a set of the population live below the poverty line. Due to the increase in food prices, this huge number of people cannot afford the necessary nutritious food. Malnutrition is the biggest problem in our country and other developing countries, and we need to solve this problem from the root. This research aims to study the demographic health, environmental, and socioeconomic factors of stunting among children's (0–59 months) using machine learning techniques. Some of the associated factors of stunting are newly determined which was not determined before. This research outcome would be valuable for the healthcare providers. They can execute appropriate inventions to improve the increase of children in the inner area of Bangladesh and other areas in the country. Governmental and non-governmental companies need to come forward to establish policies and initial steps for the benefit of the people.

Section 2 presents some related work. Section 3 explains proposed methodology. Section 4 talks about data source and data preparation how we have prepared the dataset. Section 5 is about implementations, results, and accuracy of classification. Section 6 gives conclusion and future work.

## 2 Literature Review

As per the statement of World Health Organization, 48% of children in Bangladesh suffer from malnutrition [7]. The number of stunting children under five years increased over 10 years span in Africa [8, 9]. The pervasiveness of malnutrition has three regular pointers: stunting, underweight, and wasting followed by the WHO direction and amputate periods [10]. In this paper, we have found the associated factors of stunting. In Bangladesh, physical heights are less than normal in the age ratio. It is obvious to say that these children will have negative effects of stunting until they are five years old. The main cause of severe stunting in the country is poverty and food poisoning [11, 12]. This research utilizes public data of the BDHS'2014 to find out the malnutrition (stunting) factors [13]. As determined, the major contributing factors of stunting are the family wealth index, birth order number, type of cooking fuel, type of cooking fuel, relationship with the head of the house hold, main material of the roof, drinking source of the water, the place of residence, and division. The breastfeed needs much higher rate in Bengali children than it is now, GM et al. [14]. The nutritional status of children (0–59 months) is very sensorial data of countries in several sectors like health and economic. To measure the child nutritional status, there are various methods like the Z-score method. Protein energy malnutrition (PEM) is more vital health dilemma toward Bangladesh, and it concerns the natural portion and coherent improvement of babies [15–17]. Stunting can be reduced of any nation by raising mother's education, mothers' nutrition status, and health service, Alom and Islam [18].

## 3 Proposed Methodology

In this study, the relevant information is collected from the BDHS'2014. The data collected from a survey carried out both in rural and urban areas of the nation as well as for the country as a whole. Among 17,989, separated families, interviews perfectly done 98% of all the utilized households [19, 20]. Experiment on malnutrition depends on three types underweight (weight-for-age), stunted (height-for-age), wasted (weight-for-height); however, our focus is on stunting because 36.4% children are stunted. The following Fig. 1 shows our research methodology.

In the first step, we collect data from the BDHS server, and then we, integrated the data as the data sets were in different folders. Then, we clean the data sets through SPSS (version 23.0) and separated the data sets into three categories stunting, wasting, and underweight. After that, we find the Z-score for the individual data sets. Then, we compare the Z-score ($Z > -2SD$?) to check children stunting status. Then again, we make training data sets (70%) and test data sets (30%) to apply several algorithms like decision tree, SVM, and ANN. A child is counted stunted if (s) he is higher than two standard deviations or less than ($-2$ SD) concerning the WHO reference population. In this research, the nutritional situation of children (0–59) months toward the confirmation of Z-scores mathematically expressed as:

Z-score = (specific value height or weight–Median rate of indicating population)/the standard deviation estimation of the indicating population. In the following, we briefly introduce three well-designed and popular methods used for malnutrition (stunting) prediction.

## 3.1 Artificial Neural Network

Neural network has three layers: input, outputn and hidden layer. In most casesn hidden layer consisting of units converts the input into something that output layer can use. To address complex problems, we mostly use artificial neural networks (ANNs) such as the malnutrition (stunting) prediction problem. A linear function is used in ANN to activate the rectifier activation function and sigmoid activation function and applied in hidden layers to classify malnourished child from trained data. The range of linear function is 0 and 1. The result will give the probability of being malnourished as 1 and nourished as 0.70% train data and 30% test data fit into a model which runs 1–100 epochs. Here, one forward and one backward propagation is known as epochs.

## 3.2 Support Vector Machine

For classification and regression analysis, SVM recognizes patterns and analyzes the data. To gain better performance in SVM, kernel functions have been applied [20]. Here, the train test ratio of data is 70% and 30%.

## 3.3 Decision Trees

The way of decision tree to solve a problem is tree representation. J48 version is applied in this paper for building the decision tree. Entropy (E) and information gain (I) is used to construct the tree. Percentage of train and test data is 70% and 30%. The accuracy of algorithm is 74% at depth 10.

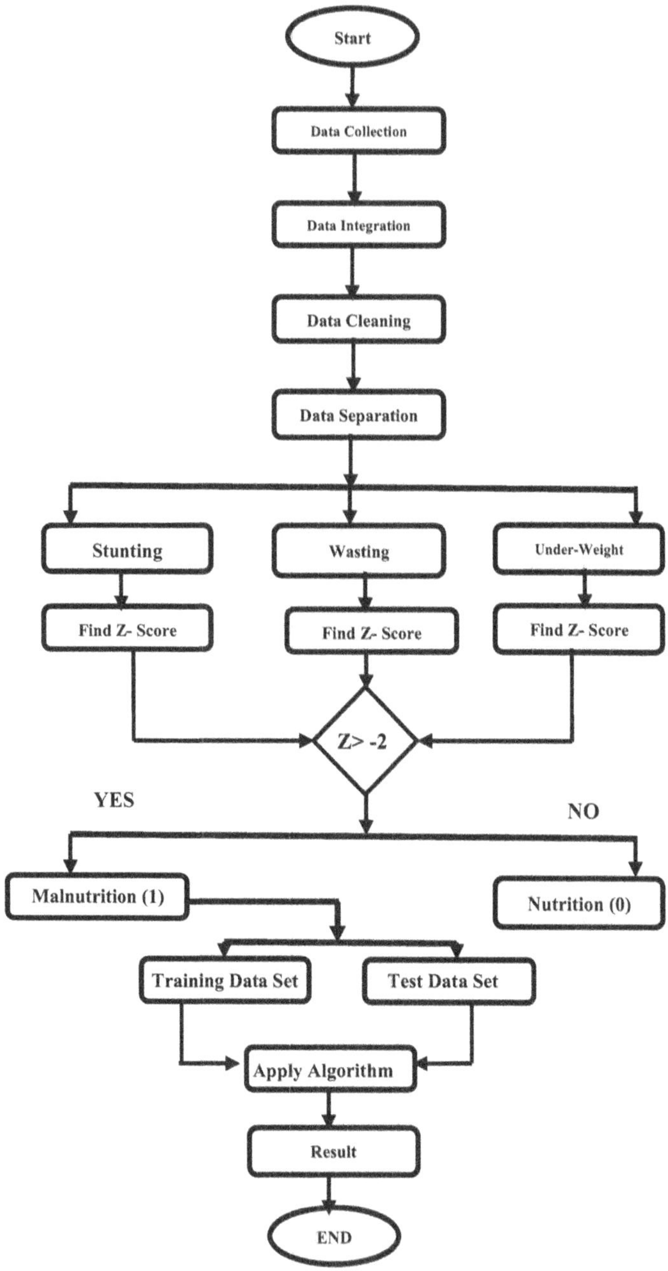


**Fig. 1.** Flowchart of working process

### 3.4 Accuracy Rule for DT, SVM, ANN

A confusion matrix is used to explain the performance of a classification model. In a confusion matrix, two-class cases: one is "0" and other "1," there are four possible outcomes of prediction, which are TN = true negatives, FN = false negatives, FP = false positives, and TR = true positives. Several performance tests such as accuracy and sensitivity are generally estimated to evaluate the classifier, as defined by,

$$\text{Accuracy} = \text{TP} + \text{TN}/(\text{TP} + \text{TN} + \text{FP} + \text{FN}) \tag{1}$$

$$\text{Sensitivity} = \text{TP}/(\text{TP} + \text{FN}) \tag{2}$$

The proportion of the cases is measured by accuracy. Sensitivity measured the fraction of negative cases those are classified as negative.

## 4 Data Source and Data Preparation

Bangladesh Demographic and Health Surveys (BDHS) is a countrywide sample survey of men and women of generative age produced to provide information on fertility and childhood mortality levels. The main key variables in our datasets are height, weight, and age. There is 7256 number of data in the dataset with the division of IR, KR, BR, HR, and PR in SAV format. IR refers to individual women's data, KR refers to children's data, BR refers to births' data, HR refers to household, and PR refers to household listing data.

To apply an algorithm, the data was not prepared. At first, the data was in different division IR, KR, BR, HR, and PR. Then, we merged the data files using statistical package for the social sciences (SPSS), PC version 23.0, tool. But the huge number of data was missing in the data set, that is why missing data was handled by the filtering procedure. Later on, the data is analyzed to find out the statistics in percentage to get how many children are suffering from stunting. Alongside, the weight of data set was on to get the accurate value.

## 5 Implementations and Results

### 5.1 Experimental Results

Table 1 presents the prevalence rate of stunting among the under-five children of Bangladesh based on BDHS 2014. It is found that 36.4% under-five children are stunted.

There is significant association with stunting status and wealth index, electricity connection at home, having mobile, mothers' education level, birth order, and age of the child. The electricity connection at home and having mobile are also included in this study to indicate economic status. It is revealed from wealth index, electricity connection at home, and having mobile that less children of upper income was suffering from stunting compared to from a lower income. Children from higher educated mother are also less in number who were suffering from stunting compared to the mother who had no education or lower level education. The number of stunned were more among the child who born as fourth or more order (Table 2).

**Table 1.** Distribution of malnutrition status (stunting) of children in percentage

| Stunting status | Frequency | Percent |
|---|---|---|
| No | 4704 | 63.6 |
| Yes | 2692 | 36.4 |

### 5.2 Accuracy of Classification

Binary classification, knowledge retrieval, and precision are the part of relevant instances in pattern identification among the retrieved examples while recall (also known as sensitivity) is the fraction of relevant examples that have been retrieved over the total number of relevant instances. Both precision and recall are consequently based on an understanding and standard of significance of three algorithms in Table 3

From 2177 trainee data, 76% precision and 84% recall is no, where 68% precision 56% recall is yes for DT. 78% precision and 86% recall is no, where 69% precision 57% recall is for SVM. 73% precision and 90% recall is no, where 70% precision 43% recall is yes for ANN.

## 6 Conclusion & Future Work

This study discussed the status of stunting in Bangladesh and found 36.4% under-five children were stunted. This study gives us a picture which shows all the major factors found for stunting like mother's highest education level, child's age, birth order number, family wealth index, currently breastfeeding, and place of residence. Stunting is not only a family problem, but it is also a major obstacle in the form of statewide skilled human resources. Malnutrition has three categories: stunting, wasting, and underweight, and in this research paper, we have only discussed stunting.

In recent years, the Government of Bangladesh is more concerned for reducing the rate of stunting. In future, we would like to work on wasting and underweight the other two indicators of malnutrition. We further would like to test our model with more algorithms and to include the data from other countries.

**Table 2.** Association between sociodemographic factors and stunting status among under-five children

| Factors | Categories | Stunting status | | Chi-square |
|---|---|---|---|---|
| | | No (%) | Yes (%) | |
| Wealth index | Ultra-poor | 49.10 | 50.90 | 325.02*** |
| | Poor | 58.80 | 41.20 | |
| | Middle | 62.30 | 37.70 | |
| | Higher middle | 68.40 | 31.60 | |
| | Rich | 79.60 | 20.40 | |
| Electricity | Yes | 68.40 | 31.6 | 130.15*** |
| | No | 55.10 | 44.90 | |
| Mobile | Yes | 65.20 | 34.80 | 108.21*** |
| | No | 45.30 | 54.70 | |
| Mother s' education | No | 49.50 | 50.50 | 276.85*** |
| | Primary | 55.00 | 45.00 | |
| | Secondary | 68.70 | 31.30 | |
| | Higher | 80.30 | 19.70 | |
| | 1 | 66.90 | 33.10 | |
| Birth order | 2 | 65.00 | 35.00% | 60.02*** |
| | 3 | 61.50 | 38.50 | |
| | 4 or more | 54.90 | 45.10 | |
| Age in month | 0–6 | 87.2 | 12.8 | 324.83*** |
| | 6–12 | 77.4 | 22.6 | |
| | 12–18 | 63.4 | 36.6 | |
| | 18–24 | 55.7 | 44.3 | |
| | 24–30 | 58.3 | 41.7 | |
| | 30–36 | 56.4 | 43.6 | |
| | 36–42 | 55.3 | 44.7 | |
| | 42–48 | 55.8 | 44.2 | |
| | 48–54 | 62.9 | 37.1 | |
| | 54–60 | 58.9 | 41.1 | |

*Note* ***, **, and * represent 1%, 5%, and 10% level of significance

**Table 3.** Confusion matrix of DT, SVM, ANN

| Algorithms | Label | Precision | Recall | f1-score | Support |
|---|---|---|---|---|---|
| DT | No | 0.76 | 0.84 | 0.80 | 1358 |
| | Yes | 0.68 | 0.56 | 0.62 | 819 |
| SVM | No | 0.78 | 0.86 | 0.82 | 1402 |
| | Yes | 0.69 | 0.57 | 0.63 | 775 |
| ANN | No | 0.73 | 0.90 | 0.81 | 1380 |
| | Yes | 0.70 | 0.43 | 0.53 | 797 |

## References


1. Toppe, J., Chomo, V.: Second international conference on nutrition (ICN2): contribution of fish to human nutrition. FAO Aquacul. Newslett. **4**(53) (2015)
2. Lartey, A.: What would it take to prevent stunted growth in children in sub-Saharan Africa? Proc. Nutr. Soc. **74**, 449–453 (2015). [CrossRef] [PubMed]
3. Seb Rengma, M., Bose, K., Mondal, N.: Socio-economic and demographic correlates of stunting among adolescents of Assam, North-East India. Anthropol. Rev. **79** (2016). [CrossRef]
4. Black, R.E., Victora, C.G., Walker, S.P., Bhutta, Z.A., Christian, P., de Onis, M., Ezzati, M., Grantham-McGregor, S., Katz, J., Martorell, R., et al.: Maternal and child undernutrition and overweight in low-income and middle-income countries. Lancet **382**, 427–451 (2013). [CrossRef]
5. Kinyoki, D.K., Kandala, N.B., Manda, S.O., Krainski, E.T., Fuglstad, G.A., Moloney, G.M., Berkley, J.A., Noor, A.M.: Assessing comorbidity and correlates of wasting and stunting among children in Somalia using cross- sectional household surveys: 2007–2010. BMJ Open **6**, e009854 (2016). [CrossRef] [PubMed]
6. Ahmed, T., Mahfuz, M., Ireen, S., Ahmed, A.M., Rahman, S., Islam, M.M., Alam, N., Hossain, M.I., Rahman, S.M., Ali, M.M., Choudhury, F.P., Cravioto, A.: Nutrition of children and women in Bangladesh: trends and directions for the future. J. Health Popul. Nutr. **30**(1), 1–11 (2012). https://doi.org/10.3329/jhpn.v30i1.11268. PMID: 22524113; PMCID:PMC3312353
7. Hong, R., Banta, J.E., Betancourt, J.A.: The relationship between household wealth inequality and chronic childhood under-nutrition in Bangladesh. Int. J. Equity Health **5**(1), 15 (2006)
8. Jahangir Alom, M.D., Islam, M.A.: Nutritional Status of Under-Five Children in Bangladesh: A Multilevel Analysis
9. Muaz, S.S.A., Hasan, M.R., Shamim, S.A., Dev, A., Kamar, S.: Nutritional status of 1–5 years children of the tea workers in Sylhet division. Bangladesh J. Child Health **34**(1), 11–16 (2010)
10. Victora, C., Adair, L., Fal, C.: Maternal and child undernutrition: Consequences for adult health and human capital. Lancet **371**, 340–357 (2008). [CrossRef]
11. WHO Multicentre Growth Reference Study Group, and Mercedes de Onis.: WHO Child Growth Standards based on length/height, weight, and age. Acta Paediatrica **95**, 76–85 (2006)
12. Green, M., Hulme, D.: From correlates and characteristics to causes: thinking about poverty from a chronic poverty perspective. World Dev. **33**(6), 867–879 (2005)
13. Motarjemi, Y., Käferstein, F., Moy, G., Quevedo, F.: Contaminated weaning food: a major risk factor for diarrhea and associated malnutrition. Bull. World Health Organ. **71**(1), 79 (1993)

14. National Institute of Population Research, Training (Bangladesh), Mitra and Associates (Firm), & Macro International.: Bangladesh Demographic and Health Survey, 2007. NIPORT (2009)
15. Hakim, M.A., Talukder, M.J., Islam, M.S.: Nutritional status and hygiene behavior of government primary school kids in central Bangladesh. Sci. J. Public Health **3**(5), 638–642 (2015)
16. Alom, J., Islam, M. A., Quddus, M.A.: Socioeconomic factors influencing the nutritional status of under-five children of Agrarian Families in Bangladesh: a multilevel analysis. Bangladesh J. Agri. Econ. **XXXII**(1 and 2), 63–74 (2019)
17. Hakim, M.A., Rahman, A.: Health and nutritional condition of street children of Dhaka city: an empirical study in Bangladesh. Sci. J. Public Health **4**(1–1), 6–9 (2016)
18. GM, R.I., Igarashi, I., Kawabuchi, K.: Inequality and mother's age as determinants of breastfeeding continuation in Bangladesh. Tohoku J. Experi. Med. **246**(1), 15–25 (2018)
19. Ahmed, S.S., Jahan, K., Arefin, S.: Nutritional status of under six children in Bangladesh: Evidence-based from nutrition survey of Bangladesh, 1995–96. Jalalabad Med. J. **2**, 17–19 (2005)
20. Kirui, C., Hong, L., Cheruiyot, W., Kirui, H.: predicting customer churn in mobile telephony industry using probabilistic classifiers in data mining. Int. J. Comput. Sci. Iss. (IJCSI) **10**(2), 165–172 (2013)

# Human Resources Information Systems and Their Impact on Employee Performance Assessment Strategy: A Practical Study on Jordan Telecom Company in the Hashemite Kingdom of Jordan

Hisham O. Mbaidin(✉)

Faculty of MIS, Abu Dhabi University, Abu Dhabi, UAE
hisham.mbaidin@adu.ac.ae

**Abstract.** The study aimed to identify the human resources information systems and their impact on the strategy of evaluating the performance of the employees of Jordan Telecom Company. The objectives of the study achieved through a questionnaire were developed for the purpose of data collection. The (SPSS.16.1) statistical software was used to analyze data. The most prominent results were: The level of importance of human resources information systems (the efficiency of human resources information systems, integration with other MIS, responsiveness) in Jordan Telecom was moderate in terms of respondents' perceptions, as well as the impact of human resources information systems in the performance assessment strategy. Human resource systems accounted for 54% of the variation in the strategy of evaluating the performance of Jordan Telecom employees. In light of the achieved results, the study recommends, the most important of which is the development of human resources information system aimed at tracking the performance of employees to improve their performance and improvement, and recommended the development of human resources management policies, so as to contribute to the detection of the potential and potential of workers.

**Keywords:** Human resources information systems · Employee performance assessment strategy

## 1 Introduction

The evaluation process is one of the important strategies and methods of performance control in accordance with predefined strategies and objectives. Its importance also lies on the constant tracking of performance stages by detecting abuses or errors that may occur. Therefore, human resource management needs to provide information of the employees performance continuously, and the criteria for the performance appraisal process. Therefore, the importance of the availability and use of specialized human resource information systems, is directly provided by providing all information related


X.-S. Yang et al. (eds.), *Proceedings of Fifth International Congress on Information and Communication Technology*, Advances in Intelligent Systems and Computing 1183, https://doi.org/10.1007/978-981-15-5856-6_6

to employees. At the beginning of the performance evaluation process, HRMS provides data about the criteria used for evaluation, which will address them to ensure that they meet the objectives set, and provides data on each factor level, to determine the required performance [13].

From this presentation, we note the extent to which the employee has benefited from the information provided by the human resources information system (HRIS), by enabling him to monitor and to assess his performance, since there are specific goals set for each employee based on his level and skills that have been identified and collected in advance, so the individual will monitor his performance and evaluate it based on the goals set for him and measure the percentage of his performance that has been achieved in the evaluation system for performance. The human resources information system examines the results presented by superiors on the employee to verify compliance with established standards and then compare these results agreed targets to determine the extent to achieve and then report them. [13], which is known as the post-performance evaluation.

## 2 The Study Problem

This study was conducted to answer the question about the impact of human resources information systems on the strategy of assessing the performance of human resources in Jordan Telecom Company and its role in the success and development of the company.

### 2.1 Study Hypotheses

In order to answer the study questions, the following hypothesis was developed:

Main Hypothesis (H01): "There is no statistically important significant impact of human resources information systems (efficiency of human resources information systems, integration with other MIS, response) in evaluating the performance of employees at the level of significance ($\alpha \leq 0.05$)."

### 2.2 The Importance of the Study

The importance of this study traces the importance of human resources in the success and development of companies and the importance of the reports provided by the HR information system in measuring the performance of employees in Jordan Telecom Company and identifying the weaknesses that companies may suffer to be processed and disposed of, and this helps companies to make appropriate and valid decisions in the evaluation process.

### 2.3 Objectives of the Study

This study aims to determine the impact of human resources information systems (HRIS) strategy in evaluating the performance of human resources and determining their role in the company's success by contributing the improvement of the employee's performance and developing them, which will positively affect the achievement of the company's main objectives.

## 3 Theoretical Framework

Measuring and evaluating performance is one of the important processes that human resources management exercises. Through measurement and evaluation, the company is able to judge the accuracy of the policies and programs it adopts, whether policies of recruitment, selection, appointment or training policy programs, development, and follow-up of its human resources. The process of measurement and evaluation is a means by which the working individual recognizes the strengths and weaknesses of his performance, especially when the results of the performance evaluation by the company, through which the individual can develop the strengths and address the weaknesses [17]. Performance evaluation is the systematic description of the strengths and weaknesses associated with the job, both individually and collectively, serving two main purposes in the companies: the development of the performance of employees in the job, in addition to providing managers and employees with the information necessary for decision-making, and the process of performance evaluation refers to that continuous function. The basic function of human resources management seeks to know the strengths and weaknesses of the overall or individual performance during a given period and to judge the performance to show progress in the work, in order to provide the substantive basis for decision-making process. [4].

One of the most important definitions addressed in the evaluation of employee performance is the evaluation of each employee of the company on the basis of the work completed within a period of time and his conduct with those who work with him [18], also known as objective measurement of employees competency and their contribution to achievement and to judge their behavior and behavior at work in a given period of time, according to specific performance criteria, and defined them [7] (Fig. 1).

The process of evaluation goes in several complex processes involving many of the forces and factors so the performance evaluators of the heads and supervisors and officials in human resources departments should take into consideration the steps mentioned by [12]:

According to Al-Mobaideen et al. [3], the study illustrates the main factors that have a direct bearing on the adoption of human resource information systems in Aqaba economic zone in Jordan. To achieve this study, four main factors were tested:

The first is TAM, the second is the information technology infrastructure, the third is the senior management support, and the fourth is the experience of individuals with computers. The study model was implemented through the use of a questionnaire that was applied to by 45 users comprising the study sample. The results of the analysis based on SPSS showed that the infrastructure has an impact. The results suggest that there are no statistically significant differences depending on the demographic variables. Based on the searcher's findings, the researchers suggested a set of recommendations for the best use of the ASESA system. A sub-model 1 has been developed based on the above to serve the original study model as the following model (Fig. 2):

According to Obeidat [10], purpose of this study is to discuss the linkage of human resources information systems and human resources management functions. The theoretical framework of the study was based on human resource management and previous research. The role of human resources contribution was measured by compliance and the level of application of human resources functions. A questionnaire was sent to five

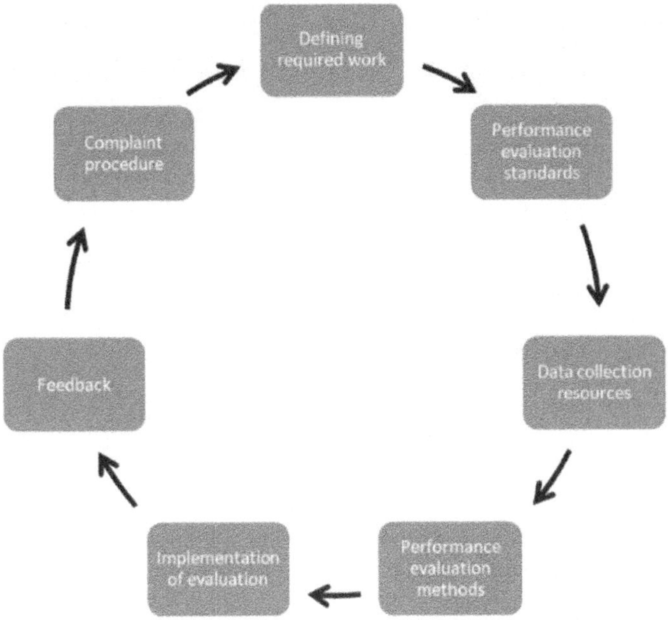


**Fig. 1.** Ways of evaluating performance. Al-Mobaideen study [3]

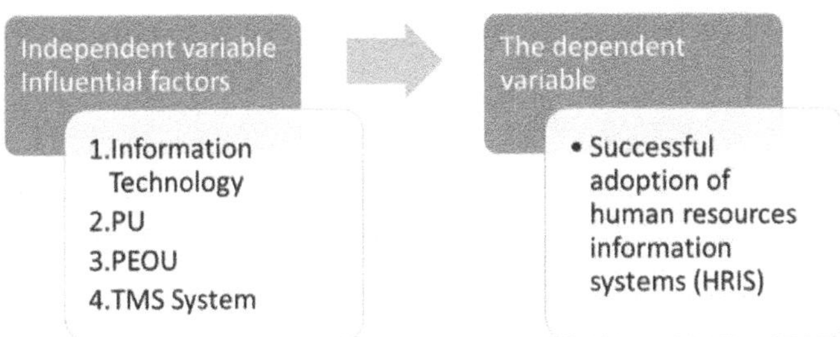


**Fig. 2.** Sub-model 1. Beadles & Others [6]

different Jordanian banks with a response rate of 61.5%. The target groups of the questionnaire were human resources departments, which included human resources managers and professionals in the Jordanian banking sector.

The functions of the human resources information systems that are studied in this paper regarding human resources management functions are especially strategic integration, forecasting, planning, human resource analysis, communication, and integration which have nothing to do with the functions, while performance development and knowledge and records management have been linked to human resource functions, which are the dimensions of human resources information systems.

Sub-model 2, illustrator of a sub-model 1, has been developed based on the above to serve the original study model (Fig. 3).

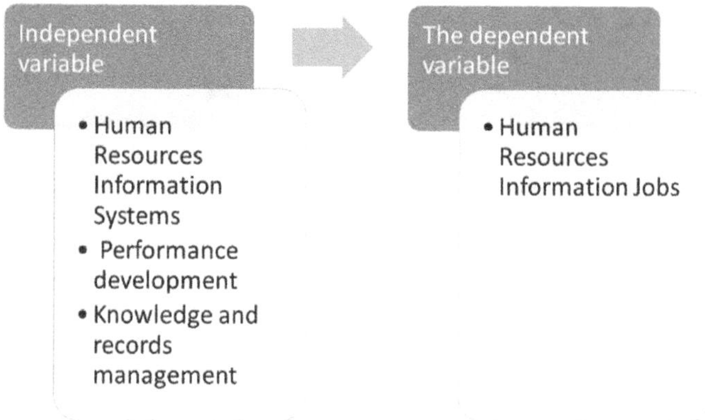


**Fig. 3.** Sub-model 2. Ball [5]

According to Stone [15], the study is to identify the factors affecting the acceptance and the effectiveness of selection of human resources systems and their impact on human resources functions. The study found that, although e-HR systems may be effective both for the organization and individuals, they may be the cause of disruption of some of the functions of organizations and individuals, such as human relations and privacy. The study recommended reducing the means of electronic inspection and supervision that limit the freedom of individuals, the use of a combination of traditional and electronic systems, and the proper and appropriate design of the system to help increase efficiency (Fig. 4).

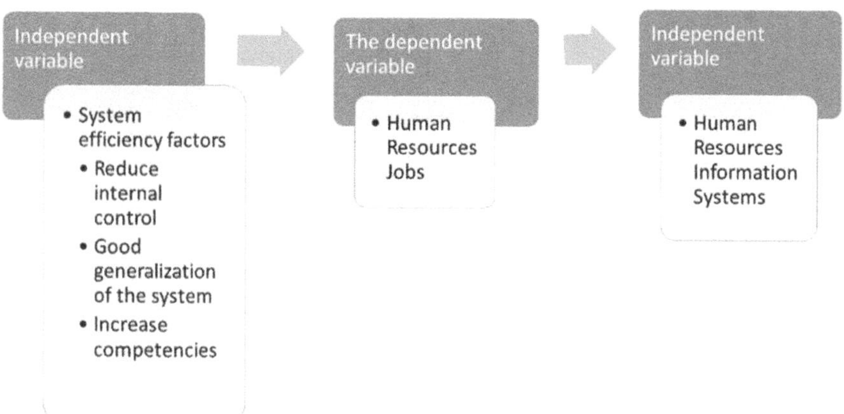


**Fig. 4.** Sub-model 3. Al-Madhoun, Abu Rahma [2]

According to Sakka [11], the study is to describe the reality of the use of the electronic human resources information system (E-HRIS) on the "Public Employees Bureau" in Gaza Strip, and the influence of this use on the performance of the chamber in four sides: (efficiency, effectiveness, quality of performance, and performance time) from the point of view of the bureau employees who use this system. The authors followed the analytical descriptive method to achieve the study objective in order to collect the research used from its initial sources via two tools. Firstly, the questionnaire was distributed to eighty-five (85) employees in the bureau, who uses the system, and secondly, individual interviews with some directors and support units in the office. The study concluded that the use of the E-HRIS led to the improvement of the performance by increasing the effectiveness, efficiency, and quality, for the functions of (recruitment, mobility, and career management processes), as well as reducing the performance of the bureau for all times jobs. The most performance aspect was improved performance time, followed by quality performance, equal efficiency, and vastly improved effectiveness of the performance of the bureau.

Finally, the study recommended the completion of computerization of the HRIS in the bureau to cover the internal and external affairs of the bureau in a wider manner than it is now, and conducting continuous assessment for the electronic system to develop the system and solve its problems and consolidating the analysis, designb and programming of information systems and government Web sites in one central location for the production of information systems and Web sites compatible with the program that can be integrated into the E-government portal later. It also recommended enacting laws to regulate electronic transactions to facilitate the transformation toward various computing transactions.

A sub-model 4 has been developed based on the above study to develop and support the main study model. As shown in the following chart (Fig. 5):

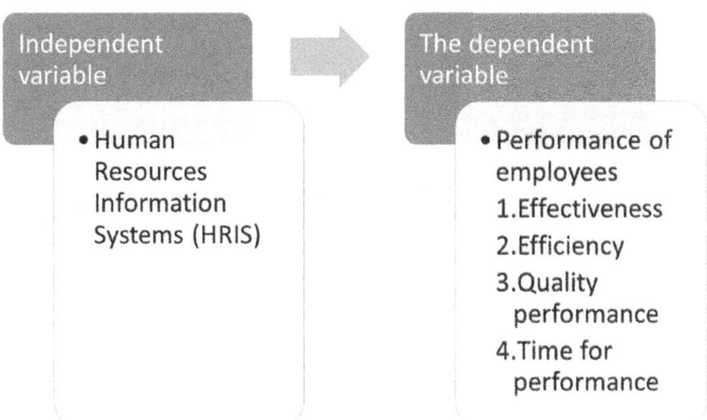


**Fig. 5.** Sub-model 4. Sourieh, Miloud [14]

According to Moayad [9], the study aims at clarifying the role of HRIS in evaluating the performance of HR. The study concluded that HRIS in the institution always needs to develop in order to fulfill the tasks that are supposed to be done and to make evaluating HR performance more effective. The two researchers mentioned a number of recommendations, including:

(A) Subjecting employees in HR department to training courses to enable them to cope with technological changes and to utilize them to perform their tasks.
(B) Establishment of a Web site for the organization to help it disseminate information about human resources such as employment announcements, advertising the results of the employment tests, and dissemination of information about the status and importance of the institution in order to attract skills.
(C) Increasing the awareness of human resources management employees about using the human resources information system and its impact on the speed and accuracy of the work carried out by them and the importance of keeping them up to date with the latest developments.

A sub-model 5 has been developed based on the above to serve the original study model. It is the following (Fig. 6):

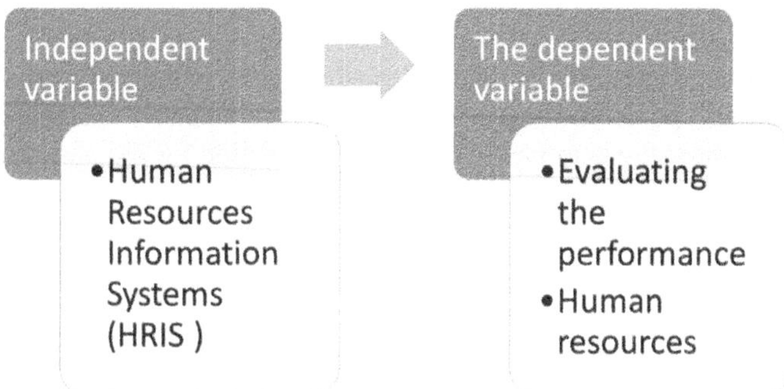


**Fig. 6.** Sub-model 5. Abu Abila [1]

According to Abu Abila [1], the objective of this study is to demonstrate the impact of HRIS on the effective performance of HRM in the Jordanian public security agency. To achieve the purpose of the study, a questionnaire was developed. A total of 450 questionnaires were distributed and analyzed. The most important results of which are:

1. The comprehension of respondents about HRIS was high, as well as their understandings of the effectiveness of HR performance.
2. There is a significant statistical impact on the application of HRIS in HRM with their dimensions in the use of the system, training, providing technical support, and supporting senior management.

The study recommended continuing work on improving and developing the management information system in the officers' affairs, personnel affairs in the Jordanian public security agency.

A sub-model 6 has been developed based on the above to serve the original study model. It is the following (Fig. 7):

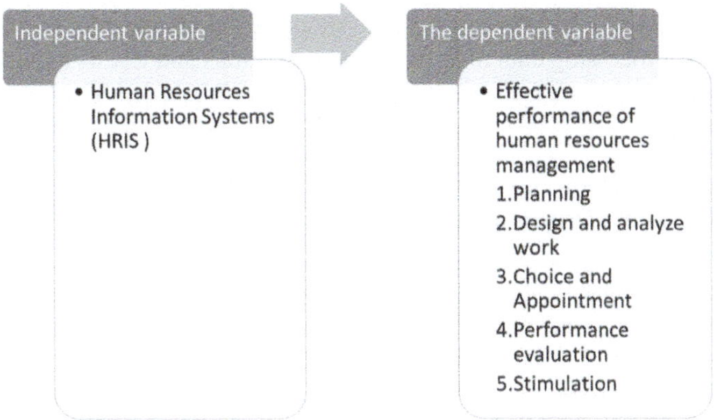


**Fig. 7.** Sub-model 6. Tartara [16]

## 4 Study Model

Based on the previous sub-models and in line with the theoretical framework discussed, in addition to the linking process that brings together human resources information systems (HRIS) with the human resource performance evaluation strategy, the study model was developed as follows (Fig. 8):

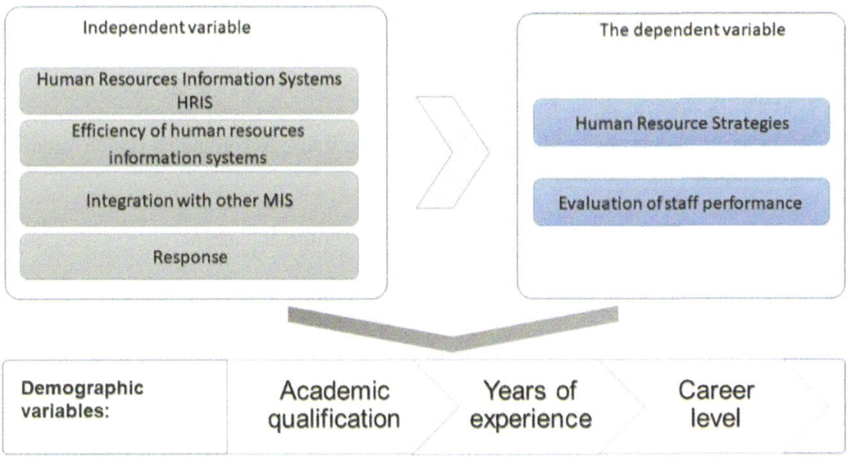


**Fig. 8.** Study model design by researcher

## 5 Data Analysis

The purpose of this study was to investigate the human resources information systems and their impact on employee performance assessment strategy of Jordan Telecom Company. Therefore, the results of this study will be presented and analyzed in this section.

The study also used descriptive analysis to describe the characteristics of the sample (SPSS.16.1) in processing data to answer study questions and test its hypotheses.

### 5.1 Study Tool

The internal consistency was calculated using the Cronbach alpha equation for the instrument in its final form as a whole and for each of its fields. Table 1 shows this.

**Table 1.** Cronbach's alpha stability coefficients for human resources information systems and human resource performance evaluation strategy

| Coefficient | Alpha |
|---|---|
| Efficiency of human resources information systems | 0.88 |
| Integration of information systems human resources | 0.85 |
| Response of human resources information systems | 0.81 |
| Human resources information systems (macro) | 0.85 |
| Performance evaluation strategy | 0.82 |

The descriptive analysis of all the variables of the study was conducted according to the responses of the sample of the study on the items in the questionnaire, where the arithmetic averages and the standard deviations were calculated. Based on this, the values of the arithmetic averages of the study were classified according to the following criterion for interpreting the data (Table 2):

**Table 2.** Descriptive analysis of all the variables of the study. According to Hatice Vatansever Bayraktar1, Kamile Özge Yilmaz1 [8]

| Arithmetic average | Level for the arithmetic mean |
|---|---|
| More than 3.5 | High |
| 3.49–2.5 | Moderate |
| 2.49–1 | Low |

Based on this criterion, if the value of the arithmetic average of the item is more than (3.5), the perception of the individuals of the sample is high. If the mean of the arithmetic mean is 2.5–3.49, (2.49) the perception of the individuals of the sample is moderate, otherwise their perception is low.

## 6 Answer Hypotheses

To answer the study question, the computation averages and the standard deviations (SD) were calculated for the perceptions of the study sample members about the level of availability of human resources information systems dimensions.

Table (3) shows that the general average of the importance of the dimensions of human resources information systems in Jordan Telecom Company was 3.48 and a standard deviation (0.77), and this means that the perception of the sample of the study about the level of importance of HR information systems in Jordan Telecom Company was moderate. By analyzing the dimensions of human resources information systems at Jordan Telecom Company, it is shown that the dimension of integrating human resources information systems ranked first with an average of (3.52) and a standard deviation (0.90), followed by efficiency of human resources information systems with an arithmetic average of (3.50), and a standard deviation (0.80), while the dimension of human resources information systems' response ranked last with an average arithmetic reached (3.41), and a standard deviation (0.75), and the following is a detailed view of the items of each dimension.

**Table 3.** Arithmetic averages and standard deviations of the perceptions of the study sample members of the level of availability of human resources information systems dimensions

| Dimensions | Arithmetic average | Standard deviation | Rank | Level of availability |
|---|---|---|---|---|
| Efficiency of human resources information systems | 3.50 | 0.80 | 2 | High |
| Integration of information systems human resources | 3.52 | 0.90 | 1 | High |
| Response of human resources information systems | 3.41 | 0.75 | 3 | Medium |
| General average | 3.48 | 0.77 | – | Medium |

## 6.1 The Efficiency of Human Resources Information Systems

The data in Table (4) indicates that the arithmetic average of the respondents' perceptions of the dimension (efficiency of human resources information systems) was high at (3.50). Item (5), which states: "The information obtained from an information system occupied the first rank with an arithmetic average of 3.62 and a standard deviation of (1.05), while item (9), which states that "there are no errors in processing and data preservation," ranked last with an average of 3.27 and a standard deviation of (1.10).

## 6.2 Integration of Human Resources Information Systems

Table (5) indicates that the arithmetic average of respondents' perceptions of the HRIS integration was high with an arithmetic average of (3.52). Item (1), which states: "The human resources information system provides integration with other activities," ranked first with an average of 3.68 and a standard deviation of 1.00, whereas item 4, which states that "HRIS is shared one database with other systems," ranked last with an average of 3.34 and a standard deviation (1.12) (Table 6).

**Table 4.** Arithmetic averages and standard deviations of the perceptions of the study sample members about the items of the dimension of efficiency of human resources information systems

| No. | Item | Arithmetic average | Standard deviation | Rank | Level of availability |
|---|---|---|---|---|---|
| 1 | The human resources information system (HRMS) includes the content to perform various tasks and functions | 3.57 | 0.99 | 2 | High |
| 2 | The human resources information system and its applications are suitable for the needs of the human Resources department | 3.55 | 0.87 | 4 | High |
| 0.3 | The human resources information system (HRIS) is easy to use | 3.54 | 0.97 | 6 | High |
| 4 | The human resources information system (HRIS) is constantly updating its data and information. | 3.55 | 1.01 | 5 | High |
| 5 | The information generated by the human resources information system (HRIS) can be thoroughly trusted for decision makers | 3.62 | 1.05 | 1 | High |

*(continued)*

Table 4. (*continued*)

| No. | Item | Arithmetic average | Standard deviation | Rank | Level of availability |
|---|---|---|---|---|---|
| 6 | The human resources information system (HRIS) is maintained regularly and continuously to ensure its continuous efficiency. | 3.43 | 0.99 | 8 | Medium |
| 7 | Equipment for the operation of the human resources information system is available accurately | 3.45 | 1.03 | 7 | Medium |
| 8 | The database used meets the needs of workers of information. | 3.43 | 1.02 | 9 | Medium |
| 9 | There are no errors in processing and saving data. | 3.27 | 1.10 | 10 | Medium |
| 10 | The human resources information system (HRIS) has helped to reduce costs resulting from traditional systems (using paper, pens, etc.) | 3.56 | 1.07 | 3 | High |
| | Total average | 3.50 | 0.80 | | High |

## 6.3 Response of Human Resources Information Systems

The data in Table (6) indicates that the mean of the respondents' perceptions of the dimension (response of HRIS) was moderate, with average (3.41). Item (1), which states: (0.93), while item (7), which states that "the system helps to determine the responsibilities of employees and their roles in the company so as not to overload the worker" got the last rank with an average arithmetic (3.21).

**Table 5.** Arithmetic averages and standard deviations of the perceptions of the study sample members on the items of the human resources information systems

| Item no. | Item | Arithmetic average | Standard deviation | Rank | Level of availability |
|---|---|---|---|---|---|
| 1 | HRMS provides integration with other activities of the company | 3.68 | 1.00 | 1 | High |
| 2 | The human resources information system assists in the preparation of reports in all directions through the exchange of information and data. | 3.52 | 0.97 | 3 | High |
| 3 | The outputs of the human resources information system (HRIS) help to provide the necessary support to different decision-making departments. | 3.56 | 1.00 | 2 | High |
| 4 | HRMS is shared in one database with other systems | 3.34 | 1.12 | 4 | Medium |
| | Total average | 3.52 | 0.90 | | High |

### 6.4 Performance Evaluation

Table (7) indicates that the mean of the respondents' perceptions of the dimension (performance evaluation) was moderate with (3.25). Item 1, which states that HRMS contributes to the improvement and development of staff performance, ranked first with an average of (3.37) and a standard deviation of 1.02. Item 4, which states that "HRMS helps to propose appropriate financial rewards for employee performance" ranked last with an arithmetic average of (3:13) and a standard deviation of (1.01).

## 7 Testing Hypotheses

The researcher conducted some tests in order to ensure the adequacy of the data for regression analysis in order to ascertain that there is no high correlation between the independent variables (multicollinearity) variance inflation factor (VIF) and tolerance test for each variable of the study variables that were used. It was ascertained also that the data follows the normal distribution by calculating the skewness coefficient considering that the data follows the natural distribution if the value of the skewness coefficient is less than (1).

**Table 6.** Arithmetic averages and standard deviations of the perceptions of the individuals of the study sample on paragraphs after the response of human resources information systems. According to Hatice Vatansever Bayraktar1, Kamile Özge Yilmaz1 [8]

| Item no. | Item | Arithmetic average | Standard deviation | Rank | Level of availability |
|---|---|---|---|---|---|
| 1 | The human resources information system (HRIS) is quick to process incoming data | 3.61 | 0.93 | 1 | High |
| 2 | The human resources information system (HRIS) helps to easily and quickly exchange information between individuals | 3.47 | 0.92 | 4 | Medium |
| 3 | HR information system provides timely information | 3.46 | 1.01 | 6 | Medium |
| 4 | The human resources information system (HRIS) provides all the outputs needed by departments and beneficiaries | 3.49 | 0.98 | 2 | Medium |
| 5 | The human resources information system (HRIS) helps the employee to view personal and functional files without any complications | 3.27 | 1.16 | 8 | Medium |
| 6 | The human resources information system responds to inquiries from other departments | 3.47 | 0.87 | 3 | Medium |

*(continued)*

**Table 6.** (*continued*)

| Item no. | Item | Arithmetic average | Standard deviation | Rank | Level of availability |
|---|---|---|---|---|---|
| 7 | The system helps identify the employees' responsibilities and roles in the company in order not to overload the worker. | 3.21 | 1.04 | 9 | Medium |
| 8 | The use of human resources information systems helps to implement the requirements of different employees at the required speed | 3.28 | 1.06 | 7 | Medium |
| 9 | Different reports resulting from human resources information systems can be prepared based on external variables. | 3.46 | 0.82 | 5 | Medium |
| | Total average | 3.41 | 0.75 | | Medium |

From Table (8), we notice that the VIF values for all variables are less than (10) and range within the (3.019–3.426). Tolerance values ranged from (0.292 to 0.331, 0.05). This is an indication of the absence of a high correlation between the independent variables (multicollinearity). It was confirmed that the data follows normal distribution by calculating the skewness coefficient, where the values were less than (1) and ranged between (0.297–0.596)

Table (9) shows the validity of the study hypotheses model, and because of the high value of ($F$) it is higher than its value at the level of ($\alpha \leq 0.05$) and the degrees of freedom (3, 96) 54%) of the total variance in the dimension (performance evaluation), and therefore, the hypotheses tested as follows:

First Hypothesis (H01): "There is no statistically significant impact at the level of ($\alpha \leq 0.05$) of HRIS (efficiency of HRIS, integration with other MIS, response) in evaluating the performance of the company's employees."

Table 10 shows the hypothesis test and multiple regression results of the analysis:

It is clear from the statistical results in Table (10), and from $t$-test values that the dimension of the response of human resources information systems had an impact on the evaluation of the performance of the employees of the company. In terms of the beta coefficients of these two dimensions as shown in the table and in terms of the increase in the values of ($t$) calculated from their tabular value at the level of significance ($\alpha \leq 0.05$), the value of ($t$) calculated was (2.939), which is a significant value at the level of significance ($\alpha \leq 0.05$). The effect of the two dimensions (human resources information systems and human resources information systems) was not shown in terms of the

**Table 7.** Arithmetic averages and standard deviations of the perceptions of the members of the study sample on the items of the dimension of evaluation of performance

| Item no. | Item | Arithmetic average | Standard deviation | Rank | Level of availability |
|---|---|---|---|---|---|
| 1 | The human resources information system (HRIS) contributes to the improvement and development of staff performance | 3.37 | 1.02 | 1 | Medium |
| 2 | HRMS contributes to determining the effectiveness of supervisors and managers in developing team members | 3.34 | 0.92 | 2 | Medium |
| 3 | HRIS reveals the capabilities of employees to be promoted to higher positions | 3.19 | 1.04 | 5 | Medium |
| 4 | HRMS helps to propose appropriate financial rewards for employee performance | 3.13 | 1.01 | 7 | Medium |
| 5 | The human resources information system (HRIS) contributes to the evaluation of human resources management programs and policies | 3.31 | 1.05 | 3 | Medium |

*(continued)*

**Table 7.** (*continued*)

| Item no. | Item | Arithmetic average | Standard deviation | Rank | Level of availability |
|---|---|---|---|---|---|
| 6 | The human resources information system (HRIS) provides assistance in establishing precise, predefined, normative performance bases. | 3.24 | 1.02 | 4 | Medium |
| 7 | The human resources information system (HRIS) helps to make the employee more responsible as a result of justice. | 3.17 | 1.20 | 6 | Medium |
| | Total average | 3.25 | 0.36 | – | Medium |

**Table 8.** Testing the amplitude coefficient of variance, permissible variation, and skewness coefficient

| Sub-dimensions | Tolerance | VIF | Skewness |
|---|---|---|---|
| Efficiency of human resources information systems | 0.293 | 3.412 | 0.304 |
| Integration of human resources information systems | 0.292 | 3.426 | 0.596 |
| Response of human resources information systems | 0.331 | 3.019 | 0.297 |

**Table 9.** Results of analysis of variance to verify the validity of the model for testing the hypotheses of the study

| Dependent variable | Coefficient of R2 determination | Degrees of freedom | Calculated F value | Level of significance F |
|---|---|---|---|---|
| Performance evaluation | 0.540 | (96.3) | 37.622 | 0.000 |

* Of statistical significance at level ($\alpha \leq 0.05$)

(Beta) coefficients of this dimension as in the table and the significance of the decrease

**Table 10.** Results of multiple regression analysis were used to test the impact of HRIS (efficiency of HRIS, integration with other MIS, response) in evaluating the performance of employees in the company

| Independent dimension | R2 selection factor | B | Standard error | Beta | Calculated t value | Level of significance t |
|---|---|---|---|---|---|---|
| Fixed regression | 0.540 | 0.313 | 0.288 | – | 1.086 | 0.280 |
| Efficiency of human resources information systems | | 0.232 | 0.157 | 0.213 | 1.475 | 0.143 |
| Integration of human resources information systems | | 0.199 | 0.134 | 0.208 | 1.492 | 0.139 |
| Response of human resources information systems | | 0.417 | 0.142 | 0.364 | 2.939 | 0.004 |

*Of statistical Significance at the level of ($\alpha \leq 0.05$)

in the value of ($t$) calculated from its tabular value at the $\alpha \leq 0.05$ level.) calculated at (1.475, 1.492), respectively. From the above, it is necessary to reject the null hypothesis and to accept the alternative hypothesis that there is a statistically significant impact at the level of ($\alpha \leq 0.05$) of human resources information systems (efficiency of human resources information systems, integration with other MIS, response) in evaluating the performance of employees in the company.

Table 11 shows the stepwise multiple regression analysis to determine the weight of each independent variable in the contribution to the arithmetical model:

Table (11) shows the order of entry of the independent variables in the regression equation, and the dimension of the response of the human resources information systems has entered the regression equation and separately interpreted 48.9% of the total variance in the evaluation of the performance of the employees of the company.

## 8 Results and Discussion

The recent study concluded a group of results that were based on the opinions of the sample members of the Jordanian Telecom Company about human resource information systems and their impact on HR strategies. In view of the responses of the sample, the following results were reached and discussed as follows:

**Table 11.** Results of stepwise multiple regression analysis to predict the performance evaluation of the company's employees through the dimensions of HRIS

| The order of entry of independent elements in the prediction equation | R2 value of the coefficient of determination | B | Calculated t value | Level of significance t |
|---|---|---|---|---|
| Invariable | | 0.416 | 1.481 | 0.142 |
| The response of human resources information systems | 0.489 | 0.504 | 3.884 | 0.000 |

Firstly, the results of the study showed that the arithmetic averages of the perceptions of the individuals of the study sample for the level of availability of the dimensions of the human resources information systems at the macro level were "medium" with an average of 3.48 and a standard deviation of 0.77. The system in any company cannot be complete even in large companies, because of the many challenges that vary in intensity and impact from one company to another. Jordan Telecom Company is one of the companies facing many difficulties and different challenges. We mention the organizational difficulties, change problems, the re-design of the organizational structure, the difficulty of building, developing. and modifying the human resources information system, the lack of specialized experts in human resource systems, the lack of acceptance of the majority of workers for technological development due to their lack of experience in these systems, and familiarity with traditional systems. We will not forget the importance of these systems in the pursuit of the development and success of these companies, which helps to collect data necessary to manage the system, maintenance, and the introduction of conversion and operation on them in order to provide the beneficiaries thereafter and provide the officials with information related to personnel working in the company in the areas of benefits, compensation, wages and insurance, training and competency development,, evaluation of performance in addition to availability of information about selection systems, assessment and analysis of activities, development of personnel capabilities, and skills and planning career courses, and showing the importance of human resources information systems in the integration with other MIS sub-systems such as marketing information system, manufacturing information system, technical operations, and conversion information systems.... And other information systems within the company to ensure the existence of an integrated macro-information system that serves decision making at the company level and helps to respond to the demands of external parties to provide the information that they need to take into account the element of confidentiality and to provide information for publication.

As for the ranking of human resources information systems axes, HRMS integration was ranked first and obtained a high arithmetic mean "3.52." This result can be explained by the fact that HRIS may provide integration with other activities of the company to assist in preparing comprehensive and integrated reports with the activities and operations of the company and to help to achieve the necessary support for different departments to make various decisions correctly.

As for the efficiency of human resources information systems, it came in the second place to obtain a "high" arithmetic average of (3.50). To explain this result, the researcher states that HRIS helps to produce accurate information that decision makers can trust to make decisions. The HR information used by JTC is comprehensive, convenient, easy to update, and continuous, as well as continuous maintenance of these systems and more importantly reducing the material costs that may result in the use of traditional systems due to the use of papers and pens and other materials.

Secondly, the results showed that the arithmetic averages of the respondents' perceptions in Jordan Telecom Company that the employee performance evaluation obtained an average of (3.25). To explain this result, the researcher attributed that the HR information system contributes to the improvement and development of the employees of Jordan Telecom, but it may suggest appropriate financial rewards for employee's performance and may not contribute to increasing the employees' sense of responsibility for their business and functions. This can be explained by the fact that staff assessment may be based on personal knowledge, friendship, and mutual interests between the subordinates and the manager. Thirdly, the results confirmed that there is an impact for resource information systems in the strategy of assessing the performance of the employees of Jordan Telecom Company. This result can be explained by the fact that the use of the HR information system in the performance evaluation of employees will contribute to improve their performance, develop, and determine the effectiveness of supervisors and managers in developing and improving their performance. This also helps to establish accurate rules and standards that are previously determined to contribute to better performance assessment.

## 9 Recommendations

Based on the findings of this research, the researcher presents the recommendations that she hopes will play a role in increasing the use of human resources information systems in human resource strategies, thus increasing the efficiency and effectiveness of the human resources department, as follows:

1. Develop the human resources information system to include strategies for assessing the performance of Jordan Telecom employees in line with their systems, policies, and objectives to track the performance of employees to develop and improve their performance.
2. Develop human resources management policies to provide important information on the level of employees' performance and help determine the factual basis from which development efforts should begin to reveal the potential and capabilities of employees.
3. Design training programs for employees of HRIS, to be trained in how to use and deal with this system to be an essential input in the process of assessing the performance of employees.

4. The researcher recommends the necessity for further studies on human resources information systems and their impact on improving business and activities of companies.

## References

1. Abu, A.: The Impact of Human Resource Information Systems on the Effectiveness of Human Resources Management Performance in the Jordanian Public Security or agency. Unpublished Master Thesis, Mutah University, Jordan (2010)
2. Al-Madhoun, Abu Rahma: The efficiency of human resources information systems in the ministries of the Palestinian National Authority in the Gaza Strip (2007)
3. Al-Mobaideen et al.: Factors influencing the successful adoption of human resource information system. In: The content of Aqaba Special Economic Zone Authority. Amman-Jordan (2013)
4. Al-mughrabi, A.: Smart Management Guide for Human Resource Development in Contemporary Organizations. Modern Library for Publishing and Distribution Mansoura, Egypt (2009)
5. Ball: The use of human resource information systems: a survey (2001)
6. Beadles & Others: The Impact of Human Resource Information Systems: An Exploratory Study in the Public Sector (2005)
7. Dajani, I.A.: The Experience of the Islamic University in Gaza in Developing the System of Performance Appraisal Management as an Input to the Development of Quality
8. Hatice Vatansever Bayraktar1 , Kamile Özge Yilmaz1, 2016
9. Moayad, A.-S.: Human Resource Management, Integrated Strategic Approach, 1st edn. Ithraa Publishing and Distribution, Amman, Jordan (2009)
10. Obeidat, B.Y.: The Relationship between Human Resource Information System (HRIS) and Human Resource Management (HRM) Functionalities. University of Jordan, Amman, Jordan (2012)
11. Sakka, A.H.H.: The Impact of the Human Resources Information System on the Performance of the Public Employees' Bureau in the Gaza Strip. Published Master Thesis, Islamic University, Gaza, Palestine (2013)
12. Shawish, Mu: Human Resources Management: Personnel Management. Dar Al Shorouk Publishing and Distribution, Ramallah (2005)
13. Soreyah, S., Tomei, M.: The Role of the Human Resources Information System in Assessing the Performance of Human Resources in the Foundation. University of Mohammed Khayder, Biskra, Algeria (2010)
14. Sourieh, M.: The Role of the Human Resources Information System in Evaluating the Performance of Human Resources in the Organization (2010)
15. Stone.: Factors affecting the effectiveness and acceptance of electronic selection systems. Hum. Res. Manage. Rev. **23** (2006)
16. Tartara: The Impact of Using Computerized Management Information Systems on Human Resources Management Functions in Public Institutions in Jordan (2006)
17. Waly, A.M.: Job Performance Assessment. Department of Planning and Follow-up, Department of Administrative Development, Development and Research Division, Jordan (2011)
18. Zewailf, M.H.: "Personnel Management" Dar Al-Safa (2003)

# Measuring Complexity of Legislation. A Systems Engineering Approach

Andres Kütt[1,2] and Laura Kask[2,3](✉)

[1] Department of Software Science, Tallinn University of Technology, Ehitajate tee 5, 19086 Tallinn, Estonia
andres.kytt@proudengineers.com
[2] Proud Engineers, Kristjani tee 5, Miiduranna 74015, Harju maakond, Estonia
[3] School of Law, University of Tartu, Näituse 20, 50409 Tartu, Estonia
laura.kask@proudengineers.com

**Abstract.** Complexity management is a well-understood concept in systems engineering with strong theoretical and practical foundations. The complexity of legal systems, however, is mainly considered in trade or tax context and remains largely qualitative in nature. Since the ability to create, develop and follow law is crucial to functioning of a society, a quantitative method for assessing the complexity of a set of laws from both its creation and consumption perspective and development of that complexity over time would be beneficial. For example, such a measure could be used to assess the sustainability of a legal system, develop "complexity budgets" for legislative texts and quantitatively measure the impact of changes. In this paper, the authors utilise a complexity measure for engineering systems in the legal context of the Republic of Estonia. A specific measure of legal complexity is developed based on ideas from systems engineering and morphological analysis. It is then applied to time series of quarterly sets of Estonian legislation from 2002 to 2019. The research shows that systems engineering approach does yield meaningful results in the legal domain and that, assuming limited cognitive capabilities, the existing trend of complexity growth is not sustainable. Policy recommendations are presented to change the trend found.

**Keywords:** Law · Complexity · Systems engineering · Estonia

## 1 Introduction

### 1.1 Background and Motivation

In his famous work, [1] showed that a human can handle $7 \pm 2$ distinct mental "chunks" at a time, that our cognitive capabilities are limited. From this fundamental limitation of ours stems, the need to assess the complexity of systems:


X.-S. Yang et al. (eds.), *Proceedings of Fifth International Congress on Information and Communication Technology*, Advances in Intelligent Systems and Computing 1183, https://doi.org/10.1007/978-981-15-5856-6_7

too complex a system stretches our processing capabilities beyond their limits. Therefore, complexity should be controlled and managed both of which assume the ability to measure complexity.

While in systems like spacecraft, oil rigs or dams, complexity is associated with a super-linearly increasing development cost [2], the impact of complexity in non-engineering systems can be less obvious. One such example is a legal system comprised of laws and regulations of a country. Here, two separate perspectives of complexity present themselves. The complexity of reading, interpreting and applying the law[1] and the complexity of developing the legal system to reflect changes in the society. Both incur non-trivial cost to either citizens or businesses or the institutions working to develop the law. For the discussion of the nature of these costs, themselves outside the scope of this article, to be effectual, an objective metric for measuring that complexity is necessary. Also, any change seeking to shift the complexity of a legal system should be based on a quantitative metric stemming from a solid theoretical foundation.

As numerous initiatives in multiple legal contexts clearly show (e.g. The European Commission's regulatory fitness and performance (REFIT) programme,[2] US H.R.946—Plain Writing Act of 2010), the fundamental problem of legal complexity described previously is clearly recognised and considered important. However, these initiatives do not appear to rely on quantitative instruments other than the number of acts.

For example, on the 19th of May 2015 the European Commission published a communication "Better regulation for better results—An EU agenda" [4] to extend the REFIT programme. The document states its goals as "simplifying the legal framework, reducing regulatory burdens across the single market and increasing regulatory predictability" (page 3) and stating one of the means of achieving that as "reducing its complexity while maintaining its level of protection". However, no specific metrics or their target values are described. A similar policy is in place in Estonia (see [5]) with the explicit title of "Plan to reduce law creation". The plan contains steps like "Assessment of the need for each legal act" and cites the need to establish a way for ministries to give a "red light" to the creation of new regulations. Again, no specific measures are described beyond the number of regulations.

That said, quantitative measures of complexity of legal systems do exist. The work of Katz et al. [3] is probably the most prominent example. That work, however, is dependent on the structure of one particular legal system, focuses on the consumer perspective of complexity and, while methodologically sound, does not utilise a previously validated metric.

It is clear, that the very definition of complexity "as that which consists of interconnected or interwoven parts; is a composite of two or more parts, is involved or intricate; or is hard to separate, analyse, or solve" [6] implies, that

[1] "With every new distinction or caveat come the costs associated with looking up, learning, and complying with relevant rules" [3].

[2] https://ec.europa.eu/info/law/law-making-process/evaluating-and-improving-existing-laws/refit-making-eu-law-simpler-and-less-costly_en.

simply counting the number of parts is not a sufficient measure of complexity. Indeed, the majority of the complexity measures listed by [7, table 2] involve either coupling or solvability in addition to the size of the system. Therefore, a useful metric of calculating complexity of a set of laws would involve either coupling or solvability. As solvability (i.e. the effort it takes to develop a system) in legal complex involves tackling fundamental philosophical questions, coupling seems a suitable candidate.

To develop such a metric, tools and approaches developed in the field of system engineering appear useful. As the concept of such prescriptive quantitative decision support methods, [8] is well-established in the field. One of the methods utilised in system engineering is the adjacency matrix describing relationships between system components. Describing a system of laws using its adjacency matrix is not a novel idea (e.g. [9]) and neither is the use of Design Structure Matrices in system architecture (e.g. [10]).

Analysis of such matrices can yield both quantitative information about the underlying system and policy recommendations for maintenance and development activities. As both would be interesting and useful in the field of law, the following fields of interest present themselves.

Firstly, the complexity measures itself. If the complexity of a legal text or system of legal texts could be objectively calculated, this would allow for benchmarking between different laws and legal systems, estimate the impact of planned changes, etc.

Secondly, changes in such a complexity measure over time allow to assess both the legal system itself and its context. If both individuals and groups have cognitive limits to the ability to handle complexity, changes in the position of a given legal system in terms of these limits appear useful:

- A relatively flat trend indicates the system to be in an equilibrium and a status quo in terms of the ability to handle the system can be assumed
- An upwards trend indicates a system that is increasingly difficult to manage as it approaches or has exceeded the abilities of its masters
- A downward trend in conjunction with the second law of thermodynamics allows to conclude that external force is applied to reduce the complexity and that the system in general is getting more manageable over time.

Thirdly, policy recommendations stemming from calculating the metric and applying the related prescriptive methods would yield policy recommendations that provide quantitative input to policies governing the given legal system.

### 1.2 Objective Statement

In this context, the authors seek to find support for the following hypothesis:

- Complexity measures used in systems engineering can be applied in the legal domain
- These measures, when applied to a temporal data series, allow to assess the sustainability of a system of legal texts in terms of complexity trends

- Applying prescriptive methods from the field of systems engineering can yield actionable policy input to the governance of a given legal system.

# 2 Material and Method

## 2.1 Complexity of Systems

The term "complexity" lacks a generally accepted definition despite being a very intuitive concept. A large number of measures have been proposed for capture our ideas about complexity [11]. This leads to the complexity metrics proposed seldom being directly computable leaving practitioners to derive specific implementations of the general ideas suitable for the particular fields. References [12, 13] are examples of such research, there are many others. Such domain-focused solutions can be hard to generalise: lines-of-code or formulae involving part mix ratios might be useful in their own context, but their applicability in a general context is questionable. The challenges in expanding the metrics beyond a given field of specialisation in turn hinder the emergence of a uniformly applicable notion of complexity and therefore also a shared set of tools and methods for its management.

In contrast to both abstract ideas and domain-specific solutions, Sinha provides an approach to complexity rooted in systems engineering, a field focused on general socio-technical systems [14]. Borrowing from the field of chemistry, a complexity index is derived that has been validated by further research (see [15]). It is based on the properties of the adjacency matrix of the system in question[3] while allowing for recursive application by the inclusion of complexity measures of subsystems. The notable feature of the measure is its focus on design, implementation and operation of system rather than their usage. While the complexity of the latter is significant in terms of user adoption and the value a system can drive, the former has a strong influence on the ability of organisations to actually deliver that value.

The index is captured in Eq. 1.

$$\begin{aligned} C(n, m, A) &= C_1 + C_2 \times C_3 \\ &= \sum_{i=1}^{n} \alpha_i + (\sum_{i=1}^{n} \sum_{j=1}^{n} \beta_{ij} A_{ij}) \gamma E(A) \end{aligned} \tag{1}$$

where $A$ is the DSM, $\alpha_i$ are the complexities of individual subsystems, $\beta_i$ are the weights of inter-subsystem relationships, $\gamma = \frac{1}{n}$ is the normalisation factor and $E(A)$ is graph energy of A as defined by Eq. 2 for eigenvalues $\sigma_i$ of A.

$$E(A) = \sum_{i=1}^{n} |\sigma_i| \tag{2}$$

This approach splits the complexity index (denoted by $C$) of a system into three aspects:

[3] More precisely, the design structure matrix (DSM) as defined in [16] is used.

- The total complexity of all subsystems ($C_1$)
- The total complexity of relationships between the subsystems ($C_2$)
- Structure of the relationships between subsystems ($C_3$).

In the context of Eq. 1, we can interpret $C_1$ and $C_2$ as the main complexity drivers with $C_3$ acting as a weighing factor either increasing or decreasing the effect of the relationships based on their structure.

Application of the formula assumes the ability to assess the complexity of subsystems, the ability to clearly define relationships between the subsystems and the ability to assess the complexity of these relationships.

## 2.2 Properties of Graph Energy

The implementations of $\alpha_i$ and $\beta_i$ capture the domain-specific aspect of the equation. $C_1$ and $C_2$ are therefore implementation-dependent while $C_3$ is generic in nature reflecting properties of the system structure. This makes $C_3$ more suitable for the derivation of specific policy recommendations: the complexity of a legal text can be estimated in a number of ways, whereas graph energy as such is relatively unambiguous.

Unfortunately, $C_3$ (or its essence, graph energy) does not lend itself easily to an analytical approach; therefore, simulations have been used extensively in its analysis. Gutman et al. [17] utilise Monte Carlo experiments to derive a relationship between the number of edges and vertices of a graph. For $N = 50$, Fig. 1a depicts the relationship between graph energy as simulated by the authors using 250 repetitions and the Erdős–Rényi G(n, M) algorithm [18] for generating random graphs. The figure shows that, for $N \geq 10$ graph energy is concave in nature with a clear maximum at $k < \frac{n(n-1)}{2}$ and that this is not true for $N < 10$ (Fig. 1b). Therefore, for large graphs a number of edges $k_i$ exist where adding edges increase $C_3$ beyond that of a monolithic system.

$k_i$ can be seen as a sort of optimum in terms of complexity as it presents the maximum number of relationships between components that still justify not building a monolithic system. It therefore presents usable policy guidance as the graph energy of a system can be calculated and compared to $k_i$. Such a policy input assumes we know the $k_i$ for a given number of vertices $n$. Analysing $\pi$-electron energies of molecules [19] derives an estimate for graph energy as a function of the number of edges and vertices (3).

$$E(A) = a\sqrt{nm}, a \approx 0.92\sqrt{2} \tag{3}$$

Solving for maximum graph energy $2n - 2$ we find that

$$2n - 2 = a\sqrt{nm} \Rightarrow 4(n-1)^2 = a^2 2nm \Rightarrow m = \frac{4(n-1)^2}{2na^2} \tag{4}$$

Equation 4 indicates a linear boundary for m. However, as McClelland operated with relatively small graphs, a more robust estimation would be needed for complex systems. Applying the same Monte Carlo approach as previously, generating random graphs with a given number of edges and computing their energies

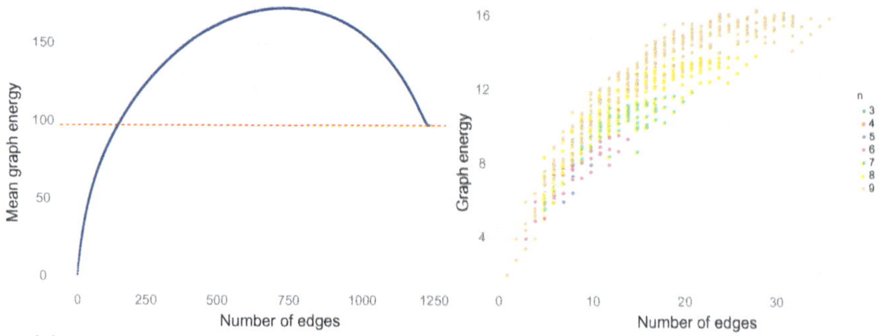


(a) Relationship between mean graph energy and number of edges in a $N = 50$ graph for 250 attempts. The red dotted line denotes the energy of a fully connected graph

(b) Relationship between graph energy and number of edges for $N \leq 10$ graph for 10 attempts

**Fig. 1.** Graph energy and number of edges for different numbers of vertices

yields the optimal number of edges for a given graph size. For $N \in [10, 500]$ and $C = 100$, this approach generates a dataset to which a linear model of $k_i \approx \lfloor 2.867n - 7.724 \rfloor$ fits with $p < 0.001$ confidence.

The coefficients do not lend themselves to an obvious interpretation but the resulting estimates are slightly larger than Eq. 3 yields confirming the need for simulations and providing a better basis for policy recommendations.

### 2.3 Method

In order to verify the hypothesis, a specific implementation of a generic complexity measure described in Sect. 2.1 was developed for both individual laws and, consisting of these, a complete set of laws for a country. That implementation was then applied to the dataset containing all parliamentary laws of the Republic of Estonia in force as of 1st of every quarter from 1st of April 2002[4] to 1st of January 2019. Quarterly resolution was chosen based on the general practice in Estonia of laws to come in force at the beginning of a quarter. A higher resolution would be unlikely to yield a statistically more significant dataset as the laws would be very unlikely to change between the points of measurement.

It is important to note Estonian legal system is much wider than this dataset covers encompassing lower level legal acts like government decrees, EU law and bilateral agreements. Also, the legal system and its complexity are influenced by the interpretation of legal acts in court practice. Analysis of that wider legal system is out of the scope of the research as a comparably long-term dataset is not available, and the results would be more difficult to interpret. Also, as Estonia is a parliamentary republic, it is natural to focus on the legal acts given by the Parliament (Riigikogu).

[4] The earliest date from which comprehensive data is available.

The dataset was captured from the Estonian State Gazette website[5] by performing an automated search for all parliamentary acts in force as of a given date, downloading the texts in XML, as supplied, and saving the result noting both the act identifier and the date filter applied.

Data capture, processing and complexity calculation were implemented as Python scripts[6] while resulting data was analysed using R [20].

The approach chosen for complexity analysis requires the ability to assess the complexity of Estonian text. This was accomplished by the use of the NLP Toolkit for Estonian language [21] for natural language processing.

For finding references between laws and between parts of individual laws, a set of regular expressions was utilised that accounts for grammar structures particular to the Estonian language.[7]

### 2.4 Acquiring and Processing the Data

Combining data scraping techniques with conventional data access methods, a set of 66 quarterly datasets was obtained. Firstly, a list of laws applicable at the beginning of the quarter was obtained. Then, the XML file containing the particular edition in force was downloaded for each law in the list. The dataset

```
Data: https://www.riigiteataja.ee/
Result: Quarterly complexity indexes of Estonian laws
for All quarters do
    Get list of all laws in effect on 1st of the quarter
    for All laws returned by query do
        if End date outside of quarter then
            if Law contains text then
                remove image content;
                save XML;
            end
        end
    end
    Remove erroneous duplicates;
    for All laws do
        calculate law complexity;
    end
    for All quarters do
        calculate quarterly complexity from complexities of individual laws of
        that quarter;
    end
end
```

**Algorithm 1:** Data processing algorithm used

[5] www.riigiteataja.ee.

[6] Available under the MIT licence at https://github.com/andreskytt/complexity.

[7] Also available as part of the Python scripts mentioned.

posed two main difficulties. Firstly, the query returned multiple redactions of the law that had to be filtered out by the processing algorithm. Secondly, certain laws were found to contain large images—mostly utilised for pre-layout of complex tables and figures (e.g. the annual state budget)—represented as text. As the images were represented as text, they were removed to prevent interference with natural language processing. Thirdly, 4 laws were found to contain no legal text whatsoever and were not included in the dataset. Finally, 12 versions of various acts were found to be duplicated, where an expert decision was needed on which particular edition of the text actually applied on a given quarter, 17 individual texts were removed by this method.

The complete algorithm of finding quarterly complexity figures for the dataset is as presented in Algorithm 1.

## 3 Theory

### 3.1 Legal Complexity

For the Estonian legal dataset, three levels of subsystems were defined for which all the assumptions of the system complexity estimation hold. Their respective complexity measures and internal relationship definitions are summarised in Table 1.

**Table 1.** Subsystem levels in Estonian legislation

| | Level | Complexity | Relationships |
|---|---|---|---|
| 1 | All parliamentary laws in effect at a given date | Using Formula 5 | Textual mentions of other laws |
| 2 | All paragraphs and sections of each given law | Using Formula 6 | Textual mention of sections and paragraphs; containment of sections in paragraphs |
| 3 | Individual sections of legal text | Using methods described in Sect. 3.2 | – |

To obtain quarterly complexity estimates (level 1 in Table 1), we must apply Formula 1 to a system of all laws and their references by defining suitable subsystem and relationship complexities. While the definition of $\alpha_i$, the complexity of subsystems, stems from the definition of complexity on level 2 presented later, $\beta_i$ presents a challenge.

Firstly, what constitutes a relationship between two laws must be defined. James [22] utilises expert assessment for relationship definition, but this approach cannot be used for the tens of thousands of references in case of laws.

Instead, textual references[8] are utilised. While this method omits many types of implicit relationships as well as indirect references, the approach is scalable as well as explicit in nature.

Secondly, a method for assessing the complexity of such a relationship must be found. A suitable estimate, $\beta_{ij}$, should reflect both the number of references between the laws and the complexity of the laws themselves as a longer legal text could, by definition, contain more references.

It must be noted that Good Legislative Practice Policy ("Hea õigusloome ja normitehnika eeskiri[9]") article 28 states that a legislative act shall not re-state something that is already stated either in the same legal act or another one. Thus, referencing other acts is compulsory and is considered a best practice in Estonia.

Two approaches present themselves. Firstly, one could argue, that complexity of a given relationship should be positively correlated with the complexities of the laws involved as more complex laws can make the references harder to find and their semantics more involved. Secondly, more complex laws might be considered to have more references between each other, and thus, the relationship complexity should be negatively correlated with the complexities of the laws involved. Neither hypothesis appears to be entirely true. The relationship between the complexity of a law and a number of references related to it indeed shows an overall positive covariance but the relationship is non-trivial in nature (see Fig. 2). Therefore, a strong positive correlation between $\beta_{ij}$ and the complexity of the laws involved would likely overemphasise the latter. Exploring alternative implementations led to signs of exponential growth indicative of an exponential relationship and culminated with Eq. 5 that. It must be noted, however, that although an implementation presented in Eq. 5 was used in the analysis, the final complexity measure is sensitive to the particular choice of $\beta_{ij}$

$$C_q^L = \sum_{k=1}^{n} C(t_k) + (\sum_{i=1}^{n}\sum_{j=1}^{n} \beta_{ij} A_{ij})\gamma E(A_q)$$
$$\text{where} \qquad \beta_{ij} = \frac{c_{ij}}{max(C(t_i), C(t_j))} \tag{5}$$

In this implementation, $\alpha_i$ are the individual complexities of laws. $\beta_{ij}$ is the relationship complexity chosen as per previous discussion and $c_{ij}$ denotes the number of references between two laws. $E(A)$ and $\gamma$ are defined as in 1.

To obtain complexities of individual laws (level 2 in table 1), the approach already used is applied to individual legal texts yielding Eq. 6. Here the system consists of sections and paragraphs rather than individual laws. While the system of laws is flat in nature, laws themselves consist of multiple layers of units, each of which can refer to each other. To capture the additional level of complexity

[8] a legal text referencing another legal text, e.g. Law A containing a text "the process of implementing this section is defined in §x in law B" is seen as referencing law B.
[9] https://www.riigiteataja.ee/akt/129122011228.

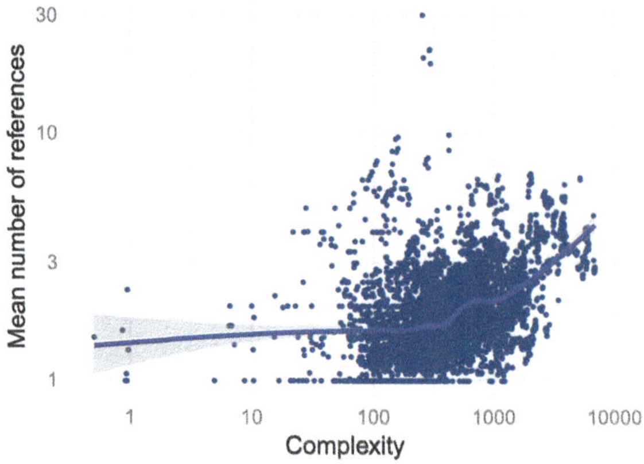


**Fig. 2.** Correlation between the complexity of laws and the mean number of relationships they have

this creates, the definition of a reference between sections can mean either a textual reference, exactly like in case of laws, or a "is a section of" relationship.

$$C_l = \sum_{k=1}^{n} C_m(t_k) + (\sum_{i=1}^{n}\sum_{j=1}^{n} \beta_{ij} A_{ij})\gamma E(A_q)$$

where (6)

$$\beta_{ij} = \frac{c_{ij}}{\max(C_m(t_i), C_m(t_j))}$$

In this instance, $\alpha_i$ are the individual morphological complexities of individual sections in the law and $\beta_i$ is chosen similarly to Eq. 5. $E(A)$ and $\gamma$ are defined as in Eq. 1.

### 3.2 Morphological Complexity

On subsystem level 3, a clear definition of subsystems and their relationships is not present. Therefore, an adaptation of Eq. 1 cannot be used and a different way of assessing the complexity of individual sections of legal text must be developed. Readability metrics like automated readability index [23] are useful, but must be calibrated to the language used. In a multi-language context, the use of Kolmogorov complexity has been proposed (see [24,25]). Kolmogorov complexity is defined as the length of a minimal program, or encoding, to be able to produce the subject message [24]. While the definition has many theoretical connotations, it is not directly computable.

Katz and Bommarito [3] utilise Shannon entropy to estimate complexity of text. Shannon entropy is roughly defined as the amount of "surprise" in communication, more formally described in Eq. 7 [26].

$$H(X) = -\sum_{i=1}^{n} P(x_i) \log_b P(x_i) \tag{7}$$

In Eq. 7, $P(x_i)$ is defined as the probability of a particular event (in our case, a sequence of words) appearing in communication. While a "bag of words" approach can be used (in [3], for example), this does not take into account the meaning encoded in word order. If that meaning is to be considered, an exponential explosion in combinations of words (for each of which, their probability needs to be estimated) occurs [27]. This makes Shannon entropy difficult to apply directly in the context of human languages.

Although it is not directly computable, Kolmogorov complexity gives rise to the idea of effective complexity [11]. To find the effective complexity of an entity, we find the shortest description of irregularities of that entity. This, of course leads to questions of defining irregularities and the algorithm of finding the shortest description. In the current context, the interest does not lie in absolute but relative complexity; i.e. our challenge is to compare the complexities of two entities and not find their absolute values. This simplifies the task considerably.

Since message encoding seeks to distil messages to their shortest possible forms, we can postulate that for the purposes of comparing effective complexities of two texts in the context of a given encoding algorithm, text A is more complex than text B if it yields a shorter encoded text. Complexity of a text t would then formally be defined as follows:

$$C(t) = \frac{|t| - |E(t)|}{|t|} \tag{8}$$

where $E(t)$ is the encoding algorithm. Montemurro et al reach a similar idea via Shannon entropy [27]. In order for the concept to be applicable for natural language, two issues must be solved. Firstly, choice of an encoding algorithm. Since Eq. 8 only requires $|E(t)|$ to define partial ordering on the set of $t$, any well-known encoding algorithm would be suitable. For this particular case, Lempel–Ziv–Markov chain algorithm [28] was chosen because of its ready availability via a Python standard library.

Secondly, the question of variability in information presentation in language must be tackled. The words "pneumonoultramicroscopicsilicovolcanoconiosis" and "silicosis" carry a very similar amount of information but are presented with two significantly different byte sequences. This is a difficult problem. On the one hand, deeply philosophical questions on meaning and information content in language appear and from the other, the semantic field of the legal language defines its own layer of meanings and complexities. To overcome these challenges, a way to represent all lemmas in a computationally identical fashion must be found. This can be accomplished, among others, via an encoding Algorithm 2.

Combining the previous arguments yields Eq. 9 for calculating morphological complexity of a given text where $\Phi(t)$ applies Algorithm 2 followed by LZMA compression to the list of lemmas extracted from the original text $t$. Note, that

```
Data: Array of lemmas
Result: Encoded byte array
c = 32
for L_i in input do
    if L_i ∉ P then
        P = P ∪ L_i
        C_i = c
        c = c + 1
        C[L_i] = C_i
    else
        C_i = C[L_i]
    end
    chr(C_i) ⇒ output
end
```

**Algorithm 2:** Encoding of lemmas

$\exists\, t|LZMA(t)| >= |t| \Rightarrow \exists\, tC_m(t) >= 1$ As no compression algorithm can guarantee reduction of the input,[10] morphological complexity can be larger than 1.

$$C_m(t) = \frac{|L(\Phi(t))|}{|\Phi(t)|} \tag{9}$$

# 4 Results and Discussion

## 4.1 Overall Trends

After downloading and processing, the dataset contained 556 distinct laws. 39% of these appear in all 66 quarters and less than 1% in one. The number of laws per quarter was found to fluctuate with no dominant trend, and the changes are presented in Fig. 3a.

The quarterly complexity indexes computed based on the previously described dataset along with a smoothed conditional mean (local polynomial regression fitting) are depicted in Fig. 3b. The figure shows a strong upward trend of complexity with growth slightly slowing after 2006. The shape of the curve was found to be dependent on the choice of $\beta_i$ in Eq. 5 but the upward trend remained very clear regardless of the particular formula utilised.

The rapid growth in complexity before 2006 can be at least partly explained by the relatively rapid growth of the legal acts in the period. This growth overlaps with the period of EU accession by Estonia on 1st of May 2004, that required the adoption of EU law into Estonian legislative framework. As a member state, that adoption process still continues but was likely to have been most intensive in the few years following the accession date.

10

$$|LZMA(x)| < |t|\forall t \Rightarrow \lim_{i\to\infty} x_i = 0 \text{ when } x_i = LZMA(x_{i-1})$$

which would violate the premise of lossless compression.

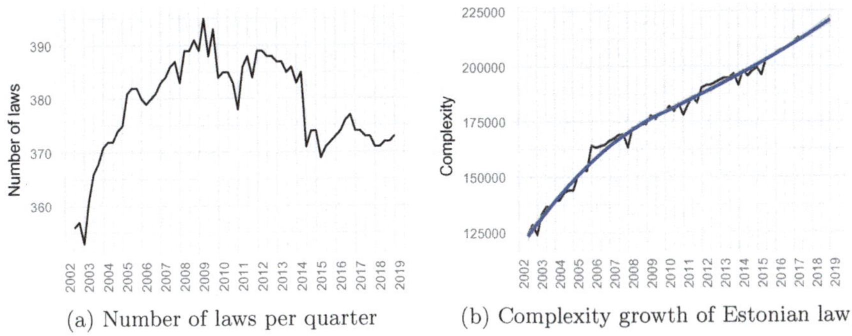


**Fig. 3.** Changes in the number of laws and complexity in Estonia between 2002 and 2019

The overall complexity growth is not necessarily indicative of an excessive legal process. Estonia is a relatively young republic having regained its independence in 1991 during the collapse of the Soviet Union. The legal system of the Soviet Union was markedly different from norms and practices in the European countries, and thus, Estonia had to redefine their legal system effectively from scratch. A process that is still going on. For example, Law of Obligations act [29] which is the cornerstone of private law was first adopted in 2001 and was enforced in 2002.

This trend is a clear indication of the number of acts being a poor proxy for the complexity of the legal system as the two variables do not exhibit covariance, especially after 2006.

To understand this growth, a deeper look into the behaviour of the key components of the complexity metric is necessary. As per Eq. 1, the three components of complexity are the total complexity of all the subsystems ($C_1$), total complexity of all the relationships between the subsystems ($C_2$) and the structure of these relationships ($C_3$). Table 2 summarises the complexity components for the individual laws. $C_1$ makes commonly up the biggest portion of the total complexity measure. C2 is generally smaller and is weighed by $C_3$ that is always smaller than one. Such a relationship supports the choice of the particular implementations of $\beta_i$ and $\alpha_i$ as the resulting complexity components are within the same order of magnitude without any of them having an overly dominant influence.

**Table 2.** Summary statistics of law complexities

| Statistic | N | Mean | St. Dev. | Min | Max |
|---|---|---|---|---|---|
| C | 25,371 | 474.456 | 653.842 | 0.502 | 6805.967 |
| C1 | 25,371 | 341.927 | 452.672 | 0.502 | 4486.265 |
| C2 | 25,371 | 184.055 | 266.334 | 0.000 | 2772.509 |
| C3 | 25,371 | 0.682 | 0.135 | 0.000 | 0.908 |

Table 3 summarises the descriptive statistics for individual complexity components of the quarterly datasets. The picture is markedly different from laws: $C_1$ is clearly dominant with $C_2$ contributing significantly less weighed by significantly smaller structural complexity. This can be explained by Eq. 5, as the complexities of individual laws (see table 2) are generally several orders of magnitude larger than the number of relationships between laws. The relatively small $C_3$ can be explained by the small $N$ and the fact that graph energy is a function of the number of vertices in a graph (Eq. 3).

**Table 3.** Summary statistics of quarterly complexities

| Statistic | N | Mean | St. Dev. | Min | Max |
|---|---|---|---|---|---|
| C | 67 | 179663.200 | 26234.740 | 124134.500 | 220633.800 |
| C1 | 67 | 179663.100 | 26234.710 | 124134.500 | 220633.600 |
| C2 | 67 | 8.256 | 1.326 | 5.205 | 10.059 |
| C3 | 67 | 0.012 | 0.001 | 0.010 | 0.014 |

Relative dynamics of the individual quarterly complexity components indicate the interplay of the various factors and thus contribute to the understanding of the overall trend of complexity growth. These are depicted on Fig. 4a. Although there are slight differences, all three complexity components seem to follow the same general pattern with most quarter to quarter changes being on the growth side but with the rate of change declining over time.

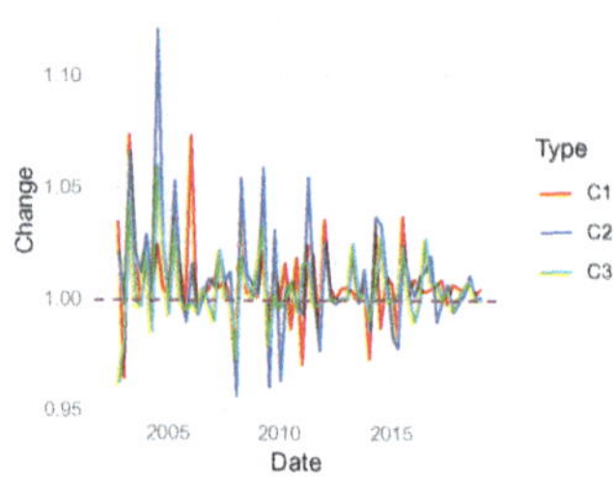


(a) Normalised quarterly complexity components with values above the unity line (highlighted) denoting an increase

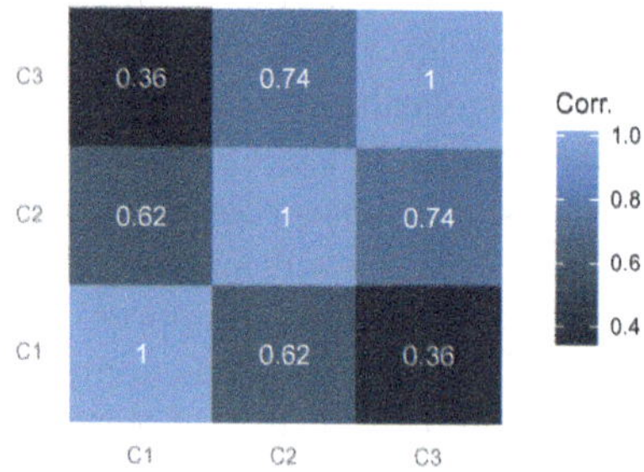


(b) Correlation between changes in complexity components

**Fig. 4.** Changes in components of quarterly complexity

Changes in $C_2$, the measure of the complexity of references, seem to exhibit the largest movement with the largest quarterly change of almost 15% between months and several spikes beyond 5%. As these spikes generally seem to coincide with changes in $C_3$ rather than $C_1$, they seem to be caused by rapid changes in the number of references between laws caused by, for example, the appearance of new laws referencing many others.

As per Fig. 4b, the changes in $C_2$ are relatively strongly correlated with changes in $C_3$. As graph energy is dependent on the number of edges in the graph, this was to be expected. The relationship is not, however, as strong as to negate the need to observe $C_2$ and $C_3$ as separate measures. $C_1$ and $C_2$, however, are relatively weakly correlated. This confirms, again, the suitability of the $\beta_i$ implementation discussed earlier in Sect. 3.1 as the impact of complexities of individual laws does not unduly interfere with the complexity measure of their relationships.

## 4.2 Complexities of Individual Laws

In addition to understanding how different types of complexity contribute to the overall system complexity, it is also important to understand the complexity of individual laws so their contribution can be determined. For this, a ratio of the complexity at the first appearance of a law to its complexity at its last appearance in the dataset was calculated. Figure 5a depicts distribution of these ratios. Almost half of the laws show no change at all while less than 10% exhibit a decrease in complexity and a significant portion showing an increase of more than 5 times. Thus, the complexity growth of the quarterly datasets seems to be, to a large extent, driven by the complexity growth of a relatively small subset of laws.

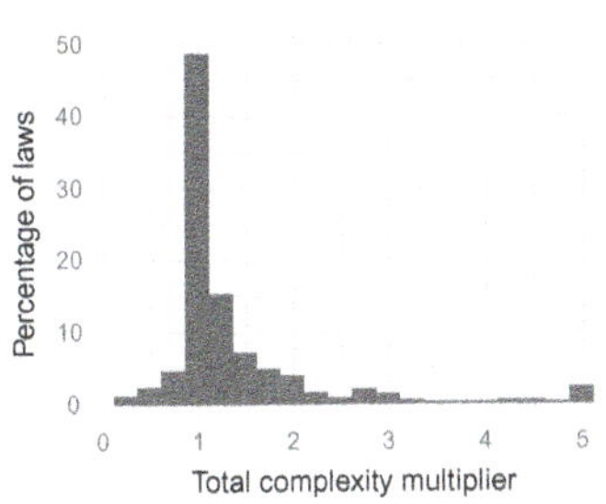


(a) Distribution of relative changes in complexities of individual laws

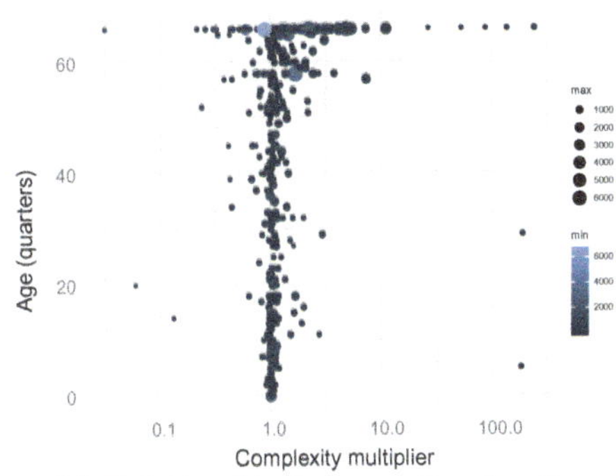


(b) Relationship between the duration a law was observed and the change in its complexities. The size and color of the points denote first and last observed complexities respectively

**Fig. 5.** Distribution of changes in complexities of laws

This hypothesis is further strengthened by Fig. 5b that shows several laws exhibiting more than ten times growth and relates the change in complexity to the duration a particular law was observed. While the expected relationship of older laws exhibiting more change is clearly evident, there are notable exceptions and significant change seems to be possible already after 3–4 years. Also, complexity growth seems to be more time-dependent than reduction. Apparently, the longer a law exists, the more complexity gets added with complexity reduction being less systemic in nature.

## 4.3 Finding the Optimum

Figure 6 depicts implementation of the previously derived theoretical optimum to both quarterly and law-level systems.

As it is evident from Fig. 6b, the number of references between laws, the edges of the graph, is significantly higher than the recommended number throughout the entire observed period. This indicates that, although the complexity metric is dominated by the complexities of individual laws (see Table 3), the structural complexity of the system of laws presents a problem in itself.

Figure 6a presents a very different picture. Here we can see that for all laws in force as of 1st of January 2019, the number of structural references in the law is significantly smaller, than the calculated optimum would suggest as most data points are located below the red unity line.

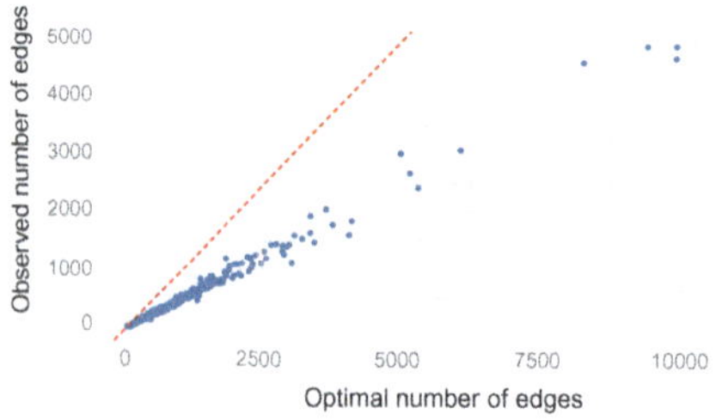


(a) Relationship between optimal and observed number of references within laws (1st quarter of 2019). The red dotted line denotes an equality line

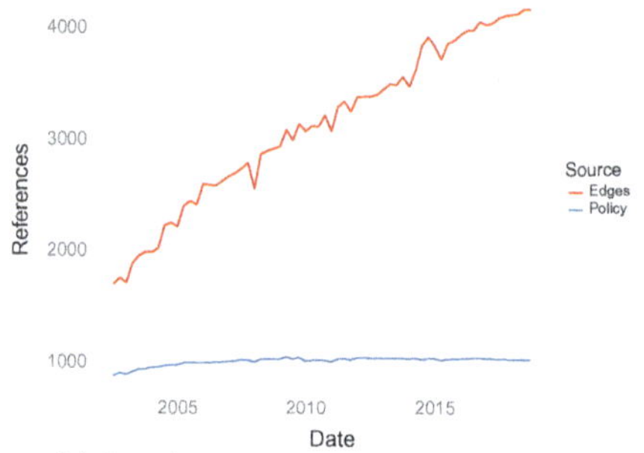


(b) Development of optimal and observed quarterly references over time

**Fig. 6.** Comparison of optimal and observed legal structure

These two findings stand seem to contradict previous findings as the complexity of individual laws does indeed contribute majority of the growth to the complexity of quarterly datasets. The dichotomy can be explained by the number of elements in both systems. This leads the quarterly dataset to have the optimal edge count up to ten times less than is the case for certain laws. Essentially, the individual laws consist of a relatively large number of relatively simple elements while the quarterly system consists of a relatively small number of relatively complex elements.

This finding further emphasises the need to consider the individual elements of complexity carefully and experiment with different implementations of $\beta_i$ and $\alpha_i$.

# 5 Conclusions

## 5.1 General

A number of conclusions can be derived from the acquired dataset by application of the complexity formula.

Firstly, it is clear that the complexity of the set of laws in Estonia is growing rapidly. Even though various choices in model options (see Sect. 4) yield different growth rates, the overall trend is very clear. Given inherent cognitive limitations of people discussed in Sect. 1, this trend is not sustainable. While its effects on society or the legal system in general are outside the scope of this paper, there are two fundamental choices: either the ability to handle complexity of the system is increased or the complexity growth is slowed. The data indicates (see Sect. 5.2), that complexity is not uniformly distributed between the laws and thus a focused action should be possible.

Secondly, basic quantitative metrics as commonly used are less suitable for assessing the complexity of a system of laws than the metrics derived from systems engineering. The number of legal texts, Fig. 3a, does not show significant growth from 2009 onwards while Fig. 3b does. Also, it was possible to both formulate Eqs. 5 and 6 and to compute on the available dataset with results that are in line with the application of the method in other domains. Thus, not only does not the data fail to disprove the first hypothesis of the system engineering methods being applicable but clear benefits over conventional metrics exist.

Thirdly, the quarterly dataset has significantly different properties from the dataset of individual laws. A significantly smaller graph energy indicates a much different structure. This was to be expected, as the structure of individual laws is by definition (see Sect. 3.1) tree-like while quarterly datasets are not. As per Fig. 5a, a relatively small portion of the laws drives a relatively large portion of complexity.

These findings allow to confirm, in Estonian context, the research hypothesis of complexity measures being both applicable and useful in a legal context. Methods derived from systems engineering can be applied in the legal domain with insightful results. Also, a clear trend emerges from the measure that allows us to draw clear conclusions about the sustainability of the system of Estonian laws.

## 5.2 Policy Recommendation

As the temporal behaviour of complexity metrics is sensitive to the particular choices made to estimate complexity of individual elements and their relations, they can only provide general policy recommendations. Although, as discussed earlier, the complexity growth is not sustainable, the urgency of the problem is difficult to estimate and solutions do not present themselves easily. Yes, quarterly complexity growth seems to stem from a relatively small set of laws but complexity of individual acts is but part of the complexity metric.

As described in Sect. 4.3, the individual laws seem to be positioned below a certain optimum whereas the quarterly datasets are located firmly above it. This clearly indicates the opportunity to merge individual laws making the individual acts more complex but reducing the structural complexity of the overall system. Of course, this change should not push the individual laws beyond their optimal structure.

**Observe and limit complexity growth** As continued complexity growth of the set of laws is not sustainable, it should be limited. Any and all changes implemented in individual laws should, on average, seek to reduce and not increase the complexity of the entire system. This can be achieved, for example, by assessing the complexity impact of changes as part of their development process using automated tools. In Estonia, control of the legal policy resides with Ministry of Justice who should adopt this "see asi, millele enne viitasime" as a practical means of measuring progress

**Stop focusing on the number of acts** The number of acts does not correlate with complexity growth observed. Therefore, it should not be used as an indicator to assess the impact of individual changes. While merging laws to reduce the overall complexity of the system would reduce the number of laws in effect, it should be made a goal in itself. Adding new acts does not necessarily have a detrimental effect on the complexity of the set of laws

**Find opportunities for merge** It would appear, the complexity of the legal system could be reduced by merging individual acts. Of course, this should not push the complexity of the new act beyond optimal doing this. Employing a clustering algorithm like [30] would be advisable. Conway's law [31] states that the structure of a system is related to the structure of the organisation producing it. It is therefore important to heed the structure of the governing institutions when performing this optimisation: one should not end up with laws simultaneously supervised by a number of ministries or agencies

## 6 Further Work

To advance the field further, more research is needed. In particular, the application of DSM for structural analysis of quarterly sets of laws should allow to further answer questions about the sources of complexity growth. In particular, work on clustering algorithms optimising for complexity of the overall system would be beneficial as current algorithms do not take the complexity aspect of the system sufficiently into account

Also, to confirm the general nature of the derived approach for complexity calculation, it should be applied to datasets in different legal and linguistic contexts and its sensitivity to different morphological complexity measures as well as $\beta_i$ implementations should be more formally investigated.

Finally, the relationship of the complexity measures developed here should be more thoroughly tested in the legal context by correlating them with expert opinions while establishing a clearer picture of the overall notion of complexity in the legal community. In particular, the validity of the theoretically developed optimal edge count measure would benefit from empirical validation in legal context.

**Acknowledgements.** The authors would like to thank professor Olivier L. de Weck for the inspiration and feedback, Kaushik Sinha for the invaluable comments and the MIT SDM program for bringing us all together.

## References

1. Miller, G.A.: The magical number seven, plus or minus two: some limits on our capacity for processing information. Psychol. Rev. **63**(2), 81 (1956)
2. Sinha, K., de Weck, O.L.: A network-based structural complexity metric for engineered complex systems. In: 2013 IEEE International Systems Conference (SysCon), pp. 426–430. IEEE, New York (2013)
3. Katz, D.M., Bommarito II, M.J.: Measuring the complexity of the law: The united states code. Fill This In (2013)
4. European Commission: Better regulation for better results - an eu agenda. http://ec.europa.eu/smart-regulation/better_regulation/documents/com_2015_215_en.pdf (2015)
5. Ministry of Justice: Õigusloome vähendamise kava. https://www.just.ee/sites/www.just.ee/files/oigusloome_mahu_vahendamise_kava_2016.pdf (2015)
6. Gove, P.B.E., et al.: Webster's third new international dictionary of the English language unabridged. BOOK, Merriam-Webster (1986)
7. Summers, J.D., Shah, J.J.: Mechanical engineering design complexity metrics: size, coupling, and solvability. J. Mech. Des. **132**(2), 021,004 (2010)
8. Selva, D., Cameron, B., Crawley, E.: Patterns in system architecture decisions. Syst. Eng. **19**(6), 477–497 (2016)
9. Liiv, I., Vedeshin, A., Täks, E.: Visualization and structure analysis of legislative acts: a case study on the law of obligations. In: Proceedings of the 11th international conference on Artificial intelligence and law, pp. 189–190. ACM, New York (2007)
10. Steward, D.V.: The design structure system: a method for managing the design of complex systems. IEEE Trans. Eng. Manage. **3**, 71–74 (1981)
11. Gell-Mann, M.: What is complexity? Complexity **1**(1), 16–19 (1995)
12. Deshmukh, A.V., Talavage, J.J., Barash, M.M.: Complexity in manufacturing systems, part 1: analysis of static complexity. IIE Trans. **30**(7), 645–655 (1998)
13. Kafura, D., Reddy, G.R.: The use of software complexity metrics in software maintenance. IEEE Trans. Software Eng. **13**(3), 335 (1987)
14. Sinha, K.: Structural complexity and its implications for design of cyber-physical systems. Ph.D. thesis, Massachusetts Institute of Technology. http://hdl.handle.net/1721.1/89871 (2014)
15. Sinha, K., Weck, O.L.: Empirical validation of structural complexity metric and complexity management for engineering systems. Syst. Eng. **19**(3), 193–206 (2016)
16. Eppinger, S.D., Browning, T.R.: Design Structure Matrix Methods and applications. MIT Press, Cambridge (2012)

17. Gutman, I., Soldatović, T., Vidović, D.: The energy of a graph and its size dependence. A Monte Carlo approach. Chem. Phys. Lett. **297**, 428–432 (1998)
18. Erdos, P., Rényi, A.: On the evolution of random graphs. Publ. Math. Inst. Hung. Acad. Sci **5**(1), 17–60 (1960)
19. McClelland, B.: Properties of the latent roots of a matrix: the estimation of $\pi$-electron energies. J. Chem. Phys. **54**(2), 640–643 (1971)
20. R Core Team: R: A Language and Environment for Statistical Computing. R Foundation for Statistical Computing, Vienna, Austria (2015). https://www.R-project.org/
21. Orasmaa, S., Petmanson, T., Tkachenko, A., Laur, S., Kaalep, H.J.: Estnltk - nlp toolkit for estonian. In: Chair, N.C.C., Choukri, K., Declerck, T., Grobelnik, M., Maegaard, B., Mariani, J., Moreno, A., Odijk, J., Piperidis, S. (eds.) Proceedings of the Tenth International Conference on Language Resources and Evaluation (LREC 2016). European Language Resources Association (ELRA), Paris, France (2016)
22. James, D.H.: Influence of system architecture changes on organizational work flow and application to geared turbofan engines. Master's thesis, Massachusetts Institute of Technology (2011)
23. Senter, R., Smith, E.A.: Automated readability index. Technical report, DTIC Document (1967)
24. Kolmogorov, A.N.: Three approaches to the quantitative definition of information. Prob. Inf. Trans. **1**(1), 1–7 (1965)
25. Juola, P.: Assessing linguistic complexity. Language complexity: Typology, contact, change pp. 89–108 (2008)
26. Shannon, C.E.: A mathematical theory of communication. Bell Syst. Tech. J. **27**, 379–423 (1948)
27. Montemurro, M.A., Zanette, D.H.: Universal entropy of word ordering across linguistic families. PLoS One **6**(5), e19,875 (2011)
28. Lempel, A.: Compression of individual sequences via variable-rate coding. IEEE Trans. Inf. Theory **24**(5), 530–536 (1978)
29. Riigikogu: Võlaõigusseadus. https://www.riigiteataja.ee/akt/122032018004 (2009)
30. Borjesson, F., Hölttä-Otto, K.: Improved clustering algorithm for design structure matrix. In: ASME 2012 International Design Engineering Technical Conferences and Computers and Information in Engineering Conference, pp. 921–930. American Society of Mechanical Engineers (2012)
31. Conway, M.E.: How do committees invent. Datamation **14**(4), 28–31 (1968)

# A Multimodal Biometric System for Secure User Identification Based on Deep Learning

Shefali Arora[1](✉), M. P. S. Bhatia[1], and Harshita Kukreja[2]

[1] Netaji Subhas Institute of Technology, Delhi, India
arorashef@gmail.com, bhatia.mps@gmail.com
[2] Indira Gandhi Delhi Technical University, Delhi, India
harshitakukreja8@gmail.com

**Abstract.** A multimodal biometric system utilizes more than one biometric modality of a person to relieve some of the shortcomings of a unimodal biometric system and improves its security. In this paper, we propose a novel deep learning approach for fusing the features extracted from the individual's face and iris (left and right) to get a more secure biometric verification system. Firstly, we extract the facial and iris features separately using various convolutional neural network (CNN) models. Further, the feature vectors of the final CNN layers of both models are fused to achieve classification of individuals with improved performance. The proposed system is tested on the CASIA-Face V5 dataset for faces and IITD iris dataset for left and right irises. The results achieved prove the superiority of the proposed multimodal system. It is efficient, reliable, and robust as compared to unimodal biometric systems.

**Keywords:** Multimodal · Information fusion · Biometric security · CNN · Pre-trained models

## 1 Introduction

As the technology is advancing in recent times, so are the threats to the privacy of the users. Conventional systems like passwords, patterns, or personal identification numbers (PINs) are failing due to a lack of uniqueness and universality among data. This, in turn, has made possible the implementation of biometric systems to eliminate the threat to user privacy and security. Biometrics is essentially the measure of human characteristics that are unique to every individual. This includes soft biometrics like the age, gender, and height of a person as well as the hard biometrics. Hard biometric traits are assumed to be present universally and can be further divided into three sub-categories to uniquely identify an individual. First, are the biological features like the DNA which is concrete in nature throughout life. Behavioral features include how a person behaves, such as his voice, signature, gait, and the keystroke. Finally, the physiological features comprise the changing biology of a person in the form of face, iris, and fingerprints. These are commonly used in biometric systems for authentication.


X.-S. Yang et al. (eds.), *Proceedings of Fifth International Congress on Information and Communication Technology*, Advances in Intelligent Systems and Computing 1183, https://doi.org/10.1007/978-981-15-5856-6_8

While the lack of integrity among the identifiers limits the unimodal biometric systems, multimodal systems overcome this limitation by fusing the extracted information from multiple unimodal systems together. This fusion can occur at different stages of recognition resulting in different accuracies for the system. Feature-level fusion involves fusing the extracted features from multiple biometrics. Decision-level fusion combines the final results of multiple classifiers. In the case of score-level fusion, the final scores of various biometric systems are fused to arrive at a decision.

In our proposed approach, we use two biometric traits, face and iris. Face is the most natural trait available and the iris is the most accurate one. A deep convolutional neural network (CNN) based on these traits is applied by using various models and feature-level fusion. We evaluate the proposed model on IITD iris dataset for the left and right irises and the CASIA dataset for faces. The resulting multimodal biometric system yields better results and outperforms the unimodal biometric system for each trait separately.

The rest of the paper is organized in the following sections. Section 2 comprises the related works on the multimodal biometric systems based on the face and iris modalities. Section 3 presents the methodology used with results achieved using the framework in Sect. 4. Finally, Sect. 5 concludes the paper.

## 2 Background

A multimodal biometric system merges two or more biometric traits like face, fingerprint, palm print, and iris together. For this, the input from a single sensor could be run on different algorithms, or more than one sensor could be used to capture single or multiple biometric traits. Multimodal systems, thus, help in improving the recognition rate of images.

There are several multimodal biometric systems that have been proposed which utilize two or more traits. Each proposed system used its own specific feature extraction and fusion method. Some of them are discussed here.

Saha et al. [1] utilized feature-level fusion of iris and retina recognition and achieved an accuracy of 98.37%. They applied principal component analysis (PCA), thereby minimizing the 'curse-of-dimensionality' limitation prevalent in their field earlier. In [2], the proposed system uses CNN along with multi-support vector machine (QG-MSVM) along with feature and decision-level fusion to create an efficient, robust, and reliable system for ECG and fingerprint recognition. The face and palm print features were extracted using methods like Gabor, Radon, Ridgelet, and Radon–Gabor, followed by feature-level fusion and SVM and SSIM classification [3]. Accuracies were increased from 92.5% for face and 76% for palm print to 98% for the fused system. Another work fuses the ear and finger knuckle print for biometric identification of a person as ear biometric is stable form the birth, hence making it unique [4]. To minimize the vulnerability of a unimodal biometric authentication, authors propose to combine the physiological and behavioral features of the face together, namely the face and facial expression [5]. The system uses facial expressions to recognize an authentic user. Another system combining physiological and behavioral traits is proposed by Bharadi et al. in [6] where they extract texture information from the iris and the dynamic pressure variation data from the online signatures. The proposed multimodal architecture based on feature vector extraction and kNN classification reached an accuracy of 90.18%, which is more than that of the individual systems, proving it to be a reliable architecture. Some of the earlier works of the kind perform score-level fusion based on symmetric sums (S-sums), which was tested on the two subsets of the NIST-BSSR1 dataset [7]. This outperforms the previous individual models. Hezil et al. [9] combine ear and palm print using feature-level fusion, and a recognition rate of 100% was obtained after experiments on IIT Delhi-2 ear and IIT Delhi palm print databases. Another multimodal biometric system combining finger vein and knuckle images using feature-level fusion is proposed, wherein they use fractional firefly (FFF) optimization. A higher accuracy of 96% is obtained [10].

A novel framework for multimodal biometric system is proposed in [11] by fusing features of face and irises of a person. A deep belief network (DBN) and a CNN-based architecture along with Softmax classifier were used to experiment on large-scale datasets like MMU1, FERET, CASIA V1.0, SDUMLA-HMT. Authors proposed a database consisting of face, fingerprint, and iris modalities of over 100 children (from 18 months old to 4 years old), with the images captured in separate sessions [12].

### 2.1 Contribution

- An efficient multimodal biometric identification system is proposed based on feature extraction for images of a person's irises and face. The framework uses deep learning and is highly flexible in terms of provision of all biometric traits or a subset of it. This helps to add security to the system as per convenience.
- An efficient use of various pre-trained CNN architectures has been employed, and the final layers involve merging of feature vectors obtained from the final CNN layer used for the classification of face and iris, respectively. Both the VGGNet and LeNet models are used for iris and face classification work for classification of users into one in $N$ classes.
- We further employ the use of an efficient deep learning architecture, which we refer to as ConvNet. It combines various CNN layers with a Softmax classifier and thus a modified and improved architecture with lesser parameters as compared to the existing pre-trained models.

## 3 Methodology

In this section, we discuss some of the pre-trained CNN architectures already being employed along with their detailed usage in our proposed framework. We also discuss the architecture of ConvNet, which we will employ in the framework of the multimodal biometric system to evaluate its performance.

### 3.1 Convolutional Neural Networks

Convolutional operations are used for linear and nonlinear image processing operations. Consider an image in the domain $I(p) = I_1(p) \ldots I_m(p)$ which is an attribute vector of m band pixel $\hat{I} = (D_I, I)$

$$p = \left(x_p, y_p\right) \in D_I \tag{1}$$

Let *A*(*p*) be a squared centered at *p* of size *LA* × *LA* such that $A \subset DI$. Let $\varphi = (A, W)$ be the filter weights *W*(*q*) associated with pixels $q \in A(p)$ such that A ⊂ DI. Let $\varphi = (A, W)$ be the filter weights *W*(*q*) associated with pixels $q \in A(p)$.

$$J_i(p) = \sum_{\forall q \in A(p)} I(q).W(q) \quad (2)$$

The process of convolution is followed by a nonlinear activation function. It extracts the maximum value out of each subregion, thus reducing the output image size. It is defined as:

$$\boldsymbol{J_i(p)} = \max(\boldsymbol{J_i(p)}, 0) \quad (3)$$

## 3.2 Pre-trained Models

CNN is a prime example of complex structures with hidden layers and is actively applied in image processing tasks. Recently, there has been an increase in the number of pre-trained CNN models available for large-scale image classification such as LeNet [13], CaffeNet [14], AlexNet [15], VGGNet [16], and ResNet [17]. LeNet is one of the first classical networks that are trained on grayscale images. With around 60,000 parameters, it uses sigmoid activation function after the pooling of the layers [18, 19]. Figure 1 shows the overall architecture of LeNet. Figure 2 depicts the architecture of pre-trained model VGGNet.

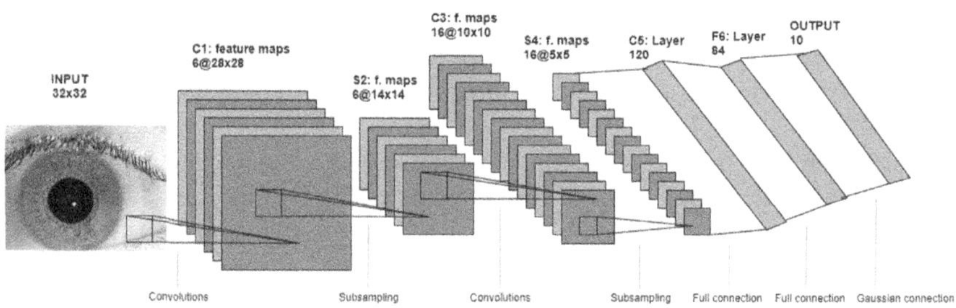


**Fig. 1.** Architecture of LeNet

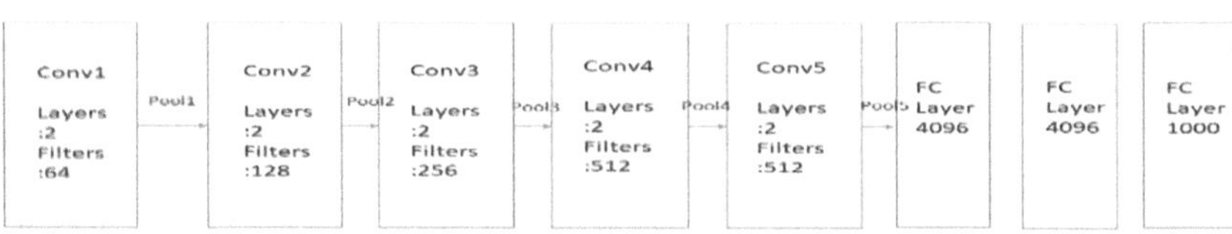


**Fig. 2.** Architecture of VGGNet

VGGNet was an improvement over models like LeNet and AlexNet with around 140 million parameters. It replaced large kernel-sized filters in AlexNet with multiple 3 × 3 kernel-sized filters one after another. In our proposed architecture ConvNet, the network consists of the following layers as depicted in Fig. 3.

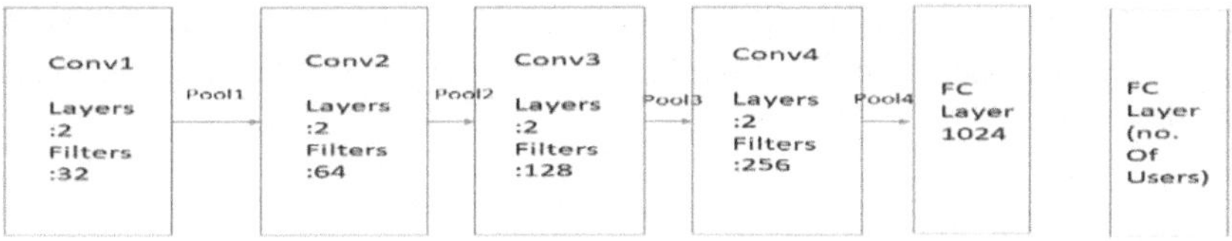


**Fig. 3.** Architecture of ConvNet

### 3.3 Proposed Framework

In this work, a multimodal biometric system is proposed based on the combination modalities of face and the two irises. The detailed framework of the proposed multimodal biometric system is illustrated in Fig. 4. Feature-level fusion is performed after feature extraction where the features from various biometric traits are merged to get a combined feature vector. In this step, the features extracted from one trait must be compatible from the features extracted from another trait for the fusion to be possible. Given F1 and F2 as two feature vectors extracted from an images with vectors V1 and V2, respectively, the fused vector will be V1 + V2 [20].

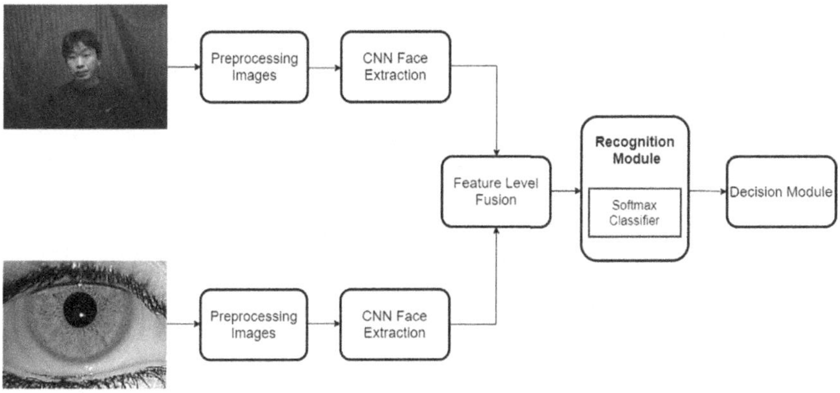


**Fig. 4.** Proposed framework

The facial and iris features are extracted using both pre-trained models and ConvNet. Feature-level fusion is done by employing different combinations of pre-trained models and ConvNet.

## 4 Experimental Setup

### 4.1 Datasets

For the purpose of facial recognition, we employ the CASIA-Face V5 database, which contains 2500 facial images. For the purpose of identification using iris, we employ the IIT Delhi iris database, which consists of around 1200 images of left and right irises of students as well as staff members. We enroll the faces and irises of around 100 users, in order to perform feature-level fusion on the vectors obtained for users during feature extraction of each.

### 4.2 Performance Metrics

The performance metrics which have been used to analyze the system are:

- False acceptance rate: It measures the possibility of the system to incorrectly accept access attempts by an unauthorized user. These users might not be enrolled in the system. A system's FAR is the ratio of the number of false acceptances to the number of attempts made at identification. The value of FAR for a biometric system should be low.
- False rejection rate: It measures the possibility of the system to reject an authorized user's attempts incorrectly.
- Accuracy: It is measured in percentage and depicts the efficiency of the biometric system in identifying users.

### 4.3 Results

The results obtained with and without feature-level fusion are illustrated in Tables 1, 2, 3. From the results, it is emphasized that a multimodal biometric system is more secure as compared to a unimodal biometric system, as it would consider the features of a user's face as well as iris. We calculate the value of FAR for the system by considering FRR to be 0 percent. Figure 5 visualizes the accuracies achieved with and without fusion of biometric modalities.

**Table 1.** Performance metrics of unimodal biometric system

| IITD iris database | Model | Accuracy (in %) | FAR (in %) @0FRR |
|---|---|---|---|
| Without fusion | VGGNet | 97.98 | 1.2 |
| Without fusion | LeNet | 89.38 | 1.5 |
| Without fusion | ConvNet | 98.99 | 1.0 |

**Table 2.** Performance metrics of unimodal biometric system

| CASIA-FaceV5 database | Model | Accuracy (in %) | FAR (in %) @0FRR |
|---|---|---|---|
| Without fusion | VGGNet | 99.79 | 0.9 |
| Without fusion | LeNet | 98.98 | 1.0 |
| Without fusion | ConvNet | 98.98 | 1.0 |

**Table 3.** Performance metrics of multimodal biometric system

| Models for fusion | Accuracy (in %) | FAR @0FRR (in %) |
|---|---|---|
| VGGNet(iris)+ Lenet(Face) | 99.42 | 0.6 |
| ConvNet(Iris)+ LeNet(Face) | 99.57 | 0.5 |
| VGGNet(Iris)+ ConvNet(Face) | 99.65 | 0.4 |
| VGGNet(Face)+ ConvNet(Iris) | 99.80 | 0.1 |

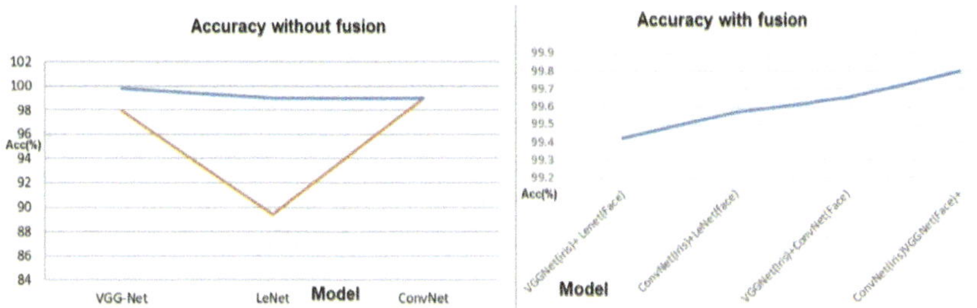


**Fig. 5.** Accuracy with and without fusion

## 5 Conclusion and Future Work

This paper presents a multimodal biometric system by fusing the features of an individual's iris and face modalities. The use of multimodal biometric system helps to overcome the limitations of a unimodal biometric system, in terms of security. Various pre-trained CNN models as well as the proposed ConvNet model are used to extract features from the two modalities. Further, various combinations are tested at the time of feature-level fusion to enhance the accuracy of authentication.

From experimental results, it is evident that the proposed multimodal systems perform better than unimodal systems. In the future, we would compare the results with score-level fusion and test the proposed system on a real database.

**Acknowledgements.** We would like to thank the Chinese Academy of Sciences as parts of research in this paper use CASIA-Face V5 [21]. We would like to thank IIT Delhi for providing the iris database [22]. We would also like to thank Hitesh Nankani, a student at NSIT, for his valuable insights.

## References


1. Saha, A., Saha, J., Sen, B.: (2019) An expert multi-modal person authentication system based on feature level fusion of iris and retina recognition. In: 2019 International Conference on Electrical, Computer and Communication Engineering (ECCE)
2. Hammad, M., Liu, Y., Wang, K.: Multimodal biometric authentication systems using convolution neural network based on different level fusion of ecg and fingerprint. IEEE Access **7**, 26527–26542 (2019)
3. Rane, M., Deshpande, P.: (2018) Multimodal biometric recognition system using feature level fusion. In: 2018 Fourth International Conference on Computing Communication Control and Automation (ICCUBEA)
4. Rajalakshmi, B., Sumathi, S.: (2018) Survey of multimodal biometric using ear and finger knuckle image. In: 2018 International Conference on Communication, Computing and Internet of Things (IC3IoT)
5. Mei Yin, D., Mukhlas, A., Wan Chik, R., Talib Othman, A., Omar, S.: (2018) A proposed approach for biometric-based authentication using of face and facial expression recognition. In: 2018 IEEE 3rd International Conference on Communication and Information Systems (ICCIS)
6. Bharadi, V., Pandva, B., Cosma, G.: (2018) Multi-modal biometric recognition using human iris and dynamic pressure variation of handwritten signatures. In: 2018 Fifth International Conference on Social Networks Analysis, Management and Security (SNAMS)
7. Cheniti, M., Boukezzoula, N., Akhtar, Z.: Symmetric sum-based biometric score fusion. IET Biometrics **7**, 391–395 (2018)
8. Ammour, B., Bouden, T., Boubchir, L.: Face–iris multi-modal biometric system using multi-resolution Log-Gabor filter with spectral regression kernel discriminant analysis. IET Biometrics **7**, 482–489 (2018)
9. Hezil, N., Boukrouche, A.: Multimodal biometric recognition using human ear and palmprint. IET Biometrics **6**, 351–359 (2017)
10. Veluchamy, S., Karlmarx, L.: System for multimodal biometric recognition based on finger knuckle and finger vein using feature-level fusion and k-support vector machine classifier. IET Biometrics **6**, 232–242 (2017)
11. Al-Waisy, A., Qahwaji, R., Ipson, S., Al-Fahdawi, S.: (2017) A multimodal biometrie system for personal identification based on deep learning approaches. In: 2017 Seventh International Conference on Emerging Security Technologies (EST)
12. Basak, P., De, S., Agarwal, M., Malhotra, A., Vatsa, M., Singh, R.: (2017) Multimodal biometric recognition for toddlers and pre-school children. In: 2017 IEEE International Joint Conference on Biometrics (IJCB)
13. Lecun, Y., Bottou, L., Bengio, Y., Haffner, P.: Gradient-based learning applied to document recognition. Proc. IEEE **86**, 2278–2324 (1998)
14. Jia Y, Shelhamer E, Donahue J, Karayev S, Long J, Girshick R, Guadarrama S, Darrell T (2014) Caffe. Proceedings of the ACM International Conference on Multimedia - MM '14
15. Krizhevsky, A., Sutskever, I., Hinton, G.: ImageNet classification with deep convolutional neural networks. Commun. ACM **60**, 84–90 (2017)
16. Simonyan, K., Zisserman, A.: (2014) Very deep convolutional networks for large-scale image recognition. CoRR. http://arxiv.org/abs/1409.1556
17. He, K., Zhang, X., Ren, S., Sun, J.: (2016) Deep residual learning for image recognition. In: 2016 IEEE Conference on Computer Vision and Pattern Recognition (CVPR)
18. Russakovsky, O., Deng, J., Su, H., Krause, J., Satheesh, S., Ma, S., Huang, Z., Karpathy, A., Khosla, A., Bernstein, M., Berg, A., Fei-Fei, L.: ImageNet large scale visual recognition challenge. Int. J. Comput. Vision **115**, 211–252 (2015)

19. CS231n Convolutional Neural Networks for Visual Recognition (2019). In: Cs231n.github.io. http://cs231n.github.io/convolutional-networks/#comp
20. Verma, S., Saravanan, C.: (2018) Performance analysis of various fusion methods in multimodal biometric. In: 2018 International Conference on Computational and Characterization Techniques in Engineering and Sciences (CCTES)
21. CASIA-FaceV5. http://biometrics.idealtest.org
22. Kumar, A., Passi, A.: Comparison and combination of iris matchers for reliable personal authentication. Pattern Recogn. **43**, 1016–1026 (2010)

# Distributed Modular Multiplication to Be Processed by a Network of Limited Resources Devices

Menachem Domb(✉)

Ashkelon Academy College [AAC], Ashkelon, Israel
dombmnc@edu.aac.ac.il

**Abstract.** Asymmetric cryptography algorithms, still considered the most robust tool available in the cryptography domain. It incorporates intensive modular exponentiation calculations, which entail considerable computing power, memory and storage space. Common IoT device is equipped with limited computing resources and so, for security purposes, it can execute symmetric and limited asymmetric cryptography. In most cases, IoT devices do not have the capacity required for executing massive modular multiplications of numbers of magnitude of 4 K bits and more. This leads to the lack of asymmetric cryptography in the IoT domain resulting with a reduction in security means to cope with the raising security challenges. The security problem increases as the number of IoT devices is expected to significantly grow soon. We propose a new implementation of asymmetric cryptography, which splits its heavy calculations into micro-processes, where each micro-process is distributed to an appropriate IoT device, connected to the network. The accepted result is transferred back to the distributing IoT. Once all the results are accumulated, a consolidation process is activated to generate the final calculation result, which then is fed into the cryptography process, to generate the Encrypted/Decrypted string, proving the ability to maintain asymmetric cryptography also at the IoT level without compromising security.

**Keywords:** RSA · IoT · Parallel/distributed processing · Exponentiation · Modular

## 1 Introduction

The term Internet of things [IoT] refers to a set of basic sensors connected to the Internet. Sensors have poor resources and security measurements and so expose the entire network to severe security risks. Commonly used security technologies require considerable computation resources, which are much beyond the capacity of existing sensors. A combination of symmetric and asymmetric encryption systems is commonly used in the industry. Symmetric cryptography requires moderate computation resources, so they are implemented successfully in IoT for all data streaming cryptography. However, the critical risk is the transmission of the symmetric key amongst the network devices. To do


X.-S. Yang et al. (eds.), *Proceedings of Fifth International Congress on Information and Communication Technology*, Advances in Intelligent Systems and Computing 1183, https://doi.org/10.1007/978-981-15-5856-6_9

so, it is common to transmit the key using public key asymmetric cryptography, such as RSA. However, RSA is based on the integer factorization problem, which entails intensive calculations of modular power, affecting its performance and cannot be executed by most IoT devices. To cope with this limitation, we propose a distributed implementation of RSA.

Our contribution in this work is the downsizing of bulk calculations into a unified implementation of RSA to comply with the embedded limitations of the IoT environment. To this end, we introduce a combination of three implementation improvements allowing suitable implementation of RSA for the IoT environment. The first improvement is allowing RSA cryptography of any length of input, the second is splitting the modular exponent calculation into micro-calculations and the third is parallel and distributed calculations. We have executed a comprehensive feasibility simulation, which proves the proposed RSA implementation is ready for IoT security implementation.

In the next section, we outline relevant related work. In Sect. 3, we describe the revised RSA implementation. In Sect. 4, we outline a feasibility experiment set-up and results and we conclude with conclusions and future work.

## 2 Related Work

RSA is based on the mathematical scheme of modular exponentiation of very large integers, which is a compute-intensive process and requires much processing power, memory and storage space. Many improvements have already been proposed. Some focus on hardware improvements and configuration changes, adding robust processes and gateways. Others propose parallel processing and distributed computing or use of alternate encryption algorithms and adding certificates. Following are selected improvements proposed in current literature. Saxena and Kapoor [1] present various parallel implementations of RSA. Kawamura et al. [2] present a parallel implementation for generating RSA keys, based on a factorisation theorem, which combines with Derome's method without Euclidean algorithm. Xian-Fu et al. [3] exploit the parallel architecture in GPU to perform RSA. It achieved 17713 and 89043 2048-bit modular exponentiation per second. It is 62.5 and 38.1 times faster than 2048-bit RSA in GTX. The overall throughput is 12% higher in random exponent bits and higher in all 1's exponent bits. Christos Stergioua et al. [4] propose an integration of IoT and cloud computing, where RSA is running on cloud providing robust asymmetric cryptography for IoT. Goyal and Vineet Sahula [5] and [6] compare the power utilization of several public key algorithms and recommend the Elliptic Curve Diffie Hellman (ECDH)/ECC algorithms. Duy An Ha et al. [7] use ECQV Implicit Certificates and DTLS Protocol to introduce a security model for IoT. It supports mutual authentication and key transmission amongst IoT devices. Fadhil and Younis [8] ran parallel RSA on a combined configuration of multi-core CPU and several single-core GPUs and achieved very significant performance improvements.

After reviewing recent related papers, it seems that the common approach is looking for an alternative to RSA with similar security level [ECC]. Our approach calls for identifying the elements causing RSA's high resource consumption, splits them into pieces and runs each piece by a selected IoT device.

## 3 The Proposed Solution

By using multi-threading, distributed processing and message passing protocols, multiple nodes can cooperate to jointly complete comprehensive and complicated tasks. We introduce three improvements to be incorporated in the implementation of RSA: (a) Splitting long input into smaller strings, which, after encryption, are concatenated to serve as the encrypted key. (b) Splitting large modular exponential calculations into smaller tasks, allowing their distribution and parallel processing by several IoT devices. The split is done according to the theorem described in [2]. (c) Smart distribution of these tasks amongst the available sensors designated to share the factorial calculation. Following, we elaborate these three steps:

a. Splitting the input message: When a sensor is about to encrypt a message, it splits the message into n pieces, $Message = Msg_1||Msg_2||Msg_3||\ldots||Msg_n$. Each piece is then sent to an available sensor, to be encrypted. The encrypted piece is then sent back to the sending sensor. Once all n pieces are in the distributor sensor, it consolidates these pieces $RSA(Message) = RSA(Msg_1)||RSA(Msg_2)||RSA(Msg_3)||\ldots||RSA(Msg_n)$. We may change the concatenation sequence and so add another security level. The new sequence may be predefined or changed dynamically based on an external parameter. Once the sub-message is encrypted, it is sent to the distributing sensor.
b. Splitting modular exponential calculations: There are several ways to split the modular exponential calculation. For simplicity, we selected the right-to-left binary calculation. Each long number is presented by its binary presentation. Each meaningful [greater than 0] exponential of 2 is calculated and modulated separately. Once all exponents are calculated--modulated, the sum of them is modulated to get the final result. Example: 56mod5 = (2^5 + 2^4 + 2^3)mod5 = (32mod5 + 16mod5 + 8mod5)mod5 = (2 + 1+3)mod5 = 1 equal to 56mod5 = 1
c. Distribution: The proposed solution assumes the existence of a local network of sensors having some processing redundancy, which its spare processing-power time, can be mapped into a firm schedule. We assign a task to a sensor that complies with the sensor capacity. To do so, a capacity and availability map is generated by a periodic process which assesses for each sensor in the network, its maximum capacity and availability. Cryptography tasks are distributed to sensors based on this capacity map. Meaning, a "stronger" sensor will get a larger number to calculate.

## 4 The Proposed Model

The model is composed of two main components, the Distributor and the Observer. The Distributor accepts the requested modular multiplication to be performed. It splits the calculation arguments into smaller pieces such that each piece will be processed by at least one available sensor. The Observer resides in each IoT device connected to the network. It acts as a listener waiting for a task to be sent to it by the Distributor and executes the calculation it is expected for performing. Figure 1 depicts the RSA-Distributor processing steps. In the first step, the module launches a heartbeat loop to ensure IoT devices connectivity and readiness to accept input and process it. The process accepts, via a GUI interface, the input for calculating its modulation.

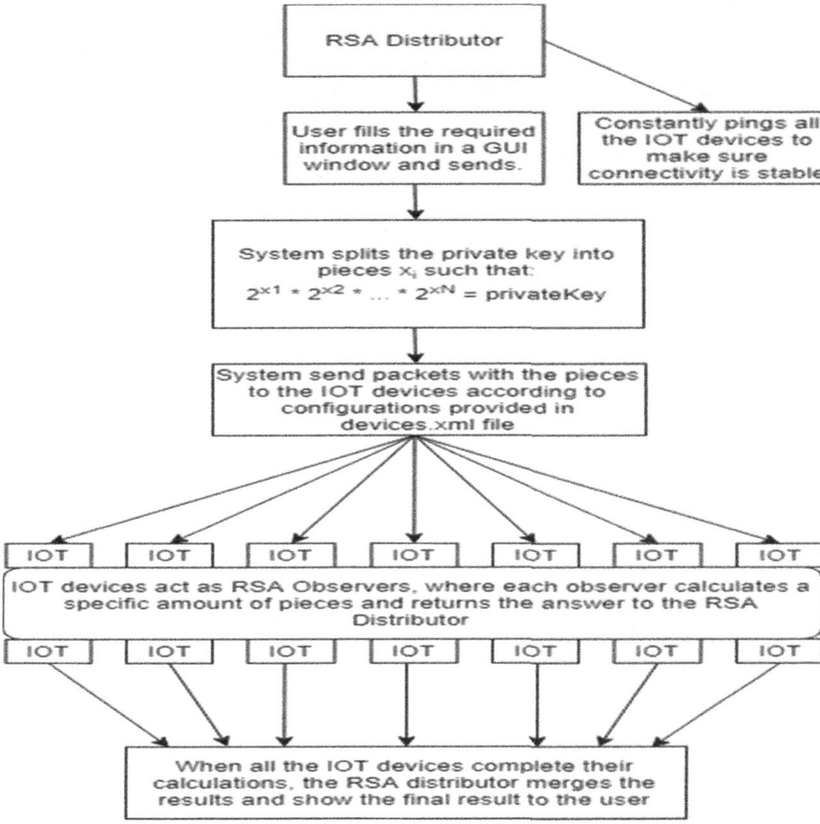


**Fig. 1.** RSA distributor

The input is translated into a binary form, which then is divided to smaller strings and are distributed in an xml–format file to the connected IoT devices. Each device calculates the exponentiation and modulation and sends the results back to the RSA-Distributor for integration and composing the result. To distribute the RSA computation, sensors supposed to communicate in a secure fashion. Hence, we assume a set-up where the sensors are securely and isolated wired-connected network, allowing safe inter-sensors messaging. We also assume that all computation devices are equipped with a reasonable proportion of computation power, memory and storage.

## 5 Experiment

Based on the described model, we have developed a distributed application incorporating most of the proposed improvements. For testing, we tested just the decryption part of RSA with the exact same data. The concept of splitting to micro-operations of the modular multiplication has already been proved mathematically [2]. The performance improvement was already proved as well by Selçuk Baktir and Erkay Savaş [9]. Therefore, in our experiment, we concentrated just on the feasibility of our proposed solution. We conducted a feasibility test using a network of four Arduino-OS devices with a basic messaging program. The test went well and proved its feasibility for IoT. To compare the performance of a non-split calculation Versus a split and distributed calculation, we used regular capacity computers connected via a long-distance network. Table 1 presents the summary results of our experiment. The left column depicts the version of RSA, standard or distributed.

**Table 1.** Experiment summary

| Elapse time | Execution time | Set-up | Version |
|---|---|---|---|
| *v* | *v* | Four-connected IoT devices | Distributed |
| *v* | 40 | 1 PC no threading | Standard |
| 2300 | 6 | 1 PC with threading | Distributed |
| 1300 | 20 | Four-connected PCs | Distributed |

The column to the right shows the four set-ups used and then the execution and elapse times columns in milliseconds. ***V*** means that the time was not collected. In line 2 we executed the standard RSA on an 8 core PC taking 40 ms. We then executed our distributed RSA using the same PC but in 16 multi-threading mode, where each thread simulates the role of an IoT device. Due to parallelism, the total execution time was dropped to 6 ms. However, creating and launching the threads took 600 ms for 994 thread, causing thread accumulation and long queues, resulting with elapse time of 2300 ms. In line 3, we executed the new decryption module using four-connected computers. While the pure execution time took about 6 ms, the network delivery caused the elapse time to be 1300 ms. As we can see, threading management and network delivery time take a big chunk of the elapse time. Although the overall RSA-decryption elapse time is below expectation, it proves the feasibility of breaking RSA into micro-tasks, distributing and parallel processing them solely by IoT devices, without any assistance of an external powerful server.

## 6 Conclusions

This work aims to enable public key cryptography [RSA] to be executed in a parallel and distributed mode by a network of small devices [IoT]. We analysed RSA components identifying the areas requiring vast processing time, memory and storage. Accordingly, we proposed three implementation improvements to RSA and show the feasibility of running RSA in the IoT framework solely by its own devices. This work can be expanded to allow RSA scaling to unlimited key lengths in multiple dimensions.

## References


1. Saxena, S., Kapoor, B.: State of the art parallel approaches for RSA public key base cryptosystem. Int. J. Comput. Sci. Appl. (IJCSA) **5**(1) (2015)
2. Kawamura, S., Koike, M., Sano, F., Shimbo, A.: Parallel computation of the generating keys for RSA cryptosystems. Electron. Lett. **32**(15), 1365–1366 (IEEE explorer)
3. Wong, X.-F., Goi, B.-M., Lee, W.-K., Phan, R.C.-W.: Performance evaluation of RSA and NTRU over GPU with maxwell and pascal architecture. J. Softw. Netw. **2017**(1), 201–220 (2017)
4. Psannisa, K.E., Kimb, B.-G., Guptac, B., Stergioua, C.: Secure integration of IoT and cloud computing. Elsevier Future Gener. Comput. Syst. **78**(3), 964–975 (2018)
5. Goyal, T.K. Sahula, V.: Lightweight security algorithm for low power IoT devices. In: 11–2016 International Conference on Advances in Computing, Communications and Informatics (ICACCI), IEEE. https://doi.org/10.1109/icacci.2016.7732296
6. Suárez-Albela, Manuel, Fraga-Lamas, Paula, Fernández-Caramés, Tiago M.: A practical evaluation on RSA and ECC-Based cipher suites for IoT high-security energy-efficient fog and mist computing devices. Sensors **18**(11), 3868 (2018). https://doi.org/10.3390/s18113868
7. Ha, D.A., Nguyen, K.T., Zao, J.K.: Efficient authentication of resource-constrained IoT devices based on ECQV implicit certificates and datagram transport layer security protocol. In: SoICT '16 Proceedings of the Seventh Symposium on Information and Communication Technology, pp. 173–179. ACM (2016)
8. Fadhil, H.M., Younis, M.I.: Parallelizing RSA algorithm on multicore CPU and GPU. Int. J. Comput. Appl. (0975—8887) **87**(6), (2014)
9. Baktir, S., Savaş, E.: Highly-parallel montgomery multiplication for multi-core general-purpose microprocessors, computer and information sciences III. pp. 467–476. Springer (2012)

# Novel Hidden Markov Scoring Algorithm for Fraudulent Impression Classification in Mobile Advertising

Iroshan Aberathne[1](✉), Chamila Walgampaya[1], and Udara Rathnayake[2]

[1] Faculty of Engineering, University of Peradeniya, Peradeniya, Sri Lanka
iroshan.du@gmail.com, ckw@pdn.ac.lk

[2] Faculty of Computing and Technology, University of Kelaniya, Kelaniya, Sri Lanka
udarar@kln.ac.lk

**Abstract.** Excessive usage of smartphones and tablets have led to drastic increase of mobile ad fraud in recent years. The fraudulent users can be either human or automated scripts with the intention of making illegal revenue or exhausting the advertiser budget are being engaged with this multimillion industry. The *ad fraud* referred to any kind of activities that are generated by a fraudulent user is a huge threat to the existence of the online advertising ecosystem. The researchers have proposed various kinds of methodologies and tools in the context of ad fraud detection and prevention. However, the fraudulent users are smart enough to bypass the significant number of existing detection and prevention systems. The combat between fraud users and researchers or solution designers in this field never ends. Thus, the novel solution of ad fraud detection and prevention techniques is needed. The proposed approach of this study to address this problem is called *hidden Markov scoring model—HMSM.* The model calculates scores for each observe/emission variable of experimental data set towards the hidden states of target variable based on hidden Markov model so that fraud impression can be classified. The experimental results show that the significance of the proposed approach to classify the fraud and non-fraud impression.

**Keywords:** Mobile advertising · Mobile click fraud · Hidden Markov model · Supervised learning

## 1 Introduction

Mobile advertising dominates a significant portion of the digital advertising market in the past few years. One of the major factors of this expansion is dramatic increase of smartphone users in the world [2]. Haider et al. [6] stated that 70% of market share will be dominated by the mobile advertising by 2019. Recent statistics show that 51% of market share already been absorbed with the worth of


X.-S. Yang et al. (eds.), *Proceedings of Fifth International Congress on Information and Communication Technology*, Advances in Intelligent Systems and Computing 1183, https://doi.org/10.1007/978-981-15-5856-6_10

more than US$200 billions [6]. The main contributors of this industry are advertisers, advertising networks (ad networks), publishers and end users. Advertiser is the one who makes a contract with ad network to publish advertisements on behalf of himself or a company. Ad network plays a broker role between advertiser and publisher. Advertiser is charged by the ad networks for publishing their advertisements. Ad networks find suitable publishers to display ads. Publisher can be a website or a mobile application which displays the advertisements to the site/app visitors [1]. The end user is an individual who surfs websites or use mobile apps where they see the ads that they may click on. During the past few years, this industry has been severely victimized by the fraudulent activities due to the availability of large sum of money. The ultimate objective of the fraudulent users is to increase his own/third-party person revenue or exhausting advertiser's budget [12].

In this study, we propose a classification algorithm based on hidden Markov model—HMM so that fraudulent impressions in mobile advertising can be classified with a higher accuracy. An impression is defined as the displaying or loading event of advertisement into an advertisement frame. This is the first time to the best of our knowledge that evaluates the applicability of HMM-based approach in mobile impression fraud detection. The remainder of this paper is organized as follows. In Sect. 2, we discuss related work with available ad fraud detection approaches. Proposed approach of this study is discussed in detail under Sect. 3. The experimental results of the model are available in Sect. 4. Conclusions and discussion are given in Sect. 5.

## 2 Related Work

The researchers have proposed and implemented a number of click or impression fraud detection tools using distinct methodologies. Xu et al. [13] have been developed a detection system where they used stepwise evaluation process including proactive functionality test at front end(i.e. user interface) backed by JavaScript and passive examination of browsing behaviour to differentiate a clickbots from human clickers. DECAF [9] proposed an offline click fraud detection approach using rule-based method to detect placement fraud by analysing the advertisement user interface status in mobile apps/pages. Not-A-Bot (NAB)[5] is a system that enables a range of verifier policies for applications that would like to separate human-generated requests from bot traffic. NAB approximately identifies and certifies human-generated activities.

Several machine learning (ML) approaches have also been experimented by researchers to improve the accuracy, performance and reliability of fraudulent clicks and impressions detection mechanisms in mobile advertising. Perera et al. [11] evaluated number of ML algorithms such as decision trees, regression trees, artificial neural networks and support vector machines on real data produced by BuzzCity Ltd. The researchers were able to identify number of different fraudulent patterns in the data set but did not focus on detecting each individual event so-called impression. Haider et al. [6] has discussed another ML-based approach which is similar to previous authors approach, but the authors were able to achieve better improvement with identifying each individual fraud impression than common pattern of fraudulent events. Botnets detection approach was proposed by Gobel et al. [4] using hidden Markov model in their study. The proposed approach has been used network traffic generated by computers to model the HMMs so that bots can be identified through measuring the distances between these HMMs.

## 3 Proposed Approach

The proposed approach is called the hidden Markov scoring model—HMSM where we evaluate a new classification approach in the context of mobile impression fraud detection. The HMSM is based on scoring approach of HMM rather than conventional probabilistic model of HMM. The proposed methodology discusses an N to N process from feature selection to target state classification via fully automated process.

### 3.1 Hidden Markov Model

Hidden Markov models (HMMs) are stochastic models which were originally introduced in statistics literature in 1957 and studied in the late 1960s and early 1970s [8]. Initially, there are two terms to be understood before moving to the discussion of HMMs. First, hidden states where possible/available states of the target/hidden variable which are not visible to the observer directly. Second, observe or emission states are outputs that are visible to the observer directly depending on the hidden states. Hidden Markov model facilitates a way of identifying most probable hidden states of a given sequence while interacting with the observe or emission variable states [8]. Thus, the HMM can be defined by transition probabilities, emission probabilities and initial state distribution, respectively. Figure 1 illustrates the sequence of hidden states(S) and respective observe states(O).

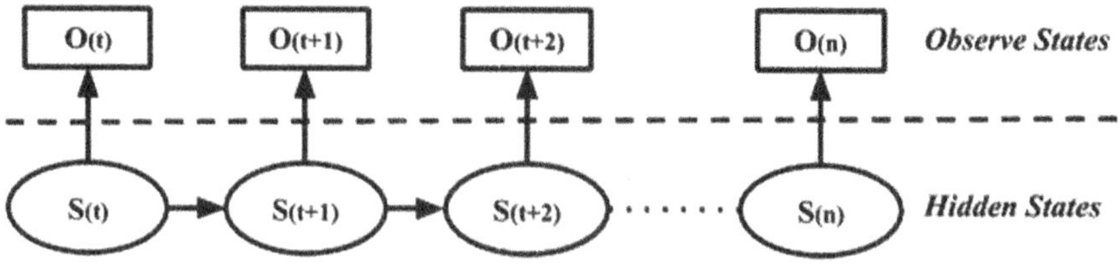


**Fig. 1.** Sequence of hidden and observe/emission states

*Transition probabilities-A*: Represents the state transition probabilities of the hidden variable conditioned on present state to a new state.

*Emission probabilities-B*: Contains the probabilities of an observe/emission variables state based on the hidden states.

*Initial state distribution-$\pi$*: Initial state distribution is a row vector representing the probability of each state of hidden variable at the beginning of the sequence. This can be seen as the prior probabilities of hidden states.

$$A_{(i,j)} = p\left(\frac{S_t = j}{S_{t-1} = i}\right), \forall_i = 1..M, \sum_{i=1}^{M} A_{(i,j)} = 1 \tag{1}$$

$$B_{(j,k)} = p\left(\frac{O_t = k}{S_t = j}\right) \tag{2}$$

$$\pi_i = p(S_1 = i), \forall_i = 1..M, \sum_{i=1}^{M} \pi_i = 1 \tag{3}$$

Here, $M$ represents total number of hidden states and $i, j$ denote the $1, \ldots, M$ index of the state while $k$ is number of possible discrete observations.

### 3.2 Data set

We used the same raw data set used by Haider et al. [6]. The data set contains a number of attributes such as *deliveryId, timestamp, clientIp, marketId, adSpaceId, accoundId, siteId, unknownDeviceId, clientVersionId, ipMarketId, ipCountyCode, ipIsp, adRelType, forcedAd, eventType, eventId, eventTimeStamp and status.* Details of the attributes can be found in [6].

**Derived Attributes** Derived attributes were introduced to the data set out of existing variables with respect to individual impression so that dimension of feature vector will be reduced. Some of derived attributes are listed below.

- *eventCount* - number of triggered events per impression
- *distEventTypes* - number of distinct event types of a given impression
- *surfTimeSec* - number of seconds users engage with an impression
- *maxEventCount* - maximum event count out of distEventTypes
- *distEventFreqGroups* - number of distinct event frequency groups

Finally, the data set is arranged in ascending order of the timestamp variable to guarantee the state transition from the previous state to the next state which is a fundamental nature of HMM.

### 3.3 Numerical Variable Transformation

Entropy-based binning is one of the supervised binning methods where it transforms numerical variables into categorical counterparts while referring the target variable [7]. The algorithm finds the best split so that the majority of the values in a bin corresponds to have the same label (target). It identifies the discretization cut-points based on maximum information gain depend on the target. Entropy-based binning may improve accuracy of the predictive models by reducing the noise or nonlinearity [3]. Since HMM basically interacts with state transition, all the numerical variables were transformed into categorical variables through entropy-based binning technique based on target variable.

### 3.4 Emission/Observe Variables Identification

Observe variables are the inputs to the proposed algorithm. Chi-square test of independence is a better option to determine the degree of relationship between two categorical variables [10]. Chi-square test statistic is compared with critical value, so that null or alternative hypothesis can be rejected. The critical value for the chi-square statistic is determined by the level of significance and the degrees of freedom. Null hypothesis-$H_0$(there is no statistically significant relationship between two variables) is rejected if the calculated chi-square test statistic is greater than the critical value.

Once numerical variables transformations are completed, all the categorical variables were evaluated with chi-square test of independence against the target variable with a significance level of 0.05. The variables which have higher chi-square test statistics than critical value were identified as the observe variables.

The final feature vector composed with nine observe variables including *adRelType, forcedAd, ipCountryCode, ipIsp, eventCount, distEventTypes, surfTimeSecCat, maxEventCount and distEventFreqGroups*. The feature vector is fed in to the proposed hidden Markov scoring model as observe variables.

### 3.5 Hidden Markov Scoring Model—HMSM Algorithm

The HMSM algorithm calculates scores for each target class based on observe variables to classify the data point in supervised learning approach. In supervised learning approach, a model should be trained first and then makes the predictions with trained parameters. Calculation of A, B and $\pi$ is referred to as training the model in HMSM. Equations (1), (2) and (3) were used to calculate the A, B and $\pi$ with the training data set.

**Scoring Approach** HMSM algorithm classifies the test data with a scoring model based on HMM. Target variable of the experimental data set has two states called OK and Fraud, where OK being the click is genuine and the Fraud being that the click is not genuine. The HMSM classifies each individual record in test data into either state.

*Feature Score*($fscore_{f,s}$): Calculate scores towards each hidden state (i.e. Fraud or OK) of target variable called $fscore_{Fraud}$ and $fscore_{OK}$ for each individual record in test data set with respect to each observe variable. Equation (4) defines the mathematical representation of $fscore_{f,s}$.

*Mean Deviation Feature Score*($mdfscore_{f,s}$): Equation (5) calculates the mean scores for each subset (i.e. $fscore_{Fraud}$ and $fscore_{OK}$) in order to calculate the deviation of the feature score from its subset mean represented in equation (6).

$$fscore_{f,s=i} = -\{log\pi_i + logA_{i,s} + logB_{s,k}\} \tag{4}$$

$$\mu_{s=i} = mean\left(fscore_{s=i}\right) \tag{5}$$

$$mdfscore_{f,s=i} = fscore_{f,s=i} - \mu_{s=i} \tag{6}$$

$$\forall_i = \{Fraud, OK\}, \forall_f = 1..F$$

*Minimum Mean Deviation Score*($mdscore_{min,s=i}$) : Select minimum mean deviation score out of $mdfscore$ for each hidden status as in equation (7).

$$mdscore_{min,s=i} = min\left(mdfscore_{s=i}\right) \tag{7}$$

HMSM identifies the most probable hidden state of a given record as the state of the maximum $mdscore_{min,s=i}$

$$hiddenState = max\left(mdscore_{min,OK}, mdscore_{min,Fraud}\right) \tag{8}$$

where $s$ = hidden states and $F$ = number of observe features.

## 4 Model Evaluation and Experimental Results

The HMSM was evaluated with 20 different test samples in order to verify the performance of the model across the data set. Accuracy, precision, recall, specificity and F-score were calculated for each individual sample as the performance measures.

### 4.1 Optimum Training Set Size

The fundamental problem of any machine learning algorithm is that finding optimum training set size to make classification or prediction. To solve this issue, HMSM was trained with 10 distinct training samples where sample size varies from 3000 to 25,000 records. Each training set is evaluated with 20 different test samples and calculated the mean value and mean of the standard deviation of performance measures to identify the optimum training set record size.

The HMSM model performs well in classifying test data when training sample size is 5000. Figure 2 illustrates the model performance with all training data sets where mean values for accuracy, recall, precision and F-score have been reached to above 80% and specificity reached to 79% at the sample size contains 5000 records. The stability of the model can be evaluated with the mean values of standard deviations of each training sample as shown in Fig. 3. The results show that the model is more stable when training data set has 5000 records where standard deviations of all the performance measures are less than 0.25 including specificity.

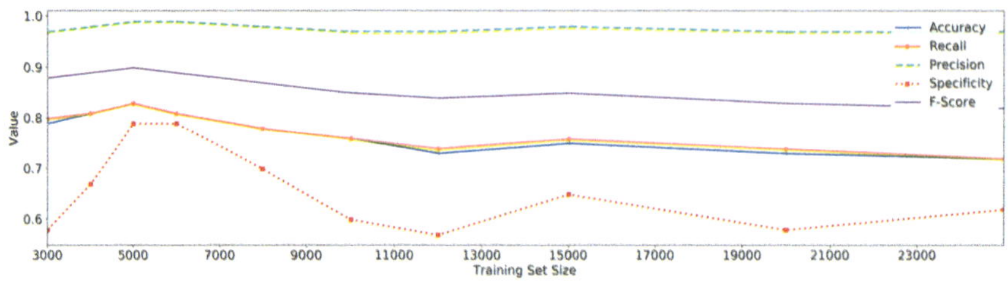


**Fig. 2.** Training set evaluation by mean

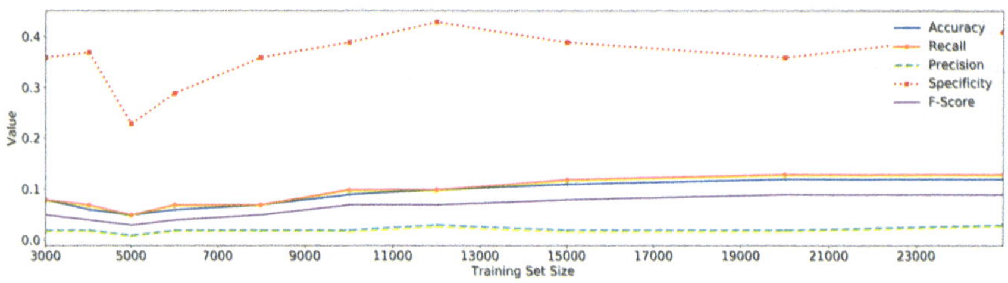


**Fig. 3.** Training set evaluation by standard deviation

### 4.2 Performance of the Optimum Model

According to the experimental results as shown in Fig. 4, the model trained with 5000 records classify test data by achieving highest performance across 20 different test samples. The optimum model achieved accuracy, precision, recall and F-score more than 75% across all samples while 15 samples perform specificity more than 65%. There is only one sample which has less than 50% of specificity.

Overall performance of the optimum model can be evaluated with the results shown in Table 1 where accuracy, recall, precision and F-score reached above 84% with the specificity of almost 80%. The lowest standard deviations of performance measures of the optimum model guaranteed that consistency or stability of the model.

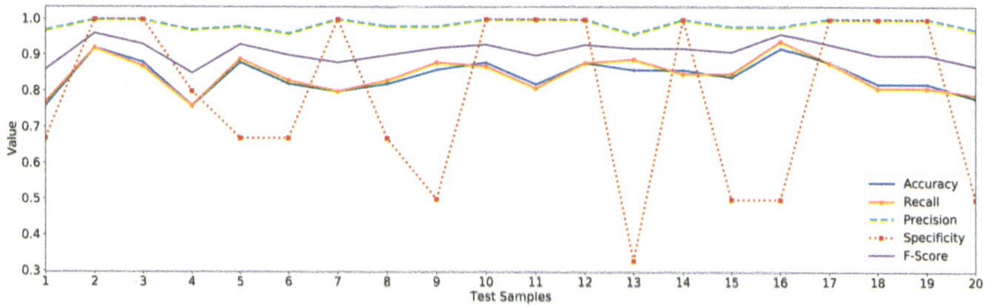


**Fig. 4.** Optimum model evaluation

**Table 1.** Consistency evaluation of optimum model

| | Mean | Std |
|---|---|---|
| Accuracy | 0.84 | 0.05 |
| Recall | 0.85 | 0.05 |
| Precision | 0.99 | 0.01 |
| Specificity | 0.79 | 0.23 |
| F-Score | 0.91 | 0.03 |

## 5 Conclusions and Discussion

The algorithm suggests the optimal size for the training data set size as 5000 records according to the 10 selected training samples started from 3000 to 25,000. Although it is generally accepted that large training data sets increase the performance of the model, the results show that the highest accuracy can be achieved by controlling the training data set size as small as 5000. This is one of the major advantages of HMSM which guarantees higher performance with smaller training data set. This concept promises that the number of training samples should not be large as much as possible and choosing the optimal training sample size helps to reduce the unnecessary computational time/power and large storage space. Therefore, the HMSM is a good alternative algorithm in supervised learning which reduces the computation time in training along with higher performance and higher learning efficiency on unseen data. Moreover, the proposed method is a good justification for the effect of reducing the sample size of the training data set. Thus, the HMSM has far reach implications in the supervised learning and is recommended for inclusion in the applications of real-time scenarios in this field.

In this paper, we proposed and studied the applicability of a new hidden Markov scoring model to classify fraudulent impressions in mobile advertising. The proposed HMSM shows significant capability as a fraudulent impression classifier with an average accuracy of 84%, average precision of 99%, average recall of 85%, average specificity of 79% and average F-score of 91% across 20 difference test samples with the standard deviation of 0.05, 0.01, 0.05, 0.23 and

0.03, respectively. The proposed HMSM can be identified as a stable classification algorithm because it performs well by identifying both positive and negative class in higher performance measures with low standard deviation. The authors plan to work with different types of feature engineering methodologies in future work so that the HMSM will be able to apply more precisely as a higher performance classification algorithm.

**Acknowledgement.** The authors of this paper would like to gratefully acknowledge the Rakin Heider who shared the label data set for this study.

## References

1. Aberathne, I., Walgampaya, C.: Smart mobile bot detection through behavioral analysis. In: Advances in Data and Information Sciences, pp. 241–252. Springer, Singapore (2018)
2. Cho, G., Cho, J., Song, Y., Kim, H.: An empirical study of click fraud in mobile advertising networks. In: 2015 10th International Conference on Availability, Reliability and Security, pp. 382–388. IEEE, New York (2015, August)
3. Dimić, G., Rančić, D., Milentijević, I., Spalević, P.: Improvement of the accuracy of prediction using unsupervised discretization method: educational data set case study. Tehnički vjesnik **25**(2), 407–414 (2018)
4. Gobel, W.: Detecting botnets using hidden Markov models on network traces (2013)
5. Gummadi, R., Balakrishnan, H., Maniatis, P., Ratnasamy, S.: Not-a-Bot: improving service availability in the face of botnet attacks. In: NSDI, vol. 9, pp. 307–320 (2009, April)
6. Haider, C.M.R., Iqbal, A., Rahman, A.H., Rahman, M.S.: An ensemble learning based approach for impression fraud detection in mobile advertising. J. Network Comput. Appl. **112**, 126–141 (2018)
7. Kotsiantis, S., Kanellopoulos, D.: Discretization techniques: a recent survey. GESTS Int. Trans. Comput. Sci. Eng. **32**(1), 47–58 (2006)
8. Li, X., Parizeau, M., Plamondon, R.: Training hidden Markov models with multiple observations-a combinatorial method. IEEE Trans. Pattern Anal. Mach. Intell. **22**(4), 371–377 (2000)
9. Liu, B., Nath, S., Govindan, R., Liu, J.: DECAF: detecting and characterizing Ad Fraud in mobile apps. In: 11th USENIX Symposium on Networked Systems Design and Implementation (NSDI 14), pp. 57–70 (2014)
10. McHugh, M.L.: The chi-square test of independence. Biochem. Med. **23**(2), 143–149 (2013)
11. Perera, K.S., Neupane, B., Faisal, M.A., Aung, Z., Woon, W.L.: A novel ensemble learning-based approach for click fraud detection in mobile advertising. In: Mining Intelligence and Knowledge Exploration, pp. 370–382. Springer, Cham (2013)
12. Wilbur, K.C., Zhu, Y.: Click fraud. Market. Sci. **28**(2), 293–308 (2009)
13. Xu, H., Liu, D., Koehl, A., Wang, H., Stavrou, A.: Click fraud detection on the advertiser side. In: European Symposium on Research in Computer Security, pp. 419–438. Springer, Cham (2014, September)

# Looking for Virtual Investors

Ion Chiţescu[1], Mădălina Giurgescu[2(✉)], and Titi Paraschiv[3]

[1] University Politehnica of Bucharest, Bucharest, Romania
ionchitescu@yahoo.com
[2] University of Piteşti, Piteşti, Romania
madalina_giurgescu@yahoo.com
[3] Military Technical Academy of Bucharest, Bucharest, Romania
titi.paraschiv@siemens.com

**Abstract.** This paper introduces a method of selecting the most probable future investment clients of a brokerage company on the capital market. The method consists in using the answers given by the virtual investors to a set of prescribed questions. Namely, the aforementioned answers are fusioned with a data mining procedure using the Choquet integrable. The scores thus obtained are classified using some preassigned thresholds, allowing to select the most probable future investment clients.

**Keywords:** Virtual investor · Data mining · Data fusion · Monotone measure · Measurable function · Choquet integral

## 1 Introduction

In this paper, we describe a mathematical model used for selecting the most probable future investment clients of a given brokerage company on the capital market. In writing this paper, we have worked in collaboration with a company named Confident Invest. This is a brokerage company on the Romanian capital market. Any company in the capital market is motivated to offer to its clients the best investment solutions. In this regard, the company wants to identify those customers who really want to invest and who have a great opening for collaboration. More specifically, a company wants to know which of its existing or potential clients really worry about finance. In this regard, Confident Invest tries to find the answers to the key question "I rarely worry about finances" by interpreting the answers to other questions the company is asking for, in a legal way. Specifically, the company realizes the risk profiles of the clients by applying a questionnaire. The risk profile of a client is made by providing 15 questions, each with 4 types of answers, ranging from the total disagreement to the absolute agreement. The answers are represented by integer numbers from 0 to 3. Thus, 0 represents total disagreement and 3 represents total agreement. Our target was


X.-S. Yang et al. (eds.), *Proceedings of Fifth International Congress on Information and Communication Technology*, Advances in Intelligent Systems and Computing 1183, https://doi.org/10.1007/978-981-15-5856-6_11

to determine the answer to the key question "I rarely worry about finances," based on the answers to the questionnaire. It is known that, as mathematical procedure, one frequently uses nonlinear integrals as fusion instruments. We shall take this into account to find the answer to the key question. Namely, this instrument should be the Choquet integral with respect to a monotone measure. The aforementioned answers are fusioned with a data mining procedure using the Choquet integral.

The scores thus obtained are classified using some preassigned thresholds, allowing to select the most probable investors. Confident Invest has given us the answers to the questions for 60 potential investors, as well as the values of preassigned thresholds to be able to calculate the target values. To calculate the values, we have created a C++ program. In the sequel, we explain how the nonlinear integral was used for data aggregation in order to solve the above-mentioned problem, and, of course, we explain the obtained results.

## 2 Preliminary Facts

Throughout the paper: $\mathbb{N} = \{1, 2, 3, \ldots\}$, $\mathbb{R}_+ = [0, \infty)$, $\bar{\mathbb{R}}_+ = [0, \infty] = \mathbb{R}_+ \cup \{\infty\}$. For any set $I$, we denote by $\mathcal{P}(I)$ the Boolean of $I$, i.e., is $\mathcal{P}(I) = \{A | A \subset I\}$.

**Definition 1.** The characteristic (indicator) function of a set $A \subset I$ is $\phi_A : I \to R_+$ given via $\phi_A(t) = 1$, if $t \in A$ and $\phi_A(t) = 0$, if $t$ is not in $A$.

For any finite set $E$, $|E|$ is the cardinal of $E$.

**Definition 2.** A measurable space is a pair $(I, \Sigma)$, where $I$ is a non-empty set and $\Sigma \subset \mathcal{P}(I)$ is a $\sigma-$algebra.

**Definition 3.** A (normalized) monotone measure is a function $\mu : \Sigma \to \mathbb{R}_+$ such that:

$$\begin{aligned} &(i)\ \ 0 = \mu(\emptyset) \le \mu(a) \le \mu(I) = 1 \text{ for any } A \in \Sigma \\ &(ii)\ \mu(A) \le \mu(B), \text{ whenever } A, B \text{ are in } \Sigma \text{ and } A \subset B \end{aligned} \tag{1}$$

Considering a measurable space $(I, \Sigma)$, a monotone measure $\mu : \Sigma \to \mathbb{R}_+$ and a $\Sigma-$measurable function $f : I \to \mathbb{R}_+$, we can define for any $a \in \mathbb{R}_+$, the level set $F_a = \{t \in I | f(t) \ge a\} \in I$. Because, for $0 \le a \le b < \infty$, one has $F_b \subset F_a$, it follows that the function $\phi : \mathbb{R}_+ \to \mathbb{R}_+$, given via $\phi(a) = \mu(F_a)$, is decreasing. So, we can compute $\int \phi dL$, where $L$ is the Lebesgue measure on $\mathbb{R}_+$.

**Definition 4.** The Choquet integral of $f$ with respect to $\mu$ is:

$$(C) \int f d\mu \stackrel{def}{=} \int \phi dL \le \infty \tag{2}$$

In case $(C)\int f d\mu < \infty$, we say that the function $f$ is Choquet integrable with respect to the monotone measure $\mu$. For generalized measure and integration theory (monotone measures, nonlinear integrals, especially Choquet integral), see [2,4,5].

Let us recall that the Choquet integral generalizes the abstract Lebesgue integral. Namely, in case $\mu$ is a classic (countably additive) measure, we have, for any $\mu$−measurable function $f : I \to \mathbb{R}_+$, the equality $(C)\int f d\mu = \int f d\mu$. For practical purpose, we recall the computation of the Choquet integral for simple functions. Consider again a measurable space $(I, \Sigma)$.

**Definition 5.** A $\Sigma$−simple (positive) function is a function $f : I \to \mathbb{R}_+$ having the form $f = \sum_{i=1}^{n} a_i \phi_{A_i}$, where $a_i \in \mathbb{R}_+$, $A_i \in I$ are mutually disjoint and $\cup_{i=1}^{n} A_i = I$.

For such a function $f$, we can reorder the values such that $0 \leq a_1 \leq a_2 \leq \cdots \leq a_n$. The reordering is unique in case all the values $a_i$ of $f$ are distinct. Then, considering that $0 \leq a_1 \leq a_2 \leq \cdots \leq a_n$ and $\mu : \Sigma \to \mathbb{R}_+$ is a monotone measure one has

$$(C)\int f d\mu = \sum_{i=1}^{n} (a_i - a_{i-1})\mu(A_i \cup A_{i+1} \cup \cdots \cup A_n) \tag{3}$$

with the convention $a_0 = 0$.

The most popular use of the previous formula is in case $I = \{1, 2, \ldots, n\}$ and $\Sigma = \mathcal{P}(I)$. Then, considering a permutation $\{1_*, 2_*, \ldots, n_*\}$ of $\{1, 2, \ldots, n\}$ such that $f(1_*) \leq f(2_*) \leq \ldots \leq f(n_*)$, we have (with convention $f(0_*) = 0$)

$$(C)\int f d\mu = \sum_{i=1}^{n} (f(i_*) - f((i-1)_*))\mu(i_*, (i+1)_*, \ldots, n_*) \tag{4}$$

# 3 Results

## 3.1 General Setting

In the sequel, $(I, \Sigma)$ will be a measurable space and $\mu : \Sigma \to \mathbb{R}_+$ will be a monotone measure.

For any $i \in I$, we shall consider a non-empty set $X_i$ and let $X \overset{def}{=} \prod_{i \in I} X_i$. Also, for any $i \in I$, let $\nu_i; X_i \to \mathbb{R}_+$ be a function. Then, for any $x = (x_i)_{i \in I}$, we can consider the function $f_x : I \to \mathbb{R}_+$, given via $f_x(i) = \nu_i(x_i)$.

Our major assumption will be that, for any $x \in X$, the function $f_x$ is Choquet integrable with respect to $\mu$. This is true, for instance, in the following two simple cases:

(a) $I$ is finite and $\Sigma = \mathcal{P}(I)$.

(b) We have simultaneously:
  - All $X_i$ are finite and equal (let $Y \overset{def}{=} X_i$ for any $i \in I$).

- All $\nu_i$ are equal (let $\nu \overset{def}{=} \nu_i$ for any $i \in I$, hence $\nu : Y \to \mathbb{R}_+$).
- It follows that, for any $x = (x_i)_{i \in I} \in \prod_{i \in I} X_i = Y^I$, the function $f_x$ takes all its values in the finite set $\nu(Y)$. We shall supplementarily assume that $f_x$ is $\Sigma$-simple (i.e., $f_x$ is $\Sigma$-measurable) for any $x \in Y^I$.

The Choquet integral $(C) \int f_x \mathrm{d}\mu$, computed for all $x \in \prod_{i \in I} X_i$, will be our data mining (fusion) instrument.

### 3.2 Prospecting Virtual Investors

We consider a brokerage company on the capital market looking for investment clients. The manager and the staff decide to search who are the most probable future investment clients among those who have shown collaboration with the company. For financial facts, see [1]. The target is to find out the answers to the key question "I rarely worry about finances." The method chosen to this end consists in asking the same set of 15 questions to all those who showed interest in the company and to compute their respective scores obtained from the answers to the aforementioned questions, via a procedure which will be described in the sequel. Each question has 4 types of answers, ranging from the total disagreement to the absolute agreement. The answers are represented by integer numbers from 0 to 3. Thus, 0 represents total disagreement and 3 represents total agreement. These scores, compared with some preassigned thresholds, will show who are the most probable virtual investors. The concrete procedure is, actually, a particular case of Sect. 3.1, point (a).
We take $I = \{1, 2, \ldots, n\}, n \in \mathbb{N}$ to be the set of the indexes assigned to the questions to which the possible investors are asked to answer. Numerical case here $n = 15$.
Our measurable space is $(I, \Sigma)$, with $I$, with $I$ as above and $\Sigma = \mathcal{P}(I)$.
We accept the existence of the monotone measure $\mu : \Sigma \to \mathbb{R}_+$. Numerical case here: $\mu$ is given as follows.

$$\mu(E) = \begin{cases} 0 & \text{if } E = \emptyset \\ 0.3 & \text{if } 0 < |E| \leq 4 \\ 0.5 & \text{if } 4 < |E| \leq 10 \\ 0.7 & \text{if } 10 < |E| \leq 14 \\ 1 & \text{if } |E| = 15 \end{cases} \tag{5}$$

We are using this measure $\mu$ because of a good previous experience with it. For any $i \in I$, $X_i$ will be the finite set of all possible answers to the question number $i$. So, every possible investor $x$ who answered to the set of questions will be identified as follows: $x \equiv (x_i)_{i \in I} = (x_1, x_2, \ldots, x_n)$. Here, for any $i \in I$, $x_i$ is the answer given by $x$ to the question number $i$.

Supplementary, with respect to the Sect. 3.1, we shall assume that any $X_i$ is ordered with the order relation $\leq_i$. This order relation puts into evidence the fact that some answers are more important (more definitory) than other answers, what concerns the decision concerning the eligibility.

On any $X_i$, we have the partial evaluation function $\nu_i : X_i \to \mathbb{R}_+$, which furnishes the marks for the answers in $X_i$. The function $\nu_i$ is monotone (another supplement with respect to the Sect. 3.1), i.e., for any $s, t \in X_i$ such that $s \leq_i t$, one must have $\nu_i(s) \leq \nu_i(t)$.
Numerical case here: for any $i \in I$, one has $\nu_i(X_i) \subset \{0, 1, 2, 3\}$. The meaning of the marks $\nu_i(t) = 0$ for "not true," $\nu_i(t) = 1$ for "partially true," $\nu_i(t) = 2$, for "generally true," $\nu_i(t) = 3$ for "always true."
*Procedure.* For any possible investor $x \equiv (x_1, x_2, \ldots, x_n)$, we form the function $f_x : I \to \mathbb{R}_+$ given via $f_x(i) = \nu_i(x_i)$. The "disponibility" of the possible investor $x$ to become a real investor is evaluated by the number

$$\nu(x) \stackrel{def}{=} (C) \int f_x(i) \mathrm{d}\mu(i) = (C) \int \nu_i(x_i) \mathrm{d}\mu(i). \tag{6}$$

The decision (prevision) concerning the possibility for $x$ to become a real investment client will be taken using the following thresholds:

$$\begin{array}{ll} \nu(x) \in [0, 0.5) & \to x \text{ is a secure investor} \\ \nu(x) \in [0.5, 1) & \to x \text{ is a potential investor} \\ \nu(x) \in [1, 2) & \to x \text{ is a neutral investor} \\ \nu(x) \in [2, 3] & \to x \text{ is a weak investor (almost} \\ & \quad \text{severely, } x \text{ will not invest).} \end{array} \tag{7}$$

## 4 Numerical Case and Conclusions

The test (questionnaire):

1. I keep all my e-mails.
2. My favorite subject in school was math.
3. I prefer to make order in the wardrobe than watching TV.
4. I prefer to work alone than to work in a team.
5. I consider myself independent.
6. I generally organize the event when I am invited to dinner or to the film.
7. I am bothering people who do not work hard.
8. I never leave anything unfinished.
9. Generally, I don't drive very fast.
10. I don't like competitive sports.
11. I don't like watching horror movies.
12. I am not eager to see new people.
13. I never get anxious when I wait for the elevator.
14. I never wear fashionable things.
15. I am never accused of having a choleric temper.

The answers of the clients:

Table 1. Answers of the clients

| Nr | 1 | 2 | 3 | 4 | 5 | 6 | 7 | 8 | 9 | 10 | 11 | 12 | 13 | 14 | 15 |
|---|---|---|---|---|---|---|---|---|---|---|---|---|---|---|---|
| 1 | 1 | 1 | 0 | 2 | 2 | 1 | 2 | 2 | 2 | 0 | 0 | 1 | 0 | 0 | 1 |
| 2 | 2 | 3 | 1 | 2 | 1 | 1 | 2 | 2 | 2 | 1 | 2 | 1 | 0 | 1 | 3 |
| 3 | 2 | 1 | 1 | 2 | 3 | 2 | 3 | 3 | 1 | 0 | 1 | 1 | 1 | 1 | 1 |
| 4 | 2 | 3 | 2 | 1 | 1 | 1 | 1 | 1 | 1 | 0 | 1 | 1 | 0 | 2 | 1 |
| 5 | 1 | 1 | 0 | 0 | 2 | 2 | 1 | 1 | 2 | 1 | 3 | 0 | 1 | 0 | 1 |
| 6 | 0 | 2 | 0 | 1 | 2 | 2 | 3 | 3 | 0 | 1 | 0 | 1 | 0 | 1 | 1 |
| 7 | 0 | 1 | 2 | 1 | 3 | 0 | 1 | 2 | 2 | 0 | 1 | 0 | 0 | 1 | 1 |
| 8 | 3 | 0 | 1 | 0 | 2 | 0 | 0 | 1 | 0 | 1 | 0 | 0 | 0 | 0 | 2 |
| 9 | 3 | 0 | 2 | 0 | 1 | 0 | 2 | 3 | 2 | 0 | 1 | 0 | 2 | 1 | 1 |
| 10 | 1 | 0 | 2 | 2 | 2 | 0 | 2 | 3 | 3 | 3 | 3 | 2 | 1 | 1 | 0 |
| 11 | 0 | 1 | 0 | 1 | 2 | 0 | 1 | 1 | 1 | 0 | 1 | 2 | 1 | 1 | 1 |
| 12 | 0 | 0 | 2 | 1 | 3 | 0 | 1 | 2 | 3 | 0 | 0 | 0 | 1 | 0 | 2 |
| 13 | 0 | 0 | 2 | 0 | 1 | 0 | 2 | 0 | 1 | 2 | 0 | 0 | 0 | 0 | 0 |
| 14 | 2 | 1 | 0 | 0 | 2 | 1 | 2 | 1 | 0 | 0 | 1 | 1 | 0 | 1 | 3 |
| 15 | 1 | 0 | 1 | 1 | 2 | 1 | 0 | 3 | 1 | 1 | 1 | 0 | 1 | 0 | 1 |
| 16 | 2 | 2 | 3 | 0 | 2 | 2 | 2 | 2 | 1 | 1 | 2 | 0 | 2 | 0 | 2 |
| 17 | 2 | 0 | 1 | 2 | 2 | 1 | 1 | 2 | 0 | 0 | 3 | 1 | 0 | 0 | 1 |
| 18 | 0 | 0 | 0 | 0 | 0 | 3 | 3 | 3 | 0 | 3 | 3 | 3 | 0 | 0 | 1 |
| 19 | 1 | 0 | 2 | 0 | 3 | 0 | 2 | 3 | 0 | 0 | 3 | 0 | 0 | 1 | 2 |
| 20 | 0 | 2 | 1 | 2 | 3 | 2 | 3 | 2 | 1 | 0 | 0 | 0 | 0 | 0 | 0 |
| 21 | 2 | 1 | 1 | 3 | 2 | 1 | 3 | 3 | 0 | 1 | 1 | 0 | 1 | 0 | 0 |
| 22 | 1 | 0 | 3 | 3 | 3 | 2 | 2 | 3 | 0 | 0 | 3 | 0 | 0 | 0 | 2 |
| 23 | 0 | 3 | 0 | 0 | 3 | 2 | 2 | 2 | 3 | 0 | 3 | 0 | 0 | 0 | 1 |
| 24 | 1 | 3 | 1 | 1 | 2 | 1 | 1 | 0 | 2 | 0 | 3 | 1 | 1 | 1 | 1 |
| 25 | 1 | 1 | 0 | 2 | 2 | 2 | 2 | 2 | 1 | 1 | 1 | 1 | 2 | 2 | 1 |
| 26 | 3 | 3 | 0 | 1 | 2 | 0 | 1 | 3 | 3 | 3 | 3 | 0 | 1 | 0 | 2 |
| 27 | 1 | 3 | 1 | 3 | 3 | 0 | 1 | 1 | 1 | 3 | 3 | 1 | 2 | 0 | 1 |
| 28 | 0 | 0 | 1 | 0 | 0 | 0 | 1 | 1 | 1 | 1 | 3 | 0 | 1 | 0 | 3 |
| 29 | 3 | 0 | 3 | 0 | 3 | 0 | 3 | 3 | 2 | 0 | 2 | 1 | 1 | 0 | 1 |
| 30 | 0 | 3 | 1 | 1 | 3 | 1 | 3 | 3 | 3 | 0 | 0 | 3 | 1 | 1 | 1 |
| 31 | 0 | 0 | 1 | 0 | 3 | 2 | 2 | 2 | 2 | 3 | 1 | 1 | 2 | 0 | 3 |
| 32 | 0 | 0 | 2 | 2 | 1 | 2 | 1 | 1 | 2 | 0 | 0 | 1 | 2 | 0 | 0 |
| 33 | 1 | 3 | 2 | 0 | 0 | 1 | 2 | 3 | 1 | 0 | 1 | 1 | 3 | 3 | 2 |
| 34 | 1 | 1 | 3 | 3 | 3 | 2 | 1 | 2 | 3 | 1 | 3 | 3 | 3 | 3 | 1 |
| 35 | 2 | 2 | 1 | 0 | 1 | 1 | 2 | 2 | 1 | 1 | 2 | 0 | 1 | 2 | 0 |

**Table 1.** Answers of the clients (*continued*)

| Nr | 1 | 2 | 3 | 4 | 5 | 6 | 7 | 8 | 9 | 10 | 11 | 12 | 13 | 14 | 15 |
|---|---|---|---|---|---|---|---|---|---|---|---|---|---|---|---|
| 36 | 0 | 0 | 2 | 0 | 3 | 1 | 1 | 2 | 1 | 0 | 0 | 0 | 1 | 0 | 0 |
| 37 | 0 | 1 | 2 | 0 | 3 | 0 | 3 | 2 | 0 | 0 | 0 | 0 | 1 | 0 | 0 |
| 38 | 2 | 2 | 2 | 2 | 2 | 1 | 2 | 2 | 1 | 1 | 3 | 1 | 1 | 1 | 1 |
| 39 | 3 | 0 | 0 | 0 | 3 | 0 | 1 | 3 | 1 | 0 | 2 | 1 | 1 | 0 | 2 |
| 40 | 3 | 0 | 2 | 0 | 1 | 0 | 1 | 3 | 1 | 0 | 2 | 0 | 2 | 0 | 2 |
| 41 | 1 | 0 | 0 | 0 | 3 | 2 | 2 | 3 | 1 | 0 | 0 | 0 | 1 | 0 | 0 |
| 42 | 1 | 3 | 0 | 2 | 2 | 1 | 1 | 1 | 1 | 1 | 1 | 1 | 0 | 0 | 0 |
| 43 | 1 | 0 | 0 | 0 | 0 | 2 | 1 | 1 | 3 | 1 | 1 | 0 | 1 | 2 | 2 |
| 44 | 2 | 0 | 2 | 1 | 2 | 0 | 1 | 2 | 1 | 0 | 3 | 0 | 2 | 1 | 1 |
| 45 | 2 | 0 | 0 | 0 | 1 | 2 | 2 | 2 | 3 | 2 | 3 | 0 | 2 | 2 | 0 |
| 46 | 1 | 3 | 0 | 1 | 2 | 2 | 2 | 3 | 1 | 0 | 3 | 0 | 2 | 0 | 1 |
| 47 | 1 | 1 | 3 | 0 | 2 | 1 | 2 | 2 | 0 | 0 | 0 | 1 | 1 | 0 | 0 |
| 48 | 0 | 2 | 1 | 0 | 1 | 2 | 0 | 2 | 0 | 3 | 3 | 0 | 0 | 0 | 3 |
| 49 | 1 | 1 | 0 | 2 | 2 | 1 | 1 | 0 | 1 | 0 | 3 | 2 | 1 | 0 | 2 |
| 50 | 0 | 0 | 0 | 0 | 0 | 1 | 3 | 3 | 2 | 0 | 0 | 0 | 0 | 0 | 0 |
| 51 | 2 | 1 | 0 | 1 | 3 | 1 | 1 | 0 | 0 | 2 | 2 | 1 | 2 | 1 | 3 |
| 52 | 2 | 1 | 2 | 1 | 2 | 0 | 0 | 2 | 1 | 1 | 3 | 1 | 2 | 0 | 2 |
| 53 | 2 | 2 | 1 | 3 | 1 | 1 | 1 | 2 | 2 | 2 | 3 | 1 | 2 | 2 | 1 |
| 54 | 0 | 3 | 1 | 0 | 1 | 0 | 2 | 1 | 0 | 1 | 0 | 0 | 0 | 0 | 0 |
| 55 | 1 | 0 | 0 | 1 | 3 | 1 | 1 | 2 | 2 | 1 | 2 | 0 | 1 | 0 | 2 |
| 56 | 0 | 3 | 3 | 0 | 2 | 0 | 1 | 3 | 1 | 0 | 3 | 0 | 0 | 0 | 2 |
| 57 | 1 | 1 | 2 | 1 | 3 | 0 | 1 | 3 | 1 | 0 | 3 | 0 | 1 | 0 | 2 |
| 58 | 0 | 0 | 0 | 0 | 1 | 1 | 1 | 0 | 2 | 1 | 3 | 2 | 0 | 0 | 1 |
| 59 | 2 | 2 | 1 | 1 | 1 | 3 | 2 | 1 | 3 | 1 | 3 | 0 | 1 | 1 | 0 |
| 60 | 2 | 2 | 2 | 0 | 0 | 1 | 2 | 3 | 2 | 1 | 0 | 0 | 1 | 0 | 1 |

To calculated the values for Table 2, we have created a C++ program (see [3]). The source code is written in C++ in the CodeBlocks development medium, 17.12 version on Windows 10 operating system, combined with GNU GCC Compiler in MinGW distribution, 6.3 version. For the matrix operations, the Eigen library version 3.3 was used (Tables 1 and 2).
The results and conclusions:

**Table 2.** Results and conclusions

| Nr | Value | Investor type |
|---|---|---|
| 1 | 0.8 | Potential investor |
| 2 | 1.5 | Neutral investor |
| 3 | 1.3 | Neutral investor |
| 4 | 1.3 | Neutral investor |
| 5 | 1.1 | Neutral investor |
| 6 | 1.1 | Neutral investor |
| 7 | 1.1 | Neutral investor |
| 8 | 1.3 | Neutral investor |
| 9 | 1.3 | Neutral investor |
| 10 | 1.3 | Neutral investor |
| 11 | 1.2 | Neutral investor |
| 12 | 1.3 | Neutral investor |
| 13 | 0.6 | Potential investor |
| 14 | 1.5 | Neutral investor |
| 15 | 1.1 | Neutral investor |
| 16 | 1.3 | Neutral investor |
| 17 | 1.1 | Neutral investor |
| 18 | 1.5 | Neutral investor |
| 19 | 1.3 | Neutral investor |
| 20 | 0.9 | Potential investor |
| 21 | 1.1 | Neutral investor |
| 22 | 1.3 | Neutral investor |
| 23 | 1.1 | Neutral investor |
| 24 | 1.3 | Neutral investor |
| 25 | 1.2 | Neutral investor |
| 26 | 1.3 | Neutral investor |
| 27 | 1.3 | Neutral investor |
| 28 | 1.5 | Neutral investor |
| 29 | 1.1 | Neutral investor |
| 30 | 1.7 | Neutral investor |
| 31 | 1.5 | Neutral investor |
| 32 | 1 | Neutral investor |
| 33 | 1.7 | Neutral investor |
| 34 | 2 | Weak investor |
| 35 | 1 | neutral investor |

**Table 2.** Results and conclusions (*continued*)

| Nr | Value | Investor type |
|---|---|---|
| 36 | 1.1 | Neutral investor |
| 37 | 1.1 | Neutral investor |
| 38 | 1.6 | Neutral investor |
| 39 | 1.3 | Neutral investor |
| 40 | 1.3 | Neutral investor |
| 41 | 1.1 | Neutral investor |
| 42 | 1.1 | Neutral investor |
| 43 | 1.3 | Neutral investor |
| 44 | 1.3 | Neutral investor |
| 45 | 1.3 | Neutral investor |
| 46 | 1.3 | Neutral investor |
| 47 | 1.1 | Neutral investor |
| 48 | 1.5 | Neutral investor |
| 49 | 1.3 | Neutral investor |
| 50 | 0.9 | Potential investor |
| 51 | 1.7 | Neutral investor |
| 52 | 1.3 | Neutral investor |
| 53 | 1.8 | Neutral investor |
| 54 | 0.9 | Potential investor |
| 55 | 1.3 | Neutral investor |
| 56 | 1.3 | Neutral investor |
| 57 | 1.3 | Neutral investor |
| 58 | 1.3 | Neutral investor |
| 59 | 1.1 | Neutral investor |
| 60 | 1.1 | Neutral investor |

We consider that this method is new and can be used to various decision situations.

## References

1. Chisholm, A.: An Introduction to Capital Markets: Products, Strategies. Participants. Wiley, Hoboken (2002)
2. Chiţescu, I., Plăviţu, A.: Computing Choquet integrals. Fuzzy Sets Syst. **327**, 46–68 (2017). https://doi.org/10.1016/j.fss.2017.02.009
3. Stroustrup, B.: The C++ Programming Language, 4th edn. Addison-Wesley Professional (2013)

4. Wang, Z., Klir, G.J.: Generalized Measure Theory. Springer, Berlin (2009)
5. Wang, Z., Yang, R., Leung, K.-S.: Nonlinear Integrals and Their Applications in Data Mining. World Scientific, Singapore (2010)

# Scalability Analysis of Low-Power Wide Area Network Technology

N. A. Abdul Latiff[1], I. S. Ismail[1](✉), M. H. Yusoff[2], A. R. Salisa[1], and J. A. Shukor[3]

[1] Faculty of Ocean Engineering Technology and Informatics, University Malaysia Terengganu, Kuala Terengganu, Malaysia
nurul_adilah@umt.edu.my

[2] Faculty of Informatics and Computing, Universiti Sultan Zainal Abidin (UniSZA), Kuala Terengganu, Malaysia

[3] School of Computing, Faculty of Engineering, University Technology Malaysia, Johor Bahru, Malaysia

**Abstract.** Low-power wide area network is a new wireless communication technology designed for low-power consumption together with long-distance communications, and LoRa technology is one of the leading technology solutions. The long-range connection between end-nodes and gateway is achievable by LoRa devices due to star-based network topology and modulation techniques used in wireless communication of the technology. One of the main features of LoRa technology is the ability to scale. Modelling and simulation can interpret the actual network behaviour of LoRa technology as accurate as possible. This paper aims to investigate the performance of the low-power wide area network technology focusing on capability of the network to scale. We model the network system based on the behaviours of the communication between the end-node and gateway. The simulation to study the scalability was done based on several parameters, such as the number of end-nodes, application time and the number of channels used by the end-node. The results show that the amount of successfully received data signal at gateway increased as the application time and channel used increased.

**Keywords:** LoRa · LPWAN · Modelling · Simulation · Scalability

## 1 Introduction

For the past few years, the advances in development of Internet-of-Things (IoT) speed up the development for new, innovative, convenience and economic benefit applications. The things are any devices holding the ability of sensing, computing and exchanging information with other devices by communicating via the Internet [1]. The technology inspired by IoT is believed to be able to enhance the effectiveness of energy consumption, resources management, productivity and environmental monitoring. The IoT has


X.-S. Yang et al. (eds.), *Proceedings of Fifth International Congress on Information and Communication Technology*, Advances in Intelligent Systems and Computing 1183, https://doi.org/10.1007/978-981-15-5856-6_12

spread its importance to various domains which previously has limitation in various possible applications, for example, real-time environmental, remote health-care, industrial control, productions systems, smart city and transportation [2].

Developers have worked on many applications for various uses in the IoT, and specific applications require specialized technology to work accordingly. Traditional short-range communication technologies such as Bluetooth and ZigBee are not ideal for applications that require long-range communication. While cellular technology can provide broader coverage than conventional radio technology, it requires much energy to operate, which is not ideal for low-power applications.

Recent advancement in applications that require smaller size device, low power consumption and cost-effective has shown positive progress in the development of a new communication technology known as the low-power wide area network (LPWAN). The new technology becomes a complement to conventional communication technology such as cellular and short-range wireless technologies by augmenting a better functionality and requirement for IoT applications. The unique features of LPWAN technology such as high coverage, low bandwidth and low power consumption, are in line with the requirements of IoT applications that only need to transmit small data sizes remotely.

LPWAN technology is designed to support billions of devices for the various applications of IoT. The technology uses a star topology architecture in which multiple end-nodes communicate directly to the gateway [3]. However, numerous end-nodes transmit the data signal to the gateway, and this causes traffic overload, and eventually, they are lost in data signal at the gateway.

The identification of the problem and further understanding of scalability for the network is necessary to improve the design of devices and gateways development. For such cause, this paper aims to analyse the performance of LPWAN in terms of scalability. The scalability model of LPWAN was designed, and simulations were run for various parameters such as the number of end-nodes, the number of channels used and the time delay between two consecutives data signal transmission. The organization of this paper is as follow: In Sect. 2, related work on the previous study was presented. The introduction of low-power wide area networks was presented in Sect. 3. Then, the proposed network model was discussed in Sect. 4. In Sect. 5, the simulation parameters were discussed, and result and discussion were presented in Sect. 6. Finally, conclude the paper in conclusion.

## 2 Related Work

Several studies on model development for LPWAN have been conducted previously for a better understanding of LPWAN's ability to scale. This section presents the previous studies on several works focusing on modelling and scalability of LPWAN.

Georgiou and Raza [4] present the study of the modelling of LPWAN to analyse the capability of the technology to scale. The authors used stochastic geometry framework to model the performance of the LoRa network that is using a single gateway. The study proposed an outage probability model which occurs at the gateway called outage condition.

Example of studies following the outage probability model by the authors in [4] was presented in [5–7]. The authors in [5] used time diversity to increase the probability of successful packet delivery from the sensor nodes to the gateways. While in [6], the authors presented the paper based on the model in [4]. It included the effect of co-SF interference and inter-SF interference as the model in [4] did not consider the interferences factor in the model. The model presented in [7] is the extending of the outage models from [4] for diversity techniques.

The study of the scalability of LPWAN was presented in [8] by Bor et al. This paper investigates the number of transmitters that LoRa network can support. The authors developed LoRa simulator called LoRaSim, which is used to study the scalability of LoRa network. Following the model and simulator from [8], several studies focus on the performance of LPWAN in terms of scalability [9–11]. The authors in [9] used LoRa communication model based on [8] to develop further an improved version of LoRaSim called EXPLoRa. Meanwhile, the authors in [11] used the model and simulator from [8] to study the performances of LoRaSim on three different simulation parameters; SF, bandwidth and coding rate.

The authors in [10] also developed a LoRa model similar to the model from [8]. The scalability of the LoRa network is studied by observing the most significant possible number of LoRa transmitter while satisfying the average packet success probability. The other model was developed by the authors in [12] to study the scalability of LoRa technology. The LoRa interference behaviour has been used in the model.

## 3 Low-Power Wide Area Network

Low-power wide area network is a wireless communication technology that enables end-nodes to communicate over long distance using low bit rates and low energy consumption [13–15]. Previous studies have shown that LPWAN technology enables the final node to communicate with gates over a distance of 3 kms for urban areas, while more than 10 kms for rural areas [16]. Additionally, in the line of sight circumstance, the last node data signal that can reach a gateway located 20 kms away [17] can still reach the gateway as far as 30 kms, as reported in [18].

The ability of end-nodes to communicate remotely with a gateway is based on two main special features of the LPWAN, the star network topology and modulation technique. The LPWAN device mostly operates in the unlicensed industrial, scientific and medical (ISM) bands at 169, 433, 868/915 MHz and 2.4 GHz [19]. However, these frequency values depend on the region in which the technology is being used. [20, 21].

Dynamic progress in LPWAN technology development has created many LPWAN-based applications and solutions in the market. The current most known LPWAN technologies are Sigfox and Semtech. The Sigfox technology uses three main components for the communication, which are ultra narrow band (UNB) radio technology, binary phase shift keying (BPSK) and Gaussian frequency shift keying (GFSK) modulation. Typically, depending on the region, the ISM band used by the technology is at 868–869 and 902–928 MHz. Sigfox devices are capable of sending small data with 12 bytes of maximum data size for uplink data while 8 bytes of downlink using the lightweight protocol. Altogether, the Sigfox frame uses 26 bytes, with 12 bytes of load data and 14 bytes for protocol overhead. This protocol overhead is smaller than conventional LPWAN technology, which applies more significant size protocol overheads to transmit data [22].

In addition to Sigfox, Semtech also developed the LPWAN technology known as LoRa technology. The technology is designed for a combination of remote, low power consumption and secure small-size data transmission. It also operates on an unlicensed SUB-GHz ISM band using a so-called chirp spread spectrum (CSS) modulation to optimize power consumption and broader communications networks. LoRa Technology uses the combination of two layers; the physical layer is known as LoRa for the connectivity and the MAC layer known as LoRaWan.

# 4 Network Model

This section describes the network model which mimic the communication between end-node and gateway for scalability study purpose. The following are the assumptions for the behaviour of the data signal from the end-node to be received by the gateway based on [4, 8, 12].

## 4.1 The Interference Conditions

Consider two or more data signal from end-node. We want to see the status of the data signal from a single end-node (reference node, Nr) whether it is successfully received by the gateway when there is one or more interference data signal from others end-node (interference node, Ni).

Let say that there are two data signals from two end-node (end-node 1 and 2) arrive at the gateway. If both of the end-node use different spreading factor or channel to each other's, then both data signal are successfully received by the gateway. These data signal conditions are said to be orthogonal to each other when the spreading factor of the end-nodes are different. For example, end-node 1 uses a spreading factor of 7, and end-node 2 uses a spreading factor of 8. The gateway will receive both the data signal. Table 1 below provides the details condition.

**Table 1.** Interference conditions of the data signal

| Spreading factor | Channel | Condition |
|---|---|---|
| Used same value | Used same value | Interference |
| Used same value | Used different value | Receive |
| Used different value | Used same value | Receive |
| Used different value | Used different value | Receive |

Interference occurs when the reference and interference node have the same spreading factor and channel. Figure 1 below illustrates all possible interference by the end-nodes. Meanwhile Table 2 below indicates the status of the reference node at the gateway. In case 2, the data signal from the reference node did not receive by the gateway. The interference node is currently being uplink to the gateway. While in case 3 and 4, the collision happens for the data signal. Both the data signal was not effectively received by the gateway. However, in case 5, both the data signal are assumed to be received by the gateway.

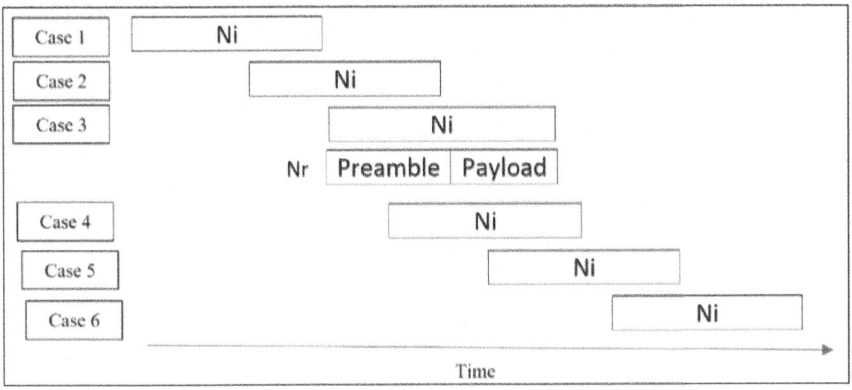


**Fig. 1.** Diagram of reference node and interference node based on time

**Table 2.** Status of reference node and interference node

| Case | Status Ni | Status Nr |
|---|---|---|
| 1 | Receive | Receive |
| 2 | Loss | Loss |
| 3 | Loss | Loss |
| 4 | Loss | Loss |
| 5 | Receive | Receive |
| 6 | Receive | Receive |

### 4.2 Spreading Factor Selection

In this study, the selection of the spreading factor for the end-node was inspired by [2]. The spreading factor was decided based on the distance between the end-node and the gateway. Typically, the end-node chose the spreading factor based on the value of RSSI and SNR at the gateway. When the distance increased as the location of the end-node located far from the gateway or the data signal is attenuated, the end-node is required to use the higher spreading factor to transmit the data signal to the gateway [12].

From the previous study, the assumption for the end-node that located far from the gateway used the SF of 12. Its data signal was assumed to be able to reach the gateway. However, if the location of the end-node located too far, the data signal cannot be received by the gateway. There should have a limit of the distance between the location of the end-node and gateway for the data signal to be successfully received by the gateway. Table 3 shows the SF for the end-node based on the distance between the end-node and gateway.

**Table 3.** SF selection of the end-node

| SF | RSSI (dBm) | Distance (km) |
|---|---|---|
| 7 | (−124, −100) | <2 |
| 8 | (−129, −124) | 2–4 |
| 9 | (−130, −129) | 4–6 |
| 10 | (−133, −130) | 6–8 |
| 11 | (−135, −133) | 8–10 |
| 12 | (−137, −137) | 10–12 |
| 12 | (−140, −137) | 12–14 |
| ~ | ~ | >14 |

## 5 Simulation

The performance on the scalability of the LPWAN is executed via simulation using MATLAB. A program is built in the MATLAB to run the simulation for the performance on the scalability of the LPWAN. Let say there are N numbers of end-nodes distributed randomly in Lx x Ly two-dimensional network field with a single gateway located at the middle of the network field.

For simplicity, the end-nodes are assumed to use specific SF based on the distance between the end-node and the gateway, *d*. Let $d(n) = (x(n), y(n))$ be the coordinate of the distributed end-nodes and $G(g) = (x(g), y(g))$ be the coordinate of the gateway location, where $n = \{1, 2, 3\ldots, N)$ and g is number of gateways. Then, the distance for the end-node j from the gateway is defined as;

$$d(G(g), d(n)) = \left[(x(g)-y(g))^2 + (x(n)-y(n))^2\right]^{1/2} \tag{1}$$

Besides that, the gateway randomly assigns the channel (CH(n)) for the end-nodes in the range of [1, CH], where CH is the total number of channels. In order to mimic the real application for the end-nodes, the starting time is assigned randomly for the end-nodes to start transmitting the packet data. Starting time (ST(n)) is randomly chosen by the end-nodes based on the range time [0, Application time]. Application time is time for end-node to transmit next data signal after complete transmitting the current data signal.

$$\mathrm{CH}(n) = \mathrm{rand}([1,\ \mathrm{CH}]) \tag{2}$$

$$\mathrm{ST}(n) = \mathrm{rand}(\left[0,\ \text{Application time}\right] \tag{3}$$

Then, the end-node start to transmit the packet data to the gateway based on its starting time. When the end-node complete transmitting the packet data, the new starting time (New_ST($n$)) for end-node is set with the combination of the starting time, time-on-air (ToA), and processing time ($PT$). Processing time is included to indicate the time for the end-node to process the data for the next transmission sequel in the range of [0,1.000 s]. Note that, the maximum time for end-nodes to process the data is assumed to be 1 s. ToA is the time for the data signal from the end-node to successfully received by gateway. However, it depends on the size of the payload, bandwidth, spreading factor and code rate used by the end-node for data transmission. Refer to [23] for more information on ToA.

$$PT(n) = [0, 1.000\text{s}] \tag{4}$$

$$\text{New\_ST}(n) = \text{ST}(n) + \text{ToA}(n) + \text{PT}(n) \tag{5}$$

At the gateway, the data signal is successfully received based on the interference conditions as discussed in the previous section. The simulation is run based on round. Each round ends if all the end-node complete transmitting the data signal to the gateway. The simulation stops when the total number of runs is reached. Then, the program calculates the percentage of received packet data (PPD). PPD is the percentage of the total received data signal at gateway over the total number of data signal transmission from end-node. Table 4 below shows the parameter used in the simulation.

**Table 4.** Parameters for the simulation

| Parameters | Values |
|---|---|
| Size of network field | 24000 m × 24000 m |
| Number of end-nodes, $N$ | 100, 200, 300, 400, 500, 600, 700, 800, 900, 1000, 1500, 2000 |
| Packet payload size | 50 bytes |
| Bandwidth | 125 kHz |
| Coding rate | 4/5 |
| Spreading factor | 7, 8, 9, 10, 11, 12 |
| Channel | 1,3,6,8 |
| Application time (s) | 600, 1200, 1800, 2400, 3600 |
| Total round | 50 |

## 6 Results and Analysis

In this work, we observe the scalability of LPWAN to investigate the total number of end-node to successfully receive in gateway. Three parameters were introduced in the simulation, such as the total number of end-node, application time and number of channels in order to investigate the effect of these parameters on the PPD. Figure 2a shows the PPD against the number of end-node with different application time for a single channel (CH1) while Fig. 2b for eight channels (CH8). Meanwhile, Fig. 3a shows the PPD against the number of end-node with a different number of CH for an application time of 600 s and Fig. 3b for an application time of 3600 s. Based on the results, higher value of PDD is preferred as more data signal from end-nodes are successfully received by the gateway.

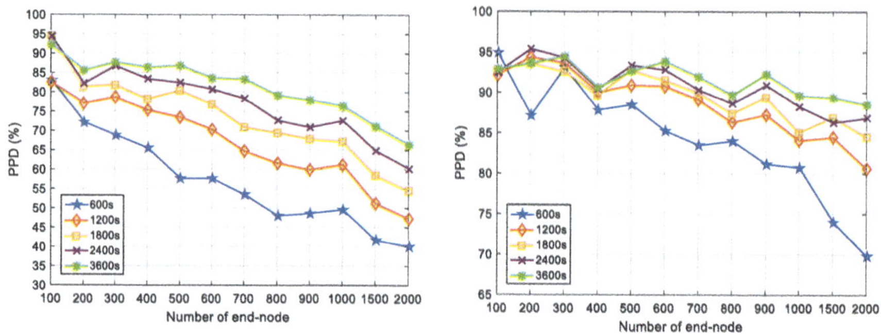


(a) (b)

**Fig. 2.** **a** PPD versus number of end-node for different application time using single channel. **b** PPD versus number of end-node for different application time using eight channels

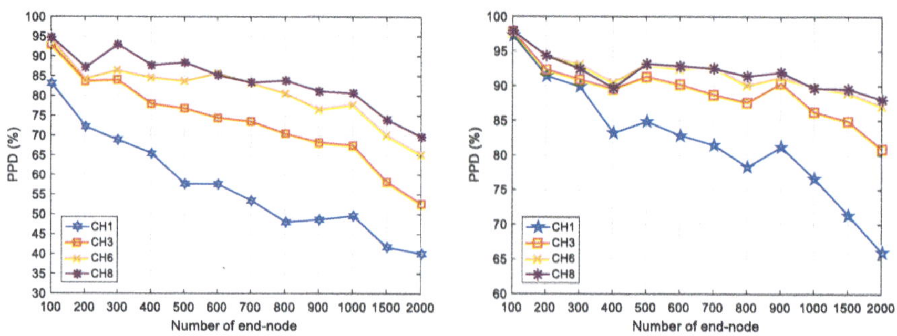


(a) (b)

**Fig. 3.** **a** PPD versus number of end-node for different channel with 600 s application time. **b** PPD versus number of end-node for different channel with 3600 s application time

Overall, the PPD decreases when the total number of end-node increases for all figures. Generally, this decrement is due to the increasing number of interruptions between the signals occurring in the gateway as the number of end-nodes increases. By referring to Fig. 2a and b, the value for PPD increases as the application time increases for all number of the end-nodes, and the start time for end-nodes is directly related to the application time. The decrement pattern of PPD for both Fig. 3a, b are similarly dissipated the different number of channels used. However, the PPD value using eight channels are higher compared to single channel while using similar application time. It can be concluded that the collision between data signals from the end-node decreases as the number of start times for the end-node to transmit data signals varies over a long period.

Meanwhile, for Fig. 3a and b, the value of PPD increases as the amount of CH increases for all number of end-nodes. The gateway randomly selected the CH used by the end-nodes. As the number of CH to be used by the end-nodes increases, the chance of signal data to collide with each other decreases, resulting in increasing the PPD values.

Limitation on the distance between the end-node and the gateway also affect the value of PPD. The end-node that is located above the distance limit that allows the data signal to reach the gateway will increase the amount of data signal loss in the gateway. The number of end-nodes located at the given distance limit increases as the number of end-nodes used in the simulation increases.

## 7 Conclusion

This paper studies the performance of LPWAN in terms of the capability to scale. A program was designed in the MATLAB platform based on the communication behaviour of the end-nodes and gateway to investigate the scalability of LPWAN technology. The limit for the communication was introduced for the data signal of the end-nodes to reach the gateway successfully. Three parameters such as the total number of end-node, application time and number of channels are used to investigate the effect on the PPD value. Based on the results, when the number of end-nodes increase, the PPD value decreases because of collision between the data signal. The increasing number of application time and CH used by the end-nodes for transmitting the data signal results in increasing the PDD value.

**Acknowledgements.** The authors would like to be obliged to University Malaysia Terengganu and Ministry of Education Malaysia for providing financial assistance under project no. 59466.

## References

1. Serdaroglu, K.C., Baydere, S.: WiSEGATE: wireless sensor network gateway framework for internet of things. Wirel. Netw. **22**, 1475–1491 (2016). https://doi.org/10.1007/s11276-015-1046-5
2. Ismail, I.S., Latiff, N.A.A., Aziemah, N.A.A.: Performance analysis of data recovery via application layer for LPWAN. In: 2019 IEEE 89th Vehicular Technology Conference (VTC2019-Spring). pp. 1–5. IEEE (2019)
3. LoRa Alliance: A technical overview of LoRa and LoRaWAN, https://www.lora-alliance.org/technology
4. Georgiou, O., Raza, U.: Low power wide area network analysis: can LoRa scale? IEEE Wirel. Commun. Lett. **6**, 162–165 (2017). https://doi.org/10.1109/LWC.2016.2647247
5. Hoeller, A., Souza, R.D., Ĺopez, O.L.A., Alves, H., De Noronha Neto, M., Brante, G.: Exploiting time diversity of LoRa networks through optimum message replication. Proc. Int. Symp. Wirel. Commun. Syst. 1–5 (2018). https://doi.org/10.1109/ISWCS.2018.8491203

6. Mahmood, A., Sisinni, E., Guntupalli, L., Rondon, R., Hassan, S.A., Gidlund, M.: Scalability analysis of a LoRa network under imperfect orthogonality. IEEE Trans. Ind. Informatics. **15**, 1425–1436 (2019). https://doi.org/10.1109/TII.2018.2864681
7. Hoeller, A., Souza, R.D., Alcaraz Lopez, O.L., Alves, H., de Noronha Neto, M., Brante, G.: Analysis and performance optimization of LoRa networks with time and antenna diversity. IEEE Access. **6**, 32820–32829 (2018). https://doi.org/10.1109/ACCESS.2018.2839064
8. Bor, M.C., Roedig, U., Voigt, T., Alonso, J.M.: Do LoRa low-power wide-area networks scale?. Proc. 19th ACM Int. Conf. Model. Anal. Simul. Wirel. Mob. Syst.—MSWiM'16. 59–67 (2016). https://doi.org/10.1145/2988287.2989163
9. Cuomo, F., Campo, M., Caponi, A., Bianchi, G., Rossini, G., Pisani, P.: EXPLoRa: extending the performance of LoRa by suitable spreading factor allocations. Int. Conf. Wirel. Mob. Comput. Netw. Commun. (2017, October). https://doi.org/10.1109/WiMOB.2017.8115779
10. Zorbas, D., Papadopoulos, G.Z., Maille, P., Montavont, N., Douligeris, C.: Improving LoRa network capacity using multiple spreading factor configurations. In: 25th International Conference Telecommunications ICT 2018. pp. 516–520 (2018). https://doi.org/10.1109/ICT.2018.8464901
11. Lavric, A., Popa, V.: Performance evaluation of LoRaWAN communication scalability in large-scale wireless sensor networks. Wirel. Commun. Mob. Comput. **2018**, 1–9 (2018). https://doi.org/10.1155/2018/6730719
12. Haxhibeqiri, J., Abeele, FVan, Van den Ablee, F., Moerman, I., Hoebeke, J.: LoRa scalability: a simulation model based on interference measurements. Sensors **17**, 1193 (2017). https://doi.org/10.3390/s17061193
13. Kim, D.H., Lim, J.Y., Kim, J.D.: Low-power, long-range, high-data transmission using Wi-Fi and LoRa. In: 2016 6th International Conference on IT Convergence and Security (ICITCS). pp. 1–3. IEEE, Prague, Czech Republic (2016)
14. Neumann, P., Montavont, J., Noel, T.: Indoor deployment of low-power wide area networks (LPWAN): a LoRaWAN case study. 2016 IEEE 12th International Conference on Wireless and Mobile Computing. Networking and Communications (WiMob), pp. 1–8. IEEE, New York, NY, USA (2016)
15. Kim, D., Jung, M.: Data transmission and network architecture in long range low power sensor networks for IoT. Wirel. Pers. Commun. **93**, 119–129 (2017). https://doi.org/10.1007/s11277-016-3482-7
16. Centenaro, M., Vangelista, L., Zanella, A., Zorzi, M.: Long-range communications in unlicensed bands: the rising stars in the IoT and smart city scenarios. IEEE Wirel. Commun. **23**, 60–67 (2016). https://doi.org/10.1109/MWC.2016.7721743
17. Pham, C.: Low-cost, low-power and long-range image sensor for visual surveillance. In: Proceedings of the 2nd Workshop on Experiences in the Design and Implementation of Smart Objects - SmartObjects'16. pp. 35–40. ACM Press, New York, New York, USA (2016)
18. Petajajarvi, J., Mikhaylov, K., Roivainen, A., Hanninen, T., Pettissalo, M.: On the coverage of LPWANs: range evaluation and channel attenuation model for LoRa technology. In: 2015 14th International Conference on ITS Telecommunications (ITST). pp. 55–59. IEEE, Copenhagen, Denmark (2015)
19. Ismail, I.S., Latiff, N.A., Rokhani, F.Z., Aziz, S.A.: A review on performances evaluation of low power wide area networks technology. Springer Singapore (2019)
20. Pop, A.-I., Raza, U., Kulkarni, P., Sooriyabandara, M.: Does bidirectional traffic do more harm than good in LoRaWAN based LPWA networks? (2017)
21. LoRa Alliance, I.: LoRaWAN$^{TM}$ 1.0.2 Regional Parameters., San Ramon, CA (2017)
22. Sigfox: Radio Technology Keypoints. https://www.sigfox.com/en/sigfox-iot-radio-technology
23. Semtech: SX1272/3/6/7/8: LoRa Modem Design Guide. https://www.semtech.com/uploads/documents/LoraDesignGuide_STD.pdf

# Enhancement of Advanced Driver Assistance System (Adas) Using Machine Learning

Santhosh Krishnarao[1], Hwang-Cheng Wang[2](✉), Abhishek Sharma[3], and Mazher Iqbal[1]

[1] Vel Tech Rangarajan Dr. Sagunthala R&D Institute of Science and Technology, Chennai, Tamil Nadu, India
santhosh310899@gmail.com, mazeriq@gmail.com
[2] National Ilan University, Yilan, Taiwan
hcwang@niu.edu.tw
[3] The LNM Institute of Information Technology, Jaipur, India
abhisheksharma@lnmiit.ac.in

**Abstract.** Most of the road accidents can be attributed to human errors. Advanced driver assistance system (ADAS) is an electronic system that guides a vehicle driver while driving. It is designed with a safe human-machine interface that is intended to increase vehicle safety and road safety. ADAS is developed to automate, adapt and enhance vehicle systems for safety and better driving. An increasing number of modern vehicles have ADAS such as collision avoidance, lane departure warning, automotive night vision, driver monitoring system, anti-lock braking system and automatic parking system. ADAS relies on input from multiple data sources like lidar, radar, and camera. This paper describes the implementation of ADAS using machine and deep learning algorithms. We implement a model which has a 360-degree camera (lens on two sides of 170 degrees each), lidar, ultrasonic sensor, and radar that provide the input for ADAS. We implement the ADAS by training this whole model using deep learning (advanced machine learning) by designing a neural network using Python in TensorFlow. Generative adversarial networks (GANs) are used in object detection when a hazed image (foggy, rainy, etc.) is detected. This reduces the sensor complexity and area in the vehicle. Results gained from the study and their implications are presented.

**Keywords:** ADAS · Human-machine interface · Lidar · Deep leaning · 360-degree camera · Generative adversarial networks

## 1 Introduction

Advanced driver assistance systems are electronic systems that assist the driver which reduces some human errors. ADAS is growing rapidly in the automotive industry. The system gets inputs from lidar, camera, radar, ultrasonic sensor, etc. The present ADAS system typically has five built-in cameras, one lidar scanner, one radar, and one ultrasonic sensor. The architecture of ADAS is complex and the cost of built-in sensors is high.


X.-S. Yang et al. (eds.), *Proceedings of Fifth International Congress on Information and Communication Technology*, Advances in Intelligent Systems and Computing 1183, https://doi.org/10.1007/978-981-15-5856-6_13

The sensors are fixed in vehicles to obtain real-time parameters. The performance of different sensors is summarized in Table 1 using various metrics.

**Table 1.** Performance of different sensors

| Parameters | Lidar | Radar | Camera |
|---|---|---|---|
| Range | H | H | VL |
| Field of view | H | L | VL |
| 3D shape | H | L | VL |
| Object recognition at long range | H | L | VL |
| Accuracy | H | L | L |
| Rain, snow, dust | H | H | L |
| Fog | M | H | L |
| Performance at nighttime | H | H | L |
| Read signs and see color | M | L | H |

From Table 1, we see that the lidar sensor performs well in nearly all aspects. Camera is well suitable for detecting traffic signs and colors, whereas radar is low in it. On the other hand, radar is used for scanning images, especially at night time or under poor weather condition with fog, rain, snow, and dust. Thus, every sensor is important, whereas the number of sensors installed differs according to the major and minor conditions.

## 2 Related Work

### 2.1 Machine and Deep Learning

Giving an automobile a set of eyes in the form of multiple cameras and image sensors is the first step, but it also will be critical for the automobile to be able to interpret those images and react accordingly. To accomplish this, machine learning and deep learning algorithms play a major role in training the models for the interpretation of images. A front-facing camera can detect pedestrians or other obstacles, and with powerful algorithms, they can assist the driver in braking. Radars in the rear side of the vehicle allow the user to park the vehicle, and the radar on the front side of the vehicle helps in achieving collision avoidance. A camera in the cockpit of the car facing the driver can identify and alert for distracted driving.

### 2.2 Generative Adversarial Networks

A generative adversarial network (GAN) [1–3] is a deep neural network that consists of two neural networks, namely generator and discriminator that compete with each other (hence adversarial). So, the generator produces fake instances from random noise and the discriminator evaluates them for authenticity, i.e., the discriminator decides whether the output of the generator is distinguishable from real instances.

The generator and discriminator are represented by $G$ and $D$. For the case of image detection, their work can be described by the formula given below:

$$\min_G \max_D V(D, G) = \mathbb{E}_{x \sim p_{\text{data}}(x)}\big[\log D(x)\big] + \mathbb{E}_{z \sim p_x(z)}\big[\log(1 - D(G(z)))\big] \quad (1)$$

Here, $x$ is a dataset, $p_{\text{data}}(x)$ is the probability that the generated data is in the dataset, $D(x)$ is the output generated by the discriminator with real data, $z$ is a noise vector, $p_x(z)$ is the probability for data generated from $z$ to be in the dataset, $G(z)$ is the output generated by the generator which is fake, and $D(G(z))$ is the output generated by the discriminator when the input is fake. Both $D(x)$ and $D(G(z))$ give a score between 0 and 1. Equation (1) may not provide a sufficient gradient for $G$ to learn well. The objective function is defined by the following equation:

$$D_G^*(x) = \frac{p_{\text{data}}(x)}{p_{\text{data}}(x) + p_G(x)}, \quad (2)$$

where $p_G(x)$ is the probability that the generated data is in $x$. $G$ tries to minimize the objective function, while $D$ tries to maximize it. Early in learning, when $G$ is poor, $D$ can reject samples with high confidence because they are different from the training data. In this case, $\log(1 - D(G(z)))$ saturates. Rather than training $G$ to minimize $\log(1 - D(G(z)))$, we can train $G$ to maximize $\log(D(G(z)))$. GANs [4] are mostly used in generating new images from the dataset. It is well suited for generating image–image translation, reconstructing a quality image from a blurred image, and so on. So, we will use GANs for object detection to remove the haze from the images detected [5].

# 3 Proposed Design

## 3.1 Vehicle Design

ADAS vehicles can be driven in two modes called self-driving mode or human driving mode. In the self-driving mode, the system detects the objects on the path and creates its path for itself using deep learning algorithms provided to it. In the human driving mode, objects are detected, and the right path is displayed to the driver in the monitor. Many applications that include ADAS features can be displayed accordingly for the safe driving of the vehicle [6–11]. In this ADAS system, we use a specially designed camera with lenses that face all sides from the center of the vehicle. The camera and a lidar scanner are fitted on the vehicle as shown in Fig. 1. Figure 2 shows in greater detail the layout of the camera and the lidar scanner. The camera has two lenses that each face 170 degrees (inclined), so that it covers all the 360 degrees from the vehicle. The two lenses are connected to a hardware module that has ADC which will convert the analog signal to digital data, and the data is then processed by image processing algorithms and finally displayed on the screen.

The vehicle is designed where the camera and lidar will be attached to the center of the vehicle at the top, so that it can cover all sides of the vehicle. The radar is fixed

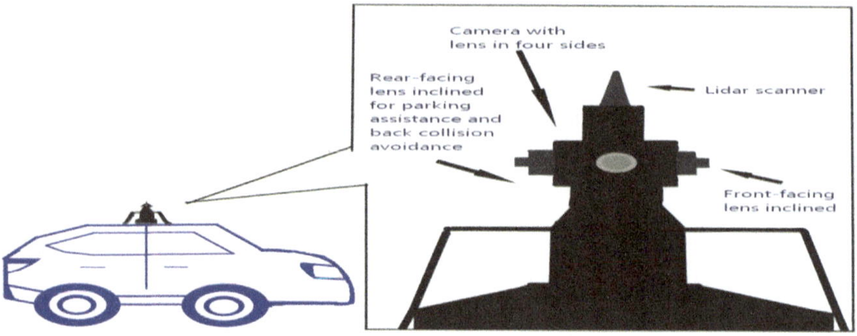


**Fig. 1.** Vehicle design with camera and lidar system

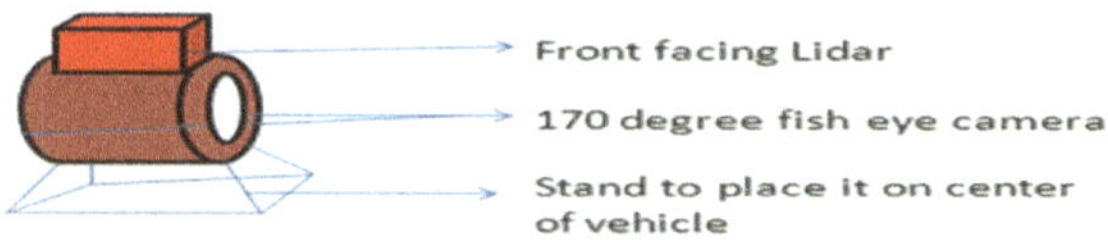


**Fig. 2.** Surround-view camera with two 170-degree cameras and lidar mounted on the camera module

in front of the vehicle for maintaining equidistance with vehicles and pedestrians. The distance $d_{\min}$ can be determined mathematically by Eq. (3) [12]:

$$d_{\min} = L + T_f\left[v_r - v_f + \rho(a_a + a_b)\right] - \frac{\rho^2 a_b}{2} + \frac{\left(T_r - T_f\right)\left(v_r + \rho_a - \left(T_f - \rho\right)a_b\right)}{2}. \tag{3}$$

Here, $L$ is the average length of the vehicles, $\rho$ is the response time of the rear vehicle, $v_r$ and $v_f$ are the velocities of the rear/front vehicle, $a_a$ and $a_b$ are the maximal acceleration/braking of the vehicle, $T_f$ is the time for the front car to reach a full stop if it would apply maximal braking, and $T_r$ is the time for the rear car to reach a full stop if it would apply maximal acceleration during the response time, and from there on apply maximal braking. This is used for the path planning of the vehicles running on the road.

### 3.2 Use of GAN in the Proposed Model

While a self-driving vehicle is in motion during extreme weather conditions like fog, rain, etc., the objects in the roads like a traffic sign, speed limit, and lanes are covered with fog or rain referred to as haze which cannot be detected by the camera. Under this situation, the objects can be easily detected by the pix-pix conditional GANs as shown in Fig. 3. During the training, the hazed images are taken, and the generator generates an image from the dataset, and the discriminator should compare the hazed image with the original image and tell us the correct sign.

The work presented in [13] proposes an AI-based test and validation process on both hardware in loop and simulation in a successive manner. The data is provided, stored, and

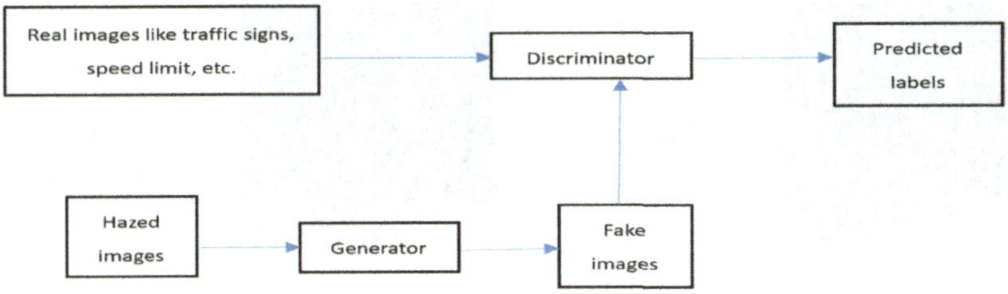


**Fig. 3.** GAN architecture for detecting hazed images used in object detection

fed again to the system in multiple test cases. This will help the system to achieve better training of an autonomous ADAS system. Deep neural networks (DNNs) also have been implemented widely in such an environment [4]. The repeated learning capabilities and power of generating augmented dataset GANs have also shown significant improvement in the ADAS scenarios. This method is well adopted in autonomous driving scenarios and has shown promising results in various environments such as fogs, nights, and variations in weather conditions [4]. The work presented in [14] shows the real-time implementation using edge computing technologies for intelligent prediction with the possibility of 5G interconnectivity to make the communication fast and reliable with the inclusion of GANs for object detection applications.

## 4 Results and Discussion

Python code is written in Anaconda, Spyder platform using a virtual environment for the simulation of a self-driving vehicle [15, 16]. This uses reinforcement learning for the training. The training is completed by performing 64 iterations. This includes path finding, object detection, obstacle avoidance, etc. After successful iterations, the simulation is done in the Kivy platform.

The simulation platform is shown in Fig. 4 with the class named vehicle. The yellow curves represent obstacles (like other vehicles, pedestrians, or buildings, as indicated by the yellow curves in Fig. 5). The vehicle has a number of sensors: a camera with lidar in red color and two radars in blue and yellow. The white rectangle stands for the vehicle. The task is to plan the shortest path along which the vehicle should go from the bottom right corner to the top left corner. In addition, the vehicle has to detect and avoid those obstacles on the road. Figure 6 shows that the vehicle has detected the obstacles, followed the curves, and eventually arrived at the destination.

The graphs shown in Fig. 7 are obtained during the training of the system where the x-axis is the time in seconds and the y-axis denotes weights for the correct move. Initially, from the first graph in Fig. 7, the model seems to be oscillating from positive to negative weights as it is in the training mode. In the second graph, the model moves from the initial point to negative weight, and then, the graph gradually increases to positive weights and maintains stability. Therefore, the model has learned from its previous errors and moves in the correct path.

It is clear from the above results that we have acquired some of the ADAS features using the proposed approach.

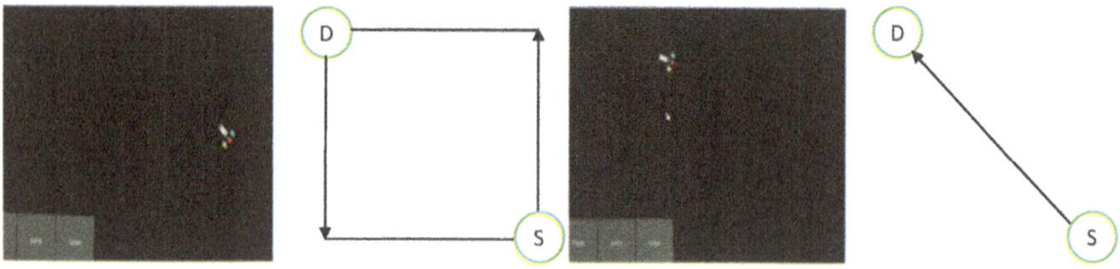


**Fig. 4.** Vehicle having three sensors is moving to the destination by traveling along the edges of the platform in the Kivy simulator. The arrow in the graph represents the path of the vehicle, S is the source, and D is the destination. The vehicle has found the shortest path

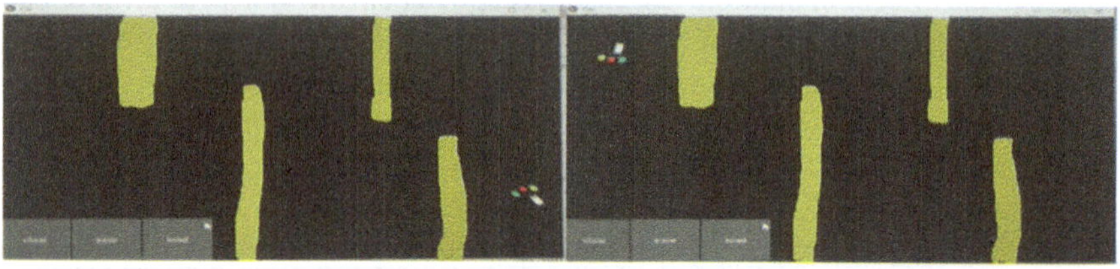

**Fig. 5.** Kivy platform with vehicle and sensors. Yellow curves denote obstacles. The vehicle moves from the bottom right corner to the top left corner

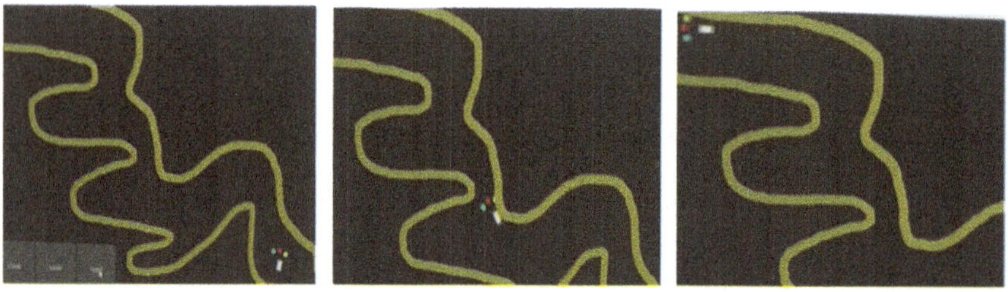

**Fig. 6.** The yellow curves define the border of the path along which the vehicle travels from the bottom right corner to the top left corner without hitting or crossing the border. The vehicle has detected the yellow curves and avoided them. Eventually, the vehicle successfully reaches the top left corner

## 5 Conclusion and Future Work

We have shown that the overall complexity in the vehicle can be reduced in the proposed model by using existing deep learning algorithms. The proposed model consists of one wide-angle camera, one lidar sensor, and two radars which can be closely monitored around the vehicle. Simulation of the proposed model was done in the Kivy platform which demonstrated desirable ADAS features such as object detection, path planning, and obstacle avoidance.

The future goal is to implement virtual reality in ADAS that helps the user to experience the real effect of driving, so that it gives an accurate path. As a result, virtual reality can convert commands to pictures to enhance user experience. Moreover, we will look

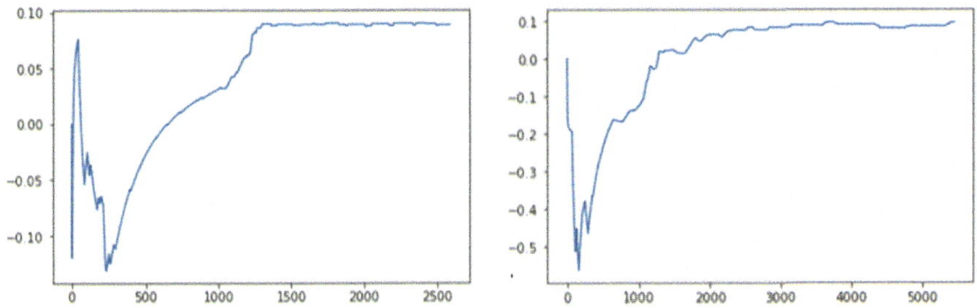


**Fig. 7.** *X*-axis denotes time in seconds, and *Y*-axis denotes weights given to the model for every right path. 0.1 is the weight given for the correct path and negative weights for incorrect movement. Initially, the model is learning to move in the correct path, and after some time, it continues to maintain in the right path

into innovative ways of exploiting machine learning algorithms to improve vehicle and road safety.

**Acknowledgements.** This work was supported in part by the TEEP@Asiaplus program sponsored by the Ministry of Education, Taiwan.

## References

1. Goodfellow, I.J.: Generative adversarial nets. In: Advances in Neural Information Processing Systems (2014)
2. Zhang, M., Zhang, Y., Zhang, L., Liu, C., Khurshid, S.: DeepRoad: GAN-based metamorphic autonomous driving system. Research Gate Publication (2018)
3. Hsieh, Y., Lin, Y.C., Shen, H.Y.: On automatic generation of training images for machine learning in automotive applications. In: IEEE International Conference on Artificial Intelligence Circuits and Systems (2019)
4. Fan, X., Wang, F., Liu, J.: Edge computing empowered generative adversarial networks for real-time road sensing. In: 26th International Symposium on Quality of Service (2018)
5. He, K., Sun, J., Tang, X.: Single image haze removal using dark channel prior. IEEE Trans. Pattern Anal. Mach. Intell. **33**(12), 2341–2353 (2011)
6. Schottle, M.: Future of driver assistance with new E/E architectures. ATZ Electron. Mag. **6**(4), 4–9 (2011)
7. Wang, J., Zhang, L., Lu, X., Li, K.: Driver characteristics based on driver behavior. In: Encyclopedia of Sustainability Science and Technology, pp. 3099–3108. Springer, Berlin (2012)
8. Popken, M., Rosenow, A., Lubcke, M.: Driver assistance systems. ATZextraworld **12**(1), 210–215 (2007)
9. Smirnov, A., Lashkov, I.: State of the art analysis of available advanced driver assistance systems. In: 17th Conference Open Innovation Association, pp. 345–349 (2015)
10. Paul, A., Chauhan, R., Srivastava, R., Baruah, M.: Advanced driver assistance systems. SAE Technical Paper 2016-28-0223 (2016)
11. Lee, J., Forlizzi, J., Hudson, S.E.: Iterative design of MOVE, a situational appropriate vehicle navigation system. J. Int. Sci. **66**(3), 198–215 (2008)
12. Yoshida, J.: Math for drive cautiously, safe distance formula. EE Times (2019)

13. Vishnukumar, H.J., Butting, B., Muller, C., Sax, E.: Machine learning and deep neural network artificial intelligence core for lab and real-world test and validation for ADAS and autonomous vehicles, AI for efficient and quality test and validation. In: Intelligent Systems Conference (2017)
14. Lee, B., Wei, Y., Guo, I.: Automatic parking of self-driving car based on lidar. In: The International Archives of the Photogrammetric, Remote Sensing and Spatial Information Sciences, vol. 42, pp. 241–246 (2017)
15. Tian, Y., Pei, K.: Deep test: automated testing of deep neural network-driven autonomous cars. In: 40th International Conference on Software Engineering, pp. 3–7 (2018)
16. Chen, S., Zhang, S., Shang, J., Chen, B., Zheng, N.: Brain-inspired cognitive model with attention for self-driving cars. In: IEEE Transactions on Cognitive and Developmental Systems, **11**(1), 13–25 (2017)

# Blockchain Applications in Logistics and Supply Chain Management: Problems and Prospects

Yulia A. Morozova(✉)

National Research University Higher School of Economics, Moscow, Russian Federation
ymorozova@ya.ru

**Abstract.** The growing interest and expectations from the blockchain applications attract many analysts to this issue. In what spheres of logistics and supply chain management blockchain is appropriate? What blockchain software solutions are available to companies now? This paper investigates the basic functionality of the existing software solutions on the market and the comparative analysis of blockchain platforms used for developing the solutions for logistics is also carried out. The main trends of blockchain applications are identified, based on the analysis of the project experience on the use of blockchain, in logistics and supply chain management, in different countries. The problems, limitations and conditions of blockchain implementation are also determined.

**Keywords:** Blockchain · Logistics · Supply chain management · Software solutions · Information technologies

## 1 Introduction

Blockchain technology is rapidly gaining popularity today, and many experts expect its significant impact on supply chains, logistics, production, market, sales and consumption of goods in the future. However, despite the great potential and interest in blockchain, the number of companies that have fully implemented and adapted blockchain technology to their needs, especially in the field of logistics, is not enough now. This paper investigates the possible limitations and problems of blockchain technology that slows down its development, as well as software solutions available on the market and the possibilities for their use.


X.-S. Yang et al. (eds.), *Proceedings of Fifth International Congress on Information and Communication Technology*, Advances in Intelligent Systems and Computing 1183, https://doi.org/10.1007/978-981-15-5856-6_14

## 2 Blockchain Platform Functionality

Blockchain technology allows transactions to be verified by a group of potentially unreliable network participants and is a distributed, unchangeable, transparent, secure and verifiable ledger. All transactions in the blockchain are available for viewing at any time and can be checked by any network node. The blockchain protocol structures information into a chain of blocks, where each block keeps a set of completed transactions at a given time. The information entered into the blockchain system cannot be changed, and data is stored without centralised control.

Cryptographic methods are used to ensure the identity of copies of the blockchain to avoid duplication of transactions and to provide access to confidential information. Public and private keys are used to ensure confidentiality and to protect information.

Two key domains of blockchain application include financial technology and the Internet of Things. Although both domains involve decentralised trusted financial transactions, the significant difference is that financial technology is particularly sensitive to security requirements with a relatively low number of transactions, while the Internet of Things is characterised by many transactions involving millions of devices. The requirements of low transaction costs and minimal transaction delays, providing real-time operation, come to the fore. More features of blockchain technology are described in [1, 29, 30].

Table 1 presents blockchain platforms, commonly used for developing applications for logistics and supply chain management.

Here, we consider basic functions of blockchain platforms that are important for logistics and supply chain management applications.

*Account registration.* Each participant of a blockchain network creates an account with an attached online wallet that allows transferring, earning and spending cryptocurrency. As a rule, the user can also withdraw cryptocurrency from the system and exchange it for another cryptocurrency at a special exchange rate, which is calculated on official verified sources. There are open (public) and closed (private, corporate) blockchain networks. Everyone can join the public blockchain network, while their computing power contributes to the network and all transaction data is available to them. Only trusted members can connect to a closed network, while access to transactions can also be limited according to confidentiality requirements.

*Consensus algorithms.* Various consensus algorithms are used to verify the correctness of transactions and determine the final state of the blockchain. The most famous of them are Proof of Work (PoW), Proof of Stake (PoS), Proof of Elapsed Time (PoET), Practical Byzantine Fault Tolerance (pBFT) and Proof of Authority (PoA) [5, 8, 9].

*Smart contracts.* A smart contract is an encrypted program code that is remotely executed in the blockchain network when the specified conditions are met. All parties of the agreement have access to the encrypted information. The result of the smart contract execution is confirmed by all network participants. Smart contracts can be created to conduct a transaction, to provide a service, to buy goods or sell them. Some platforms allow writing your own code, other platforms offer ready smart contract codes. The most famous is the solidity smart contract, offered by the Ethereum platform.

**Table 1.** Comparison of blockchain platforms

| Platform | Cryptocurrency | Consensus algorithm | Transaction fee | Type | Smart contracts |
|---|---|---|---|---|---|
| Ethereum [5] | ETH | PoW | Yes | Public | Solidity |
| Ripple [24] | XRP | RPCA | Yes | Public | No |
| Hyperledger Sawtooth [8] | No | pBFT, PoET, Devmode, RAFT | No | Private | DAML |
| Hyperledger Fabric [9] | No | pBFT, Kafka, SOLO | No | Private | Go.Java |
| R3 Corda [4] | No | RAFT, pBFT | No | Private | Java, Kotlin |
| Stellar [28] | XLM | SCP, pBFT | Yes | Public | JavaScript, Python, Golang, PHP |
| IBM Blockchain Platform [10] | No | pBFT, XFT | No | Private | JavaScript, Java, Go.Java |
| Microsoft Azure Blockchain Workbench [16] | No | Aura PoA | No | Private | JavaScript, Rust, C, C ++ |
| Quorum [22] | ETH | RAFT, iBFT, Clique PoA | Yes | Private | Solidity |

## 3 Analysis of Blockchain Applications in Logistics and Supply Chain Management

In order to reveal features of blockchain applications in logistics and supply chain management, we analysed 21 blockchain projects that are being conducted in such countries as Australia, Bulgaria, Denmark, Finland, France, Hong Kong, Russia, Singapore, Sweden, United Kingdom and USA (Fig. 1). Some projects are initiated by large corporations (IBM [11], Russian Railways [19] and Gazprom Neft PJSC [7]), and some projects are start-ups conducted by a small group of developers.

Figure 2 shows that the most popular platforms for development of applications in logistics and supply chain management are Ethereum and Hyperledger Fabric. Other group included R3 Corda, Stellar, TBSx3 [31], Yojee [33] and some original applications, partly based on Ethereum. Ethereum is the most popular platform because it is open sourced. However, it has limitations that enforce the development of new platforms.

Figure 3 presents popular functions, implemented in blockchain applications, for logistics and supply chain management. The most popular function is tracking that gives a great advantage of supply chain and logistic operation visibility.

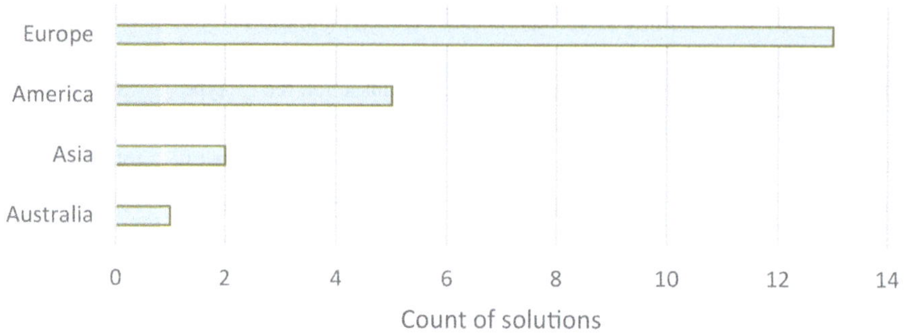


**Fig. 1.** Blockchain solutions in logistics and supply chain management by subcontinents

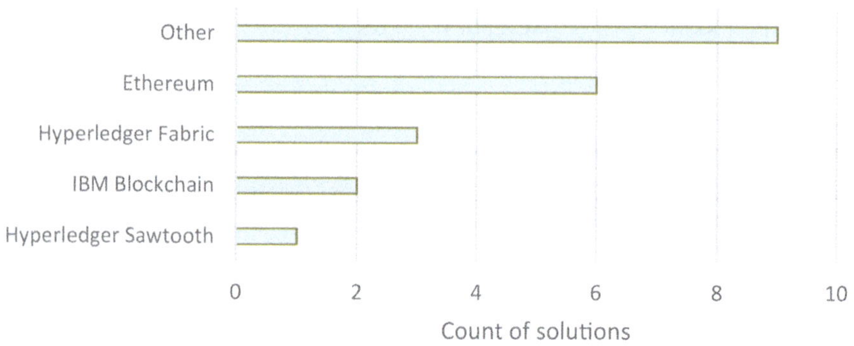


**Fig. 2.** Blockchain solutions in logistics and supply chain management by platform

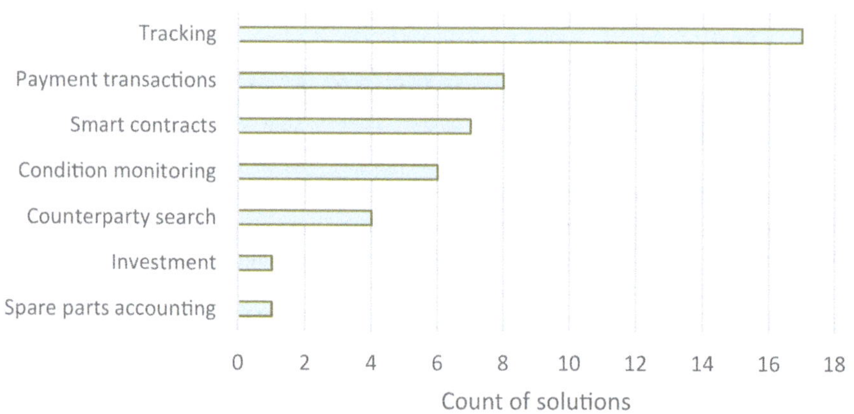


**Fig. 3.** Functionality of blockchain solutions in logistics and supply chain management

Based on an analysis of blockchain applications, we outline the following main areas of its use in logistics and supply chain management.

*Simplification of workflow and mutual settlements between contractors.* The execution of some logistics operations, for example, transportation of refrigerated cargo in containers, requires the coordination of many participants, and sometimes, the order processing process can take several days. Carrying out these operations through the blockchain saves time and reduces costs [3, 12, 15, 18, 32].

*Food quality control, falsification prevention, certification guarantee of goods (equipment, vehicle).* The product information, including the date of entry, name, qualitative and quantitative characteristics, components and other data, that identifies the product, is registered in the blockchain database and is available for every participant in the logistics system, for checking the product of interest in the system. Data entries cannot be changed or deleted, which guarantees the certification of goods [2, 6, 11, 20, 31].

*Tracking and condition monitoring of the goods during transportation.* State sensors embedded into containers for transportation and storage of goods [14, 17, 26, 27], register data on the current state of the goods, temperature conditions, location of the container or vehicle, thereby forming up-to-date information on all processes in the logistics system, which is available to all contractors involved in the transaction. This information can also be used by insurance companies to investigate cases of damage of goods.

*Electronic platform for the provision of logistics services.* Blockchain provides a service [12, 25, 27] that allows you to place orders for delivery, transportation, to find a supplier, a carrier, to track the process of fulfilling an order and to carry out settlements on the services provided.

*Attracting investment and establishing contacts between suppliers and customers.* Blockchain can be used as a tool to attract investment in production and to establish direct contacts with consumers, thereby forming more efficient supply chains [13].

*Managing IoT devices.* The development and expansion of the use of unmanned vehicles (drones, unmanned cars, robots) will require a powerful network infrastructure and servers to manage these devices. A distributed blockchain network may be a good solution to this problem [21, 23].

## 4 Limitations and Problems of Using Blockchain Technology

In addition to the advantages and prospects of using the blockchain, we should note the problems and limitations that currently exist and can be significant obstacles to its distribution.

As the blockchain system develops, the need for memory resources will increase, as the technology involves the complete copying of all data, including the history of all changes, at each network node involved in data processing.

Also, one of the biggest drawbacks of the blockchain is that platforms usually take quite a while to register transactions. There are ways to get around this limitation, for example, using transactions outside the chain, but, in most cases, writing data to the blockchain does not happen instantly.

Blockchain does not guarantee complete transparency. Moving data to the blockchain can help increase the transparency of the company, but this does not make it completely "open". The degree of transparency largely depends on what information the company wants to share with other participants.

The blockchain deals with virtual objects, i.e., ensures the relevance and correctness of information about the object, yet it cannot interfere with their physical damage and theft.

All the data stored in the public blockchain is available to anyone who wants to read it. This means that the company will not be able to store data on the blockchain in such a way that only authorised users can see it. There are possible solutions for controlling access to data based on the blockchain, but they are difficult to implement and cannot guarantee complete confidentiality of data.

Global supply chains operate in a complex environment, in which participants must act in accordance with the law, including customs restrictions, maritime laws and regulations, commercial codes, laws regarding property rights and others. Currently, the legislation of most countries does not have normative acts regulating the use of blockchain technology. The widespread adoption of blockchain technology, with these limitations, can be quite a challenge.

The implementation of the blockchain involves the involvement of all counterparties in the network, some of which may not be ready for this, which will require long negotiations.

Blockchain corporate networks, with a limited number of participants, who have access to data, according to some experts, are less decentralised, which means they are more vulnerable to hacker attacks.

Many supply chain partners, located in developing and least developed countries, do not have enough level of computerisation, and are far from ready to implement blockchain. However, without their participation, it is difficult to realise the full potential of the blockchain in the supply chain.

## 5 Conclusion

In conclusion, we note that the approach, using a distributed blockchain-based registry, is not a panacea for all ills and, in our opinion, will continue to compete with the model based on a traditional database. The low speed of data updating, the duration of transactions, the rapid growth of requirements for available memory and other technical and organisational problems suggest that there must be reasons for choosing a system implementation based on the blockchain. In general, the use of blockchain is advisable under the following conditions:

- many participants are involved in the process;
- there is a need for a common database shared with other participants in the process;
- there is a conflict of interest and distrust of the participants of each other;
- there are differences in the methods and rules for controlling the behaviour of participants;
- there is a need for an objective, unchanged data registry;
- transaction rules rarely change.

In addition, undoubtedly, the limiting factor is that the number of qualified specialists in the field of blockchain is still very small, especially for the blockchain integrated with the concept of the "Internet of Things", which is so important for the logistics industry. Currently, the lack of competent specialists is impeding the widespread adoption of blockchain technology.

## References

1. Antonopoulos, A.M., Wood, G.: Mastering Ethereum: Building Smart Contracts and DApps. O'Reilly, USA (2018)
2. Bext360. https://www.bext360.com. Accessed 20 Nov 2019
3. CargoCoin. https://thecargocoin.com. Accessed 20 Nov 2019
4. Corda. https://docs.corda.net. Accessed 20 Nov 2019
5. Ethereum. https://www.ethereum.org, last accessed 2019/11/20
6. Everledger Diamond Platform. https://diamonds.everledger.io. Accessed 20 Nov 2019
7. Gazprom Neft PJSC. https://www.gazprom-neft.ru/press-center/news/1814734. Accessed 20 Nov 2019
8. Hyperledger Sawtooth. https://sawtooth.hyperledger.org. Accessed 20 Nov 2019
9. Hyperledger. https://www.hyperledger.org. Accessed 20 Nov 2019
10. IBM Blockchain Platform. https://www.ibm.com/blockchain/platform. Accessed 20 Nov 2019
11. IBM Food Trust. https://www.ibm.com/ru-ru/blockchain/solutions/food-trust. Accessed 20 Nov 2019
12. IMMLA International multimodal logistic application. https://immla.io/wp-content/uploads/2018/05/Booklet_Immla_en.pdf. Accessed 20 Nov 2019
13. Krasnushkina, N.: Blockchain from the plow. How to raise a Russian village using cryptocurrency. https://www.kommersant.ru/doc/3658179. Accessed 20 Nov 2019
14. Kshetri, N.: Blockchain's roles in meeting key supply chain management objectives. Int. J. Inf. Manage. **39**, 80–89 (2018)
15. Marine Transport International. https://www.marinetransportint.com/comship. Accessed 20 Nov 2019
16. Microsoft Azure. https://azure.microsoft.com/en-us/solutions/blockchain. Accessed 20 Nov 2019
17. Namiot, D.E., et al.: Blockchain transport applications. Int. J. Open Inf. Technol. **12**(5), 130–134 (2017)
18. OpenPort. https://openport.com/technology/blockchain-protocol. Accessed 20 Nov 2019
19. Poletaev, E.: Blockchain will fight with counterfeit. https://www.rzd-partner.ru/zhd-transport/news/s-kontrafaktom-poboretsya-blokcheyn. Accessed 20 Nov 2019
20. Provenance. https://www.provenance.org/business/platform. Accessed 20 Nov 2019
21. Pustisek, M., Kos, A.: Approaches to front-end IoT application development for the Ethereum blockchain. Procedia Comput. Sci. **129**, 410–419 (2018)
22. Quorum. https://www.goquorum.com. Accessed 20 Nov 2019
23. Reyna, A., et al.: On blockchain and its integration with IoT challenges and opportunities. Future Generation Comput. Syst. **88**, 173–190 (2018)
24. Ripple. https://www.ripple.com. Accessed 20 Nov 2019
25. Shipit.to. https://shipit.to/en. Accessed 20 Nov 2019
26. Smart Containers. https://www.smartcontainers.ch. Accessed 20 Nov 2019
27. Smartlog. https://smartlog.kinno.fi. Accessed 20 Nov 2019
28. Stellar. https://www.stellar.org/developers/guides. Accessed 20 Nov 2019

29. Swan, M.: Blockchain: Blueprint for a New Economy. O'Reilly, USA (2015)
30. Tapscott, D., Tapscott, A.: Blockchain Revolution: How the Technology Behind Bitcoin Is Changing Money, Business, and the World. Portfolio Penguin, Canada (2018)
31. Tbsx3. https://tbsx3.com. Accessed 20 Nov 2019
32. Tradelens. https://www.tradelens.com. Accessed 20 Nov 2019
33. Yojee. https://yojee.com. Accessed 20 Nov 2019

# Augmented Reality Storytelling Teachers and Preschool Children Experience

Faiz bin Meor Othman, Wan Adilah Wan Adnan, and Zan Azma Nasruddin(✉)

Faculty of Computer and Mathematical Sciences, Universiti Teknologi MARA (UiTM), Shah Alam, Malaysia
faizmeorothman@yahoo.com, {adilah,zanaz}@tmsk.uitm.edu.my

**Abstract.** This study aims to use AR technology to develop an AR-based learning of a digital storybook for preschool children aged 6 years old as to motivate their reading. This application will show digital storytelling based on book titled "The Three Bears" by Emma Bailey focusing on pages 4 and 5. User testing with both teachers and preschool children was conducted to examine the potential of the AR approach in motivating them learning by using both qualitative and quantitative interview in order to measure their learning experience. The findings indicate that the AR-based storybook approach does affect their enjoyment, engagement and motivation. A suggestion for further research is to embed elements of surprise to prevent boredom in an AR-based storybook to preschool children to increase their engagement in reading.

**Keywords:** Augmented reality storytelling · Learning experience · Teacher · Preschool children

## 1 Introduction

Technology has been evolved constantly which is taking the education in more attractive new ways to the student. Usage of augmented reality (AR) technology is possible due to its attractiveness [1]. People begin to accept different types of human–computer interactions with the emergence of AR browser and smartphones [2].

Today, education usually involves student to learn and understand certain concept or scenario that no longer exist such as learning about planets, microorganisms or history. The common way to teach certain concept is through the usage of relatable analogy. By utilizing analogy method, a concept is described in terms of generally observable reality since students can relate and imagine the concept to experience [3].

Generally, AR interface provide students smooth interaction by combining both real and virtual world in the learning environment. This approach has a significant attractiveness and effectiveness when teaching and learning session is conducted. In a way, AR changes how students interact with computer-generated virtual space as it overlays and uses an interface onto the physical world [4]. The usage of AR most


X.-S. Yang et al. (eds.), *Proceedings of Fifth International Congress on Information and Communication Technology*, Advances in Intelligent Systems and Computing 1183, https://doi.org/10.1007/978-981-15-5856-6_15

likely makes students become more motivating, engaging and supporting the creativity of students in a restricted school environment, in ways that otherwise it could not be possible [5].

The ability to read and write is important to learn in early age, preferably during their preschool year. Reading and writing skills are basic skills that will be used for the rest of their life, especially in education. In addition, literacy can help develop how to shape children to think, feel and speak [6]. Preschool literature programme can create positive attitude as it has influence on emotional sources and cognitive enrichment [7, 8].

An interview was conducted with a kindergarten teacher who has 17 years of experience teaching kindergarten children. She stated that some children took up to 6 months to learn reading due to lack of motivation to learn. They struggle to remember the alphabet and word pronunciation.

The aim of this paper is to examine the potential of AR approach in motivating children's learning. User testing for both teachers and children were conducted. In the next section, we will discuss the learning theories, augmented reality, digital storytelling and AR book.

## 2 Literature Review

A. Learning theory

Preschool education usually done in classroom settings. AR technology can help students construct knowledge for themselves as they are learning individually [9]. AR can assist in students' self-expression. In this paper, learning theories chosen are constructivist learning theory and Vygotsky's social cultural theory.

- *Constructivist learning theory*

Constructivist learning theory acknowledges complexity and uniqueness of a learner. This theory emphasizes on understanding the learning environment in order to motivate, engage and enjoy to the learner [10].

- *Vygotsky's social cultural theory*

Vygotsky believes that learning started form birth and continues forever as long as that person lives. When children try to understand something, they often self-talk to themselves to sort everything that happens in their mind. Vygotsky zone proximal development (ZPD) can be understood easily as a teacher and a learner both work together on a particular task that is difficult [11].

B. Augmented reality

The main purpose of AR is to simplify user's need for information by bringing it to their surrounding immediately. Furthermore, user has access to both virtual and physical world setting. This can increase user's awareness and interaction in real world with the help of virtual environment [12]. With the rapid growth of mobile devices development, AR is becoming common and more research is being made in various areas such as education, training and advertising [13].

There are two techniques involved, which are marker-based AR and markerless AR. Marker AR can be different images that act as a marker. This marker is detected by a camera and used with software as the location for virtual assets placed in a scene [14]. Markerless AR can be referred to tracking without AR marker. The tracking is done by using GPS or geolocation to locate user viewpoint [15]. A study involved AR indoor decoration system by [16] compares both marker and markerless AR based on their technicality.

The main use of AR is not as replacement of education way but as an additional tool to improve content, instructions and as supplement instructions in both traditional classroom and special education classroom as AR can extend content to the world outside classroom [17]. Several amount of literature regarding AR applications in educational domain are studied in [18–20].

### C. Digital storytelling

Digital storytelling is a combination of storytelling with digital media that may include multimedia elements, which are text, audio, graphics, animation and video. These multimedia elements are combined using software usually to tell a story from a specific character point of view with a chosen theme [21]. Typically, these stories last for a few minutes and have various uses including retelling of historical event, instruction on particular topic and personal stories [22]. Digital storytelling is very mobile since it can be experienced at anywhere and anytime by anyone. It is also a form of nonlinear information transmission which are not addressed or delivered to a particular audience [23]. Seven elements of digital storytelling are often being used as reference before any digital stories are being worked on. The elements are point of view (author perspective), dramatic question (answered question by story's ending), emotional content (issues that speaks to audience), get of your voice (a way to personalize the story to help the audience understand the context), the power of soundtrack (music), economy (enough content without information overload) and pacing [24].

An innovative way can be used by teachers to deliver more instructional content, idea presentation, increase understanding, review material and procedure illustration. Aside from teachers, students also get benefits from digital storytelling where they can share their experience, reflect their understandings and report their findings [25].

A study has been conducted in Crotia that focused on investigating how digital storytelling affects preschool education in computer and mathematical literacy [26]. There are total 55 students involved in the experiment where 29 students in experimental group and 26 students in the control group. The finding proves that digital storytelling method is age appropriate and flexible. Furthermore, children show enthusiasm to complete all four stories with higher motivation.

### D. AR book

Today learners are "digital learners" and do not prefer to read hard copy books mainly since they can search keywords rather than read the whole book in order to find what they need. The hard copy books are not so attractive for the new generation learners as they prefer interactive books which have more 3D effects and more visual effects [27]. The intervention of AR technology through the addition of Multimedia elements in a book could bring a tremendous impact on learning and learner's attitude towards reading. In this context, AR is being implemented in the form of markers based on existing storybook [27].

## 3 Methodology

User testing with 4 teachers and 15 preschool children aged 6 years old from Tadika Didik Kreatif was conducted to examine the potential of the AR approach in motivating children's learning.

The title of AR-based storybook developed for user testing is "The Three Bears" by Emma Baily focusing on pages 4 and 5. The software used are blender, unity 3D and Adobe Photoshop Creative Cloud 2019.

Figures 1 and 2 show screenshots of The Three Bears AR Storybook application, and AR book features are integrated in this application.

The features were applied in the development as in Figs. 1 and 2 including audio, text, graphic, animation and 3D models as suggested by [27].

## 4 Results and Analysis

The results were gathered based on learning experience from teachers and preschool children. Both give positive feedbacks while using AR storytelling.

- **Teachers Experience Using AR**

The teachers expressed their interest in the AR tools. They seem to point out that children get attention easier and help them understand more easily through the use of animated 3D objects in the AR book. The following are parts of expressions quoted from the teachers.

Expression by Teacher 1:

> "Children likes character movement. When the character moves in the storybook, it helps students to understand more easily. For example, they get to see the character opens the door. Some of the children think the character just pushing the door while other may think the character touching the door knob. Reading with animated 3D object is more exciting than just looking at a picture."

Expression by Teacher 2

> "AR has lots of features. They combined a lot of things (words, colours, audio) all in one. Other medium usually limited to passive learning such as books that have no movement. Excitement will arise with the usage of AR."

**Fig. 1.** Main menu screenshot

Expression by Teacher 3:

> "Traditional education style does not work anymore. A new modern way of education is needed. Back in the day, we just learn to read by sitting face to face with teachers. No gadgets, no smartphone, no laptops. That hardly works now since children are exposed to technology and they use it all the time."

Expression by Teacher 4:

> "This is nice. The animated 3D character follows the story and her action mirroring the book."

Based on the observation and interview, AR can potentially be a key component in future learning in preschool education to support teaching and learning environment. All teachers explained that AR technology is an interesting tool to be used in classroom as to initiate an active participation of both teachers and children. Thus, this learning environment is aligned to the Vygotsky's social cultural theory.

**Fig. 2.** Application's screenshot

AR is different from other instructions as it is combined with a lot of different features including photograph, text and animated 3D objects [28]. AR method of teaching is different than the traditional way of teaching since kids nowadays respond to mobile gadgets, especially smartphones and laptops [29, 30].

- **Preschool Children Experience Using AR**

The excitement shows when they were told that they will be reading using gadget instead of whiteboard. They reacted to the 3D objects that pop up when they direct the smartphone's camera to the storybook. A child was so excited as she pushed her friends away to have a good look at the AR application.

She stated:

"Wow, something pop out of the book! the character is walking around!"

The preschool children show disbelieved when they see the 3D objects pop out on the screen. Almost all of them say, it is magic that happened on the smartphone. Based on their expression, it seems like they really enjoy using the AR storybook. They seemed amazed when the 3D objects popped out of the screen.

The preschool children seem motivated as they interact with AR storybook through observing on how active they are using the application. Each of them wanted to use the smartphone and the AR storybook. They even waited patiently for their turn. Their positive attitude towards AR storybook shows they want to learn more about it. Their curiosity to see what happened next was also one of the factors contributing to the motivation. While using the application,

A child stated:

> "I only use my father's phone to watch Upin Ipin and play games. I do not know that I can read this way".

The participants seem to show positive reactions and speak their mind when being asked open-ended question. Their motivation and engagement when using AR application are really great.

Furthermore, a questionnaire is used to measure preschool children's emotion towards AR storybook. Figure 3 shows the graphical view of the results.

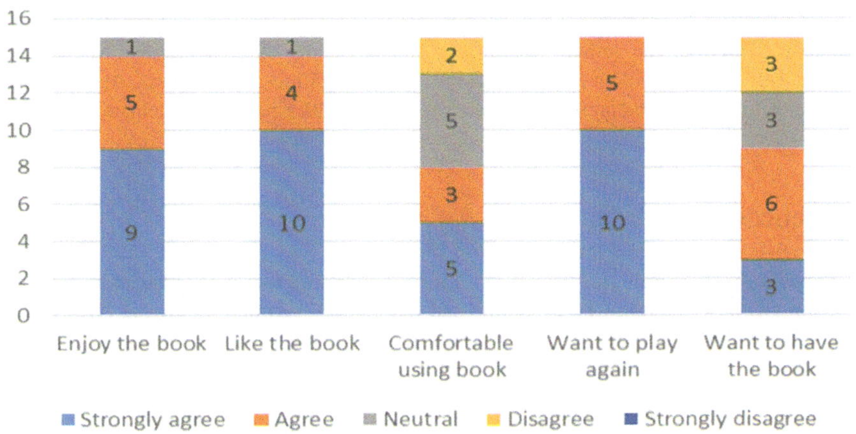


**Fig. 3.** Result of the questionnaire

The first question asked if participants enjoy the book. Nine of them selected strongly agreed, five agreed, and one was neutral. The second question is related to how they like the book. 10 of them strongly agreed, four agreed, and one was neutral. The third question asked how comfortable participant used the book. For this question, it was shown almost half of them did not comfortable using this book. This due to the book was old, and they were afraid to damage the book since they were not familiar with the researcher's book. For fourth question, the participants were asked whether they wanted to play again. Ten of them strongly agreed while five of them agreed. Most of them were excited to see the 3D object popped out of the book on top of the AR marker. Question five asked participant if they wanted to have the book. Three of them strongly agreed, while six of them agreed, three of them neutral and three of them disagreed.

The results of the questionnaire were shown as positive feedbacks. It seems like colourful, 3D animated and unique interface by mixing both real and virtual world attract the attention for the preschool children. Most of them enjoy using AR storybook. This application can be fun and engaging tool as they are excited to use the application over and over again.

The findings from user testing on preschool children's experience using AR support the constructivist learning theory which argue that learning environment can influence learners' motivation and engagement. Some studies related on the emotional include in [31, 32].

## 5 Conclusion

Findings from interview with teachers and preschool children indicate that the AR-based storybook approach does affect their enjoyment, engagement and motivation. The animated 3D models with narration in AR becomes the main attraction for them. Teachers expressed that a new trend of teaching tool should be implemented in the classroom. For the future recommendation, the interaction using AR should include more emotion elements such as surprise, angry and others.

**Acknowledgements.** The authors would like to thank Faculty of Computer and Mathematical Sciences, Universiti Teknologi MARA, Malaysia for the facilities and supports. Also, it is great appreciation to teachers and preschool children at Tadika Didik Kreatif to participate in this study.

## References

1. Piovesan, S., Passerino, L., Pereira, A.: Virtual reality as a tool in the education. In: IADIS International Conference on Cognition and Exploratory Learning in Digital Age (CELDA 2012), pp. 295–298 (2012)
2. Alkhamisi, A.O., Monowar, M.M.: Rise of augmented reality current and future application areas. Int. J. Internet Distrib. Syst. **1**(4), 25–34 (2013)
3. Christou, C.: Virtual reality in education. In: Affective, Interactive and Cognitive Methods for E-Learning. June 2010, pp. 228–243 (2015)
4. Kesim, M., Ozarslan, Y.: Augmented reality in education: current technologies and the potential for education. Procedia Soc. Behav. Sci. **47**, 297–302 (2012)
5. Persefoni, K., Tsinakos, A.: Use of augmented reality in terms of creativity in school learning. CEUR Workshop Proc. **1450**, 45–53 (2015)
6. Abdul Razak, F.H.A., Salleh, K., Azmi, N.H.: Children's technology: how do children want it? In: Proceeding IVIC 2013, Third International Visual Informatics Conference on Advances in Visual Informatics, vol. 8237, pp. 275–284. ISBN: 978-3-02957-3 https://doi.org/10.1007/978-3-319-02958-0_26
7. Rohaya, D., Rambli, A., Matcha, W., Sulaiman, S., Nayan, M.Y.: Design and development of an interactive augmented reality edutainment storybook for preschool. IERI Procedia **2**, 802–807 (2012)
8. Dibrova, A.: AR Books and Pre-school Children's Engagement (2016)
9. Harasim, L.: Constructivist learning theory. Learn. Theory Online Technol. 61–79 (1991)

10. Olusegun, S.: Constructivism learning theory: a paradigm for teaching and learning. Int. J. Res. Method Educ. **5**(6), 66–70 (2015)
11. Vygotsky, L.: Vygotsky's social constructivists theory of learning the zone of proximal development. Impact contructivism Education Language Discourse Mean. **5**(2002) (1978)
12. Carmigniani, J., Furht, B.: Handbook of Augmented Reality. Springer, New York (2011)
13. Liang, S., Roast, C.: Five features for modeling augmented reality. Commun. Comput. Inf. Sci. **434**(434), 607–612 (2014). PART I
14. Shetty, M., Lasrado, V., Sc, M.M.: Marker based application in augmented reality using android. Int. J. Innov. Res. Comput. Commun. Eng. An ISO Certif. Organ **3297**(7), 146–151 (2007)
15. Clarke, S.K.: Markerless augmented reality for visualization of 3D objects in the real world. Department of Science and Technology, Linköping University, Norrköping, Sweden (2014)
16. Cheng, J.C.P., Chen, K., Chen, W.: Comparison of marker-based and markerless ar: a case study of an indoor decoration system. In: Lean and Computing in Construction Congress Proceedings of the Joint Conference on Computing in Construction, vol. 1, pp. 483–490 (2017)
17. Antonioli, M., Blake, C., Sparks, K.: Augmented reality applications in education. J. Technol. Stud. **40**(2), 96–107 (2014)
18. Radu, I.: Why should my students use AR? A comparative review of the educational impacts of augmented-reality. In: ISMAR 2012-11th IEEE International Symposium on Mixed and Augmented Reality Science Technology Paper, pp. 313–314 (2012)
19. Radu, I.: Augmented reality in education: a meta-review and cross-media analysis. Pers. Ubiquitous Comput. **18**(6), 1533–1543 (2014)
20. Santos, M.E.C., Chen, A., Taketomi, T., Yamamoto, G., Miyazaki, J., Kato, H.: Augmented reality learning experiences: survey of prototype design and evaluation. IEEE Trans. Learn. Technol. **7**(1), 38–56 (2014)
21. Robin, B.: The power of digital storytelling to support teaching and learning. Digit. Educ. Rev. **30**, 17–29 (2016)
22. Robin, B.R.: The educational uses of digital storytelling. In: Society for Information Technology and Teacher Education International Conference, pp. 709–716 (2011)
23. Fasi, M.: Digital Storytelling in Education (2009)
24. Wang, S., Zhan, H.: Enhancing teaching and learning with digital storytelling. Int. J. Inf. Commun. Technol. Educ. **6**(2), 76–87 (2010)
25. Preradovic, N.M., Lesin, G., Boras, D.: Introduction of digital storytelling in preschool education: a case study from Croatia. Digit. Educ. Rev. **30**, 94–105 (2016)
26. Ivanova, G., Yuksel A., AIvanov, A.: Augmented reality textbook for future blended education. In: Proceedings of International Conference E-Learning, vol. 14, pp. 130–136 (2014)
27. Gopalan, V. et al.: A Review of the Features of Augmented (2015)
28. Sakamat, N., Sabri, S.N., Diah, N.M.: Multimedia elements of digital storytelling for Dyslexic children. Sci. Res. J. **14**(2), 35–48 (2017). ISSN: 1675-7009
29. Zamin, N., Khairuddin, F.A., Ramli, D.R.A., Ibrahim, E.N.M., Soobni, M.S.A.: Building empathy in young children using augmented reality. Int. J. Eng. Technol. (UAE) **7.28**, 191–192 (2018)
30. Lokman, A.M., Aziz, A.A.: Feasibility study on the readiness, suitability and acceptance of M-learning AR in learning history. In: AIP Conference Proceedings, vol. 1705(1), p. 20009 (2016)
31. Redzuan, F., Khairuddin, A.A., Daud, N.A.: Emotional augmented reality-based mobile learning design elements: a Kansei engineering approach. Indonesian J. Electri. Eng. Comput. Sci. **14**(1), 413–420. http://doi.org/10.11591/ijeecs.v14.i1

32. Mahmood, F., Adnan, W.A.W., Noor, N.L.M., Saman, F.M., Nasruddin, Z.A.: User perception towards cultural based E-Government portal design. In: APIT 2019 Proceedings of the 2019 Asian Pacific Information Technology Conference, pp. 65–69. ISBN: 978-1-4503-6621-2 (2019). https://doi.org/10.1145/3314527.3314539

# TSP Algorithm for Optimum Path Formulation of AUV for Data Collection in Underwater Acoustic Sensor Network

S. Ansa Shermin, Aditya Malhotra, and Sarang Dhongdi(✉)

BITS Pilani, K K Birla Goa Campus, Goa, India
{p20180017,f20150460,sarang}@goa.bits-pilani.ac.in

**Abstract.** Underwater acoustic sensor network (UASN) marks a new era in ocean exploration, enabling various scientific, military and commercial applications. This paper describes one of the scientific applications named coral reef monitoring in a desired region of interest in the Arabian Sea. The events of coral bleaching that leads to massive destruction of coral reef is a current alarm faced by most of the buildups. UASN can be deployed to monitor the environmental parameters of various such regions. In this paper, a number of nodes are deployed in the form of clusters at various regions. An autonomous underwater vehicle (AUV) is used to collect data by visiting the clusters/cluster-heads periodically. Travelling salesman problem (TSP) is used to find optimum tour for the AUV in the data collection phase. The application of TSP in data gathering helps the data collection by forming minimal tour for the vehicle. This paper provides the implementation of protocol stack along with detailed results and analysis using an underwater network simulator (UnetSim).

**Keywords:** Coral reef monitoring · UASN · TSP · AUV · Clusters · Data gathering

## 1 Introduction

The downswing of enthralling buildups of attols and coral reefs sway the core existence of Lakshadweep archipelago. The encircling buildups in the Arabian Sea endow the coral island with immense resources and provide an inherent defence system. But their upheaval impacts the aqua biodiversity as well as the economy, demanding a real-time ocean column monitoring and sensing [1]. The design and implementation of reliable networking paradigm for a three-dimensional mobile underwater acoustic sensor network (3D UASN) emerged out as a major breakthrough. This paper presents architecture and protocol stack for a 3D UASN. Network topology developed for the requirement is as follows: The region of interest (ROI) assumes to be cuboid structure where nodes are


X.-S. Yang et al. (eds.), *Proceedings of Fifth International Congress on Information and Communication Technology*, Advances in Intelligent Systems and Computing 1183, https://doi.org/10.1007/978-981-15-5856-6_16

deployed and a mobile entity known as an AUV is assigned for collecting the data from the deployed nodes. Designed network is more suited for sparse and long term deployment. Data is gathered from localized regions with the aid of a mobile entity. Here, nodes are deployed in the locations predefined by sink node and these localized groups form the clusters. A probabilistic cluster-head (CH) selection technique is used to select CH among cluster nodes. Meanwhile, sensed data is sent to CH using a TDMA-based MAC protocol and an AUV collects data from CH. As this problem is quite similar to the mission planning of autonomous vehicles, the travelling salesman problem (TSP) is used for the formulation of optimal tour [2]. TSP algorithm is developed in this paper for planning trajectory for AUV to collect data from CH and to relay data to sink node. The application of TSP in path formation of AUV has reduced delay in data aggregation by forming minimal tours for the vehicle. The network is simulated using UnetSim simulator.

This paper is organized as follows: Sect. 2 discusses the background, motivation and related works. In Sect. 3, the proposed system model and architecture are described, while Sect. 4 presents and discusses obtained results. Finally, Sect. 5 concludes the paper.

## 2 Background and Motivation

This section gives an insight about ROI and provides whys and where offs of considering this application under surveillance using UASN deployment. Coral reefs are one of the most ancient marine ecosystems housing myriads of flora and fauna. These buildups play vital roles in island formation as well as tsunami prevention. The coral reefs across Indian costs are distributed across Lakshadweep Island, Gulf of Kutch, Gulf of Mannar and Andaman Islands. However, the elixir's biodiversity is nearing extinction due to natural and anthropogenic causes [3]. Global warming, increased presence of carbon dioxide and other toxic dissolvent triggered a phenomenon called coral bleaching, resulting in massive destruction of coral reefs. Lakshadweep Isle being more prone in India having a probable chance for even creating first climate change refugee inhabitants. However, the deployment of relentless monitoring and sensing underwater networks could aid the incumbent bodies to take preventive measures against the malice.

Lakshadweep, the smallest Indian union territory inheriting a coral origin is situated in the Arabian Sea between 080 00' N and 120 30' N latitude and 710 00' E and 740 00' E longitude and at a distance of 220–440 km from the west coast of India. The riveting charm of isle is due to the presence of coral reefs found mostly in following islands like Agatti, Kadmat, Kavaratti, Bangaram and Kalpeni. Among these, Kavaratti Island was chosen here. Kavaratti the capital of Lakshadweep around 3.63 $km^2$ has a lagoon on west coast, with coral structure spreading around 4.9 $km^2$ [4]. The objective is to deploy UASN for in situ monitoring and to collect high quality data of water temperature, salinity, pH, dissolved carbon dioxide, other solvents/pollutants, etc., required for understanding the perilous impacts on coral diversity [5].

## 2.1 Related Works

Variants of TSP are used along literature to solve vehicle routing problem in wireless sensor network as well as underwater network. In [6], TSP algorithm is used to plan the trajectory of AUV resulting in optimum tour length for AUV, but as number of tour point increases the complexity of TSP increases, so different heuristics like genetic algorithm, ant colony optimization, one-opt neighbours, etc., are used for effective solutions. A ferry-based data gathering technique is adopted in [7] to gather data in a large scale deployment, which uses genetic algorithm to solve tours obtained by TSP. For large scale networks, data gathering using a single AUV will increase data latency. Multiple AUVs are used in partitioned network in [8] and it uses the concept of multiple travelling salesman problem.

This work presents a data collection scheme using a data collector whose traversal is determined by TSP algorithm, apart from the above-mentioned literature, the simulation platform is setup using an underwater simulator. It permits the agent implementation to be shared between deployment and simulation environment. Once the protocol is developed and tested in simulation, it is ready for deployment and at sea-testing with the help of UnetStack-compatible modems.

# 3 System Model and Architecture

The deployment aims monitoring of coral reefs in lagoons of Kavarathi Island spreading across five square kilometre and having an approximate depth of 200 m from sea level. ROI assumes to be a cuboid structure of dimension 5000 m × 5000 m × 200 m. The application is suited for sparse and long term deployment. The network architecture is shown in Fig. 1. Nodes are deployed across five localized regions predefined by sink node. These regions are defined such that each assumes to be a cuboid of dimension 525 m × 525 m × 200 m and form five clusters. In each cluster, sensor nodes are randomly distributed and nodes are connected via acoustic links. The clusters formed are isolated from each other. It is assumed that nodes are location aware and self-configurable, and communication range of node is twice the sensing range.

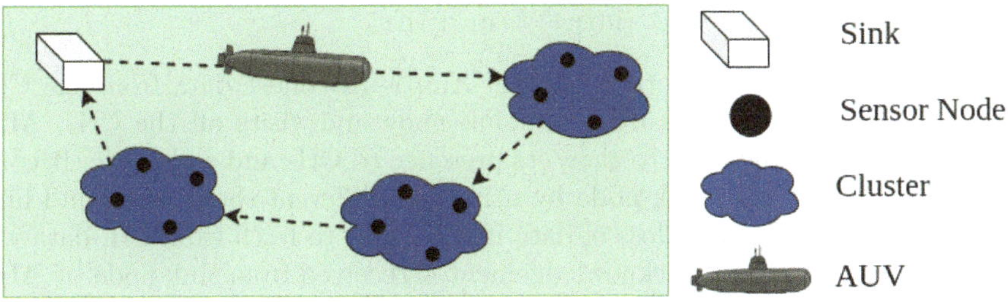


**Fig. 1.** Network architecture

A data gathering scheme with minimal collection points and a mobile data collector could optimize the tour of mobile collector. The collection points are reduced by selecting a CH among nodes in cluster. Data is collected using an AUV such that the path chosen by the AUV is the shortest. This is explained in two phases.

***Phase 1. Cluster-head selection phase***: A probabilistic CH selection is initiated by assigning probabilistic values (weights) to each node. Nodes broadcast their assigned weights using a power level that enables transmission among all nodes in their individual clusters. Nodes in respective clusters receive weights of their neighbours, and every node can select node with higher weight as its CH. The selected CH broadcasts a CH message. The message includes CH ID and schedule for cluster members to send their data to CH. Data collection starts after CHs are selected. A TDMA-based MAC protocol is chosen for sending sensed data from sensor nodes to their respective clusters. Soon after CHs are selected, a message is broadcasted by CH to all member nodes in clusters initiating them to send their sensed data to respective CHs. Each node receiving the message will send data at time slots allotted to them after a specified delay given in (1).

$$D = t_{\mathrm{PDT}} * (1 + \mathrm{rem}(A\%10)) \tag{1}$$

A delay of D ms is introduced to make underwater acoustic environment intact. Where $D$ is delay in ms, $A$, the address of node, $rem(I\%J)$ refers to remainder of $I$ when divided by $J$. Multiplying factor of $t_{\mathrm{PDT}}$ (milliseconds) is packet delivery time. Packet delivery time $t_{\mathrm{PDT}}$ can be calculated as follows:

Consider, $S$ as size of packet to be transmitted (in bits), $B$ the bit rate in bps, $c$ is Average propagation speed of acoustic signal in water, i.e., 1500 m/s, and $R$ is the distance of communication.

Then, Packet transmission time $t_{\mathrm{PTT}}$ is

$$t_{\mathrm{PTT}} = S/B \tag{2}$$

Propagation time $t_{\mathrm{PT}}$ is

$$t_{\mathrm{PT}} = R/c \tag{3}$$

Packet delivery time $t_{\mathrm{PDT}}$ is

$$t_{\mathrm{PDT}} = t_{\mathrm{PTT}} + t_{\mathrm{PT}} \tag{4}$$

***Phase 2. Data Collection phase***: Aim is to collect data from the CHs using an AUV. AUV starts from the sink node and visits all the CHs. AUV broadcasts a message signal to show its presence to CHs and sink node. It communicates with CH and sink node by using two different short range and high data rate channel. To avoid loss of data if AUV fails to fetch from CH, data will be stored in them until an acknowledgement is received from sink node by AUV in next visit.

The tour of AUV is described by travelling salesman problem, it should start from sink node, after covering all CHs once should return back to sink.

A mathematical formulation for this is described in Sect. 3. To initiate communication between AUV and sensor node, AUV will first send a control message to respective CH indicating its presence, and the CH starts to send data. The network can be formulated as a graph. Let $Q = (C, E)$ be a directed graph, where $C$ represents set of points including the sink node and all CHs, $E$ the set of edges connecting them. So, $C$ contains the elements from 0 to $m$, where element zero represents sink and the elements from 1 to $m$ represents set of all CHs. The points in the graph is represented by the coordinates ($p_k$, $q_k$, $r_k$) and $k = 0, \ldots, m$. The edge between the two points in the graph is represented by $e^{kh}$, i.e. the edge between the CHs 1 & 2 is denoted by $e^{12}$. A cost factor $\xi_{kh}$ is defined for each $e^{kh}$ that defines distance between $k$ and $h$. Some weighing other than Euclidean distance is defined. They are as follows:

***(a) Risk Cost***: A risk cost is assigned to each cluster based on the fact some regions in ROI are more prone to bleaching or any hazardous impact. More the impact larger the cost.

The cost factor is defined as $R_i \in (100, 1000)$.

***(b) Inter Sink Cluster-head Distance***: The distance from sink node to any CH matters during the time AUV decides which node to start with and prefer a node with a smaller tour, so the CH nearer to the sink node is given a greatest weight defined by

$$\xi_d(k) = 1/S(C(k) : K \neq 0, C(0)) \tag{5}$$

$S(C(k), C(o))$, is the distance between the corresponding CH and sink node.

***(c) The Cluster-Head Adjacency***: The CHs which are adjacent to each other is given a largest weight given by

$$\xi_a(k) = 1/\min(s(c(k), c(h))k, h = 1, 2 \ldots m, k \neq h \tag{6}$$

Then the total weight $W_{tot}$ is the sum of both,

$$W_{\text{tot}} = \left( \sum_{k=1}^{m} \xi_d(k) + \xi_a(k) \right) * R_i \tag{7}$$

The total cost between two points in the graph is reciprocal of Euclidean distance between them which got multiplied by the total weight is given by

$$\text{Total cost } = 1/\left(\xi_{\text{kh}} * W_{\text{ tot }}\right) \tag{8}$$

Based on obtained cost, TSP algorithm is run to determine the traversal of AUV. To initiate the communication between the AUV and sensor node, the AUV will first send a control message to the respective CH indicating its presence, and CH starts to send the data, the time the AUV stays in each CH is termed as the brake time and is given by Eq. (9).

$$Bt(k) = M(s) \times T_b + T_d \tag{9}$$

where $M(s)$ memory size of node in bits, $T_b$ time taken for a bit to get transmitted and $T_d$ the guard time. The time taken by the AUV to cover the entire network is predefined, the AUV must be back to the sink node after a particular interval of time, and this was done to prevent unexpected delays caused by any kind. The total brake time of the AUV is given by

$$\text{Tot } Bt(k) = \sum_{k=1}^{m} Bt(k) \tag{10}$$

Total distance is given in (11)

$$D_{\text{tot}} = \sum_{k=1}^{m} \sum_{\substack{h=1 \\ k \neq h}}^{m} S(C(k,h)) + S(C(m), C(0)) \tag{11}$$

$$T_{\text{tot}} = D_{\text{Tot}}/V \tag{12}$$

Total travel time covering the CHs is given in (12) where $V$ is the velocity of AUV. Where $S(C(k,h))$ denotes the tour from node $k$ to node $h$ and $S(C(m), C(0))$ distance from the $m$th CH node to sink node. Travel time of AUV is the sum total of the brake time and travel time from one CH node to other.

$$\text{Travel Time } = T_{\text{tot}} + Bt(k) \tag{13}$$

### 3.1 Mathematical Formulation

*Problem statement: Design data gathering strategy to optimize AUV traversal.*

Integer linear programming (ILP) for travelling salesman problem is adopted for above constraint. It is applicable to mission planning of AUV. This is applied for optimizing tour of AUV. AUV is deployed in network to gather data from CHs. A brief overview is illustrated in Fig. 1. The mobile entity should start from the sink node has to cover all CHs only once using a minimal tour and then return back to sink [9].

First a decision variable is defined as follows

$$a_{\text{kh}} = \begin{cases} 1, & \text{if the edge } e^{kh} \text{ is on the tour of AUV} \\ 0, & \text{otherwise} \end{cases}$$

To achieve primary constraints of minimizing tour of AUV, an objective function is defined in (14), and (15 and 16) implies all CHs in network are visited by AUV.

Minimize,

$$\sum_{k=1}^{m} \sum_{h=1}^{m} \xi_{kh} a_{kh} \tag{14}$$

Subject to

$$\sum_{\substack{k=1 \\ k \neq h}}^{m} a_{kh=1} \quad h = 1 \ldots m \tag{15}$$

$$\sum_{\substack{h=1 \\ k\neq h}}^{m} a_{kh=1} \quad k = 1\ldots m \tag{16}$$

But these constraints result in various sub-tours for AUV. To eliminate sub-tours of AUV using the Miller–Tucker–Zemlin (MTZ) formulation, a dummy variable $T_k$ is introduced such that:

$$T_1 = 1 \tag{17}$$

$$2 \leq T_K \leq m \tag{18}$$

$$T_K - T_h + 1 \leq (m-1)\,(1 - a_{kh})\,\forall k \neq 1, \forall h \neq 1 \tag{19}$$

And (18) excludes sub-tour of AUV as for $(k, h)$ forces $T_h \geq T_k + 1$, when $a_{kh} = 1$.

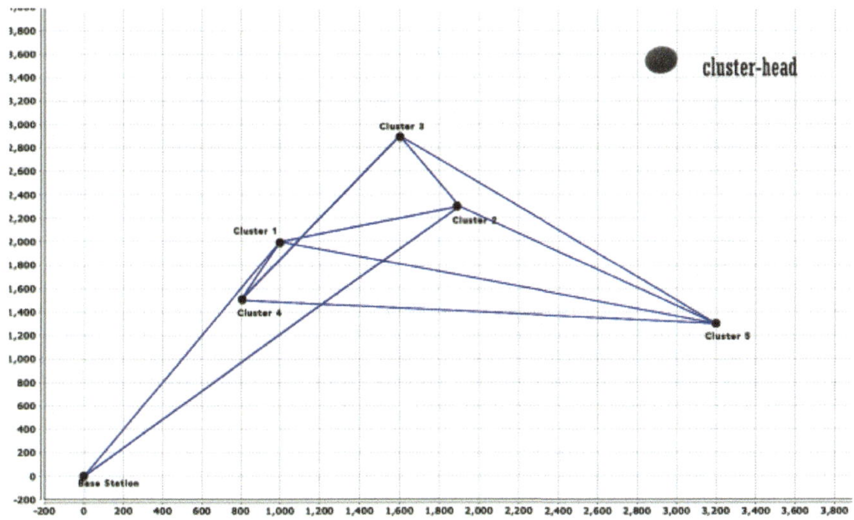


**Fig. 2.** AUV tour in UnetSim

## 4 Results and Discussion

Unet simulator is used for the implementation of proposed protocols [10]. The parameters of network architecture and simulation set-up are provided in Table 1. The events and time taken for the clustering process are shown in Table 2. Simulation results show the packet delivery ratio (PDR) of 100 %. All packets generated in the network are received by the sink node as the packets are transmitted by the single hop between the nodes and the cluster-heads. The bit rate

**Table 1.** The parameters of network architecture and simulation set-up

| Network architecture parameters | | Simulation parameters | |
|---|---|---|---|
| Total depth | 200 m | Frequency of operation (f) | 25 kHz |
| | | Transmission range | 750 m |
| No. of clusters | 5 | Source level (SL) | 135 dB re $\mu$PA |
| No. of nodes in each cluster | 4 | Average propagation speed of acoustic signal in water (c) | 1500 m/s |

**Table 2.** Schedule of the data collection in a cluster

| Event | Time (μs) |
|---|---|
| CH broadcast | 60 |
| Member node's transmission time to CH | 96 |
| AUV control message transmission time to CH | 594 |
| CH data transmission time to AUV | 600 |

**Table 3.** Size of the packets

| | |
|---|---|
| Size of the packet sent by the cluster member to CH | 16 bytes |
| size of the packet sent by the CH to the AUV | 80 bytes |

**Table 4.** Path traversed by AUV

| Next location | Sink (0 m, 0 m, 0 m) |
|---|---|
| CH 2 | (1000 m, 2000 m, −200 m) |
| CH 5 | (1900 m, 2300 m, −200 m) |
| CH 3 | (1600 m, 2900 m, −200 m) |
| CH 4 | 800 m, 1500 m, −200 m) |
| CH 1 | 3200 m, 1300 m, −200 m) |
| Location | (0 m, 0 m, 0 m) |

**Table 5.** Distribution of costs among the CHs

| Cluster | Sink | 1 | 2 | 3 | 4 | 5 |
|---|---|---|---|---|---|---|
| Sink | 0 | 10 | 15 | 20 | 40 | 60 |
| 1 | 10 | 0 | 35 | 25 | 45 | 65 |
| 2 | 15 | 35 | 0 | 30 | 15 | 23 |
| 3 | 20 | 25 | 30 | 0 | 15 | 15 |
| 4 | 40 | 45 | 15 | 15 | 0 | 67 |
| 5 | 60 | 65 | 23 | 15 | 67 | 0 |

used is 2400 bps. As the data is collected at regular intervals size of packets remains constant despite time in which the data is collected. Table 3 illustrates the packet transmission and reception of network. The path of AUV assigned to collect data is determined by TSP algorithm. Table 4 illustrates the distribution of inter-edge cost among various CHs that determines the TSP algorithm. Then, TSP is applied to costs distributed among CHs to find optimal tour. Figure 2 gives AUV traversal in UnetSim. The AUV visits the CHs based on optimal path found using TSP, here, AUV starts from sink node and traverse in order of CHs (3, 5, 2, 4 and 1), then, traces back to sink. The location of nodes and sequence of visit by AUV using TSP algorithm is shown in Table 5. From results, it is inferred that the application of TSP has optimized path of AUV.

## 5 Conclusion and Future Work

This paper discusses an application-based network scenario for 3D UASN. Here, isolated clusters are deployed along the desired region of interest. The data is gathered from CH nodes which are selected using a probabilistic clustering approach. The sensed data is sent to the AUV using a TDMA-based MAC protocol. A mobile entity (AUV) whose tour is optimized using a TSP is used for collecting the data from CH nodes. This results in the minimal tour of the AUV reducing the latency in the network. Simulation of the proposed network and protocols is performed using UnetSim simulation platform for an ideal simulation environment. For future work, multiple AUVs can be used in partitioned networks, and their tours can be formulated using MTSP techniques. Further, simulation parameters can be modified to evaluate the working of the proposed protocol in more realistic and dynamic ocean environment.

## References

1. Verstelle, J.T.: The growth rate at various depths of coral reefs'in the dutch east Indian archipelago. Treubia **14**(1), 117–127 (2016)
2. Kara, I., Bektas, T.: Integer linear programming formulations of multiple salesman problems and its variations. Eur. J. Oper. Res. **174**, 1449–1458 (2006)

3. Arthur, R.: Patterns of benthic recovery in the Lakshadweep Islands. Ten years after bleaching-facing the consequences of climate change in the Indian Ocean. CORDIO Status Report, 39–44 (2008)
4. Gopinath, A., Nair, S., Kumar, N., Jayalakshmi, K., Pamalal, D.: A baseline study of trace metals in a coral reef sedimentary environment, Lakshadweep Archipelago. Environ. Earth Sci. **59**, 1245–1266 (2010)
5. Suzuki, T., Kato, K., Makihara, E., Kobayashi, T., Kono, H., Sawai, K., Yamashiro, H.: Development of underwater monitoring wireless sensor network to support coral reef observation. Int. J. Distrib. Sens. Netw. **10**, 189643 (2014)
6. Ghoreyshi, S., Shahrabi, A., Boutaleb, T., Khalily, M.: Mobile data gathering with hop-constrained clustering in underwater sensor networks. IEEE Access **7**, 21118–21132 (2019)
7. Alnuaimi, M., Shuaib, K., Alnuaimi, K., Abdel-Hafez, M.: Data gathering in delay tolerant wireless sensor networks using a ferry. Sensors **15**(10), 25809–25830 (2015)
8. Van Le, D., Oh, H., Yoon, S.: HiCoDG: a hierarchical data-gathering scheme using cooperative multiple mobile elements. Sensors **14**(12), 24278–24304 (2014)
9. Pataki, G.: Teaching integer programming formulations using the traveling salesman problem. SIAM Rev. **45**(1), 116–123 (2003)
10. Chitre, M.A., Bhatnagar, R., Soh, W.: UnetStack: an agent-based software stack and simulator for underwater networks. 2014 Oceans - St. John's, pp. 1–10. IEEE, Canada (2014)

# Preprocessing Improves CNN and LSTM in Aspect-Based Sentiment Analysis for Vietnamese

Duy Nguyen Ngoc[1,2,3(✉)], Tuoi Phan Thi[1,2], and Phuc Do[2,4]

[1] Ho Chi Minh City University of Technology (HCMUT), Ho Chi Minh City, Vietnam
{1680475,tuoi}@hcmut.edu.vn, duynn@ptithcm.edu.vn
[2] Vietnam National University, Ho Chi Minh City, Vietnam
phucdo@uit.edu.vn
[3] Posts and Telecommunications Institute of Technology, Ho Chi Minh City, Vietnam
[4] University of Information Technology, VNU-HCM, Ho Chi Minh City, Vietnam

**Abstract.** The deep learning method has achieved particularly good results in many application fields, such as computer vision, image processing, voice recognition, and signal processing. Recently, this method has also been used in the field of natural language processing and has achieved impressive results. In this field, the problem of categorizing subjective opinions which is an individual's thinking or judgment of a product or an event or a cultural and social issue. Subjective opinions have received attention from many producers and businesses who are interested in exploiting the opinions of the community and scientists. This paper experiments with the deep learning model convolution neural network (CNN), long short-term memory (LSTM), and the boxed model of CNN and LSTM. Training data sets comprise reviews of cars in Vietnamese. Cars are objects with a significant number of specifications that are provided in user reviews. The Vietnamese opinion set is preprocessed according to the method of aspect analysis based on an ontology of semantic and sentimental approaches. A Vietnamese corpus experiment with CNN, LSTM, and CNN + LSTM models are used to evaluate the effectiveness of the data preprocessing method that was used in this paper. To assess the validity of the test models with the Vietnamese opinion set, the paper also tests the sentiment classification with the English Sentence Collection Stanford Sentiment Treebank (SST).

**Keywords:** Corpus · Deep learning · Classification · CNN · Convolution neural network · LSTM · Long short-term memory · Sentiment analysis

## 1 Introduction

Everyone has the opportunity to express their thoughts and assessments about a product, an event, or other problems in the world due to the development of the Internet. Such user opinions are increasing in the Internet environment. This is a particularly useful resource


X.-S. Yang et al. (eds.), *Proceedings of Fifth International Congress on Information and Communication Technology*, Advances in Intelligent Systems and Computing 1183, https://doi.org/10.1007/978-981-15-5856-6_17

for individuals and organizations who want to exploit the opinions of the community, as well as individuals who consider or want to buy a product. There is a large demand for the exploitation of this resource. With a substantially large number of opinions, the classification to exploit them using scientific methods represents an indispensable requirement. Many methods have been proposed and tested for the problem of sentiment analysis of each opinion. Machine learning methods are highly popular in this field, such as support vector machine, Naïve Bayes, and conditional random fields [1, 2]. The results have been highly promising. These machine learning methods are based on a sentiment dictionary with a predefined weight for each item. An important feature of a sentiment dictionary is that it was built for a specific language and certain topics. Therefore, if a dictionary is not sufficiently good or used for an inappropriate topic, it can affect the quality of the sentiment analysis.

Deep learning has only relatively recently been applied to natural language processing but has demonstrated good results, such as Kalchbrenner et al. in [3] achieving an accuracy of 86.8% and Li et al. in [4] producing an accuracy of 89.81%. The deep learning method is not a requirement for sentiment dictionary. Building a sentiment dictionary is a difficult task, especially for the less popular languages in the world. Methods that need a sentiment dictionary are a major obstacle to less popular languages in the world. Therefore, the approach of the deep learning method brings substantial opportunities for less popular languages such as Vietnamese to develop the sentiment analysis field in natural language processing quickly. Vietnamese has not yet had a sentiment dictionary commonly for research. For less popular languages when using deep learning methods, the important focus is testing, which requires substantial time to identify the specific characteristics of these languages. From there, it is possible to develop solutions to exploit deep learning methods for this field in an effective manner.

Studies on deep learning methods for the field of sentiment analysis in Vietnamese, such as [5–7] also obtained good results. However, the number of such studies has not yet to reach a sufficient level. The Vietnamese corpus is not yet rich in the topic. This chapter develops the method of [5] in terms of enriching the corpus and ontology. In addition, this research conducted additional tests of some deep learning methods in the sentence- and aspect-level analyses. From there, some characteristics can be drawn from using the deep learning method for sentiment analysis in Vietnamese as well as a more detailed evaluation of the method proposed in [5].

The rest of the paper is organized as follows: Sect. 2 reviews the related works. Section 3 gives an overview of the deep learning models used in testing and our preprocessing data method. Section 4 discusses the configuration of the deep learning system and the experimental results. Finally, Sect. 5 presents conclusions.

## 2 Related Works

In the form of sentence-level sentiment classification, Kalchbrenner et al. [3] used the dynamic convolution neural networks model to model sentences by a Dynamic K-Max Pooling operator as a nonlinear sampling function to test sentiment classification using the SST corpus. The accuracy of [3] was 86.8%. Wang et al. [8] also tested on the SST opinion set but by using a combination of CNN and RNN. The accuracy of [8] was

89.95% on SST with three sentiment labels. The accuracy of [8] was 51.5% only if Wang et al. used SST with five sentiment labels. For Pang and Lee's film rating, 2005, the accuracy of [8] was found to be 82.28%.

In the form of aspect-level sentiment classification, Li et al. [4] used the CNN to obtain an accuracy of 89.81%. Corpus used in the experiment is Chinese blog sites. Meanwhile, Dhanush et al. [9] obtained an accuracy of only 76.1% results when using CNN to classify the sentiment of restaurant reviews.

## 3 Approach Method

As mentioned in the previous sections, this research will use some models of deep learning methods to implement sentiment classification. In the following sections, the paper presents an overview of the LSTM and CNN models, as well as the combined model of CNN and LSTM for sentiment classification testing.

### 3.1 Deep Learning Model for Experiments

**Convolution Neural Network**. A convolution neural network is first used in the field of digital signal processing. Based on the principle of information conversion, scientists have applied this technique to digital photo and video processing. The CNN model as Fig. 1. The layers are linked together through a convolution mechanism. The next layer is the cumulative result from the previous layer. As a result, we have local connections. Hence, each neural in the next layer is generated from the filters imposed on a local data area of the previous neural layer. As a result, we have local connections.

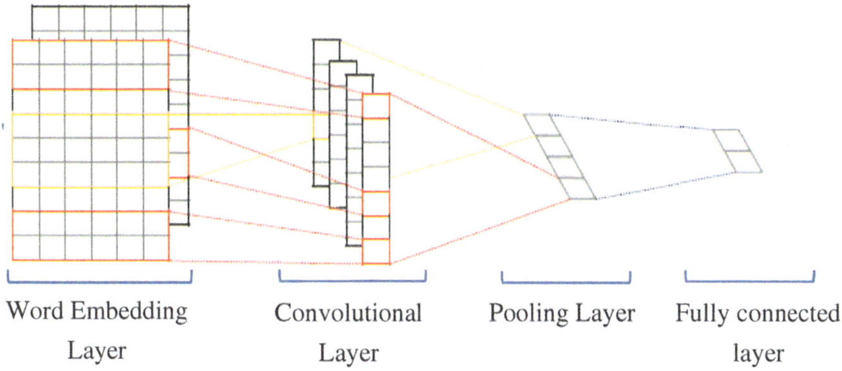


**Fig. 1.** Basic convolutional neural network model [12]

- Word Embedding Layer: This class includes matrices of size $n \times k$, representing sentences with $n$ words, each representing a $k$-dimensional vector.

This layer encodes every word in the selected sentence into a word vector. Let $l \in R$ be the sentence length, $|D| \in R$ is the vocabulary size and $W^{(l)} \in R^{k \times |D|}$ is the matrix embedded vectors from $k$ dimensions. The $i$ word in the sentence is transformed into a $k$ dimensional vector $w_i$ using Eq. (1):

$$w_i = W^{(l)} x_i \tag{1}$$

where $x_i$ is a one-hot vector representation for the $i$ word one-hot vector.

We can use the tool word2vec or Glove to create the matrix for this layer.

- Convolution Layer: This layer uses convolution to process data by sliding a fixed size slide (also called a kernel) on the input data matrix to obtain a refined result.
- Pooling Layer: This layer is responsible for summarizing the result vector of the convolution layer and retaining the most important vectors.
- Fully Connected Layer: In a convolution neural network, there are one or more fully connected layers after the convolution layer. This layer is simply a traditional neural network that uses the remaining vectors in the upper layers as input to produce the final result through training.

**Long short-term memory**. Long short-term memory (LSTM) has four interaction layers and two status signals: hidden state and cell state [10], as shown in Fig. 2.

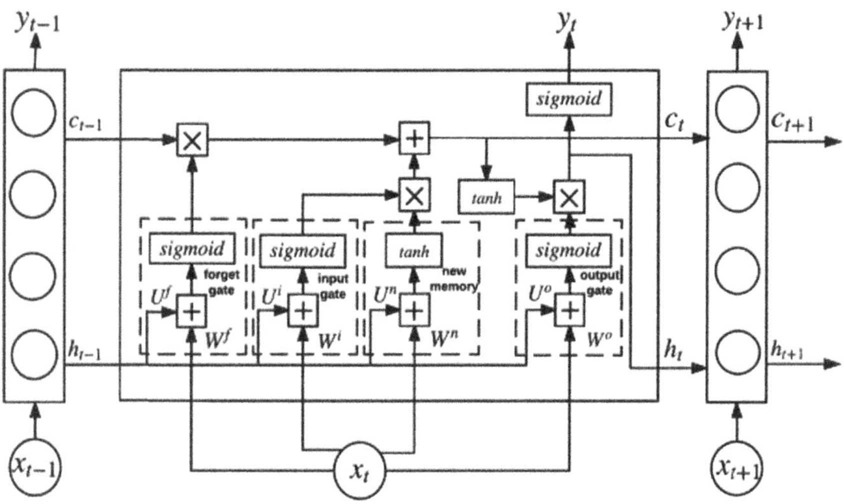


**Fig. 2.** Long short-term memory model [10]

At the time $t$, the LSTM decides that information poured into the cell state based on the sigmoid function or the $\sigma$, floor, forget gate. This function takes $h_{t-1}$ from the previous hidden layer and input signal $x_t$ in the present to create a number in [0, 1] as in formula (2). Whether the new information is saved to the cell state or not is based on the calculation at the input port with the sigmoid function as formula (3). A vector of new candidate values $\tilde{C}_t$ is created through the tan $h$ layer using formula (4). The cell state at time t is calculated by (5) based on the previous cell state $C_{t-1}$, candidate $\tilde{C}_t$ and $f_t$ function. The $f_t$ function can control the slope that passes through it and explicitly deletes or updates memory. The LSTM network decides the output based on the cell state. Through the sigmoid function, the value of the cell state is calculated by the formula (6) and reached at the output gate. This value is via the tan $h$ function and multiplied by the output of the sigmoid port to obtain the $h_t$ value via formula (7).

$$ft = s(Wfxt + Ufht - 1) \tag{2}$$

$$i_t = s\left(W^i x_t + U^i h_{t-1}\right) \tag{3}$$

$$\tilde{C}_t = \tan h(W^n x_t + U^n h_{t-1}) \tag{4}$$

$$o_t = \sigma(W^0 x_t + U^0 h_{t-1}) \tag{5}$$

$$C_t = f_t * C_{t-1} + i_t * \tilde{C}_t \tag{6}$$

$$h_t = o_t * \tan h(C_t) \tag{7}$$

**CNN–LSTM**. The convolution layer of CNN creates a feature map vector. The number of feature vectors is equal to the number of filters used during the convolution. In the pooling layer, the best feature map values from each class will be chosen to obtain the most important feature of the comment. The feature vectors over a fully connected CNN network generate a set of parameters at the output of CNN. The LSTM model uses output parameters of CNN to carry out the process of classifying comments. This combined model is shown in Fig. 3.

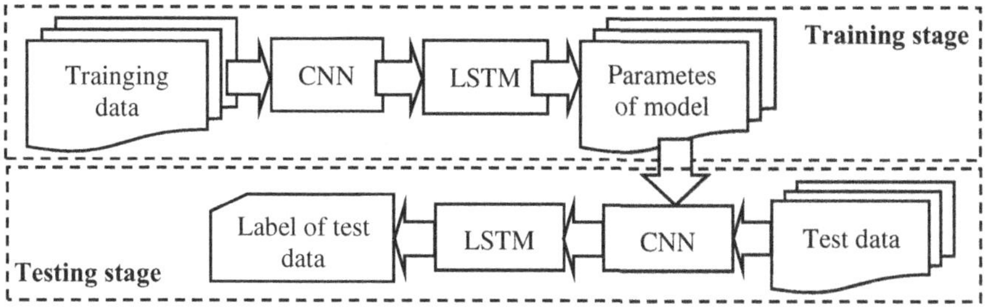


**Fig. 3.** CNN-LSTM model

## 3.2 Corpus

**English corpus**. Stanford Sentiment Treebank is a corpus of film reviews in English [11]. Opinions are categorized using five labels: very positive, positive, neutral, negative, and very negative. To have similarities with the Vietnamese corpus, this paper will carry out an empirical classification of sentiments using three labels; the very positive and the positive will be combined into one positive label; very negative and negative classes into one negative label.

**Vietnamese corpus**. The Vietnamese corpus is a collection of documents that record the review of car objects (Car Opinions in Vietnamese—COV). This topic has attracted the attention of many people in large countries such as the USA and Germany where car ownership is particularly popular, and Vietnam where such ownership is popular in some large cities.

This corpus was built and processed according to the method presented in the paper [5]. Collected opinions that have been processed by this method comprise a so-called standardized opinion set (COVn). Opinions that have not been processed by this method are referred to as raw comments (COVr). The pretreatment method according to [5] can be summarized as follows.

*Creating a set of aspects for the objects*. The aspects of cars are technical characteristics that often used by manufacturers when introducing products, or customers are interested in and assessing products. Each aspect has an official name that is commonly used by manufacturers and a semantic equivalent. They are the common name that users use when expressing their opinions in the Internet environment. The aspect entities collect by two methods as statistical and word2vec tool.

*Creating the structure of a semantic and sentiment vocabulary hierarchical tree*. Both aspect and sentiment entities are divided into three tiers in the Semantic and Sentiment Vocabulary Hierarchical Tree (SSVHT) ontology. Aspect set refers to words/phrases used by the manufacturer or user to refer to the components or specifications of a car.

*Labeling sentences with the semantic and sentiment on aspects of object*. The labeling of semantics and sentiments of the sentence according to the aspect of the car is done manually. The labeling process is divided into two stages. The first stage carries out the labeling according to the aspect of the sentence. The next stage is the sentimental labeling as positive, neutral, and negative. The process of labeling will remove sentences without Vietnamese accents, without sentiment, or not the same subject. Semantic labels are nouns or noun phrases belonging to the aspect set of cars.

*Standardizing sentiment aspects of the object*. Sentiment sentences in the corpus will be re-formulated into the simply structured sentence

$$N + A \tag{8}$$

$$N + P + A \tag{9}$$

$$N + A + P \tag{10}$$

where:

N noun representing an aspect.
R adverb or adverbs
A sentiment complements (adjectives or verbs).

Based on the SSVHT ontology, the sentences are normalized using the formulas (8), (9), and (10). The features of the corpus after the above processing shown in Table 1.

**Table 1.** Organization of the corpus

| Features of corpus | Quantity |
|---|---|
| Sample of cars | 102 |
| Opinions | 3,214 |
| Training sentences | 6,517 |
| Testing sentences | 1,130 |
| Sentiment labels | 3 (positive, neutral, negative) |
| Sentences labeled with positive polarity | 2,134 |
| Sentences labeled with neutral polarity | 2,055 |
| Sentences labeled with negative polarity | 2,328 |

### 3.3 Experiment Configuration

**System configuration.** Table 2 presents the components of the system used in the experiment of this paper.

**Table 2.** Configuration of the convolution neural network

| Component of system | Version name |
|---|---|
| Operating system | Ubuntu 16.04 LTS |
| Programminng language | Python |
| Framework deep learning | Theano–Keras |

**Deep learning models.**

- CNN: Activation function is sigmoid ($1/(1 + e^{-x})$), embedding word size is 300, number of filters is 300, filter size is 3, and dropout is 0.5.
- LSTM: Activation function is sigmoid ($1/(1 + e^{-x})$), dropout is 0.2, embedding word size is tested at 300, and number of filters tested at 100 and 200.
- CNN-LSTM: The configuration parameters of CNN and LSTM when combined are the same as when operating independently.

## 4 Experiment

This chapter attempts to change the number of filters in the LSTM model. The experiment results with the SST, COVr, and COVn corpus are shown in Figs. 4, 5, 6 and 7. The best values were obtained during the testing of three sets of comments by the LSTM, CNN, and CNN + LSTM model according to the number of iterations shown in Table 3.

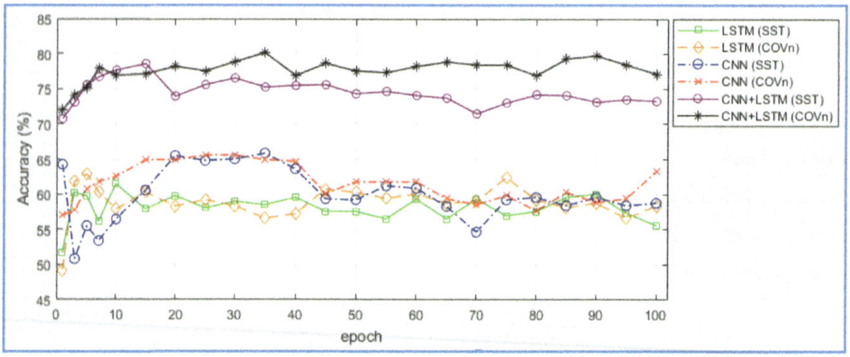


**Fig. 4.** Accuracy of experimenting on SST and COVn with LSTM has 200 filters

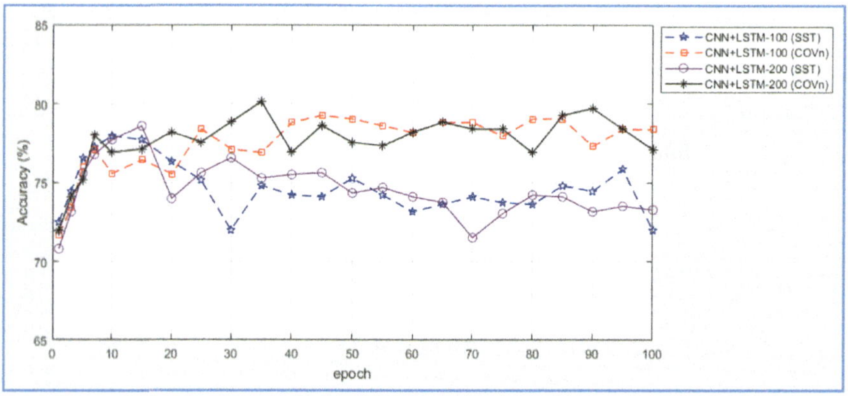


**Fig. 5.** Accuracy of experimenting on SST and COVn with LSTM has 100 and 200 filters

The LSTM has 200 filters and CNN with 300 filters.

From the experiment results shown in Figs. 4, 5, 6, 7, and Table 3, the following statements can be made:

- Both CNN and LSTM models operate independently, obtaining a fairly low accuracy when they analyze aspect-level sentiment. The CNN and LSTM models operating independently demonstrated an accuracy that fluctuated significantly when the number of iterations (epoch) was low (less than 50 times). Within 100 iterations, the best results showed low iterations for both SST and COV opinion sets. By increasing the number

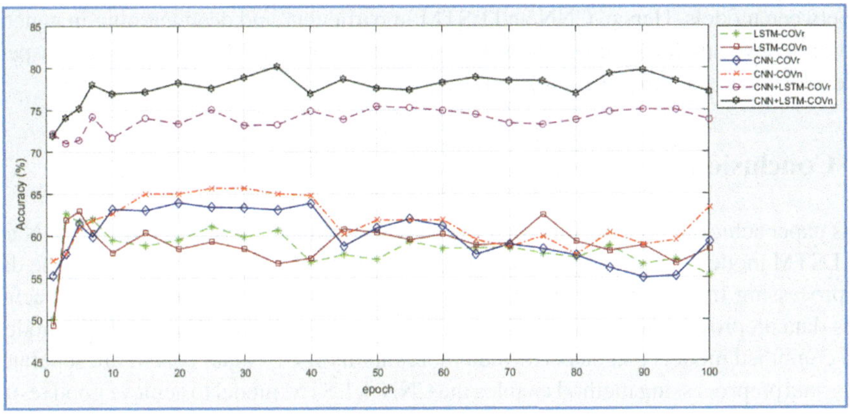


**Fig. 6.** Accuracy of experimenting on COVr and COVn with LSTM has 200 filters

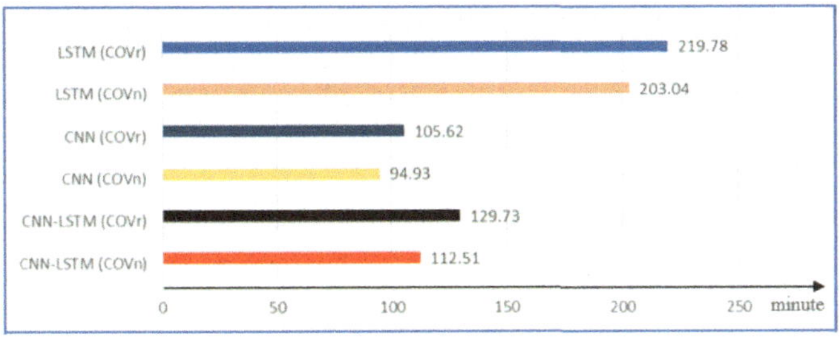


**Fig. 7.** Execution time of 100 epoch of the model

**Table 3.** Best accuracy results of each corpus

| Data | Accuracy (%) | Epoch |
|---|---|---|
| CNN + LSTM (200 filter–COVr) | 75.39 | 50 |
| CNN + LSTM (200 filter–COVn) | *80.13* | 35 |
| CNN + LSTM (200 filter–SST) | 78.61 | 15 |

of iterations and filters, the result was more stable and reliable. The accuracy of the combined CNN and LSTM model is better than the accuracy of each model when they run independently, as shown in Table 3. This accuracy improved by approximately 10% compared to when these models are operated separately.

- The aspect-level data preprocessing method for comments based on SSVHT ontology improves accuracy in both CNN, LSTM, and CNN + LSTM models. In particular, the CNN + LSTM model has an accuracy difference of up to 4% when it was trained by the preprocessed dataset compared to the raw dataset.
- Although the results obtained on SST and COV tests differed with respect to the accuracy, there were many similarities in the results achieved in each model as well as

between models. Hence, CNN and LSTM in particular, and deep learning in general, do not have any language barriers as well as the subject of the sentence- and aspect-level sentence sentiment classification.

## 5 Conclusion

This paper achieved good results for aspect-level sentiment analysis using the CNN and the LSTM model, particularly the model combining CNN + LSTM by helping the data preprocessing method based on an ontology of semantic and sentimental approaches. This data preprocessing method helps the learning process of CNN and LSTM models, and combined models can be performed more rapidly if CNN and LSTM are separated. This data preprocessing method enables the CNN + LSTM model to achieve good results with a small corpus. The importance of this method is in understanding the subject of the corpus. The data preprocessing method was tested in the aspect-level sentiment classification. Next, the authors will test document-level sentiment classification as well as improve the method so that we can obtain good results for that problem.

The test results of this paper on the SST corpus in English and COV corpus in Vietnamese demonstrate that the deep learning method does not meet the limitations for different languages. This feature of the deep learning method is of high significance to the less popular languages in the world, such as Vietnamese. It will help these languages to develop the sentiment analysis field in natural language processing more rapidly.

## References

1. Balazs, J.A., Velásquez, Juan D.: Opinion mining and information fusion: a survey. Inf. Fusion **27**, 95–110 (2016). https://doi.org/10.1016/j.inffus.2015.06.002
2. Moraes, R., Valiati, J.F., Neto, W.P.G.: Document-level sentiment classification: an empirical comparison between SVM and ANN. Expert Syst. Appl. **40**(2), 621–633 (2013). https://doi.org/10.1016/j.eswa.2012.07.059
3. Kalchbrenner, N., Grefenstette, E., Blunsom, P.: A convolutional neural network for modelling sentences. In: Proceedings of the 52nd Annual Meeting of the Association for Computational Linguistics, Maryland, USA, pp. 655–665 (2014). https://doi.org/10.3115/v1/p14-1062
4. Li, Q., Jin, Z., Wang, C., Zeng, D.D.: Mining opinion summarizations using convolutional neural networks in Chinese microblogging systems. Knowl. Syst. **107**, 289–300 (2016). https://doi.org/10.1016/j.knosys.2016.06.017
5. Ngoc, D.N., Thi, T.P., Do, P.: A data preprocessing method to classify and summarize aspect-based opinions using deep learning. In: Nguyen, N., Gaol, F., Hong, T.P., Trawiński, B. (eds.), Intelligent Information and Database Systems. ACIIDS 2019. LNCS (LNAI), vol. 11431, pp. 115–127. Springer, Cham (2019). https://doi.org/10.1007/978-3-030-14799-0_10
6. Vo, Q.H., Nguyen, H.T., Le, B., Nguyen, M.L.: Multi-channel LSTM-CNN model for Vietnamese sentiment analysis. In: Proceedings of the 9th International Conference on Knowledge and Systems Engineering (KSE), pp. 24–29. Hue, Vietnam (2017). https://doi.org/10.1109/kse.2017.8119429
7. Pham, T.-H., Le-Hong, P.: The importance of automatic syntactic features in Vietnamese named entity recognition. In: Proceedings of the 31st Pacific Asia Conference on Language, Information and Computation, pp. 97–103. Cebu, Philippines (2017)

8. Wang, X., Jiang, W., Luo, Z.: Combination of convolutional and recurrent neural network for sentiment analysis of short texts. In: Proceedings of COLING 2016, the 26th International Conference on Computational Linguistics: Technical Papers, pp. 2428–2437. Osaka, Japan (2016)
9. Dhanush, D., Thakur, A.K., Diwakar, N.P.: Aspect-based sentiment summarization with deep neural networks. Int. J. Eng. Res. Technol. **5**(5), 371–375 (2016). https://doi.org/10.17577/IJERTV5IS050553
10. Zhang, L., Wang, S., Liu, B.: Deep learning for sentiment analysis: a survey. Wiley Interdisciplinary Rev. Data Mining Knowl. Discov. **8**(4), e1253 (2018). https://doi.org/10.1002/widm.1253
11. Socher, R., Perelygin, A., Wu, J., Chuang, J., Manning, C.D., Ng, A., Potts, C.: Recursive deep models for semantic compositionality over a sentiment treebank. In: Proceedings of the 2013 Conference on Empirical Methods in Natural Language Processing, pp. 1631–1642. Washington, USA (2013)
12. Kim, Y.: Convolutional neural networks for sentence classification. In: Proceedings of the 2014 Conference on Empirical Methods in Natural Language Processing, EMNLP, pp. 1746–1751. Doha, Qatar (2014). https://doi.org/10.3115/v1/D14-1181

# Improving Soft Skills in Agile Software Development by Team Leader Rotation

Jose Libreros[1](✉), Ivan Viveros[2], Maria Trujillo[1], Mauricio Gaona[2], and David Cuadrado[3]

[1] Multimedia and Computer Vision Group, Universidad del Valle, Cali, Colombia
{jose.libreros,ivan.viveros}@correounivalle.edu.co
[2] Doctoral Studies in Computer Science Group, Universidad del Valle, Cali, Colombia
{maria.trujillo,mauricio.gaona}@correounivalle.edu.co
[3] Truora Inc., San Francisco, CA, USA
dcuadrado@truora.com

**Abstract.** New agile techniques have brought advantages over traditional techniques. However, agile techniques fail to solve soft skills problems. People with high technical knowledge still find difficulties to adapt to industrial work environments due to the lack of soft skills. At the individual level, specific skills are not perfected, questioned or transmitted. Team members with weaknesses in leadership and communication do not establish a regular channel for dissemination and resolution of disputes. Soft skill problems must be resolved quickly, at least, at the beginning of a project. Thus, people with high potential can lead to having a high impact on development iterations. This paper presents an agile technique taking into account weaknesses in soft skills. The proposed technique is based on rotating the leader team according to an iteration. Experimental evaluation shows that team members reach minimum communication soft skills after a short time. At the same time, teams gain more cohesion and better knowledge of everyone.

**Keywords:** Soft skills · Agile practices · Software development · Developers · Agile projects

## 1 Introduction and Related Work

Soft skills are not developed due to how curricula of universities are structured. According to enterprises, communication, leadership, collaborative environment and other soft skills are a weak point that professionals present. Involve people with no experience and lack in soft skills is an open challenge [1]. Enterprises prefer new professionals with high level of soft skills and lack of technical abilities rather than high-level technicians and scientists with low level of soft skills [2].


X.-S. Yang et al. (eds.), *Proceedings of Fifth International Congress on Information and Communication Technology*, Advances in Intelligent Systems and Computing 1183, https://doi.org/10.1007/978-981-15-5856-6_18

Andrews et al. [3] have characterised employability into abilities. Business-specific issues as hard business-related knowledge and skills; interpersonal competencies as soft business-related skills; and work experience and work-based learning are the three most general and significant areas in employability. Moreover, European Skills, Competences, Qualifications and Occupations bring the set of skills for different positions places at a high value on soft skills [4].

Agile techniques have introduced more agility of development. Additionally, recent graduate professionals do not have enough experience in agile developments. Nevertheless, software projects often do not comply with traditional standard measurements of success, namely time, cost and specifications. Time expend to develop software, meetings such as sprint review, retrospective, may be reduced by having better communication techniques, leadership and adequate channel for problem disruption [5].

Related works have proposed different activities involving rotation. Zurita et al. [6] have proposed rotation of team members among teams. They agree with many authors that communication, collaboration, creativity, critical thinking and problem-solving are the most important skills for success of professionals.

Vinot et al. [7] have proposed an analysis of skills of software development team. Additionally, effect of team's abilities on quality of final software product has been reported. Results imply that inexperienced software team members have responsibility of fulfilling other responsibilities that are poorly adapted to their skill level. Therefore, it has a significant adverse effect on quality of software products.

Jia et al. [8] have presented a classification of soft skills requirements for mobile developers. They enhance four soft skills: abilities such as responsibility, the initiative in meeting challenging work, positive attitude to work and coding habit. According to their research, mobile developers should have a stronger ability to fast learning. This can be extended to other types of teams such as back-end, front-end and DevOps with better cohesion teams [9].

A way to strengthen communication abilities and leadership of team members is to adapt to environments where all team members must be leaders. This brings the benefit of team members gaining experience in soft skills.

In this paper, an agile technique, so-called Team Leader Rotation, is presented. The proposed technique aims to overcome soft skills. Experimental evaluations have shown that using Team Leader Rotation promote the development of soft skills.

## 2 Team Leader Rotation

Team Leader Rotation is an agile practice with the aim of integrating professionals among different roles and responsibilities. In agile techniques philosophy, a team is homogeneous and with no hierarchy. However, many team members need a reference person that has more knowledge. In addition, in a given iteration, some team members fail to convey their concerns or controversies, and those controversies are never escalated to the relevant people. These communication failures can be improved by having a team leader, especially when the work is remotely done, or face-to-face sessions are few. Figure 1 shows the steps as Team Leader Rotation Flowchart. Some observations can be applied:

1. Scrum master or project facilitator can moderate the decision of who will be the next team leader.
2. When team members have been all team leaders, a round begins again.

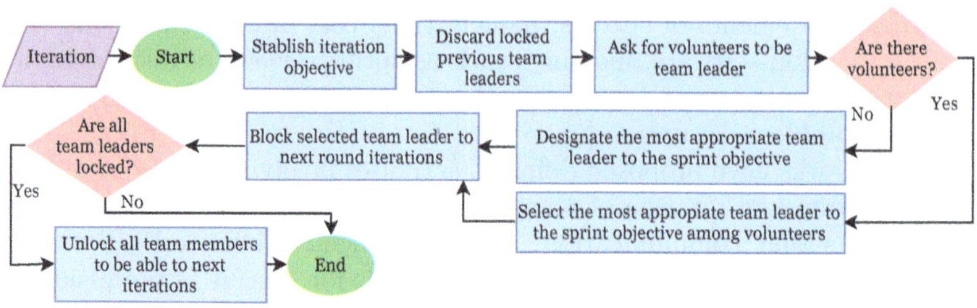


**Fig. 1.** Team Leader Rotation flowchart

# 3 Evaluation

## 3.1 Research Questions

Qualitative assessment method was used. This tests the hypothesis that teams show better behaviour when the team leader is designated in determined iterations. Through evaluation of 11 soft skills, and the conclusions that are obtained, the following research questions will be answered. This method is aligned with evaluation methods proposed by Creswell [6,10]:

1. According to team members, to what extent does Team Leader Rotation contributes to the development of creativity?
2. Which is the perception team members get about the contribution of Rotation of Team Leader to promote collaboration?
3. Were team members engaged in the project and proposed topics?
4. Did the results clarify power relationships?
5. Does team members recommend Team Leader Rotation?

## 3.2 Study Cases

Team Leader Rotation was applied into two project groups. First one is a project in coordination between the University and a specific enterprise. The second project is an academic project course. Having two study cases allows to item's responses be consistent across constructs. Test are determined by scores stable over time and teams [10]. Two cohorts have been selected. Each cohort is composed of six sprints, each week-long sprint. This is to show changes between initial and final iterations. Behaviour is extracted as follows. Positive and negative mentions that express strengths or weaknesses of a specific skill in a given sprint have been extracted from the Sprint Review Document at each respective project.

*Univalle*-**Truora project** In this study case, team members are a group of nine students of fifth semester of computer science program. Those students were involved in a project about face and OCR recognition. Project is conducted by Truora, an information security enterprise, and a research group of the *Universidad del Valle*. The main topic of this project is images processing. Students have prior knowledge about numerical analysis and methods, essential to image processing during the fifth semester. However, image processing is not a mandatory course of the computer science program. Some students have shown some vague knowledge on image processing, from what was learned in the course of analysis and numerical methods. Therefore, everyone has needed to receive training in image processing, machine learning and computer vision. The project had covered that formation. All students have finished secondary school and entered the university immediately. At the time of their hiring in the project, they have not had any previous experience in software projects.

**Software Engineering Trends, Graduate Course Project** Second study case describes an academic project with six master students and one undergraduate student. Five of the six master students are computer science professionals. The last master student is a mechanical engineer. The undergraduate student is enrolled in a computer science bachelor. The project was about a web application with a multitenant layer. Only two of the six team members had knowledge about multitenant and technologies associated with the implementation. At the beginning, many of the shortcomings were related to lack of experience of technologies to be used. One of these professionals has worked in the banking sector with experience in back-end development. Other students have had professional experience developing web applications, but oriented to one client.

## 4 Experimental Results

Table 1 shows the percentage of positive and negative opinions and comments on retrospectives with keywords related to each category. Description of each category is as follows [8]:

1. *Communication skills*: the ability to transmit anything effectively to others.
2. *Team level*: satisfaction as a team regarding the experience gained.
3. *Analytical and problem-solving skills*: ability to overcome problems, solve controversies or know how to escalate.
4. *Responsibility*: assessment of the importance of individual work, and helping others, as part of the entire team and the team responsibility.

5. *Fast learning*: regarding the level of interest in learning and reply knowledge. Additionally, the ability of self-learning in a short time.
6. *Initiative in meeting challenging work*: set challenges, know that it has a leader who can help directly, or escalate the concerns.
7. *Working independently*: complete tasks independently. In addition, to learn, find new solutions, new approaches and help in resolution of disputes using independence.
8. *Positive attitude to work*: e.g. enthusiastic, patient and hard.
9. *Organizational skills*: the ability to make team members work efficiently.
10. *Reading*: the ability to read documents, papers, journals and new algorithms.

**Table 1.** Soft skills behaviour according to retrospective skill keywords mentions

| Skill | Cohort 1 | | Cohort 2 | |
|---|---|---|---|---|
| | Positive (%) | Negative (%) | Positive (%) | Negative (%) |
| Communicational skills | 0.00 | 17.78 | 15.56 | 13.33 |
| Team level | 11.11 | 0.00 | 46.67 | 0.00 |
| Analytical and problem skills description | 6.67 | 4.44 | 4.44 | 0.00 |
| Responsibility | 6.67 | 2.22 | 20.00 | 2.22 |
| Fast learner | 75.56 | 0.00 | 100.00 | 0.00 |
| Initiative in meeting challenge work | 11.11 | 33.33 | 44.44 | 17.78 |
| Time management | 15.56 | 55.56 | 15.56 | 33.33 |
| Working independently | 15.56 | 0.00 | 82.22 | 0.00 |
| Positive attitude | 88.89 | 8.89 | 100.00 | 0.00 |
| Organisational skills | 8.89 | 8.89 | 17.78 | 17.78 |
| Reading | 66.67 | 0.00 | 100.00 | 0.00 |

## 5 Discussion

Three types of improvements emerge from Fig. 2. We noticed a substantial increment of positive comment after rotating team leaders. Team Leader Rotation contributes to improve soft skills and motivation of team members. Also, level of team is evidenced in the work developed by each team. Second, responsibility for tasks developed. That is, since team has a leader that encourages responsibility. From the beginning of projects, team members have shown skills of adaptation and rapid appropriation of knowledge. This ability is related to being motivated by a leader to explain and clarify doubts quickly.

The presence of a leader has a significant improvement in the skill of meeting challenge work. Basically, leaders concern and request for review of proposals are rapidly escalated to product manager and other teams with more experience. Meeting challenge work is enhanced when that leader focuses on subject in

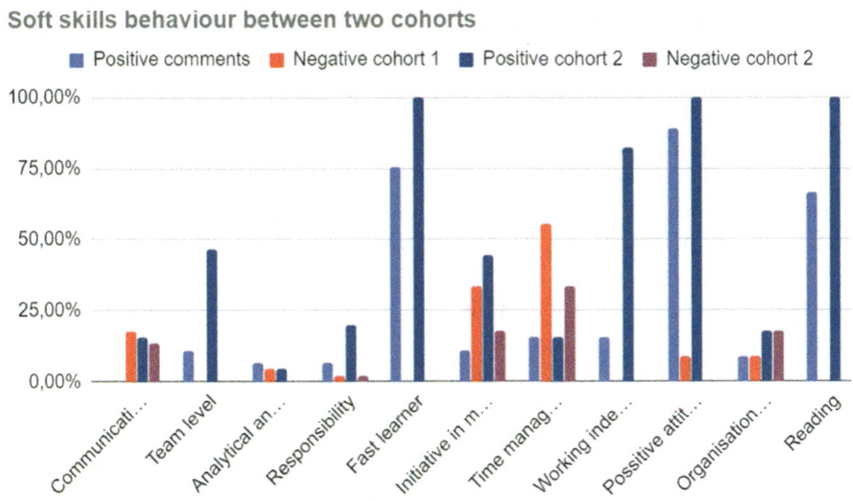


**Fig. 2.** Soft skills behaviour according to retrospective skill keywords mentions—percentage

which he feels identified, following the line of proposal presented in this article. This need for initiatives in meetings in cohort 1 is evident, based on negative comments in retrospective. In general, percentage of negative comments have decreased. It is evident that there has been a good appropriation and interest of the team and scrum master to improve evaluated soft skills. Thus, another advantage of having team leaders emerges. Independent work has been another benefited area. In retrospectives of the first cohort, they were hardly named. However, in the retrospectives of the second cohort is mentioned the satisfaction for the work done autonomously.

There was a decrease in positive comments in analytical and problem skills description skill. However, it is noted that percentage of negative observations decreased to zero in the second cohort. Having a leader who acts as a receiver of concerns, controversies have been resolved quickly. However, at the beginning of the project, motivations to be team leader must be instilled. It is necessary with teams with no experience. Thus, dispute resolution is a skill that must be worked on and improved in teams with little or no experience in projects.

As opportunities for improvement, in addition to the one mentioned above, organisational skills are a pending task. There is an increment of negative comments about having a better time organisation. This opens possibility of examining other alternatives that increase satisfaction of team concerning this field.

From the discussion carried out and conclusions obtained, research questions can be answered with qualitative evaluation.

1. *Contribution to the development of creativity*: According to perception of members of teams, the ability to read, initiative in meetings, self-learning and reduction of negative comments on dispute resolution are evidently driven by

a creative desire of team members. By having a leader who identifies with the particular objective of each sprint, there is more confidence and dissemination of knowledge.

2. *Promotion of collaboration*: It has been evidenced in development of the projects, that there are volunteers who are offered to be team leaders in an iteration, respecting the rule of finishing the whole round. This makes them feel identified and appropriate with this new practice. In addition, in other teams of the organisation, this practice is carried out –in the case of the academy-industry project– with satisfactory results in terms of collaboration and communication. This has contributed these teams being more experienced.
3. *Proposing research topics, schedules and engagement with the project*: As having a leader who represents a team in an iteration, the team have topics to learn. This has allowed them to contribute to the project, within the accompaniment they have with experienced people. In addition, it has been shown that team members feel more confident about themselves. This since in each iteration they have a reference to address their concerns –which is at the same time a partner.
4. *Elucidating of power relationships*: Implementing Team Leader Rotation makes team members clear about the power relations in the team, where leadership does not imply a hierarchical structure. Thus, philosophy of agile methodologies is respected. Team members have expressed intentions to recommend this practice for future projects.

In the case of applying Team Leader Rotation to an agile technique, Scrum is considered in this analysis. The Scrum Master is a specialist who knows and fully understands Scrum. Scrum master helps the team training it and making sure that all Scrum processes are implemented correctly. Additionally, Product Owner is part of the team [11]. A Team Leader who knows the topic of sprint, who is a member of the team, does care about technical aspects, communication and dispute resolution. Thus, Team Leader complements the work of the Scrum Master, development-oriented and soft skills of each team member. With respect to Product Owner, some aspects of the implementation can be quickly solved using the Team Leader as a channel for dissemination of concerns and controversies.

## 6 Final Remarks

Software development has been growing rapidly over the last years. Growing of software development has created new job positions and careers. New computer scientists leave universities with high technical knowledge. Nevertheless, employers have proposed higher requirements for both hard skills and soft skills of team members used in software engineering practice.

A new agile practice is proposed in this paper in order to improve soft skills to inexperienced teams. Two projects have been considered as study cases. First one is an university-enterprise project. Second one is an academic project. Qualitative evaluation of this practice has been proposed. Evaluation is composed of five research questions to probe the hypothesis of effectiveness of Team Leader Rotation practice in strengthening soft skills. Qualitative evaluation was performed through measurement of 11 soft skills identified [8].

Results suggest that 7 soft skills were improved by using Team Leader Rotation as a new agile practice to software development. Other two practices had improvement but with a low rate. Results reported here provide some good ideas to be applied in theory and practice. Two soft skills, however, have not been improved by using this practice. Thus, this practice has some limitations. Strengthening soft skills in inexperienced teams remains an open problem. Limitations have been found in strengthening communication skills, and the capacity of problem disruption. Moreover, Team Leader Rotation allows having a person in charge of problems of team members not in the area of methodology. Thus, Team Leader complements the work of the project facilitator role in any agile software development methodology. It is necessary to improve rotation of Team Leader with accompaniment of more experienced partners at the organisation. This could be achieved by rotating team members among different teams. Research could be directed in that way.

## References

1. Hollins, P., Humphreys, S., Yuan, L., Sleightholme, G., Kickmeier-Rust, M.: The "Water Cooler" game. In: Pivec, M., Grundler, J. (eds.) Proceedings of the 11th European Conference on Games Based Learning, ECGBL 2017. pp. 262–271. Academic Conferences and Publishing International Limited (2017)
2. Valentin, E., Carvalho, J.R.H., Barreto, R.: Rapid improvement of students' soft-skills based on an agile-process approach. In: 2015 IEEE Frontiers in Education Conference (FIE), pp. 1–9. IEEE, New York (2015)
3. Andrews, J., Higson, H.: Graduate employability, 'soft skills' versus 'hard' business knowledge: a European study. Higher Educ. Europe **33**, 411–422 (2008)
4. European Commission, ESCO: European Classification of Skills/Competences, Qualifications and Occupations (2013). http://bookshop.europa.eu/en/esco-europeanclassification-of-skills-competences-qualifications-andoccupations-pbKE0313496
5. Udayakumar, R., Kaliyamurthie, K.P., Khanaa, Geo, A.V.: Mechanism to ensure success in software projects. World Appl. Sci. J. **29** (2014)
6. Zurita, G., Baloian, N., Pino, J.A., Boghosian, M.: Introducing a collaborative tool supporting a learning activity involving creativity with rotation of group members. J. UCS **22**, 1360–1379 (2016)
7. Vinod, V., Dhanalakshmi, J., Sahadev, S.: Software team skills on software product quality **8**, 8–13 (2009)
8. Jia, J., Chen, Z., Du, X.: Understanding soft skills requirements for mobile applications developers. In: 2017 IEEE International Conference on Computational Science and Engineering (CSE) and IEEE International Conference on Embedded and Ubiquitous Computing (EUC), vol. 1, pp. 108–115. IEEE, New York (2017)
9. Al Shatti, L.: Investigating the impact of teamwork quality on project effectiveness in a multiple-project management setting (2018)

10. Creswell, J.W.: Research Design: Qualitative, Quantitative, and Mixed Methods Approaches. University of Nebraska-Lincoln, Los angeles (2009)
11. Schwaber, K., Beedle, M.: Agile Software Development with Scrum, vol. 1. Prentice Hall, Upper Saddle River (2002)
12. Zykov, S.: Software development crisis: human-related factors' influence on enterprise agility. In: Vallverdú, J., Mazzara, M., Talanov, M., Distefano, S., Lowe, R. (eds.) Advanced Research on Biologically Inspired Cognitive Architectures, pp. 126–148. IGI Global, Hershey, PA (2017)
13. Grenning, J.W.: Agile requirements, estimation and planning-iteration zero. In: Embedded Systems Conference, Boston, MA (2012)

# Holistic Factors that Impact the Under-Representation of Women in ICT: A Systematic Literature Review

Lomé Spangenberg and Hendrik Willem Pretorius(✉)

University of Pretoria, Pretoria, South Africa
Henk.Pretorius@up.ac.za

**Abstract.** An under-representation of women in the Information and Communication Technology (ICT) industry exits. Current research tends to focus on either social aspects (social construction) or physical aspects as cause for this phenomenon. Consequently, there is a lack of a holistic perspective of factors that causes the under-representation of women in ICT. This research provides a holistic perspective of factors that causes the under-representation of women in the ICT industry. This research was performed by conducting a systematic literature review that considered 89 articles to identify factors that cause the under-representation of women in ICT. The identified factors were classified as: organizational, economical and socio-psychobiological. The under-representation of women in ICT can now be better addressed by holistically considering this classification of factors to increase female participation in ICT.

**Keywords:** ICT · Women in ICT · Systematic literature review

## 1 Introduction

For decades, women of all ethnicities have been trying to empower each other and prove their worth in society. For the most part of history, it has been a long and tiresome journey to ensure that future generations of women will be able to work alongside men without being solely defined by their gender [1, 2].

There are still women who are facing gender discrimination in the modern day and age [2, 3]. This is mainly caused by the different societies that these different women are a part of. Gender essentialism is an active element that contributes to the under-representation of a specific gender in certain career industries such as women in the ICT industry [4–7].

Current factors that causes the under-representation of women in ICT are classified into two schools of thought. First, there is inherent differences between genders that are responsible for the under-representation of women in the ICT industry. Secondly, that the male-domination in the ICT industry is a cause of social construction making the industry more suitable for men and less suitable for women [6].


X.-S. Yang et al. (eds.), *Proceedings of Fifth International Congress on Information and Communication Technology*, Advances in Intelligent Systems and Computing 1183, https://doi.org/10.1007/978-981-15-5856-6_19

The systematic literature review performed in this research provides a holistic classification of factors which influence career selection and career choices of women toward the ICT industry. Insight is given into the struggles that women face and need to overcome to be equally represented in the ICT industry.

## 2 Literature Background

This section briefly discusses characteristics of the ICT industry, thought schools, gender development theories, and computer phobia, which has been shown to be responsible for the under-representation of women in the ICT industry.

### 2.1 Characteristics of the ICT Industry

The ICT industry expanded rapidly over the past 20 years into an extensively innovative, competitive, and demanding industry [8]. Powerful economies strive to be in possession of technologies that would give them an economic advantage, implicating that ICT is a developmental instrument [9, 10].

The ICT industry can be summarized as a demanding, masculine, ever-changing, moving industry that interferes with work--life balance. [11–14]. The industry has been shown to cause burnout and leads to high levels of stress [15]. These intimidating characteristics discourage women from entering, or staying in the ICT industry, even in cases where anti-discrimination legislation (e.g., The Promotion of Equality and Prevention of Unfair Discrimination Act, 2000[1]) gives preference to women who applies for occupations in the ICT field, with the aim to represent gender fairly in the ICT industry [13].

As the ICT industry expands at an accelerated rate, new and flexible skill sets are required. With new skills sets in demand, new employment opportunities will present themselves [16, 17]. One of the characteristic of the ICT industry is that it requires both flexible hard- and soft-skill sets. Hard-skills refer to a person's ICT technical ability and understanding, whereas soft-skills refer to collaborative interactions and focusses more on the interpersonal and team work [18, 19]. People can often progress to reach the executive level of the organization by relying on hard-skills, but often fail due to their lack of soft-skills [18].

Finally, the ICT industry is characterised as an information society is characterized by an increase of information and theoretical knowledge that influences our daily lives and aids in our socio-economic evolution [20, 21]. ICT diffusion has played a large role in creating this information society [21].

### 2.2 Thought Schools Regarding the Under-Representation of Women in ICT

This section briefly discusses the two schools of thought when dealing with the under-representation of women in ICT.

[1] Comprehensive South African discrimination law that prohibits unfair discrimination (also by gender) by the government and by private organisations and individuals and forbids hate speech ad harassment.

**Although improvement is seen in some parts of the world, there is still an under-representation of women in the ICT industry** [7, 15, 22, 23]. Two main schools of thought exist to think about the causes for the under-represented of women in the ICT industry [15, 24, 25]. The first school blames social factors and social construction for the under-representation of women in the ICT industry [24]. The second school of thought blames physical/biological differences between the genders for the under-representation of women in the ICT industry [25]. However, both social and physical factors contribute to a person's perception of careers choice.

Most articles tend to focus only on either the social aspects or the physical aspects and therefore lacks a holistic perspective of the situation. This literature review however provides a holistic perspective and classification that include both these schools of thought in the hope of providing a better understanding of the under-representation of women in the ICT industry. Both social and physical aspects play an important role in determining why women are so under-represented in the ICT field.

## 2.3 Gender-Based Theories for ICT

There are many gender-based theories that may help us to understand why the ICT sector is mostly male dominated. These theories might provide new and unexplored insights to better understand gender-based career perceptions within adults and also why there is an under-representation of women in the ICT industry. However, this does not mean that the concept of gender and its associated roles are entirely derived from these theories.

### 2.3.1 Cognitive Developmental Theory

Stereotypical knowledge and behavior concerning gender can be a factor when exploring cognitive theories. However, it is not the sole factor that determines a child's gender preference and gender behavioral patterns.

The cognitive developmental theory is structured on "gender constancy," which is a term used to describe the point in a child's life where they are able to identify themselves with a particular sex and comprehend, in its most basic form, that this is not going to change [26, 27]. Children will instinctively begin to shape their behavior and opinions on gender norms by formulating and managing genders categorically [27, 28]. They do this by using their physical gender traits to categorize information that is being presented to them during their everyday lives and environment [29].

The way children categorize information can be interpreted from a social constructive perspective. This suggests that children formulate their own perspective of certain gender-based roles, such as the mother conducting house-related chores, whereas the father's role is a more technical role [30]. For the purpose of this systematic literature review, the cognitive developmental theory will be discussed as a socio-psychobiological aspect.

### 2.3.2 Gender Schema Theory

The gender schema theory differs from the cognitive developmental theory in that, it focuses more on the processing of information. It discards gender constancy and tends to focus more on the basic understanding that children have, such as gender identification [31, 32].

This theory not only acknowledge that children play a core part in their own gender development, but suggests that they create their own psychological connotations between gender information, which then influences their behavioral patterns [28, 33]. It was found that children who started developing their skills to label genders before reaching the age of 28 months were far more likely to play with same-sex toys than those that developed these skills later on [31, 32].

Children that distinguishes between types of toys that they perceive to be gender specific are often influenced by advertisements or simply different parenting styles. However, companies are becoming more concerned with equality and are striving to improve on their advertisements and store layouts in order to create and demonstrate a gender-neutral society [34]. For this systematic literature review, the gender schema theory will be discussed as a socio-psychobiological aspect.

#### 2.3.3 Extremist Views

The future of women in the ICT industry can be predicted by exploring two extreme visions. Firstly, the utopian vision describes women being best suited for the digital age [35]. Women are considered as hybrid workers in this scenario, that will essentially lead the ICT industry to new heights by introducing their inherent "soft" skills to this industry [36].

Secondly, the dystopian vision describes that women will be at a disadvantage as genetics will play a key role in women being overpowered in the male dominated ICT industry [35]. This scenario is supported by the gender stereotype that men possess the logical aspects necessary to thrive in technology-based industries that women do not.

These two visions are merely theoretical versions of what the future technological industry could be. Both are influenced by gender stereotypes and what the interpreter believes to be the best-suited skills for this industry, whether it be "soft" or logical skills. The existing under-representation of females in the ICT industry places to some extent strain on the economy due to losing important and precious skills that women can bring to this industry [37].

#### 2.3.4 Technophobia

Computer use between genders differ. Women tend to send more emails and perform task and study-related functions, whereas men are more likely to intensively use the Internet and perform personal-related functions [38, 39]. Women also tend to show more anxious behavior than men, when interacting with new technology [40, 41]. When exposed and introduced to new technology, men are more likely to respond with an optimistic and confident attitude where women will indicate a less positive attitude toward the technology [42, 43].

It has been found that the amount of anxiety experienced over time relating to computers reduces significantly for men whereas their female counterparts demonstrated little improvement [44, 45]. Women tend to have less experience with completing intricate computerized tasks as well as programming, and it has also been found that females receive less encouragement regarding these fields from friends and family [43, 46].

Gender-related technophobia can be a cause of gender stereotypical behavior that is directly or indirectly presented to children at a young age. The lack of computer interaction and encouragement that female experience can be one of the factors that aids their anxiety and lack of self-confidence when introduced to new technology at a later stage in life.

## 3 Research Method

The following section will briefly discuss the process that was used during the systematic literature review to answer the following research question:

*What holistic factors impact the under-representation of women within the ICT sector*?

### 3.1 Search Terms Used

("Gender (NEAR/2) Information and Communication* Technolog*" OR "Femal* (NEAR/2) Information and Communication* Technolog*" OR "Wom* (NEAR/2) Information and Communication* Technolog*") AND (Defin* OR sector* OR industry* OR physical* OR review OR Differenc*)

### 3.2 Selection Criteria

Table 1 presents the selection criteria (what was included and excluded) for the literature review.

**Table 1.** Selection criteria for the literature review

| Selection criteria | |
|---|---|
| *Inclusion criteria* | *Exclusion criteria* |
| 1. The ICT sector | 1. All other industries except the ICT sector |
| 2. Articles that explain the relationship between genders and the ICT industry | 2. Employment losses due to artificial intelligence |
| 3. English articles | 3. Articles published before 1990 |
| 4. Information relating to developed countries | 4. All languages except English |
| 5. Articles between 1990 and 2019 | 5. Articles that were not peer reviewed |
| 6. Qualitative research | |

### 3.3 Source Selection

The following databases were used in the literature review: Google Scholar, MEDLINE, WorldCat, WorldCat.org, ERIC, and Academic Search Complete.

WorldCat returned 30 297 results and by adding the 72 hand-chosen results from Google Scholar, the total number of results equaled 30 369. The amount of records left after duplications were removed equaled 26 120. We excluded a total of 25 984 irrelevant articles. During the eligibility phase, we selected only 67 qualitative articles. We added an additional 22 articles from Google Scholar to end with a total of 89 qualitative articles for this systematic literature review. A list of all the articles consulted during the systematic literature review was carefully documented, using Microsoft Excel.

## 4 Analysis of Findings

This systematic literature review examined what influences the under-representation of women in the ICT industry. After exploring 89 relevant articles on this topic, three distinctive categories were identified. These categories include the following: organizational aspects, economical aspects, and socio-psychobiological aspects.

### 4.1 Organizational Aspects

Technology alone can do a little for an organization. However, if that technology is strategically introduced and integrated into an organization, it can cut expenses and improves business functions [47, 48]. Organizational aspects such as culture and ethics can play a role in how women perceive and experience the ICT industry [49–51].

#### 4.1.1 Organizational Culture

The culture of an organization is defined as the collective principles and standards that are practiced in an organization [49]. The culture refers to how people think and act as a group to create the beliefs and value system of that organization. When the rapid evolution of the ICT industry began, it was the norm to associate the ICT industry with masculinity [50]. Male-dominated industries tend to focus on results, whereas feminine industries focus on social interactions [51].

During an interview with fifteen females from the ICT industry, the females suggested that they need additional communication skills to be able to interact with their male co-workers [13]. They needed to converse in a manner that exhibited logical skill sets [13]. Women tend to be more emotionally involved and sensitive, which makes communication in a male-dominated environment challenging.

It is also no secret that the ICT industry is a demanding and rapidly changing industry associated with burn-out and stress. The extensive hours and work that is demanded of employees result in exhausted personnel that struggle to keep up and properly recuperate [52]. The demands and extensive workhours, means that the line between a person's personal life and work life can become distorted [53]. This is difficult for women to handle since they have responsibilities outside of work such as managing a household and caring for children.

## 4.2 Economical Aspects

Productivity is the power that controls an economy and innovation determines that productivity [54]. The ICT industry impacts an economy by supporting economic development, economic growth, and employment. It is therefore important that women enter this industry.

### 4.2.1 Development

Nearly 30 percent of all resources in the public and private sectors invest in ICT research and development [10]. However, three factors are important to successfully take advantage of ICTs in the public and private sectors. These three factors include infrastructure, expert employees, and finances to be able to incorporate the necessary ICTs [55].

### 4.2.2 Economic Growth

When the implementation and investment in ICTs are done successfully, it provides an organization with a competitive lead in the market [10]. Furthermore, the correlation between ICTs and economic growth is usually positive. In other words, to be able to compete in the global market, it would be wise to make use of available ICTs in order to reap the worth of this resource [56–58].

India, for example, has developed itself as the leading country in exporting software and services related to ICT. They have increased their ICT export software and services with more than twelve times during the years 2000--2013 [59]. In general, one could argue that ICTs cause a positive relationship with economic growth. However, to accurately determine this correlation, both ICT and socio-economical aspects should be considered [60].

### 4.2.3 Employment

The ICT revolution requires new and improved skills that involve greater problem-solving, innovation, and soft skills [61]. There was 210% increase in ICTs and 69% increase in ICT companies in the USA in 2011, as opposed to the 1980s [62]. This demonstrates the rapid growth of the technology industry and with that growth comes new employment opportunities that need to be filled. Employment opportunities that women should also take advantage of.

## 4.3 Socio-Psychobiological Aspects

Sociology is the study of the society, social interaction, relationships, and every day culture [63]. Sociological factors are socially accepted and cultural norms that may influence gender stereotypes. On the other hand, psychobiology provides a greater comprehension of both psychology and biology. Psychobiology deals with humans on a biological level but incorporates a person's thought process [64]. By combining these sciences, it provides a better understanding of gender development behavior. This section presents various sub-categories of socio-psychobiological aspects that influence the under-representation of women in the ICT industry.

### 4.3.1 Gender Stereotypes

The over-representation of males in the ICT industry provokes negative stereotypes related to gender, for example, that females might not possess the necessary skills for these types of industries [65–67]. The fact that such stereotypes exist, means that society is struggling to see that in fact, maybe discrimination against females in the ICT and related industries exist [67]. This suggests that stereotypical behavior masks against a certain gender exist. This then leads to a continued over-representation of men in the ICT field. Women are often less willing to work in environments where they feel that they are being discriminated against [67]. Perhaps if women receive more encouragement to join such occupations and they are assured that they will only be assessed against their skills and not their gender, it will increase female participation.

As previously discussed, two extreme visions can be identified which can be associated with gender stereotypes. The first being, the utopian view. This view suggests that women have inherent soft skills that will aid them in dominating the IT industry. It describes women as being the ultimate hybrid worker [35]. The second being, the dystopian view. This view on the other hand, predicts that women will remain the gender that is at a disadvantage. The ICT industry is and will continue to be a masculine-dominated industry in this scenario of dystopia [35].

### 4.3.2 Gender Encouragement

Women tend to be more drawn to IT courses if they have been exposed to female educators in their early lives [68]. It can be assumed that the views of other female and parental figures in a young female's life plays a significant part in their understanding and acceptance of the ICT industry [43]. Encouragement is the first step in the right direction to possibly increase the representation of women in the male-dominated ICT industry.

Encouragement programs do in fact exist that are focused on encouraging potential female participation in fields such as ICT. Examples of these types of programs include the following. Firstly, students from Massachusetts Institute of Technology who are female provides encouragement seminars with the assistants of the "Women's Initiative Program" to female school pupils [69]. Secondly, certain webpages are demonstrating and presenting their female employees to the public in hope of encouraging young women who aspire to become IT professionals [69]. Lastly, certain ICT organizations are offering more female courses in which these women have the chance to be introduced to and engage with other women that are currently in ICT-related occupations [69]. These types of initiatives aim to increase female participation and decrease the over-representation of men in these types of industries. However, female idols are not the only aspect to inspiring young women, it is also important that these idols do in fact present a good image that could make these young women feel more at ease [69]. This suggests that the fact that an idol is female is only one part of the equation, it is important that they do not represent a stereotypical image of female IT professionals [69]. They should rather aim to inspire all types of young women to join these types of fields and not just the stereotypical "geek." It is essential to encourage young females to partake in

IT careers not only to provide the industry with a fresh viewpoint but to make high-end occupations accessible to more women [69].

### 4.3.3 Gender Programs

During 2004, programs such as the LDC Initiative has begun to focus their efforts on sustaining and improving academies [8]. These types of initiatives are finding ways in which they can empower and increase female participation in the ICT industry. A positive correlation exists between women's participation in this industry and their encouragement to related fields. Unfortunately, not many initiatives or programs are set in place to increase female involvement in ICT-related sectors [70]. However, a research network formed in 2004 named "Gender Research in Africa into ICTs for Empowerment" in short GRACE, aimed to empower females in the ICT industry and promote equality [71]. The GRACE Network provides ample mentor programs for women. Twelve women from Nairobi that have ICT-related occupations were interviewed. These women all participated in the mentoring program. Their support structure has aided them in numerous ways. Firstly, by developing and comprehending the necessary skillsets that they require in ICT [70]. Secondly, they acquired better self-efficacy [70]. Thirdly, they developed a more in-depth comprehension of occupational possibilities in the ICT industry [70]. Fourthly, they received important and beneficial information about the ICT industry [70]. Lastly, they acquired a valuable support structure to assist them on their journey [70].

### 4.3.4 Gender Development

As previously discussed, gender development is explained by various theories that can be applied to young children. Such theories include cognitive perspective theories and gender schema theories. The cognitive theory relies on a concept known as "gender constancy." It suggests that a child will gradually come to the realization that they belong to a certain gender and allows them to start behaving accordingly [27]. It is believed that this phase occurs between the ages of three and seven [27, 72]. Gender constancy is divided into three important categories. The first being, identity. This suggests that the child can acknowledge which gender they themselves and the people surrounding them belong to [72]. Secondly, stability. The child will comprehend that if they are a certain gender, it will be the same gender once they have grown-up [72]. Lastly, consistency. It will be clear to the child that in spite of personal preference, their sex will not be altered [72].

The gender schema theory suggests that once a child acknowledges their sex, they will start to make psychological connotations regarding different sexes [33]. Gender labelling can be influenced by any person or gender specific organization that a child comes in contact with [32]. This gender-related labelling plays a substantial role various aspects of a person's life, for example, the type of toys that children prefer [33]. Gender-related labelling also plays a role in how different genders perceive the ICT industry and may contribute to the under-representation of women in the ICT industry.

### 4.3.5 Gender Self-Efficacy

Differences in the relationship between gender and technology was more prevalent in the 1980s than today [73]. During the 1990s, only five percent of users on the Internet were predicted to be female but of late this, number seems to surpass that of men, especially regarding social media [39]. Female users tend to limit their technology usage to explore and sustain contacts, whereas males, on the other hand, focus on task-related actions [74]. Even though more women are participating in technology-related activities, they are still under-represented in ICT careers. Women often experience more anxiety than men, when interacting with new technology [41]. Asian and Western cultures were compared to see if there would be a difference in gender technological usage. It was found that both cultures propose that men's confidence levels were higher than their female counterparts when interacting with technology [75]. It would seem that the reason for this under-representation is the female's self-efficacy combined with a lack of curiosity [43]. Even so, it is suggested in some studies that there were no differences recorded or observed between the two genders [41]. The results of these studies indicate that gender cognitive perspectives of the subjects and not necessarily biology play a greater role in gender-technology usage [45].

### 4.3.6 Sexual Dimorphism

Men have physical characteristics that differ from women [76]. However, cognitive factors (involving the brain) can also be taken into consideration when distinguishing between the genders [77]. Research indicates that the female brain causes women to possess greater social and recall abilities, whereas the male brain causes men to possess greater visual-spatial and motoring abilities [78]. Neuroimaging is used to determining these differences between genders in one's brain. Women were found to have a smaller crania[2] with their gray matter[3] percentage higher than men's [79]. Men on the other hand demonstrated a higher volume in their white matter[4] [78].

Empathy is defined as the response to the state of another individual's emotions when witnessed [80]. Research on neuroimaging suggest that there are gender differences in the neuro-functional processes related to both cognition and emotions in certain areas of the brain [81, 82]. Women tend to showcase higher levels of empathy when they are observing another individual's state of emotions [80]. On the other hand, men depend more on their cognitive abilities to decide on how to express empathy toward another person [80]. These results provide some insight to why men are deemed to be more suited for industries such as ICT which is regarded as task orientated and logical. On the other hand, this may also provide some insight into what women with inherent soft-skills, could potentially offer to the ICT industry.

The ICT industry has been previously seen as a logical career, but more recently, it started to recognize the need to include emotional and inter-personal aspects as well such as customer and social collaboration skills [83, 84]. Social interactions can be divided

[2] Part of the skull that encloses the brain.

[3] Responsible for muscle control and sensory perception such as seeing and hearing, memory, emotions, speech, decision making, and self-control.

[4] The tissue through which messages pass between different areas of gray matter.

into two main categories, namely agentic and communal. Men tend to excel in the agentic aspect which focusses on tasks, logic, dominance, and competitiveness [85, 86]. On the other hand, women are more prone to surpass men in the communal aspect which refers to the nurturing side of a person [86]. These behaviors creates gender segregation.

The research suggests that if these traits were widely considered to be accurate, it would suggest that certain career paths are more acceptable for a specific gender than other careers [87]. Therefore, if someone were to be more communal in nature, which in this case is associated with the female gender, they should choose a profession that relates to caregiving or social interactions such as providing services to others [87]. On the other hand, if a person were to be more agentic in nature, which in this case is associated with being male. It is suggested that they should consider entering any trade that is task orientated, logical, and competitive in nature [87].

## 5 Conclusion

This systematic literature review has recognized a total of 89 articles that provided a holistic perspective on factors that affect or contribute to the under-representation of females in the current ICT field. The data obtained from these articles were evaluated and grouped into three distinct categories into answer the research question: What holistic factors impact the under-representation of women within the ICT sector?

The categories were identified as: organizational, economical, and socio psychobiological factors. The theories and concepts within the categories are not necessarily linked to each other which could be expanded in further research. The categories that emerged were chosen and used to create a holistic perspective of factors that cause the under-representation of women in ICT.

This review contributes to the body of knowledge by presenting a holistic perspective of factors that explain the under-representation of women in the ICT, not focusing on a single set or group of aspects or factors.

## References

1. MacLean, N.: The hidden history of affirmative action: working women's struggles in the 1970s and the gender of class. Feminist Stud. **25**(1), 42 (1999)
2. Peters, J.S., Wolper, A.: Women's rights, human rights: international feminist perspectives. Routledge (2018)
3. Ferree, M.M., Hess, B.: Controversy and coalition: the new feminist movement across four decades of change. Routledge (2002)
4. Crump, B.J., Logan, K.A., McIlroy, A.: Does gender still matter? a study of the views of women in the ICT industry in New Zealand. Gender Work Org. **14**(4), 349–370 (2007)
5. Charles, M., Grusky, D.B.: Occupational Ghettos: The Worldwide Segregation of Women and Men, vol. 200. Stanford University Press Stanford, CA (2005)
6. Michie, S., Nelson, D.L.: Barriers women face in information technology careers: self-efficacy, passion and gender biases. Women Manage. Rev. **21**(1), 10–27 (2006)
7. González, A.M., Revelles, B., Almeda, E., Bosch, N.V., García, J.S.: Women in ICT: opportunities for their inclusion in an international labor market. In: IFIP World Information Technology Forum, pp. 171–177. Springer (2016)

8. Kramer, W.J., Jenkins, B., Katz, R.S.: The Role of the Information and Communications Technology Sector in Expanding Economic Opportunity. Kennedy School of Government, Harvard University, Cambridge, MA (2007)
9. Avgerou, C.: The link between ICT and economic growth in the discourse of development. In: Organizational Information Systems in the Context of Globalization, pp. 373–386. Springer (2003)
10. Tarutė, A., Gatautis, R.: ICT impact on SMEs performance. Proc. Social Behav. Sci. **110**, 1218–1225 (2014)
11. Houghton, K., Miller, E., Foth, M.: Integrating ICT into the planning process: impacts, opportunities and challenges. Aust. Plan. **51**(1), 24–33 (2014)
12. Danner, R.B., Pessu, C.O.: A survey of ICT competencies among students in teacher preparation programmes at the University of Benin, Benin City, Nigeria. J. Inf. Technol. Educ. Res. **12**(1), 33–49 (2013)
13. Smith, L.: Working hard with gender: gendered labour for women in male dominated occupations of manual trades and information technology (IT). Equal. Diver. Incl. Int. J. **32**(6), 592–603 (2013)
14. Tennakoon, K.U.S., Da Silveira, G.J., Taras, D.G.: Drivers of context-specific ICT use across work and nonwork domains: a boundary theory perspective. Inf. Organ. **23**(2), 107–128 (2013)
15. Pretorius, H.W., Mawela, T., Strydom, I., Villiers, C.D., Johnson, R.D.: Continuing the discourse of women in information technology: a South African perspective. Gender Technol. Develop. **19**(3), 346–369 (2015)
16. Terzi, N.: The impact of e-commerce on international trade and employment. Proc. Social Behav. Sci. **24**, 745–753 (2011)
17. Holtgrewe, U.: New new technologies: the future and the present of work in information and communication technology. New Technol. Work Employ. **29**(1), 9–24 (2014)
18. Deepa, S., Seth, M.: Do soft skills matter? implications for educators based on recruiters' perspective. IUP J Soft Skills **7**(1), 7 (2013)
19. Saad, M.S.M., Majid, I.A.: Employers' perceptions of important employability skills required from Malaysian engineering and information and communication technology (ICT) graduates. Glob. J. Eng. Educ. **16**(3), 110–115 (2014)
20. Webster, F.: Theories of the Information Society, 3rd edn. Routledge, London and New York (2014)
21. Vu, K.M.: ICT as a source of economic growth in the information age: empirical evidence from the 1996–2005 period. Telecommun. Policy **35**(4), 357–372 (2011)
22. Olatokun, W.M.: Availability, accessibility and use of ICTs by Nigerian women academics. Malaysian J. Lib. Inf. Sci. **12**(2), 13–33 (2017)
23. Mueller, R.E., Truong, N.K., Smoke, W.: Underrepresentation of women in Canada's information and communication technology sector: what can we learn from a Canadian survey of adult skills? Can. Publ. Policy **44**(S1), S73–S90 (2018)
24. Nyborg, K., Anderies, J.M., Dannenberg, A., Lindahl, T., Schill, C., Schlüter, M., Adger, W.N., Arrow, K.J., Barrett, S., Carpenter, S.: Social norms as solutions. Science **354**(6308), 42–43 (2016)
25. Papineau, D.: There are no norms of belief. The Aim of Belief. Oxford University Press, United States of America (2013)
26. Ruble, D.N., Taylor, L.J., Cyphers, L., Greulich, F.K., Lurye, L.E., Shrout, P.E.: The role of gender constancy in early gender development. Child Dev. **78**(4), 1121–1136 (2007)
27. Slater, A., Bremner, J.G.: An introduction to developmental psychology. Wiley (2017)
28. Martin, C.L., Ruble, D.N., Szkrybalo, J.: Cognitive theories of early gender development. Psychol. Bull. **128**(6), 903 (2002)
29. Martin, C.L., Ruble, D.: Children's search for gender cues: cognitive perspectives on gender development. Curr. Dir. Psychol. Sci. **13**(2), 67–70 (2004)

30. Alesina, A., Giuliano, P., Nunn, N.: On the origins of gender roles: women and the plough. Q. J. Econ. **128**(2), 469–530 (2013)
31. Bussey, K., Bandura, A.: Self-regulatory mechanisms governing gender development. Child Dev. **63**(5), 1236–1250 (1992)
32. Weisgram, E.S., Fulcher, M., Dinella, L.M.: Pink gives girls permission: exploring the roles of explicit gender labels and gender-typed colors on preschool children's toy preferences. J. Appl. Devel. Psychol. **35**(5), 401–409 (2014)
33. Bjorklund, D.F., Causey, K.B.: Children's thinking: cognitive development and individual differences. Sage Publications (2017)
34. Sweet, E.V.: Boy builders and pink princesses: gender, toys, and inequality over the twentieth century. University of California, Davis (2013)
35. Moore, K., Griffiths, M., Richardson, H., Adam, A.: Gendered futures? Women, the ICT workplace and stories of the future. Gender Work Org. **15**(5), 523–542 (2008)
36. Woodfield, R.: Woman and information systems development: not just a pretty (inter) face? Inf. Technol. People **15**(2), 119–138 (2002)
37. Papastergiou, M.: Are computer science and information technology still masculine fields? High school students' perceptions and career choices. Comput. Educ. **51**(2), 594–608 (2008)
38. Imhof, M., Vollmeyer, R., Beierlein, C.: Computer use and the gender gap: the issue of access, use, motivation, and performance. Comput. Hum. Behav. **23**(6), 2823–2837 (2007)
39. Herring, S.C., Stoerger, S.: 29 Gender and (A) nonymity in computer-mediated communication. In: The Handbook of Language, Gender, and Sexuality, 567 (2014)
40. Chua, S.L., Chen, D.-T., Wong, A.F.: Computer anxiety and its correlates: a meta-analysis. Comput. Hum. Behav. **15**(5), 609–623 (1999)
41. Cai, Z., Fan, X., Du, J.: Gender and attitudes toward technology use: a meta-analysis. Comput. Educ. **105**, 1–13 (2017)
42. Broos, A.: Gender and information and communication technologies (ICT) anxiety: male self-assurance and female hesitation. CyberPsychol. Behav. **8**(1), 21–31 (2005)
43. Beyer, S.: Why are women underrepresented in Computer Science? Gender differences in stereotypes, self-efficacy, values, and interests and predictors of future CS course-taking and grades. Comput. Sci. Educ. **24**(2–3), 153–192 (2014)
44. Todman, J.: Gender differences in computer anxiety among university entrants since 1992. Comput. Educ. **34**(1), 27–35 (2000)
45. Powell, A.L.: Computer anxiety: comparison of research from the 1990s and 2000s. Comput. Hum. Behav. **29**(6), 2337–2381 (2013)
46. Busch, T.: Gender differences in self-efficacy and attitudes toward computers. J. Educ. Comp. Res. **12**(2), 147–158 (1995)
47. Nograšek, J., Vintar, M.: E-government and organisational transformation of government: black box revisited? Gov. Inf. Quart. **31**(1), 108–118 (2014)
48. Ma, J.X., Buhalis, D., Song, H.: ICTs and Internet adoption in China's tourism industry. Int. J. Inf. Manage. **23**(6), 451–467 (2003)
49. Tong, C., Tak, W.I.W., Wong, A.: The impact of knowledge sharing on the relationship between organizational culture and job satisfaction: the perception of information communication and technology (ICT) practitioners in Hong Kong. Int. J. Human Res. Stud. **5**(1), 19 (2015)
50. Iniesta-Bonillo, M.A., Sánchez-Fernández, R., Schlesinger, W.: Investigating factors that influence on ICT usage in higher education: a descriptive analysis. Int. Rev. Publ. Nonprofit Market. **10**(2), 163–174 (2013)
51. Bankole, F.O., Bankole, O.O.: The effects of cultural dimension on ICT innovation: empirical analysis of mobile phone services. Telematics Inform. **34**(2), 490–505 (2017)
52. Schaufeli, W.B., Taris, T.W.: A critical review of the job demands-resources model: implications for improving work and health. In: Bridging Occupational, Organizational and Public Health, pp. 43–68. Springer (2014)

53. Ninaus, K., Diehl, S., Terlutter, R., Chan, K., Huang, A.: Benefits and stressors—perceived effects of ICT use on employee health and work stress: an exploratory study from Austria and Hong Kong. Int. J. Qual. Stud. Health Well-being **10**(1), 28838 (2015)
54. Cardona, M., Kretschmer, T., Strobel, T.: ICT and productivity: conclusions from the empirical literature. Inf. Econ. Policy **25**(3), 109–125 (2013)
55. Manochehri, N.N., Al-Esmail, R.A., Ashrafi, R.: Examining the impact of information and communication technologies (ICT) on enterprise practices: a preliminary perspective from Qatar. Electron. J. Inf. Syst. Develop. Countries **51**(1), 1–16 (2012)
56. Ishida, H.: The effect of ICT development on economic growth and energy consumption in Japan. Telematics Inform. **32**(1), 79–88 (2015)
57. Kiveu, M., Ofafa, G.: Enhancing market access in Kenyan SMEs using ICT. Global Bus. Econ. Res. J. **2**(9), 29–46 (2013)
58. Sassi, S., Goaied, M.: Financial development, ICT diffusion and economic growth: lessons from MENA region. Telecommun. Policy **37**(4–5), 252–261 (2013)
59. Jorgenson, D.W., Vu, K.M.: The ICT revolution, world economic growth, and policy issues. Telecommun. Policy **40**(5), 383–397 (2016)
60. Jin, S., Cho, C.M.: Is ICT a new essential for national economic growth in an information society? Gov. Inf. Quart. **32**(3), 253–260 (2015)
61. Redecker, C., Johannessen, Ø.: Changing assessment—towards a new assessment paradigm using ICT. Eur. J. Educ. **48**(1), 79–96 (2013)
62. Hathaway, I.: Tech starts: high-technology business formation and job creation in the United States. Ewing Marion Kauffman Foundation Research Paper (2013)
63. Turner, B.S.: The Cambridge Dictionary of Sociology. Cambridge University Press, U.K (2006)
64. Rossi, A.: Gender and the Life Course. Routledge, Amherst Massachusetts, United States (2018)
65. Good, C., Aronson, J., Harder, J.A.: Problems in the pipeline: stereotype threat and women's achievement in high-level math courses. J. Appl. Devel. Psychol. **29**, 17–28 (2008)
66. Cundiff, J.L., Vescio, T.K., Loken, E., Lo, L.: Do gender—science stereotypes predict science identification and science career aspirations among undergraduate science majors? Soc. Psychol. Educ. **16**(4), 541–554 (2013)
67. Cundiff, J.L., Vescio, T.K.: Gender stereotypes influence how people explain gender disparities in the workplace. Sex Roles **75**(3–4), 126–138 (2016)
68. Zhang, W.: Why IS: understanding undergraduate students' intentions to choose an information systems major. J. Inf. Syst. Educ. **18**(4) (2007)
69. Cheryan, S., Drury, B.J., Vichayapai, M.: Enduring influence of stereotypical computer science role models on women's academic aspirations. Psychol. Women Quart. **37**(1), 72–79 (2013)
70. Nakafeero, A., Okello, D., Comfort, K., Mandlate, J., Kyomuhendo, G.B., Yitamben, G., Tafnout, M., Sifuna, O., Hassanin, L., Macueve, G.: African women and ICTs: investigating technology, gender and empowerment. Zed Books Ltd. (2013)
71. Keifer-Boyd, K.: Book review: African women & ICTs: investigating technology, gender and empowerment. J. Int. Women's Stud. **12**(1), 212–217 (2011)
72. Halim, M.L., Ruble, D.N., Tamis-LeMonda, C.S., Zosuls, K.M., Lurye, L.E., Greulich, F.K.: Pink frilly dresses and the avoidance of all things "girly": children's appearance rigidity and cognitive theories of gender development. Dev. Psychol. **50**(4), 1091 (2014)
73. Popovich, P.M., Gullekson, N., Morris, S., Morse, B.: Comparing attitudes towards computer usage by undergraduates from 1986 to 2005. Comput. Hum. Behav. **24**(3), 986–992 (2008)
74. Kimbrough, A.M., Guadagno, R.E., Muscanell, N.L., Dill, J.: Gender differences in mediated communication: women connect more than do men. Comput. Hum. Behav. **29**(3), 896–900 (2013)

75. Yau, H.K., Cheng, A.L.F.: Gender difference of confidence in using technology for learning. J. Technol. Stud. **38**(2), 74–79 (2012)
76. Dixson, A.: Primate sexuality. In: The International Encyclopedia of Human Sexuality, pp. 861–1042 (2015)
77. Oliveira-Pinto, A.V., Santos, R.M., Coutinho, R.A., Oliveira, L.M., Santos, G.B., Alho, A.T., Leite, R.E., Farfel, J.M., Suemoto, C.K., Grinberg, L.T.: Sexual dimorphism in the human olfactory bulb: females have more neurons and glial cells than males. PLoS ONE **9**(11), e111733 (2014)
78. Ingalhalikar, M., Smith, A., Parker, D., Satterthwaite, T.D., Elliott, M.A., Ruparel, K., Hakonarson, H., Gur, R.E., Gur, R.C., Verma, R.: Sex differences in the structural connectome of the human brain. Proc. Natl. Acad. Sci. **111**(2), 823–828 (2014)
79. Sacher, J., Neumann, J., Okon-Singer, H., Gotowiec, S., Villringer, A.: Sexual dimorphism in the human brain: evidence from neuroimaging. Magn. Reson. Imaging **31**(3), 366–375 (2013)
80. Schulte-Rüther, M., Markowitsch, H.J., Shah, N.J., Fink, G.R., Piefke, M.: Gender differences in brain networks supporting empathy. Neuroimage **42**(1), 393–403 (2008)
81. Azim, E., Mobbs, D., Jo, B., Menon, V., Reiss, A.L.: Sex differences in brain activation elicited by humor. Proc. Natl. Acad. Sci. **102**(45), 16496–16501 (2005)
82. Hofer, A., Siedentopf, C.M., Ischebeck, A., Rettenbacher, M.A., Verius, M., Felber, S., Fleischhacker, W.W.: Gender differences in regional cerebral activity during the perception of emotion: a functional MRI study. Neuroimage **32**(2), 854–862 (2006)
83. Kelan, E.K.: Emotions in a rational profession: the gendering of skills in ICT work. Gender Work Org. **15**(1), 49–71 (2008)
84. Valenduc, G.: Not a job for life? Women's progression, conversion and dropout in ICT professions. Int. J. Gender Sci. Technol. **3**(2), 483–500 (2011)
85. Delfgaauw, J., Dur, R., Sol, J., Verbeke, W.: Tournament incentives in the field: gender differences in the workplace. J. Labor Econ. **31**(2), 305–326 (2013)
86. Collins, B.J., Burrus, C.J., Meyer, R.D.: Gender differences in the impact of leadership styles on subordinate embeddedness and job satisfaction. Leadership Quart. **25**(4), 660–671 (2014)
87. Wood, W., Eagly, A.H.: Two traditions of research on gender identity. Sex Roles **73**(11–12), 461–473 (2015)

# The Design of an Effective Extreme_Controller_Mechanism Scheme for Software-Defined Cognitive Radio Network

Brian Sibanda(✉) and Mthulisi Velempini

University of Limpopo, Polokwane, South Africa
bsib1234@gmail.com, mthulisi.velempini@ul.ac.za

**Abstract.** Security is a major concern in Software Defined Cognitive Radio Network (SDCRN). SDCRN is an integration of the Software Defined Network (SDN) with the Cognitive Radio Network (CRN). SDN is an architecture developed to reduce the complexity of the network, whereas CRN is a technology developed to allow radios to learn and adapt to their environment. However, the architecture and technology are susceptible to a number of malicious attacks such as Distributed Denial of Service (DDoS) and Primary User Emulation (PUE), respectively. The DDoS and PUE attacks could be launched onto the SDCRN with the intention of disrupting service. The design of effective security schemes that enhance maximum protection of SDCRN from these malicious attacks is a sought after solution in network security. Hence, this research study proposes a security mechanism that addresses the effects of DDoS and PUE attacks in SDCRN.

**Keywords:** Denial of service · Primary user emulation · Software-Defined Cognitive Radio Network

## 1 Background

The integration of Software-Defined Network (SDN) with Cognitive Radio Network (CRN) has improved the functionality and the management of networks resulting in efficient spectrum usage. However, the SDN and CRN are prone to Distributed Denial of Service (DDoS) and Primary User Emulation (PUE) attacks, respectively [1, 2]. The effects of these two attacks are likely to be compounded in Software-Defined Cognitive Radio Network (SDCRN). A number of studies have been conducted in SDN and CRN. To the best of our knowledge, the effects of these attacks in an SDCRN have not been investigated. This research proposes an Extreme_Controller_Mechanism (XCM) that addresses the effects of these attacks in an SDCRN.

XCM is an effective, efficient and lightweight security scheme which utilises less memory and processing time designed to curb these attacks in SDCRN. In addition, the XCM security scheme can be used as a benchmark for future schemes in SDCRN.


X.-S. Yang et al. (eds.), *Proceedings of Fifth International Congress on Information and Communication Technology*, Advances in Intelligent Systems and Computing 1183, https://doi.org/10.1007/978-981-15-5856-6_20

The study investigates the attributes of the attacks which can be detected. An effective technique that addresses the DDoS and PUE attacks in an SDCRN is desirable. Lastly, it optimises the DDoS and PUE security for efficient memory and processor utilisation.

## 2 Related Work

The SDCRN provides better network management and efficient spectrum usage. However, it is susceptible to DDoS [1, 3] and PUE [2, 4]. A number of studies [1, 2, 4–12] address DDoS and PUE in SDN and CRN, respectively.

A threshold-based method which prevents DDoS attacks in SDN was proposed in [6]. In [7], the DDoS attack was mitigated using a source-based Internet Protocol (IP) filtering technique. However, the method is efficient when the malicious traffic is light. In [5], a filtering scheme which detects and addresses the effects of DDoS attacks in SDN is proposed. The simulation results show that the scheme is effective. In [8], the SDN-Oriented DDoS blocking scheme was proposed. The results show that the DDoS can be prevented. In [1], a machine learning technique was observed to be effective in mitigating the effects of DDoS attacks in SDN. Also, a lightweight method in [9] results shows that the scheme is efficient in detecting the DDoS. The method used Self Organising Maps, an unsupervised artificial neural networks to prevent DDoS.

In [4], an energy localisation and variance mechanism were proposed. The results show that the PUE attack can be detected and controlled effectively in CRN. In [2], a filter-based technique was employed in addressing the effects of the PUE. The filter-based results show that the scheme outperformed the technique called the RSS-based localisation in terms of probabilities of miss detection and detection. However, the filter-based technique failed to identify the initial coordinates of the primary user (PU) and an attacker which is close PU. In [10], the PUE attack in Cooperative Spectrum Sensing (CSS) in CRN was investigated and the study observed that the PUE attack has a negative effect on the operation of CSS. This results in the poor utilisation of vacant spectrum bands when malicious users (MUs) broadcast false reports. A Probability Density Function-based scheme was in [11]. The study observed that as the number of MUs increases, the probability of false alarms also increases. Research revealed that encryption-based schemes can be effective in detecting PUE. An Advanced Encryption Standard scheme was implemented in [12], and the results show that it can detect the PUE.

We propose a security scheme that mitigate the effects of the DDoS and PUE in SDCRN. The XCM scheme detects the DDoS and PUE attacks based on the duration of the connection and src_bytes (number of data bytes sent). The security mechanism incorporates the use of neural network (NN) concepts which is one of the machine learning techniques designed to mitigate the effects of the attacks in SDCRN. The increasing relevance and application of this technique in the prevention of security attacks in networking environments presents NN as a security design choice [1, 13]. Furthermore, it is noted in [1, 13] as a better prevention mechanism as compared to signature techniques.

# 3 Methodology

The study employed three tools, namely the Objective Modular Network Testbed (OMNeT++), Octave, and Matrix Laboratory (MATLAB) to evaluate the scheme. OMNeT++ was used as the main tool for the study. The selection of OMNeT++ was largely based on its support of SDNs. It is also regarded as one of the best-known simulators used for distributed protocols in wireless networks [14, 15].

Octave and MATLAB were used as supporting evaluation tools. Octave was used in the training and testing of the NN whilst MATLAB was used in the confirmation of all results. The selection of OCTAVE was based on the fact that it is open-source software. MATLAB is a widely used simulation tool [16, 17].

In an endeavour to design an effective security mechanism scheme that can mitigate the effects of the DDoS and PUE, a network model was designed. The network model was used in DDoS, PUE, and normal traffic generation. Algorithms B1 and B2 illustrate the generation and detection of the DDoS and PUE attacks, respectively.

**Algorithm B1**: Generate and detect a DDoS Attack in SDCRN Simulation Tool: OMNeT++

1. Supposedly, we have three networks, namely N1, N2, and N3. We establish connection between the networks.
2. N1 consists of 1 user: Host 5; N2 consists of 2 users: Host 1 and Host 4; N3 consists of 1 user: Host 6.
3. Host 1, Host 5 (Victim), and Host 6 were all good users whilst Host 4 was a bad user (DDoS attacker).
4. Host 4 launches a DDoS attack overwhelming the controller. This makes Host 5 to fail to send and receive data from any users in Networks, N2 and N3.
5. Host 4 was used to generate the traffic for DDoS attack whilst Host 1 was used to generate the traffic for a normal user.

**Algorithm B2**: Generate and detect a PUE attack in SDCRN
Simulation Tool: OMNeT++

1. We have a licensed band that has three channels: f1, f2, and f3.
2. We assume that a primary base station (BS) is not using any of the channels to transmit to PU. Therefore, channels f1, f2, and f3 are idle.
3. This allows Host 2, Host 3, and Host 7 to use the idle channels {f1, f2, f3} to transmit.
4. However, in the presence of a PUE (Host 8) mimics, the primary signal of f1 misleading Host 2 and Host 7 into vacating channel f1.
5. Host 8 was used to generate the traffic of a PUE attack whilst Host 3 was used to generate non malicious traffic.

## 3.1 Proposed XCM Scheme

The XCM detects and prevents the DDoS and PUE attacks in SDCRN. The scheme is built based on the multi-layer feed-forward NN. The incorporation of NN is inspired by the works such as [18–20]. In [18], NN has been reported as a successful mechanism in detection and prevention of security attacks. Studies in [19, 20] affirm that NN is more effective in the detection of attacks in different networking environments. It is also a good modeling tool of complex tasks.

The generated traffic dataset was trained in OCTAVE in order to teach the XCM scheme to detect and prevent similar attacks in future in the same environment. Four attack outcomes were expected to be produced during the classification phase by the XCM scheme which are: DDoS, PUE, DDoS, and PUE as well as neither DDoS nor PUE attacks as depicted in Fig. 1.

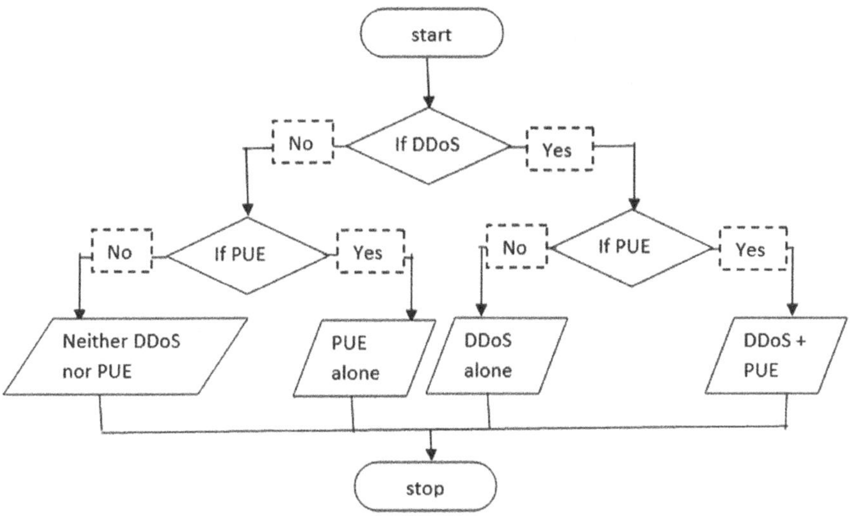


**Fig. 1.** Flowchart on the classification of attacks

After the training of the NN, we implemented the XCM scheme using the test dataset generated in a similar SDCRN platform in MATLAB with the intention of evaluating its performance in terms of detection time, detection rate, false positive, false negative, memory and processor utilisation.

The performance of the XCM scheme was evaluated using these six metrics because they were regarded as the most effective performance evaluation metrics [21–25].

i. Detection Time---time taken to detect an attacker A lesser detection time is required for a scheme to be classified as effective [21].
ii. Detection Rate---positive detection of malicious traffics [22]. In [22], the higher the detection rate the scheme is classified to be efficient. The formula given (Formula 1) is adapted from [22].

$$\text{Detection Rate} = (\text{True Positive})/(\text{False Negative} + \text{True Positive}) \times 100\% \quad (1)$$

iii. False Positive---the amount of malicious traffic flagged as non-malicious [23]. In [21, 23], the lower false positive rate, the scheme is considered to be performing well. In [23], a formula used to calculate false positive rate for any attack is given in Formula 2.

$$\text{False Positive Rate} = (\text{False Positive})/(\text{False Positive} + \text{True Negative}) \times 100\% \tag{2}$$

iv. False Negative---the amount of traffic that are incorrectly detected and dropped [21]. A scheme that incurs a lower false negative is deemed to a better performing scheme [21, 24].
v. Memory Utilisation---the utilisation of the RAM. In [25–27], it was noted that a scheme is supposed to have a lower memory utilisation for it to be regarded as a good scheme, as lower memory utilisation infers the scheme is lightweight as it consumes less space.
vi. Processor Utilisation---processing time consumed by the computer during the execution of the program [25]. A lower processor utilisation is ideal [25, 27].

## 4 Results and Analysis

The algorithms described in algorithms B1 and B2 were used in the generation of datasets. The duration and src_bytes network attributes were extracted for the DDoS, PUE, and normal traffic. The choice of these two network attributes was influenced by work in [19]. They are regarded as the effective attributes that can be used to detect any type attack. The datasets were then trained using a multi-layer feed-forward NN. Figure 2 depicts the structure of the multi-layer feed-forward NN which was implemented.

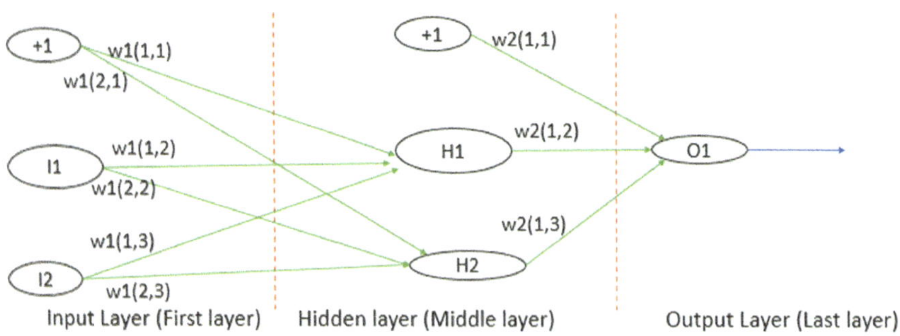


**Fig. 2.** Structure of the nodes in a multi-layer feed-forward NN

After the NN has managed to learn and produced correct results, we generated a similar dataset from the same SDCRN environment and implemented the XCM scheme using the new dataset. We then evaluated the performance of our scheme based on

**Table 1.** Statistical results for our proposed XCM scheme performance metrics

| Performance metric | Threshold value |
|---|---|
| Detection time ($\mu$s) | <5 |
| Detection rate (%) | >99 |
| False positive (%) | <1 |
| False negative (%) | <1 |
| Memory utilisation (%) | <2 |
| CPU utilisation (%) | <2 |

detection time, detection rate, false positive, false negative, memory, and CPU utilisation. Table 1 shows the statistical results of the performance of the XCM scheme.

The results show that results of our scheme are consistent with DDoS attack results in SDN [5] and PUE attacks results in CRN [4]. Our study is designed to detect and prevent both attacks in an SDCRN; we therefore conclude that our scheme performs better than schemes optimised for either PUE in CRN or DDoS in SDN. Hence, our proposed XCM scheme is able to address the effects of the DDoS and PUE attacks in an SDCRN.

## 5 Conclusion

In this study, we proposed an effective XCM scheme that addresses the effects of DDoS and PUE attacks in an SDCRN-integrated environment. The analysis shows that our XCM scheme took less time (approximately less than 5 $\mu$s) to detect the DDoS and PUE attacks in SDCRN-integrated environment. The scheme also managed to yield a higher detection rate (above 99%) of the two attacks in an SDCRN-integrated environment. In addition, the scheme has 1% (0.01) false positive and negative rates in the presence of DDoS and PUE attacks. The analysis further shows that our scheme is lightweight since it consumes less resources (approximately less than 2%) in terms of memory and processor time. The analytical results attests that our XCM scheme can effectively detect and prevent the DDoS and PUE attacks in SDCRN-integrated environment.

## References

1. Ashraf, J., Latif, S.: Handling intrusion and DDoS attacks in software defined networks using machine learning techniques. In: 2014 National Software Engineering Conference, pp. 55–60. Rawalpindi
2. El Mrabet, Z., Arjoune, Y., El Ghazi, H., Abou Al Majd, B., Kaabouch, N.: Primary user emulation attacks: a detection technique based on Kalman filter. J. Sensor Actuators Netw. **7**(26), 1–14 (2018)
3. Ahmad, I., Namal, S., Ylianttila, M., Gurtov, A.: Security in software defined networks: a survey. IEEE Commun. Surv. Tutor. **17**(4), 2317–2346 (Fourthquarter 2015)

4. Sultana, R., Hussain, M.: Mitigating primary user emulation attack in cognitive radio network using localization and variance detection. In: Proceedings of First International Conference on Smart System, Innovations and Computing, 9 January 2018. Available: SpringerLink, https://link.springer.com. Accessed 10 May 2018
5. Huang, X., Du, X., Song, B.: An effective DDoS defense scheme for SDN. In: 2017 IEEE International Conference on Communications, pp. 1–6. Paris (2017)
6. Hyun, D., Kim, J., Hong, D., Jeong, J.: SDN-based network security functions for effective DDoS attack mitigation. In: 2017 International Conference on Information and Communication Technology Convergence, pp. 834–839. Jeju (2017)
7. Dao, N.N., Park, P., Park, M., Cho, S.: A feasible method to combat against DDoS attack in SDN network.In: 2015 International Conference on Information Networking, pp. 309–311. Cambodia (2015)
8. Lim, S., Ha, J., Kim, H., Kim, Y., Yang, S.: A SDN-oriented DDoS blocking scheme for botnet-based attacks. In: 2014 Sixth International Conference Ubiquitous and Future Networks, pp. 63–68. Shanghai (2014)
9. Braga, R., Mota, E., Passito, A.: Lightweight DDoS flooding attack detection using NOX/OpenFlow. In: IEEE Local Computer Network Conference, pp. 408–415. Denver (2010)
10. Kadu, R.D., Karde, P.P., Thakare, V.M.: Performance of CSS cognitive radio networks under primary user emulation attack. In: 2017 International Conference on Trends in Electronics and Informatics, pp. 19–24. Thirunelveli (2017)
11. Sharma, H., Kumar, K.: Primary user emulation attack analysis on cognitive radio. Indian J. Sci. Technol. **9**(14), 1–6 (2016)
12. Alahmadi, A., Abdelhakim, M., Ren, J., Li, T.: Defense against primary user emulation attacks in cognitive radio networks using advanced encryption standard. IEEE Trans. Inf. Forensics Secur. **9**(5), 772–781 (2014)
13. Trung, P.V., Huong, T.T., Tuyen, D.V., Duc, D.M., Thanh, N.H., Marshall, A.: A multi-criteria-based DDoS-attack prevention solution using software defined networking. In: 2015 International Conference on Advanced Technologies for Communications, pp. 308–313. Ho Chi Minh City (2015)
14. Abdulqadder, I.H., Zou, D., Aziz, I.T., Yuan, B., Li, W.: SecSDN-cloud: defeating vulnerable attacks through secure software-defined networks. IEEE Access **6**, 8292–8301 (2018)
15. Diaz, A., Sanchez, P.: Simulation of attacks for security in wireless sensor network. Sensors **11**, 1–27 (2016)
16. GNU Octave, gnu.org, 2018. Available: https://www.gnu.org/software/octave/. Accessed on 28 Sept 2018
17. Amjad, M., Rehmani, M.H., Mao, S.: Wireless multimedia cognitive radio networks: a comprehensive survey. IEEE Commun. Surv. Tutor. **20**(2), 1056–1103 (Secondquarter 2018)
18. Naik, A., Ahmad, S.W.: Data mining technology for efficient network security management. Int. J. Comput. Sci. Trends Technol. **3**, 7–12 (2015)
19. Diro, A.A., Chilamkurti, N.: Distributed attack detection scheme using deep learning approach for internet of things. Future Gener. Comput. Syst. **82**, 761–768 (2018)
20. Bienstock, D., Munoz, G., Pokutta, S.: Principled deep neural network training through linear programming, vol. 1810, pp. 1–26 (2018)
21. Chen, S., Zeng, K., Mohapatra, P.: Hearing is believing: detecting mobile primary user emulation attack in white space. In: 2011 Proceedings IEEE INFOCOM, pp. 1–5. Shanghai, China (2011)
22. Oo, M.M., Kamolphiwong, S., Kamolphiwong, T., Vasupongayya, S.: Advanced support vector machine (ASVM)-based detection for distributed denial of service (DDoS) attack on software defined networking. J. Comput. Netw. Commun. **2019**, 1–12 (2019)

23. Chen, Z., Jiang, F., Cheng, Y., Gu, X., Liu, W., Peng, J.: XGBoost classifier for DDoS attack detection and analysis in SDN-based cloud,. In: 2018 IEEE International Conference on Big Data and Smart Computing, pp. 251–256. Shanghai China (2018)
24. Benabbou, J., Elbaamrani, K., Idboufker, N.: Security in OpenFlow-based SDN, opportunities and challenges. Photon Netw. Commun. **37**(1), 1–23 (2018)
25. Hameed, S., Khan, H.A.: SDN based collaborative scheme for mitigation of DDoS attacks. Future Internet **10**, 1–18 (2018)
26. Ghanem, W.R., Mohamed, R.E., Shokair, M., Dessouky, M.I.: Particle swarm optimization approaches for primary user emulation attack detection and localization in cognitive radio networks. In: 2018 35th National Radio Science Conference, pp. 1–16. Cairo, Egypt (2018)
27. Pimpalkar, A.S., Patil, A.R.B.: Defense against DDoS attacks using IP address spoofing. Int. J. Innov. Res. Comput. Commun. Eng. **3**, 1919–1926

# A Smart Ontology for Project Risk Management Based on PMI's Framework

Wiem Zaouga[1(✉)] and Latifa Ben Arfa Rabai[1,2]

[1] SMART Laboratory, Université de Tunis, ISG Tunis, 2000 Bardo, Tunisia
wiemzaouga89@gmail.com, latifa.rabai@gmail.com
[2] College of Business, University of Buraimi, P.C. 512, Al Buraimi, Oman

**Abstract.** Whenever Project Risk (PR) exists, there is complexity. The difficulty to make concerned decisions related to Project Risk Management (PRM) increases project complexity and even its failure. In order to assist practitioners and professionals to better study the potential impacts of their decisions and assess the PR as precisely as possible, this chapter put forward an ontological approach based on OWL ontology with SWRl rules, that provides the project team clear guidelines to effectively manage PR, and then make the appropriate decisions based on the right recommendations. This approach takes advantages of ontology semantic strengths as it represents a unified PRM knowledge relying on PMI's frameworks. As well, through SWRL reasoning rules, the proposed ontology generates recommendations by which a team member ask for risk-related request more targeted. The proposed ontological approach was evaluated, in term of content and structure, achieving promising results based on the F-measure metric.

**Keywords:** Domain ontology · SWRL rule · Project risk management · PMBOK fifth best practice · PMI's standard for project risk management

## 1 Introduction

Project management, as a well-defined approach dealing with strategic change, contributes to create a strategic value chain with which company respond to change and gain a competitive edge [1]. To carry out project, various parameters and often, various stakeholders must be considered by the managers so as to better study the potential impacts of their decisions and assess the risks as precisely as possible [2]. Whenever Project Risk (PR) exists, there is complexity. Thus, PRM is an integral part of project management since it can assess risks and envision the required mitigations.

In the academic literature, PRM is considered as a necessity to follow standards, audits, and regulations [3] and provides benefits when it is developed in compliance with best practice to make decisions openly and without bias [4]. For that aim, we have selected PMI's frameworks by combining PRM knowledge area recognized in "Pro-


X.-S. Yang et al. (eds.), *Proceedings of Fifth International Congress on Information and Communication Technology*, Advances in Intelligent Systems and Computing 1183, https://doi.org/10.1007/978-981-15-5856-6_21

ject Management Body of Knowledge PMBOK 5th best practices" and PMI's practice standard for PRM, based on the following raisons [3, 4]:

- PMI's frameworks describe in detail the artifacts and T&T for each process.
- PMI's standard for PRM is globally applicable and consistently applied.
- Finally, PMI is incorporated in the ISO/PC236 and ISO/TC258 committee, so ISO and PMI's PRM standard are closely aligned.

Although PRM frameworks and risk models are enough mature to provide a systematic approach for managing risks, [5–7] pointed out a lack of a common vocabulary, which involves incomplete understanding of PRs and inconsistent communication of risk-related concepts. Hence, ontology, as a formal explicit way for representing the concepts and relations of a shared conceptualization [8], has been introduced.

In this context, this work explores the field of PRM in conjunction with ontological approach in attempts to develop an ontological model that provide the project team clear guidelines to effectively manage risks, and then make decisions based on the right recommendations. The proposed approach takes advantages of the ontology strengths as it represents a unified PRM knowledge relying on PMI's frameworks. Also, it exploits lessons learned stored in the ontology for decision support purposes. Finally, through SWRL reasoning rules, this ontology can generate recommendations, by which a team member ask for risk-related request more targeted.

This chapter is structured as follows. In Sect. 2, the related work is briefly reviewed. In Sect. 3, a detailed description of the ontology development process, its different steps with their outputs are presented. We develop the SWRL reasoning rules in Sect. 4. Finally, in Sect. 5, we conclude with the main findings and the future directions of this work.

## 2 Related Work

When considering the application and studies of ontology related to PRM, pertinent examples have been proposed especially for construction project. With regard to the knowledge conceptualization, Wang and Boukamp [9] have defined an ontological model to structure the knowledge about activities, job steps and hazards, and then to enhance access to a company's Job Hazard Analysis (JHA) knowledge using reasoning mechanism that identifies the safety rules for given activities. In [7], risk-vulnerability ontology has been developed to systematically assess risk-related variables that may lead to cost overrun. In this model, the experiences of domain experts related to risks were stored in lessons learned concept. To overcome the lack of formality in construction RM, risk-oriented ontology was introduced by [10] who seek to integrate risk knowledge domain with the risk monitor object domain in an effort to unify the generic concepts, the relations as well as the inferring rules. Further research has been driven for modeling ontology-based-semantic retrieval system, which correlate the PR domain ontology and the documents' content of previous projects, to provide a complete glossary of concepts with their interrelations and to facilitate the usage of previous projects' experiences [11]. Nonetheless, there are others research trend shifting toward IT project. In [1], FLEX-INET project sought to develop a reference ontology that supports risk assessment for

GNP in an interoperable manner by means of network of products or PSS. Also, in [12], authors put forward enterprise ontology for the assessment of procurement risks in supply chains in a way to implement an Early Warning System (EWS) for monitoring suppliers.

All these efforts have proved the potential benefits of ontology related to PRM, from which, three main items can be deduced (i) most of these ontologies do not address the overall PR process whereas their scope is only limited to one process (risk analysis and assessment are the most cited), (ii) none of them focus to conceptualize PRM knowledge with respect to standards and best practices, (iii) rather than a conceptual framework, ontology provides powerful querying and reasoning tools successfully exploited by Decision System (DS); whil, ontology and DS are handled separately in the PRM and even the examples of PR Ontology-Based DS are quite limited.

In this paper, we propose an ontological approach intending to combine and extend the previous studies in several ways. First, the knowledge extracted complies with PMI's frameworks. Second, it incorporates "lessons learned" as a concept into the proposed ontology to better manage risks through experiences' reusing and sharing. Third, a set of inferences rules will be developed to guide a team member, at each PR process, by the most appropriate recommendations for a specific case.

## 3 Toward a Unified PRM Domain Ontology

We adopt an ontology-based approach, PRM-ontology, allows modeling the PRM knowledge and reasoning over it to provide a set of recommendations. The PRM knowledge is formalized by means of ontology into OWL axioms and SWRL rules. Formally, Ontology "$O$" is a quadruple $O = (C, R, I, A)$ where $C$: collection of concepts, $R$: set of relationships (attributive/cross tree), $I$: Instances for concepts and $A$: formal Axioms (constrain the concept behaviors). As well, we adopt the Methontology presented by [13] as the most mature ontology development methodology with enough details of techniques and activities over the ontology life cycle. The design process proposed by Methontology involves five main steps (see Fig. 1).

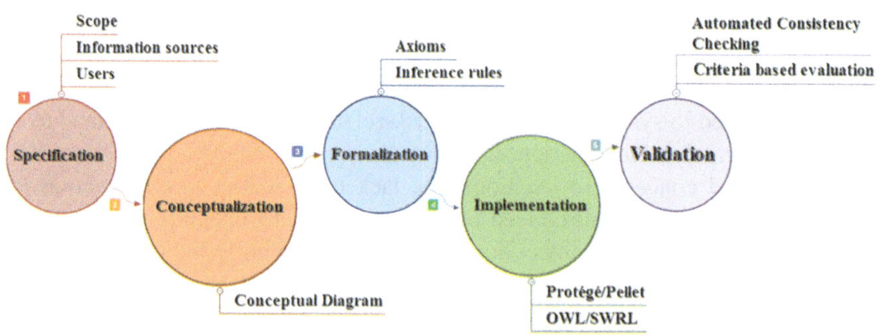


**Fig. 1.** Ontology design process of methontology

### Step 1: Specification

To develop domain ontology, the scope, knowledge sources, and the intended users are defined first, using a set of competences questions (CQs). These CQs include, but are not limited to: what is the domain that will cover?, what's the scope of the ontology?, what are the knowledge sources?, and who will use this ontology?

The answers related to these CQs are in order:

- The domain considered is PRM knowledge area of PMBOK 5th Guide.
- The scope of the ontology is to standardize the description of concept related to PRM knowledge and then facilitates the knowledge reasoning.
- The sources of knowledge are PMBOK 5th best practices and PMI's standard for PRM.
- The project team members who are our potentials users as they seek even better ways to understand and be able to deal with risk.

### Step 2: Conceptualization

Based on these informal CQs and their respective answers, we have elicited information about the main concepts, instances, and their properties, to be formally represented in the PRM-ontology. In this context, a model with 20 classes, 21 subclasses, 26 object properties, 21 data type properties, and 132 instances has been obtained. As a result, the conceptual model of our domain ontology is depicted in Fig. 2.

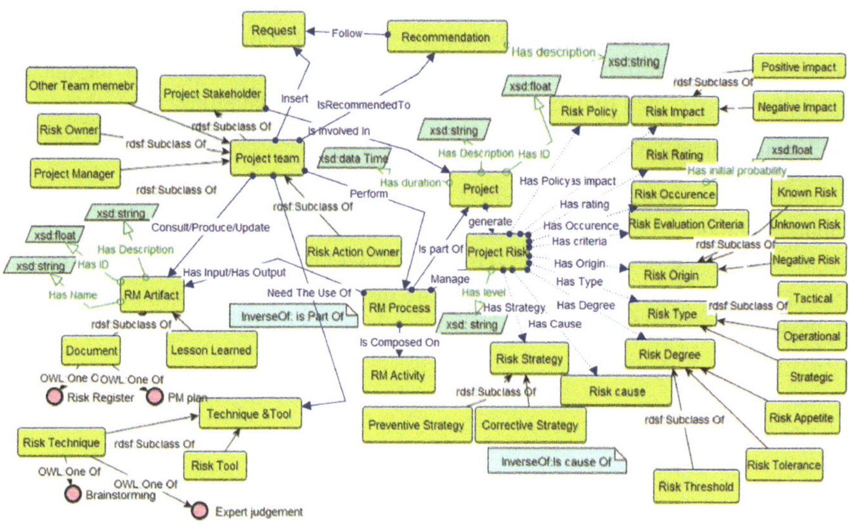


**Fig. 2.** An excerpt of Graffoo diagram summarizing the PRM-ontology class hierarchy

Considering the description of concepts, the PMI [14, 15] defines risk as uncertain event or condition that, if it occurs, has a positive or negative impact on at least one project objective. For each Project Risk, the PMI specifies the Risk Policy, Risk Cause,

Risk Rating, Risk Evaluation Criteria, and Risk Occurrence. It defines Risk Type that can be Operational, Tactical, or Strategic; Risk Strategy that can be Corrective or Preventive Strategy. Even, Risk Degree that can be Risk Appetite, Risk tolerance, or Risk Threshold. Also, Project risk has its origins: known, unknown or negative. The PMI defines the Project Stakeholders which can be Project Manager, Risk Manager, Risk Owner, Risk Action Owner or Other Team member. For each team member, this ontology defines the RM Artifacts to consult the RM Process and RM Activity to perform and the needed T&T to produce/update RM Artifacts as outputs. Finally, this ontology collects a set of recommendations from a single request with respect to each team member profile.

Rather than the concepts taxonomy, the proposed ontology establishes a set of properties. Each property has a domain and a range. Table 1 presents some properties of our domain ontology.

**Table 1.** Some defined properties for PRM-ontology

| Property | Type domain | Range |
|---|---|---|
| Has cause | Object project risk | Risk cause |
| Has level | Data-type project risk | String |

As well, we have defined the following object properties: "rdfs: Subclass Of" specify a parent/child relationship between two classes. The causality relationships between Project Risk and risk RiskCause are modeled explicitly with the relation "Has Cause." The pair-wise inverse relationships "IsComposedOn" and its inverse "Is Part Of" are used to identify the RM Activity that are parts of a specific RM Process. For the datatype, properties are associated to data values as string, float and data time.

**Step 3: Formalization**

The resulting conceptual diagram (Fig. 2) is converted into a formalized model or semi-computable model. As illustrated below, some terminological (TBox) and assertional (ABox) axioms are defined where TBox describes the domain structure in terms of classes and properties and ABox defines the instances described in the TBox. These axioms are formalized by means of description logics (DLs) [16]:

**//—Axioms used to describe classes—//**

RM Artifact ⊆ ⊤ (∃ID. float) ∩ (∃ Name. string) ∩ (∃ Description. string) ∩ (≥ 1 IsConsultedBy ∩ IsProducedBy. Project team ∩ IsUpdatedBy.Project Team ) ∩ (≥ 1 Is the Input Of ∩ Is the OutputOf. RM Process)
Document ⊆ RM Artifact
Lesson Learned ⊆ RM Artifact
Document ∩ Lesson Learned ⊆ ⊥ //- disjoint classes-//
(….)

**//—Axioms used to express properties—//**

Is Composed Of − ⊑ Is PartOf
IsCauseOf − ⊑ HasCause

Consult − ⊑ IsConsultedBy

NeedTheUseOf Risk Tool ⊑ NeedTheUseOf

(….)

**//—Individual assertions—//**

**Step 4: Implementation by OWL**

Based on the definition of classes and their associated properties given above, the PRM domain knowledge is implemented in protégé5.5, using Ontology Web Language (OWL2). In this way, concepts, relationships, and attributes were defined as "classes," "object properties," and "data properties," respectively. Also, axioms were modeled by means of OWL restrictions as illustrated in Fig. 3.

**Step 5: Evaluation**

Ontology evaluation is the task of measuring the ontology quality with respect to given criteria. Thus, it is important to select the suitable evaluation criteria and approaches, since some may not fit well to the ontology scope and its domain [17]. In this perspective, criteria-based approach was selected to verify our ontology in term of consistency, completeness, expandability, and coverage using Pellet v.2.2 [16] reasoning engine:

- *Consistency*: our ontology is consistent as it does not contain or allow circular class hierarchies, inconsistent axioms, and redundant naming schemes.
- *Expandability*: refers to the ontology's ability to expand itself. Our proposed ontology is built to be scalable since it can extend other domains such as construction RM, IT RM based on the generic conceptual model (Fig. 3).
- *Completeness*: PRM-ontology is complete since all the relevant concepts mentioned in PMI framework's related to PRM Knowledge area are defined and covered with their instances and their properties (domain → range).

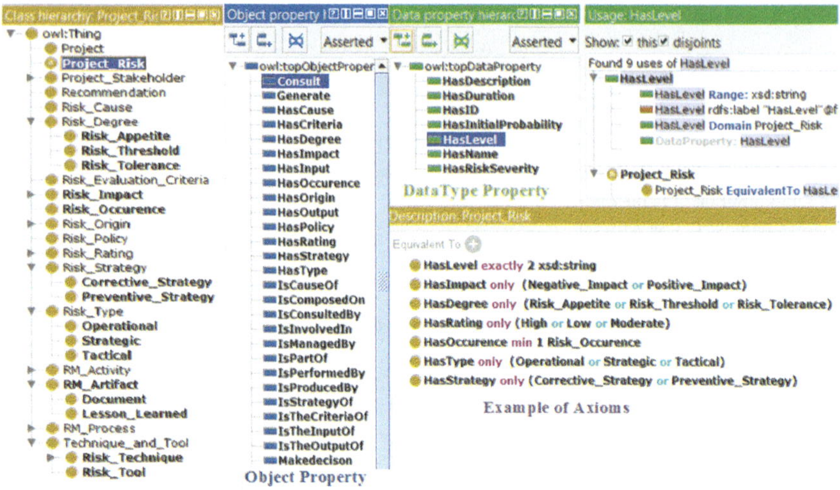


**Fig. 3.** Fragment of the ontology in Protégé 5.5 OWL screenshot

- *Coverage*: to ensure the ontology coverage during the development process; the mainly concepts and attributes used in PMI's Framework's were manually extracted and organized.

While coverage can be assessed objectively based on data-driven approach in which our proposed ontology is compared with respect to a corpus of information, PMI's standards for PRM, using various measures, extensibility, and completeness are difficult to quantify as the criteria-based evaluation was a subjective approach. According to [18], we consider counting the number of overlapping terms, vector space similarity, and structural fit by clustering and mapping terms, and conditional probability. We will only focus on the first two measures as our corpus is not annotated:

- *Number of Overlapping Terms*: measure the overlap between the retrieved terms in the ontology and the corpus to which it is compared.
- *Vector Space Similarity* (*VSS*): measures the similarity score using the cosine angle. Where $O$ Ontology, $C$, the Corpus, and n represents the numbers of concepts in $O$ and $C$; we obtain a VSS measure as:

$$\text{Similary}\,(O, C) = \cos(\theta) = \frac{\sum_{i=1}^{n} O_i * C_i}{\sqrt{\sum_{i=1}^{n} (O_i)^2} \cdot \sqrt{\sum_{i=1}^{n} (C_i)^2}} \tag{1}$$

We have also adopted the information retrieval measures such as:

- *Precision*: the percentage of ($O$) concepts that overlap with the ($C$):

$$\frac{|O \cap C|}{|O|} \tag{2}$$

- *Recall*: the percentage of the ($C$) terms that overlap with the ($O$):

$$\frac{|O \cap C|}{|C|} \tag{3}$$

- *F-Measure*: the harmonic mean that combines both the values of precision and recall. It attains its best score at 1 and its worst value at 0.

$$\text{F-Measure} = \frac{2}{\frac{1}{\text{Recall}} + \frac{1}{\text{Precision}}} \tag{4}$$

The evaluation results demonstrate that the proposed ontology fits well with the PMI's standard as presented in Table 2.

**Table 2.** Results of data-driven approach for measuring the coverage criterion

| \|O\| | \|C\| | \|O ∩ C\| | VSS | Precision | Recall | F-measure |
|---|---|---|---|---|---|---|
| 783 | 1836 | 659 | 0.197 | 0.841 | 0.358 | 0.52 |

- The overlapping terms (659) present 35.6% of PRM-ontology coverage in PMI's standard for PRM.
- The VSS measure puts greater weighting for terms that appear more often. A high similarity score tells us that the ontology covers words that appear more frequently. VSS value shows that all terms frequently and infrequently appeared are considered.
- The proposed ontology has a precision score of 0.841, a recall score of 0.358; thus, we obtain an average result of F-measure score of 0.502 in [0, 1].

## 4 SWRL Reasoning Rules

Besides the knowledge explicitly defined in our ontology using RDF/OWL, additional implicit knowledge which typically are in the form of rules is needed. Thus, the PRM-ontology will be extended with reasoning rules, constructed based on our interpretations of both PMBOK 5th and PMI's RM standard. Then will be used by the ontology to obtain the corresponding recommendations which are expressed in the rule-based format. Thus, Semantic Web Rule Language (SWRL) is used [10]. In order to illustrate how rules are codified, we take the rule_10 as an example to infer the steps followed when a known Project Risk "PR" is allocated to a single Risk Owner "x" so as to plan strategies and actions with the equired artifacts and the needed T&T. The rule_10 is defined as:

ProjectRisk(?PR) ^(Is AllocatedTo max 1 RiskOwner) (?x)^Consult (? x, RMPlan)
^Document ( RMPlan)^ Consult (? x, RiskRegister) ^ Document ( RiskRegister)
Consult(? x, ? L1)^ lesson_Learned (?L1) ^IsRecommendedTo(? x, Recommendation5)
Recommendation(?Recommendation5) ^NeedTheUseOfTool (?x, DecisionSupportTool)
^RiskTool (DecisionSupportTool)
^IsrecommendedTo(? x, ? Recommendation6)^NeedTheUseOfTechnique( ?x, StrategyIm plementationTechnique) ^RiskTechnique
(StrategyImplementationTechnique)^ IsRecommendedTo(? x, Recommendation7) ^Need TheUseOfTool (?x, Contingent Response Strategy) ^RiskTool ( Contingent Response Strategy) ^IsRecommendedTo(?x, Recommendation8)^IsPartOf(? activity5 , ? P5)^ RM

In order to execute these rules, Drools engine, a powerful open-source OWL 2 rule engine, is applied. In this context, 22 SWRL rules are exported as shown in Fig. 4.

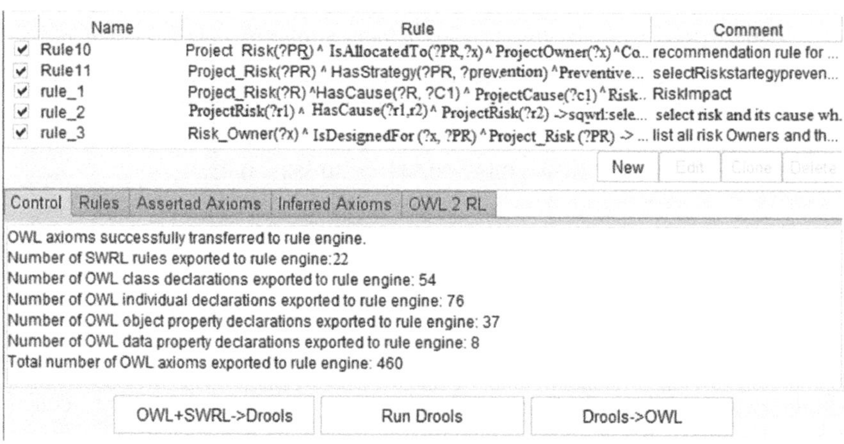


**Fig. 4.** Screenshot of SWRL rules editor and drool engine interaction

## 5 Conclusion

Project Risk Management remains a knowledge intensive domain, which requires semantic approach such as domain ontology and reasoning tools such as SWRL Rules. The singularity of our work is, firstly, the implemented approach relies on knowledge which complies with best practice and standard. Then, it incorporates "lessons learned" as a concept into the PRM-ontology to exploit PR past experiences; and finally, this work provide a team member, at each PR process, the appropriate decision according to its request. Based on these findings, an extension of this work is needed to implement an ontology-Based Decision Support System as a reliable support tool to enhance the efficiency of PR processes in a great extent.

## References


1. Palmer, C., et al.: An ontology supported risk assessment approach for the intelligent configuration of supply networks. J. Intel. Manuf. **29**(5), 1005–1030 (2018)
2. Manotas-Niño, V., et al.: Towards a model of integration between risk management and lesson learning system for project management. In: 2015 International Conference on Industrial Engineering and Systems Management (IESM). IEEE (2015)
3. Pittl, B., Fill, H.-G., Honegger, G.: Enabling risk-aware enterprise modeling using semantic annotations and visual rules (2017)
4. El Yamami, A., et al.: Representing IT projects risk management best practices as a metamodel. Eng. Technol. Appl. Sci. Res. **7**(5), 2062–2067 (2017)
5. de Gusmão, C.M.G., de Moura, H.P.: Multiple risk management process supported by ontology. In: International Conference on Product Focused Software Process Improvement. Springer, Berlin, Heidelberg (2006)
6. Dikmen, I., Talat Birgonul, M., Fidan, G.: Assessment of project vulnerability as a part of risk management in construction. In: Proceedings of Joint 2008 CIB W065/W055 Symposium (2008)
7. Fidan, G., et al.: Ontology for relating risk and vulnerability to cost overrun in international projects. J. Comput. Civil Eng. **25**(4), 302–315 (2011)
8. Studer, R.V., Benjamins, R., Fensel, D.: Knowledge engineering: principles and methods. Data Knowl. Eng. **25**(1–2), 161–197 (1998)
9. Wang, H.-H., Boukamp, F.: Ontology-based representation and reasoning framework for supporting job hazard analysis. J. Comput. Civil Eng. **25**(6), 442–456 (2011)
10. Zhong, Botao, Li, Yong: An ontological and semantic approach for the construction risk inferring and application. J. Intell. Rob. Syst. **79**(3-4), 449–463 (2015)
11. Wang, H., et al.: Research on ontology-based knowledge presentation and reasoning in civil aviation emergency decision. Comput. Eng. Sci. **33**(4), 129–133 (2011)
12. Emmenegger, S., Laurenzini, E., Thönssen, B.: Improving supply-chain-management based on semantically enriched risk descriptions. In: KMIS (2012)
13. Fernández-López, M., Gómez-Pérez, A., Juristo, N.: Methontology: from ontological art towards ontological engineering (1997)
14. http://dinus.ac.id/repository/docs/ajar/PMBOKGuide_5th_Ed.pdf
15. http://www.innovativeprojectguide.com/documents/PMIPracticeStandardforProjectRisk Management.pdf

16. Sirin, E., et al.: Pellet: a practical owl-dl reasoner. Web Semantics: science, services and agents on the World Wide Web **5**(2), 51–53 (2007)
17. Guo, B.H.W., Goh, Y.M.: Ontology for design of active fall protection systems. Autom. Const. **82**, 138–153 (2017)
18. Brewster, C., et al.: Data driven ontology evaluation (2004)

# Developing an Integrated IoT Blockchain Platform: A Demonstrator

Fazel Naghdy(✉), Golshah Naghdy, and Samaikya Malreddy

School of Electrical, Computer and Telecommunications Engineering, University of Wollongong, Wollongong, NSW, Australia
fazel@uow.edu.au

**Abstract.** Ensuring the integrity and security of an Internet of Things (IoT) system, and maintaining and updating the firmware driving its devices, becomes critical as the number of nodes, sensors, actuators and control loops increases. The security of IoT can be addressed by deploying blockchain technology. The feasibility of building a blockchain-based IoT on a low-cost microcontroller is demonstrated. This is achieved by implementing an Ethereum node on Raspberry Pi 3 controlling an LED. The LED data of turning ON and OFF by the user are collected and stored on a smart contract. The interaction with the smart contract is performed by a DApp called, Status, developed based on Light Ethereum Subprotocol (LES). The feasibility of the concept was successfully demonstrated, showing the potential of blockchain technology in developing more secure IoT systems.

**Keywords:** IoT · Blockchain · Light Ethereum · DApps · Status · Go · Geth

## 1 Introduction

Distributed control and sensing, deployed in large-scale automation such as smart cities and environmental monitoring and control, is reaching to a new stage of development and complexity with the pervasive and large-scale connectivity offered by the Internet.

The emerging concept of Internet of Things (IoT) refers to an ensemble of smart objects that communicate and share data through the Internet [1]. Each object represents a node with sensing, actuation, computation and communication elements as well as security and privacy. The unique characteristics of IoT are that the Internet provides a seamless and efficient communication between various devices and clients on the system. Otherwise, similar concepts using a private network have been around under different titles in different engineering disciplines such as distributed control or wireless sensor and actuator networks [2].

The major tasks associated with the Internet of Things (IoT) such as ensuring the integrity and security of the whole system, maintaining and updating the firmware driving the devices, increasing with the number of nodes, sensors, actuators and control loops


X.-S. Yang et al. (eds.), *Proceedings of Fifth International Congress on Information and Communication Technology*, Advances in Intelligent Systems and Computing 1183, https://doi.org/10.1007/978-981-15-5856-6_22

and managing billing services between devices can become potentially a barrier to the expansion of IoT device ecosystem and its sustainability.

The IoT is currently a major area of research and development while various concepts and technologies are rapidly deployed in developing major IT infrastructures [3, 4]. It is expected that the combined markets of the IoT will reach to $520B, more than double of $235B spent in 2017 [5]. It is expected that the number of devices interconnected within Internet will reach over 50 billion by 2020 [6]. In the survey conducted in [6], the customers identify three most significant issues in the IoT as security, integration with the existing technology and uncertain return on investment. Other studies also show that concerns about security are major hindrances in the expansion and further development of IoT [7].

The security of IoT can be addressed by deploying blockchain technology acting as an immutable ledger where each new block is connected to the previous block creating a chain-like, decentralized and distributed structure in a peer-to-peer network [8]. In such a system, no single node controls the network. In blockchain, the distributed database ledger of block records and grows when a new block is built by miners resisting to modification of data intrinsically among all the built nodes participating in a distributed blockchain network. A transaction referred to an agreement or operation between parties of two/peer-to-peer through a consensus protocol changes the network states. The ledger holds the batches of transactions that cannot be altered retrospectively, without the centralized trusted third party safeguarding the history immutability, privacy, fault-tolerant distribution and transaction automation.

In blockchain, the data are tracked in secured cryptocurrencies, hashing and distributed consensuses. Amongst different platforms, Ethereum has proved to have a less transaction time of 12–16 s, rather than 10 min of bitcoin [9]. Ethereum is a distributed, public, open-source platform featuring smart contracts functionality under blockchain-based computation. It sets the shared computing system with rules by writing smart contracts to define the IoT behavior through private or public key infrastructure authentication used in decentralized applications (DApps). Overall, Ethereum is a much more feasible platform to build decentralized applications in IoT and to run blockchain smart contracts.

The main objective of the study reported in this paper is to explore how effectively a blockchain-based IoT can be developed on a low-cost microcontroller. This is demonstrated by implementing an Ethereum node on Raspberry Pi 3 controlling an LED. The LED data of turning ON and OFF by the user are collected and stored on a smart contract. The interaction of the user with the smart contract is performed by a DApp called, Status, developed by Status.im [10], based on Light Ethereum Subprotocol (LES). Using Status, a mobile device becomes a light node on the Ethereum network, providing access to Ethereum's entire ecosystem from anywhere. This offers a powerful tool to introduce blockchain technology into new areas such as sensor/actuator networks.

During the course of the chapter, the IoT and blockchain technologies are introduced and a background on the application of blockchain in IoT is provided in Sect. 2. The overall design of the system and its major characteristics are described in Sect. 3. This is followed by describing the steps taken to build the demonstrator in Sect. 4. Some conclusions on the work are finally drawn in the Sect. 5 of the chapter.

## 2 Deployment of Blockchain in IoT

The security of IoT is critical because the devices operating within it generate, process and interchange a large amount of sensitive and critical data. Various security features in nodes, network and data processing required data encryption, identity authentication, secure multiparty computation and cloud computing. The crucial technologies currently deployed operate by adopting hop or end-to-end encryption mechanism, TLS/SSL or IPSec communication security mechanism, protecting sensor data and symmetric and asymmetric cryptographic algorithms [11]. The general architecture of IoT and the security requirements at each level are illustrated in Figs. 1 [12].

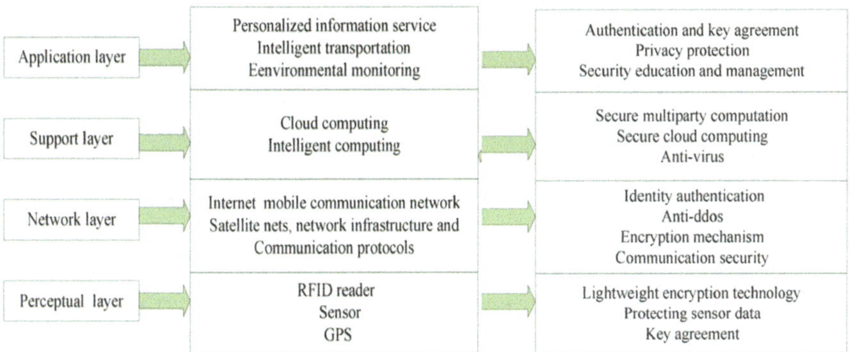


**Fig. 1** Architecture of IoT and its security requirements [12]

Blockchain has potential to effectively address the security issues in IoT by providing decentralized security and privacy. However, there are major challenges associated with deploying blockchain in IoT as it introduces high level of energy, delay and computational overheads—well above what the IoT devices can accommodate or manage. The IoT devices are generally low-powered and resource-constrained, consequently requiring a more specific approach to the blockchain design. Overall, the blockchain-based IoT faces three main challenges of scalability, power and processing time and amount of memory required to store data. The scalability challenge of the blockchain-based IoT is due to a need to store an ever-growing blockchain on resource-constrained devices [13]. This limitation is further intensified by the fact that some low-powered IoT devices require less than 10 KB of data memory and less than 100 KB of program memory, whereas the memory requirements of a blockchain node are in the order of multiple gigabytes.

In a system proposed by Cha et al. [14] a privacy-aware blockchain gateway acts as a mediator between users and the IoT nodes. The user has access to the data produced by the IoT devices through the gateway rather than directly from nodes. This ensures the security of sensitive data. On average, the proposed gateway requires 13 s for a data query to process a smart contract at the server side. Further research into shortening this time is needed to for more efficient means of protecting data on the device end.

In the work conducted by Biswas et al. [14], the scalability of ledger and the rate of transaction execution in blockchain when applied in IoT are studied. A large number of

IoT devices generally produce a very high rate of transactions, well beyond the capability of current blockchain solutions to manage. In order to address this challenge, a local peer network is proposed to bridge the gap between IoT and blockchain. The local peer network restricts the number of transactions which enters the global blockchain by implementing a scalable local ledger, without compromising on the peer validation of transactions at local and global level.

In a major study conducted by Lo et al. [15], various methods proposed in the literature to integrate blockchain and IoT are reviewed and systematically analyzed. The study, based on 35 published papers, shows that the majority of the research works are either at the conceptual level or at an early stage of development. Among the reviewed papers, 19 deploy blockchain in IoT for data storage, 10 as an access control mechanism, six as a means of platform connector and one as an incentive distributor. Ethereum has been used as a platform in 16 of the studies against four using bitcoin network, four multichain, three Hyperledger and one Monax. The proposed methods address the challenges associated with implementing blockchain in the IoT. The paper also provides some insights on how the developed solutions and proposed research methodologies can be improved.

## 3 Overall Design

The system developed in this study is a Light Ethereum Subprotocol (LES) node implemented on a Raspberry Pi 3—a low-cost but versatile microcontroller with multiple digital and analog ports and standard utilities. A simple IoT is created by interfacing an LED to Raspberry Pi, running smart contracts executed within the Ethereum-based decentralized application Status App. The LED status is read by Raspberry Pi and updated in smart contract. The change in the status of LED as requested by the user is applied by the Raspberry Pi. When a user requests to turn the LED light ON or OFF, the smart contract takes the data and the Raspberry Pi drives the LED to turn it ON or OFF accordingly. The workflow of the designed simple IoT is illustrated in Fig. 2.

Status is a mobile operating system built for Ethereum developed by Status.im with the aim of providing Ethereum capabilities within a smartphone. Status was introduced in 2016 at Devcon in Shanghai and has gone through major developments since then.

Status runs a full implementation of Geth Ethereum. Within Status, the DApps such as Gnosis, Aragon, Etherisc, Uport, Ethlance, CryptoKitties, Bancor, Peepeth and others are available within the app and run on the mobile phone. The chat feature of Status is built on top of Whisper, an open-source IOS and Android mobile app. Whisper is an anonymous social media, allowing users to post and share photo and video messages anonymously. The overall architecture of a typical Status network with Ethereum IoT node driven from a mobile platform is illustrated in Fig. 3.

In order to verify and demonstrate the feasibility of communicating from Status mobile platform running Light Ethereum Subprotocol (LES) to an IoT device using blockchain technology, an IoT system based on Raspberry Pi 3 with a microSD card of 32 GB was deployed. The Raspberry Pi microcontroller was interfaced to an LED. The protocol for communication was to turn the LED ON and OFF from the mobile device using status by sending a text message using Status and blockchain. The simple

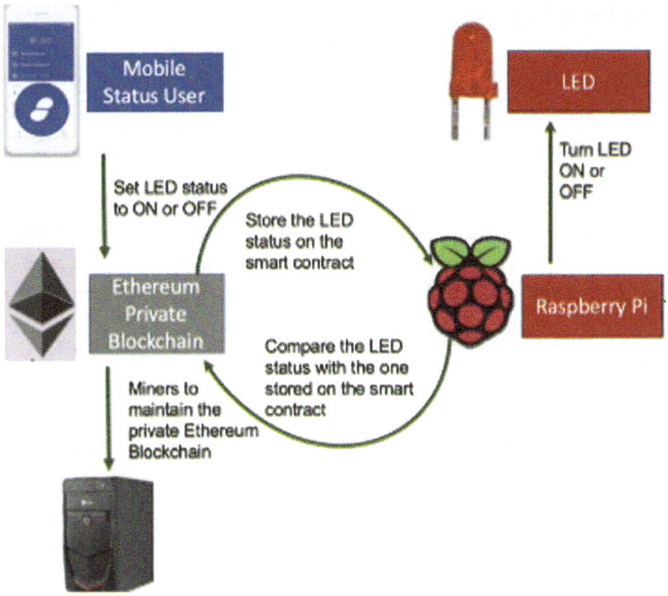


**Fig. 2** Workflow of simple IoT

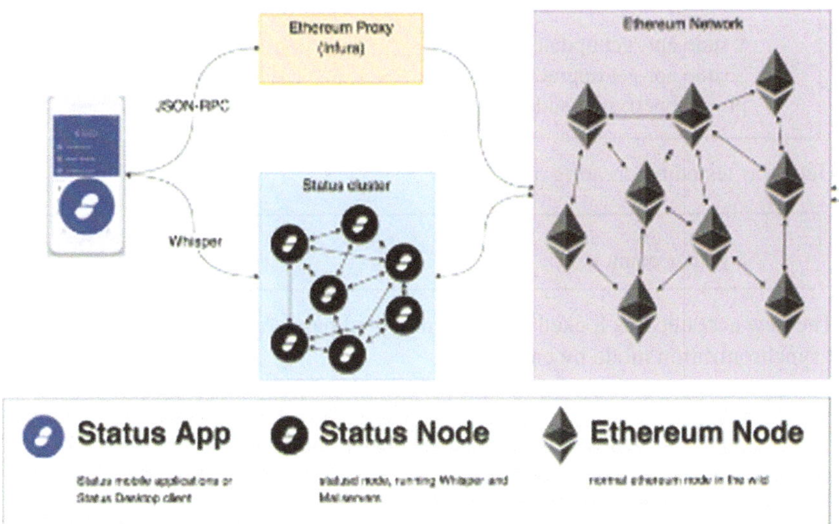


**Fig. 3** Overall architecture of status network with Etherium IoT node [16]

IoT device was chosen to ensure that the work in the first stage of its development was focused on the blockchain-based secure communication.

# 4 Platform Development

The major steps taken to build the demonstrator are briefly outlined in this section.

## 4.1 Building an Ethereum Node

Raspbian desktop that is a Unix-like operating system for Raspberry Pi was installed on Raspberry Pi to create a simple and powerful computational environment. The next step was to configure the Raspberry Pi as an Ethereum node. This was achieved by implementing and running Geth that is the command-line interface for running a full Ethereum node implemented in Go [17].

## 4.2 Installing Geth

Using Geth, it is possible to conduct blockchain-associated activities such as mining real ether, performing transactions between different nodes, creating contracts and send transactions and exploring block history. After start-up, Geth either connects to the exiting live blockchain or create its own according to its setup. In this application, Geth was configured as a Light Ethereum client to make it compatible with Status. Private user accounts were created on Geth.

Geth installation and running was performed by the following commands

```
$ sudo apt -get update
$ sudo apt-get upgrade -y
pi@rasbberrypi~$sudo apt-get dist-upgrade
```

Geth was tested by creating an account:

```
$ geth account new
```

The new account was locked with an encrypted password. The node was started in light synchronization mode by entering:

```
$ geth –syncmode light –cache 64 –maxpeers 12
```

## 4.3 Installing Go

Go is an open-source programming language designed for simple, reliable and efficient software. Compared to other programming languages, it is quite expressive, concise, clean and efficient. It provides concurrency mechanisms that can be deployed to effectively write code for multicore and networked machines. It is also ideal to construct flexible and modular programs due to its novel type system. Go compiles fast, is statically typed and though a compiled language, feels like an interpreted language [18].

Go for Ethereum node implementation was installed and copied to bin by entering:

```
$ mkdir sam
$ cd sam
$ git clone -b release/1.7 https://github.com/ethereum/go-ethereum.git
$ cd go-ethereum
$ make
$ sudo cp build/bin/geth/usr/local/bin/
```

## 4.4 Connecting to JavaScript Console

The node running on the Raspberry Pi was connected to JavaScript console by entering

```
pi@raspberrypi~$ geth --autodag console
>eth.accounts
>eth.syncing
```

This creates the command-line interface to check the eth accounts and eth syncing. At this point, the Ethereum is installed as a light client node on Raspberry Pi and can synchronize with the live chain (mainnet).

## 4.5 JavaScript Run-Time Environment

JavaScript was used in the Raspberry Pi to monitor and alter the status of LED and access the smart contract functions. The LED status was logged into the smart contract every 15 s.

In order to provide a run-time environment to run JavaScript programs, Node.js was downloaded and installed on the Raspberry Pi. Since the microcontroller was a Raspberry Pi 3, ARMV71 was deployed as the Node.js.

## 4.6 Smart Contract

The main function of smart contract is to log the desired status of the LED. At each sampling interval, the Raspberry Pi compares the current status of LED with the desired status. If there is a difference, the change is applied to LED and the status of LED is changed in the ledger.

The smart contracts running within Ethereum were developed and compiled using Solidity compiler [19]. An online Solidity compiler called Remix was used to compile the smart contract switching the LED ON and OFF [20].

## 5 Conclusions

In this chapter, a simple IoT based on blockchain technology was developed successfully by deploying Raspberry Pi 3 Model B, Status Dapp, an LED and Ethereum private blockchain. The primary aim was to change the status of the LED according the command issued from a user under privacy protected and secure blockchain. The LED data were stored by the smart contract, showing the reaction of user manipulation in the Status DApp under high protection of Ethereum. This simple demonstrator shows the potential of blockchain to develop more secure IoT system.

The focus of the chapter has been on the proof-of-concept and demonstrating that the current available technologies can be effectively integrated to produce IoTs with much higher security. The actual measurement of the overheads associated with using blockchain and its impact on the real-time operation of an IoT system with a higher number of nodes and more complexity is the focus of the future work.

## References

1. Weber, R.H.: Internet of things—new security and privacy challenges. Comput. Law Security Rev. **26**, 23–30 (2010)
2. Salarian, H., Chin, K.W., Naghdy, F.: Coordination in wireless sensor-actuator networks: a survey. J. Parallel Distrib. Comput. **72**(7), 856–867
3. Lampropoulos, G., Siakas, K., Anastasiadis, T.: Internet of things in the context of industry 4.0: an overview. Int. J. Entrepreneurial Knowl. **7**(1), 4–19 (2019). https://doi.org/10.2478/IJEK-2019-0001
4. Mccann, J.A., Picco, G.P., Gluhak, A., Johansson, K.H., Törngren, M., Gide, L.: Connected things connecting Europe. Commun. ACM **62**(4), 46–51. https://search-ebscohost-com.ezproxy.uow.edu.au/login.aspx?direct=true&db=bth&AN=135580877&site=eds-live. Accessed on 23 July 2019
5. Bosche, A., Crawford, D., Jackson, D., Schallehn, M., Schorling, C.: Unlocking opportunities in the internet of things. https://www.bain.com/insights/unlocking-opportunities-in-the-internet-of-things/. Accessed on 07 Aug 2018
6. Roman, R., Zhou, J., Lopez, J.: On the features and challenges of security and privacy in distributed internet of things. Comput. Netw. **57**(10), 2266–2279 (2013)
7. Ali, S., Bosche, A., Ford, F.: Cyberseurity is the key to unlocking demand in the internet of things. https://www.bain.com/insights/cybersecurity-is-the-key-to-unlocking-demand-in-the-internet-of-things/. Accessed on 13 June 2018
8. Papadodimas, G., Palaiokrasas, G., Litke, A., Varvarigou, T.: Implementation of smart contracts for blockchain based IoT applications. In: Proceedings of the 2018 9th International Conference on the Network of the Future, NOF (2018)
9. Xu, Q., He, Z., Li, Z., Xiao, M.: Building an ethereum-based decentralized smart home system. In: IEEE 24th International Conference on Parallel and Distributed Systems (ICPADS) (2018)
10. Information on the operating system is available at status.im
11. Suo, H., Wan, J., Zou, C., Liu, J.: Security in the internet of things: a review. In: Proceedings of International Conference on Computer Science and Electronics Engineering (2012)
12. Yang, G., Xu, J., Chen, W., Qi, Z.H., Wang, H.Y.: Security characteristic and technology in the internet of things. J. Nanjing Univ. Posts Telecommun. (Nat. Sci.) **30**(4) (2010)
13. Padma, M., Kasiviswanath, N., Swathi, T.: Blockchain for iot application: challenges and issues. Int. J. Recent Technol. Eng. **7**(5C), 34–37 (2019)

14. Cha, S., Chen, J., Su, C., Yeh, K.: A blockchain connected gateway for BLE-based devices in the internet of things. IEEE Access **6**, 24639–24649 (2018)
15. Biswas, S., Sharif, K., Li, F., Nour, B., Wang, Y.: A scalable blockchain framework for secure transactions in IoT. IEEE IoT J. **6**(3), 4650–4659 (2019)
16. Lo, S.K., Liu, Y., Chia, S.Y., Xu, X., Lu, Q., Liming, Z.: Analysis of blockchain solutions for IoT: a systematic literature review. IEEE Access **7**, 58822–58835 (2019)
17. Retrieved from https://discuss.status.im/t/use-dappnode-to-host-a-status-node/597
18. Geth can be downloaded from https://github.com/ethereum/go-ethereum/wiki/geth
19. Go can be downloaded from https://golang.org/
20. Information can be retrieved from https://solidity.readthedocs.io/en/v0.4.24/

# Impacts of the New General Data Protection Regulation for Small- and Medium-Sized Enterprises

Ralf Christian Härting(✉), Raphael Kaim, Nicole Klamm, and Julian Kroneberg

Business Administration, Aalen University of Applied Sciences, 73430 Aalen, Germany
Ralf.haerting@hs-aalen.de, Raphael.kaim@hs-aalen.de, kmu@kmu-aalen.de, j.kro10@gmx.de

**Abstract.** The European General Data Protection Regulation (GDPR) implies a lot of new regulations. The implementation of these new regulations is a major challenge for many small- and medium-sized enterprises (SMEs). Therefore, we investigated which factors influence the implementation of the GDPR in already existing business models for SMEs. Our model is focusing on already existing business models for SMEs. It is based on empirical data from German experts. It has been developed by using clustering and qualitative content analysis. Important influencing factors that are now open for quantitative verification are know-how, expenditure of time, uncertainty, costs, provision of information and process adaption.

**Keywords:** Data protection · GDPR · Impacts · Qualitative study · SMEs

## 1 Introduction

The European General Data Protection Regulation (GDPR) was adopted by a large majority on 14 April 2016. The GDPR replaces the Data Protection Directive from 1995 [1]. The GDPR became binding on 25 May 2018 after a transitional period of two years [2]. By 25 May 2018, all companies must adapt their business processes to the new legal situation. The aim of the new regulation is the modernisation of the European data protection law. A balance between economic and consumer interests in times of progressive digitalisation is to be created. Citizens should be given greater transparency and more control over their data [3]. In addition, the GDPR creates a future-oriented legal framework for data processing companies, as well as innovative business models [4].


X.-S. Yang et al. (eds.), *Proceedings of Fifth International Congress on Information and Communication Technology*, Advances in Intelligent Systems and Computing 1183, https://doi.org/10.1007/978-981-15-5856-6_23

In general, this regulation has six principles that are addressed. Those are data minimisation, purpose limitation, accuracy, storage limitation, confidentiality, fairness and lawfulness and lastly integrity [5].The new regulation prescribes a new framework of how the personal data of citizens of the European Union has to be used [6]. The GDPR creates a uniform level of data protection throughout Europe and increases customer rights by expanding the scope of rights for data subjects, including data portability an access [7]. This removes competitive barriers and market access barriers because of initially different data protection regulations [8]. However, the implementation of the new regulations of the GDPR can be very challenging, especially for SMEs. Before the 25 May 2018, many SMEs have been concerned that they will miss the deadline and that they are not able to prove the compliance of the GDPR processes. There are new regulated fines for breaches. The fines could be up to EUR 20 Million [9]. The research target of this paper is to identify impacts of the implementation of new GDPR. Moreover, constructs are determined, which influence the dependent variable in a positive or negative way. Therefore, we formulated the following research question: Which impact has the implementation of the GDPR in already existing business models for SMEs? To answer this research question, a qualitative study was implemented in which German experts in the field of data protection and data security in SMEs were interviewed. This research method will allow a better understanding of the challenges posed by the new general data protection regulation for SMEs. Thereby, the paper is structured as follows. First, the research design is presented and the data collection as well as the methodology is described. After this, an overview about our main results of the evaluation is given. In the following, a model of the impacts of the implementation of the GDPR in already existing business models for SMEs is developed. Finally, a conclusion with limitations and an outlook for further research are given.

## 2 Research Design

In this section, we give an overview about our research design. First, we define the overall group of experts and give some facts about the selected interview partners. In the following, we describe the used questionnaire. Then we would like to present the methodology used for our analysis.

In order to conduct our qualitative research, we looked for experts in the field of data protection or data security. One specific selection criterion is that the experts have to be responsible in the field of data protection in their respective company. CEOs, managing directors or IT managers therefore come into consideration. We addressed our survey also to data protection officers. In addition, the experts should have a long-time experience in the field of data protection or data security, which they gained through scientific work or working experience. Furthermore, we searched only for experts, who work in SMEs. We considered SMEs with up to 500 employees or an annual turnover of 50 million Euro. As a result, we conducted 13 interviews with experts in the field of data protection or data security. After completion of the survey phase, it became clear that the

**Table 1.** Interviewed experts

| Expert No. | Job position/responsibility | Size of enterprise |
|---|---|---|
| 1 | CEO | 5 |
| 2 | Commercial management | 230 |
| 3 | Data protection officer | 14 |
| 4 | CEO | 1 |
| 5 | CEO | 3 |
| 6 | CEO | 19 |
| 7 | Data protection officer | 500 |
| 8 | CEO | 8 |
| 9 | Company officer with statutory authority | 15 |
| 10 | CEO | 45 |
| 11 | CEO | 33 |
| 12 | Executive secretary (data protection officer) | 21 |
| 13 | CEO | 27 |

majority of the experts came from smaller companies with up to 50 employees. The experts thereby represent enterprises in different industries like producing enterprises and commercial enterprises. Table 1 gives an overview about our 13 interviewed experts with their job position and the size of enterprise.

### 2.1 Questionnaire

The questionnaire consists of eight open-structured questions and was carried out exactly once for each expert. The questionnaire was distributed a few days before the interview, so the experts had the opportunity to prepare and examine topics in their own company. The experts were then interviewed either in person or by telephone. The expected duration of twenty-five minutes was communicated in advance. That gave the experts ample time to respond. The survey was conducted in Germany, considering that the survey was conducted in German. The answers are transcribed and thereafter evaluated.

### 2.2 Methodology

The methodological procedure of this research consists of a qualitative survey. The subsequent evaluation of the answers was conducted using a hybrid method. The analysis method consists of two different methods: a cluster analysis according to Everitt et al. and a summary and structured content analysis according to Mayring [11]. The method is depicted in Fig. 1.

Forming clusters is the first part of the analysis. A cluster is either defined by the homogeneity of its items, or by the external isolation from other items. The purpose of clustering in this research is to divide the main sample into different smaller samples. Each sample is analysed individually. The content of the questionnaire is based on different types of impacts as expected by the authors. These expected impacts are formulated in the individual questions. Therefore, the clusters in this research are built by separating the questions. The answers to the individual questions resemble each other. Each question is different regarding their kind of impact on the topic of GDPR [10]. This kind of clustering differs from the clustering technique used in studies with greater sample sizes. Clusters in this kind of studies are formed within the experts. Such a division would be disadvantageous, as it is a homogenous group of experts.

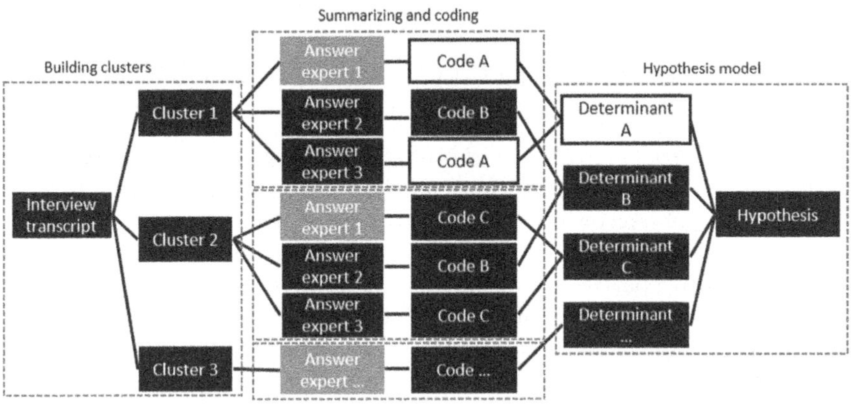


**Fig. 1.** Research model

Creating summaries and codes is the second part of the analysis. Both of those techniques are part of the qualitative content analysis according to Mayring. Their aim is to gather different statements from a defined amount of data and to interpret them [11]. This summary is not available for all of the answers. Therefore, it is an optional step to improve the comparability of the answers. The corresponding texts are at first with the exception of technical vocabulary paraphrased. Subsequently, the texts are reduced. Filler words or explanations as well as digressions are deleted. This forms the basis for the second technique of qualitative content analysis, the structured content analysis. It deals with the development of codes that allow a cross section through the entire material [12]. The coding procedure is linear in the first phase. Each answer of an expert in a cluster is provided with one to four codes. This means that the first answer will only contain new codes. From the second answer onwards, the codes of all previous answers are taken into account. This is resulting in a match with regard to the topic and its form. The answer receives the same code. In the second phase, after a cluster has been fully coded, the codes of a cluster are checked. Possible overlaps are eliminated or codes formulated in more detail. This procedure also allows codes to be present in several clusters. Codes can be found in cluster overlaps due to a weakness of the cluster technique. According to Everitt, instead of discovering structures, an investigator can impose a structure onto the data [10]. Therefore, in combining the techniques of clusters and structured content analysis, we created a more objective view on our data.

These codes form the basis to develop the hypothesis model. A code named by several experts has to be regarded as relevant. The constructs of the hypothesis model are formed from these relevant codes. This is achieved by forming thematic groups that clearly distinguish themselves from each other. The constructs and the dependent variable form the hypothesis model, which is explained in the following part of this paper.

## 3 Evaluation

In the following section, we summarise the most important key messages from our surveys.

*Question 1: Which guidelines of the GDPR have already been implemented? Which guidelines of the GDPR have not been implemented so far? Is there a lack of know-how?* According to the status of implementation, three experts said that they have already implemented the GDPR guidelines. Nine experts said that they have partially implemented the GDPR, and one expert said that he has nothing implemented of the GDPR so far. Five experts, of those that have not fully implemented the GDPR, told us that the know-how is missing to implement the GDPR guidelines completely. Also five experts said that they have not completely implemented the GDPR about of other reasons.

*Question 2: Which are the three biggest challenges in the implementation of the GDPR?* With eight mentions, "time" was the most frequent answer. Four experts called the costs, associated with the implementation of the GDPR, as a challenge. Three experts questioned the relevance of the GDPR topic and three further called the provision of information a challenge. Two experts found the technical expertise as well as the legal knowledge as a challenge and inter alia one expert called the complexity as a major challenge.

*Question 3: Do customers claim their rights from the GDPR? If yes, which rights?* Relating to this question, two of our experts said that they had customer requests about the GDPR, namely customer information requests. One of these two experts in addition had a customer request for customer data cancellation.

*Question 4: Does the implementation of the GDPR improve the quality of the customer relationship? Does the GDPR lead to the demolition of customer contacts?* On this question, with eight experts the majority of them stated that the GDPR has no impact on customer relations. However, two experts had the opinion that there is a tendency to customer relationship improvement and two said the GDPR leads to a tendency to deterioration. In addition, an expert was sure that it leads to a deterioration of the customer relationship.

*Question 5: Do you have costs of implementation, costs of opportunity or a loss in revenue by implementing the GDPR?* Almost all experts answered that they have mostly implementation costs and opportunity costs because of the GDPR. Three experts said they have running costs for external data protection officers and two experts answered that they have a loss in revenue because of this new regulation. In contrast to those statements, only one expert said that he had no costs because of the GDPR. But the reason for that is that he has not implemented the GDPR so far.

*Question 6: How does the GDPR impact the quality of customer data?* Three experts said that the quality would decrease due to the implementation of the GDPR, because a lot of data has to be deleted from the CRM that will miss in future conversations and in the customer relationship. In contrast, four experts stated that their data quality would increase because of the necessary database cleaning, which will leave them with correct data only of customers interested in the company. Six experts do not see any change in the quality of customer data at all.

*Question 7: What uncertainties does the GDPR implicate?* All experts said that they have legal uncertainty through these new regulations. Nobody knows exactly what they have to do and what they are allowed to do. That is for the inconsistent information, which has missing meanings, the lack of interpretation by courts and also the topicality of the information. In addition, seven experts told us that they have uncertainties because of the warnings. Three experts said that they have uncertainties in communication, especially with their customers and suppliers.

*Question 8: How does the GDPR impact the digitization of your company?* Five experts expect a slower digitization, mostly because of the changing and elaborating processes. Only one expert said that the GDPR will accelerate the digitalization. Five experts told us that the GDPR has no impact on digitalization.

## 4 Hypotheses Model

After evaluating our empirical data with our hybrid method, we elaborated six different constructs. These influence the dependent variable. The whole model was developed based on the collected and evaluated data. The hypotheses model including all single items are depicted in Fig. 2. In the following, we describe the different constructs which influence our dependent variable.

The first construct is **know-how** and consists of the lack of technical know-how and the lack of legal know-how in the companies. Expert 9 said that the technical aspect on the IT implementation is a major challenge. Expert 8 mentioned that the GDPR is too complicated and added that the guidelines are not transparent enough. There is also an **expenditure of time** that results from the procurement of information regarding the GDPR and the time for the implementation to fulfil the GDPR guidelines. As an example expert 1 reported: "The biggest challenge is to find enough time to implement these guidelines". Another very important construct is the **uncertainty**. In SMEs, these consist of legal uncertainties, missing explanation to the information, the fear of financial

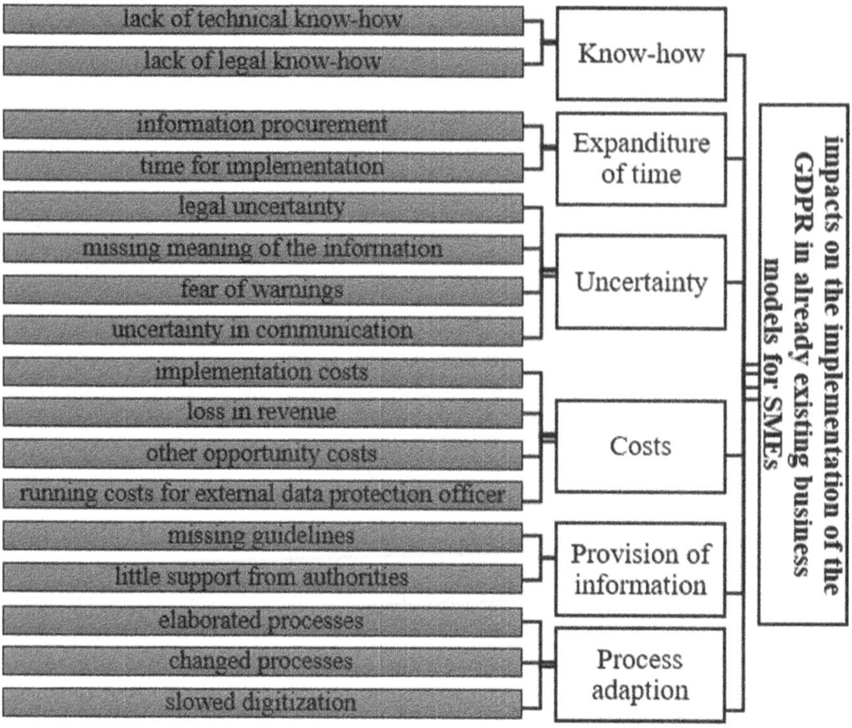


**Fig. 2.** Hypotheses model including items, constructs and the dependent variable

penalties and the uncertainties in communication with partners or customers. Uncertainty exists in the communication because it is not clear which data is allowed to be used. Expert 9 mentioned uncertainties about the possible high fines for violations of the guidelines. This is consistent with the findings of Kaim et al. in which uncertainties about economic policies have steadily increased in recent years [12].

There are additional costs, which enterprises have to face. For example, costs of implementation to fulfil the guidelines of the GDPR, the loss in revenue and other opportunity costs. Moreover, some companies have to face the running costs for an external data protection officer. Expert 9 said, while the implementation takes a lot of time, he could not do other work, which might be a value-adding activity. Another issue is the **provision of information** because there are some missing guidelines. Additionally, there is only little support from the responsible authorities. Expert 13 mentioned that there is simply too little information material and help.

The last construct is the **process adaption**. This variable stands for all the elaborated processes and the changed processes. On top of that, the digitisation is slowed down. Expert 10 mentioned that the GDPR creates some elaborate processes, which are needed to select and delete data. Expert 2 said that other work processes are superseded, and expert 3 added that the digitalisation is slowed down. He also stated that legal compliance of processes always has to be taken into account.

All the constructs have a negative impact on the dependent variable "impacts on the implementation of the GDPR in already existing business models for SMEs". Therefore, six new hypotheses can be established:

1. The missing know-how has a negative impact on the implementation of the GDPR in already existing business models for SMEs.
2. The high expenditure of time has a negative impact on the implementation of the GDPR in already existing business models for SMEs.
3. The big uncertainties have a negative impact on the implementation of the GDPR in already existing business models for SMEs.
4. The high costs have a negative impact on the implementation of the GDPR in already existing business models for SMEs.
5. The insufficient provision of information has a negative impact on the implementation of the GDPR in already existing business models for SMEs.
6. The difficult process adaptation has a negative impact on the implementation of the GDPR in already existing business models for SMEs.

## 5 Conclusion

We developed a research model to investigate the impacts of implementing the GDPR in already existing business models of SMEs. We identified know-how, expenditure of time, uncertainty, costs, provision of information and process adaption as influencing aspects. For all constructs, we assume a negative impact. These findings will help practitioners directing their capacities when implementing these and similar regulations. Also the makers of regulations get an impression, where they can support companies with problems. Our research contributes to evaluate the impacts of the GDPR guidelines. With a sample size of 13 experts, we received different opinions and experiences. A quantitative model validation is needed to prove or disprove the hypotheses we set up. However, there are still questions open for further research. What other challenges do SMEs have to face regarding the new regulations? How expensive will it be for the companies? How much effort do SMEs have to implement the guidelines and how aware are they with the whole topic?

The study has several limitations. The analysis of the interviews showed mainly negative influences of the GDPR. Although some interviewees also mentioned positive aspects, our qualitative research model is focused only on German experts and therefore be biased, as the GDPR includes parts of the old German Federal Data Protection Act.

## References


1. de Hert, P., Papakonstantinou, V.: The new general data protection regulation: still a sound system for the protection of individuals? Comput. Law Secur. Rev. **32**(2), 179–194 (2016)
2. The European Parliament and the Counsel: Regulation (EU) 2016/679 of the European Parliament and the Counsel of the protection of natural persons with regard to the processing of personal data and on the free movement of such data, and repealing Directive 95/64/EC (General Data Protection Regulation). Official Journal of the European Union, L 119 (2016)
3. Sobolewski, M., Mazur, J., Palinski, M.: GDPR: a step towards a usercentric internet? Intereconomics **52**(4), 207–213 (2017)
4. Chen, J.: How the best-laid plans go awry: the (unsolved) issues of applicable law in the general data protection regulation. Int. Data Privacy Law **6**(4), 310–332 (2016)
5. Goddard, M.: The EU General Data Protection Regulation (GDPR): European regulation that has a global impact. Int. J. Market Res. **59**(6), 703–705 (2017)
6. Constantinides, P., Henfridsson, O., Parker, G.G.: Introduction-platforms and infrastructures in the digital age. Inf. Syst. Res. **29**(2), 381–400 (2018)
7. Todorovic, I., Komazec, S., Krivokapic, D., Krivokapic, D.: Project management in the implementation of General Data Protection Regulation (GDPR). Euro. Project Manage. J. **8**(1), 55–64 (2018)
8. The Federation of German Industries (BDI). https://english.bdi.eu/article/news/the-eus-general-data-protection-regulation-answers-to-the-most-important-questions/. Last accessed 16 May 2019
9. Jackson, O.: Many Small Firms are Still Unprepared for GDPR. International Financial Law Review, London (2018)
10. Everitt, B.S., Landau, S., Leese, M., Stahl, D.: Cluster Analysis, 5th edn. Wile, Chichester (2011)
11. Mayring, P.: Qualitative Content Analysis. Forum Qualitative Soc. Res. **1**(2), Art. 20 (2000)
12. Kaim, R., Härting, R.C., Reichstein, C.: Benefits of agile project management in an environment of increasing complexity—a transaction cost analysis. Intell. Decis. Technol. **2019**, 195–204 (2019)

# Innovative Classroom Activity with Flipped Teaching for Programming in C Course—A Case Study

Shikha Maheshwari[1(✉)], Suresh Kumar[1], Naresh Kumar Trivedi[1], and Vijay Singh Rathore[2]

[1] CUIET, Chitkara University, Punjab, India
{shikha.maheshwari,suresh.kaswan, nareshk.trivedi}@chitkara.edu.in
[2] IIS University, Jaipur, India
vijaydiamond@gmail.com

**Abstract.** The concept of teaching through the flipped classroom technique is always considered as one of the most important in higher education in the Indian context. It enables undergraduates to gain fundamental learning even before they attend the class and also facilitate educators to manage discussions among the undergraduates during the classroom interaction hours. Likewise, the undergraduates share their insights gained during such discussions in the classroom. It is seen that the majority of the Engineering courses are lacking in implementing the blended learning approach into their curriculum. The case study presented in this research paper is an attempt to prepare undergraduates before they attend their class to provide further speculation in the subject matter. The educator disseminated the learning course material for the chosen programming subject for five topics even before the beginning of the regular classes in a semester and interacted with the undergraduates during the hours meant for classroom interaction in the semester. From the observations, it is found that undergraduates have actively participated in the flipped classroom approach and also lead to improved learning outcomes of the undergraduate over the span of time during the semester.

**Keywords:** Flipped classroom · Engineering courses · Innovative teaching–learning methods

## 1 Introduction

The present education system is the outcome of the continuous evolutions that happened in India for the past ten years with educators from all states actively working to identify and implement various innovative techniques for 21st centennial undergraduate students. However, educators of engineering educational institutes are still delivering their instructions through obsolete approaches such as chalk and talk, etc. even for problem-solving


X.-S. Yang et al. (eds.), *Proceedings of Fifth International Congress on Information and Communication Technology*, Advances in Intelligent Systems and Computing 1183, https://doi.org/10.1007/978-981-15-5856-6_24

technical concepts and trying to induce critical thinking among undergraduates [1]. As per Bloom's taxonomy, it is widely known that the key factor in Engineering Education is critical thinking, analysis, evaluate, and create. Many experienced educators are still allotting home assignments with a student-centred approach applied in later stages. However, with the exceptionally student-centric subject matter, less self-motivated undergraduates are facing numerous challenges. Many educators have acknowledged that during classroom instruction delivery engineering undergraduates are satisfied with the blended approach of learning [2]. One of the effective methodologies to enable the undergraduate student to attend the class lecture with an object-oriented approach and comprehend the concept without any intervention of the educator is to share the course material beforehand. Generally, the educator attempts to explain the concept, clarify the doubts, if any during the classroom interactions and make an attempt to identify the level of understanding of the undergraduates by asking few questions orally [3]. However, in a flipped classroom approach, the undergraduate will interact with the educator through the online content available with them and effectively raising the questions on their comprehension. By addressing these questions, peer undergraduates comprehend the general concepts of the subject, and henceforth the learning in the classroom can be seen as student-centric. Yusong Li and Tareq Daher [4] have presented a case study for the flipped classroom for some topics in civil engineering. During their study, it was found that the majority of the undergraduates are satisfied with the flipped classroom approach.

The motivation for the educator is (a) plan for interactive lectures (b) distribute the course content in advance through an online medium (c) encourage undergraduates to become self-learners and assist them in practicing the subject [5].

One of the noted advantages of the flipped classroom approach is that the educators need not clarify his 60-min long interaction either through chalk and talk method or utilizing a power-point/video lecture [5]. Undergraduates attend the lecture in their classrooms with various questions and educators are responsible for making them understand the concept based on the questions asked by the undergraduate [6]. Another advantage is that even undergraduates can also interact during the non-teaching hours for clarifying their doubts [7].

## 2 Methodology

For implementing the flipped classroom approach, various class notes, power-point presentation, and videos for four important topics in programming in C language course was prepared:

1. History of Programming Languages
2. Strategy for Designing Algorithms
3. Evolution of Data Types
4. Relational and Logical Operators.

The short videos of a maximum of 4–5 min and power-point presentation with a maximum number of 20 slides for each selected topic have been prepared. This course content was distributed among undergraduates one week before the actual commencement of the classes. Simultaneously, a set of assessment tools is adopted in the present studies which are given below:

### 2.1 Assessment Tools

For assessing the student, the following principles are implemented:

1. **Messy Point Markup**: Explaining the most confusing part sub-topic the course content provided to the students to enable them in engaging on an understanding of the content and its learning outcome.
2. **Summary Generation**: Students will be asked to summarize the topics in the content provided in order to assess their understanding of the topic.
3. **Concept Mapping**: This will enable the educator how well the students learned the fundamental concepts and understood the programming concepts provided in the course content.
4. **Formative Assessment**: For each topic covered in the course content shared with the students, a formative assessment is conducted, and results were analyzed.

Every set of 35 students among three batches participated in the flipped classroom teaching activity. Initially, the educator distributed expected learning outcomes of the subject and conducted the pre-test for each batch separately for identifying whether all undergraduates meet the pre-requisites for the course or not [5]. The following are the student learning outcomes list for programming in C course.

On the completion of the course, the students should be able to:

**SLO1**: Analyze the problem statement.
**SLO2**: Choose the appropriate C programming constructs to solve the problems.
**SLO3**: Demonstrate the advantages and disadvantages of specific techniques to be used.
**SLO4**: Differentiate between the efficient and inefficient way of programming.
**SLO5**: Determine and demonstrate bugs in a program and recognize the needed basic operations.
**SLO6**: Formulate new solutions for programming problems or improve existing code to program effectively.

The student learning outcomes have been assessed and evaluated, and the results are shown in Table 1.

From Table 1, it is observed that the success rate of the students is high and pass percentage in the subject is around 92%, and many students got more than 75% in the assessment conducted. Also, it is found that 65% of the students secured more than 60% marks.

A student survey has been conducted to get their perceptions on the experience of the flipped classroom for the topics described above. The questionnaire mainly consisted of understanding the concepts, classroom management of the educator, student engagement, and problem-solving skills gained. For conducting this survey, a rubric has been prepared and circulated among the students. After conducting the survey, the students

**Table 1.** Result analysis using innovative methods for SLOs

| Student learning outcome | No. of students with % marks | | | | | |
|---|---|---|---|---|---|---|
| | <40 | 40–50 | 60–70 | 70–80 | 80–90 | >90 |
| SLO 1 | 3 | 17 | 22 | 10 | 8 | 3 |
| SLO 2 | 2 | 15 | 24 | 14 | 6 | 2 |
| SLO 3 | 2 | 13 | 25 | 12 | 7 | 4 |
| SLO 4 | 1 | 12 | 29 | 12 | 7 | 2 |
| SLO 5 | 6 | 11 | 21 | 14 | 8 | 3 |
| SLO 6 | 1 | 12 | 25 | 12 | 9 | 4 |

have been asked to compare the advantages of the traditional teaching approach, i.e., chalk and talk and flipped classroom teaching approach. It is found that around 76% of the students are in the opinion to continue for the flipped classroom approach as compared to the traditional teaching methods. Also, this approach helped in identifying the slow learners (<60% in the pre-test) and fast learners before the commencement of the course delivery, and appropriate content followed by its assessment is carried out for the slow learners (Fig. 1).

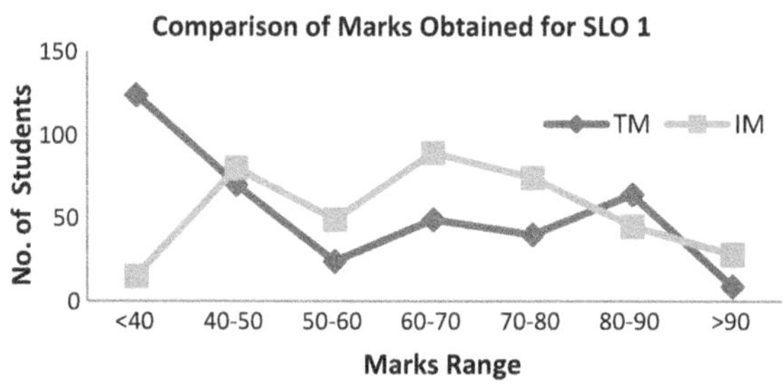


**Fig. 1.** Comparison of grades between innovative methods (IM) and traditional methods (TM)

Pre-test survey conducted for the undergraduates indicated that students are not having the awareness related to the flipped classroom module. But still, they believed that it will be one of the improvements that can be done to the teaching–learning process. It is observed from the feedback of the students during the classroom that more than 95% of the students watched the videos and followed the power-point presentations without any interruption due to the availability of the internet facilities. More than 70% of the students understood the concepts without any help and hence they were able to discuss it in the classroom.

Similarly, attainments for other SLOs were compared using the same basis as that of SLO1. As inferred from Fig. 2, each SLO has shown improvement in their attainment using innovative methodologies as compared to traditional teaching methodologies.

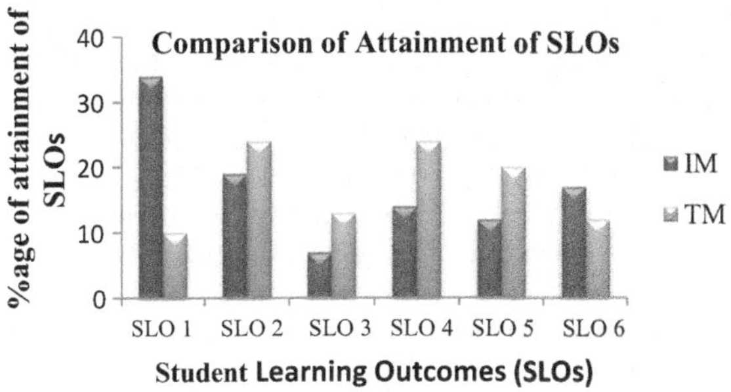


**Fig. 2.** Comparison of Attainment of SLOs

Comparisons were made based on the formative test scores conducted both for flipped module and traditional teaching technique (previous batch) proved that the students enrolled with the flipped module have a better score (class average percentage). The positive finding is attributed that the students have additional opportunities to re-visit the concepts during the classroom interaction with the educator.

## 3 Concluding Remarks

The present case study suggested that the flipped module is beneficial to the students for the selected specifically practical subjects and the topics. Even though the educator is not proficient in implementing the flipped module teaching, the student survey suggested that more than 70% of the students agreed for the classroom management, better understanding and better student engagement including enhancements in problem-solving skills along with critical thinking. Most of the students are interactive in the messy point markup and summary generation for the topic. The educator needs to engage more classes for enabling them to understand the concepts and problem-solving techniques for the selected topics. Overall the students were of the opinion that they could devote more time for learning for the topics given when compared to classroom engagement. The educator purposefully implemented the flipped classroom module for programming in C course for the undergraduates to continue the technology-based learning in HEI experienced by them during their $K - 12$.

## References

1. Prince, M., Felder, R.M.: Inductive teaching and learning methods; definitions, comparisons and research basis. J. Eng. Educ. **95**, 123–138 (2006)
2. Felder, R.M., Brent, R., Prince, M.: Effective instructional development strategies. J. Eng. Educ. **100**, 89–122 (2011)
3. Hotle, S.L., Garrow, L.A.: Effects of the traditional and flipped class rooms on undergraduate student opinions and success. J. Prof. Issues Eng. Educ. Pract. **142**, 1–10 (2016)
4. Li, Y., Daher, T.: Integrating innovative classroom activities with flipped teaching in a water resources engineering class. J. Prof. Issues Eng. Educ. Pract. **143**, 1–6 (2017)

5. Rathore, V.S, Rajawat K, Maheshwari S. (2018) Impact of open-ended project-based collaborative learning model for teaching undergraduate-level database management systems. In: Third International Congress on Information and Communication Technology, ICICT 2018, London, Springer AISC, vol. 797. Springer, Singapore. https://doi.org/10.1007/978-981-13-1165-9; Softcover ISBN—978-981-13-1164-2; eBook ISBN—978-981-13-1165-9
6. Mason, G.S., Shuman, T.R., Cook, K.E.: Comparing the effectiveness of an inverted class room to a traditional class room in an upper division engineering course. IEEE Trans. Educ. **56**, 430–435 (2013)
7. Francis, R., Shannon, S.J.: Engaging with blended learning to improve students' learning outcomes. Eur. J. Eng. Educ. **38**, 359–369 (2013)

# Development of Remote Monitoring and Control System for MMSU i4.0 Platform: Energy Self-sufficient Small-Scale Smart Laboratory Using MQTT Protocol

Vladimir P. Ibañez(✉), Willen Mark D. Manzanas, Thomas D. Ubiña, and Shirley C. Agrupis

National Bioenergy Research and Innovation Center, Mariano Marcos State University, City of Batac, Ilocos Norte, Philippines
ibanezvp@gmail.com, wmanzanas@gmail.com, tndconcepts@yahoo.com, shirleyagrupis@yahoo.com

**Abstract.** Automation is one of the inventions of this generation, and it is widely used not only here in the Philippines but all over the world. Amid all these technological advancements and innovations, electrical energy plays a vital role in our lives. It powers our devices, machinery, equipment, modes of transportation, and many other things that we need and use. To date, solar energy is one of the most abundant renewable energy sources that have been gaining attention in the past few years. This study generally aims to develop a remote monitoring and control system for the Mariano Marcos State University Industry 4.0 (MMSU i4.0) Platform, particularly for an energy self-sufficient small smart laboratory. By using solar energy, the implementation of the Maximum Power Point Tracking (MPPT) technique needed to supply a small-scale smart laboratory integrating the use of the Internet of Things (IoT) in controlling and monitoring devices. The system controlled through a web application and mobile application to lessen the human effort in manual switching and monitoring tasks. It can print weekly and monthly power consumption of the smart small-scale laboratory. Since it is automated, users can monitor the actual voltage and current, the temperature of the power source, and the total generated power from renewable energy. It will improve the level of comfort by controlling and managing laboratory devices automatically and also applicable to enhance buildings, houses, and other work areas cleanly, thus helping the environment.

**Keywords:** Renewable energy application · MQTT · MPPT · i4.0 · Sustainable development


X.-S. Yang et al. (eds.), *Proceedings of Fifth International Congress on Information and Communication Technology*, Advances in Intelligent Systems and Computing 1183, https://doi.org/10.1007/978-981-15-5856-6_25

# 1 Introduction

Solar energy is one of the most important renewable energy sources that have been gaining attention in the past few years. Solar energy is abundant; it has the highest availability compared to other energy sources. Solar energy is clean and free of emission since it does not produce pollutants. The investment of installing solar panels in our home is worth it for, in the long run, the cost of consumed electricity from the grid is much more expensive. An MPPT system is known for optimized energy harnessing from renewable energy sources; however, it is not readily available in the locality. Hence, developing one is worth exploring. Also, one recent innovation under solar energy is the usage of the Internet in controlling devices in a particular establishment or buildings and or most commonly known the Internet of Things (IoT). It allows the user to control potentially any device with their laptop, cellular phone, or tablets as long as it has an internet connection. While there are many ways to communicate with machines, Message Queuing Telemetry Transport (MQTT) is one of the most reliable protocols used in communication to connect monitoring devices and send data to a remote server. As these innovations are increased, consumption of energy in various buildings like schools is growing drastically due to the excessive use of devices. Monitoring and controlling the usage of tools help to control energy usage, prevent energy waste, and save cost to benefit the school and the world. With the help of this, it could control and monitor the lighting system, energy usage of the outlets in buildings even they are not near the vicinity, giving them an easier and less human effort to monitor using mobile phones, laptops, tablets, and computers.

IoT-based laboratory automation powered by a solar panel is not implemented yet in the Ilocos region, which in fact, this system could help them save energy, cut down energy costs, enhance security, and improve the standard of living.

This study generally aims to develop a remote monitoring and control system for the Mariano Marcos State University Industry 4.0 (MMSU i4.0) Platform, particularly for an energy self-sufficient small smart laboratory. Specifically, it aims to: (a) design, fabricate, and test an MPTT system for the self-sufficient energy laboratory; (b) develop a mobile and web applications for remote monitoring and control; and (c) evaluate the success rate performance of the developed remote and monitoring system

It will improve the level of comfort by controlling and managing laboratory devices automatically and also applicable to enhance buildings, houses, and other work areas in a clean way, thus helping the environment. The research output would benefit the school by helping to lessen the expenses on electricity bills.

# 2 Methodology

## 2.1 Development Procedure

This study adopted the developmental research design enriched by a descriptive approach. Some aspects and contents need to be defined in detail; hence, an illustrative method was also used in the study. The study observed the development cycle in Fig. 1. The development starts with the feature conceptualization, then hardware and software design based on the desired features of the whole system. The hardware is fabricated at the same time the program was evaluated before integration. Then testing the system and the hardware before the final assembly. Several iterations were done until the results of the testing are acceptable.

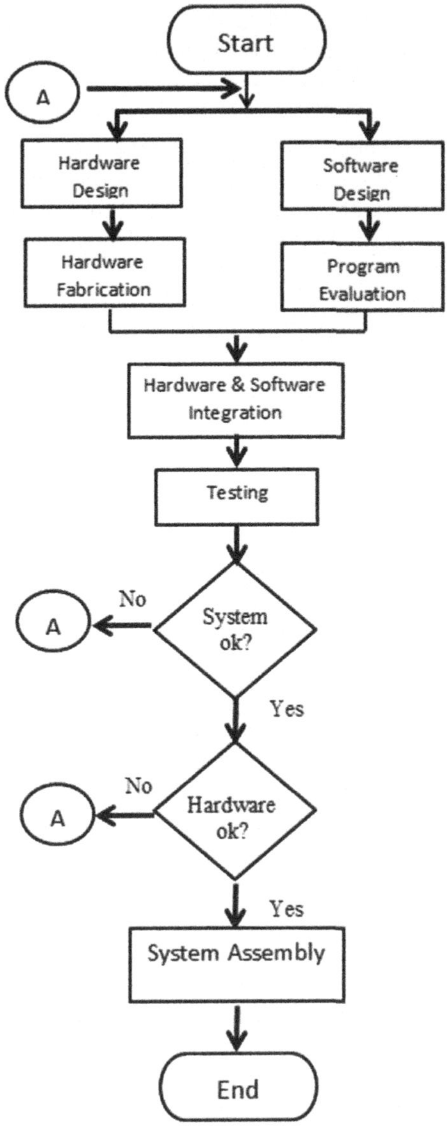


**Fig. 1.** System development flowchart

### 2.2 Testing and Data Gathering Procedure

For the MPPT system, the actual implementation was done to determine its effectiveness in harnessing the maximum possible energy. The system was installed in a 500 Wp solar power system to supply a small smart laboratory to make it energy self-sufficient. Monitoring all the input and output (voltage and current) from 7 am to 5 pm by the two identical clamp meters, multi-testers, and cameras. An advantage of using the camera as a recording device for the data gathering is the inclusion of the actual time and date in the recording.

The developed android application and web application were expected to control lights, outlets, cameras, view power consumption and can leave an important message to the user. The response of the electrical devices was recorded to the broker. The success rate was calculated by dividing the number of successful attempts ($N$) by the total number of trials ($T$), multiplied by 100%, as shown below:

$$\text{Success Rate} = (N/T) * 100; \tag{1}$$

where: $N$ = Number of successful attempts; $T$ = Total number of trials.

## 3 Results and Discussion

The power graph harnessed by the MPPT system shown in Fig. 2 was gathered on April 28, 2019, during a cloudy day. The vertical value is the power measured in Watts (W), and the horizontal is the time from 7:00 am to 5:00 pm. The total energy harnessed by the system is about 1.8 kWh or 72% energy conversion during a cloudy day from the 500 Wp solar panel.

The system architecture of the developed remote monitoring and control system is shown in Fig. 3. It consists of solar panels, Maximum Power Point Tracking (MPPT) Charge Controller Prototype, batteries, off-grid inverter, DC circuit breaker, AC circuit breaker, MQTT systems, and other electrical laboratory equipment; Watt-hour meter, Manual Switches, Raspberry Pi for the repository of data (local and cloud), smartphone and laptop for controls, and the laboratory devices or loads.

A PV System produces a DC voltage that is fed to the system together with the sensors that measure the input and output voltage and current. The LCD will display the power produced by the solar panel from the charge controller. The controllers take specific actions based on the measured voltage and current received and will do particular work based on the power received. A buck converter serves the MPPT as a power controller, and it is a specific type of switching regulator that steps down the input voltage to a lower level output and regulates the energy to charge the battery.

The power from the battery and the MPPT charge controller goes to the off-grid inverter, which converts direct current (DC) to alternating current (AC), fed into the Small-Scale Smart Lab 220 V electrical system. In case of insufficient supply from renewable energy, the utility grid and the battery will serve as a backup.

The android and web application Graphical User Interface (GUI) shown in Fig. 4 contains buttons to turn on and off all devices, "LIGHTS" to view added lights, "OUTLET" to see all outlets, a recent activity container that shows the current controlled device

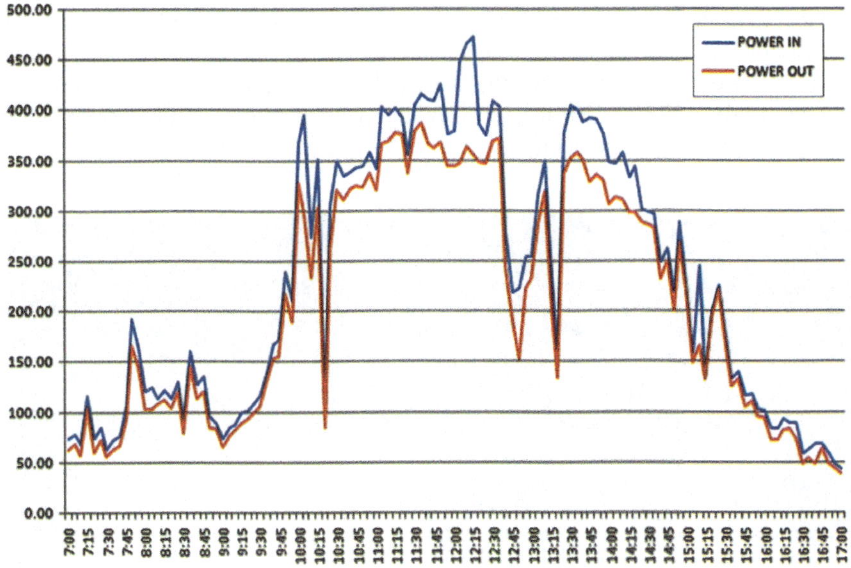


**Fig. 2.** MPTT system power harnessing graph

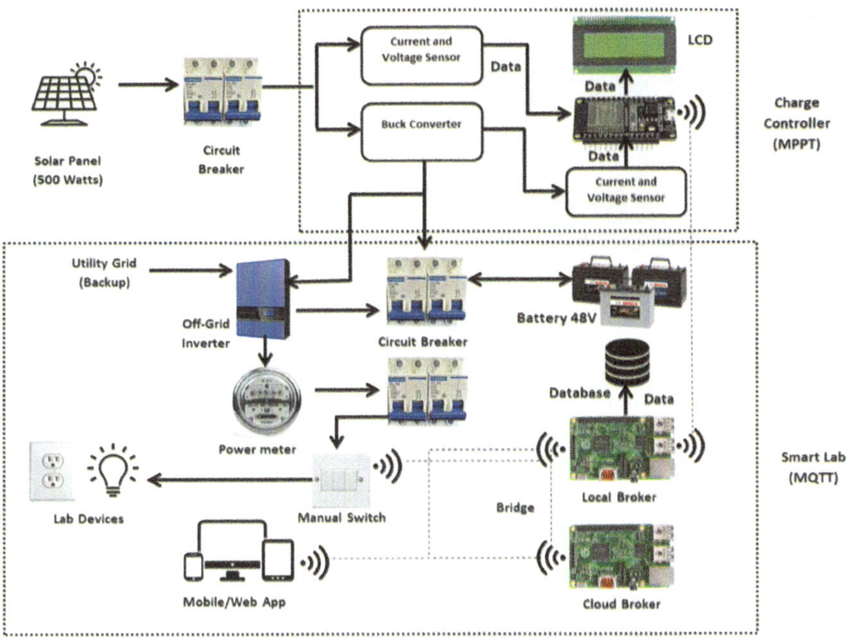


**Fig. 3.** System architecture

and message, a connection status is placed in the upper left corner of the application in which you can monitor your connectivity with the broker, on the other side of the top panel the nickname of the user is displayed. A user settings icon, and a navigation panel at the bottom, "CHAT" to move to the chat page same with "METER" that redirects you to the Power Consumption Graph, "MPPT" for the graphs and readings of the solar panel and lastly is the "CAMERA" to view the page of camera. Sliding right and left moves you from on-page to another.

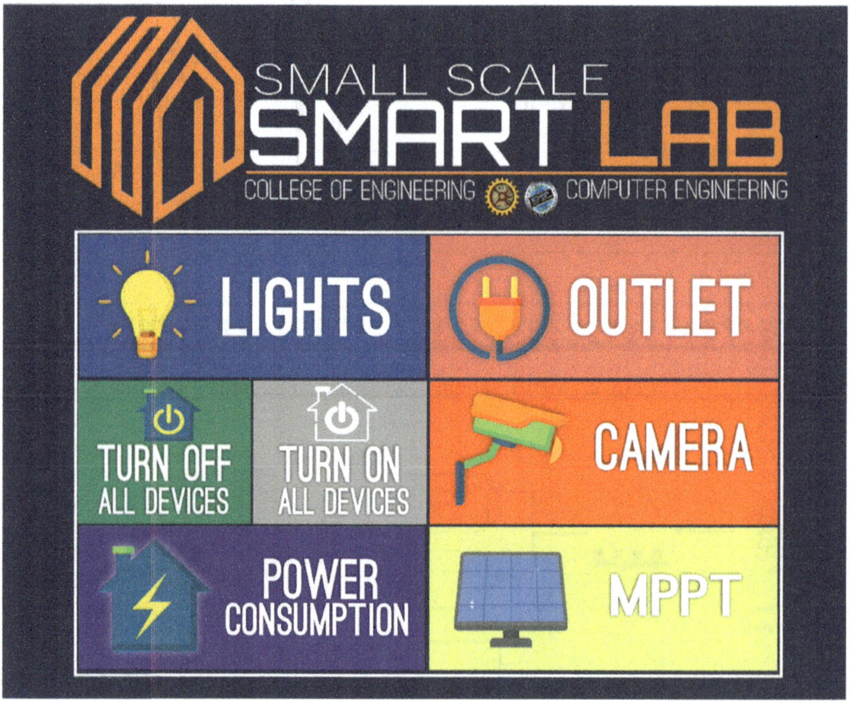


**Fig. 4.** Android and web application controller

The power meter module shows the graph of the power consumption and the small-scale smart lab and its previous use daily. Also, users can view the power consumption table and download the excel file for reporting. The MPPT Monitor module can display the current temperature, voltage, and current of the MPPT side, and the user can also retrieve old readings. The camera module can display the surveillance camera in the small-scale smart lab, and at the same time, the user can turn on and off the outlet where the camera is plugged in.

Table 1 shows that the developed remote monitoring and control system can control connected lights and outlets remotely. Monitoring the activities in the smart laboratory was achieved as well as monitoring of energy harnessed and consumed in a remote location.

**Table 1.** Summary of test result on the features of the develop remote monitoring and control system

| Items | Device | Test step | Success rate (%) |
|---|---|---|---|
| 1 | Lights | On/off | 100 |
| 2 | Outlets | On | 100 |
| 3 | Camera | On | 100 |
| 4 | Power consumption monitoring | Read | 100 |
| 5 | Messaging | Send/receive | 100 |

## 4 Summary of Findings

The MPP varies, based on the iridium and cell temperature, as well as the weather condition. Hence, appropriate algorithms were utilized to track the MPP. However, an Adoptive Perturb and Observe method is used in this study because it has faster dynamics and improved stability compared to traditional Perturb and Observe. The said algorithm was setup employing numerical simulations and experimental tests, confirming the effectiveness of the method. With this method, the energy harnessed during a cloudy day is 1.8 kWh from 500 Wp solar panel.

The mobile application and web application developed to monitor and control the lighting system, outlets, devices installed in the Smart Laboratory, and even determine actual power consumption. The software application lessens the human effort in manual switching and monitoring tasks. It can print weekly and monthly power consumption of the small-scale laboratory. Since it is solar-powered, users can monitor the actual voltage, the temperature of the power source, and current of the small scale laboratory. The researchers have conducted manual testing, functionality testing, white box testing, black-box testing, and acceptability testing. All the testing results showed a favorable implication to overall performance and acceptability to the user with a 100% success rate.

## 5 Conclusion

The design and fabrication of an MPPT system harnessed 1.8 kWh from 500 Wp solar panel on cloudy weather. The MPPT system is about 72% energy conversion and is sufficient to supply the load requirement of a small smart laboratory. The remote monitoring and control system is useful in controlling connected lights and outlets in a remote location. Monitoring of the energy harnessed and consumed, as well as the activities inside the laboratory, was easily achieved. Generally, test results showed a 100% success rate in all of the desired features of the system.

## References


1. Amanse, J., Bermejo, C.V., Imperial, C.P.: Work-Flow Based Platform for a Home Automation System. Ateneo de Manila University (2017)
2. Alberto, K.M., Cortez, C.D., Santos, J.L., et al. Development of multi-home alarm system based on GSM technology. Int. J. Electron. Electr. Eng. **4** (2016)
3. Alimed, E.S.A., Mohammed, Z.K.A.: Internet of things Applications, Challenges and Related Future Technologies, Sudan, Africa (2017
4. Gupta, N., Kaduskar, V.P., et al.: IOT Based Lab Automation System. Pune, India (2017)
5. Int. J. Adv. Res. Comput. Commun. Eng. ISO 3297:2007 Certified **6**(3) (2017)
6. Iyer, K.R., Jaikar, S.P.: A survey of messaging protocols for IOT systems. Int. J. Adv. Manag. Amity University Mumbai
7. Liu, Z., Lu, T., Yang, B.: Laboratory instruments management system based on IOT. Jiujiang University, Jiujiang, Jiangxi, China (2015)

# Intelligent Search for Strategies to Minimize the Risks of Internet Communication of Teens and Youth

Elena Brodovskaya, Tatyana Vladimirova, Anna Dombrovskaya(✉), Natalya Leskonog, Alexander Ognev, Lyubov Shalamova, and Yulia Shchegortsova

Moscow Pedagogical State University, 119991 Moscow, Russia
brodovskaya@inbx.ru, {tn.vladimirova,yum.shchegortsova}@mpgu.su, an-doc@yandex.ru, leskonog@mail.ru, altognev@mail.ru, fsh57@mail.ru

**Abstract.** This article raises issues of children and young people communication safety in the Internet space. The materials of the article contain the results of an intellectual search conducted by the authors to compile a list of the most common risks of Internet communication faced by Russian children, or teachers and counselors working with children, including using the Internet, note the growth of these risks. Research design combines the following methods: social media analytics (provides content, dynamic, structural and discourse characteristics of relevant social media streams, including automatical metrics: tag cloud, audience activity markers, online opinion leaders, and so on) and eye-tracking for measuring perception of Internet memes by teens and youth. The article presents descriptions of markers that allow to identify these risks, possibilities of neurovisual correction of these risks, as well as makes recommendations for teachers and counselors, schools and universities, departments and agencies for social work with teens and youth to minimize them.

**Keywords:** Internet communications · Online risks · Internet communication risks · Data mining · Social media analytics · Digital markers · Neurovisual techniques to counter destructive network communities · Teens · Youth

## 1 Introduction

In the modern world, the concept of the young man social life environment which creates sets of the conditions attending his daily activity, forms certain opportunities for expansion of communication processes, search of various information, leisure filling by various forms of self-realization, and significantly extended [3]. First of all, Internet becomes an integral part of this environment. With its help the new communities constructed on the principles of open communications, virtual images of personal perception are formed.


X.-S. Yang et al. (eds.), *Proceedings of Fifth International Congress on Information and Communication Technology*, Advances in Intelligent Systems and Computing 1183, https://doi.org/10.1007/978-981-15-5856-6_26

At the same time, as M. Castels noted "through communication, human consciousness interacts with its social and natural environment. This process of communication takes place in accordance with the structure, culture, organization and technology of communication in the society" [2].

Since modern gadgets, which accompany a child since his/her birth, become an integral and prestigious element of social life, children perceive them as necessary means of life support.

In addition, the Internet provides fast access to large amounts of information, instant speed of its distribution. It reveals almost unlimited opportunities to the user: which makes it extremely attractive to children and young people.

Competent strategies of Internet behavior are formed as a person gains life experience, there are ideas about the ways of behavior in the particular situation, which children still have minimal. Russian children and teenagers involve unlimited and uncontrolled use of Internet content by children, which threatens their personal security [1, 7, 8].

The problem of Internet communication risks impact on the children and young people socialization begins to draw the researchers' attention. It should be recognized that researchers all around the world slightly earlier than in Russia began to study it.

The impact of online risks on children, protection from them and prevention is considered by Brown [4]. The research of European scientists in the field of Internet safety of young media users within the framework of the pan-European project "EU Kids Online" and the national project "British Children on the Internet" (UK Children Go Online) under Professor Livingstone [6]. For example, the study found that users with higher levels of digital competence face more online risks than their less convergent peers.

Important for this work are Milosevic's research [8] on studying the actions of social networks, including Facebook, Twitter, Snapchat and Instagram, to overcome the problem of cyber intimidation by young users. Very notable is the study of Frumkin [5], who studied cyberbullying using experimental methods, where the level of predisposition to the role of the cyberbullying victim served as a dependent variable.

As a part of the personal-oriented approach, the authors of the article initiated a study that allowed to formulate a list of risks faced by schoolchildren, with which children turn to teachers and counselors for help, or teachers and counselors themselves note these risks, being involved in Internet communication with students. The study was conducted by the Federal Coordination Center for Training and Support of Senior Personnel (hereinafter FCC), working on the basis of the Moscow State Pedagogical University during 2018–2019 and showed the existence of a whole list of topical threats, independent of the Russian Federation region where children, teachers, counselors live. Among the most common risks on which we focused the continuation of the study noted: cyberbullying; pushing for suicidal behavior; distribution of extremist online content, terrorist content and involvement in extremist, terrorist activities (school shooting).

## 2 Methodology

The methodology of social media flows analysis presenting digital markers of Internet risks data for schoolchildren involved implementation of the following research procedures algorithm: justification of groups selection in social media as objects of message content qualitative analysis (20 online communities have been selected, and 1800 documents have been analyzed); developing of the digital markers dictionary for automated information flows accumulation; automated upload of Internet content using created marker dictionaries and online service for monitoring social media (software) IQBuzz, a total of 953,000 messages have been uploaded (together for all types of streams about Internet risks), depth of accumulation 2016–2018; the interpretation was performed according to the criteria specified by the program: weight of information flows; dynamics of information flows; opinion leaders' publication activity and its audience.

## 3 Results

### 3.1 Cyberbullying/Harassment on the Internet

Cyberbullying is understood by the authors as a systematic and targeted negative online impact on the social media users for the purpose of causing psychological trauma and/or introducing depression.

Total number of social media documents accumulated according to these digital markers—280,000, source of unloading—blog hosting "VKontakte," depth of unloading—period from 1.01.2017 to 10 October 2018.

The dynamic characteristics of the resulting arrays of relevant documents are shown in Fig. 1. According to these data, a remarkable intensification of the analyzed flow occurred at the end of February—the beginning of March 2018, and the maximum extremum is at the end of September 2018. These peaks have substantial intersections with offline events concerning cyberbullying. At the beginning of this spring, there was an active discussion in traditional and new media of the intensive growth of network groups with a destructive focus: pushing to harm health and life, immoral and criminal social actions. It was during this period that the number of reports attempting to make sense of the factors, causes, triggers of cyberbullying, and ways to prevent and counter this type of Internet risk increased dramatically. At the end of September 2018, the surge in reports of cyberbullying could be explained by the indirect impact of the Kerch tragedy, which significantly affected social media content, including information flows about cyberbullying.

On the one hand, the digital environment becomes a space of harassment, moral damage, and mental damage, while the social media space provides opportunities to express the reaction of cyberbullying victims, as well as to find like-minded and sympathizers.

There is a complex correlation between the emergence of depression in adolescents and young people, threatening suicide attempts and involvement in communication through social media. This connection is expressed in the fact that depressive states did not end in suicidal actions most often in those members of the younger generation who had the opportunity to be constantly online, and, on the contrary, removal from communication in the network as a whole more often caused attitudes for suicide.

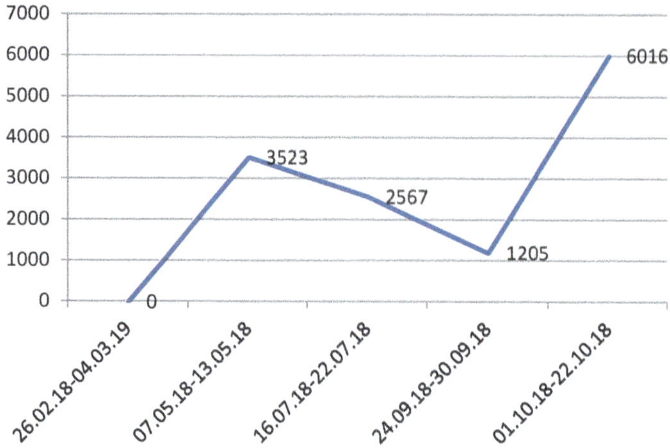


**Fig. 1.** Dynamics of information flows on cyberbullying in social media in the period from 26.02.2018 to 22.10.2018

### 3.2 Pushing for Suicidal Behavior

Pushing for suicidal behavior in a network environment as a research case was interpreted as a targeted psychological influence on the user in order to form a behavioral setting on committing suicide.

By means of the listed digital markers 102 802 messages of their blog hosting "VKontakte" from March 1, 2016 till November 1, 2018 were accumulated.

Analysis of the names of influential blogs and online groups reflecting conversations about suicide shows that the most numerically voluminous audience has a block of priest Vladislav (see Table 1) providing spiritual assistance in difficult life situations. The remaining groups have from hundred to four hundred among their user audience. Group names are entirely based on the names of their creators and do not carry a serious semantic load. These groups are not so-called death groups that stimulate and encourage adolescents to commit suicide. Real "death groups"—closed communities, accounting and unloading of messages of which online service for monitoring social media does not carry out.

The socio-demographic analysis of the suicide report's authors indicates the prevalence of girls among them, users aged 15–17 and most often residents of large cities.

The dynamics of the social and media flow about suicide, presented in Fig. 2, shows that this flow is unstable, has many peaks of different intensity, and falls to virtually zero. Most of the relevant messages are recorded in the blog hosting "VKontakte." The peak in March 2016 is due to a case that took control of the Civic Commission on Human Rights: "On March 23, 2016, the Civil Commission on Human Rights was approached by Andrei Vershinov from the city of Arzamas, who reported that on March 10, 2016 his aunt Elena Vershinova (both names changed) committed suicide allegedly after having taken an antidepressant on the appointment of a psychiatrist" (http://www.cchr.ru/news/

**Table 1.** Publication activity of network leaders forming suicide attitudes

| Name | Blog hosting | Audience |
|---|---|---|
| Alexander Konstantinopolsky (id399034192) | VKontakte | 483 |
| Danil Tukabaev (id391676907) | VKontakte | 112 |
| Nasibat (nasibat_maromedova) | Instagram | 259 |
| Zinaida Pershina (id1444926) | VKontakte | 347 |
| Vitaly Ivanov (id201457128) | VKontakte | 81 |
| Emergency spiritual help! (priest_vladislav) | Instagram | 19,189 |
| Total | | 20,481 |

464.htm). Although the woman was not in suicidal groups, a surge in social media activity triggered this due to the active circulation of reports of this suicide online.

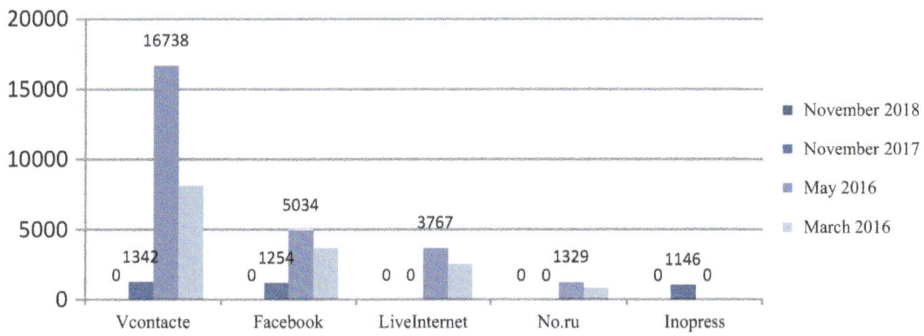


**Fig. 2.** Dynamics of social and media reports of suicides between March 2016 and November 2018, in thousands

Taking into consideration the fact that messages from closed "death groups" were not included in the unloading, and the statistics presented in Fig. 2 (measured in thousands of documents) reflect only an open online discourse about suicide, it is obvious that there is a very wide interest in the topic of stimulating suicidal behavior in social media.

### 3.3 Distribution of Online Content of Extremist and Terrorist Content, Involvement in Extremist, Terrorist Activities (School Shooting)

This type of Internet risk is interpreted by the team of authors from the point of view of mobilizing users through online propaganda of radical, extremist attitudes: readiness to participate in actions of civil disobedience, terrorist actions. An integral part of the risk type is school shooting, which can be reduced to motivation in the network space for the organization of terrorist acts in educational institutions. School shooting served as a specific case for exploring the process of distributing online content of extremist and terrorist content.

570,000 messages were unloaded in the Russian-language and English-language segments of social media: “Facebook,” “VKontakte,” “YouTube,” “Odnoklassniki,” “Twitter,” “Instagram” for the period from January 8, 2018 to October 22, 2018.

Figure 3 shows a surge in the intensity of flows about school shooting in social media in connection with the tragic accident in Kerch in September 2018. An alarming signal is the fact that the main contexts analysis of this stream messages indicates attempts to glorify the teenager who carried out this terrorist act and romanticize his motives.

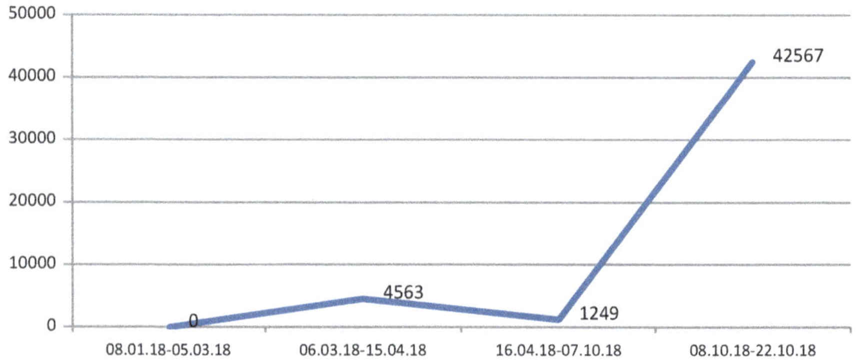


**Fig. 3.** Dynamics of social and media flow reflecting digital traces of school shooting between January 2018 and October 2018 in blog hosting and microbligs: “Facebook,” “VKontakte,” “YouTube,” “Odnoklassniki,” “Twitter,” “Instagram”

Analyzing the publication activity of the most influential network groups associated with school shooting (see Table 2), note that the largest user audience has a community “Natural selection,” and it features 20,027 participants. It is a group that designs and distributes ideas of necessity of natural selection, rejection of values of humanity and mercy, realization of the principle “Survives the strongest.”

**Table 2.** - Publication activity of network leaders forming school shooting installations

| Name | Blog hosting | Subscribers |
|---|---|---|
| Natural selection | VKontakte | 20,027 |
| Word of serial killers | VKontakte | 281,718 |
| Word of criminals | VKontakte | 45,579 |

“The World of Maniacs and Serial Killers” group generally semantically and discursively repeats the contours of ideas circulated in the “Natural selection” community, spreading the values of mass murder in order to uphold their principles and achieve their goals. Participants of this network group deny, subvert the importance of morals and morality in human society. This group has a very wide audience (281,718).

In another rather representative online community—"World of Crimes"—there is a discussion of various ways of killing people, options for committing crimes. Despite the absence of explicit calls for school shooting or other crimes in this group, the very fact of active communication on these topics can serve as a trigger for the formation of the user's position on the organization of socially dangerous actions.

Analysis of the tag cloud typical for the school shooting flow revealed several semantic word usage groups associated with a particular discussion context. The first group of words: "Crimea 'college' placel" indicates the intense circulation in social media of reports about the Kerch episode of school shooting, which literally blew up the network environment in late September—early October 2018. The second group of tags: "murder 'life' childl" concerns an indication of the social essence of school shooting. And the third group of the most common in the analyzed stream of word usage: "weapons 'explosion' murderl" concerns the technological aspect of the organization of school shooting.

## 4 Conclusions and Discussion

Summarizing the results of the social media analysis, it should be noted that the wide scale of distribution of flows, which reflect markers of Internet risks of the younger generation in the global network, as well as their coverage of a huge adolescent user audience, has been revealed. Common to all the streams studied was the extremum of graphs of their dynamics converging at a point dated to the end of the first decade of October 2018 and related to the tragic accident in Kerch. For the flow of school shooting, this event served as a trigger for a sharp increase in the number of reports on the topic of mass killings, including the reason for romanticization and glorification of the teenager who carried out the attack. "Suicidal Stream" was also stimulated by this tragedy, because it served for network leaders, who form in their user audience installations for suicide, some "model" of termination of life. The cyberbullying stream also experienced the impact of the Kerch attack, as the opinion centers of the stream were aimed at exploiting the idea of teenager harassment as the main reason for his mass murder.

Significant measures to prevent the involvement of adolescents in online school shooting communities are their belief in the need to have an opinion, to critically sense, to reflex from various sides the information coming from the network; discuss what consciousness manipulation technologies are used in online groups mobilizing extremism and terrorism, and the benefits that network leaders derive from their criminal activity.

It is obvious that in conditions of internment risks it becomes extremely important to manage effectively strategies of child behavior by teachers, parents, and other participants of educational environment, to build in children resistance to emerging internment threats.

The criteria for selecting the relevant content may be its entertainment (what can shift attention), logic (what meets common sense), and normative acceptability (what meets the rules of the reference group). Opposition to such content should consist in encouraging alternative creative actions through the formation and awakening of positive meanings, the popularization of acceptable standards of behavior and role models.

The effectiveness of countering destructive network communities in all three areas mentioned above can be facilitated by neurovisual techniques such as the use of well-known warning symbols, the management of discourse by changing attention concentrator objects, the targeted use of verbal and graphic metaphors. Thus, our eye-tracking studies with the help of eye trackers show that additional mobilization in the perception of graphic images can be provided by the use of familiar symbol-warnings. The appearance of a red line on the computer screen for 7–10 s in combination with the inscription of the type "Beyond the line it is impossible!" significantly reduced the time of respondents to consider high-risk zones and increased the time of view stay in safe zones. For example, when looking at a person on a computer screen standing on the edge of a cliff, on the ledge of a high-rise building, on a rock, the time of sight in places of possible fall after the demonstration of a red line inadvertently decreased on average more than twice (567 respondents participated in experiments). At the same time, the time of sight in safe places of such images increased more than one and a half times.

According to the authors, the effectiveness of large-scale counter-risks of children and young people Internet communication can be increased by training teachers and counselors to digital navigation of teenagers, which includes both the ability to identify risks in the process of Internet communication with students, and to minimize them in the course of educational and out-of-school activities. The continuation of the study may involve the development and testing of a digital navigation model for teachers, the improvement of neurovisual techniques to counter destructive network communities.

**Acknowledgement.** The study carried out with the support of the state task budget of Moscow Pedagogical State University concerning to "All-Russian school counselors."

## References

1. Brodovskaya, E., Dombrovskaya, A., Nechayev, V.: Perception of the freedom borders and security in the Internet space by Russians. Moscow Univ. J. **4**, 38–50 (2015)
2. Castels, M.: Power of Communication. Moscow (2016) [In Rus.]
3. Soldatova, G., Zotova, E., Lebesheva, M., Khepnikov, V.: Internet: opportunities, competencies, security. In: Methodological manual for employees of the general education system. Google, Moscow, 165 p. (2013) [In Rus.]
4. Brown, J.: Online Risk to Children: Impact, Protection and Prevention, 1st edn, p. 244. John Wiley & Sons Publ, Ltd (2017)
5. Görzig, A., Frumkin, L.: Cyberbullying experiences on-the-go: When social media can become distressing. Cyberpsychol. J. Psychosoc. Res. Cybersp. 7(1), (2013). Last access at http://dx.doi.org/10.5817/CP2013-1-4. 20 Oct 2019
6. Livingstone, S.: Online risk, harm and vulnerability: reflections on the evidence base for child internet safety policy. ZER: J. Commun. Stud. **35**, 13–28 (2013)
7. Kowalski, R., Limber, S.: Agatston Cyberbullying: Bullying in the Digital Age, 2nd edn, p. 294. Wiley, New York (2012)
8. Milosevic, T.: Protecting children online?: Cyber bullying policies of social media companies (Information Society Series) Hardcover—February 2. The MIT Press, 296 p (2018)

# Sec-IoT: A Framework for Secured Decentralised IoT Using Blockchain-Based Technology

Muhidul Islam Khan[1(✉)] and Isah A. Lawal[2]

[1] Tallinn University of Technology, Ehitajate tee 5, 19086 Tallinn, Estonia
mdkhan@ttu.ee
[2] Noroff University College, Kristiansand, Norway
Isah.Lawal@noroff.no

**Abstract.** Blockchain technology has been used recently as a secure method for authenticating digital information in many applications. Inspired by the success of the technology, we envision the potential of the blockchain for secured communication in a decentralised Internet of things (IoT). In this paper, we envisage a framework for a secured IoT and describe the infrastructure and mechanism of the entire system. Also, we provide solutions to overcome some of the limitations of blockchain technology including miner selection and reaching consensus, for a decentralised IoT by incorporating a learning-to-rank method for node selection. We also contemplate using hybrid consensus algorithm in the blockchain to detect faulty node and to improve the node convergence.

**Keywords:** Decentralised IoT · Blockchain technology · Learning-to-rank

## 1 Introduction

Internet of things (IoT) is a network of devices connected via the Internet. Several applications have been deployed over IoT space leveraging the ubiquitousness of sensors, actuators, radio-frequency identification (RFID) and communication protocols [14]. The classical centralised communication method commonly used for IoT is problematic for secured IoT because it is prone to failure [9]. A decentralised communication method typical of peer-to-peer networks can, however, improve the reliability of IoT. But security is a serious concern for decentralised communication. This is because IoT devices are used in many cases to collect and transmit sensitive data, keeping these data secure and accessible to only authorised entities is vital for many IoT applications. Recently, blockchain technology appeared as a trusted peer-to-peer communication method in many applications


X.-S. Yang et al. (eds.), *Proceedings of Fifth International Congress on Information and Communication Technology*, Advances in Intelligent Systems and Computing 1183, https://doi.org/10.1007/978-981-15-5856-6_27

[3,5,6]. The technology is capable of handling an enormous amount of data in a secured manner using a distributed ledger technique. Data communication in the blockchain is done by creating blocks, where each block represents a single data transaction. Each transaction is verified using a mechanism known as mining, and the devices involved in the process are called the miners [7]. Moreover, blockchain ensures security via cryptography, whereby blocks are encrypted using hashing techniques [8]. In this paper, we adopt the blockchain technology for secured communication in a decentralised IoT. We propose a generalised security framework for IoT with consideration to important functionalities and requirements of the IoT such as an adaptive way to select the miners, encryption standard and consensus algorithm. Specifically, we incorporate a learning-to-rank method for miner selection in the blockchain for the decentralised IoT. We also propose a hybrid consensus algorithm to hasten convergence and to detect compromised miners.

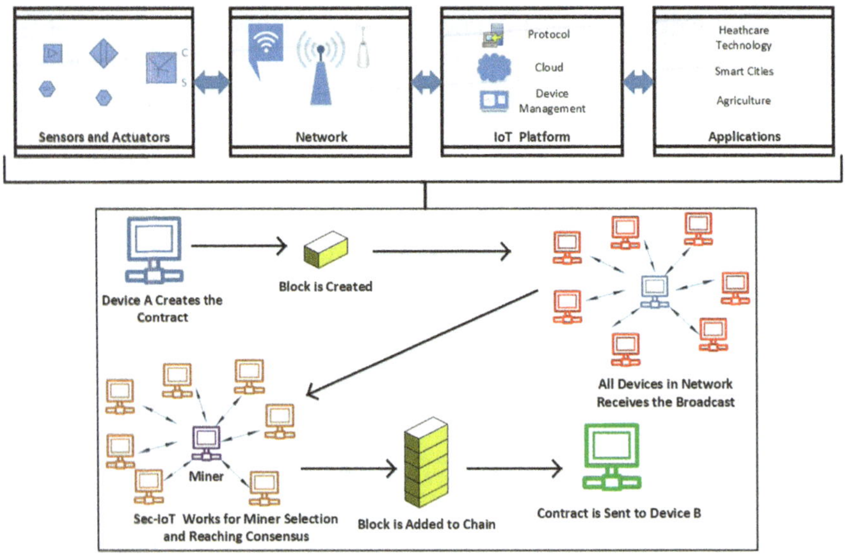


**Fig. 1.** Blockchain-based secured decentralised communication in IoT. **Top:** A generic IoT setup consisting of the following components; sensors/actuators, network, platform, and applications. **Bottom:** The blockchain mechanism applicable to any of the IoT components

## 2 The Sec-IoT Framework

Figure 1 shows the proposed framework. We consider a generic IoT setup with several IoT devices (i.e. sensors, actuators) for any particular application (i.e.

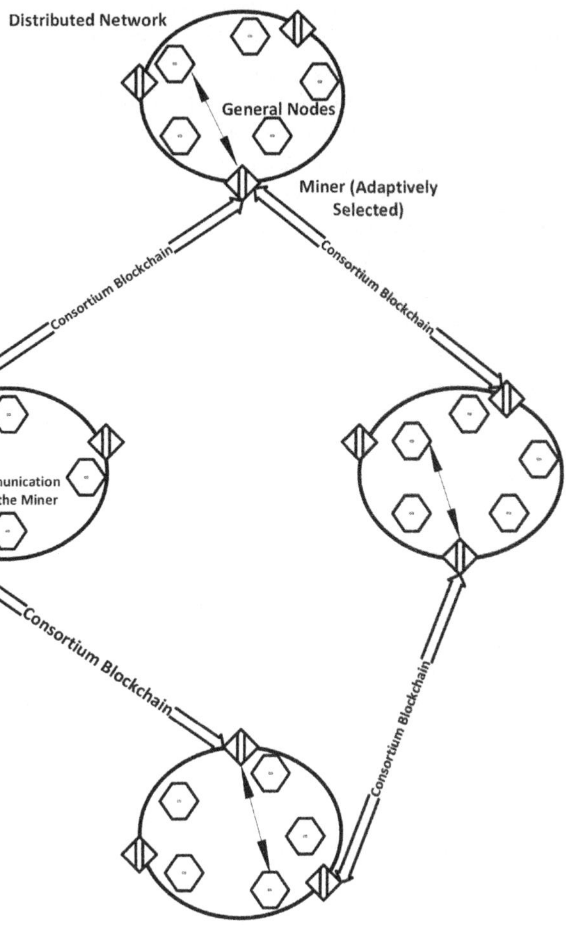


**Fig. 2.** Consortium blockchain mechanism. The circles represent four networks consisting of general nodes (diamond shape) and miners (rectangular shape). The arrows represent communication with the general node and the miners, respectively

healthcare, smart cities, agriculture, etc.) which are connected through a communication technology (i.e. Wi-Fi, Bluetooth, WiMAX, LTE-advanced, etc.) using IoT-based protocol, cloud infrastructure and device management. The devices (also referred to as nodes) are distributed in terms of computing and data processing. The nodes are grouped into different networks with each network comprising of some special nodes, called the miners. These nodes have the most computing power, faster hash rate generation and lower block propagation delay. Communication between two different miners is achieved through the consortium blockchain strategy shown in Fig. 2. A consortium blockchain is a concept that incorporates the efficiency and privacy of private blockchain while leveraging the decentralised governance of public blockchain [4]. To transfer data from one node to another, the sending node must initiate the communication by creating a contract. The contract then undergoes a two-phase verification process. Firstly, a miner is selected from the sending node network. This miner is responsible for generating an eligible signature for the contract and forwarding the signed contract to other miners in the network. Secondly, a consensus process whereby the miners examine the authenticity of the signed contract is executed. If the signature is genuine, a block is created for the signed contract. Figure 3 illustrates the process of reaching consensus among the miners in a network. The newly created block is added to a chain of other blocks and delivered to the receiving node. The miners then record the new blockchain information in a shared record system that is continuously updated, after each block verification process [16]. Because the record system is distributed across all the miners, ensures its authenticity over time. The processing power of miners is very important for the verification of contracts in blockchain. Thus, in the IoT network, miners need to be selected appropriately to ensure adequate processing capability. To improve the block verification process, we introduce a method for selecting the best miner which incorporates a learning-to-rank machine learning method. We also introduce a hybrid consensus strategy for aiding decision-making processing. The following sections discuss the miner selection method and the hybrid consensus technique.

## 2.1 Machine Learning-Based Miner Selection

Miner selection is the process of choosing the best node from a set of nodes in a network, used to generate a legitimate block in the blockchain technology. This process is time-consuming especially for IoT networks with many nodes. To simplify the selection process, we introduce a node ranking procedure using the learning-to-rank (LTR) technique [11]. The LTR method has been used previously in product rating [12]. The algorithm takes as inputs the node's properties such as computing power, hash generation rate and block propagation delay and then finds the optimal sorting of the entire nodes. The node with the most desirable properties is ranked top and can be selected as a miner. The LTR is used to learn a scoring function by mapping the attributes of the nodes to real-valued scores from previously labelled node data. The scoring function is used to sort and rank the nodes as follows:

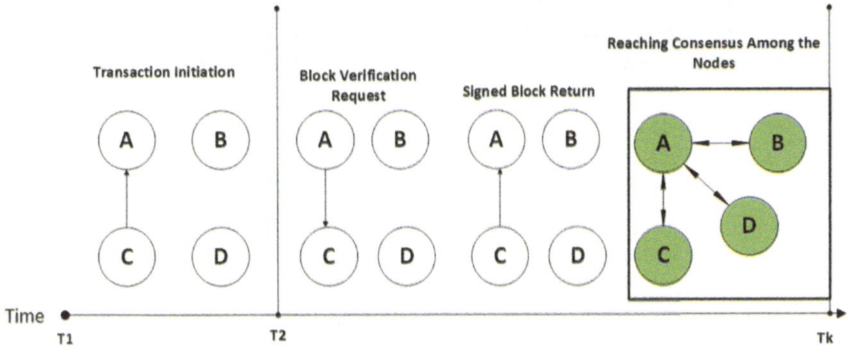


**Fig. 3.** Illustration of the data transfer process using blockchain technology. A sending node $C$ initiates a data transfer request at the time $T1$. At the time $T2$, a selected miner, $A$ examines the transfer credentials, creates a contract and then requests the sending node to sign and return the contract. The miner then broadcasts the signed contract to other miners in the network. At the time $Tk$, all the miners have reached a consensus to accept the signed contract. A block is created for the signed contract and added to the blockchain

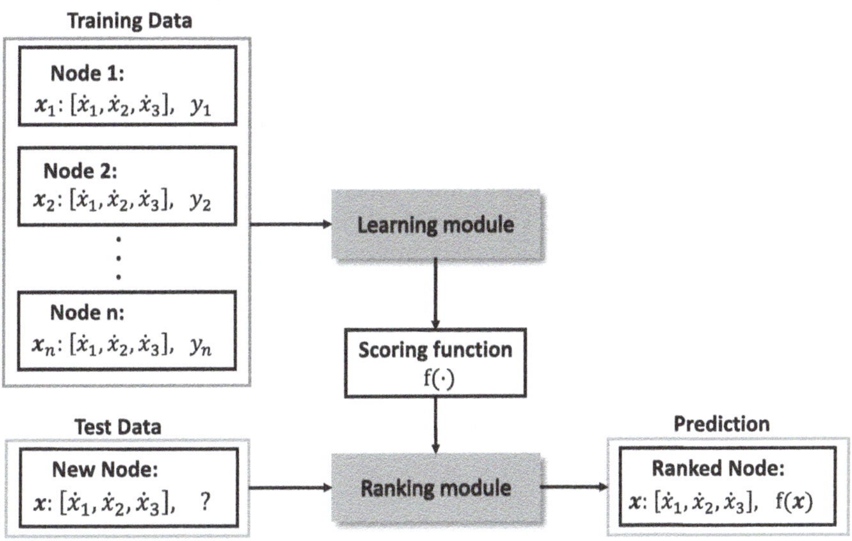


**Fig. 4.** A simple illustration of the LTR ranking procedure. The scoring function can be learned using ListNet [2] or DeepRank [13]. The ranking module is simple sort algorithm

Let $\tau = \{(\boldsymbol{x}_1, y_1), \ldots, (\boldsymbol{x}_n, y_n)\}$, be a set of data describing $n$ nodes in a decentralised IoT network, where $\boldsymbol{x}_i$ is the $i^{th}$ node and $y_i$ is the corresponding real-valued score associated with the $\boldsymbol{x}_i$ used to rank the node in the past. Each $\boldsymbol{x}_i \in \Re^d$ is represented by a set of attributes (feature vector) of dimension $d$. $\boldsymbol{x}_i = [\dot{x}_1, \dot{x}_2, \ldots, \dot{x}_d]$, where $\dot{x}_i$ is the $i^{th}$ attribute of the node, such as computing power. The goal is to exploit $\tau$ to create a scoring function $f(\cdot)$, such that for any given new node $\boldsymbol{x}_j$, it will generate a score $f(\boldsymbol{x}_j)$. The scoring function can be created using publicly available tools such as ListNet [2] and DeepRank [13]. Thus, for any given $j$th IoT network with $n$ nodes, we can obtain a sorted list of scores $s^j = [f(\boldsymbol{x}_1^j), \ldots f(\boldsymbol{x}_n^j)]$ corresponding to the rank of the nodes. Figure 4 illustrates the setup for the learning-to-rank procedure. The creation of the scoring function can be conducted offline and then deploy online for evaluating and ranking the nodes.

### 2.2 Hybrid Consensus Algorithm

In blockchain technology, a consensus algorithm must fulfil the following objectives: gathering all the miners' agreement, collaboration, participation and maintaining equal rights to participating miners. In addition to the aforementioned objectives, it is important to be able to detect faulty miners for IoT networks. For this reason, we introduce a hybrid consensus algorithm consisting of binary and average consensus mechanisms [10]. In binary consensus, every miner in the network can hold two status, zero and one. One being functional and zero being faulty. Before two miners communicate and run the set of rules, they compare their current state and assume a new state if their observation is different [1]. In this way, a rogue miner can be detected easily and removed. With average consensus, two miners can reach an agreement by averaging the values of their initial states [15]. While the binary consensus mechanism helps with the detection of faulty miners, the average consensus mechanism helps to converge the decision of the miners faster. Algorithm 1 illustrates the hybrid consensus algorithm in a decentralised IoT. The algorithm commences with the creation of a new contract. The participating miners proceed with the consensus process by checking the convergence rate and the possibility of a faulty miner. If the convergence rate is smaller than a predefined threshold, then the average consensus mechanism is activated. However, if there exist some faulty miners, which may disrupt the process of reaching consensus, then the binary consensus mechanism is activated to detect and remove the affected miners.

## 3 Discussion

In a decentralised IoT, where there is no central control, it is difficult to provide security. Also, the limitations of different communication standards make the working condition of the decentralised IoT challenging. For example, GSM is power-hungry and expensive and Bluetooth technology has a very limited range. Thus, blockchain technology which offers practical solutions to some of the aforementioned limitations is considered appropriate for implementing decentralised

**Algorithm 1** Hybrid consensus algorithm for decentralised IoT

```
procedure TO REACH CONSENSUS
    Initiate participating miners
    Initiate average and binary consensus mechanism
    Set consensus criteria (e.g. Convergence rate)
    loop:
    Create a new signed contract
    if convergence rate < threshold then
        Invoke the average consensus mechanism
        Broadcast to all participating miners
    if a faulty miner is suspected then
        Invoke the binary consensus mechanism
        Broadcast to all participating miners
    Check if consensus is reached:
    Create a new block for the contract
    Update all the miners' record with the new block information
    Go to loop.
```

IoT. In this paper, we present a framework for secured communication in a decentralised IoT using consortium blockchain, which incorporates machine learning-based miner selection techniques and hybrid consensus mechanism. Our study helps to provide some pointers to future research directions for implementing blockchain technology for IoT. In the following sections, we discuss the research directions.

### 3.1 Blockchain Mechanism

Blockchain mechanisms including public, private and consortium pose challenges for many IoT applications. A public blockchain incorporates openness, but it is limited in security and poor data transfer time. Private blockchain provides distribution but requires authorisation from some private nodes. Though data transfer time is reduced in a private blockchain, there is no transparency and openness. Consortium blockchain, on the other hand, overcomes the limitations of both public and private blockchain.

### 3.2 Cryptographic Algorithm

For secured data transfer, encryption is an important issue to consider during nodes communication. Generally, blockchain uses hash functions for encrypting data which is computationally complex. However, for IoT, the cryptographic mechanism should be computationally less complex. An investigation into the various cryptography algorithms is necessary to ensure a computationally efficient encryption algorithm for IoT applications.

### 3.3 Miner Selection

The selection of miners with good processing capability, hash generation rate and suitable block propagation delay is quite a challenge for many blockchain applications. Because of the roles the miners play in reaching consensus and creating blocks for the blockchain, it is important to adaptively select only those with the best characteristics for the mining task. Machine learning can be used to provide an intelligent way to select a miner. However, with the different machine learning techniques available, choosing a suitable method for a given blockchain-based IoT application can be difficult.

### 3.4 Consensus Algorithm

A consensus algorithm is helpful for nodes to decide any particular data communication. In a decentralised IoT, since there is no central control, this is even more pertinent. Various consensus algorithms exist for distributed node settings. However, there are limitations when it comes to their adaptation to blockchain-based IoT. The future research direction is to investigate and to get the empirical evaluation for using different consensus algorithms in various settings of blockchain-based IoT.

## 4 Conclusion

There will be billions of devices soon connected using various communication standards for different IoT applications. For scalability, robustness and proper distributed access to information, the decentralised IoT is the way to go. However, the lack of a central entity makes the security aspect of the decentralised IoT more challenging. Blockchain is a promising technology that can be used to improve security in IoT. In this paper, we present a framework for secured communication in a decentralised IoT network. We discuss the overall blockchain mechanism including the consensus strategy used, the cryptographic algorithm for security and the miner selection method. We also highlight some research directions for the implementation of blockchain technology for other IoT applications. The proposed framework is a work in progress, and future work will involve experimentation that will evaluate the performance of the various components of the framework.

## References


1. Al-Nakhala, N., Riley, R., Elfouly, T.M.: Binary consensus in sensor motes. In: Proceedings of the International Wireless Communications and Mobile Computing Conference, pp. 1337–1342 (2013)
2. Cao, Z., Qin, T., Liu, T.Y., Tsai, M.F., Li, H.: Learning to rank: From pairwise approach to listwise approach. In: Proceedings of the International Conference on Machine Learning, pp. 129–136 (2007)

3. Cong, L., He, Z.: Blockchain disruption and smart contracts. Rev. Finan. Stud. **32**(5), 1754–1797 (2019)
4. Dib, O., Brousmiche, K.L., Durand, A., Thea, E., Hamida, E.: Consortium blockchains: overview, applications and challenges. Int. J. Adv. Telecommun. **11**(1), 51–64 (2018)
5. Dorri, A., Kanhere, S.S., Jurdak, R., Gauravaram, P.: Blockchain for IoT security and privacy: The case study of a smart home. In: Proceedings of the International Conference on Pervasive Computing and Communications Workshops, pp. 618–623 (2017)
6. Elisa, N., Yang, L., Chao, F., Cao, Y.: A framework of blockchain-based secure and privacy-preserving e-government system. Wireless Networks, pp. 1–11 (2018)
7. Fanning, K., Centers, D.: Blockchain and its coming impact on financial services. J. Corporate Account. Finan. **27**(5), 53–57 (2016)
8. Fisher, J., Sanchez, M.: Authentication and verification of digital data utilizing blockchain technology. U.S. Patent Application 15/083,238 (2016)
9. Fortino, G., Guerrieri, A., Russo, W., Savaglio, C.: Integration of agent-based and cloud computing for the smart objects-oriented IoT. In: Proceedings of the International Conference on Computer Supported Cooperative Work in Design, pp. 493–498 (2014)
10. Kraft, D.: Difficulty control for blockchain-based consensus systems. Peer-to-Peer Network. Appl. **9**(2), 397–413 (2016)
11. Li, H.: A short introduction to learning to rank. Trans. Inf. Syst. **94-D**(10), 1854–1862 (2011)
12. Liu, T.Y.: Applications of Learning to Rank, pp. 181–191. Springer, Berlin (2011)
13. Pang, L., Lan, Y., Guo, J., Xu, J., Xu, J., Cheng, X.: Deeprank: A new deep architecture for relevance ranking in information retrieval. In: Proceedings of the ACM Conference on Information and Knowledge Management, pp. 257–266 (2017)
14. Perera, C., Zaslavsky, A., Christen, P., Georgakopoulos, D.: Sensing as a service model for smart cities supported by internet of things. Trans. Emerg. Telecommun. Technol. **25**(1), 81–93 (2014)
15. Xiao, L., Boyd, S., Kim, S.J.: Distributed average consensus with least-mean-square deviation. J. Parallel Distrib. Comput. **67**(1), 33–46 (2007)
16. Zhang, A., Lin, X.: Towards secure and privacy-preserving data sharing in e-health systems via consortium blockchain. J. Med. Syst. **42**(8), 140– (2018)

# Methodology to Build Radio Cartography of Wi-Fi Coverage

Dhouha Kbaier Ben Ismail[1(✉)] and Deep Singh[2]

[1] Cyber Research Institute, University of Wolverhampton, Wolverhampton, UK
D.Kbaier@wlv.ac.uk
[2] Satisnet, Luton, UK

**Abstract.** In the modern world, wireless local area networks (WLANs) have seen major growth and almost become a necessity for organisations regardless of its size due to its flexibility, easy installation and cost-efficiency compared to traditional wired LANs. On the other hand, Wi-Fi cartography is being utilised to enhance wireless network deployment, optimise Wi-Fi infrastructure through performing a radio frequency (RF) spectrum analysis (2.4 and 5 GHz) and identifying crucial points to improve the performance of voice and data communications. This paper presents a methodology to build radio cartography of Wi-Fi coverage. Furthermore, simulations are run and Wi-Fi cartography results performed in several areas at the University are presented. A thorough examination and critical analysis of the simulated areas are discussed. The authors reflect upon propagation environments, access points (APs) detected with respective standards, techniques and speeds supported, latency, packet loss and quality of service parameters.

**Keywords:** Access point · Cartography · Data rate · Heatmap · LAN · Packet loss · Received signal strength indicator · Wi-Fi

## 1 Introduction

Wi-Fi, an abbreviation of wireless fidelity [1], is a set of standards for wireless local area networks (WLANs) and based on IEEE 802.11 specifications. Wi-Fi technology uses radio waves to provide wireless network connectivity and provide wireless connectivity to the device by emitting frequencies between 2.4 and 5 GHz. Wi-Fi is being used in various applications such as residential Wi-Fi, enterprise Wi-Fi, Wi-Fi hotspots, Wi-Fi on trains and so forth. Wi-Fi radio cartography has been used in several corporate and urban wireless settings to enhance the wireless network deployment and infrastructure by performing a radio frequency (RF) spectrum analysis to identify propagation of Wi-Fi network signal and signal quality study of all existing access points to identify the areas with best and least signal strength. Considering all the benefits of Wi-Fi cartography, there are no available instructions and guidelines to aid public in building their own cartography. The main motivation of this paper is to share the methodology used to build


X.-S. Yang et al. (eds.), *Proceedings of Fifth International Congress on Information and Communication Technology*, Advances in Intelligent Systems and Computing 1183, https://doi.org/10.1007/978-981-15-5856-6_28

the radio cartography and critically analyse the simulation results run at the University. This paper will identify the areas in the University that are most exposed to interferences, white zones, areas with strong and weak signal strengths. The simulations are run in selected areas that are generically accessible by everyone in the University such as receptions, restaurants, café and bar, IT suite, where users are usually understood to use Wi-Fi facilities in such environments.

The paper is structured as follows. Section 2 provides some practical strategies to achieve cartography at any desired infrastructure. Tools, applications, hardware and software requirements are discussed as well as tips for users interested in building cartography. Then, the authors investigated several areas at the University to build a radio cartography of Wi-Fi coverage. Simulation results for different locations at the University are available in both Sects. 3 and 4 where the authors analyse several network performance parameters and generate a signal strength Heatmap of Wi-Fi coverage. Finally, Sect. 5 draws some conclusions.

## 2 Practical Suggestions/Strategies to Generate a Wi-Fi Cartography

There are several tools around the market that allow one to achieve the cartography. Some options to choose from are Acrylic Wi-Fi Heatmaps, SolarWinds, NetSpot, Ekahau, etc. All the mentioned tools offer similar functionalities and features. The techniques discussed here are universal and can be applied to any application being used.

First, there are number of tools and applications needed as follows:

- **Acrylic Wi-Fi Heatmaps**---Allows to perform a Wi-Fi site survey and collects parameters such as signal levels, AP information, bandwidth and latency.
- **Acrylic Wi-Fi Professional**---Using this tool along with Heatmaps works flawlessly, this tool defines the number of clients connected to the network, it identifies AP transmission speeds and it performs the analysis of wireless traffic.
- **Floor Blueprint/Floorplan**---To be able to use any application, users are required to have a floor plan or the blueprint which the application would use to map the data and cartography upon.
- **Iperf Server Application**---In order to obtain parameters such as latency, packet loss and bandwidth, an Iperf server needs to be set up on local LAN.
- **TracePing**---This is a mobile application, which is another option for users to collect real time like latency, jitter, packet loss, round-trip time (RTT). At times, tools such as Acrylic Wi-Fi do misbehave and can display incorrect packet loss and latency results. Therefore, in such cases, this tool is extremely beneficial to collect correct data.

Below are the system requirements for application:

- Windows Vista (32-bit, 64-bit) or above
- Memory—4 GB recommended
- Processor 1.5 GHz or above
- Disc space over 500 Mb.

Finally, external hardware is also needed:

- Wireless card (supporting monitor mode)---An external wireless card is required to perform an active survey, this is important in collecting active data such as bandwidths, latency and packet loss.
- Airpcap cards (Professional Wi-Fi Hardware)—This is professional hardware used for an optimal measurement and can collect additional data such as radio emissions.

The authors investigated several areas at the University. The areas were selected based upon their accessibility and usage of Wi-Fi to students, staff and guests. Hence, the areas analysed were receptions at the Park Square and Campus Centre. The café and restaurant areas were examined on both campuses. Further, the bar, informal meeting and social area at the Campus Centre building were investigated. On the other hand, the library IT suite was also examined to determine the Wi-Fi coverage and outline any differences. The following sections critically analyse the output of the simulations.

## 3 Wi-Fi Cartography at the University: Park Square Campus-Reception and Café

This section analyses the results acquired when cartography is performed at the Park Square campus's reception and lounge area. Figure 1 displays the floor map of the Park Square buildings reception and café area where the cartography took place.

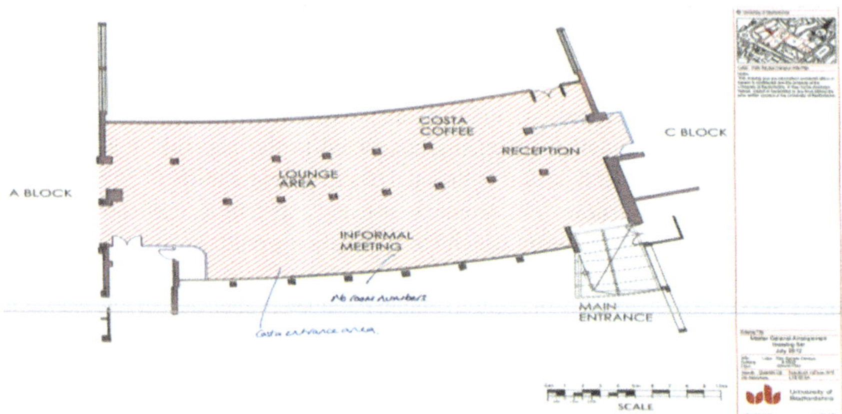


**Fig. 1.** Floor map of Park Square's reception and café

The graph in Fig. 2 summarises the networks, MAC addresses, and physical access points (APs) grating network accesses at this location. The graph represents that Park Square reception and café area detected 8 networks, 27 APs and 68 basic service set identifiers (BSSIDs). As one would note, that the number of BSSIDs can exceed the number of physical APs because a single AP can propagate more than one network by using different MAC addresses. Also, some APs can belong to the same network.

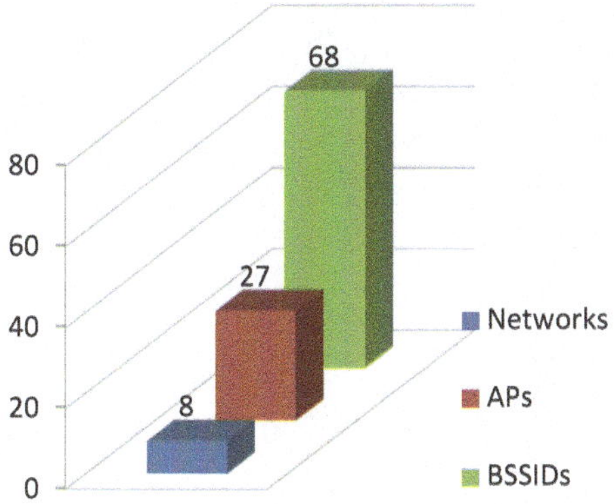


**Fig. 2.** Networks, APs and BSSIDs at Park Square's Café

Figure 3 represents the BSSIDs detected operating on 2.4 GHz channel. There are 14 BSSIDs detected on channel 1, 26 by channel 6 and 26 by channel 11. Other BSSIDs which are relatively small in number were detected on channel 3 and 4, respectively. The 2.4 GHz channel is the most widely used frequency. This frequency can penetrate through solid objects and offers a larger range than the 5 GHz channel. On the other hand, as this channel is used by almost every device, e.g. cordless phones, automatic garage doors and other home appliances, therefore, this also has made the channel overcrowded [2]. Indeed, the IEEE 802.11 WLAN standard sets the attributes for the different channels that may be used. These attributes allow diverse Wi-Fi devices to communicate with each other effectively. The graph in Fig. 3 submits that the number of BSSIDs had intervals of 5 channels in between to prevent channel overlap and reduce interference with adjacent channels. However, a few BSSIDs were found on channels 3 and 4 that overlapped with other channels and interfered with adjacent channel signals [3, 4]. These BSSIDs detected on channel 3 and channel 4 do not belong to the eduroam (educational roaming) network and are not coded as intranetwork, they are simply from another network.

The maximum data rates of the detected BSSIDs are available in Fig. 4. The graph shows that 64 BSSIDs offered a maximum data rate of 216.7 Mbps, one BSSID of 144.4, one of 54 and two BSSIDs of 270 Mbps. The 54 Mbps data rate is supported by the IEEE 802.11 g standard which is a popular wireless standard today. This standard offers transmission over distances of 150 feet and speeds up to 54 Mbps compared to the 11 Mbps for the 802.11b standard [5]. IEEE 802.11n standard offers 144.4 and 216.7 Mbps. This standard makes use of multiple antennas to increase its data rate, also referred to as multiple-input multiple-output (MIMO) [6]. This technique uses multiple radio and antennas to transmit multiple data streams, known as spatial streams. The default configuration of 802.11n uses 20 MHz channel width with 800 ns guard interval between the transmitted signals. 802.11n can offer up to 144.4Mbps when using 20 MHz channel width, 400 ns guard interval and 2 streams. Further, when 802.11n uses 3 streams on the 20 MHz channel, the maximum bandwidth that could be achieved is 216.7 Mbps. Figure 4 also shows two BSSIDs supporting a maximum data rate of 270Mbps. This bandwidth is achieved by 802.11n standard when operating on 40 MHz bandwidth, a guard interval of 800 ns and with 2 streams. The standard 802.11n uses the

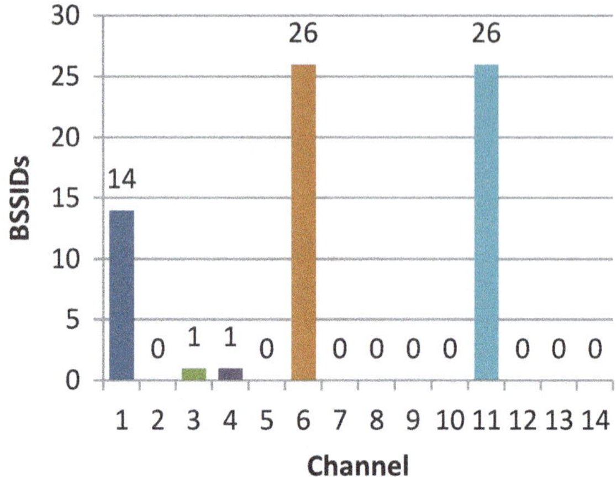


**Fig. 3.** Number of BSSIDs detected on the 2.4 GHz band

orthogonal frequency-division multiplexing (OFDM) technique with the integration of MIMO. MIMO technology is frequently integrated with the OFDM access technique to improve its performance [7]. This integration improves the robustness in fading channels when the cyclic guard interval is included. Therefore, the MIMO-OFDM system is now being adopted by several applications, for example, 802.11n [6, 8].

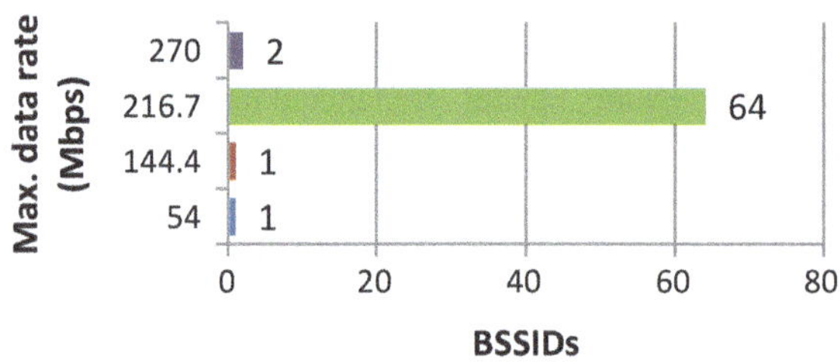


**Fig. 4.** Maximum data rates for the detected BSSIDs

Figure 5 represents the cartography of Park Square's reception and café area. The figure displays the signal strength at each area on the floor map. The signal strength translates to how efficiently the network is reaching the surveyed area, indicating how the eduroam network is received at each location. Signal strength values range from 0 to −100 dB, being −100 dB the worse performance.

Figure 6 provides a grid for Park Square's reception and café. It divides the location into several areas in order to simulate parameters such as the best received signal strength indicator (RSSI), average RSSI, APs detected and maximum data rate at each point on the map of Fig. 5.

Table 1 displays the best and average RSSI for all the cell areas of Fig. 6, alongside the number of APs detected. The measured maximum data rate offered by a single AP at each cell was 216.7 Mbps. As it can be analysed from the grid in Fig. 6 that two

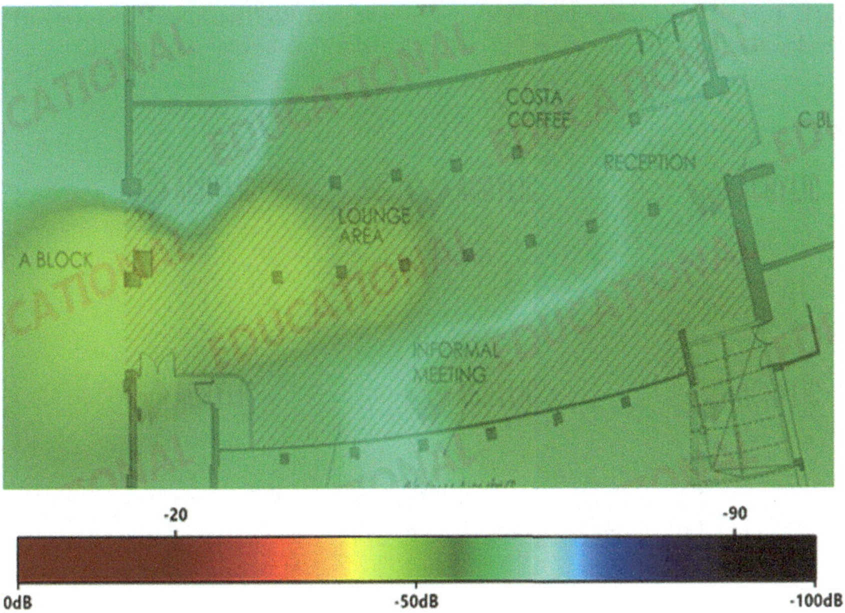


**Fig. 5.** Signal strength map of the Park Square location

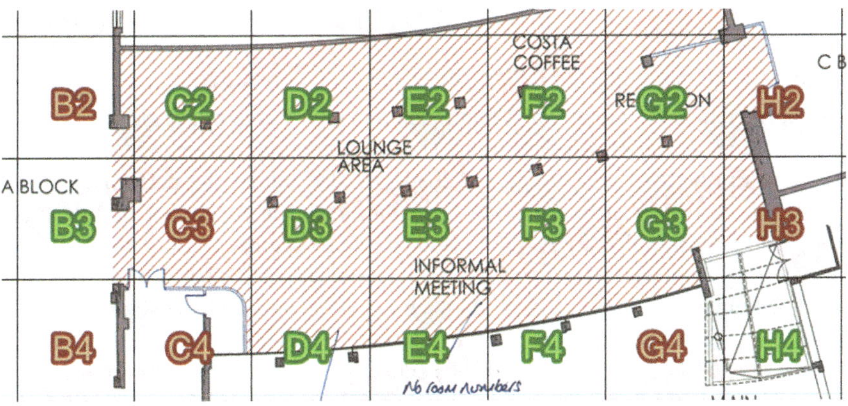


**Fig. 6.** Detailed grid of the Park Square location

contrasting areas can have the same average RSSI, however, they have an important difference in the simulated best RSSI such as for cell G2 and cell B3. G2 defines the area near Park Square reception and B3 is located near the café exit. Table 1 shows that G2 has an average RSSI of −68 dB and the best RSSI of −59 dB. On the other side, the B3 area has respective average RSSI, however, a significant difference was realised in the best average RSSI of −49 dB. Although these figures may not display substantial differences at first sight, such differences are quite significant in wireless communication.

**Table 1.** Best and average received signal strength indication (RSSI) at different cell areas in Park Square location, alongside the number of AP's detected within the eduroam network

| Areas | Best RSSI (dB) | Average RSSI (dB) | Detected APs |
|---|---|---|---|
| C2 | −60 | −74 | 9 |
| D2 | −58 | −74 | 9 |
| E2 | −59 | −71 | 8 |
| F2 | −57 | −71 | 9 |
| G2 | −59 | −68 | 6 |
| B3 | −49 | −68 | 7 |
| D3 | −49 | −65 | 6 |
| E3 | −52 | −65 | 6 |
| F3 | −58 | −72 | 10 |
| G3 | −55 | −68 | 7 |
| D4 | −58 | −69 | 6 |
| E4 | −56 | −69 | 8 |
| F4 | −56 | −69 | 5 |
| H4 | −56 | −70 | 4 |

This analysis shows that the signal at the A block (area B3) is ten times better than the signal received near the reception. The propagation environments are the most common phenomena that affect the wireless signal such as reflection, refraction, diffraction and absorption. Further, network usage and load can also hinder the performance of the network [9]. As it is can be specified from the area near reception, this location is surrounded by glass walls, doors, pillars and people frequently visiting the reception or the Costa Coffee situated right opposite causing congestion at peak periods. Hence, the propagation environment is not ideal. As the glass doors and walls will cause refraction, this likely causes the signal to change the direction of travelling from sender to receiver and causes lower data rates and lessens the capacity. Further, at peak times the network usage and load also increase due to people trying to access the Wi-Fi while waiting to be served at the café or their break. Therefore, as more clients utilise the network bandwidth, the less the channel can share and thus the bandwidth requirements increase. On the other hand, the A block zone (B3 in Fig. 6) has more open space, not very congested and is surrounded with fewer obstacles unlike the zone near reception. Therefore, the client/user has a direct view of the AP with less interference and obstacles.

Table 2 displays the absorption rate of different materials. For example, the glass wall and doors at the G2 location have an absorption rate of 6 dB, in addition, combined with an increase in network load and usage, this is bound to affect the wireless signal near reception. While A Block offers better propagation conditions, the results show a significant improvement in received signal strength at point B3 than compared to G2 as previously analysed in Table 1. It is important to notice that an absorption of -3 dB effectively demonstrates that the bandwidth is reduced to half [11]. These absorptions also come from OFDM which is one of the drawbacks of this access technique.

**Table 2.** RF absorption rates by common materials (CWNA certified wireless network administrator official study guide)

| Material | Absorption rate (dB) |
|---|---|
| Plasterboard/drywall | 3–5 |
| Glass wall and metal frame | −6 |
| Metal door | 6–10 |
| Window | 3 |
| Concrete wall | 6–15 |
| Block wall | 4–6 |

## 4 Wi-Fi Cartography at the University: Simulations at Campus Centre-Reception and Restaurant

This section reports the results acquired when cartography is performed at the Campus Centre reception and restaurant area. Figure 7 represents the cartography of the Campus Centre reception and restaurant area that has been simulated. The figure displays the signal strength at each area on the floor map. The RSSI translates to how efficiently the network is reaching the surveyed area, indicating how the eduroam network is received at each location.

Figure 8 shows the numbers of APs granting access to the eduroam network that were detected within the surveyed area. Table 3 displays the best and average RSSI for all the cell areas, alongside the number of APs detected. The measured maximum data rate offered by a single AP at each cell was also 216.7 Mbps.

Table 3 establishes the best RSSI at the survey location was measured at point H4 which is a restaurant area. Oppositely, a very poor signal was received at point E4 which is a sitting area located next to the entrance of the building. At the restaurant area, the RSSI value is −53 dB whereas at point E3 the RSSI value is −65 dB. This is due to different environmental characteristics. The E4 zone

is surrounded by two concrete walls and a glass wall with two pillars. Further, just opposite of the area, there is a staircase, whereas at the H4 location, the area is well-spaced and without any containment. Although this area is covered by fewer APs compared to the E4 location as shown in both Fig. 8 and Table 3, however, it provides greater signal strength against any other areas. Referring to absorption values in Table 2, the glass wall and walls absorb 6 dB minimum of the connection at the location E4. Moreover, the glass wall will cause refraction causing the signal to change the direction of travelling from sender to receiver and lessens the network capacity. In comparison to point E4,

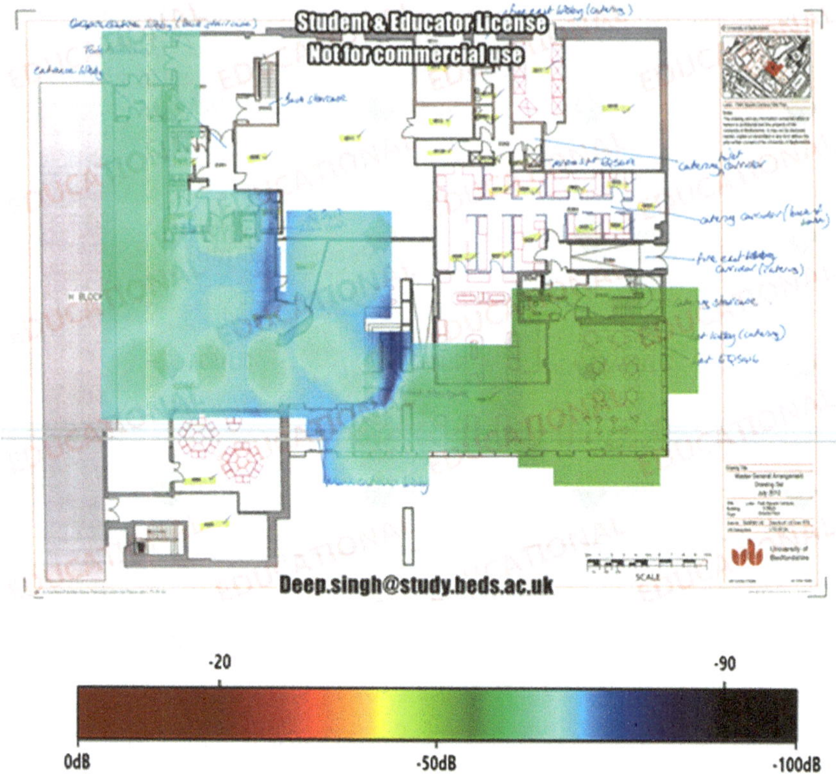


**Fig. 7.** Signal strength Heatmap of the Campus Centre reception and restaurant location

the location H4 provides a better propagation environment, henceforth, the RSSI values display a significant improvement.

Further analysis was carried on the packet loss; H4 location encountered no packet loss whereas at the location E3 the packet loss rate is calculated at 29%. This represents an issue since the performance of any applications are degraded, the network load is increased, as the packets that are lost need to be retransmitted to the users which effectively increases the network usage, and quality of service (QoS) for demanding applications is not met [12]. For example, the packet loss should be less than 1% and no more than 5% for VoIP and video streaming applications, respectively. Since the E4 location suffers from a high packet loss, any user utilising video conference application or streaming video will notice a deficient performance. Although E4 confronted high packet loss, most other areas at the Campus Centre location offered higher bandwidth and no packet loss, which meets QoS requirements for both VoIP and Video traffic. Therefore, users generically should obtain optimum performance on high demanding applications such as VoIP and video conference even at the peak of times [13].

The signal strength Heatmap was also performed in other areas such as the bar and café in the Campus Centre as well as the IT suite in Park Square campus. Although this location provides great wireless performance for the users nowadays, however, this was

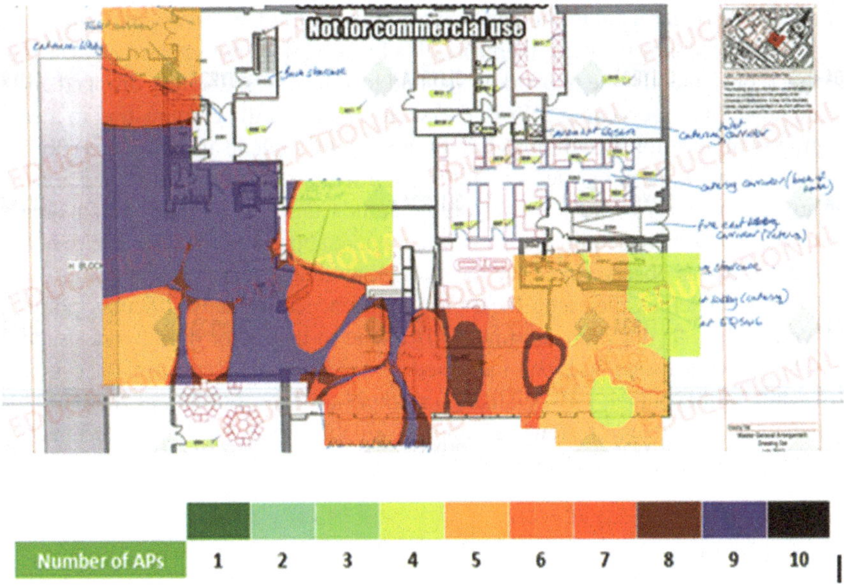


**Fig. 8.** Number of APs detected at the Campus Centre reception and restaurant location

**Table 3.** Best and average received signal strength indication (RSSI) at different cell areas in Park Square location, alongside the number of APs detected within the eduroam network

| Areas | Best RSSI (dB) | Average RSSI (dB) | Detected APs |
|---|---|---|---|
| B0 | −65 | −71 | 5 |
| B1 | −63 | −69 | 7 |
| C2 | −65 | −72 | 9 |
| B3 | −62 | −69 | 9 |
| C3 | −67 | −74 | 11 |
| E3 | −67 | −76 | 5 |
| G3 | −53 | −67 | 5 |
| C4 | −63 | −67 | 6 |
| D4 | −63 | −70 | 12 |
| E4 | −65 | −73 | 6 |
| F4 | −56 | −71 | 10 |
| G4 | −53 | −70 | 5 |
| H4 | −53 | −66 | 7 |

not the case last year. Due to the new library built by the University, more users have transitioned to the new location; therefore, this location does not see as many users as it once did. Consequently, the network usage and load have significantly dropped and the APs are now able to serve more Wi-Fi users. However, it will be interesting to measure cartography of new library buildings first floor as many students inclined to use this area a lot for their work, group projects and it is often seen to be a very congested area.

Cartography analysis was also performed at the Campus Centre restaurant area and simulation results were acquired at peak time, where approx. 60–80 users were noted at the location. It was observed that the location granted access to 6 access points outside peak periods, whereas the same location granted access to only 2 access points (as displayed by Fig. 9) during peak time. This is due to the factor of network usage and load. The network usage and load increased at this time as several users accessed the Wi-Fi while at their lunch for socialising or other purposes. Further, as there are more simultaneous users there are fewer channels available for users. This reflects also the principle of first in, first out (FIFO) that takes place.

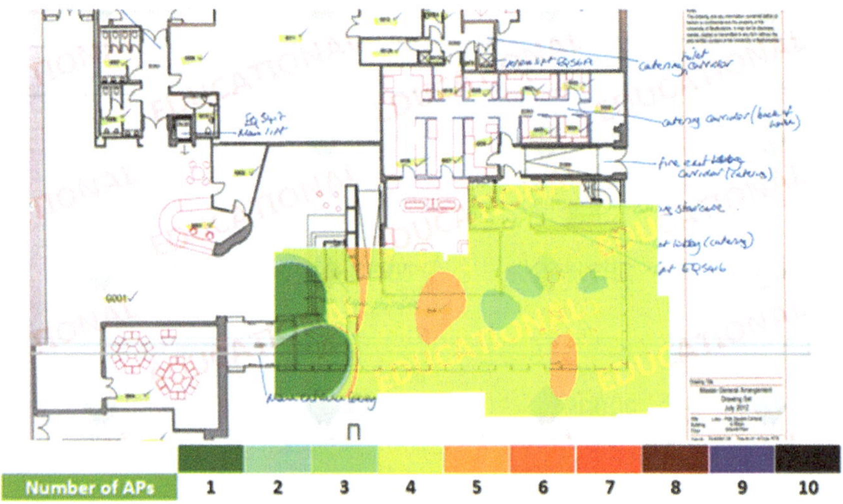


**Fig. 9.** Number of APs detected at the Campus Centre restaurant location during peak time

## 5 Conclusion

To conclude, the University provides a satisfactory Wi-Fi performance across its campus. Park Square's reception location providing the best coverage, particularly the A Block area due to the ideal propagation environments discussed. Oppositely, a specific area at Campus Centre's restaurant faced high packet loss, high latency and low signal strength contributing to poor Wi-Fi performance, while areas such as IT suite showed great improvements if compared to last year due to the number of users transitioning to the new library. However, in the future, the new library's Wi-Fi cartography should be analysed to determine its Wi-Fi performance. The authors provided practical suggestions of tools, applications, hardware and software requirements. These strategies are universal and thus can be used to achieve cartography at other infrastructures.

**Acknowledgements.** The authors would like to thank Miss Nici Wilkinson, Admin and Planning Manager at the University, for providing the floor plans of the University that enabled them to carry out the simulations.

## References

1. WiFi Alliance. http://www.wi-fi.org/certified-products-search
2. CISCO: [online] (2017). Available at https://www.cisco.com/c/dam/en/us/products/collateral/wireless/aironet-3600-series/white-paper-c11-713103.pdf. Accessed 14 Nov 2019
3. Chieochan, S., Hossain, E., Diamond, J.: Channel assignment schemes for infrastructure-based 802.11 WLANs: a survey. IEEE Commun. Surveys Tutorials 2010 **12**, 124–136 (2017)
4. Szigeti, T., Hattingh, C.: Quality of service design overview, section QoS requirements of data. In: End-to-End QoS Network Design: Quality of Service in LANs, WANs, and VPNs (2004)
5. Vassis, D., Kormentzas, G., Rouskas, A., Maglogiannis, I.: The IEEE 802.11 g standard for high data rate WLANs. IEEE Netw. **19**(3), 21–26 (2005)
6. Paul, T.: Wireless LAN comes of age understanding the IEEE 802.11n amendment. IEEE Circ. Syst. Magaz. **8**(1), 28–54 (2008)
7. Paul, T.: Wireless LAN Comes of Age: Understanding the IEEE 802.11n Amendment. [online] IEEE Xplore (2017). Available at http://0-ieeexplore.ieee.org.brum.beds.ac.uk/stamp/stamp.jsp?arnumber=4460436. Accessed 21 Jan 2020
8. Ganesan, R., Gunturi, S.: Robust channel estimation for 802.11n (MIMO-OFDM) systems. In: International Conference on Signal Processing and Communications (SPCOM), 22–25 July 2014
9. Jee, R.: The 802.11n MIMO-OFDM standard for wireless LAN and beyond. Wireless Personal Communications, 37(3–4):445-453
10. Jee, R.: The 802.11n MIMO-OFDM standard for wireless LAN and beyond. Wirel. Personal Commun. **37**(3–4), 445–453 (2006)
11. Ogunjemilua, K., Davies, J.N., Grout, V., Picking, R.: An Investigation into Signal Strength of 802.11n WLAN. In: Proceedings of the Fifth Collaborative Research Symposium on Security, E-Learning, Internet and Networking SEIN 2009, 26–27 November 2009, pp 191–204 held in Darmstadt, Germany, Plymouth University (2009)
12. Research Gate. (2013). [online]. Available at https://www.researchgate.net/post/Please_tell_me_Why_the_cutoff_frequency_is_taken_for_3dB_and_not_other_values_like_1_or_2_db. Accessed 14 Nov 2019
13. Noction. (2015). Network latency and packet loss effects on performance|Noction [online]. Available at https://www.noction.com/blog/network_latency_packet_loss_effects. Accessed 21 Jan 2020

# Digital Transformation: The Evolution of the Enterprise Value Chains

Rui Ribeiro(✉)

ECATI—Escola de Comunicação Arquitetura Artes e Tecnologias de Informação, COPELABS—Cognitive and People Centric Computing Labs, Universidade Lusofona de Humanidades e Tecnologias, Lisbon, Portugal
rui.ribeiro@ulusofona.pt

**Abstract.** Market competition today has given a massive advantage for those who can better use data. The main reason is that they understand how to develop medium long-term correct strategies, with short-term pragmatic operational approaches. The challenge is not on how to get or how to produce data, but on how to use it and transform it oriented to get business focus and business value. For that reason, it is fundamental to understand the details on how companies apply the well-known Ackoff's DIKW hierarchy (data, information, knowledge, and wisdom) in their value chains. This ability to transform data into wisdom in a real-time mode is pressuring companies to transform themselves, and that is an internal process change, commonly known as digital transformation. In reality, companies are turning their value chains since they understand the power and the use of information systems as a strategic value-added decision-making tool. That means it is more than a simple technological support activity like Michael Porter's value chain defined initially. This paper, being a conceptual paper, intends to present the evolution that enterprise value chains need to face and the need they have to include information systems as a core activity nowadays. Doing so, they will achieve the best data to wisdom transformation in a real-time loop and continuous mode, leveraging their value chains to higher probabilities of optimization of company's market value. Those who address digital and information systems strategies will be able to get faster optimization of their market value functions.

**Keywords:** Digital transformation · Information · Strategy plan · Value chain · Information systems · Enterprise strategy

## 1 Introduction

This conceptual paper intends to present that digital transformation is also a value chain transformation. Companies, with technology, are able today to address the data transformation into information, knowledge, and wisdom (DIKW) layers in a real-time loop and continuous knowledge improvement. The ones that have a structured and innovative information systems and digital strategies will be able to achieve business efficient mod-


X.-S. Yang et al. (eds.), *Proceedings of Fifth International Congress on Information and Communication Technology*, Advances in Intelligent Systems and Computing 1183, https://doi.org/10.1007/978-981-15-5856-6_29

els to the referred DIKW cycle. That means they will optimize their internal operation functions, referred to as optimum market value function. So, the goal of this paper is to present today's evidence that digital and information systems (DIS) are a core activity inside companies, side-by-side with logistics, operations, sales and marketing and post-services. DIS are proactive wheels for the enterprise business strategy. Companies that understand these new trends of digital and information systems strategies are the ones that will be more prepared to face future market competitions.

This paper is structured in four main sections, which firstly presents the importance of leveraging data to wisdom and the ability to achieve the optimum company's value if the right processes are deployed. After, it explained the importance of information systems in today's competitive market position, meaning that information systems are no longer the old information technology support activity, but are a crucial core activity on the delivery of a product and service to customers like operations or logistics are. The third structured section presented the importance of having a strategy inside a company, from the top management to the operational implementation. These three sections will be crucial to understand the importance of having specific digital and information systems strategies in this new core activity of the enterprise value chains.

## 2 From the Information Age to the Information Prediction

In 1985, Michael Porter referred that "the information revolution is sweeping through our economy" [1]. Today, after more than 30 years of this statement, we read it and think "it is obvious," but at that time, like today, the majority of the enterprises' decision-makers tend to forget the new challenges. If in the decades of 1980 and 1990, companies were starting to face the challenge to create data and get information, today, data and information are already considered commodities, meaning that decision-makers need to upgrade their information and digital ecosystems to acquire knowledge and wisdom, in order to be able to take the lead and be able to get a competitive advantage. This is known by the famous DIKW—data, information, knowledge, and wisdom pyramid—supposedly identified by Ackoff [2].

There is a set of discussions about the relevance and the real applicability of the DIKW hierarchy to enterprise methodologies, like the ones referred by David Weinberger in his HBR article [3] saying that "knowledge isn't determined by information, because it is the knowledge that decides which information should be important" or by Jennifer Rowley when in her research of several study analysis concludes that the main pain point to adopt the hierarchy is that there is "less consistency in the description of the processes that transform elements lower in the hierarchy into those above them" [4]. Although these skeptics on the DIKW process, the majority of those researches do not refer that the DIKW hierarchy should be considered as a spiral of knowledge, meaning that the pyramid is, in fact, a virtuous cycle of growing knowledge (Fig. 1—DIKW virtuous cycle searching for the optimum market value function), where the predictive models should be applied in the real-world environment and next collect the data from what really happens, which will be transformed into information, that will be analyzed to get new knowledge and generate new predictive wisdom. Basically, the next figure concept tries to represent the knowledge management spiral to be considered and the mathematical function model that is always searching for the "perfect" function, named as "optimum market value function" (OMV function) with the exact reactions of the real world.

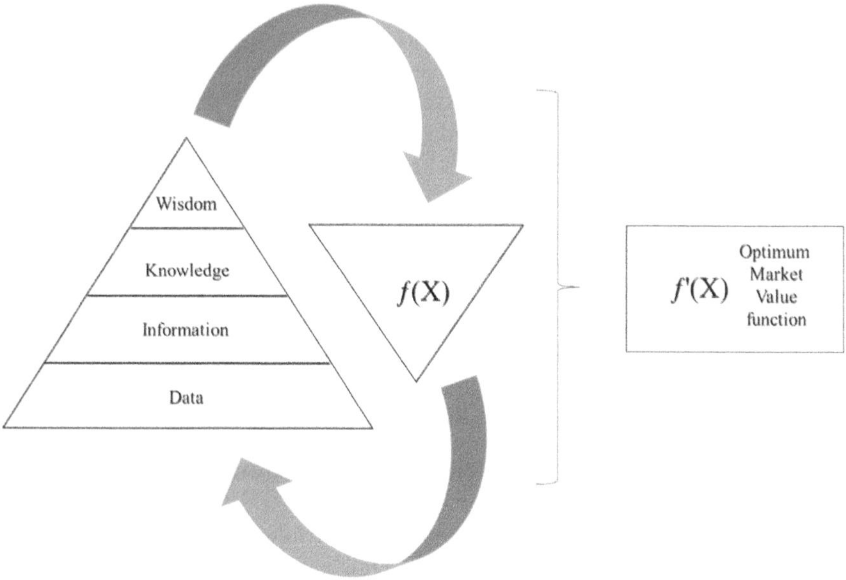


**Fig. 1.** DIKW virtuous cycle searching for the optimum market value function

The reality is that the DIKW hierarchy, which is a knowledge management process basis, has surpassed the enterprise strategy concepts, being today a reality with the emerging of new technologies and business trends based on big data [5], Internet of things (IoT), and artificial intelligence.

There are several cases in the real business world several, even on the primary sector, which is supposedly an analogic industry, but the transformation of this industry is searching to achieve exponential efficiencies with deployments of "connected farms," "soil-moisture sensors," "autonomous tractors," and "stock and animal wearables" [6]. What are these changes? Basically, with IoT, several sensors are data producers that bring information to the farm management, which across time, the knowledge is created and, with technology, the predictability will be achieved. That is only valid because there is a business goal, namely for controlling diseases, that will be more accurate in artificial insemination or milk quality. The real examples of the DIKW virtuous cycle applications are, for instance, developed by Fujitsu, the Japanese technology firm, and its customers. And this is not recent, because Fujitsu started this business and technological deployment, with a Microsoft cloud infrastructure [7], in a Japanese farm in 2013 [8], where the sensors were pedometers attached to cows delivering the steps given by cows (data), which permitted to see the visualization of the cows position (information) and understand the normal and the unnormal behavior of the cows, which led to accurate detection of estrus (knowledge). This technology implementation (IoT + Big Data +

Artificial Intelligence together) permitted them to be 70% more productive than before because they have got +95% of estrus detection accuracy versus 56% previously and a 67% pregnancy rate versus 39%. But this case gives us the fundamental vision to achieve the optimum market value function (OMV function) which was that after a set of cycles of data--information--knowledge, they have found the wisdom ambition, which was to get an high level of accuracy that making the insemination prior to the optimum point (16 h after the first detection of estrus) they probably would get a female, but the insemination after the optimum point of insemination they probably would get a male. This is an enormous capacity for business planning, which is to be able to control their operations and fit the future contracts, imagine an agreement to sell meat or milk within one year. Basically, that farm will be able to generate what they want in their production environment, according to their sales orders. It is the applicability of just-in-time fulfillment operations in an industry that was miles away from this concept. This is a dream come true for any business, which is to control almost all the variables in their operations. The challenge is to keep moving on the DIKW virtuous cycle, maintain the enterprise ethics, and to be able to control all the value chain.

Examples like this one are starting to happen in other industries, like health care, government, logistics, upon others with proven business values, not only on efficiency, and also on effectiveness [9]. The investments on these set of technologies to create the wisdom level are enormous, from big companies small and medium businesses. The movement to align the DIKW hierarchy with technologies, in order to achieve new business innovations and get competitive advantages are being made from big companies like Coca-Cola, GE, Rio Tinto, Bank of America to small companies like Pulmonary Apps (Healthcare), Team Companies (HR Services), Harps Food Stores (Retail), and many others [10, 11].

If we analyze the B2C market, companies would search and invest in the identification of the set of variables that influence their customers. Basically, knowing that the leading companies' goal is to generate value for its stakeholders, bring true advantages to higher productivity to make money to their stockholders. The challenge would be to have the "perfect" OMV function to control their customers as "puppets."

This is a process of creating the culture for an enterprise intelligence, which is the business transformation of understanding of the power that data can deliver, as a value added for the business.

A real known example of the "*puppetization*" is the basis of a well-known public scandal about the data breaches of Facebook and used by Cambridge Analytica [12]. The underneath concept of what happened is, in reality, nothing new and should not be a surprise because since the beginning of the creation of the business intelligence concepts this was the goal [13]. The title of the book of Liautaud is an example of that: "e-Business Intelligence: turning information into knowledge into profit", and it was written in 2000. So, and knowing that Business Intelligence is a set of "mathematical models and analysis methodologies that exploit the available data to generate information and knowledge useful (…)" [14], a company makes profits selling more, and it only sells to customers, meaning that the mathematical models and analysis methodologies are in fact the first steps to achieve the OMV function, meaning that there was since the early stage of the decision support systems [15] a goal to get the control of all the variable of the customer and the internal processes.

## 3 The Digital Transformation of Porter's Value Chain

Michael Porter in his book "Competitive Advantage: creating and sustaining superior performance" (Porter [16]) has introduced an important concept, that has established a standard in the economics science and which explains the systematic structure of the set of activities needed to deliver products or services value added by enterprises: the value chain. At that time, porter also understood the information value has a fundamental infrastructure to achieve efficient models, between the several flows that optimize the support and core activities of the enterprise. Although, today the Digital Transformation paradigm and the achievement of technology maturity higher levels, the importance of Information Systems has growth to be a fundamental source of sustainability and acceleration in the value chain like described by Jim Collins when he refers that the Technology Accelerators are one of the explosions in the flywheel momentum [17], which means that the technology is in between of the processes and the several resources of the companies.

Knowing that information systems are different from information technology because it is a set of three main components—people, processes, and information technology—[18] and understanding the value of the information within today's business world, also predicted by Michael Porter and other authors, and the confirmation that technology is a fundamental accelerator for efficient and sustainable companies, means that the information systems are more than the technology development within the support activities of the value chain. In this twenty-first century, information systems need to be understood as a primary activity on the daily basis of the company and essential for the product or service delivery (Fig. 2—Second-generation value chain). This recognition as a primary activity is the assurance that there is a specific and critical area within the company that is always searching for the right fit between people, processes, and technology. This will assure today's and near future competitive advantage.

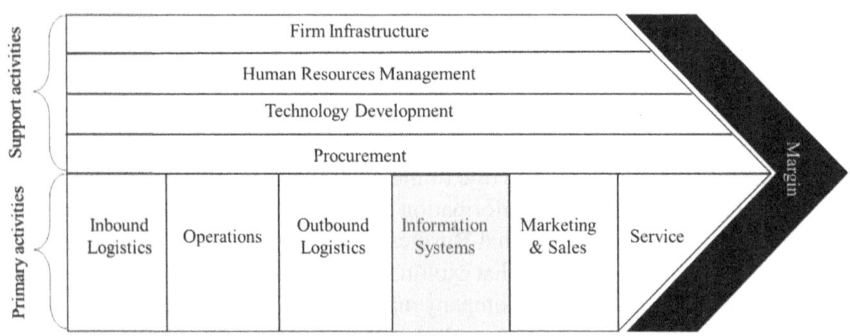


**Fig. 2.** Second-generation value chain

Several authors analyzed that information technology, by their own, suffer from the "productivity paradox" [19], which means that they did not improve business productivity and pointed out a set of mismanagement activities, neither their usage would bring real advantages to higher productivity [20]. Additionally, Nicholas Carr, in his article "IT Doesn't Matter" [21], touched the pain point, when he understands that Information Technology is a commodity for enterprises, like water and electricity. In 2003, when Carr wrote this article, he knew that the competitive advantage was not on technology, but the way that companies used it and leveraged it as a strategical issue. This is an example that the competitive advantage is not on information technologies, but on information systems.

Companies need to have this information systems management that has a focal point in the company, to assure that the DIKW virtuous cycle, previously explained, is a strategical and core activity in the value chain. The appearance of roles like chief information officer (CIO) and chief digital officer (CDO), in the organizational structure of several companies, is example of that value chain change. They are not the chief technology officer (CTO), which manages only the technology infrastructure (the commodity factor), meaning that they assure that the technology is working in the expected delivery of the operational flows within the company. A CTO is not expected to be a strategic engine, but an operational engine aligned with the day-to-day support of the enterprise infrastructure, as viewed in Porter's value chain (first generation). CIO and CDO are engines for enterprise intelligence, meaning that their roles are more aligned with the strategic future and long-term movement of the enterprise.

Information systems need today to be understood as a strategical influencer of the company, meaning it is not the "old" IT of a company, or in terms of an analogy with the "well-known" ecologic food chain, popularized by Charles Elton in 1927 [22] in his book "Animal Ecology," information systems have climbed up from the producers and decomposers to the top carnivores, meaning that they were seen as the technical support and are (should) now a fundamental strategic decision-maker in the "decision chain." This happens due to the fact we are now in the world that data are transformed in information, where the business transactions are no longer made in traditional ways (supplier + distributer + customer), but are a set of possible networked transactions [23], in other words, the knowledge and the ability to understand that the information systems are a strategic competence and a primary activity to achieve a sustainable company with competitive advantages. For that reason, information systems are the engine that can join business, technology, and information. Information systems (people, processes, and technology) ensure that the long-term strategy and goals are present in the overall decisions.

It is also relevant to confirm that in IBM CIO Study 2009 [24], it is referred that "*CEOs are more dependent on CIO insights to innovate and improve these processes*," confirming the climb up on the "food chain" of the enterprise decision-maker, which in practical terms is the recognition of the importance of the constant and fundamental work made by the information systems, to leverage data and analytics and creating efficiency in new or existing processes within the company.

## 4 The Importance of a Strategy

It is a fact that marketplaces today are living very competitive and saturated times with products, offerings, and other services to a global consumer base, meaning that no business can afford to be without detailed information about: their business (financials, accounting, revenues, costs, among others), their customers (profiles, psychographics, new habits), or changes that are impacting their business future. That is a primary reason to develop a strategy, in order to assure that managers can have a decision support guideline and rules to create the path of their business scope and growth, even knowing that they will make decisions under partial ignorance [25]. Defining a strategy is to create a business approach to set goals and policies which need to be carried out to achieve competitive positioning in the environment of the company [26].

One way to understand the concepts of the word strategy is to know the origin of the word. Strategy derives from the Greek word *stratēgos,* which is formed by two words: Stratos (army) and Ago (ancient Greek for leading). This meant that *stratēgos* referred to the military general during the age of Athenian Democracy, which was able to understand the environment and lead the troops in the field. So, it is easy to understand, that a corporate strategy is the set of guidelines with a scope under the internal and external environment of the company, knowing which resource capability the company has, in order to be able to allocate and reallocate them in the field, according to with a set of values, expectations, and goals to be achieved in the long term [27].

It can then be resumed that strategy is the pattern of resource allocation decisions made throughout an organization. These decisions encapsulate both desired goals and beliefs about what is acceptable and, most critically, unacceptable means for achieving them. One of the most critical decisions, when a company is defining its strategy, is to know what are the paths that doesn't want to go, because de way to achieve its goals there will be several temptations to change, several opportunities to allocate new resources that aren't at the core values.

A strategy always starts with a vision, which determines the target to be achieved, or at least, the field of opportunity to be surpassed is. This vision milestone definition is the trigger for the strategy process definition, which can be resumed in three main steps: analysis, choice, and implementation [27].

The analysis can be described as the process of internal and external understanding of the environment, the culture, the capabilities, and beliefs.

The choice is the generation of paths to achieve the vision, meaning that, for example, we have several route options to go from city A to city B (by car, by train, by air, by bicycle, or by other mobility way) and each option has an economic cost (Fig. 3—Economic function cost of strategy route options), where it needs to be understood which are the variables that need to be considered (for example time to achieve, toll costs, comfort of the travel, among others) and the weight of each variable. This symbolic exercise is the same work that a company needs to do when it is analyzing the several routes it has when it knows what wants to be and to achieve. This means options like acquisitions, like technology investments, integrating more human resources, installing factories, and so, are examples of variables that need to be analyzed and weighted to compound the economic function cost of each route option and choose the best one to implement.

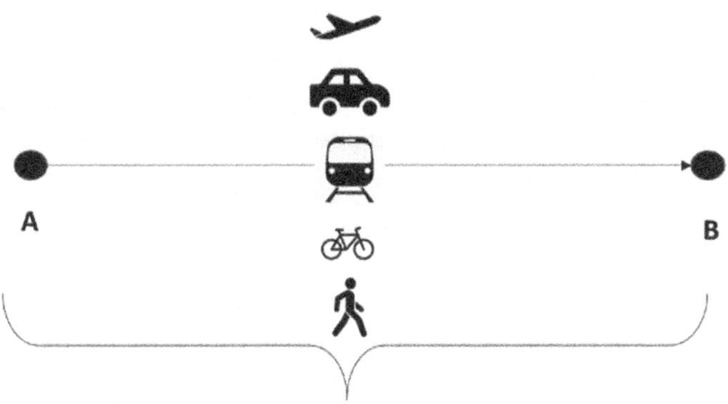


$$f(x) = a_1 * p1 + a_2 * p2 + a_3 * p3 + \cdots$$

$$\sum_{1}^{n} p_n = 100\%$$

$$a_n - variable\ economic\ vaue$$

**Fig. 3.** Economic function cost of strategy route options

The strategy implementation is the operational deployment details of the strategic route option that has been chosen, which is generally known as the action plan of the strategy. This implementation plan needs to have the set of milestones and the key performance indicators that will guide the confirmation acknowledge, during those years of implementation, of the control range if the company is being of the right path for the 3–5 years of the strategy definition plan. The milestones and key performance indicators will deliver the BHAG—Big Hairy Audacious Goals, introduced by Jim Collins and Jerry Porras [28], which means that a strategy will need to push for the overall company to achieve more, to higher its limits, and like those authors also refer the company has to leverage and develop their strengths to surpass to other levels of competitive capabilities.

## 5 Digital and Information Systems Strategy

Once Stephen Hawking said: "The world has changed far more in the past 100 years that in any other century in history. The reason is not political or economic but technological—technologies that flowed directly from advances in basic sciences." This is an enforcing of what is described in the previous sections of this article, that the business and the way companies are doing businesses is changing, and it is changing fast and each year is being faster. There are several pieces of evidence that technology adoption is pushing much faster the product or service consumptions than ever. An amazing infographic created by Nicholas Felton for the NY Times [29] is an example of an artifact where it can be seen the speed of adoption of a set of everyday commodities like the electricity that took almost 30 years to reach 50% of the US households and Internet took around ten years, or that the telephone took more than 30 years to achieve those

50% of US households and the cell phone took just five years. But when we enter in the digital world, the speed of the production of digital data growth starts to be even faster transforming the rates in exponential increases, because all the numbers "explode," meaning around 50× growth rates between 2010 and 2020 and 90% of that data have been produced in the last two years [30].

These fastest times are consequently introducing changes in the marketplace, where the disappearance and rise of new companies are astonishing. As can be seen in "Fig. 4—Average company lifespan on S and P 500 Index—adapted from Innosight, Credit Suisse and other public sources," the marketplace change in the S and P 500 index has been enormous. The lifespan of companies in that index in 1958 was 61 years and in 2012 was 14 years, with an estimated lifespan of 10 years in 2027 [31–33]).

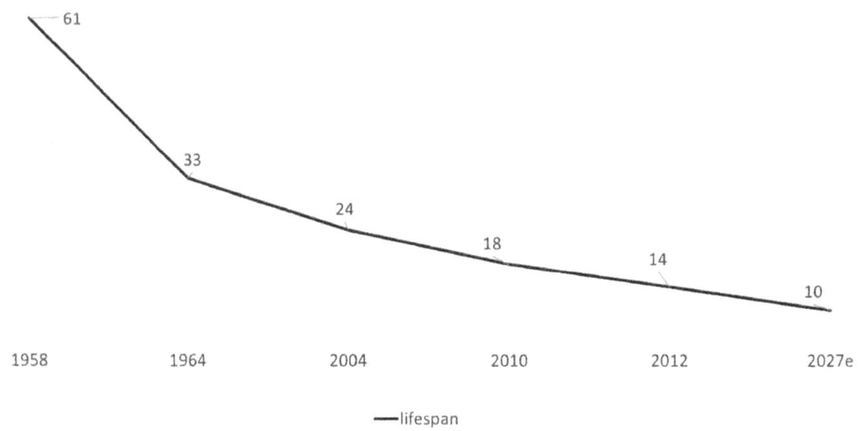


**Fig. 4.** Average company lifespan on S and P 500 Index-adapted from Innosight, Credit Suisse and other public sources

The shrink of companies' longevity puts pressure on the overall stakeholders, namely investors, which need to rethink even their valuation methods, portfolio sizing and capabilities new and traditional variables for a company to have a sustainable competitive advantage. This is quite important because, for instance, Credit Suisse February 2017 Report identifies a quite good correlation between longevity and their innovation index [32].

The common challenge identified by several studies is the power of the present and near-future digital disruption that is changing markets. That power is pushing operations to be efficient as they have never been, which means that technology needs to be present and correctly deployed because only technology can help to re-engineer the existing processes in companies and can understand more deeply new digital interactions within the various channels.

Market globalization and its global competition creation are a daily unknown challenge and a run for a wheel of continuous innovation speed, pressuring the product life cycles and existing business models. Basically, the term digital transformation is transversal to all industries value chains. And companies and leaders that understand it are the ones that will prevail.

Several research analysts, like MIT Center for Digital Business, like McKinsey, Gartner or IDC, have common conclusions, namely that best digital companies are 26% more profitable than their industry competitors and generate 9% more revenues. Recently, a new pattern recognition of those studies has found that companies that are effectively using the new data science capabilities are more productive and profitable in 5–6% than its peers [31, 34, 35, 36].

Gartner predicts for 2020 that 75% of the businesses will be digital or are preparing to become digital [37]. This understanding of becoming digital is a strategic first movement for companies to redesign their business strategy and culture. Basically, the appearance of a digital strategy plan, besides the business and information systems plans, is starting to be understood as an essential tool for companies' future sustainability. These digital strategy plans are the first steps for companies to know how to convert digital value propositions of their businesses into revenue-generating digital offers [38]. Today, this means to create hybrid solutions (digital and physical) able to engage new experiences to customers. These experiences are able to create unique emotional relations and new views of loyalty between the customer and the company in a continuous operation. Today, methodologies of design thinking to understand customer needs and expected experiences with a constant business application development of the business and information strategies, based in DevOps methodologies, are essential to create competitive advantages and future value added for the stakeholders. The next figure represents the fit of these overall strategies.

The previous figure represents, in fact, the reality of the new culture that has appeared: Digital strategy as the center of an enterprise intelligence (Fig. 5—Digital strategy has the center of an enterprise intelligence). It links technologies, namely the new disruptive trends of analytics, IoT, 3D printing, blockchain, natural user interfaces, artificial intelligence with customer experience, business strategy, and information systems strategy.

The digital strategy will deliver digital governance, which is fundamental to define a clear responsibility of aligning digital strategy, policies, and digital standards. Digital governance, when effective defined and implemented, will, for sure, help the development of digital and agile business and sustainable company [39].

## 6 Future Research

This article permitted to identify a set of paths for a more in-depth understanding of the digital business impacts, meaning that it is a basis for future researches like:

- Analysis of correlations of companies P and L growth, the existence of digital strategy, presence of formal CIO and CDO
- Analysis of digital transformation (un)successful cases and the presence of the CIO or CDO roles
- Analysis of digital framework strategy definitions, governance and implementation methodologies, to identify additional optimization processes.

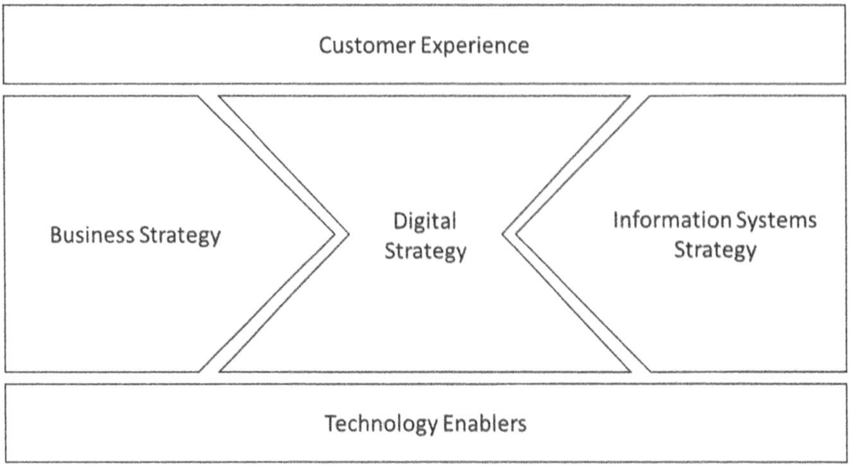


**Fig. 5.** Digital strategy has the center of an enterprise intelligence

## 7 Conclusions

With this conceptual article, it can be understood the impact that information systems have on the knowledge and business management inside companies, namely the creation of capabilities of leveraging data analytics able to understand deeper the value chain. Basically, these analytics intend to automate the data, information, knowledge, and wisdom (DIKW) hierarchy, potentiate that managers and decision-makers within companies are capable of maximizing their business value generation. This means that, if a company is considered as a set of processes, a company by itself is a business model function where this DIKW analytics, virtuous, continuous, and interactive cycle of adopting strategic or tactic actions in the market, will analyze the impacts and reactions of the market. A manager or decision-maker is then searching for its optimum market value function (OMV function).

The digital transformation of a company is, in fact, a process re-engineering inside the companies, adding technologies that will create a vision of "real time" everywhere in the value chain and will run as an automation authority, creating additional efficiencies and competitive advantages (OMV function optimization).

This trend of digital transformation, to get the OMV function optimization, is changing the way that companies understand its core strategic activities. What Michael Porter [1] has structured as a set of inbound logistics, operations, outbound logistics, sales and marketing and post-sales services activities, is changing with the need to incorporate information systems as a core and strategic activity of the enterprises value chains, because information systems are more than the "old and only" information technologies. In reality, information systems are the set of people, processes, and technologies! Information systems are fundamental to address the strategic optimization of the automation and efficiency transformation within a company. This means that when a company is building a business strategy, needs today not only to create, but fundamentally to assure the right alignment with Digital and Information Systems strategies.

It is a fact that the speed of new technologies' appearance can create market disruptions and add continuous pressures to existing markets' *status quo*. The most adaptive and digital companies, according to several studies, are 26% more profitable than their industry competitors and generate 9% more revenues, which represents an output of the intentions of this article's conceptual terms. The future challenges are to understand the several steps and variables that comply with a set of best practices strategies that can transform the duo "data and information" basis in a digital and business optimized company.

## References


1. Porter, M.E., Millar, V.E.: How Information Gives You Competitive Advantage. Harv. Bus. Rev (1985)
2. Ackoff, R.L.: From Data to Wisdom. J. Appl. Syst. Anal. **16**, 3–9 (1989)
3. Weinberger, D.: The Problem with the Data-Information-Knowledge-Wisdom Hierarchy. Harv. Bus, Rev (2010)
4. Rowley, J.: The wisdom hierarchy: representations of the DIKW hierarchy. J. Inf. Sci. **33**, 163–180 (2007). https://doi.org/10.1177/0165551506070706
5. Rothberg, H.N., Erickson, G.S.: Big data systems: knowledge transfer or intelligence insights? J. Knowl. Manag. Kempston **21**, 92–112 (2017). https://doi.org/10.1108/JKM-07-2015-0300
6. Kite-Powell, J.: Fujitsu Bets on Connected Cow Technology to Transform Farms One by One [WWW Document]. Forbes (2017). https://www.forbes.com/sites/jenniferhicks/2017/08/31/fujitsu-bets-on-connected-cow-technology-to-transform-farms-one-by-one/. Accessed 11.11.18
7. O'Reilly: "Connected Cows?"—Joseph Sirosh (Strata + Hadoop 2015) (2015)
8. Gallagher, S.: The Internet of Cows: Azure-powered pedometers get dairies mooovin' [WWW Document]. Ars Tech. (2015). https://arstechnica.com/information-technology/2015/04/the-internet-of-cows-azure-powered-pedometers-get-dairies-mooovin/. Accessed 11.11.18
9. AIG. IoT Case Studies: Companies Leading the Connected Economy (2016)
10. Uzialko, A.C.: 6 Incredible Ways Businesses are Using Artificial Intelligence Today [WWW Document] (2016). https://www.businessnewsdaily.com/9542-artificial-intelligence-businesses.html. Accessed 11.11.18
11. Pyatakova, S.: 4 Real-life IoT Examples Proven to Transform Business [WWW Document] (2018). https://www.itransition.com/blog/4-real-life-iot-success-cases-proven-to-transform-business. Accessed 11.11.18
12. Granville, K.: Facebook and Cambridge Analytica: What You Need to Know as Fallout Widens. Times, NY (2018)
13. Liautaud, B.: E-Business Intelligence: Turning Information into Knowledge into Profit. McGraw-Hill Inc, New York, NY, USA (2000)
14. Vercellis, C.: Business Intelligence: Data Mining and Optimization for Decision Making. Wiley (2011)
15. Power, D.J.: A Brief History of Decision Support Systems [WWW Document] (2018). http://dssresources.com/history/dsshistoryv28.html. Accessed 11.14.18
16. Porter, M.E.: Competitive Advantage: Creating and Sustaining Superior Performance, 1st edn. Free Press, New York (1998)
17. Collins, J.: Good to Great: Why Some Companies Make the Leap and Others Don't, 1st edn. Harper Business, New York, NY (2001)
18. Bourgeois, D.T.: Information Systems for Business and Beyond. Lulu.com, S.l (2014)

19. Brynjolfsson, Erik: The Productivity Paradox of Information Technology: Review and Assessment. ACM, Commun (1993)
20. Brynjolfsson, Erik, Hitt, Lorin M.: Beyond the Productivity Paradox: Computers are the Catalyst for Bigger Changes. ACM, Commun (1998)
21. Carr, N.: IT doesn't matter. Harv. Bus. Rev (2003)
22. Elton, C.S.: Animal Ecology, Reprint. University of Chicago Press, Chicago (2001)
23. Andal-Ancion, A., Cartwright, P.A., Yip, G.S.: The Digital Transformation of Traditional Business. MIT Sloan Manag. Rev (2003)
24. IBM Corporation: The new voice of the CIO: Implications for the C-Suite [WWW Document]. URL undefined (2009). Accessed 12.26.18
25. Ansoff, H.I.: Corporate Strategy: An Analytic Approach to Business Policy for Growth and Expansion, Later Printing. McGraw-Hill, London (1965)
26. Porter, M.E.: Competitive Strategy: Techniques for Analyzing Industries and Competitors, 1st edn. Free Press, New York (1998)
27. Johnson, G., Scholes, K.: Exploring Corporate Strategy. Prentice Hall (1993)
28. Collins, J., Porras, J.I.: Built to Last: Successful Habits of Visionary Companies, 3rd edn. Harper Business, New York (1994)
29. Nicholas Felton: How Americans Spend their Money [WWW Document] (2008). https://archive.nytimes.com/www.nytimes.com/imagepages/2008/02/10/opinion/10op.graphic.ready.html. Accessed 2.19.19
30. Peter Ffoulkes: The Exponential Growth of Data. insideBIGDATA (2017)
31. Anthony, S.D., Viguerie, S.P., Schwartz, E.I., Landeghem, J.V.: Corporate Longevity Forecast: Creative Destruction is Accelerating. Innosight (2018)
32. Mauboussin, M.J., Callahan, D., Majd, D.: Corporate Longevity: Index Turnover and Corporate Performance, GLOBAL FINANCIAL STRATEGIES. Credit Suisse (2017)
33. Mochari, I.: Why Half of the S&P 500 Companies Will Be Replaced in the Next Decade [WWW Document]. Inc.com (2016). https://www.inc.com/ilan-mochari/innosight-sp-500-new-companies.html. Accessed 2.22.19
34. CapGemini, MIT Center for Digital Business: The Digital Advantage: How Digital Leaders Outperform their Peers in Every Industry. Capgemini Worldw (2012)
35. Court, D., Perrey, J., McGuire, T., Gordon, J., Spillecke, D.: Big Data, Analytics, and the Future of Marketing & Sales. McKinsey (2015)
36. Kasturi, V.: Digital speed: 5 strategies to rapidly deliver business value from digital transformation [WWW Document]. CIO (2018). https://www.cio.com/article/3267981/digital-transformation/digital-speed-5-strategies-to-rapidly-deliver-business-value-from-digital-transformation.html. Accessed 2.22.19
37. Mohammed Hashim: Art of Digital Jujutsu. In: ResearchGate. Presented at the Dell EMC World (2016)
38. Ross, J.: Digital Is About Speed—But It Takes a Long Time. MIT Sloan Manag. Rev (2018)
39. Welchman, L.: Managing Chaos: Digital Governance by Design, 1st edn. Rosenfeld Media, Brooklyn, New York (2015)

# Improving In-Home Appliance Identification Using Fuzzy-Neighbors-Preserving Analysis Based QR-Decomposition

Yassine Himeur[1(✉)], Abdullah Alsalemi[1], Faycal Bensaali[1], and Abbes Amira[2]

[1] Department of Electrical Engineering, Qatar University, Doha, Qatar
{yassine.himeur,a.alsalemi,f.bensaali}@qu.edu.qa
[2] Institute of Artificial Intelligence, De Montfort University, Leicester, UK
abbes.amira@dmu.ac.uk

**Abstract.** This paper proposes a new appliance identification scheme by introducing a novel approach for extracting highly discriminative characteristic sets that can considerably distinguish between various appliance footprints. In this context, a precise and powerful characteristic projection technique depending on fuzzy-neighbors-preserving analysis based QR-decomposition (FNPA-QR) is applied on the extracted energy consumption time-domain features. The FNPA-QR aims to diminish the distance among the between class features and increase the gap among features of dissimilar categories. Following, a novel bagging decision tree (BDT) classifier is also designed to further improve the classification accuracy. The proposed technique is then validated on three appliance energy consumption datasets, which are collected at both low and high frequency. The practical results obtained point out the outstanding classification rate of the time-domain based FNPA-QR and BDT.

**Keywords:** Appliances identification · Feature extraction · Time-domain descriptor · Dimensionality reduction · FNPA-QR · Bagging decision tree

## 1 Introduction

Various studies claim that domestic buildings are responsible of up to 40% of the total energy use around the globe in the past decades [1]. A large part of this energy is wasted because of inappropriate usage habits and absence of data highlighting the consumption rate of each device in residential area. In addition, individuals have not fully harnessed possibilities to reduce their wasted energy. Consequently, creating an energy efficiency lifestype begins with (i) pro-


X.-S. Yang et al. (eds.), *Proceedings of Fifth International Congress on Information and Communication Technology*, Advances in Intelligent Systems and Computing 1183, https://doi.org/10.1007/978-981-15-5856-6_30

viding them with fine-grained power use profiles at real-time, (ii) helping them comprehend their consumption behavior and (iii) encouraging them to make energy saving actions through allowing tailored recommendations [2]. Therefore, it would be of great positive potential if smartphones can alert users with notifications about power consumption (PC) level of each appliance and provides appliance usage status [3,4]. Currently, sub-meters and smart sensors are the candidate solutions to collect this kind of data. The first one provides aggregated consumption profiles without permitting to pick up individual loads, while the second category is very costly to be deployed in residential buildings [5–7].

In that direction, energy disaggregation for appliance identification becomes increasingly a hot area of research. Individual consumption footprints can be picked up by sampling the whole PC signal through making use of a non-intrusive load monitoring (NILM) scheme [8]. The proposed framework focuses on the design of an efficient appliance identification system using a robust feature extraction scheme based on dimensionality reduction. First, time-domain (TD) descriptors are implemented to capture relevant features from PC footprints. They can pick up step-changes in PC signals generally occurring when an appliance is switched on/off. Second, a fuzzy-neighbors-preserving analysis based QR-decomposition, namely FNPA-QR, is introduced to reduce the dimension of TD signatures. In this context, as the PC signals have an important variance nature, the extracted TD data from the PC signals are usually scattered freely within the initial feature ensemble. Third, a bagging decision tree (BDT) classifier is designed to robustly identify each appliance category. The BDT exploits multiple weak machine learning models to develop a nearly optimal classifier that outperform other classification models, especially when it is applied on TD features.

The remaining of this paper consists of the following parts. The proposed appliance identification system is explained in Sect. 2. Section 3 presents the obtained empirical results with a set of comparisons at different levels of the proposed appliance recognition architecture. Finally, Sect. 4 summarizes the main outcomes of this framework.

## 2 Proposed System

In this section, contributions of this framework are highlighted through explaining in details the main modules proposed to design a robust in-home appliance recognition architecture. Figure 1 portrays the block-diagram of the proposed system.

### 2.1 Feature Extraction Using TD Descriptors

In order to select the best TD descriptor for the appliance recognition purpose, five TD descriptors are firstly considered and their performance are compared. The best descriptor is then implemented to design the appliance identification

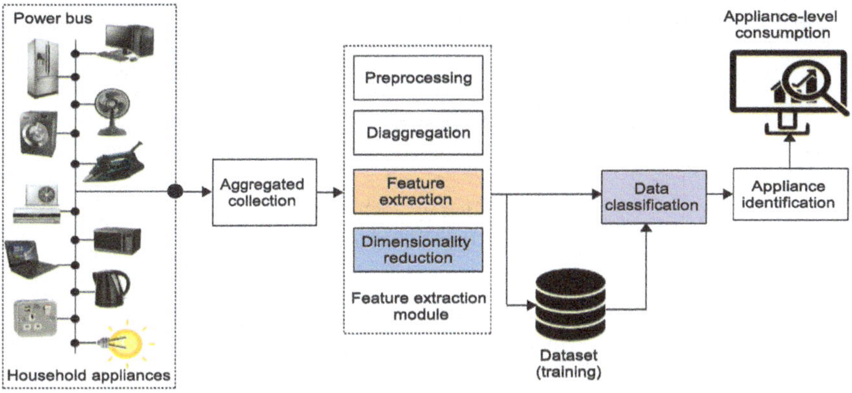


**Fig. 1.** Block diagram of the proposed in-home appliance identification system

system. In order to describe the TD feature descriptors used under our framework, let us consider a sampled version of the energy consumption signal denominated as: $y[i]$, with $i = 1, 2, \dots, M$, of length $M$ and collected at the sampling frequency $f_y$, in order to extract the TD features, a windowing process is applied on $y$ where a window length $N$ is employed and the TD property $Y(k)$ of each window $k$ $(k = 1, 2, \dots, K)$ is then collected as follows:

- Root mean square feature (RMSF)

$$Y_{\text{RMS}}(k) = \sum_{i=1}^{N} \sqrt{\tfrac{1}{N}(y_i^2)} \tag{1}$$

- Mean absolute deviation feature (MADF)

$$Y_{\text{MAD}}(k) = \sum_{i=1}^{N} \tfrac{1}{N} \left| y_i - \mu \right| \tag{2}$$

where $\mu$ represent the central tendency,

- Integrated absolute magnitude feature (IAMF)

$$Y_{\text{IAMF}}(k) = \frac{1}{N} \sum_{i=1}^{N} \tfrac{y_i^2}{2} \text{sgn}(y_i) + \mu \tag{3}$$

- Waveform length feature (WLF)

$$Y_{\text{WLF}}(k) = \log\left( \sum_{i=1}^{N-1} \left| y_{i+1} - y_i \right| \right) = \log\left( \sum_{i=1}^{N-1} \left| \Delta y_i \right| \right) \tag{4}$$

- Slope sign change feature (SSCF)

$$S_{\mathrm{SSC}}(k) = \sum_{i=2}^{N-1} f[(y_i - y_{i-1}) \times (y_i - y_{i+1})] \tag{5}$$

where

$$f(y) = \begin{cases} 1 \text{ if } y \geq \text{threshold} \\ 0 \text{ otherwise} \end{cases} \tag{6}$$

### 2.2 Dimensionality Reduction Using FNPA-QR

The FNPA-QR is applied on the TD signals, which is a variant of fuzzy-linear discriminant analysis (FLDA). The latter has been used to study the class relationship between samples, however the main drawback of FLDA is related to the fact that it could not find out the regional geometric structure of samples. Accordingly, in discriminant analysis, the regional arrangement is more prominent in comparison to the global one [9]. In addition, the discrimination ability between samples pertaining to different groups can be improved if the local structure is preserved. To that end, the FNPA-QR introduces a novel feature projection scheme that can map the samples into a new subspace by analyzing the adjacent patterns. Consequently, it makes adjacent samples with the same label more close and in contrast, turns the adjacent coefficients with different labels to be far away. The main steps to perform the FNPA-QR approach are summarized.

1. Set the features data using a matrix structure $Y(i,j)$, where $i$ is the index of feature vectors and $j$ is the number of patterns in each vector and set the number of reduced samples $r$;
2. Estimate the within-class-scatter (WCS) array $Y_W$ as:

$$Y_W = \left(YDY - YWY^T\right) = YL_1Y^T \tag{7}$$

   where $L_1 = D - W$ represents the Laplacian array as described in [10,11]. $D$ constitutes a diagonal array that its coefficients are derived by summing the WCS array $W$. $W$ is the array of the WCS patterns;
3. Estimate the between-class-scatter (BCS) array $Y_B$ as:

$$Y_B = (MEM^T - MBM^T) = ML_2M^T \tag{8}$$

   where $M$ is the mean matrix of total patterns. $L_2 = E - B$ and $E$ represents a diagonal array; its inputs are column sums of $B$. $B$ is the array of the BCS patterns;
4. Estimate the transformation matrix $\mathbf{H}_{\text{FNPA-QR}}$ as follows

$$\mathbf{H}_{\text{FNPA-QR}} = \text{argmax trace}\left(\frac{\mathbf{H}^T Y_B \mathbf{H}}{\mathbf{H}^T Y_W \mathbf{H}}\right) \tag{9}$$

5. Calculate the matrix $Q$ using $H = Q \times R$, where $R$ represents an upper-triangular array and $Q$ defines an orthogonal array;
6. Make $\mathbf{H}_{\text{FNPA-QR}} = Q$ and project the feature matrix with the transformation matrix as follows:

$$Y'(i,r) = Y(i,r) \times H_{\text{FNPA-QR}} \tag{10}$$

### 2.3 Bagging Decision Tree (BDT) Classifier

The idea behind BDT algorithm relies on the fact that the bagging process can trigger unstable weak classifiers to generate nearly optimal classifier. Instead of training the weak classifier on the global data set, each one is trained on a bootstrap set that is derived from the ensemble set. This makes the samples distribution along the training similar to the initial distribution. Consequently, the individual classifiers in a bagging ensemble can accurately classify the samples. Finally, a majority vote process is performed to normalize the performance of the BDT model and to additionally improve the classification accuracy. Figure 2 explains in details the steps required to implement the BDT classifier.

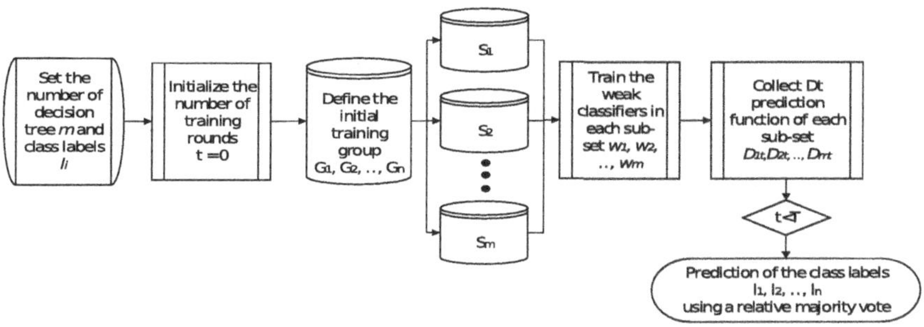


**Fig. 2.** Flowchart of a the BDT algorithm

## 3 Experimental Results

The performance of the developed appliance in-home identification architecture is evaluated using three different energy consumption datasets. The ACS-f2 is the second version of the appliance consumption signature repository. It encompasses the electricity consumption footprints of various electrical device categories using 0.1 Hz sampling frequency [12]. The GREEND: collects daily PC data of more than 35 domestic appliances deployed in 8 different households for a period ranging from 6–12 months. Additionally, consumption fingerprints are gathered using 1 Hz sampling frequency [13]. The WITHED: captures electricity consumption signatures for up to 47 appliance classes for a short period of 5 sec and using a 44,000 Hz sampling rate. Under this framework, 11 appliance classes are employed to validate the proposed technique [14].

First of all, the performance of the different TD descriptors using different window lengths is investigated with regard to the BDT classifier. The window lengths considered in the evaluation are selected according to the length of power consumption signals in each database. The accuracy outputs are then plotted in Fig. 3. It is clearly shown that RMSF descriptor can slightly outperform other descriptors, especially under window lengths of 2048 and 3072 for both GREEND and WHITED, and under a window length of 128 for ACS-F2. Furthermore, is has better performance stability than the other descriptors. Therefore, in the rest of this paper, the results of the proposed scheme are collected with reference to the RMSF descriptor using window lengths of 128 for ACS-F2 and 2048 for both GREEND and WHITED.

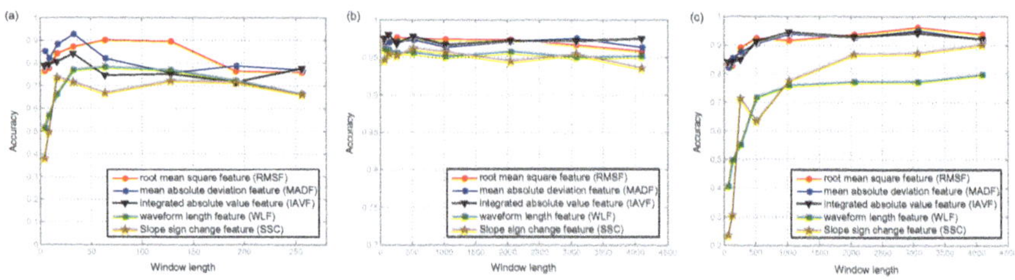


**Fig. 3.** Accuracy performance of different TD descriptors for the **a** ACS-F2, **b** GREEND and **c** WHITED datasets

### 3.1 Comparison Versus Other Dimensionality Reduction

The performances the FNPA-QR system are validated with reference to other feature projection schemes, including principal component analysis (PCA), linear discriminant analysis (LDA) and FLDA for appliance identification. Table 1 depicts the accuracy, F-score and computational complexity of the proposed technique based on FNPA-QR in comparison to other dimensionality reduction approaches. The outputs are collected using MATLAB 9.4 (R2018a) executed on a desktop that has a Core i7-3770S processor, 16 GRAM and 3.1 GHz. It is witnessed that the accuracy and F-score of the FNPA-QR are highly improved in comparison to other feature projection schemes. However, the time complexity is increased and this is due to the fact that FNPA-QR uses a fuzzy process along with the QR decomposition.

### 3.2 Comparison with Other Classifiers

Under the classification stage, in addition to the developed BDT classifier, other machine learning (ML) models were also deployed in the experiments including: support vector machine (SVM), deep neural networks (DNN), K-nearest neighbors (KNN), decision tree (DT). These models were proceeded with respect to

**Table 1.** Performance comparison of FNPA-QR versus other dimensionality reduction techniques

| Dataset | Performance | PCA | LFDA | FLDA | FNPA-QR |
|---|---|---|---|---|---|
| ACS-F2 | Accuracy | 0.78 | 0.83 | 0.92 | **0.99** |
| | F-score | 0.76 | 0.82 | 0.92 | **0.98** |
| | Time complexity (s) | **0.035** | 1.81 | 0.71 | 0.85 |
| GREEND | Accuracy | 0.86 | 0.92 | 0.95 | **1** |
| | F-score | 0.85 | 0.92 | 0.94 | **1** |
| | Time complexity (s) | **1.6** | 34.3 | 15.2 | 17.5 |
| WHITED | Accuracy | 0.83 | 0.91 | 0.94 | **1** |
| | F-score | 0.83 | 0.9 | 0.93 | **1** |
| | Time complexity (s) | **0.071** | 6.8 | 3.8 | 3.4 |

**Table 2.** Accuracy of the appliance identification system based on FNPA-QR with reference various ML classifier

| ML | Classifier | ACS-F2 | | GREEND | | WHITED | |
|---|---|---|---|---|---|---|---|
| algo | Parameters | accuracy | F-score | Accuracy | F-score | Accuracy | F-score |
| SVM | Linear Kernel | 0.94 | 0.93 | 0.95 | 0.95 | 0.98 | 0.98 |
| SVM | Quadratic kernel | 0.92 | 0.91 | 0.93 | 0.93 | 0.9 | 0.86 |
| SVM | Gaussian kernel | 0.94 | 0.93 | 0.95 | 0.95 | 0.93 | 0.9 |
| KNN | K = 1/Euclidean distance | 0.97 | 0.97 | 0.98 | 0.98 | 0.96 | 0.93 |
| KNN | K = 10/Weighted euclidian dis | 0.96 | 0.95 | 0.98 | 0.98 | 0.95 | 0.92 |
| KNN | K = 10/Cosine dist | 0.94 | 0.94 | 0.96 | 0.96 | 0.92 | 0.87 |
| DT | Fine, 100 splits | 0.98 | 0.97 | 0.99 | 0.99 | 0.93 | 0.91 |
| DT | Medium, 20 splits | 0.93 | 0.93 | 0.96 | 0.93 | 0.94 | 0.9 |
| DT | Coarse, 4 splits | 0.92 | 0.89 | 0.94 | 0.92 | 0.91 | 0.88 |
| DNN | 50 hidden layers | 0.96 | 0.95 | 0.98 | 0.98 | 0.97 | 0.96 |
| EBT | 30 learners, 42 k splits | **0.99** | **0.98** | 1 | 1 | 1 | 1 |

different classification parameters. Table 2 provides the obtained results in terms of the accuracy and F-score. It is shown that the BDT classifiers surpasses other ML algorithms for all the datasets examined under this framework.

## 4 Conclusion

This paper presented a robust non-intrusive appliance identification system using TD descriptors. More specifically, our focus was on the use of the FNPA-QR as a dimensionality reduction module. FNPA-QR played an important role since it (i) reduced the amount of data samples in the feature vectors, (ii) decreased the distance between data points of the same appliance category and increased the distance among the samples from dissimilar classes. Moreover, a BDT classifier was designed that further enhanced the identification accuracy. Consequently, the results of the evaluation were promising, since they showed that using the proposed TD descriptor based FNPA-QR and BDT classifier, it was possible to reach an optimal identification accuracy and also outperform other dimensionality reduction techniques and ML models.

**Acknowledgements.** This paper was made possible by National Priorities Research Program (NPRP) grant No. 10-0130-170288 from the Qatar National Research Fund (a member of Qatar Foundation). The statements made herein are solely the responsibility of the authors.

## References

1. Rashid, H., Singh, P., Stankovic, V., Stankovic, L.: Can non-intrusive load monitoring be used for identifying an appliance's anomalous behaviour? Appl. Energy **238**, 796–805 (2019)
2. Alsalemi, A., Sardianos, C., Bensaali, F., Varlamis, I., Amira, A., Dimitrakopoulos, G.: The role of micro-moments: a survey of habitual behavior change and recommender systems for energy saving. IEEE Syst. J. **13**(3), 3376–3387 (2019)
3. Alsalemi, A., Bensaali, F., Amira, A., Fetais, N., Sardianos, C., Varlamis, I.: Smart energy usage and visualization based on micro-moments. In: Bi, Y., Bhatia, R., Kapoor, S. (eds.), Intelligent Systems and Applications, pp. 557–566. Springer International Publishing, Cham (2020)
4. Sardinos, C., Varlamis, I., Dimitrakopoulos, G., Alsalemi, A., Bensaali, F., Amira, A.: I want to ...change micro-moment based recommendations can change users. In: Proceedings 8th International Conference on Smart Cities and Green ICT Systems (2019), pp. 1–10
5. Li, D., Dick, S.: Residential household non-intrusive load monitoring via graph-based multi-label semi-supervised learning. IEEE Trans. Smart Grid **10**(4), 4615–4627 (2019)
6. Alsalemi, A., Ramadan, M., Bensaali, F., Amira, A., Sardianos, C., Varlamis, I., Dimitrakopoulos, G.: Endorsing domestic energy saving behavior using micro-moment classification. Appl. Energy **250**, 1302–1311 (2019)
7. Alsalemi, A., Bensaali, M.F., Amira, A., Sardianos, C., Varlamis I., Dimitrakopoulos, G.: Towards domestic energy efficiency: using micro-moments for personalized behavior change recommendations. In: 8th Global Conference on Global Warming (GCGW-2019), Doha, Qatar (2019)
8. Welikala, S., Dinesh, C., Ekanayake, M.P.B., Godaliyadda, R.I., Ekanayake, J.: Incorporating appliance usage patterns for non-intrusive load monitoring and load forecasting. IEEE Trans. Smart Grid **10**(1), 448–461 (2019)
9. Cai, D., He, X., Zhou, K., Han, J., Bao, H.: Locality sensitive discriminant analysis. In: Proceedings of the 20th International Joint Conference on Artificial Intelligence (2007), pp. 708–713
10. He, X., Yan, S., Hu, Y., Niyogi, P., Zhang, H.K.: Face recognition using Laplacianfaces. IEEE Trans. Pattern Anal. Mach. Intell. **27**(3), 328–340 (2005)
11. Khushaba, R.N., Kodagoda, S., Liu, D., Dissanayake, G.: Muscle computer interfaces for driver distraction reduction. Comput. Methods Programs Biomed. **110**(2), 137–149 (2013)
12. Ridi, A., Gisler, C., Hennebert, J.: Acs-f2—a new database of appliance consumption signatures. In: 6th International Conference on Soft Computing and Pattern Recognition (SoCPaR) (2014), pp. 145–150

13. Monacchi, A., Egarter, D., Elmenreich, W., D'Alessandro, S., Tonello, A.M.: Greend: An energy consumption dataset of households in Italy and Austria. In: IEEE International Conference on Smart Grid Communications (SmartGridComm) (2014), pp. 511–516
14. Kahl, M., Haq, A.U., Kriechbaumer, T., Jacobsen, H.-A.: Whited-a worldwide household and industry transient energy data set. In: 3rd International Workshop on Non-Intrusive Load Monitoring (2016)

# The Use of Two-Dimensional Landmark-Based Geometric Morphometrics to Assess Spinal and Vertebral Malformations in Individuals with Spinal Cord Injuries: A Pilot Study

Maria Auxiliadora Marquez[1], Giovanni Galeoto[2](✉), Anna Berardi[3], Marco Tofani[4], Massimiliano Mangone[5], and Paolo Colangelo[6]

[1] CPO—Ostia Paraplegic Center, Rome, Italy
[2] Department of Public Health and Infectious Diseases, Sapienza University of Rome, Rome, Italy
giovanni.galeoto@uniroma1.it
[3] Department of Human Neurosciences, Sapienza University of Rome, Rome, Italy
[4] Neurorehabilitation Unit, Bambino Gesù Children's Hospital, Rome, Italy
[5] Department of Anatomy Histology, Forensic Medicine and Orthopedics, Sapienza University of Rome, Rome, Italy
[6] Research Institute on Terrestrial Ecosystems, National Research Council (CNR), Monterotondo, Rome, Italy

**Abstract.** The purpose of this study was to use geometric morphometrics (GMM), a technique suitable for the study of complex anatomical objects, to investigate spinal cord injury (SCI). Eight individuals with SCIs who underwent radiologic evaluation of their lumbar column in a lateral seated position prior to recruitment were included in the study. Each individual was assessed with rasterstereography using the Formetric 4D, and the results were compared with an X-ray of the column analysed through two-dimensional landmark-based GMM. A principal component analysis (PCA) was performed to describe shape variation. Subsequently, the correlation between the Formetric 4D indexes and the shape of the first principal component axis (PC1) was measured with Spearman's rank correlation coefficient. Thin-plate spline deformation grids were used to describe shape changes in the morphospace depicted by the PCA and to describe shape changes predicted by linear regression. Through the analysis of human X-ray plates, we highlighted the ability of GMM to describe the shape of the column and to evaluate spinal and vertebral malformations. This pilot study is the first step for using a GMM approach to investigate human spinal cord abnormalities. These results provide clinicians and researchers a new method to evaluate bone structures that could provide important information about the development and progression of various deformities in the future.

**Keywords:** Column · Deformity assessment · Geometric morphometrics · Rasterstereography · Spinal cord injury · Technology · X-ray


X.-S. Yang et al. (eds.), *Proceedings of Fifth International Congress on Information and Communication Technology*, Advances in Intelligent Systems and Computing 1183, https://doi.org/10.1007/978-981-15-5856-6_31

# 1 Introduction

Geometric morphometrics (GMM), a method for the statistical analysis of shape variation and its covariation with other variables, [1] quantifies in two or three dimensions geometric features of anatomical structures. [2–6].

In GMM, shape is represented by the Cartesian coordinates of anatomical landmarks. Generalized Procrustes analysis (GPA) is a method of statistical analysis that is used to compare the shapes of objects, or the results of surveys, interviews, or panels. GPA removes the effects of non-shape variation (size), and GMM quantifies the variation in the shape of anatomical objects despite their scale, position and orientation. [1, 2, 5, 7, 8] Furthermore, the analysis of shape coordinates can be directly visualized as transformation of 2D or 3D landmark configurations that can be interpreted directly and graphically [1, 7]. This analytical power is not achieved by traditional morphometrics, and in recent decades, GMM has become a state-of-the-art complex biological shape analysis method. Because of its evident advantage in the analysis of structures such as bones, GMM is a potentially useful technique for spinal deformity assessment.

Patients with SCIs use wheelchair in most of their daily activities [9]. Periodic assessment and monitoring of the patient sitting posture are necessary in order to prevent the development and worsening of deformities and to avoid joint or tendon retractions [10]. The minimization of complications and optimization of outcomes are some of the essential goals of patient treatment. A recent study reported a spinal deformity incidence of 21% among adult patients with SCIs [11].

The present study aimed to use GMM to describe vertebral and cord shapes from X-ray plates of individuals with SCIs. Using GMM, the authors collected anatomical landmarks chosen to describe not only single lumbar vertebrae but also the entire shape of this spinal segment. This method allowed the description of shape changes among patients with different degrees of spinal cord deformations and the comparison of the results obtained with those from the rasterstereography indexes.

# 2 Materials and Methods

This study was developed by rehabilitation professionals, anthropologists and clinical engineers from Sapienza University of Rome and from ROMA—Rehabilitation and Outcome Measures Assessment Association. The research group has carried out of many outcome measures in Italy [12–29].

## 2.1 Participants and Spinal Deformity Assessment

Participants were recruited in June 2018 from Ostia Paraplegic Centre. In order to be included in the study, participants had to be older than 18 years old, have a spinal injury at the lumbar spine level, hold a radiologic evaluation of their lumbar column in a lateral seated position prior to recruitment, not have pressure injuries or be pregnant during the study. Eight subjects agreed to participate, firmed the informed consent and completed the study.

Every person included in the study was informed about procedures and purposes of the study during an interview conducted as preliminary step. In order to evaluate spinal deformities in people with SCIs, authors used Formetric 4D [4]. This tool uses rasterstereography and by analysing the spinal column, it facilitates clinical practice [6–9]. As a result of this optical scan, the system detects anatomical landmarks. The data obtained are analysed by the software which reconstructs the shape of the column, and the position of the basin in three dimensions; it creates 12 images in 6 s and calculates an average value.

## 2.2 Geometrico Morphometric

Sixteen anatomical landmarks of the lumbar vertebrae were collected on X-ray plates using the TPSdig2 software (see Fig. 1). Landmark coordinates were aligned using a generalized Procrustes analysis (GPA) and subsequently used to perform statistical analyses. A principal component analysis (PCA) was performed to describe shape variation among the investigated patients. Subsequently, the correlation between the Formetric 4D indexes and the shape described by the first principal component axis (PC1) was tested using Spearman's correlation coefficient. The statistical significance was set for p-values less than 0.05. Finally, thin-plate spline deformation grids [1] were used to describe shape changes in the morphospace depicted by the PCA and to describe shape changes predicted by linear regression.

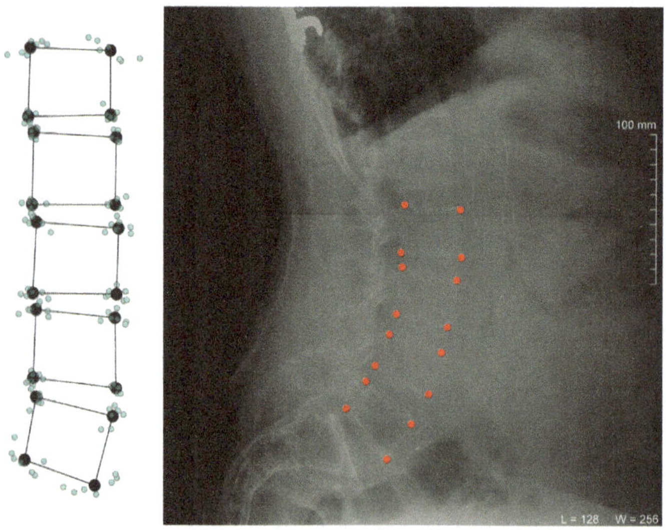


**Fig. 1.** Sixteen landmarks were collected on X-ray plates (right) on eight participants. Landmark configuration of each patient was subsequently aligned using GPA (left)

Statistical analysis was performed using the *R* statistical environment. Shape analysis, including the GPA, PCA and thin-plate spline, was performed using the geomorph software package.

## 3 Results

Eight subjects completed the study. The mean (standard deviation) age was 45.22 (12.21) years, and they spent a mean (SD) 11.67 (2.45) hours in their wheelchairs per day. The demographic characteristics of the participants are summarized in Table 1.

**Table 1.** Demographic characteristics of the eight participants

| | Mean | Median | Standard deviation |
|---|---|---|---|
| Age | 45.22 | 52.00 | 12.21 |
| Mean hours on wheelchair everyday | 11.67 | 10.00 | 2.45 |
| Years from injury | 23.14 | 26 | 14.21 |
| | | Frequence | Percentage |
| Gender | Male | 4 | 44 |
| Lesion level | C7 | 1 | 12.5 |
| | L2 | 1 | 12.5 |
| | L3 | 1 | 12.5 |
| | T10 | 1 | 12.5 |
| | T12 | 2 | 25 |
| | T3 | 1 | 12.5 |
| | T4 | 1 | 12.5 |

According to the PCA, the eight patients analysed were aligned in the morphospace along the PC1, which explained 77.14% of the results, according to the degree of lordosis. The deformation grids associated with each patient can be observed in Fig. 2. Along the second PC axis, the minor shape deformation associated with a single vertebra can be observed (see Fig. 2). Because the PC1 reflected global spine deformation, we used the PC1 scores for subsequent statistical analyses. There were no significant correlations between the Formetric 4D indexes and the PC1 except for the lumbar arrow (see Table 2).

Linear regression (see Fig. 3) showed a positive relationship between the lumbar arrow and shape, which suggests that smaller lumbar arrow values correspond to a return of normal lumbar lordosis, whereas larger lumbar arrow values correspond to an accentuated lordosis (see Fig. 3).

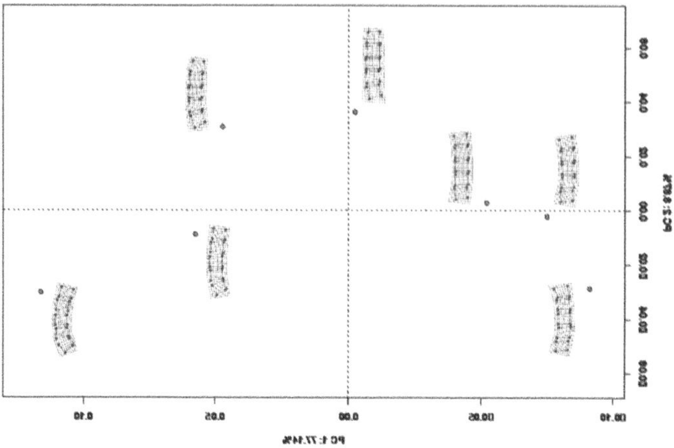

**Fig. 2.** Principal component analysis of the landmark coordinates of seven patients with spinal injuries. Patients are represented by black points. Each point is associated with a thin-plate spline deformation grid of the lumbar spinal segment

## 4 Discussion

This pilot study shows the potential use of landmark-based shape analysis in the study of spinal deformities. A comparison with the rasterstereography measured with Formetric 4D, which was analysed using Spearman's rho, showed one positive correlation with the lumbar arrow. Among other measurements performed by the Formetric 4D, such as prominent vertebra, the cervico-thoracic ratio, trunk length, and pelvic inclination, the lumbar arrow was the most likely measurement to correlate with a lateral X-ray plate of the lumbar column. The lumbar arrow clinically measured the distance (arrow) of lordosis from a plumb line that touches the apex of the lumbar lordosis from the occipital tuberosity. GMM is able to compute shape deformation of a lumbar segment and of single lumbar vertebrae. This allows a quantitative and qualitative assessment of spinal deformation. Positive linear correlation between the lumbar arrow and the shape confirm that GMM correctly identifies the degree of deformation in the segment of the spine. Furthermore, the local shape change associated with each vertebra suggests that GMM could also provide information on single element deformation in relation to a specific global deformation of the whole spinal segment.

**Table 2.** Spearman's rank correlation coefficient between the results of the Formetric 4D analysis and geometric morphometrics of lateral X-Ray plates of the column of patients with spinal cord injuries

| | Spearman's Rho | *p* value |
|---|---|---|
| Pelvic inclination [°] | 0.48 | 0.274 |
| Pelvic inclination [mm] | 0.51 | 0.247 |
| Twist Emibacini [°] | −0.57 | 0.200 |
| Kyphotic apex [mm] | 0.61 | 0.167 |
| Thoraco-lumbar inversion Point [mm] | 0.57 | 0.200 |
| Lordotic apex [mm] | 0.21 | 0.662 |
| Lumbo-sacral inversion reversal point [mm] | 0.32 | 0.498 |
| Cervical arrow (Stagnara) [mm] | −0.57 | 0.200 |
| Lumbar arrow (Stagnara) [mm] | 0.79 | 0.048 |
| Cervico-thoracic inversion-Thoraco-lumbar inversion [°] | −0.30 | 0.518 |
| Thoraco-lumbar inversion-Lumbo-sacral inversion | 0.39 | 0.396 |
| Surface rotation (Rms) [°] | 0.00 | 1.000 |
| Surface rotation (Amplitude) [°] | −0.25 | 0.595 |
| VPDM—lateral deviation (+Max) [mm] | −0.11 | 0.840 |
| VPDM—lateral deviation (−Max) [mm] | −0.36 | 0.444 |
| VPDM—lateral deviation (Amplitude) [mm] | 0.43 | 0.354 |

*VP* prominent vertebra; *DM* average point of the line that connects right and left lumbar dimple

A potential extension of this method is the analysis of large populations to study spinal and vertebral deformations according to different categories of potential risk.

In Conclusion, this study provides information that allows the use of GMM in clinical and research settings. This pilot study is the first step for the use of a GMM approach to investigate human spinal cord abnormalities. Finally, this research lays the basis for the use of this method in other fields of medicine. Specifically, these results provide clinicians and researchers a new method to evaluate bone structures, which could provide important information about the development and progression of various deformities in the future.

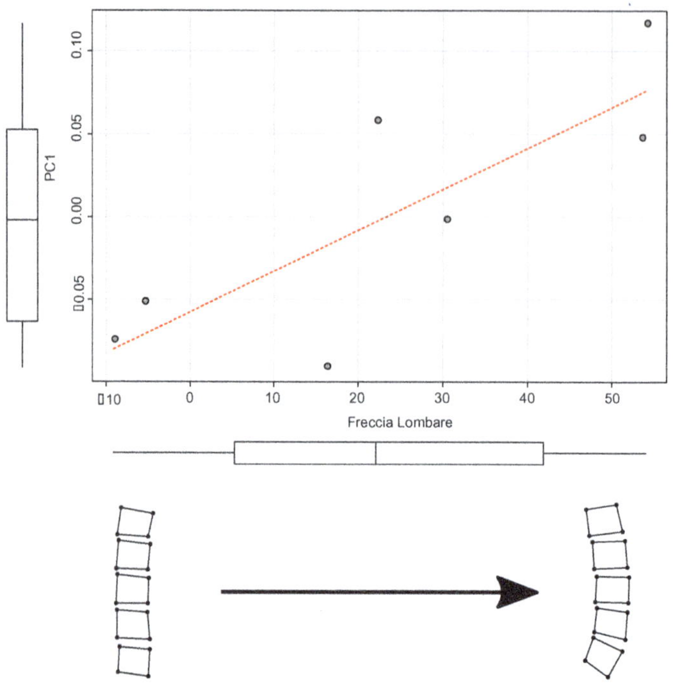


**Fig. 3.** Linear regression analysis between the lumbar arrow (*x* axis) and shape (*y* axis). The linear model shows a positive correlation between the two variables. Shape changes associated with the minimum and maximum value of the lumbar arrow suggest a spinal deformation from rectification (minimum) to accentuated lordosis (maximum)

## References

1. Bookstein, F.L.: Morphometric tools for landmark data geometric and biology. Cambrige University, Cambrige (1991)
2. Rohlf, F.J., Marcus, L.F.: A revolution in morphometrics. Trends Ecol. Evol. **8**, 129–132 (1993)
3. Bookstein, F.L.: Biometrics, biomathematics and the morphometric synthesis. Bull. Math. Biol. **58**, 313–365 (1996)
4. Adams, D.C., Rohlf, F.J.: Ecological character displacementin Plethodon: biomechanical differences found from a geometricmorphometric study. Proc. Natl. Acad. Sci. **97**, 4106–4111 (2000)
5. Mitteroecker, P., Gunz, P.: Advances in geometric morphometrics. Evol. Biol. **36**, 235 (2009)
6. Klingenberg, C.: Evolution and development of shape: integrating quantitative approaches. Nat. Rev. Genet. **11**(9), 623–635 (2010)
7. O'Higgins, P.: The study of morphological variation in the hominid fossil record: biology, landmarks and geometry. J. Anatomy **197**(1), 103–120 (2000)
8. Zelditch, M.L., Swiderski, D.L., Sheets, H.D.: Geometric morphometrics for biologists: a primer. Academic Press (2012)
9. Sonenblum, S.E., Sprigle, S.H., Martin, J.S. Everyday sitting behavior of full-time wheelchair users. J Rehabil Res Dev (2016)

10. Mattie, J., Borisoff, J., Miller, W.C., Noureddin, B.: Characterizing the community use of an ultralight wheelchair with "on the fly" adjustable seating functions: a pilot study. PLoS One (2017)
11. Yagi, M., Hasegawa, A., Takemitsu, M., Yato, Y., Machida, M., Asazuma, T.: Incidence and the risk factors of spinal deformity in adult patient after spinal cord injury: a single center cohort study. Eur. Spine J. **24**(1), 203–208 (2010)
12. Dattoli, S., Colucci, M., Soave, M.G., De Santis, R., Segaletti, L., Corsi, C. et al.: Evaluation of pelvis postural systems in spinal cord injury patients: Outcome research. J. Spinal Cord. Med. 1–13 (2018)
13. Covotta, A., Gagliardi, M., Berardi, A., Maggi, G., Pierelli, F., Mollica, R. et al.: Physical activity scale for the elderly: Translation, cultural adaptation, and validation of the Italian version. Curr. Gerontol. Geriatr. Res. (2018)
14. Galeoto, G., Colucci, M., Guarino, D., Esposito, G., Cosma, E., De Santis, R., et al.: Exploring validity, reliability, and factor analysis of the quebec user evaluation of satisfaction with assistive technology in an italian population: a cross-sectional study. Occup. Ther. Health Care. **32**(4), 380–392 (2018)
15. Galeoto, G., Sansoni, J., Valenti, D., Mollica, R., Valente, D., Parente, M., et al.: The effect of physiotherapy on fatigue and physical functioning in chronic fatigue syndrome patients: a systematic review. Clin. Ter. **169**(4), e184–e188 (2018)
16. Berardi, A., Dhrami, L., Tofani, M., Valente, D., Sansoni, J., Galeoto, G.: Cross-cultural adaptation and validation in the Italian population of the wolf motor function test in patients with stroke. Funct. Neurol. **33**(4), 229–253 (2018)
17. Berardi, A., Biondillo, A., Màrquez, M.A., De Santis, R., Fabbrini, G., Tofani, M. et al.: Validation of the short version of the Van Lieshout Test in an Italian population with cervical spinal cord injuries: a cross-sectional study. Spinal Cord. (2018)
18. Galeoto, G., Scialpi, A., Grassi, M.L., Berardi, A., Valente, D., Tofani, M. et al.: General sleep disturbance scale: translation, cultural adaptation, and psychometric properties of the italian version. CRANIO®, 1–9 (2019)
19. Savona, A., Ferralis, L., Saffioti, M., Tofani, M., Nobilia, M., Culicchia, G. et al.: Evaluation of intra-and inter-rater reliability and concurrent validity of the Italian version of the Jebsen–Taylor Hand Function Test in adults with rheumatoid arthritis. Hand Ther. **24**(2), 48–54 (2019)
20. Ruggieri, M., Palmisano, B., Fratocchi, G., Santilli, V., Mollica, R., Berardi, A., et al.: Validated fall risk assessment tools for use with older adults: a systematic review. Phys. Occup. Ther. Geriatr. **36**(4), 331–353 (2018)
21. Galeoto, G., Iori, F., De Santis, R., Santilli, V., Mollica, R., Marquez, M.A. et al.: The outcome measures for loss of functionality in the activities of daily living of adults after stroke: a systematic review. Topics in Stroke Rehabilitation (2019)
22. Berardi, A., Galeoto, G., Tofani, M., Mangone, M., Ratti, S., Danti, A., Sansoni, J., Marquez, M.A.: The validity of rasterstereography as a technological tool for the objectification of postural assessment in the clinical and educational fields: pilot study. In: Methodologies and Intelligent Systems for Technology Enhanced Learning, 9th International Conference, Workshops, pp. 55–62. https://doi.org/10.1007/978-3-030-23884-1_8
23. Puledda, F., Viganò, A., Celletti, C., Petolicchio, B., Toscano, M., Vicenzini, E., et al.: A study of migraine characteristics in joint hypermobility syndrome a.k.a. Ehlers-Danlos syndrome, hypermobility type. Neurol. Sci. **36**(8), 1417–1424 (2015)
24. Nobilia, M., Culicchia, G., Tofani, M., De Santis, R., Savona, A., Guarino, D.: Italian version of the Jebsen-Taylor hand function test for the assessment of hand disorders: a cross-sectional study. Am. J. Occup. Ther. **3**(3), 7303205080p1–7303205080p6 (2019)

25. Galeoto, G., Iori, F., De Santis, R., Santilli, V., Mollica, R., Marquez, M.A., et al.: The outcome measures for loss of functionality in the activities of daily living of adults after stroke: a systematic review. Top Stroke Rehabil. **26**(3), 236–245 (2019)
26. Tofani, M., Galeoto, G., Cazzetta, D., Berardi, A., Sansoni, J., Valente, D.: Validation of the pediatric evaluation of disability inventory in an italian population with autism spectrum disorder: a cross-sectional study. La Clinica Terapeutica **170**(6), e460–e464 (2019)
27. Galeoto, G., Turriziani, S., Berardi, A., Sansoni, J., Santilli, V., Mascio, M. et al.: Levels of cognitive functioning assessment scale: Italian cross-cultural adaptation and validation. Ann. di. Ig. (2020)
28. Berardi, A., Galeoto, G., Guarino, D., Marquez, M.A., De Santis, R., et al.: Construct validity, test-retest reliability, and the ability to detect change of the Canadian occupational performance measure in a spinal cord injury population. Spinal Cord. Ser. Cases **5**, 52 (2019)
29. Berardi, A., Saffioti, M., Tofani, M., Nobilia, M., Culicchia, G., Valente, D., et al.: Internal consistency and validity of the Jebsen-Taylor hand function test in an Italian population with hemiparesis. NeuroRehabilitation **45**(3), 331–339 (2019)

# Impact of Dimensionality on the Evaluation of Stream Data Clustering Algorithms

Naresh Kumar Nagwani(✉)

Computer Science and Engineering, National Institute of Technology Raipur, Raipur, India
nknagwani.cs@nitrr.ac.in

**Abstract.** Handling stream data is a tedious task. Recently numerous techniques are presented for analysing stream data. Stream data clustering is one of the important tasks in stream data mining. A number of application programming interfaces (APIs) are available for implementing the stream data clustering. These APIs can handle the stream data of any dimension. The objective of the presented paper is to explore the impact of dimensionality over the existing standard data stream clustering algorithms. Selected standard data stream clustering algorithms are compared for different dimensions of stream using six performance parameters, namely adjusted Rand index, Dunn index, entropy, F1 measure, purity and within cluster sum of square measure.

**Keywords:** Data stream · Stream clustering · Clustering evaluation · Stream processing

## 1 Introduction and Related Work

The sources of stream data are increasing in the modern world of Internet and vast amount of stream data is generated every hour. Analysing these vast amounts of continuous upcoming data is a big challenge. Stream data mining is the recent focus area where numerous inventions are going on to explore the possibility of better analysing the streamed data. Although a number of technologies and API's (Application Programming Interfaces) are available today to analyze the streamed data still there is a chance of improvements on it. Dimensionality of the streamed data is one of the key inputs which can impact the stream mining techniques. In this study, stream clustering task is focused for analysing the behaviour of the stream clustering algorithms for higher dimensionality of streamed data. For smaller dimension of the streamed data existing data stream clustering algorithm can perform better but it will be interesting to observe the behaviour of such algorithms with higher dimensionality of the streamed data. The presented work can be helpful to suggest the knowledge miner for selecting and improving the stream clustering task to accommodate the real-world streamed data consisting of higher dimensions.


X.-S. Yang et al. (eds.), *Proceedings of Fifth International Congress on Information and Communication Technology*, Advances in Intelligent Systems and Computing 1183, https://doi.org/10.1007/978-981-15-5856-6_32

Some of the existing works related to evaluation of stream data clustering algorithms are presented in the earlier works but the dimensionality impact still need some attention. Evaluation of some of the stream clustering algorithms are evaluated and compared using different performance parameters in different datasets by Mansalis et al. [1]. Various issues and challenges in applying stream clustering algorithms are also addressed in the paper. Some of the data stream methods and algorithms are reviewed and evaluated on standard synthetic data streams and real-life data streams by [2]. Density micro-clustering and density grid-based clustering algorithms are covered in the comparative analysis. It was presented that a single algorithm cannot satisfy all the performance measures. The objective of the present study is to evaluate the standard stream clustering algorithms for the streamed data having relatively higher dimensions.

## 2 Stream Data Clustering Algorithms

In this work, five standard stream clustering algorithms available in R statistical package [3] are used for the experiments and study. These algorithms are BICO [4], DBSTREAM [5], DSTREAM [6, 7], Sample (reservoir sampling-based clustering algorithm) [8–10], Window [11] with the horizon parameter as 20 and 100. All these algorithms are available in the stream package of *R* [12, 13].

BICO [4] is based on k-means clustering algorithm and performs fast computation of coresets in a data stream. It is derived from the popular clustering algorithm Balanced Iterative Reducing Clustering using Hierarchies (BIRCH). Instead of actual data points clustering is applied on the clustering features which store triplet information, i.e. number of data points, linear sum of data points and sum of squared data points. DBSTREAM [5] clustering algorithm is density-based stream clustering algorithm; in this algorithm, data points are assigned to micro-clusters using suitable radius and implements shared density-based re-clustering. The DBSTREAM algorithm verifies the newly arrived data point in the incoming stream using the pre-defined threshold value, and merging of data points takes place if the threshold value of dissimilarity value is less than any existing micro-clusters. DSTREAM (D-Stream) algorithm is grid-based clustering algorithm; it estimates the density by counting the data points in a cell. DSC_DStream is a variation of the D-Stream algorithm implemented by Chen and Tu [6]. Adjacent dense cells are merged for re-clustering to form macro-clusters. The concept of attraction between grids cells can be utilized alternatively for re-clustering [7]. Sample clustering algorithm extracts a fixed-size sample using reservoir sampling from a data stream. It takes the parameter *k* that represents the number of data points to be selected as sample from the stream. DSC_Sample is a clustering interface based on DSO_Sample data stream operator. Reservoir sampling [8] is used to select the number of representative points from the stream. Unbiased sampling is used for all data points in the algorithm by McLeod and Bellhouse [9]. To evolve data streams, biasing of samples towards recent data points can be more appropriate. Aggarwal et al. [10] have proposed another algorithm for biased sampling. Window-based clustering algorithms use a sliding window of fixed amount for focusing on the most recent data in the stream. It is driven by two parameters, namely horizon and lambda. Horizon represents the window length and DSC_Window represents a clustering interface. It is based on sliding window using a user supplied number known as window length and the dampened window models [11] pointing to most recent data points.

## 3 Experiments and Result Analysis

All the experiments are performed using stream API available in *R* [12, 13]. The description of dataset, setting of the algorithm parameters and analysis of the results of the experiments are presented in this section.

### 3.1 Dataset

For all the experiments, the stream data is generated using the stream package in *R*. Gaussians stream generated function from stream package is used for experiments. Gaussians generates [14] the data of randomly created fixed clusters with random multivariate. It uses the parameters number of clusters to be created, dimensions, matrix of mean values for all dimensions, list of covariance matrices, probability vector for presenting the likelihood of data points between cluster centres for minimize overlapping. Noise and noise probability is also selected between 0 and 1. Uniform distribution function is used for noise over a noise range. The stream generator is used with the number of clusters as 3 for different values of the dimensions. The values of the dimensions are taken as 2, 3, 4, 5, 10, 20 and 50. Noise probability of 5% (i.e. noise = 0.05) is taken while generating the stream data with the Gaussians generator. A sample of stream generated with three clusters and three dimensions using Gaussians generator is presented in Fig. 1.

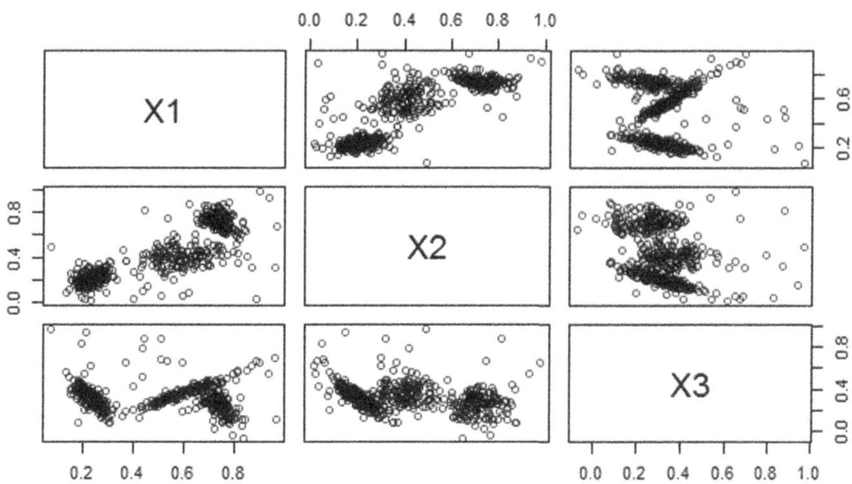


**Fig. 1.** Generated sample of 3D stream for the experiments

### 3.2 Performance Parameters

Six performance parameters are considered for evaluating the performance of clusters generated by stream clustering algorithms. The parameters selected are adjusted Rand index (cRand), Dunn index, entropy, F1 measure (F1), purity [15] and sum of squared measure within cluster (SSQ). Rand index is derived using the false positives and negatives. It measures the number of correct measures by the total measures. In this paper, cRand variation is used which is called as adjusted Rand index. The maximum value of adjusted Rand index is 1 when pair of clusters is same. It can have negative values also if the clusters are not identical. Dunn index is the measure of quality of clusters generated by clustering process. A higher Dunn index represents a better cluster. It is the ratio of minimum separation with the maximum diameter. Distribution of cluster membership is measured by entropy parameter. Less value of entropy indicates a good clustering result. F1 measure is the combined measure (harmonic mean) of the precision and recall. It is calculated using the values available in confusion matrix. Purity indicates whether the data instance is falling under the correct cluster or not. The ideal value of purity approaches to 1 when the clustering is perfect. SSQ parameter represents the within cluster sum of square measure. It indicates within cluster variation. A good clustering should have less within cluster variation, and hence, less value of SSQ is desired.

### 3.3 Algorithms Settings

All the experiments presented in this study are performed in *R*. All the experiments are performed on the same input data stream generated by Gaussians stream generated by varying the number of dimensions. Before analysing and applying the algorithms on the input stream, parameter values for the algorithms are required to be set. The parameters for BICO algorithm are selected as follows, and the value of cluster centre is taken as 3 as the same number of clusters is specified in the generated input stream, coreset size of 10, number of random projects for nearest neighbours is taken as 10 and number of iterations specified for offline clustering is selected as 10. For the algorithm DBSTREAM, the value of radius of micro-cluster is taken as 0.45, and remaining parameters are taken with the default value provided in the API. For DSTREAM algorithm size of grid cell is taken as 0.1 and density threshold used to detect dense grid is taken as 1.2. For sampling-based stream clustering algorithm (SAMPLE) number of points to be sampled is taken as 20. For the windows-based algorithm and its variation, the window length (horizon) is taken as 20 and 100, respectively, for analysing the impact of dimensionality of the streamed data. The decay factor for damped window model is taken as default, i.e. 0.

After setting the parameter values for the selected algorithm, clustering of streamed data is performed by varying the dimensions of the data. A sample output of clustering of three-dimensional streamed data using selected algorithms is presented in Fig. 2. The evaluation of performance parameters for the 5000 instances for three-dimensional streamed data using selected algorithms is presented in Fig. 3.

## 4 Result Analysis

After making the environmental set-up ready for the experiments, the Gaussians stream is generated by varying the dimension parameter and the same set of algorithms is

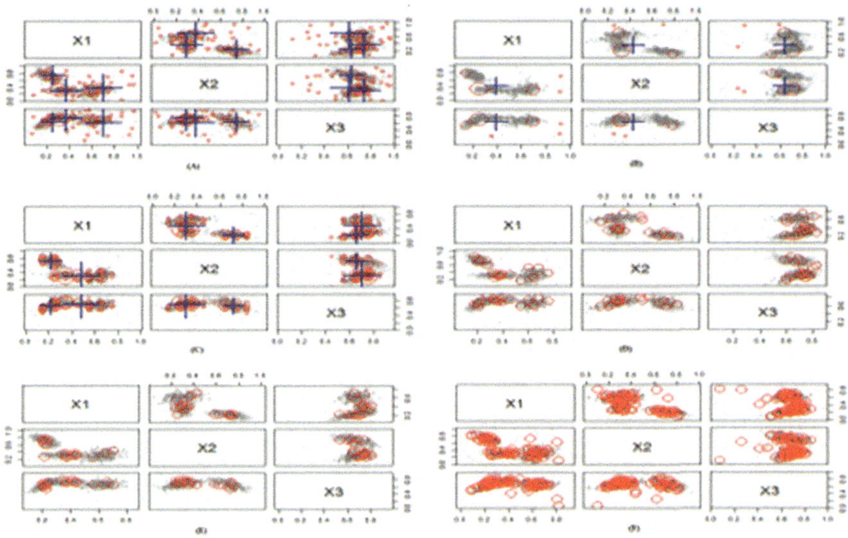


**Fig. 2.** Clusters generated for the 3D Sample stream using algorithms **a** BICO, **b** DBSTREAM, **c** DSTREAM, **d** SAMPLE, **e** Window with 20 horizon and **f** Window with 100 horizon

applied and clusters are evaluated. The performance parameters are tracked for different dimensions of the stream for all the algorithms, and the results are presented in Fig. 4. The analyses of the experiments are summarized in Tables 1 and 2. Out of the six selected algorithms best algorithm(s) against the performance parameters is presented in Table 1. The impact of dimensionality of stream for the algorithms is presented in Table 2. Table 2 summarizes about the algorithms which can accommodate higher dimensions of stream data using the selected performance parameter for example Window-based stream clustering algorithm with window size of 100 performs still better when the dimensions of the streams are increased for the all the performance parameters but SSQ.

## 5 Conclusion and Future Scope

Five standard stream data clustering algorithms, namely BICO, DSTREAM, DBSTREAM, Sample and Window-based clustering (with 20 and 100 horizons) are selected for clustering a commonly used stream generated using Gaussians stream generator in *R* and various performance parameters are evaluated by varying the dimension of the stream. It is observed that as the dimension of the streamed data increases algorithms indicate a slight drop in the performance parameters but overall most of the algorithms are stable and acceptable for high dimensional stream. Only the Window-based stream clustering algorithm performs better when the dimension of the streamed data is increased using all the performance parameters except the SSQ. For the performance parameter SSQ, algorithm DSTREAM performs better as compared to the other algorithms when the dimensionality of the streamed data is increased. The best performing algorithms using various performance parameters are tabulated in Table 1, similarly, the algorithms which accommodate the high dimensional streams and perform

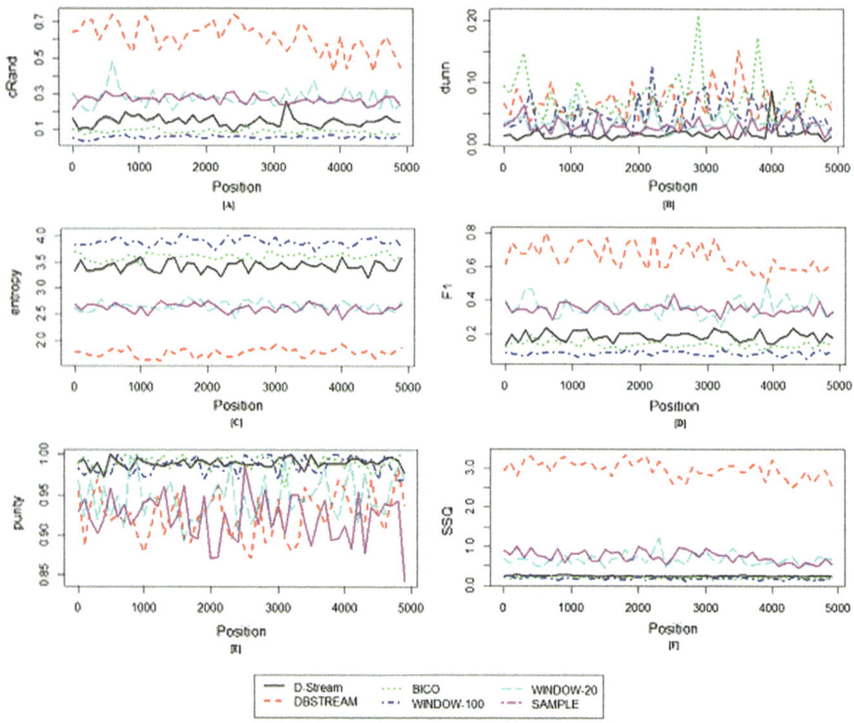


**Fig. 3.** Performance parameters evaluation over different positions of 3D Stream **a** cRand, **b** Dunn, **c** entropy, **d** F1, **e** purity and **f** SSQ

better are tabulated in Table 2. Future work related to the presented work can be making fine-tuning of the selected clustering algorithms so that approximately same number of clusters can be generated and more accurate comparisons can be made. Another work can be, the presented study can also be performed on some real-world datasets with different dimensions for better validation of the presented study.

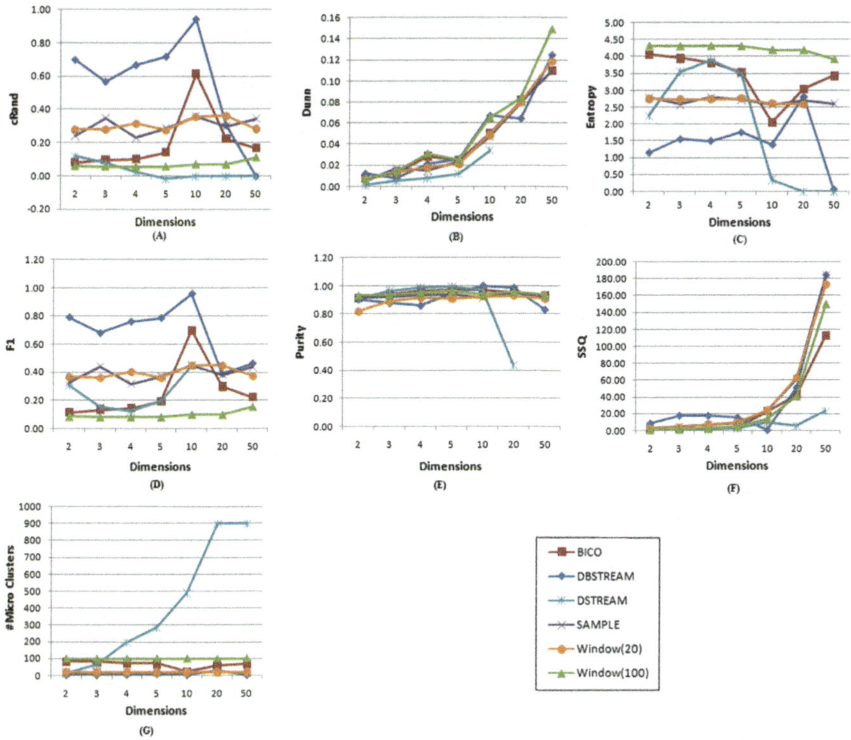


**Fig. 4.** Performance parameters for different stream clustering algorithms for varying dimensions **a** cRand, **b** Dunn, **c** entropy, **d** F1, **e** purity, **f** SSQ and **g** number of micro-clusters generated

**Table 1.** Performance parameterwise best algorithm

| Parameter | Best algorithm |
|---|---|
| cRand | DBSTREAM |
| Dunn | Window(100) |
| Entropy | DBSTREAM |
| F1 | DBSTREAM |
| Purity | BICO, DBSTREAM, SAMPLE, Window(100), Window(20) |
| SSQ | DSTREAM, BICO, Window(100) |

**Table 2.** Impact of dimensionality over clustering algorithms

| Algorithm | cRand | Dunn | Entropy | F1 | Purity | SSQ |
|---|---|---|---|---|---|---|
| BICO | | √ | | | √ | |
| DBSTREAM | | √ | √ | | | |
| DSTREAM | | | √ | | | √ |
| SAMPLE | | | | | | |
| Window(20) | √ | √ | √ | | √ | |
| Window(100) | √ | √ | √ | √ | √ | |

## References


1. Mansalis, S., Ntoutsi, E., Pelekis, N., Theodoridis, Y.: An evaluation of data stream clustering algorithms, statistical analysis and data mining: the ASA. Data Sci. J. **11**(4), 167–187 (2018)
2. Kokate, U., Deshpande, A., Mahalle, P., Patil, P.: Data stream clustering techniques, applications, and models: comparative analysis and discussion. Big Data Cogn. Comput. **2**(4), 32 (2018)
3. Stream: Infrastructure for Data Stream Mining in R. Available at https://cran.r-project.org/web/packages/stream/index.html, last accessed 2019/06/20
4. Fichtenberger, H., Gillé, M., Schmidt, M., Schwiegelshohn, C., Sohler, C.: BICO: BIRCH meets coresets for k-means clustering, In: European Symposium on Algorithms, pp. 481–492. Springer, Berlin, Heidelberg (2013)
5. Hahsler, M., Bolaños, M.: Clustering data streams based on shared density between micro-clusters. IEEE Trans. Knowl. Data Eng. **28**(6), 1449–1461 (2016)
6. Chen, Y., Tu, L.: Density-based clustering for real-time stream data. In: Proceedings of the 13th ACM SIGKDD International Conference on Knowledge Discovery and Data Mining, pp. 133–142. San Jose (2007)
7. Tu, L., Chen, Y.: Stream data clustering based on grid density and attraction. ACM Trans. Knowl. Disc. Data (TKDD) **3**(3), 12 (2009)
8. Vitter, J.S.: Random sampling with a reservoir. ACM Trans. Math. Softw. (TOMS) **11**(1), 37–57 (1985)
9. McLeod, A.I., Bellhouse, D.R.: A convenient algorithm for drawing a simple random sample. J. Roy. Stat. Soc. Ser. C Appl. Stat. **32**(2), 182–184 (1983)
10. Aggarwal, C.C.: On biased reservoir sampling in the presence of stream evolution. In: Proceedings of the 32nd International Conference on Very Large Data Bases, pp. 607–618. Seoul, Korea (2006)
11. Zhu, Y., Shasha, D.: Statstream: statistical monitoring of thousands of data streams in real time. In: VLDB'02: Proceedings of the 28th International Conference on Very Large Databases, pp. 358–369. Morgan Kaufmann (2002)
12. Hahsler, M., Bolanos, M., Forrest, J.: Introduction to stream: an extensible framework for data stream clustering research with R. J. Stat. Softw. **76**(14), 1–50 (2017)
13. Hahsler, M., Bolanos, M.: Stream: infrastructure for data stream mining. R package version 1.3-1. https://CRAN.R-project.org/package=stream (2019)
14. Jain, A.K., Dubes, R.C.: Algorithms for Clustering Data, vol. 6. Prentice Hall, Englewood Cliffs (1988)

15. Christopher, D., Manning, P.R., Schütze, H.: Introduction to Information Retrieval. Cambridge University Press. Available at https://nlp.stanford.edu/IR-book/html/htmledition/evaluation-of-clustering-1.html, last accessed 2019/06/19

# Command Pattern Design for Web Application

Sagar Gupta(✉)

SAP Labs, Bangalore 560066, India
Sagar.gupta01@sap.com

**Abstract.** In recent times, Web application development has become essential for the IT industry. In the past, Web applications were used primarily to provide a graphical view of data stored in backend systems, and business logic execution or any form of data manipulation was programmed in the backend systems. Web applications were treated as dummy client interfaces because client machines were not proficient in performing complex computational operations or executing business operations such as data manipulation and analysis. Now, as client machines have become more advanced and capable of performing complicated business operations, application developers can make client interfaces more intelligent. Enabling client interfaces to handle such complex business processes is of great importance for some business scenarios, like in the case of collaborative user interface designing where data representation and data insertion are not the sole purpose of the Web applications. Adapting design patterns like "Command Pattern" (Long in IEEE, 2017 [1]; Betts et al. in Avail Maintainability Windows Azure, 2013 [2]) can make the Web client smarter and more efficient. This pattern reduces the burden on the backend systems when performing all kinds of business operations [3]. This design pattern provides flexibility to the developers for converting all user actions executed in the user interface (UI) into simple, discrete commands. In this paper, a detailed approach is discussed on how to apply the pattern in more optimized manner by which one can then submit commands individually in an asynchronous manner to the backend system for processing. Handling of these user actions through individual asynchronous commands discussed in the paper makes the UI non-blocking. This also provides flexibility in performing some of the UI changes locally until a response is received from the backend. This paper details the architecture for using "Command Pattern" and design approach for any Web application development and best practices that improve the user experience.

**Keywords:** Command pattern · Non-blocking UI · Collaborative Web application space · Server-side events


X.-S. Yang et al. (eds.), *Proceedings of Fifth International Congress on Information and Communication Technology*, Advances in Intelligent Systems and Computing 1183, https://doi.org/10.1007/978-981-15-5856-6_33

## 1 Introduction

Data representation is always considered an essential aspect of business administration, as it provides a holistic view of business to users. To efficiently represent data digitally, a new user interface known as graphical user interface (GUI) was introduced in the late 60s. GUI allows users to interact with electronic devices through graphical icons and visual indicators like secondary notations, typed command labels, and text navigation. This new way of data representation gave birth to some commonly accepted standards like hypertext markup language (HTML) for creating Web pages and Web applications to fulfill the needs of remote users who want to view data. With Cascading Style Sheets (CSS) and JavaScript, these three technologies form a triad of cornerstone technologies for the World Wide Web. This technology stack enables Web application developers to transform data into various formats depending upon the user's choice like bar graphs, histograms, and pivot tables. Since all these data transformation techniques are specific to the user's choice, the life cycle of such requests needs to be handled on the client's machine. In an ideal scenario, every client's machine sends a request to backend systems to load the specific information or data that the user requested. Once the data is loaded completely, based on the user's choice, the front-end logic uses HTML, CSS, and JavaScript to render the data into UI elements like graphs, pie charts, lists, and tables.

The problem with legacy Web applications in managing UI based on backend data processing results in a high amount of data transfer between client and server [4]. Any request submitted to the backend always treats the whole UI state as one object. This UI state is transmitted to the backend. This request to the backend server results in a new data state, in case a success or a failure message is sent when the processing of the request results in business or technical issues. To reduce this data transfer volume, the UI should transmit only specific changes to the backend. For this, the UI state needs to be broken into various expressive sub-parts. For example, if a user is altering his or her profile, the UI state can be broken into personal information (including name, age, gender, and address), professional information (including company information, work experience, and skill set), or general information. Whenever a user changes any of these fields, instead of transmitting the complete user profile, only the specific change can be transmitted. The success or failure of the request modifies the state of only this attribute in the UI.

Creation of such atomic requests for processing might lead to issues such as concurrency or atomicity of the data between the UI and backend. As stated earlier, the backend can result in failure that further raises concerns on what should happen to any subsequent changes performed by the user without waiting for a server response. With the proposed architecture, the UI design will become robust for handling event-specific consistency for the server response. The proposed design can also help to update related information on the UI based on the server response. For example, if a user changes the address that is related to the current location of a user. In this case, if the address modification is successful, the current location should be changed automatically. Another example is when a user updates his or her marital status, and spouse information must be updated automatically. If the request results in failure, the spouse details should roll back to the previous state. The paper also explains the usage of some best practices for handling concurrent requests with shorter data fetching cycles from the backend system that keep the UI in sync with the backend system.

## 2 Architecture

For a robust and resilient UI for modern Web applications, understanding how to efficiently handle data and errors is important. A common mistake by many programmers is not handling failure properly. Consider an interaction between the server and the client through a dedicated channel. Through this channel, the UI receives different types of information packets, which can be of the following three types:

a. Success
b. Error
c. Completion.

Each of these types of information packets represents a form of data. If we treat all three packets—including errors [4]—as data, the only concern left for the design pattern applied in the UI is to respond to this data.

As discussed, with *command pattern* [4], all the user's actions are discrete commands fired to the backend system for processing. The backend can return data in any mentioned type. Now, after receiving the data, the *command pattern* decides how the UI state should respond to it (Fig. 1).

*Command pattern* architecture consists of three main modules:

i. Controller
ii. Command processor
iii. Query processor.

- **Controller**: The *controller* acts as a central processing unit (CPU) for this architecture. It monitors all the UI changes made by the user. A user's action sets the context of the *controller* that is required to understand if the request requires backend processing or just data reload. Based on this context, the *controller* either passes the request to the *command processor* or the *query processor* for backend execution. When the *controller* receives data from the backend in the form of a response, the *controller* verifies if it is success data, error data, or a completion flag. After identifying the response state, the *controller* begins data parsing to update the data model.

- **Command processor**: This module manages the execution of all user commands. When a user performs a UI change, the *controller* observes this change and depending on the nature of the change, it may result in data change commands. Once the command is created, the *command processor* receives the command. This command gets inserted into a queue of commands for processing. This queue is scanned by the *QueueExecutor*, which fires the command to the backend.
  Once the *command processor* receives the response of the command, the response is sent to the controller, which updates the data model accordingly. The data model acts

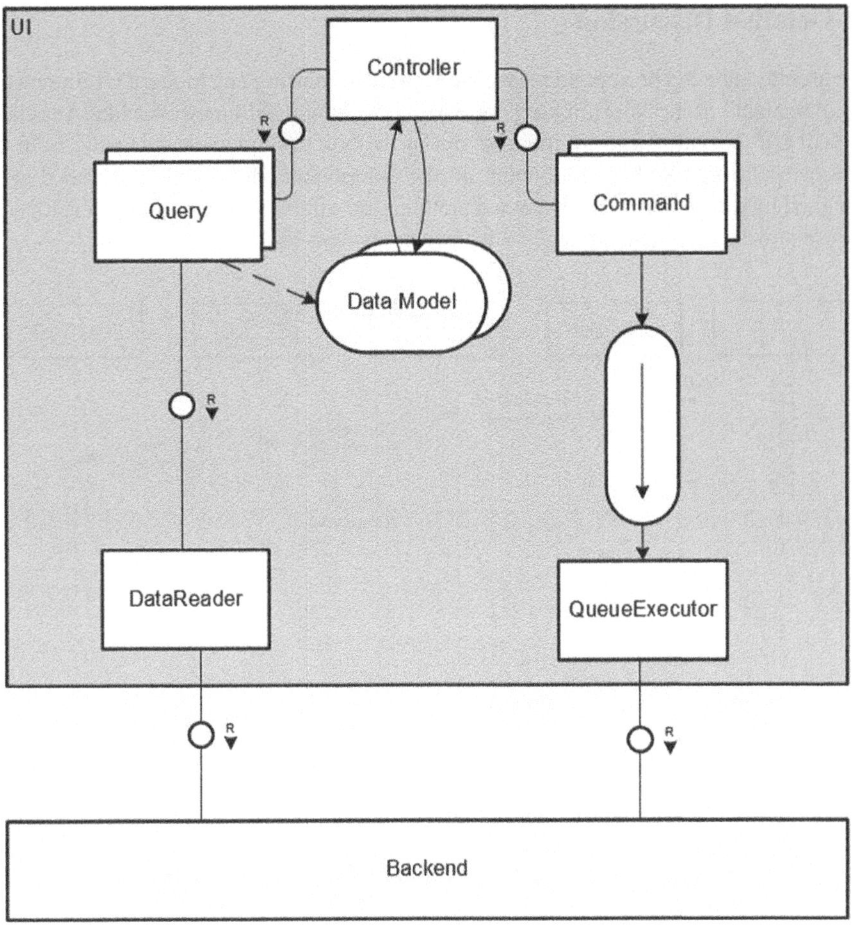


**Fig. 1.** Command pattern architecture

as a primary cache for the UI state, and all the UI elements are bound to this model. Any change in the model results in a change of UI state, but the controller limits the change to the specific element rather than re-rendering the whole UI.

- **Query processor**: The other aspect in the UI is to keep the state in sync with the backend. So, the query executor module performs the *data model* update whenever a state change is required or if the UI requires a complete refresh of the data in sync with backend data. The module also provides the functionality to reload specific granular-level information. For example, if the user wants to reload only information about work experience from the backend, the query executor reloads the data from the backend system and changes the data model with fresh information.

## 3 Detailed Discussion

The architecture can be applied to any Web application using any modern UI framework like AngularJs, ReactJs [5], jQuery, or SAP UI5. Some UI frameworks like AngularJs or SAP UI5 provide basic components out of the box to implement this pattern. In the case of traditional Web development, all the components discussed above need to be designed by developers. The discussed architecture offers flexibility to the developer to choose any development strategy to achieve equivalent results (Fig. 2).

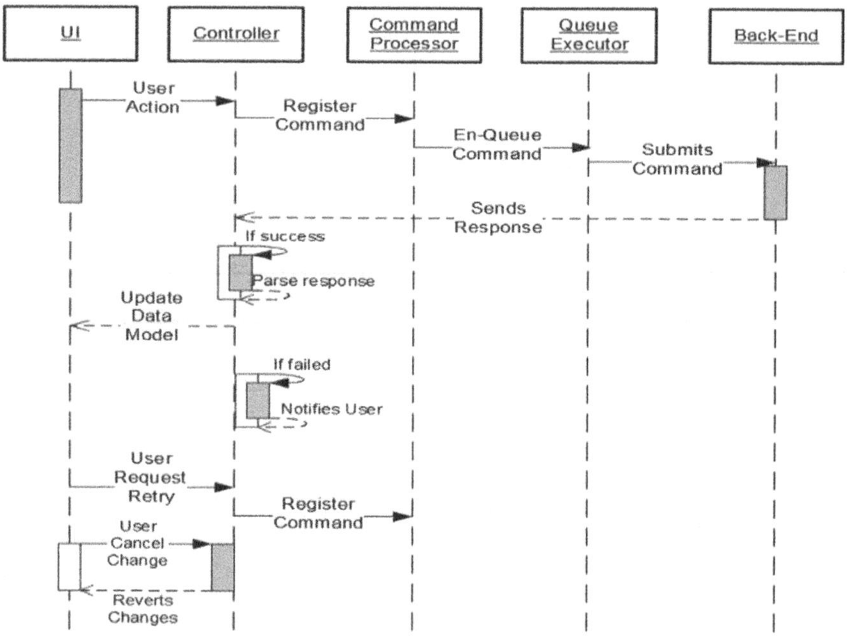


**Fig. 2.** Sequence diagram for command executor

When the *controller* receives any user action, it examines the type of action required, if the action requires any data update, or if it needs only a refresh of data. If the action needs backend processing, then the *controller* decides which part or sub-part of the current UI state demands to be part of the backend request. Once the data change to the UI element is recorded, the change is then passed to the *command processor*. The data change is then appended to the payload of the command that is bound for handling data changes of that UI element. Once the command formulation is successful, the controller pushes it to the queue for processing.

After obtaining a response from the backend, the *controller* assesses the state of response before parsing. It commonly checks the status code set by the backend to highlight the result. These status codes are standard response codes like HTTP codes or some pre-configured values that convey the result of the execution in the backend, such as success, failure, or bad request. If the response code evaluates into success, the response gets parsed by the *controller* and the response is registered in the *data model*.

If the request results in failure, the *controller* notifies the user of the failure and provides an option to either retry the action or cancel the changes. If the user wishes to cancel the action, then the *controller* reverts the changes in the UI based on the old state in the *data model* [6]. If the user wishes to retry, the command again pushes to the command queue.

For the second scenario, if the *controller* senses the user action needs only a data refresh from the backend, the *controller* sends the request to the *query executor* [6] to reload the specific information from the backend. For example, if a user wants to reload the profile picture when it has not been rendered properly in the UI (Fig. 3).

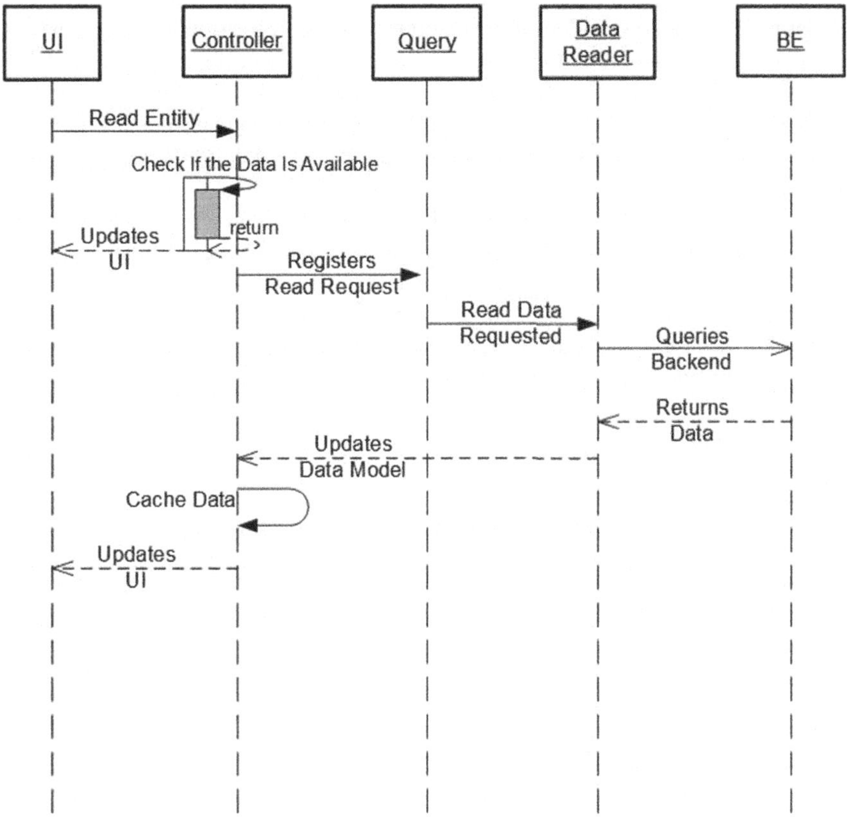


**Fig. 3.** Sequence diagram for query processor

When the UI requests data reload, the query executor tries to respond first with the locally cached data. If the data is not available in the *data model*, the *DataReader* submits a new request to the backend for supplying the required data. Once the data gets delivered from the backend, the *DataReader* [6] passes it to the *controller* for examination before rendering the data in the UI. The *query processor* also stores a local copy of data in the *data model* for future reference.

## 4 Results and Discussions

There are many implementation strategies already available for *command pattern* focusing only for backend. The discussed architecture details architecture for UI with some additional features like handling local changes for UI till the response from backend is received compared to the classical implementation of the design pattern. This architecture also overcomes the drawback of the classical implementation of the design pattern is which it needs extensive data fetching from the backend.

As mentioned above, every single user action is converted into a specific command that is submitted to the backend system. Keeping the UI state in sync with the backend involves a lot of data fetching [7]. This is required because a command can only provide information about the success or failure related to the specific action. If the action results in some other data changes in the backend, the changes are never visible in the UI. For such complex commands [2], therefore, the UI state should be refreshed completely. Also, there are some user actions like reload or refresh of Web page data that get loaded again by the pattern, as the classical implementation never handles the refresh or reload scenarios.

The query executor feature provides clear handling of data changes through commands and data retrieval. In an enterprise application, some background activity or job can also produce data changes. This feature enables the backend to transmit such updated data in an asynchronous way to the UI. This technique of data communication is known as server-side events. These data changes in the backend are delivered to the UI without explicitly submitting a request for data retrieval.

## 5 Comparison

We have compared the proposed solution with the existing industry known Web-development frameworks such as UI5, ReactJs (including fulx/redux) [5], and AngularJs (Fig. 4).

## 6 Benefits

The discussed approach helps application developers segregate [1] the concerns of data updates and data retrieval. Additionally, because of the segregation in handling commands and queries, some UI changes requested by the user can be performed locally in the UI. This handling of UI updates without waiting for the backend response reduces latency and makes the UI non-blocking. Once the backend sends the response to the UI after completing the processing, the data model can be updated asynchronously.

Handling UI changes locally provides a new dimension to developers making the UI experience seamless and non-blocking. It also helps to provide a collaborative working environment to multiple users. Under this approach, the UI handles all the user changes locally first and then transmits them to the backend for actual processing [3]. Later, when the backend system replies with the response, the user changes can be updated in the data model for future reference.

| Main Driver | AngularJs | ReactJs | UI5 | Proposed Solution |
|---|---|---|---|---|
| Transferring the discrete changes made on the UI to the backend | ✗ | ✗ | ✗ | ● |
| ODATA Support | ✗ | ✗ | ● | ● |
| Data Read from backend without UI intervention | ✗ | ✗ | ✗ | ● |
| Combine smaller UI commands /actions into bigger change | ✗ | ● | ✗ | ● |
| Perform local changes for the UI actions | ✗ | ● | ✗ | ● |
| Native support for Undo-Redo operations | ✗ | ✗ | ✗ | ● |
| Native support to load master data or staticdata at load time | ✗ | ✗ | ● | ● |
| Intelligence to handle UI actions locally based on the intent of the user's action | ✗ | ● | ✗ | ● |
| Handling UI changes in blocking and non-blocking mode | ✗ | ✗ | ✗ | ● |

**Fig. 4.** Comparison study between frameworks

Distributing user actions into specific commands also allows deferring some operations. For example, if a user chooses a profile picture, there is a high chance that the user may upload multiple images before finalizing the profile picture. If we upload all the images to the server, this results in high data volume being transmitted between the UI and the backend. Also, if the user wants to revert the changes, the backend always needs to store the old profile image.

## 7 Conclusion

The above-discussed design pattern helps developers create a robust and resilient solution. Implementing the discussed strategy helps reduce the amount of data traffic between the UI and backend. Changes in the UI can now be handled in a non-blocking manner without waiting for a backend response. Applying non-critical UI changes locally without waiting for the backend response enhances the user experience and provides a collaborative working environment to all business users. The discussed solution also reduces the chances of deadlock situations.

This solution also reduces the overall turnaround time for a Web application, as backend changes can now occur in an asynchronous mode without making the UI wait for a response. By using the query executor feature, server-side events [7] can also be easily integrated with the UI. This feature helps the Web page automatically obtain updates from a server without requesting them.

## References

1. Long, Z.: Improvement and implementation of a high performance CQRS architecture. In: 2017 International Conference on Robots and Intelligent System (ICRIS), IEEE (2017)
2. Betts, D., Dominguez, J., Melnik, G. et al.: Exploring CQRS and Event Sourcing: A journey into high scalability. Avail. Maintainability Windows Azure (2013)
3. Sugrue, J.: Command Pattern Tutorial with Java Examples. [Online]. Available https://dzone.com/articles/design-patterns-command
4. w3sDesign.com.: The Command design pattern - Problem, Solution, and Applicability. [Online]. Available http://w3sdesign.com/?gr=b02&ugr=proble
5. Wheeler, K.: Getting to Know Flux, the React.js Architecture. [Online]. Available https://scotch.io/tutorials/getting-to-know-flux-the-react-js-architecture
6. Debski, A., Szczepanik, B., Malawski, M., Spahr, S., Muthig, D.: A scalable, reactive architecture for cloud applications. IEEE Softw. **35**(2)
7. Bidelman, E.: Stream Updates with Server-Sent Events. Online (https://www.html5rocks.com/en/tutorials/eventsource/basics/)

# Lean Thinking Simulation Model to Improve the Service Performance in Fast Food

Diana Sandoval[1](✉), Manuela Palomares[1], Jose Rojas[1], Pablo Mendoza[1], and Carlos Raymundo[2]

[1] Business Management Engineering, Peruvian University of Applied Sciences, Lima, Peru
dianasandovm@gmail.com, manuelapalomaresf@gmail.com, {jose.rojas,pablo.mendoza}@upc.edu.pe

[2] Department of Investigation, Peruvian University of Applied Sciences, Lima, Peru
carlos.raymundo@upc.edu.pe

**Abstract.** The fast food industry, specifically the service department, has experienced sustainable economic growth and has evolved in the application of substantial methods, however, it has presented a slowdown due to deficient operation level provided in the customer service area. A large number of customers do not make the purchase owing to a series of unmet trends that face the consumer behaviour which leads to significant economic losses and inefficient service. They have therefore focused their efforts on finding impulse mechanisms through allowing them to migrate to less costly processes and/or to achieve better utilisation of available resources without success. This research inquires into the effectiveness of the Lean Thinking Simulation (LTS) model, which consists in the development of a set of methodological phases and the adaptation of the technological support termed as Digital Change to improve the performance of customer service in Peruvian fast food. The main result of this practical study was defined by a Dashboard in real-time, and as a first approximation of the model, a 17.03% improvement can be shown in the performance of customer service on the fast food selected.

**Keywords:** Customer service · Discrete events simulation · Fast Food · Lean thinking · LTS model · Simulation model

## 1 Introduction

In recent years, Peru has been ranked as the second-best country for its economic growth and quality of services because it has an investment rate of 4% of annual GDP owing to the flow of 62% of Peruvians, who spend about 25% of their wages motivated by gastronomic culture according to the Doing Business ranking in Latin America. In the services sector, the main sub-sector is the foodservice industry, led by the fast food industry, which generates an annual turnover of S/.585 million among fast foods, nevertheless, there has been a deceleration as a result of their deficient service level provided in the customer service area giving rise to high migration of its customers to direct competitors.


X.-S. Yang et al. (eds.), *Proceedings of Fifth International Congress on Information and Communication Technology*, Advances in Intelligent Systems and Computing 1183, https://doi.org/10.1007/978-981-15-5856-6_34

In this context, the problem is affecting 49.5% of Peruvian fast foods, in the city of Lima, as they are short-staffed and overburdened the existing staff, this extends the customers waiting time during the purchase process due to poor service performance. Because of this, they have focused their efforts on finding impulse mechanisms among them are staff turnover measurement and employ of quality tools such as Service Performance (SERVPERF), Quality Function Deployment (QFD) and statistical analyses that evaluate customer satisfaction surveys. While it is ascertained that there is only 79% reliability in the responses, it can be asserted that until to date this problem that causes customer dissatisfaction, concretely during periods of increased public influx in a Peruvian fast food has not been tackled.

Consequently, we seek to reduce customer churn as a result from deficient service performance. For this purpose, the Lean Thinking Simulation (LTS) model is proposed to improve the performance of the customer service, which is based on a continuous improvement cycle that uses the simulation of discrete events as a tool under a strategic implementation of Digital Change. In this way, Peruvian fast food has been chosen in a very popular area of Lima, with 68.33% of all Peruvian fast foods, where it aims to demonstrate the effectiveness of the proposed model to remain in the competitive market through improving the performance of customer service during the purchase process.

In view of the above, the revised literature for the sustenance of the proposed model is presented in accordance with the typologies categorised according to the techniques, Lean Thinking and Simulation of Discrete Events. Followed by the contribution section which explains in general and detailed the proposed model and its components applied to the elected Peruvian fast food. Continuing validation, which describes the test stage along with its initial results and gets a subsequent result that is evaluated through KPIs. Finally, the findings are projected as a result of the implementation of the model at issue.

## 2 Literature Review

Simulation models exposed in the fast food industry examine worker capacity and customer behaviour during the purchase process. Then, [1, 2, 3, 4] utilise a simulator to visualise the current process that the customer experiences during the peak hours of public influx through the record of random times under parameters that are set according to fast food. Then, they linked the waiting time problem as a variable to the low performance of workers, in order that they increased the number of its and/or their labour capacity, achieving a 15–20% reduction in waiting times and an increase in customer satisfaction.

According to [5, 6], Simulation models applied in the service department use simulation as validation and quality tools such as Total Quality Management (TQM), which enables them to perform continuous long-term improvement that is important to customers, thereby increasing workers operational capacity through the Plan-Do-Check-Act (PDCA) and verifying their performance through KPIs. However, they do not analyse the productivity and labour capacity that afflicts them during peak hours, due to a lack of knowledge of demand.

Lean Thinking models associated with the service industry focus on tangible and transactional deliverables from customer experience. Furthermore [7], this is possible by applying the five Lean Thinking principles, (i) specifying value from the end-customer perspective, (ii) identifying value flow, (iii) influencing value to reduce processing and design time, (iv) providing what the customer wants only when he wants it, and (v) pursuing perfection by constantly eliminating waste layers as part of the continuous improvement process. Thus, they measured the performance and capacity of workers through Dashboards and programmes that strengthen their skills development [8, 9, 10, 11].

In addition to these models integrating Lean Thinking and simulators identify inefficiencies in the service sector. To this end, [12, 13, 14, 15, 16] analyse the initial state of the production process through the Value Stream Mapping (VSM), that locates waste types to improve worker performance in the activities and to follow a customer-oriented organisational model. Continue [17, 18, 19, 20] with a simulator and tools, such as SAP Success Factors-Performance and Goals, which track worker performance based on the fulfilment of the objectives of the executive and SAP S/4HANA, which contributes to the virtual monitoring of product stock at times of higher influx, resulting in reduced customer waiting times and activities that do not add value to the process.

## 3 Research Method

The main objective of this investigation is to define and evaluate the applicability of LTS model to improve service performance in fast food. In view to achieve this goal, the similarities in using Lean tools for problem identification could be identified among the revised literature. However, the development of the models found is mostly focused on the elaboration processes neglecting the customer service area, where problems have been encountered directly influence in the purchasing behaviour.

### 3.1 Description of the Model

The LTS model is based on a four-phase Lean Thinking improvement cycle, which uses simulation as a tool to enable service performance by companies in the fast food industry that can be easily adapted to the constant change in consumer behaviour. The first phase is a situational diagnosis, which presents a combination of the value stream mapping tools and the Arena Rockwell Software simulator that are geared towards visual process management in the flow of material and information, which determines the problem area of the purchasing process by evaluating the workload, operational capacity, sales, customer output, and system duration variables. The process at this stage focuses on the identification of problems that interfere with the optimal development of the purchasing process, which consists of ordering counter, kitchen and dispatch areas referred to as order service; preparation service and delivery service, respectively.

The second phase, systemic analysis, has as its main function to identify the changes in the specific area in problem through the Flow Chart and to estimate with the simulator the impact of improvements that would eliminate these wastes on the process of the specific area in order to address a more efficient workplace.

The third phase, strategic implementation represents the execution of the proposed three-week model in the days and hours of higher influx, whose information will be channelled and processed through Tablets, whose technological support has been termed as Digital Change, which is comprised of SAP SuccessFactors—Performance and Goals, SAP S/4HANA, and the simulator that will result in a Dashboard in real-time, which is a control panel to enables real-time evaluation of the performance of the staff support performance during the procurement process of the areas to guide the strategy towards achieving the stated objectives.

The fourth and final phase of management control aims to measure the performance of the results achieved by the Dashboard in real-time and the Enhancement Simulator through KPIs that demonstrate whether the solutions implemented in the fast food meet the requirements identified in the first and second phases as it is shown in (Fig. 1).

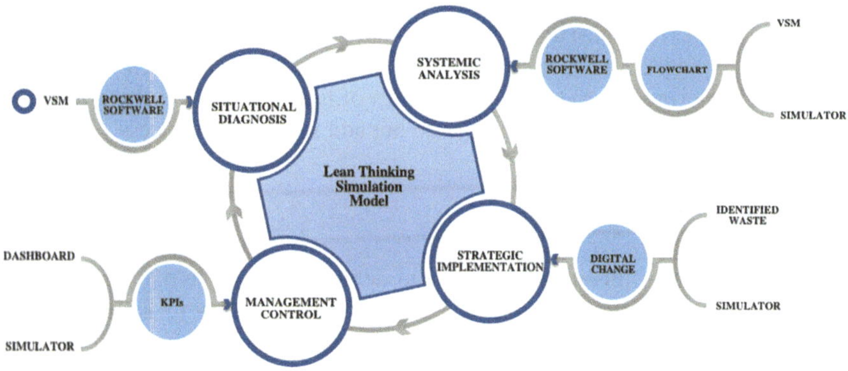


**Fig. 1.** LTS model

### 3.2 Components

This introduces three new components as shown in TOC., the integration with the simulation that consists of the inclusion of the value stream mapping tools, the Flow Chart and the simulator that allow you to visualise the entire process and determine the problem area of the purchasing process by evaluating the operational capability, sales, customer output; adapting the simulation to Lean Thinking, whose phases were reduced to four; and the Digital Change application that combines traditional Dashboard and simulator to obtain an in-real-time Dashboard that evaluates service performance.

To measure the progress of the LTS model through KPIs, the productivity, quality of service, operational capability, cycle time, and average wait time metrics are used to enable an evolutionary analysis of the performance of the affected area (Table 1).

## 4 Validation

### 4.1 Description of the Stage

The model was applied to a Peruvian fast food service, where customer churn was identified for maximum waiting time when placing an order has significant economic losses

**Table 1.** KPIs of the LTS model

| Metrics | Description |
|---|---|
| Productivity | Percentage of sales made |
| Service quality | Percentage of service effectiveness |
| Operational capacity | Work in process |
| Cycle time | Duration of purchase process |
| Average waiting time | Average length of customer wait |

for the organisation, being the branch understudy the one that presents the highest percentage of affectation of the problem in 22%. Thus, we saw the opportunity to improve this process that has been operating traditionally for many years without applying new advantages to its customers. In this way, the investigation took place during peak hours which are parameterized on Friday and Saturday from 20:30 to 23:30 h according to the information provided by the organisation. Therefore, the procurement process established by the fast food was then described, consisting of the Ordering Counter, Kitchen and Dispatch areas, whose activities begin with the customers' attention, the registration of the order, followed by the preparation and assembly of the order to settle with the delivery of the order.

### 4.2 Results

A notable improvement in the study fast food can be seen during the validation period. Thus, KPIs used to measure the model during its implementation demonstrate a reduction in operational capacity to 75.66% of customer workers who worked above its capacity, which is 85% as per the ILO (International Labour Organization). As a result, there was an impact on the other areas of the process and increased by 17.03% the service performance, thereby it is representing the recovery of 102 customers churn. In view of this result, it has been identified that the simulator obtained a 22.58% reduction of the problem thus having a margin of error of ±5.55%. However, the outcome of the implementation may be increased because it was found that, over the hours, the operators became more despite having little training time, which is the result of the use of a model of continuous improvement that was based on Lean Thinking in this research. In this regard, Table 2 shows the significant improvements that have been generated from the application of the LTS model in Peruvian fast food. Hence, the proposed model achieved its purpose.

## 5 Conclusions

The LTS model achieved the purpose of recovery customers churn due to the low level of service performance in Peruvian fast food. This was possible through the realisation of the four model phases that integrated the Arena Rockwell Software simulation tool into the Lean Thinking methodology, whose tools such as the Flow Chart and VSM

**Table 2.** Results of the LTS model through KPIs

| Metrics | Actual situation | Improve situation |
|---|---|---|
| Productivity (%) | 40.55 | 57.58 |
| Service quality (%) | 60.02 | 73.05 |
| Operational capacity (%) | 92.69 | 75.66 |
| Cycle time | 13.271 min | 11.008 min |
| Average waiting time | 3.052 min | 2.532 min |

enabled a comparison between the technique and methodology in order to obtain results closer to reality. Thus, Digital Change was able to be implemented in the establishment, which obtained information on new customer service time-frames and waiting times that resulted in an in-real-time Dashboard that allowed a comparison between the current situation and the situation of improvement of the establishment where improvements in customer service performance could be quantified.

In accordance with the proposed model, improvement in service performance could be validated. This required a three-week implementation in which it was executed under the Lean Thinking methodology. In addition, Arena Rockwell Software's discrete systems simulator was used in conjunction with technological support, Digital Change, which provided an opportunity to make a comparative chart to actually assess the impact that the model would have on the fast food chosen. This achieved an 17.03% improvement in the performance of customer service, which recovery compared to customer churn due to excessive waiting times. This result reflects improvements in 57.58% in productivity, 73.05% of service quality and 75.66% in operational capacity, while cycle time had a notable reduction to 11,008 min as well as the average waiting time that ended in 2532 min.

## References

1. De Vries, J.: Worth the wait? How restaurant waiting time influences customer behavior and revenue. J. Oper. Manag. (2018). https://doi.org/10.1016/j.jom.2018.05.001
2. He, H., Hu, Z.: Analysis of fast food service capability based on flexsim modeling and simulation. IOP Conf. Ser. Mater. Sci. Eng. https://www.doi.org/10.1088/1757-899X/394/5/052005
3. Koide, T., Kobayashi, T., Kikuda, M.: A multi-agent simulation study for congestion miniga-tion in a university campus restaurant. Trans. Eng. Technol. (2019). https://doi.org/10.1007/978-981-13-2191-7_8
4. Singh, J., Singh, H., Singh, D., Singh, K.: Evaluating lean thinking using value stream mapping in manufacturing industry—a case study. Int. J. Prod. Qual. Manag. **22**(1), 101–116 (2017). https://doi.org/10.1504/IJPQM.2017.085849
5. Li, G., Field, J.M., Davis, M.M.: Designing lean processes with improved service quality: an application in financial services. Qual. Manag. J. **24**(1), 6–19 (2017). https://doi.org/10.1080/10686967.2017.11918497
6. Vadhvani, M.D., Bhatt, M.G.: Application of lean principles in indian service sector: exploratory analysis. Productivity **58**(3), 331–339 (2017)

7. Wijnhoven, F., Beckers, D., Amrit, C.: Reducing waste in administrative services with lean principles. In: Thirty Seventh International Conference on Information Systems (2016)
8. Amaro, P., Alves, A.C., Sousa, R.M.: Lean thinking: a transversal and global management philosophy to achieve sustainability benefits. Lean Eng. Glob. Dev. 1–31 (2019). https://doi.org/10.1007/978-3-030-13515-7_1
9. Baag, P.K., Kavitha, P., Sarkar, A.: Application of lean tool for process improvement of bank branches. IIM Kozhikode Soc. Manag. Rev. **8**(2), 131–142 (2019). https://doi.org/10.1177/2277975219836502
10. Martins, A.F., Alves, A.C., Leão, C.P.: Development and implementation of dashboards for operational monitoring using participatory design in a lean context. Adv. Intell. Syst. Comput. **621** (2018). https://doi.org/10.1007/978-3-319-61121-1_21
11. Mohammad, A.: Approaching the adoption of lean thinking principles in food operations in hotels in Egypt. Tourism Rev. Int. **21**(4), 365–378 (2017). https://doi.org/10.3727/154427217X15094520591349
12. Ghiyasinasab, M., Lehoux, N., Ménard, S., Cloutier, C.: Using lean techniques and simulation to improve the efficiency of engineered wood production: a case study in a small factory. Ind. Eng. Manag. **7**(269) (2018)
13. Greasley, A., Smith, C.M. Using activity-based costing and simulation to reduce cost at a police communications centre. Policing Int. J. Police Strat. Manag. **40**(2), 426–441 (2017). https://doi.org/10.1108/PIJPSM-03-2016-0044
14. Nguyen, Q., Nisar, T.M., Knox, D., Praka, G.: Understanding customer satisfaction in the UK quick service restaurant industry: the influence of the tangible attributes of perceived service quality. Br. Food J. (2018). https://doi.org/10.1108/BFJ-08-2017-0449
15. Singh, J., Singh, H., Singh, A., Singh, J.: Managing industrial operations by lean thinking using value stream mapping and six sigma in manufacturing unit: case studies. Manag. Decis. (2019). https://doi.org/10.1108/MD-04-2017-0332
16. Tamás, P., Illés, B.: Examining the integration possibilities for lean tools and simulation modeling. Solid State Phenom. **261**, 516–522 (2017). https://doi.org/10.4028/www.scientific.net/SSP.261.516
17. Lunsford, D., Phillips, P.: Tools used by organizations to support human capital analytics. Perform. Improv. **57**(3), 6–15 (2018). https://doi.org/10.1002/pfi.21767
18. McDonald, K., Fisher, S., Connelly, C.E.: e-HRM systems in support of smart workforce management: an exploratory case study of system success. Electron. HRM Smart Era, 87–108 (2017). https://doi.org/10.1108/978-1-78714-315-920161004
19. Poschadel, F.: SAP S/4HANA: performance management in real time? In: Buttkus, E.M., Eberenz, R. (eds.) Performance Management in Retail and the Consumer Goods Industry. pp. 137–151 (2019). https://doi.org/10.1007/978-3-030-12730-5_9
20. Ziebell, R.C., Albors-Garrigos, J., Schoeneberg, K.P., Marin, M.P.: Adoption and success of e-HRM in a cloud computing environment. Int. J. Cloud Appl. Comput. **9**(2), 1–27 (2019). https://www.doi.org/10.4018/IJCAC.2019040101

# PUEA Impact on Sensing and Throughput in RF Powered Cognitive Radio Networks

Avik Banerjee(✉) and Santi P. Maity

Indian Institute of Engineering Science and Technology, Shibpur, Howrah 711103, India
avikbanerjee@ieee.org, santipmaity@it.iiests.ac.in

**Abstract.** This paper explores primary user emulation attack (PUEA) impact on spectrum sensing (SS), sum secondary user (SU) throughput and energy harvesting (EH) in cognitive radio network (CRN). Cognitive radio (CR) system model consists of a set of transmit-receive node pairs, one fusion center (FC) and one PUEA node. At the initial time slot of the frame, simultaneous EH and spectrum sensing (SS) are done through power splitting (PS) mode. Then based on SS decision, at the FC, CR transmit nodes either perform EH or transmit data in time division mode. Closed form expressions of the optimal sensing duration and transmit power for each SU are found while maximizing the sum SU throughput under the constraints of SS reliability, energy causality on each SU, interference at PU and outage probability on individual SU link. Simulation results show that ∼20% increase in PUEA power enhances the throughput ∼9.76% and residual energy ∼0.31% while meeting the detection and false alarm probabilities 0.95 and 0.05, respectively.

**Keywords:** Cognitive radio · PUEA · Energy harvesting · Residual energy · Secondary outage probability

## 1 Introduction

Recently energy harvesting (EH) [1,2] enabled cognitive radio (CR) [2,3] addresses spectrum scarcity problem with promise of green communication. One of the basic CR operation is spectrum sensing (SS)/cooperative SS (CSS) [2,3] which experiences primary user emulation attack (PUEA) [2,3] where some external (outside CR system) node mimics primary user (PU) signal to avail PU spectrum for its own use. Increase in PUEA power has two adverse effects, one in SS performance [2,4] and the second one is interference to secondary user (SU) receiver. However, PUEA interference being a radio frequency (RF) signal, can also be used by the SUs to perform EH [2,5]. Hence, the role of PUEA power on SS, SU throughput and EH are the focus of this present problem.


X.-S. Yang et al. (eds.), *Proceedings of Fifth International Congress on Information and Communication Technology*, Advances in Intelligent Systems and Computing 1183, https://doi.org/10.1007/978-981-15-5856-6_35

A frame structure for the CR system is suggested that keeps provision of simultaneous EH and SS using power splitting (PS) mode followed by reporting the local SS decision to the fusion center (FC). Non-linear EH (NL-EH) model is used for harvesting of energy using both PU and PUEA signals. SUs transmit data during PU's non-transmission phase on equal slot in TDMA. The objective is to maximize the sum secondary throughput of the network under the constraints of meeting the sensing reliability of PU, individual energy causality for each secondary transmitter, interference at PU receiver (during missed detection of PU) and individual secondary outage probability. The expressions of the optimal sensing duration and SU transmit power for data transmission are derived in closed form. The main contributions of the present work are given below:

1. The expressions of total residual energy and individual secondary outage probability are derived analytically that build various design parameters.
2. A global maximum point on the sum secondary throughput can be achieved at the optimal values of sensing duration and SU transmit power.
3. It is possible to achieve enhanced (gain) sum secondary throughput and total residual energy values with the increase in PUEA transmit power.

The organization of the paper is given as: The system model is described in Sect. 2. NL-EH and residual energy calculation are done in Sect. 3. In Sect. 4, the problem formulation and the proposed solution are provided. Section 5 presents the numerical results and the conclusion of the paper is given in Sect. 6.

## 2 System Model

The system model, shown in Fig. 1, comprises of a primary transmitter ($PT$)-receiver ($PR$) pair, '$K$' pairs of secondary transmitter ($ST_i$)-receiver ($SR_i$) that are governed by a fusion center (FC). The SU network utilizes the frequency band of PU opportunistically, through CSS. The frame structure, as shown in Fig. 2, is repetitive (say $N_c$ number of cycles). Each frame begins with a simultaneous sensing and EH over the time slot $\tau_s$ where each SUs simultaneously performs the local SS and RF-EH from the transmitted signal of both $PT$ and PUEA using PS mode at $\rho_i$:$(1-\rho_i)$. The symbol $\rho$ denotes the PS ratio. During the reporting sub-slot $\tau_{ri} = (1-\alpha)\tau/K$, for $i = 1, 2, \ldots K$, each $ST_i$ sends its local binary SS result to the FC where a global decision of CSS is made. Based on CSS decision, each $ST_i$ either performs EH or transmits data to its respective $SR_i$ on equal time sharing over duration $(T_f - \tau)/K$. The duration of $\tau_s$ and $\tau_r$ (reporting slot) are $\alpha\tau$ and $(1-\alpha)\tau$, respectively, where $0 < \alpha < 1$. Here PUEA interfering power is used by STs to perform EH during both $\tau_s$ and $(T_f - \tau)$. $ST_i$ utilizes a part of the harvested energy ($P_i$) in data transmission such that the total interference power must be within the limit $I_{th}$ to $PR$ for missed detection on PU transmission. Global channel state information (CSI) of all secondary nodes [1] and transmit power of PU and PUEA nodes [3,6] are available to the FC.

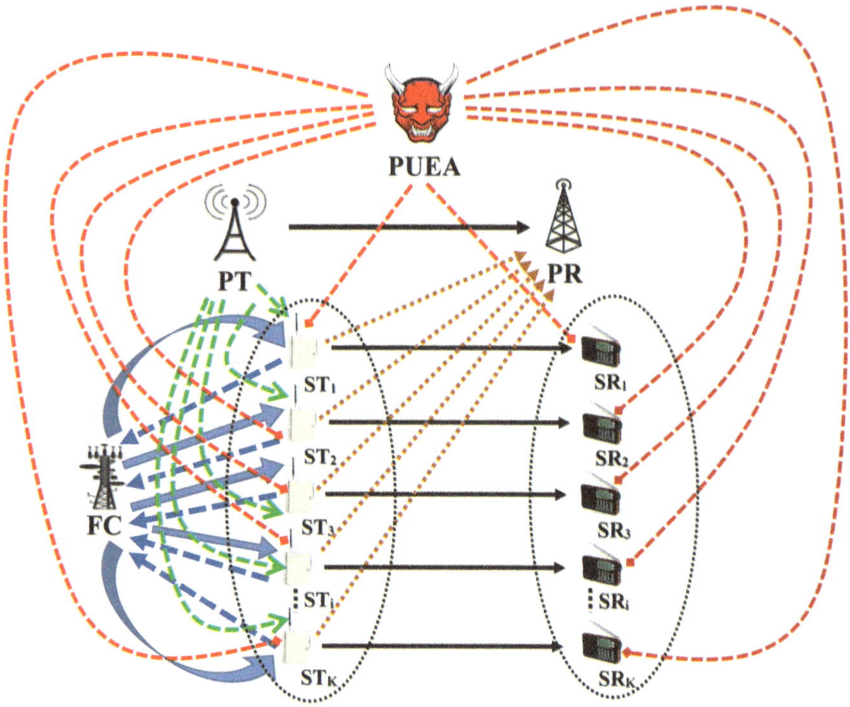


**Fig. 1.** System model

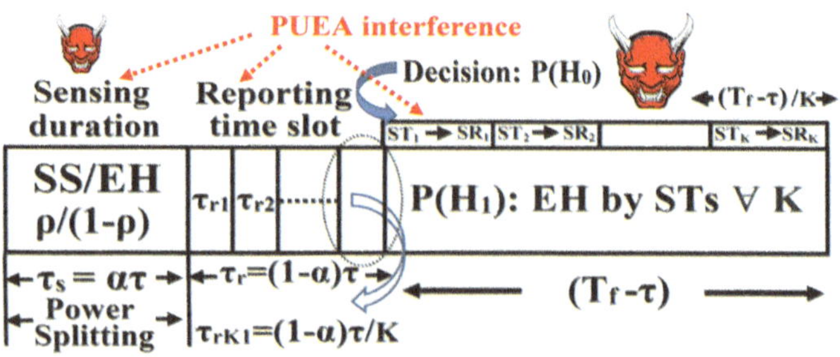


**Fig. 2.** Frame structure of CRN with PUEA interference

## 2.1 Signal and Channel Modelling

During $\tau_s$, the $n$-th ($n = 1$ to $N$) sample of the received signal at $ST_i$ is

$$y_{r_i}(n) = \underbrace{\phi h_{ps_i} x_p(n)}_{\textbf{PU signal component}} + \underbrace{h_{ms_i} x_m(n)}_{\textbf{PUEA signal component}} + \underbrace{v_i(n)}_{\textbf{noise component}} \tag{1}$$

Sampling frequency $f_s$ and sample number $N$ relate as $N = \tau_s f_s = \alpha\tau f_s$. The probabilities for transmission and non-transmission states of $PT$ are $\mathcal{P}(\mathcal{H}_1) = \mathcal{P}(\phi = 1)$ and $\mathcal{P}(\mathcal{H}_0) = \mathcal{P}(\phi = 0)$, respectively. PS device, at each $ST_i$, splits this received signal for EH and SS purposes as $y_{rh_i}(n) = (1-\rho_i)y_{r_i}(n)$ and $y_{rs_i}(n) = \rho_i y_{r_i}(n)$, respectively [1]. The symbols $x_p(n)$ and $x_m(n)$ denote the circularly symmetric complex Gaussian (CSCG) signal with $P_p$ and $P_m$ as transmit power of $PT$ and PUEA, respectively for zero mean and variances $E[|x_p(n)|^2] = P_p$ and $E[|x_m(n)|^2] = P_m$. The symbols $v_i(n)$ and $nr_i(n)$ denote the additive noise signal at the receiver of $ST_i$ and $SR_i$, respectively and are assumed to be independent and identically distributed (i.i.d.) CSCG random sequences with zero mean and variances $\sigma^2_{ns_i}$ and $\sigma^2_{nr_i}$, where $E[|v_i(n)|^2] = \sigma^2_{ns_i}$ and $E[|nr_i(n)|^2] = \sigma^2_{nr_i}$, for $i = 1$ to $K$.

Wireless channels are assumed to be independent to each other and undergo quasi-static Rayleigh fading with the coefficient $h_m \sim \mathcal{CN}(0, d_m^{-\psi_m})$, '$d_m$' and '$\psi$' denote the distance between the nodes and path loss exponent, respectively. The mean channel power gain of $|h_m|^2$ is given by $\frac{1}{\lambda_m} \triangleq d_m^{-\psi_m}$ where $m \in \{pp, ps_i, pr_i, spr_i, ms_i, mr_i, sr_i,\}$. It is worth mentioning that the mean and the variances of the channel coefficients associated with PUEA i.e. $h_{ms_i}$ and $h_{mr_i}$ are assumed to be available to $ST$s [6].

## 2.2 Cooperative SS

$ST_i$ uses an energy detection for the samples of $PT$ i.e. $y_{rs_i}(n)$ with the test statistics $T_{st_i} = \frac{1}{N}\sum_{n=1}^{\alpha\tau f_s} |y_{rs_i}(n)|^2$. The probability of detection ($p_{d_i}$) and probability of false alarm ($p_{f_i}$) at $ST_i$ can be expressed respectively as [1]

$$\begin{aligned} p_{d_i} &= \mathcal{Q}\left(\left(\frac{\zeta_i}{\rho_i(d_{ms_i}^{-\psi_{ms_i}} P_m + \sigma^2_{n_{si}})} - \gamma_i - 1\right)\frac{\sqrt{\alpha\tau f_s}}{(\gamma_i + 1)}\right) \\ p_{f_i} &= \mathcal{Q}\left(\left(\frac{\zeta_i}{\rho_i(d_{ms_i}^{-\psi_{ms_i}} P_m + \sigma^2_{n_{si}})} - 1\right)\sqrt{\alpha\tau f_s}\right) \end{aligned} \tag{2}$$

where $\mathcal{Q} = \frac{1}{2\pi}\int_x^\infty e^{\frac{-r^2}{2}} dr$ and $\zeta_i$ is the local detection threshold used at $ST_i$. The signal-to-noise plus interference (SINR) ratio at $ST_i$ is $\gamma_i = (d_{ps_i}^{-\psi_{ps_i}} P_p)/(d_{ms_i}^{-\psi_{ms_i}} P_m + \sigma^2_{n_{si}})$. For simplicity in mathematical analysis, the distances from the $ST_i$ to $PT$ and PUEA are assumed to be of much larger than the distances among the ST nodes, i.e. $d_{ps_i} = d_{ps}$, $d_{ms_i} = d_{ms}$, $\psi_{ps_i} = \psi_{ps}$, $\psi_{ms_i} = \psi_{ms}$ $\zeta_i = \zeta$, $\gamma_i = \gamma$, $\sigma^2_{ns_i} = \sigma^2_{ns}$, $\rho_i = \rho$, $p_{d_i} = p_d$, $p_{f_i} = p_f$, $\forall i = 1$ to $K$ [1].

During $\tau_s$, the energy consumed by $ST_i$ for local SS is given by

$$E_s = \rho\big[\mathcal{P}(\mathcal{H}_1)(d_{ps}^{-\psi_{ps}} P_p + d_{ms}^{-\psi_{ms}} P_m + \sigma_{ns}^2) + \mathcal{P}(\mathcal{H}_0)(d_{ms}^{-\psi_{ms}} P_m + \sigma_{ns}^2)\big]\alpha\tau = \rho\alpha\tau\mathcal{S} \tag{3}$$

where $\mathcal{S}_1 = \mathcal{P}(\mathcal{H}_1)(d_{ps}^{-\psi_{ps}} P_p + d_{ms}^{-\psi_{ms}} P_m + \sigma_{ns}^2)$, $\mathcal{S}_2 = \mathcal{P}(\mathcal{H}_0)(d_{ms}^{-\psi_{ms}} P_m + \sigma_{ns}^2)$ and $\mathcal{S} = \mathcal{S}_1 + \mathcal{S}_2$. During reporting, each $ST_i$ forwards its SS decision to the FC sequentially (duration of $\frac{(1-\alpha)\tau}{K}$) in an error free common control channel. FC uses "OR" fusion rule in order to combine the binary SS decisions. The cooperative probability of detection ($P_d$) and false alarm ($P_f$) are [1]

$$P_d = 1 - (1 - p_d)^K, P_f = 1 - (1 - p_f)^K \tag{4}$$

The required time for broadcasting the CSS decision from the FC to all $ST$s, being negligible over sensing and reporting slots, is not considered in analysis. Energy consumption for reporting SS decision is also neglected.

## 3 NL-EH and Residual Energy Calculation

The harvested power at $ST_i$ using non-linear (NL) model is given as [1,7]

$$\Phi_{\text{EH}}^{\text{NL}} = \frac{\frac{M}{1+\exp(-a(P_\xi - b))} - \frac{M}{1+\exp(ab)}}{1 - \frac{1}{1+\exp(ab)}} \tag{5}$$

where $P_\xi$ represents RF power used for EH and $M$, $a$ and $b$ are same in [1,7].

$ST_i$ performs EH during $\tau_s$ (PS mode) and $(T_f - \tau)$ after CSS under $\mathcal{P}(\mathcal{H}_1|\mathcal{H}_1)$ and $\mathcal{P}(\mathcal{H}_1|\mathcal{H}_0)$, with harvested energies $E_{\mathcal{H}_1}^{NL} = \Phi_{\text{EH}_1}^{\text{NL}}\alpha\tau$ and $E_{\mathcal{H}_2}^{NL} = \Phi_{\text{EH}_2}^{\text{NL}}(T_f - \tau)$, respectively ($\forall$ i=1 to $K$). $\Phi_{\text{EH}_1}^{\text{NL}}$ and $\Phi_{\text{EH}_2}^{\text{NL}}$ are evaluated by putting $P_\xi$ (in Eq. (5)) as $(1-\rho)\mathcal{S}$ and $\mathcal{B}_{11}$, respectively, where $\mathcal{B}_{11} = P_d\mathcal{S}_1 + P_f\mathcal{S}_2$.

The energy consumption for SS by $ST_i$ is $\rho\alpha\tau S$ as discussed before. Now based on $\mathcal{P}(\mathcal{H}_0|\mathcal{H}_0)$ and $\mathcal{P}(\mathcal{H}_0|\mathcal{H}_1)$, energy required at $ST_i$ for transmitting data to $SR_i$ during $\frac{(T_f-\tau)}{K}$ is

$$\begin{aligned} E_{tr,ST_i} &= \frac{(T_f - \tau)}{K} P_i\Big[\overbrace{\mathcal{P}(\mathcal{H}_0)(1 - P_f)}^{\mathcal{P}(\mathcal{H}_0|\mathcal{H}_0)} + \overbrace{\mathcal{P}(\mathcal{H}_1)(1 - P_d)}^{\mathcal{P}(\mathcal{H}_0|\mathcal{H}_1)}\Big] \\ &= (T_f - \tau)P_i\mathcal{C} \end{aligned} \tag{6}$$

where $\mathcal{C} = \frac{1}{K}\Big(\mathcal{P}(\mathcal{H}_0)(1 - P_f) + \mathcal{P}(\mathcal{H}_1)(1 - P_d)\Big)$. So residual energy ($E_{res}$) is

$$\begin{aligned} E_{res} &= \sum_{i=1}^{K}\left(E_{\mathcal{H}_1}^{NL} + E_{\mathcal{H}_2}^{NL} - E_s - E_{tr,ST_i}\right) \\ &= \sum_{i=1}^{K}\left\{\Big(\Phi_{\text{EH}_1}^{\text{NL}} - \rho\mathcal{S}\Big)\alpha\tau + \Big(\Phi_{\text{EH}_2}^{\text{NL}} - P_i\mathcal{C}\Big)(T_f - \tau)\right\} \end{aligned} \tag{7}$$

# 4 Problem Formulation and Proposed Solution

$ST_i$ transmits data to $SR_i$ for the slot $\frac{(T_f-\tau)}{K}$ under $\mathcal{P}(\mathcal{H}_0|\mathcal{H}_0)$ and $\mathcal{P}(\mathcal{H}_0|\mathcal{H}_1)$. PU transmitter is assumed to be located far apart from $SR_i$ and hence its interference at the latter node, during missed detection i.e. under condition $\mathcal{P}(\mathcal{H}_0|\mathcal{H}_1)$, is not considered in the analysis. The effect of additive noise at $SR_i$, being much smaller than interference power, is also not considered. Thus the signal-to-interference ratio (SIR) at $SR_i$ is expressed by $\gamma_{sr} = \frac{P_i|h_{sr_i}|^2}{P_m|h_{mr_i}|^2}$.

**Secondary Outage Probability**: Data transmission over the $i^{th}$ SU link experiences an outage if $\mathcal{R}_{sr_i}$ falls below target rate $\mathcal{R}_{th}$. Hence secondary outage probability ($\mathscr{P}_{s,out_i}$) for the $i$-th link is in Eq. (8), where $\mathcal{R}_{sr_i} = \frac{(T_f-\tau)}{T_f}\mathcal{C}\log_2(1+\gamma_{sr_i})$, $\gamma_{th} = 2^{\frac{\mathcal{R}_{th}T_f}{(T_f-\tau)\mathcal{C}}} - 1 \approx \frac{\mathcal{R}'_{th}T_f}{(T_f-\tau)\mathcal{C}}$, $\mathcal{R}'_{th} = 0.693\mathcal{R}_{th}$ [7] (details omitted due to space constraint).

$$\mathscr{P}_{s,out_i} = P_r(\mathcal{R}_{sr_i} < \mathcal{R}_{th}) = 1 - \left[\frac{P_i d_{sr_i}^{-\psi_{sr_i}}}{P_m\gamma_{th}d_{mr_i}^{-\psi_{mr_i}} + P_i d_{sr_i}^{-\psi_{sr_i}}}\right] \tag{8}$$

## 4.1 Secondary Sum Throughput

The frame structure, being repetitive, spectrum efficiency (SE) of $ST_i$ node includes an average value on large set of instantaneous channel gains, and is represented by $\mathcal{R}_{sr_i}^{avg} = \frac{(T_f-\tau)\mathcal{C}}{T_f}\log_2\left(1+\frac{P_i d_{sr_i}^{-\psi_{sr_i}}}{P_m d_{mr_i}^{-\psi_{mr_i}}}\right)$ [1,8]. Here the mean channel power gain of $|h_{sr_i}|^2$ and $|h_{mr_i}|^2$ are calculated as $E[|h_{sr_i}|^2] = d_{sr_i}^{-\psi_{sr_i}}$ and $E[|h_{mr_i}|^2] = d_{mr_i}^{-\psi_{mr_i}}$, respectively ($\because$ $h_{sr_i} \sim \mathcal{CN}(0, d_{sr_i}^{-\psi_{sr_i}})$, $h_{mr_i} \sim \mathcal{CN}(0, d_{mr_i}^{-\psi_{mr_i}})$). The symbol $E[.]$ denotes the expectation operator. Sum secondary throughput for $K$ number of $ST_i - SR_i$ links is given by

$$\sum_{i=1}^{K}\mathcal{R}_{sr_i}^{avg} = \sum_{i=1}^{K}\frac{(T_f-\tau)\mathcal{C}}{T_f}\log_2\left(1+\frac{P_i d_{sr_i}^{-\psi_{sr_i}}}{P_m d_{mr_i}^{-\psi_{mr_i}}}\right) \tag{9}$$

## 4.2 Problem Formulation and Optimal Solution

The optimization problem can mathematically be stated as mentioned below

$$\begin{aligned}
&\max_{\tau,P_i}\sum_{i=1}^{K}\mathcal{R}_{sr_i}^{avg} \equiv \max_{\tau,P_i}\sum_{i=1}^{K}\frac{(T_f-\tau)\overline{\mathcal{C}}}{T_f}\log_2\left(1+\frac{P_i d_{sr_i}^{-\psi_{sr_i}}}{P_m d_{mr_i}^{-\psi_{mr_i}}}\right)\\
&s.t.\ (i)\ P_d = \overline{P}_d, P_f = \overline{P}_f \implies \gamma\sqrt{\alpha\tau f_s} + (\gamma+1)\mathcal{Q}_{pd} - \mathcal{Q}_{pf} = 0\\
&\quad (ii)\ E_{\mathcal{H}_1}^{NL} + E_{\mathcal{H}_2}^{NL} \geq E_s + E_{tr,ST_i}\\
&\quad\quad \implies \left(\Phi_{\mathrm{EH}_1}^{\mathrm{NL}} - \rho\mathcal{S}\right)\alpha\tau + \left(\Phi_{\mathrm{EH}_2}^{\mathrm{NL}} - P_i\overline{\mathcal{C}}\right)(T_f-\tau) \geq 0\\
&\quad (iii)\ \mathcal{P}(\mathcal{H}_1)(1-\overline{P}_d)P_i d_{spr_i}^{-\psi_{spr_i}} \leq I_{th}
\end{aligned}$$

$$(iv)\ \mathscr{P}_{s,out_i} \leq \theta_s \implies 1 - \left[ \frac{P_i d_{sr_i}^{-\psi_{sr_i}}}{P_m \gamma_{th} d_{mr_i}^{-\psi_{mr_i}} + P_i d_{sr_i}^{-\psi_{sr_i}}} \right] \leq \theta_s$$
$$\implies \theta_s^{'} - \frac{P_m \mathcal{A}_1 \mathcal{R}_{th}^{'} T_f}{P_i (T_f - \tau)\overline{\mathcal{C}}} \geq 0 \quad (10)$$

SS reliability constraint is given in Eq. (10) (i), where $\overline{P}_d$ and $\overline{P}_f$ are the thresholds for detection and false alarm probability, respectively. Equation (10) (ii) denotes the energy causality constraints for $ST_i$, where $P_d = \overline{P}_d$ and $P_f = \overline{P}_f$ are used in the expressions of $\mathcal{B}_{11}$ to get $\overline{\mathcal{B}}_{11} = \overline{P}_d \mathcal{S}_1 + \overline{P}_f \mathcal{S}_2$. Interference power constraint by $ST_i$ at $PR$ under $\mathcal{P}(\mathcal{H}_0|\mathcal{H}_1)$ is mentioned in Eq. (10) (iii). The individual secondary outage constraint is given in Eq. (10) (iv), where $\theta_s$ denotes the secondary outage probability threshold, $\theta_s^{'} = \frac{\theta_s}{(1-\theta_s)}$, $\mathcal{A}_1 = \frac{d_{mr_i}^{-\psi_{mr_i}}}{d_{sr_i}^{-\psi_{sr_i}}}$ and $\gamma_{th} \approx \frac{\mathcal{R}_{th}^{'} T_f}{(T_f - \tau)\overline{\mathcal{C}}}$ [7].

The optimal solutions to Eq. (10) i.e. $P_i^*$ and $\tau^*$ are found using Lagrange multiplier and Karush-Kuhn-Tucker (KKT) conditions [7] and are given as

$$P_i^* = \frac{\mathcal{Y}_{11} + \sqrt{\mathcal{Y}_{11}^2 - 4\mathcal{X}_{11}\mathcal{Z}_{11}}}{2\mathcal{X}_{11}}, \tau^* = \frac{(b_{11} - \overline{\mathcal{C}} P_{\text{mean}}^*) T_f}{(b_{11} - \overline{\mathcal{C}} P_{\text{mean}}^*) - a_{11}\alpha} \quad (11)$$

where, $\mathcal{X}_{11} = \mathcal{B}_1 E_{11} \left( \frac{I_P K \gamma_Q}{\alpha f_s} - \frac{a_{11}\alpha \mathcal{A}_1 P_m}{\theta_s^{'} E_{11} \ln 2} \right)$, $\mathcal{Y}_{11} = P_m \mathcal{A}_1 K (T_f - \frac{\gamma_Q}{\alpha f_s})^3$, $\mathcal{Z}_{11} = P_m \mathcal{A}_1 \mathcal{Y}_{11}$, $I_p = \frac{I_{th}}{\mathcal{P}(\mathcal{H}_1)(1-\overline{P}_d) d_{spr_i}^{-\psi_{spr_i}}}$, $\gamma_Q = \left[ \frac{\mathcal{Q}_{pf} - (\gamma+1)\mathcal{Q}_{pd}}{\gamma} \right]^2$, $E_{11} = (a_{11}\alpha - b_{11}T_f)\frac{\gamma_Q}{\alpha f_s} + b_{11}T_f$, $a_{11} = \Phi_{\text{EH}_1}^{\text{NL}} - \rho \mathcal{S}$, $b_{11} = \Phi_{\text{EH}_2}^{\text{NL}}$. The value of $\tau^*$ is evaluated by taking the average value of $P_i^*$ denoted as $P_{\text{mean}}^*$ for $K$ number of $ST$s to maintain synchronization among the latter. Furthermore, $a_{11}$ and $b_{11}$ in $\tau^*$ is evaluated similarly by taking the average values on parameters.

## 5 Results and Discussion

The numerical results of sum SU throughput ($\sum \mathcal{R}_{sr}^{avg}$) are done by averaging over 10,000 Monte Carlo simulations. Here $\{d_m^{\min}, d_m^{\max}\}$ and $\{\psi_m^{\min}, \psi_m^{\max}\}$ represent the minimum and maximum range of the distance and the path loss exponents, respectively between two nodes. Simulation parameters are set as $N_c$ =10,000, $\overline{P}_d$= 0.95, $\overline{P}_f$ = 0.05, $\mathcal{P}(\mathcal{H}_1)$ = 0.3, $\mathcal{P}(\mathcal{H}_0)$ = 0.7, $T_f$ = 100 ms, $f_s$ = 10 kHz, $\alpha$ = 0.3, $\rho$ =0.1, $P_p$ =40 mW, $P_m$ =35 mW, $K$ =12, $I_{th}$ = 12 mW, $\mathcal{R}_{th}$ = 0.03 mbps/Hz, $\theta_s$ =0.05, $d_{ps}$ = 2.5 m, $\psi_{ps}$ = 4, $d_{ms}$ = 2.2 m, $\psi_{ms}$ = 4, $d_{sr}^{\min}$=2.2 m, $d_{sr}^{\max}$ = 2.4 m, $\psi_{sr}^{\min}$=2.7, $\psi_{sr}^{\max}$ =3.1, $d_{spr}^{\min}$=2.1 m, $d_{spr}^{\max}$ =2.6 m, $\psi_{spr}^{\min}$ = 3.3, $\psi_{spr}^{\max}$= 3.5, $\sigma_n^2$ = 0.45 mW, $\sigma_{ns}^2$ =0.55 mW, $M = 24\ mW$, $a = 1500$, $b = 0.0022$ as reported in [1,7].

Figure 3a shows the variation on $\sum \mathcal{R}_{sr}^{avg}$ with $\tau$ and a gain ~2.31% in $\sum \mathcal{R}_{sr}^{avg}$ is found for $\rho = 0.2$ compared to $\rho = 0.1$. Increase in $\rho$ value requires decrease in $\tau$ value to meet $P_d = \overline{P}_d$ and $P_f = \overline{P}_f$ and increases the required data

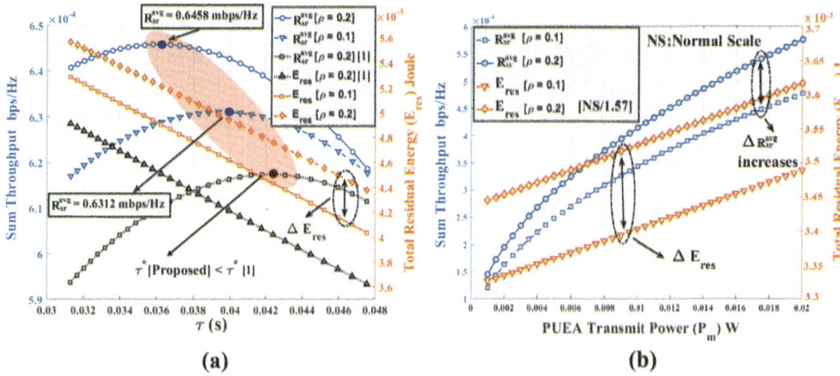


**Fig. 3.** $\sum \mathcal{R}_{sr}^{avg}$ and $E_{res}$ versus **a** $\tau$ and **b** $P_m$

transmission slot $(T_f - \tau)$ of all $STs$ to meet $\mathcal{R}_{th}$ and $\mathcal{R}_s$ values. It is also observed that at $\{P_{\text{mean}}^* = 0.73\text{mW}, \tau^* = 36.3\text{ms}\}$ and $\{P_{\text{mean}}^* = 0.67\text{mW}, \tau^* = 40\text{ms}\}$, the maximum values of $\sum \mathcal{R}_{sr}^{avg}$ are 0.6458 and 0.6312 mbps/Hz, at $\rho = 0.2$ and $\rho = 0.1$, respectively. The variation of $E_{res}$ is also shown with $\tau$, where the former is calculated using $P_d = \overline{P}_d$ and $P_f = \overline{P}_f$ in Eq. (7). It is seen that $E_{res}$ value for $\rho = 0.2$ is ~8.25% greater than $\rho = 0.1$ at $\tau = 46.4$ms. Increase in $\rho$ value decreases $\tau$ and enhances $(T_f - \tau)$ slot (earlier mentioned) which reduces the required $P_{\text{mean}}$ value. Performance comparison on $\sum \mathcal{R}_{sr}^{avg}$ and $E_{res}$ is shown with [1]. The maximum value of $\sum \mathcal{R}_{sr}^{avg}$ is observed to be ~4.5% more whereas the value of $E_{res}$ at $\tau = 36.3s$ is ~14.79% more for $\rho = 0.2$ over [1]. It is also observed that $\tau^*$ values for both $\rho = 0.1$ and $\rho = 0.2$ are less over [1], hence enhanced transmission slot $(T_f - \tau)$ reduces $P_{\text{mean}}$ value and consequently increase in $E_{res}$ value is seen over [1].

The variation on $\sum \mathcal{R}_{sr}^{avg}$ and $E_{res}$ with $P_m$ for $\rho = 0.1$ and $\rho = 0.2$ is shown in Fig. 3b. When $P_m$ value is increased, the interfering signal at FC and to each $SR_i$ (see Eq. 9) also increases. Increased $P_m$ needs increase in the value of $\tau$ (to maintain SS reliability) which reduces $(T_f - \tau)$ slot. To compensate the increased interfering effect at $SR_i$ in this reduced transmission slot, the value of $P_{\text{mean}}$ increases significantly (this increased power is supported from the energy causality in EH) leading to an overall enhancement on $\sum \mathcal{R}_{sr}^{avg}$ value of ~9.76% for $\rho = 0.2$. However, increase in $P_m$ value (~20%) also increases further the scope of EH from PUEA which increases $E_{res}$ value of ~0.31% for $\rho = 0.2$.

## 6 Conclusions

To conclude, the presence of PUEA enables (i) a unique maximum point for sum secondary throughput ($\sum \mathcal{R}_{sr}^{avg}$) to achieve at some optimal values of individual transmit power and the sensing duration, (ii) a gain ~ 4.5% in the maximum value of $\sum \mathcal{R}_{sr}^{avg}$ and ~14.79% is achieved for the total residual energy ($E_{res}$) at $\tau = 36.3$s over [1], (iii) an increase in power splitting ratio ($\rho$) value increases both $\sum \mathcal{R}_{sr}^{avg}$ and $E_{res}$ values through an increased value in data transmission slot and reduced value of $P_{\text{mean}}$, (iv) increase in PUEA transmit power ($P_m$) increases SS duration, at the same time also enhances $\sum \mathcal{R}_{sr}^{avg}$ and $E_{res}$ values through enhanced scope of EH. In brief, $\sim 20\%$ increase in PUEA value enhances both $\sum \mathcal{R}_{sr}^{avg}$ $\sim$9.76% and $E_{res}$ $\sim$0.31% for $\rho = 0.2$ at $\overline{P}_d = 0.95$ and $\overline{P}_f = 0.05$.

## References

1. Banerjee, A., Maity, S.P., Das, R.K.: On throughput maximization in cooperative cognitive radio networks with eavesdropping. IEEE Commun. Lett. **23**(1), 120–123 (2019). Jan
2. Banerjee, A., Maity, S.P.: On residual energy maximization in cooperative spectrum sensing with PUEA. IEEE Wirel. Commun. Lett. **8**(6), 1563–1566 (2019)
3. Chen, C., Cheng, H., Yao, Y.: Cooperative spectrum sensing in cognitive radio networks in the presence of the primary user emulation attack. IEEE Trans. Wirel. Commun. **10**(7), 2135–2141 (2011). July
4. Banerjee, A., Maity, S.P.: On optimal sample checkpoint for energy efficient cooperative spectrum sensing. Digital Signal Process. **74**, 56–71 (2018)
5. Alahmadi, A., Fang, Z., Song T., Li, T.: Subband PUEA detection and mitigation in OFDM-based cognitive radio networks. IEEE Trans. Infor. Forensics Security **10**(10), 2131–2142 (2015)
6. Karimi, M., Sadough, S.M.S.: Efficient transmission strategy for cognitive radio systems under primary user emulation attack. IEEE Syst. J. **12**, 3767–3774 (2018). Dec
7. Banerjee, A., Maity, S.P.: On residual energy maximization in cognitive relay networks with eavesdropping. IEEE Syst. J. **13**(4), 3836–3846 (2019). Dec.
8. Banerjee, A., Paul, A., Maity, S.P.: Joint power allocation and route selection for outage minimization in multihop cognitive radio networks with energy harvesting. IEEE Trans. Cogn. Commun. Netw. **4**, 82–92 (2018). March

# The Design and Performance Evaluation of 4-SSB Using Hilbert Transform with SISO Turbo and Shadow Equalizer Toward 5G Communication Networks and Beyond

Alhassani Mohammed Mustafa[1(✉)], Quang N. Nguyen[1(✉)], Gen-Icchiro Ohta[2], and Takuro Sato[1]

[1] Graduate School of Fundamental Science and Engineering, Waseda University, Tokyo, Japan
mohammed@akane.waseda.jp, quang.nguyen@aoni.waseda.jp
[2] Yokosuka Telecom Research Park, Inc, Yokosuka, Japan

**Abstract.** As the Hilbert transform is known as the main cause of intersymbol interference (ISI) in single-sideband generation (SSB), the design and performance evaluation of relevant equalizers over Hilbert transform with different numbers of taps is critical to minimize the ISI effect. In this paper, we develop a mathematical model of Hilbert transform with the different number of taps with the relevant equalizers over four single-sideband (4-SSB), given that 4-SSB modulation can carry a double amount of information while using half of the bandwidth compared to the conventional modulation of SSB. As expected, the evaluation results demonstrate that the soft-input soft-output (SISO) Turbo equalizer has degraded the ISI issue with small number of tap since the Hilbert transform is showed to work empirically in the case of small tap number. Similarly, the Shadow equalizer verifies the benefits and robustness of the Hilbert transform effect with the lowest number of taps, particularly Bit Error Rate (BER) performance is converged when the number of taps is greater than 15. This demonstrates a new insight into applying the SISO Turbo and Shadow equalizer using the novel concept of 4-SSB modulation for data transmission in the case of coded and uncoded wireless environments, respectively, toward 5G communication networks and beyond.

**Keywords:** Soft-input soft-output (SISO) equalizer · Shadow equalizer · Inter-Symbol interference (ISI) · Four Single-Sideband (4-SSB) · Hilbert transform · Turbo coding · 5G

## 1 Introduction

The new wireless technology requires data transmission with a high data rate and the quality of service to match the huge content demands. In this context, the single-sideband (SSB) technology [1] has demonstrated unique merit with the capability to transmit the


X.-S. Yang et al. (eds.), *Proceedings of Fifth International Congress on Information and Communication Technology*, Advances in Intelligent Systems and Computing 1183, https://doi.org/10.1007/978-981-15-5856-6_36

information through transmitter using only half of the spectrum bandwidth, i.e., save the power compared to the equivalent schemes. However, SSB is an amplitude modulation which is affected by the sensitivity of the random nature of the wireless channel environment. Hence, SSB endures the symbol intersymbol interference (ISI), especially in high data rate and high-order modulation like quadrature amplitude modulation (QAM) and phase-shift keying (PSK).

In SSB, the signal is generated using the Hilbert transform [2]. Currently, most of the research dealt with the non-ideal Hilbert transform in infinite impulse response (IIR), whereas the finite impulse response (FIR) is studied for the practical system implementation. The SSB is validated for optical communication to reach a high data rate in gigabit per second. Recently, the most common SSB technology is twin SSB to double the amount of information on both sides of the spectrum, instead of only one side as of the conventional SSB. Twin SSB, however, is mainly applied for wired networks using the optical modulator in which the coefficient of channel fading $h$ is equal to one.

Also, given that the spectrum bandwidth is limited and bounded by the government regulations, we need a new kind of modulation to match the 5G specification. Currently, there are several existing modulation technologies that realize the high data rate to guarantee the quality of service, e.g., the non-orthogonal multiple access (NOMA) and the orthogonal multiple access (OMA) [3]. The main concept of these technologies is provided the many users with the same allocation frequency which implying the save the power spectrum.

To exploit the advantage of SSB, several researches started improving the capacity of the SSB scheme by combining it with quadrature phase-shift keying (QPSK) to increase the capacity of the channel due to the fact that SSB requires the orthogonal basis. For instance, the authors in [1] applied widely minimum mean square error (MMSE) algorithm in Turbo equalizer to compensate ISI. The result showed good performance by successfully transmit data over QPSK with SSB. However, the research has a limitation, which is difficult to be applied in 4G and beyond systems because the data rate does not match the requirements of modern wireless systems.

To improve the SSB capability, Ohta et al. [4] proposed the new modulation scheme by combining the four independent signals to produce the four single-sideband (4-SSB). This modulation can carry a double amount of information while using half of the bandwidth compared to the conventional modulation of SSB. Specifically, 4-SSB enables the same feature as of Twin SSB [5] and is applicable to the wireless communication systems in which Twin SSB does not work because the channel fading is randomly changed in the dynamic wireless environment, which is presented by Gaussian distribution. To decrease the ISI effect, the widely MMSE Turbo equalizer is applied to keep the quality of the received signal [6].

The 4-SSB technique is also improved by applying the OFDM for increasing the capacity which matches the requirements in 4G [7]. However, on this day, 5G has been launched and it enables ten times higher data rate compared to 4G. For that reason, the 4SSB over OFDM scheme was improved by applying new Turbo equalizers called Shadow equalizer and SISO equalizer to increase the data rate for the high-order modulation like 16-QAM and 64-QAM [8, 9].

Since the Hilbert transform with finite impulse response (FIR) has not been considered in previous 4-SSB studies, in this research, we study the effect of Hilbert transformer in 4-SSB using relevant equalizers specified by Verizon 5G Technical Forum (V5GTF) prototype [10] for 5G radio specification toward 5G and beyond to improve the receiver signal performance with decreased complexity. We also find out that SISO Turbo equalizer is a feasible approach for 4-SSB modulation toward 5G and beyond since it performs well even with a small number of taps, as shown in the evaluation section (Sect. 5).

## 2 Four Single-Sideband (4-SSB) System Model

In this section, we introduce the brief concept of the 4-SSB system model. In general, the 4-SSB system is modulated and demodulated based on the conventional SSB generation by using the Hilbert transform to be applicable to the high-order M-QAM (Quadrature Modulation). The 4-SSB signal is generated by using the four independent discrete signals denoted by $u$, $v$, $p$ and $r$ as the modulation scheme sequence generation as 1 follows [4]:

$$S_{4\mathrm{SSB,I}} = u - \hat{v} + p + \hat{r}, \tag{1}$$

$$S_{4\mathrm{SSB,Q}} = -\hat{u} - v + \hat{p} - r, \tag{2}$$

where $S_{4\mathrm{SSB,I}}$ and $S_{4\mathrm{SSB,Q}}$ denote the In phase 4-SSB and Quadrature modulation phase 4-SSB, respectively, and $\hat{u}$, $\hat{v}$, $\hat{p}$ and $\hat{r}$ denote the Hilbert transforms of original discrete signals.

The two complex signals of $d_1$ and $d_2$ are combined in 4-SSB to increase the channel capacity and data rate of the system. The demodulation process of 4-SSB is a critical procedure for the successful reception of the two complex data $d_1$ and $d_2$ and can be modeled as follows:

$$d_{1,\mathrm{LSB}} = \frac{1}{2}\left[u + \hat{v} + j(u + v)\right], \tag{3}$$

$$d_{2,\mathrm{USB}} = \frac{1}{2}\left[p + \hat{r} + j(r - \hat{p})\right], \tag{4}$$

where $d_{1,\mathrm{LSB}}$ denotes the lower sideband of 4-SSB and $d_{2,\mathrm{USB}}$ denotes the upper sideband of 4-SSB.

## 3 The Hilbert Transform Approach

In general, the Hilbert transform is mainly used for signal processing applications like Fourie, $Z$ and Laplace, Hartley, and Wavelet transform [11–14]. In this research, the modulation and demodulation procedure of the 4-SSB scheme for the Hilbert transform is conducted in the same way as Eqs. (1) and (2) in which SSB is generated mainly through using the Hilbert transform. Also, to be applied in the time domain, the Hilbert transform only changes the frequency sign rather than changing the signal domain. This characteristic of the Hilbert transform will be described in the next section.

### 3.1 Hilbert Transform with a Mathematical Model

Together with the Fourier transform, Hilbert transform realizes a well-known transform for the analytical spectral signal. The mathematical model of the Hilbert transform is performed by using the feature of the Fourier transform.

To enable various applications, we use the Hilbert transform for the 4-SSB modulation and demodulation process to enhance the received signal performance and compensate ISI at the same time by using the different type of equalization process.

In particular, the mathematical model for the Hilbert transform can be described by considering a Fourier transform signal pair as follows:

$$u(t) \overset{F}{\Leftrightarrow} U(f), \tag{5}$$

where () is a unit step function. $U(f)$ can be represented in the frequency domain as a complex number with a real part and imaginary part as:

$$U(f) = A(f) + jB(f). \tag{6}$$

$U(f)$ is then an analytical signal of complex Fourier signal of $u(t)$. By using the kernel of integral information for signal $u(t)$, the Hilbert transform can be defined in the time domain as follows:

$$v(t) = H[u(t)] = \frac{1}{\pi t} * u(t), \tag{7}$$

$$u(t) = H[u(t)]^{-1} = -\frac{1}{\pi t} * v(t) \tag{8}$$

() and () realize the pair of Hilbert transform and can be denoted as the following transformation:

$$u(t) \overset{H}{\Leftrightarrow} v(t). \tag{9}$$

In the next section, we describe the Hilbert transform with infinite impulse response FIR.

### 3.2 Hilbert Transform with Finite Impulse Response (FIR)

Since modern communications include the built-in digital signal processors, all of the signals must be converted to a digital signal before they are initially processed. This process is known as an analog to digital converter (A/D).

The Hilbert transform is mainly performed in the digital domain called the discrete Hilbert transform or Finite impulse response FIR. Figure 1 illustrates two types of Hilbert transform: one includes the infinite delay in the continuous-time domain, whereas the other is the finite or discrete-time domain (FIR). The 4-SSB system is designed to use Hilbert transform as FIR to realize a practical wireless system. By default, the filter of the Hilbert transform includes 21 taps in transmission and

receiver, as suggested by our prior study [7]. In this research, to deduce the highly feasible and efficient system, we evaluate the optimized number of taps to gain high performance with low complexity using different types of equalization like SISO with Turbo equalizer and Shadow equalizer.

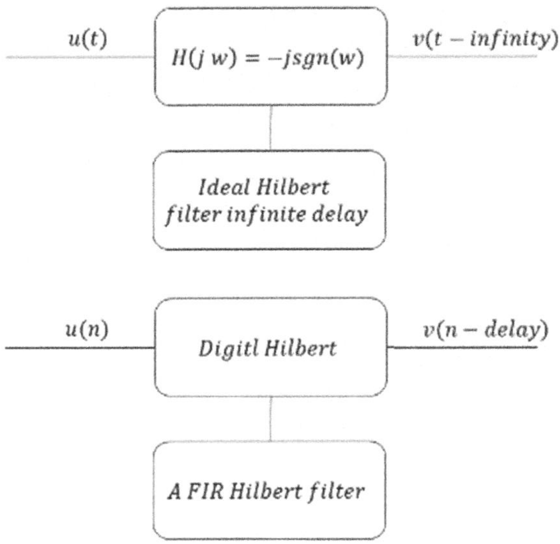


**Fig. 1.** Structure of two types of Hilbert transform (infinite and finite impulse response)

## 4 Mathematical Model of Hilbert Transform with Taps Over 4-SSB

As can be seen in Fig. 1, the ideal Hilbert transform is $() = -()$ where $()$ is the sign function of the frequency domain. Then, if the frequency is in a positive band, the sign will become negative and vice versa, as shown in Fig. 2. The impulse response in each tap of Hilbert transform in the time domain can be represented as the following equation:

$$h = \frac{1}{\pi t}. \tag{10}$$

The cascade tap results in higher delay, but there is a trade-off between the optimized number of tap and efficient performance of 4-SSB in a wireless channel environment.

### 4.1 Hilbert Transform Model with Odd and Even Number of Taps

In discrete Hilbert transform, the impulse response in odd and even number of taps can be represented by the notation $g_{HT}$. Also, the discrete signal can be represented with discrete Hilbert transform as the following equation:

$$\hat{x}[n] = \sum_{k=\frac{-N}{2}+1}^{\frac{N}{2}} g_{HT}[k]x[n-k] \tag{11}$$

Let $N$ be the number of taps in the Hilbert transform, the Hilbert impulse response with respect to the tap number $N$ can be expressed as:

$$N \text{ even: } g_{\mathrm{HT}}[k] = \begin{cases} \frac{2}{N}\sin^2\left(\frac{\pi.k}{2}\right)\cot\left(\frac{\pi.k}{N}\right), & k \neq 0 \\ 0 & k = 0 \end{cases} \tag{12}$$

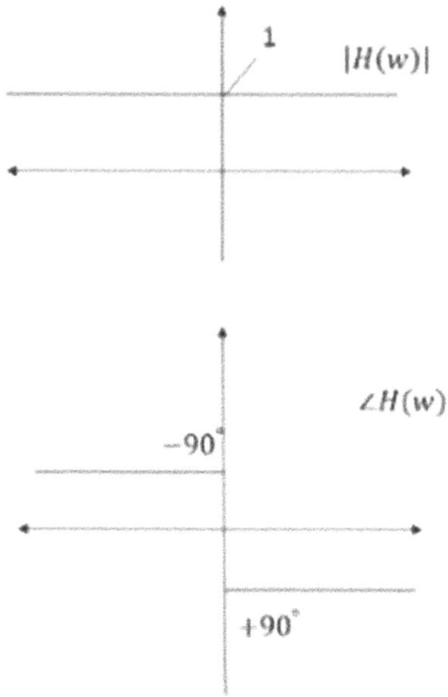


**Fig. 2.** Hilbert transform impulse in infinite

$$N \text{ odd}: g_{\mathrm{HT}}[k] = \begin{cases} \frac{2}{\pi.k} & k \text{ odd} \\ o & k \text{ even} \end{cases}; \tag{13}$$

Note that due to the Hilbert transform properties in the discrete domain, the performance of the proposal remains unchanged in 4-SSB modulation and demodulation processes, irrespective of the number of taps (odd or even). The USB and LSB in 4-SSB are then represented in all discrete domain as the following expressions:

$$x_{\mathrm{SSB}}[n] = \sum_k h_{\mathrm{SSB}}[k].x[n-k], \tag{14}$$

$$h_{\mathrm{SSB}}[k] = \begin{cases} \pm j.g_{\mathrm{HT}}[k], & k \neq 0 \\ 1 \pm j.g_{\mathrm{HT}}[k], & k = 0 \end{cases} \tag{15}$$

Equation (14) is a discrete convolution with channel impulse response of Hilbert transform. As the Hilbert transform in the continues time domain was mentioned in previous sections, Eqs. (14) and (15) form the mathematical model of Hilbert transform, which is applicable in the 4-SSB model for the performance evaluation.

## 5 Performance Evaluations and Discussion

### 5.1 The Hilbert Transform Performance Using Turbo Equalizer with SISO in AWGN

In this section, the effect of the number of Hilbert transform taps in odd is described. It can be deduced from Eq. (13) that half of the coefficient of the impulse response is equal to zero (i.e., is even and differs from zero). Figure 3 shows that the Hilbert transform does not work well because to realize the benefits of Hilbert transform, a high number of taps is needed to make an appropriate approximation. Also, when the number of taps is increased to 11, the performance of 4-SSB is increased as the SNR reaches 12 dB. When the number of taps is increased to 21, the proposal achieves the best performance under the Additive white Gaussian noise AWGN.

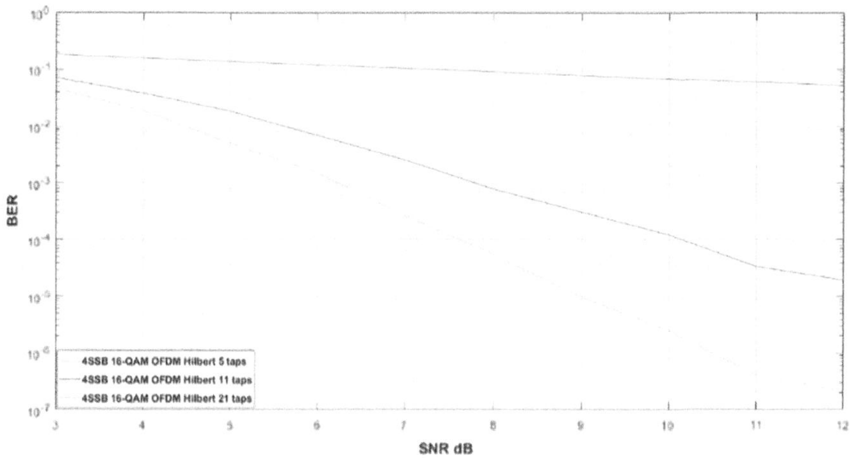


**Fig. 3.** Performance evaluation of the Hilbert transform with taps in SISO Turbo equalizer

### 5.2 The Hilbert Transform Performance Using Shadow Equalizer in AWGN

Figure 4 shows the performance of the Hilbert transform tap evaluation via the simulation of Shadow equalizer in AWGN when the SNR range is taken from 0 dB to 30 dB. As can be seen, the Hilbert transform with the Shadow stills performs well even with the smallest number of taps. Although SISO equalizer shows a higher BER performance when the number of taps is not greater than 11, we can observe that Shadow equalizer achieves a higher and more stable performance for the 4-SSB received signal when the number of taps is greater than 15, provided that Shadow equalizer is run under the uncoded system compared to the SISO coded system.

## 6 Conclusion

In this paper, the Hilbert transformer tap effects in the 4-SSB model system are studied. Two types of equalizers, including SISO and Shadow equalizers were evaluated. A new design of Hilbert transforms in 4-SSB-based wireless communications and its performance evaluations of the effect as one step to verify the proposed scheme for the 5G network is proposed. However, some necessary work remains regarding using OFDM technology, e.g., synchronization in the up and downlink channels. Also, the other parameters will be evaluated to further prove the robustness of Hilbert transform with the Hilbert filter, such as the number of the taps of the equalizer.

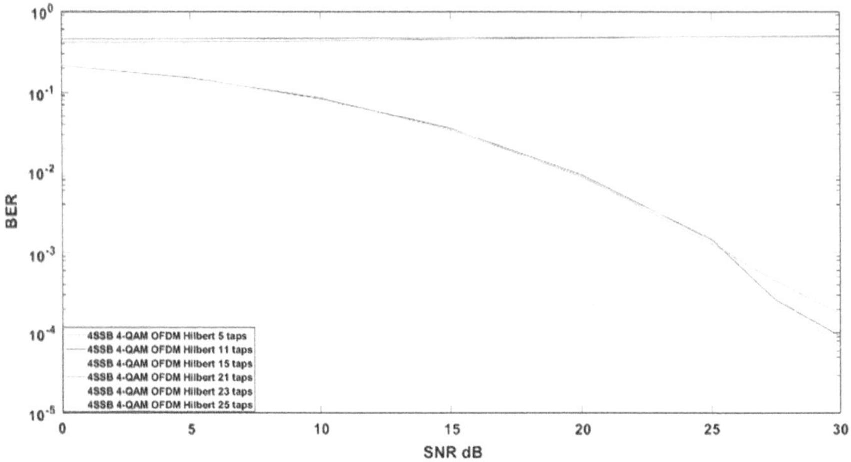


**Fig. 4.** Performance evaluation of the Hilbert transform with taps in Shadow equalizer

## References


1. Pitakdumrongkija, B., Suzuki, H., Suyama, S., Fukawa, K: Single sideband QPSK with turbo equalization for mobile communications. In: 2005 IEEE 61st Vehicular Technology Conference, Stockholm, vol. 1, pp. 538–542 (2005). https://doi.org/10.1109/vetecs.2005.1543349
2. Hahn, S.L.: Hilbert Transforms in Signal Processing (1996)
3. Zhang, D., Liu, Y., Ding, Z., Zhou, Z., Nallanathan, A., Sato, T.: Performance analysis of non-regenerative massive-MIMO-NOMA relay systems for 5G. IEEE Trans. Commun. **65**, 4777–4790 (2017)
4. Ohta, G., Nanri, M., Uesugi, M., Sato, T., Tominaga, H.: A Study of New Modulation
5. Shi, J., Zhou, Y., Xu, Y., Zhang, J., Yu, J., Chi, N.: 200-Gbps DFT-S OFDM using DD-MZM-based twin-SSB with a MIMO-Volterra equalizer. IEEE Photon. Technol. Lett. **29**, 1183–1186 (2017)
6. Tuchler, M., Singer, A.C.: Turbo equalization: an overview. IEEE Trans. Inf. Theor. **57**(2), 920–952 (2011)
7. Jiang, Y., Zhou, Z., Nanri, M., Ohta, G.I., Sato, T.: Performance evaluation of four orthogonal single sideband elements modulation scheme in multi-carrier transmission systems. In: 2011 IEEE Vehicular Technology Conference (VTC Fall), pp. 1–6 (2011)
8. Mustafa, A.M., Nguyen, Q.N., Sato, T., Ohta, G.: Four single-sideband M-QAM modulation using soft input soft output equalizer over OFDM. In: Proceedings of the 2018 28th International Telecommunication Networks and Applications Conference (ITNAC), pp. 1–6. Sydney, Australia, 21–23 Nov 2018
9. Alhasani, M.M., Nguyen, Q.N., Ohta, G.-I., Sato, T.: A novel four single-sideband M-QAM modulation scheme using a shadow equalizer for MIMO system toward 5G communications. Sensors **2019**, 19 (1944). https://doi.org/10.3390/s19081944
10. Verizon 5G TF: Air Interface Working Group. Verizon 5th Generation Radio Access; Multiplexing and Channel Coding (Release 1); Verizon Wireless. NY, USA (2016)
11. Tsai, L.-C., Hsue, C.-W.: Dual-band bandpass filters using equal length coupled-serial-shunted lines and Z-transform technique. IEEE Trans. Microw. Theor. Tech. **52**(4), 1111–1117 (2004). https://doi.org/10.1109/TMTT.2004.825680

12. Nahin, P.J.: Behind the laplace transform. IEEE Spectrum **28**(3), 60 (1991). https://doi.org/10.1109/6.67288
13. Averchenko, A.P., Zhenatov, B.D.: Hartley transform as alternative to fourier transform in digital data processing systems. In: 2014 Dynamics of Systems, Mechanisms and Machines (Dynamics), pp. 1–4. Omsk, 2014. https://doi.org/10.1109/dynamics.2014.7005633
14. Bruce, A., Donoho, D., Gao, H.: Wavelet analysis [for signal processing]. IEEE Spectrum **33**(10), 26–35 (1996). https://doi.org/10.1109/6.540087

# E-learning Course for Healthcare Professionals: Continuing Education for Idiopathic Scoliosis

Donatella Barbina[1], Giovanni Galeoto[2](✉), Debora Guerrera[1], Alessadra Di Pucchio[1], Pietro Carbone[1], Valter Santilli[3], Anna Berardi[4], Donatella Valente[4], and Alfonso Mazzaccara[1]

[1] National Institute of Health—Training Office—Istituto Superiore di Sanità, Rome, Italy
[2] Department of Public Health and Infectious Diseases, Sapienza University of Rome, Rome, Italy
giovanni.galeoto@uniroma1.it
[3] Department of Anatomy, Histology, Forensic Medicine and Orthopedics, Sapienza University of Rome, Rome, Italy
[4] Department of Human Neurosciences, Sapienza University of Rome, Rome, Italy

**Abstract.** The objective of this study is to analyze the results of the e-learning course "Health promotion, prevention, diagnosis and treatment of idiopathic scoliosis," directed to the Italian healthcare professionals involved in idiopathic scoliosis management. The e-learning course, based on the problem-based learning methodology, was directed to health professionals previously enrolled to the University of Rome's research project "Prevention, diagnosis and treatment of idiopathic scoliosis: evaluation of skills acquired in health professionals through distance learning." This study is focused on the analysis of the participant's data. A significant improvement in knowledge ($t = 17.2$; $p < 0.001$) was found in participants' scores from pre- and post-test. 80.1% of participants passed the final certification test. The final satisfaction questionnaire showed a high level of satisfaction among participants. Health professionals involved in idiopathic scoliosis management need adequate training, based on methods appropriate to their professional context. The e-learning course's results are positive and in line with the project's task; moreover, this confirms the validity of an active didactic approach such as the problem-based learning.

**Keywords:** Scoliosis · Education · Distance · Problem-based learning · Education · Medical · Continuing

## 1 Introduction

Idiopathic scoliosis (IS), when untreated, may lead to severe trunk deformities, which limit the capacity and functional biomechanics of the chest, exercise capacity, general fitness and ability to work, and all factors related with impairment on quality of life


X.-S. Yang et al. (eds.), *Proceedings of Fifth International Congress on Information and Communication Technology*, Advances in Intelligent Systems and Computing 1183, https://doi.org/10.1007/978-981-15-5856-6_37

[1]. Being crucial early detection and correct initial management in effective IS treatment, an overarching goal is health professionals' training on the main concepts in IS and in basic criteria for diagnosis and treatment. Since continuing medical education (CME) is mandatory in most countries and in Italy, in addition to traditional educational approaches, the use of e-learning has steadily increased among health professionals worldwide [2].

Starting from 2004, the e-learning working group of the Italian National Institutes of Health (NIH) [3–5] is delivering e-learning courses on the platform https://www.eduiss.it (EDUISS), assessing the quality of e-learning programs paying attention to several key factors, such as navigability, multimedia approach, and degree of interactivity. A specificity of the e-leaning methodologies developed and mainly adopted in EDUISS is an innovative ways to reproduce problem-based learning (PBL), originally devised for classroom learning, in the e-learning context, using the best available Web tools [3–5].

To address healthcare workers' training needs on IS topics, the Department of Anatomical, Histological, Medical-Legal Sciences and Locomotor System's—University of Rome "Sapienza" organized a national congress and, in co-organization with the e-learning group of the training office of the NIH of Italy developed an e-learning course on "Health promotion, prevention, diagnosis and treatment of idiopathic scoliosis." The congress and the e-learning course are part of the research project "Prevention, diagnosis and treatment of idiopathic scoliosis: evaluation of skills acquired in health professionals through distance learning." The research project was carried out by health professionals at Sapienza University in Rome and by Rehabilitation and Outcome Measures Assessment Association (R.O.M.A). The research group has carried out many outcome measures in Italy [2, 6–15]

The e-learning course was delivered from November 22, 2017 to March 22, 2018.

The purpose of this study was to analyze the results of the CME e-learning course, focusing on the knowledge acquired by the participants and their level of satisfaction.

## 2 Materials and Methods

### 2.1 Design

A one-group pre-post study design was used: participants were assessed before the training (pre-training—T0) at the beginning of the course and again after the training (post-training—T1). Moreover, after receiving the post-test's feedback on the specific learning objective to be reviewed before taking the final certification test, participants at the end of the course filled in a knowledge final evaluation test and the satisfaction questionnaire (one-group post-test-only design).

The cohort of participants included in the study comprised healthcare professionals who completed the course: (1) filled out the two self-assessment test (T0 pre-test, T1 post-test), (2) passed the knowledge final certification test with a score at least of 75% of correct answers, and (3) filled out the satisfaction questionnaire.

The study excluded participants who enrolled the course but did not complete it for voluntary withdrawing or that not started it at all (dropout) and participants that failed in passing the knowledge final test (not passed).

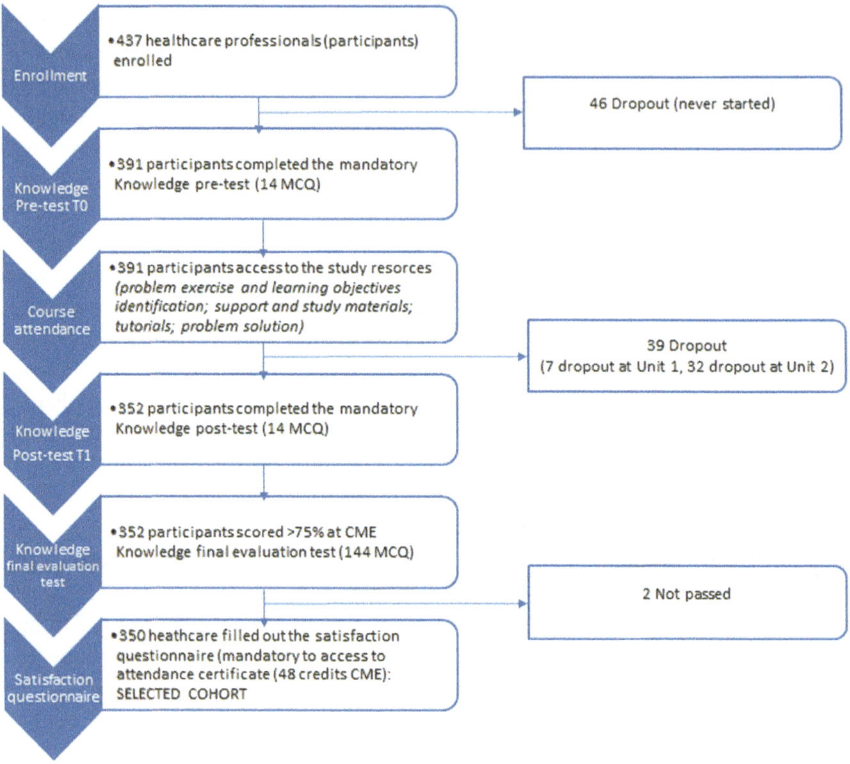


**Fig. 1.** Flowchart of cohort selection

Applying these criteria, the cohort comprised 350 participants, as reported in Fig. 1.

The course was developed by experts with different backgrounds: experts who provided the scientific contents (from University of Rome "Sapienza," Department of Anatomical, Histological, Medical-Legal Sciences, and Locomotor System) and experts in the e-learning methods and technological aspects (training office of the Italian NHI).

The course was direct to health professionals involved in all the IS aspects: physicians, occupational therapists, orthopedic technicians, physiotherapists, therapists of neuro and psychomotricity of the developmental age and Nurses.

Specifically, to the course could enroll participants previously included into the research project "Prevention, diagnosis and treatment of idiopathic scoliosis: evaluation of skills acquired in health professionals through distance learning."

The learning method selected to develop the e-learning course was based on the integration of problem-based learning (PBL), an active learning methodology, with the e-learning tools available on the platform EDUISS.

In summary, PBL is a methodology that challenges participants to "learn to learn" by working in small groups to solve real-world problems that mirror their working context, with the goal of developing life-long learning skills [3]. This way, the process of working toward the understanding or solving of a problem leads to the participants' learning along a seven steps cycle.

The learning management system (LMS) used for the learning project was Totara Learn 11 (based on a Moodle extension), that offered all the technical resources to reproduce the selected methodological approach (PBL). Indeed, various LMS tools have been used for the implementation of the course structure. The PBL cycle has been set up using the platform tools such as feedback, Web pages, quiz and other learning tools, such as Shareable Content Object Reference Model (SCORM).

The access to the activities was set up in order to schedule the learning path according to the conditions that the previous activity was completed. Only upon successfully completing all the course activities an attendance certificate was provided for each participant. The tutors were available by e-mail to clear up any doubts regarding the course contents, methodological aspects, and technological issues.

The course was structured into two learning units, both reproducing an entire PBL cycle.

The general objectives of the e-learning course were to update the participants about the main concepts in IS and to guide healthcare professionals about the clinical practice for IS. The two units and the related specific learning objects were:

1. Idiopathic scoliosis concepts and epidemiology
2. Evaluation and treatment of idiopathic scoliosis.

Participants were expected to spend 48 h to complete the course: they could access the course at any time; however, 20 weeks was the time of fruition.

The course was developed within the Continuing Medical Education (CME) Italian regulation and delivered through the e-learning platform of the Italian NHI—EDUISS (https://www.eduiss.it).

### 2.2 Study Instruments and Data Collection

The e-learning course included three types of activities for participants' assessment and evaluation and data collection:

- Pre-post test. Two self-assessment knowledge tests were required to be filled out at the beginning (pre-test—T0) and at the end of the course (post-test—T1). Pre-post test was composed of the same set of 14 multiple-choice questions (MCQ) to establish a baseline before training and to compare the score gained with the post-test one.
- Final certification test (CME final test). At the end of each unit, participants were required to pass a MCQ test, consisting in 72 questions on knowledge for a total of 144 MCQs (0 points for wrong answers—1 point for correct answer). The passing mark was set at 75% of the scored questions answered correctly. Three passing attempts were allowed. Time to complete the test was 150 min.
- Satisfaction questionnaire. At the very end of the course, participants which successfully completed the final MCQ test were also required to fill in the satisfaction questionnaire (SQ). The SQ consisted in 18 closed questions. The closed questions

were Likert type---scale: from 1 to 5 (1 = I do not agree at all, 2 = not I agree; 3 = neither agree nor disagree; 4 = I agree; 5 = are decidedly agree).

### 2.3 Ethical Considerations

According to institutional regulation in Italy, ethics committee approval was not required for this type of interventional study: participants were informed about methods and aims of the study and agreed to the use of anonymous data in accordance with Italian Data Protection Regulations.

### 2.4 Statistical Analysis

Descriptive analysis of participants' characteristics was performed. Values are presented as mean, median, standard deviation (SD), and frequency (%).

Differences between each one person's pre-test and post-test scores have been calculated. Comparisons between pre- and post-test were performed using the paired samples student's T test, assuming that values at T0 were lower than T1.

Mean, median, and standard deviation (SD) are presented for participants' final certification test score.

Satisfaction questionnaire (SQ) results were analyzed and represented by a radar chart, grouping the 18 questions under three headings: learning methodology (7 questions/radials), contents (8 questions/radials), and platform (3 questions/radials). The scale was from 1 to 5, where 5 expresses the highest degree of satisfaction (supplementary data are presented as mean and SD for each item and for the SQ total mean score).

Analyses were performed using the statistical software Statistical Package for Social Science (SPSS) version 25.0 (IBM, Corp., Armonk, NY).

## 3 Results

### 3.1 Participant Characteristics

A total of 437 health professionals enrolled to the course, demographic and professionals data of completers are presented in Table 1.

### 3.2 Pre-post Test Analysis Results

Significant differences in the pre-post self-assessment test were found: the mean of percentage of total correct answers was $8.2 \pm 3.2$ (from 0 to 14) in pre-test and $11.4 \pm 2.7$ (from 1 to 14) in post-test. There is strong evidence ($t = 17.2$ $p < 0.001$) that the teaching intervention improves marks.

At post-test (T1), 72.3% of participants had improved their results, 16% had invariant condition, and 11.7% of participants reported worst results.

**Table 1.** Demographic and professionals data of completers ($N = 350$)

| | Gender | | Total sample |
|---|---|---|---|
| | Female | Male | |
| Gender *N* (%) | 182 (52%) | 168 (48) | 350 (100) |
| Age, mean (range) | 46 (27–63) | 50.8 (26–79) | 48.3 (26–79) |
| *Professional profiles (CME) N (%)* | | | |
| Physician | 86 (24.6) | 99 (28.3) | 185 (52.9) |
| Physiotherapist | 74 (21.1) | 42 (12.0)) | 116 (33.1) |
| Orthopedic technician | 12 (3.4) | 25 (7.1) | 37 (10.6) |
| Occupational therapists | 5 (1.4) | 1 (0.3) | 6 (1.7) |
| Nurse | 4 (1.1) | 0 (0.0) | 4 (1.1) |
| Therapist of neuro and psychomotricity of the developmental age | 1 (0.3) | 1 (0.3) | 2 (0.6) |
| *Employment status* | | | |
| Employee | 112 (32.9) | 107 (30.6) | 219 (62.6) |
| Private contractor | 4 (1.1) | 11 (3.1) | 15 (4.3) |
| Self-employed | 65 (18.6) | 49 (14.0) | 114 (32.6) |
| Unoccupied | 1 (0.3) | 1 (0.3) | 2 (0.6) |

*Note 1* SD, standard deviation

### 3.3 Final Certification Test

The final certification test scores were calculated on a scale ranging from 0 to 100 and expressed in percentage. For completers, results on the final certification test ranged from 77 to 100% and the mean score of the cohort at the end of the course training was 93.3% (SD ±5.0). The majority of participants (72%) scored over 90% of correct answers.

### 3.4 Level of Satisfaction

The majority of participants (91.7%) expressed high level of satisfaction on the learning methodology adopted, as well as on the adequacy of the contents and on the e-learning platform functioning.

## 4 Discussion

The primary goal of rehabilitation for idiopathic scoliosis is to reduce the progression of curves thereby decreasing the risk of secondary impairment, including back pain, breathing problems, and cosmetic deformities, and improve the quality of life [16–19]

In line with the aims of the research project "Prevention, diagnosis and treatment of idiopathic scoliosis: evaluation of skills acquired in health professionals through distance learning" an e-learning course was provided free of charge by the ISS e-learning platform and addressed to all the HCWs involved in the IS management.

A total of 437 participants registered to attend the course; 230 (53%) were females; the average age of participants was 49 years.

A very high percentage (350; 80.1%) of participants completed successfully the course (completers); the dropouts (participants that enrolled the course but not completed it for voluntary withdrawn or not started at all) were 85 (19.5%); only 2 (0.5%) failed in passing the final test (not passed). This represents a very satisfying result, considering the expected level of completion in distance training context, as descripted in literature [20] and found out in ISS's previous e-learning experiences [21–30]. Moreover, the high level of completion could be also explained by the preliminary selection of the participants, who were already involved as participants in other training activities within the research project.

The completers' data show a prevalence of females (182; 52%); the mean age was 48.3 years (standard deviation SD $\pm$10.7).

The general objectives of the e-learning course consisted in training the HCWs on the topics related to IS promotion, prevention and treatment. Coherently with these objectives, the most represented professions among completers were: physicians (24.6%), physiotherapists (21.1%), and orthopedic technicians (3.4%).

The majority of participants worked as employee in the Italian National Health Service (32.9%), likely the workers mainly involved in the management of IS. The self-employed and private contractors, who in Italy represent a significant part of the operators involved in the treatment of IS, were the 19.7%. As expected, the category of unoccupied was the less represented.

Statistical analysis on pre-test and post-test data ($N$ questions = 10) shows in all the questions a significant improvement in participants learning level after having completed the course. The improvement calculated on each item/question shows some differences in percentage improvement between the different questions. It could be assumed that not all the questions presented the same level of difficulty or that the associated specific objectives were not easily reachable with the corresponding study materials. Although the completion was mandatory, the aim of the two learning activities was mainly formative, since participants could assess their own improvement in knowledge comparing by themselves the scores obtained pre and post having completed the course. In any case, further analysis is required.

The satisfaction questionnaire analysis shows a high overall liking from the majority of participants (91.7%), on the learning methodology adopted, as well as on the adequacy of the contents and on the e-learning platform functioning.

## 5 Conclusions

The results obtained are positive. This study highlights the importance of training healthcare professionals in the IS management and treatment.

We found a statistically significant increase in participants' knowledge after having attended the course and a good average score at the final certification test.

The result of the satisfaction questionnaire also indicates a high level of appreciation on the learning methodology, contents, and e-learning platform functioning

The high rate of completion, as well as the level of satisfaction expressed by participants, represents important aspects of our study, reflecting the interest shown by the participants to the training project, confirming their committing to IS management and treatment.

Our global results show that e-learning can provide statistically relevant knowledge gains in healthcare professionals in IS management, underlying the importance of e-learning as a appropriate technology for training in health sector and confirm the importance of an active approach, involving and motivating participants.

Also, it indicates future directions and developments. In the next courses, it will be proposed to participants a questionnaire on perceived knowledge, attitudes, and practice behaviors (KAPb) at three time points: baseline, post-training, and at 6 months post-training, respectively.

Furthermore, it will prosecute the direction of an even more faithful reproduction of the PBL cycle in e-learning, with active involvement of the participants.

## References


1. Negrini, S., Donzelli, S., Aulisa, A.G., Czaprowski, D., Schreiber, S., de Mauroy, J.C., et al.: 2016 SOSORT guidelines: orthopedic and rehabilitation treatment of idiopathic scoliosis during growth. Scoliosis Spinal Disord. **13**, 3 (2018)
2. Berardi, A., Dhrami, L., Tofani, M., Valente, D., Sansoni, J., Galeoto, G.: Cross-cultural adaptation and validation in the Italian population of the wolf motor function test in patients with stroke. Funct. Neurol. **33**(4), 229–253 (2018)
3. Barrows, H.S., Tamblyn, R.M.: Problem-based Learning: an approach to medical education. Springer Publishing Company, New York (1980)
4. Barbina, D., Guerrera, D., Mazzaccara, A.: Study of the effectiveness of an e-learning course to high interactivity. Rome Mondo Digitale **13**, 33 (2014)
5. Barbina, D., Bonciani, M., Guerrera, D., Mazzaccara, A., Guerra, R.: Case study—National Health Institute, Italy. In: Cole, J., Henrick, G., Cole, J. (eds.) Moodle 2.0 for Business: Beginner's Guide. Implement Moodle in Your Business to Streamline Your Interview, Training, and Internal Communication Processes. Birmingham, pp. 161–164. Packt Publishing (2011)
6. Dattoli, S., Colucci, M., Soave, M.G., De Santis, R., Segaletti, L., Corsi, C. et al.: Evaluation of pelvis postural systems in spinal cord injury patients: Outcome research. J. Spinal Cord. Med. 1–13 (2018)
7. Covotta, A., Gagliardi, M., Berardi, A., Maggi, G., Pierelli, F., Mollica, R. et al.: Physical activity scale for the elderly: translation, cultural adaptation, and validation of the Italian version. Curr. Gerontol. Geriatr. Res. (2018)
8. Galeoto, G., Colucci, M., Guarino, D., Esposito, G., Cosma, E., De Santis, R., et al.: Exploring validity, reliability, and factor analysis of the quebec user evaluation of satisfaction with assistive technology in an italian population: a cross-sectional study. Occup. Ther. Health Care. **32**(4), 380–392 (2018)
9. Galeoto, G., Sansoni, J., Valenti, D., Mollica, R., Valente, D., Parente, M., et al.: The effect of physiotherapy on fatigue and physical functioning in chronic fatigue syndrome patients: a systematic review. Clin. Ter. **169**(4), e184–e188 (2018)

10. Berardi, A., Biondillo, A., Màrquez, M.A., De Santis, R., Fabbrini, G., Tofani, M. et al.: Validation of the short version of the Van Lieshout Test in an Italian population with cervical spinal cord injuries: a cross-sectional study. Spinal Cord. (2018)
11. Galeoto, G., Scialpi, A., Grassi, M.L., Berardi, A., Valente, D., Tofani, M. et al.: General sleep disturbance scale: translation, cultural adaptation, and psychometric properties of the Italian version. CRANIO®, 1–9 (2019)
12. Savona, A., Ferralis, L., Saffioti, M., Tofani, M., Nobilia, M., Culicchia, G., et al.: Evaluation of intra-and inter-rater reliability and concurrent validity of the Italian version of the Jebsen-Taylor Hand Function Test in adults with rheumatoid arthritis. Hand Ther. **24**(2), 48–54 (2019)
13. Ruggieri, M., Palmisano, B., Fratocchi, G., Santilli, V., Mollica, R., Berardi, A., et al.: Validated fall risk assessment tools for use with older adults: a systematic review. Phys. Occup. Ther. Geriatr. **36**(4), 331–353 (2018)
14. Galeoto, G., Iori, F., De Santis, R., Santilli, V., Mollica, R., Marquez, M.A. et al.: The outcome measures for loss of functionality in the activities of daily living of adults after stroke: a systematic review. Topics in Stroke Rehabilitation (2019)
15. Puledda, F., Viganò, A., Celletti, C., Petolicchio, B., Toscano, M., Vicenzini, E., et al.: A study of migraine characteristics in joint hypermobility syndrome a.k.a. Ehlers-Danlos syndrome, hypermobility type. Neurol Sci. **36**(8), 1417–1424 (2015)
16. Berardi, A., Galeoto, G., Tofani, M., Mangone, M., Ratti, S., Danti, A., Sansoni, J., Marquez, M.A.: The validity of rasterstereography as a technological tool for the objectification of postural assessment in the clinical and educational fields: pilot study. In: Methodologies and Intelligent Systems for Technology Enhanced Learning, 9th International Conference, Workshops, pp. 55–62 https://doi.org/10.1007/978-3-030-23884-1_8 (2019)
17. Weinstein, S.L., Ponseti, I.V.: Curve progression in idiopathic scoliosis. J. Bone Joint Surg. Ser. A **65**(4), 447–455 (1983)
18. Ascani, E., Bartolozzi, P., Logroscino, C.A., et al.: Natural history of untreated idiopathic scoliosis after skeletal maturity. Spine **11**(8), 784–789 (1986)
19. Thiese, M.S.: Observational and interventional study design types; an overview. Biochem Med (Zagreb). **24**(2), 199–210 (2014). https://doi.org/10.11613/BM.2014.022
20. Anwer, S., Alghadir, A., Shaphe, M.A., Anwar, D.: Effects of exercise on spinal deformities and quality of life in patients with adolescent idiopathic scoliosis. Biomed. Res. Int. **2015**, 123848 (2015). https://doi.org/10.1155/2015/123848
21. Barbina, D., Carbone, P., Guerrera, D., Mazzaccara, A.: Il problem-based learning in salute pubblica con moodle: esperienze e prospettive. In: Proceedings Moodle Moot Conference Italia, pp. 75–83 (2017)
22. Calabrò, G.E., Mazzaccara, A., Barbina, D., Carbone, P., Guerrera, D., Federici, A., Ricciardi, W., Boccia, S.: Distance learning training in genetics and genomics practices for Italian physicians. Eur. J. Pub. Health **27**(3), 337–338 (2017)
23. Vaona, A., Banzi, R., Kwag, K.H., Rigon, G., Cereda, D., Pecoraro, V., Tramacere, I., Moja, L.: E-learning for health professionals. Cochrane Database Syst. Rev. **1**, CD011736 (2018). https://doi.org/10.1002/14651858.cd011736.pub2
24. Walsh, K.: The future of e-learning in healthcare professional education: some possible directions. Ann Ist Super Sanità **50**(4), 309–310 (2014)
25. Cook, D.A., Hatala, R., Brydges, R., Zendejas, B., Szostek, J.H., Wang, A.T., et al.: Technology-enhanced simulation for health professions education: a systematic review and meta-analysis. JAMA **306**, 978–988 (2011)
26. Nobilia, M., Culicchia, G., Tofani, M., De Santis, R., Savona, A., Guarino, D.: Italian Version of the Jebsen-Taylor hand function test for the assessment of hand disorders: a cross-sectional study. Am. J. Occup. Ther. **73**(3), 7303205080p1–7303205080p6 (2019)

27. Galeoto, G., Iori, F., De Santis, R., Santilli, V., Mollica, R., Marquez, M.A., et al.: The outcome measures for loss of functionality in the activities of daily living of adults after stroke: a systematic review. Top Stroke Rehabil. **26**(3), 236–245 (2019)
28. Tofani, M., Galeoto, G., Cazzetta, D., Berardi, A., Sansoni, J., Valente, D.: Validation of the pediatric evaluation of disability inventory in an italian population with autism spectrum disorder: a cross-sectional study. La Clinica Terapeutica **170**(6), e460–e464 (2019)
29. Galeoto, G., Turriziani, S., Berardi, A., Sansoni, J., Santilli, V., Mascio, M. et al.: Levels of cognitive functioning assessment scale: Italian cross-cultural adaptation and validation. Ann. di. Ig. (2020)
30. Choi, H.J., Park, J.H.: Testing a path-analytic model of adult dropout in online degree programs. Comput. Educ. **116**, 130–138 (2018)

# Organizational, Technical, Ethical and Legal Requirements of Capturing Household Electricity Data for Use as an AAL System

Sebastian Wilhelm[1(✉)], Dietmar Jakob[1], Jakob Kasbauer[1], Melanie Dietmeier[1], Armin Gerl[2], Benedikt Elser[1], and Diane Ahrens[1]

[1] Deggendorf Institute of Technology, Technology Campus Grafenau, Hauptstraße 3, 94151 Grafenau, Germany
{sebastian.wilhelm,dietmar.jakob,jakob.kasbauer, melanie.dietmeier,benedikt.elser,diane.ahrens}@th-deg.de

[2] Chair of Distributed Information Systems, University of Passau, Innstraße 41, 94032 Passau, Germany
armin.gerl@uni-passau.de

**Abstract.** Due to demographic change, elderly care is one of the major challenges for society in near future, fostering new services to support and enhance the life quality of the elderly generation. A particular aspect is the desire to live in one's homes instead of hospitals and retirement homes as long as possible. Therefore, it is essential to monitor the health status, i.e. the activity of the individual. In our data-driven society, data is collected at an increasing rate enabling personalized services for our daily life using machine-learning and data mining technologies. However, the lack of labeled datasets from a realistic environment hampers research for training and evaluating algorithms. In the project BLADL, we use data mining technologies to gauge the health status of elderly people. Within this work, we discuss the challenges and caveats both from a technical and ethical perspectives to create such a dataset.

**Keywords:** Data models · Privacy · Elderly care · Electricity load profiles · Household electricity data · Ethics · Home appliance · Smart metering · Data recording · Data transmission · Ambient-assisted living · Human activity detection · Data recording architecture

## 1 Introduction

Due to a steadily rising life expectancy and a decreasing birth rate, the German society is aging. As a result, the number of people requiring or depending on care increases. Most affected people have the desire to lead a self-determined life


X.-S. Yang et al. (eds.), *Proceedings of Fifth International Congress on Information and Communication Technology*, Advances in Intelligent Systems and Computing 1183, https://doi.org/10.1007/978-981-15-5856-6_38

into old age [1–3]. However, at the same time, as the demand for care increases, the number of specialists in the elderly care system is shrinking. Therefore, professional circles and the media are already talking about a care crisis [1]. New technical support and assistance systems can help to enable the elderly and those in need of care to live as long as possible in their own homes. Furthermore, the assistance systems can support and unburden the nursing staff [2].

As part of the research project *BLADL—Better life in old age through digital solutions* we develop a prototype for a system analyses electricity consumption for cues if a person is in need of assistance. To detect such cues, we use machine learning to identify individual (household) devices by analysing the total household power consumption. Based on the usage of these devices, the system learns regular daily appliance usage patterns. Deviations from these patterns can serve as an indicator for a person in need of care [4]. Recording and evaluating our approach will be done in 21 private households.

This work discusses the complexity of requirements—organizational, technical, ethical and legal—that have to be considered generating a dataset enabling the development of appropriate algorithms, thus striving to support scientifical progress in this domain. The system is designed to be unintrusive while still creating a high quality dataset. The challenges of a real-world measurement set-up as well as the diversity of the test households complicates the data recording.

## 2 Power Patterns and Related Work

### 2.1 Electrical Power Consumption Patterns

Electrical appliances produce distinctive temporal patterns in the power consumption timeseries. The electrical power varies over time, depending on the device's internal operation mode. Each operation mode powers a different electrical consumer (e.g. heating element vs motor) and consequently the electrical load changes. A further cause for recurring patterns is thermal threshold switches in heating and cooling appliances. For example, a refrigerator activates its compressor when the internal temperature exceeds a threshold temperature. Due to a steady heating rate across the thermal insulation, the internal temperature will cross the threshold in regular time intervals. Consequently, the compressor will turn on and off in a recurring timely pattern.

The peak power consumption and the temporal pattern in the power consumption timeseries are distinctive features for a specific electrical device [5,6]. Figures 1 and 2 visualize the load profiles of a baking oven and a dishwasher. Although the peak power on both devices is at a similar high level, the course differs. For the oven, we have regular power peaks after an initial heating phase. The dishwasher also has a heating phase and a high power consumption at the end of the rinsing process (drying). In between, a (almost) constant, low power consumption is observed.

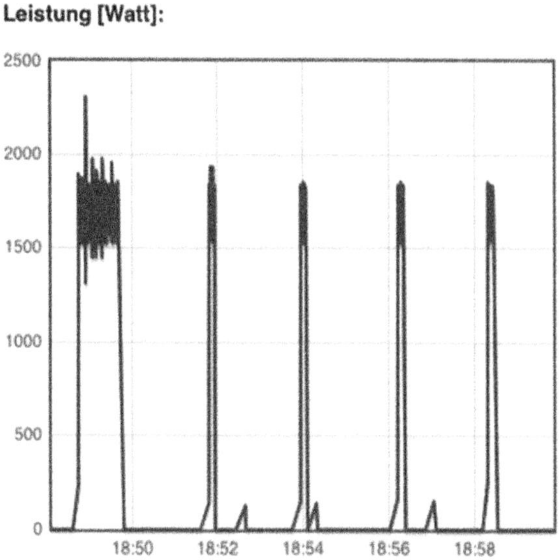


**Fig. 1.** Load profile of a baking oven. Reprint of [6]

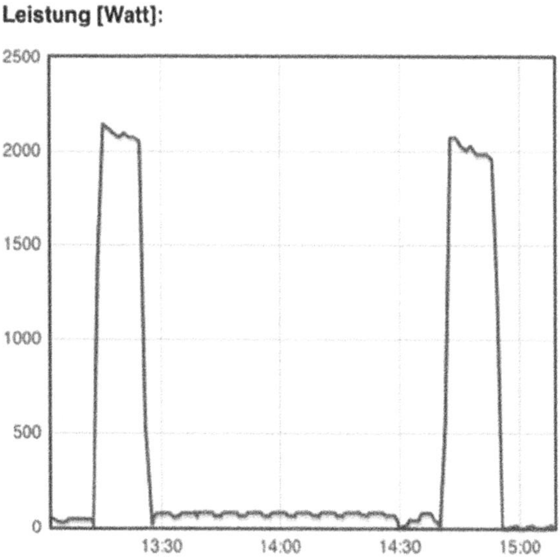


**Fig. 2.** Load profile of a dishwasher. Reprint of [6]

## 2.2 Previous Research

Greveler et al. [7] have already validated the recognition of individual devices via characteristic signatures in the total electricity consumption of a household. In a laboratory experiment, the household appliances—refrigerator, kettle, water heater, light bulb, energy-saving lamp, coffee pad machine, stove, microwave, oven, washing machine, dishwasher and TV—were identified based on the power consumption. In addition to the laboratory experiments, numerous other research project have already investigated the electricity consumption behaviour in private households (also with extended sensors). Table 1 shows an overview of already existing datasets and briefly presents their properties. The resulting datasets differ basically in the number of households investigated, the number and type of sensors used, the resolution of the data, the duration of the test phases in the individual households and the measured features [8].

Based on the information on the activity of individual (household) devices, activities of the residents in the household can be detected (*HAR Human Activity Detection*) [17]. For example, the following assumption can be made: While the stove is active, the resident cooks. Information about the residents' activities now helps to identify potential emergencies. Clement and Kabitzsch [17] divide the approaches to detect emergencies into two types:

- **Inactivity detection**:
  Inactivity detection is a very basic approach to identify cases where people are inactive for an unusually long time. In each case, the cummulated duration of the inactivity $t_{inact}$ is measured. Inactivity ends with any input signal that indicates a human action. Increasing the duration of an inactivity $t_{inact}$ above a certain *threshold* value, this is an indication of a need for assistance. Figure 3 illustrates such an inactivity profile.

  The inactivity threshold can be adjusted for individual households and time intervals (for example, during the day/at night) [18]. The cummulative inactivity monitoring can detect situations in which the inhabitants have fallen down or show no activity, e.g. caused by illness [17].
  However, this approach is coarse grained: A running TV does not necessarily show that the resident is actively using the TV. This work focusses on direct interactions with a device (e.g. power on/off). Another hurdle is that some devices switch off automatically after a certain time. Accordingly, power signal not caused by user interaction must not be identified as human activity.
- **Detection of Activity of Daily Living (ADLs)**:
  It is important to recognize the habits of a person and determine deviations. This approach is mostly based on the results of clinical investigations, which have shown that human activities follow a periodic cycle governed by the biological rhythm of the individuals. This cycle lasts about 24 h [19]. Based on historical data of a person's daily behaviour (training record), a typical behavioural profile of a person is created. If the activity on a certain day deviates too much from the typical day, then an emergency situation can be assumed. However, there is a natural margin of deviation between two regular days, which can also be observed from the training data [19–21].

The pyramid in Fig. 4 visualizes the layers of knowledge extraction from power data to person wellbeing.

**Table 1.** Existing datasets for energy consumption in households. Extended reprint form [8]

| Dataset | Location | Duration | #Households |
|---|---|---|---|
| ACS-F1 [9] | Sqitzerland | 1 h session (2 sessions) | N/A |
| AMPds [10] | Greater Vancouver | 1 year | 1 |
| BLUED [11] | Pittsburg, PA | 8 days | 1 |
| GREEND [8] | Austria, Italy | 1 year (2–5 month completed) | 9 |
| HES | UK | 1 month (255 houses) 1 year (26 houses) | 251 |
| iAWE [12] | India | 73 days | 1 |
| IHEPCDS[a] | France | 4 years | 1 |
| OCTES[b] | Finland, Iceland, Scotland | 4–13 month | 33 |
| REDD [13] | Boston, MA | 3–19 years | 6 |
| Sample dataset[c] | Austin, TX | 7 days | 10 |
| Smart* [14] | Western Massachussets | 3 months | 1 Sub-metered + 2 (Agg + Sub) |
| Tracebase [15] | Germany | N/A | 15 |
| UK-DALE [16] | UK | 499 days | 4 |
| **BLADL** | Germany | 21 days | 21 |

**Table 1.** (*continued*)

| Dataset | #Sensors (per house) | Features | Resolution |
|---|---|---|---|
| ACS-F1 [9] | 100 devices in total (10 types) | I, V, Q, f, varphi | 0.1 Hz |
| AMPds [10] | 19 | I, V, pf, F, P, Q, S | 0.016 Hz |
| BLUED [11] | Aggregated | I, V, switch events | 12 Khz |
| GREEND [8] | 9 | P | 1 Hz |
| HES | 13–51 | P | 0.0083 Hz |
| iAWE [12] | 33 sensors (10 appliance level) | V, I, f, P, S, E, varphi | 1 Hz |
| IHEPCDS[a] | 3 circuits | I, V, P, Q | 0.016 Hz |
| OCTES[b] | Aggregated | P, Energy price | 0.14 Hz |
| REDD [13] | 9–24 | Aggregate: V, P Submetered: P | 15 Khz (aggr.), 0.33 Hz (sub) |
| Sample dataset[c] | 12 | S | 0.016 Hz |
| Smart* [14] | 25 circuits, 29 appliance monitors | P, S (circuits), P (submetered) | 1 Hz |
| Tracebase [15] | 158 devices in total (43 types) | P | 0,1–1 Hz |
| UK-DALE [16] | 5 (house 3) 53 (house 1) | Aggregated P, Sub P, switch status | 16 Khz (aggr.), 0.16 Hz (sub) |
| **BLADL** | 13 | Aggregated P, Sub V, P | 1 Hz (aggr.), 7 Hz (sub) |

[a] https://tinyurl.com/IHEPCDS
[b] http://octes.oamk.fi/final/
[c] https://www.pecanstreet.org/project/consortium/

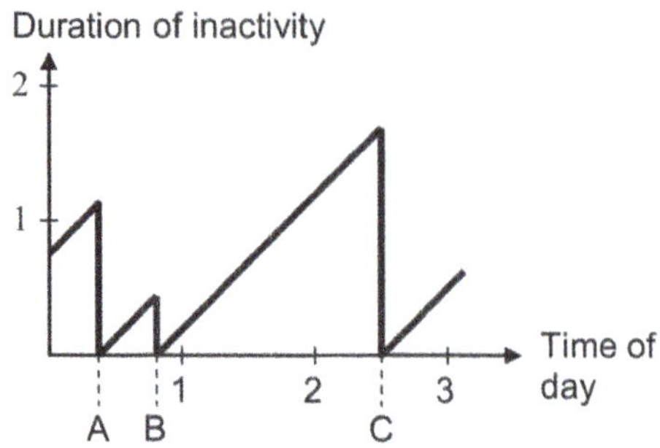


**Fig. 3.** Principle of an inactivity profile. User activity signals (A, B, C) reset the cummulative inactiviy profile. Reprint of [18]

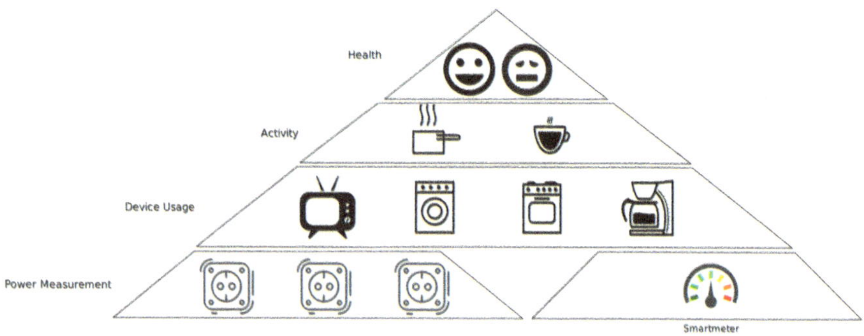


**Fig. 4.** Layers of knowledge extraction. The lower layers process data with a high volume and little information. Whereas the top layer contains less data with a higher information content

## 3 Selection of Measuring Instruments

First, we have to identify active devices in a household based on the total electricity consumption. For this purpose, it is necessary to analyse the total electricity load of the household using smart meters. Smart meters measure the effective electrical power per phase conductor in a household ($L_1$, $L_2$, $L_3$). The advantage of measuring the power per phase is that individual devices are usually connected to a single phase. When two devices consume power at the same timepoint, the devices can be separated if the devices are connected to different phases.

According to Revuelta et al., measurement frequencies from 1 Hz are suitable for detecting electrical signatures [22]. Currently, in the Federal Republic of Germany, legislature only requires a resolution of at least 0.001 Hz and an optical (data) interface [23, Sect. 60 Abs. 3].

As part of an analysis on the market available digital ammeters, only the Smart Meter *Q3M* of the company EasyMeter GmbH, Bielefeld, with the associated Smart Meter Gateway *Discovergy Meteroit* of the company Discovergy GmbH, Heidelberg, could be identified as suitable. This smart meter gateway allows the sampling of the measured values in a resolution of 1 Hz via a REST-API. These smart meter gateways transmit the real-time power consumption of a household to the IT infrastructure of the manufacturer. However, when using these electricity meters, special attention should be noted to the privacy policy of Discovergy GmbH. The Company reserves the right to disclose household consumption data for marketing, promotional and product development purposes.[1]

In order to obtain label data for device identification, it is necessary to record the separate power consumption of individual household devices. For this purpose, we use single-device-power-meters (intermediate power socket) which are connected directly to the device to be measured. These connectors must also be able to provide measurement data in a temporal resolution of at least 1 Hz [22].

Another requirement to the corresponding hardware is to network the devices wirelessly (for example via wireless LAN). After a market research, the adapter `ALL3075v3` from ALLNET GmbH, Germering, could be identified as suitable for this purpose. These measurement sockets provide a local network interface (API) for reading the measured values. An automated data transfer to server of the manufacturer or other third parties does not take place. The individual electricity meters achieve a measurement samplerate of 7 Hz.

Formally, we record the following variables:

$$\begin{aligned} household &= (address, statistic\text{-}information) \\ measurement_{device} &= (household, time, current, voltage, power, device) \\ measurement_{total} &= (household, time, energy, power_1, power_2, power_3) \end{aligned}$$

whereas:

$$\begin{aligned} time &= \text{NTP Synchronized UNIX Timestamp (in ms)} \\ current &= \text{Electric Current (in A)} \\ voltage &= \text{Voltage (in V)} \\ power &= \text{Electric Power (in W)} \\ device &= \text{The electrial appliance name / ID} \\ energy &= \text{Cumulative Energy consumption (in kWh)} \\ power_1 &= \text{Power over Line Conductor 1 (in W)} \\ power_2 &= \text{Power over Line Conductor 2 (in W)} \\ power_3 &= \text{Power over Line Conductor 3 (in W)} \end{aligned}$$

[1] see: https://discovergy.com/privacy Version: 18-April-2018.

## 4 Data Transmission and Storage

To perform the data evaluation and data processing of the measured variables, the measurement data must be available in a structured databse. For this purpose, it is necessary to collect the individual measurement data via software/scripts and to store it centrally in a database. In the scope of this project, the data is stored exclusively on internal servers for reasons of data protection and data security. It is stored on its own virtual machine with a PostgreSQL database for archiving the measurement data. The corresponding server is only accessible from the intranet or via VPN. However, data can also be stored on other persistent servers or cloud environments, provided that data protection and data security requirements can be guaranteed. In the following, the readout of the measured data and the transmission of the measured values of the smart meters and the individual electricity meters to the internal server is briefly outlined.

- **Smart Meter**:
  The manufacturer provides a REST API from which the measurement data of all smart meters connected to a user account[2] can be retrieved. It is possible to assign several counters to one account. The data are maintained in a resolution of 1 Hz even over a longer, but not exactly defined period. In order to be able to view high-resolution data for data analysis over as long a period as possible, the measured data are regularly read out and stored on an internal server. This is done via a Python script, which runs automatically on the virtual machine.
- **Single electricity meters**:
  The `ALL3075v3` single electricity meters allow the readout of measurement data in a resolution of 7 Hz via an open API of the devices. However, this API is only available within the household network to which the devices are connected. Therefore, it is necessary to use single board computers in the individual households, which read out the data of the single electricity meters and transfer them to the internal servers. *ROCK Pi 4* from Radxa are used as single board computers. The advantage of *ROCK Pi 4* over conventional Raspberry Pi models is that they are equipped with an eMMC memory, which promises higher reliability. In addition, the *ROCK Pi 4* can be upgraded with an Real Time Clock which is required for time synchronization of the measurement data. The corresponding single-board computer regularly calls the API of the individual electricity meters, extracts the relevant measured variables and sends them compressed to the internal server via a VPN tunnel using the AMQP protocol. At the server, the data is persistently stored in an SQL database.

Figure 5 shows the software architecture for data transfer.

To ensure the highest possible data quality and prevent data loss, we took various measures. Because Internet outages can occur frequently in private households, a fallback Internet connection via LTE was maintained during the test

[2] Discovergy user account.

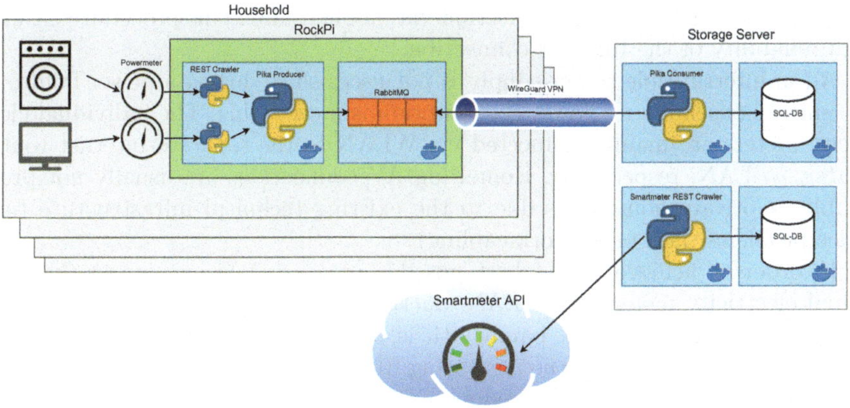


**Fig. 5.** Software architecture data for acquisition

phase using an additional router. Furthermore, the single powermeter data is cached on the single board computer and transferred to the internal server using RabbitMQ.[3] The RabbitMQ-queue ensures that even in the case of connection loss, server failures or power outages already collected data is not lost because the previous recordings remain in the transmission-queue. The content of the transmission-queue will eventually be sent to the storage server.

For data analysis, the data of the smart meters and the individual electricity meters must be combined on a common time axis and synchronized in time. To ensure that the independently collected data (smart meters / single stream meters), each with its own clock, fits together in the analysis, both the smart meter data and the single-board computers are synchronized with an NTP server. The single-board computer is also equipped with a battery-backed Real Time Clock.

## 5 Networking in Private Households

As part of the research project, 21 private households will be equipped with the appropriate measuring equipment (see Fig. 6). The networking of smart meters and the Internet router of the household $router_{hh}$ is wired via LAN or HomePlug AV connections. Alternatively, the connection between the smart meter and the $router_{hh}$ can also be established via a WLAN bridge. The data throughput as well as any packet losses play a subordinate role here, since the smart meter only transmits approx. 18 MB of data per day as well as integrated data buffering for disconnections.

The networking of the individual electricity meters is realized via a separate Internet router $router_{add}$, which is implemented in the home network of $router_{hh}$

[3] https://www.rabbitmq.com.

and has a fall back Internet connection via mobile radio. This procedure ensures high reliability of the Internet connection.

An uninterruptible power supply is not used, since during a power failure in the entire household, no power consumption is measurable. The individual electricity meters are mainly connected via WLAN to the separate Internet router $router_{add}$. LAN, respectively, HomePlug AV connections are usually not practicable in private households due to the existing technical infrastructure (e.g. space problems, lack of network cabling).

In order to achieve the highest possible measuring frequency on the individual electricity meters via WLAN, packet losses and disconnections must be minimized. At least WLAN bandwidth is required for the data stream of the measured values, otherwise the measuring frequency drops. For this purpose, the separate router $router_{add}$ is placed as possible at an equidistant position to the respective individual electricity meters. Due to the different building constructions (1-story, multi-story) and external interference factors (e.g. water pipes, radio signals) perhaps additional WLAN repeaters connected to $router_{add}$ via a MESH network may be required. The single-board computer for transmitting the single-current measured values to the internal servers is integrated in the network of the separate router $router_{add}$ by means of a LAN connection. Figure 6 shows the cross-linking of the components schematically.

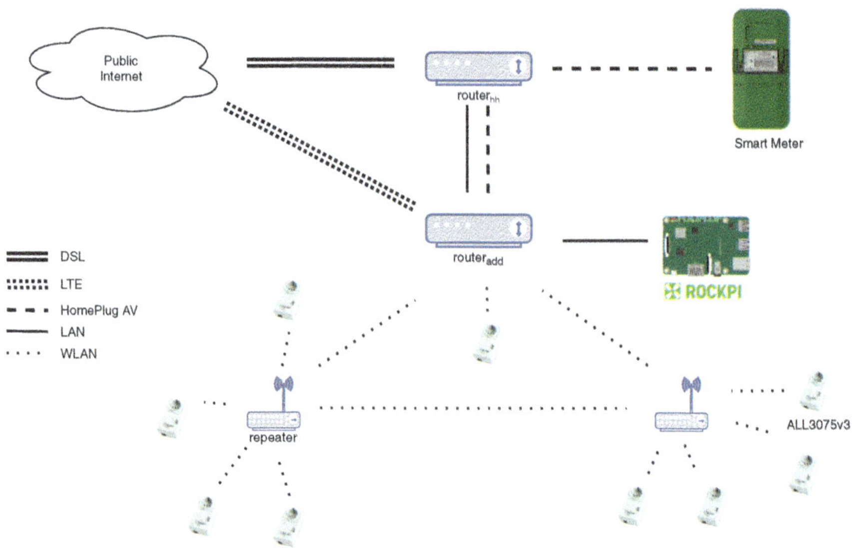


**Fig. 6.** Schematic representation of the networking of trade fair equipment in private households

## 6 Evaluation of Measurement Data

During the sensor recording periods, we monitor the data ingress rate to assess if the recording set-up is still operational. When a household setup fails, we detect a drop in data rate.

### 6.1 Samplerate

The wallsocket powermeters provide their measurements via a WLAN interface. A varying WLAN throughput affects the samplerate. Figure 7 visualizes the samplerate of the individual wallsocket powermeters. There is a correlation between the samplerates. We assume that the cause of the correlation is the varying WIFI throughput because the WIFI signal strength is a common factor of all wallsocket powermeters. In this household we tested three different locations for the WIFI Access Point and Repeater, therefore the samplerate pattern changes three times.

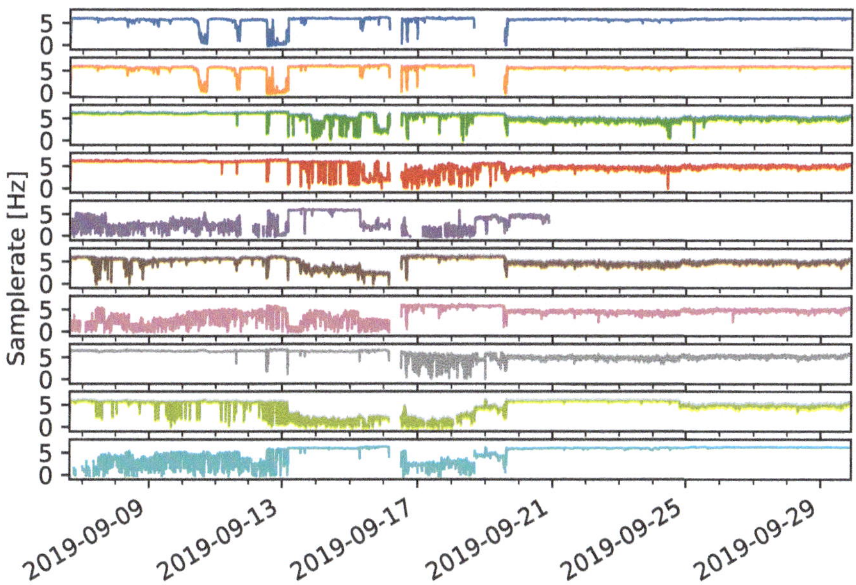


**Fig. 7.** Samplerate per Wallsocket Powermeter over time. One powermeter per row. Too little WIFI bandwidth reduces the samplerate. The WIFI throughput affects all powermeters and therefore there is a correlation between the individual samplrates. In the shown timespan, we tested three locations for the WIFI-Repeater and the correlated samplerate change at three timepoints. The gaps are caused by WIFI-Failures or measurement-instrument failures

### 6.2 Comparing the Power Timeseries of Smartmeter with Wallsocket Powermeter

Figure 8 simultaneously shows two power-measurement timeseries caused by a single coffeemachine. The pattern of both measurement devices is similarly shaped and in temporal proximity. The similarity of both patterns demonstrates that a pattern-recognition algorithm can detect the coffeemachine's pattern in the household's power-timeseries using the wallsocket powermeter-timeseries as training-data.

## 7 Ethics and Privacy

Gaining special importance in this project are the aspects of ethics and privacy since sensitive data concerning the highly personal sphere of life of the test persons are collected and processed.

### 7.1 Ethics

In order to comply with ethical norms and values as well as a maximum of process transparency for the test participants, an ethical reflection on the project (especially for the data collection and further processing) was created in advance based on a guideline of the Ethics Commission DGP e.V. [24].

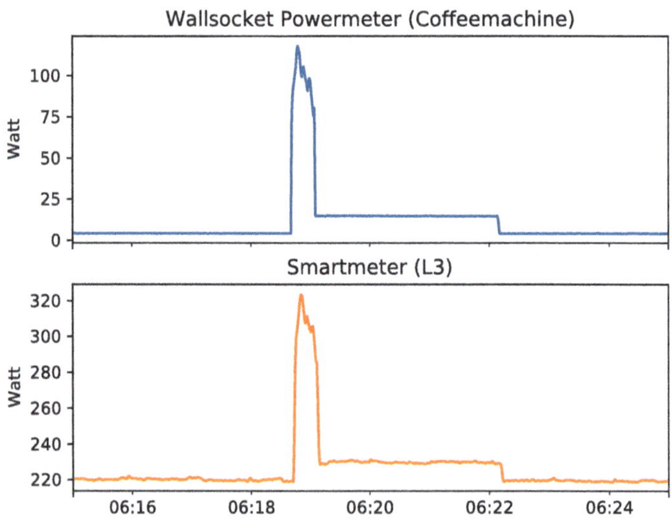


**Fig. 8.** Power-timeseries-pattern generated by the coffeemachine is also visible in the household power-pattern

**7.1.1 Procedure for Selecting Participants** The main target groups of participants are people between the ages of 55 and 80, who live in their own home, who can take care of themselves and do not need any nursing assistance. We exclude persons from the participants if we cannot determine whether the person is able to give their informed consent. If the mental, physical or mental abilities of a participant deteriorate significantly during the project, the project will be terminated for this participant. There are no plans to involve vulnerable people in the project.

**7.1.2 Possible Risks for Participants and Preventive Measures** We identified that the following risks for the participants can occur during the data recording.

- **Increased electricity costs**

  *Probability of occurrence:* high

  *Description:* In order to carry out the project, various technical devices are used in the subscriber's home (for example routers), which consume electricity themselves.

  *Preventive measures:* The participants will receive an allowance.

- **Reduced available Internet bandwidth in the home network**

  *Probability of occurrence:* high

  *Description:* The measurement data is sent in real time via the household Internet connection. This creates a continuous stream of data that occupies the available data rate for other purposes.

  *Preventive measures:* The data transmission was developed algorithmically in order to send as few data packets as possible. In addition, during the test phase with the single-powermeters, a fallback option is introduced into the household via LTE in order to compensate for possible bottlenecks.

- **Faulty switching of single current measuring devices**

  *Probability of occurrence:* medium

  *Description:* The individual current measuring instruments used can switch off the connected household appliances. Faulty switching or incorrect operation may cause the connected device to stop functioning.

  *Preventive measures:* The measuring instruments have an analogy on / off switch and can be switched independently by the participants. They will be instructed. In addition, it is always possible to remove the meter.

- **Health effects of high-frequency radiation**

  *Probability of occurrence:* low

  *Description:* The participants are exposed to additional high-frequency radiation by WLAN. The self-claimed electrical radiation hypersensitivity can lead to health problems for participants.

  *Preventive measures:* Only devices approved for the German market are used. An additional wireless network will only be set-up for a period of about 1 month. If health problems arise due to the alleged hypersensitivity, the project will be stopped immediately.

- **Risk of fire and short circuit due to technical devices**

  *Probability of occurrence:* low

  *Description:* For the project, additional technical devices are used in the household (for example, FRITZ! Box, measuring instruments). These devices can generally pose a fire and short-circuit hazard..

  *Preventive measures:* Only CE-certified and approved for the German market technical devices are used. Participants are required to report any abnormalities immediately. Technical data sheets / manuals are handed out.

- **Cyber-attack**

  *Probability of occurrence:* low

  *Description:* The participants can be spied on by a cyber-attack, or household appliances can be switched off by cyber-attacks.

  *Preventive measures:* All common safety measures have been taken to minimize this risk. The individual electricity meters are also only accessible in the intranet of the home network. The data transfer always takes place via an encrypted VPN connection.

- **Psychological burden of supervision**

  *Probability of occurrence:* low

  *Description:* The participants can feel supervised by the project and thus be psychologically stressed.

  *Preventive measures:* It is made clear that the project does not intend to monitor individuals. If psychological problems occur and become known, the project will be stopped immediately.

- **Mental stress situation due to additional work**

  *Probability of occurrence:* low

  *Description:* The participants are required to check the function of the equipment required for the project. Especially in the case of an error, this can lead to a stress situation among the participants.

  *Preventive measures:* Participants will receive contact information on employees of *Deggendorf Institute of Technology* who will assist in case of a fault. If an ongoing stress situation is detected among the participants, the project is terminated.

**7.1.3 Ensuring Informed Consent** Participation in the project is voluntary. The participants will receive a comprehensive and easy-to-understand information sheet in advance, in which the entire project (project framework, test procedure, safety precautions and contact persons) will be explained. Participants have the opportunity to ask questions about the project at any time (also in advance) and receive information from the staff of the *Deggendorf Institute of Technology*. The participant agrees to participate by signing a corresponding data usage contract.

### 7.2 Data Protection

In order to ensure the highest level of data security and data protection, all data related to the project and collected, processed and analysed are stored exclusively on the servers of the *Deggendorf Institute of Technology*. All measured data are saved by default pseudonymized (see Table 2). This means that the measurement data can only be assigned to a specific household using the meter number. This assignment (see Table 3) is saved separately and can only be viewed by the employees directly involved in the project and named in the data usage contract. The data usage contract constitutes the contractual basis for data collection and processing for the participants and also includes their data subject rights under the GDPR [25].

**Table 2.** Sample data record for total current meter readings

| MeterID | Time | Energy | $Power_1$ | $Power_2$ | $Power_3$ |
|---|---|---|---|---|---|
| cb2dcf3a83a | 1563681798323 | 135238596018000 | 58800 | 83080 | 57280 |
| cb2dcf3a83a | 1563681804323 | 135238594912000 | 59120 | 83080 | 56660 |
| cb2dcf3a83a | 1563681802323 | 135238593807000 | 59430 | 82910 | 56360 |

**Table 3.** Sample data record of the test logging for the assignment of the measured values

| MeterID | Household |
|---|---|
| cb2dcf3a83a | John Doe; 123 Main St Anytown |

At the latest upon completion of the research project, all measurement data will be anonymized, so that no conclusions can be drawn on individual households (see Table 4). The anonymized data can be used for further research projects.

**Table 4.** Sample data set from Table 2 in anonymous form

| ID | Time | Energy | $Power_1$ | $Power_2$ | $Power_3$ |
|---|---|---|---|---|---|
| 001 | 1563681798323 | 2211000 | 58800 | 83080 | 57280 |
| 001 | 1563681804323 | 1105000 | 59120 | 83080 | 56660 |
| 001 | 1563681802323 | 0 | 59430 | 82910 | 56360 |

A general publication of the data in a not completely anonymized form takes place at no time. It is always possible for the participants to drop out of the project. If a participant drops out, all measurement data will be anonymized and the personal data will be deleted.

## 8 Discussion and Further Work

The aim of the research project *BLADL—Better Life in Old Age through digital solutions* is to use data analysis on daily power consumption behaviour to detect deviations in daily inhabitant behaviour, which can serve as indicators of a need for assistance [4].

Through direct citizen participation, data can be collected under real-world conditions. This dataset allows a realistic development of intelligent algorithms. Using the data collected, the research project BLADL develops and evaluates algorithms for the identification of individual (household) devices by analysing the total electricity consumption, determining the (in-)activity profiles of individual households and identifying a need for assistance. The high-resolution dataset can also be used for further research and development projects in a wide variety of disciplines (geragogics, dementia research, social science, electrical engineering or computer science).

The scalability of data collection is limited by the equipment, installation effort and time requirement for each individual household.

## 9 Conclusion

This paper highlights the organizational, technical, ethical and legal requirements needed to generate a dataset of labeled electricity data. The Sects. 3, 4 and 5 showed the technical implementation of data recording, transmission and storage. Section 5 deals with the complexity of integrating sensors into the inhomogeneous networks of private households. A further aspect of data collection is the ethical issue, which has been explained in Sect. 7.1. One result of the ethical reflection is that vulnerable persons should not be included in data collection. The reason for the exclusion is that it cannot be ensured that vulnerable persons can give informed consent to the procedure. The data collected in this work are highly sensitive, since the recordings allow a reconstruction of the daily activities of a private household. Data protection and data security must therefore be explicitly taken into account in the dataset creation and storage process (Sect. 7.2).

**Acknowledgements.** This work was funded by the *Bavarian State Ministry of Family Affairs, Labour and Social Affairs* within the project *Digitales Dorf—BLADL/Besser Leben im Alter durch digitale Lösungen.*

## References

1. Fischer, Florian, Krämer, Alexander (eds.): eHealth in Deutschland. Springer, Berlin, Heidelberg (2016)
2. Hoffmann, S.: Technik die unser Leben vereinfacht (2016) (Date last accessed 23-August-2019)
3. Paulus, W.: Selbständig zuhause leben im Alter: Auf dem Weg zu einer integrierten Versorgung. Technical report, Institut Arbeit und Technik (IAT), Westfälische Hochschule, Gelsenkirchen (2015)
4. Ahrens, D.: Digitales Dorf—BLADL Besser Leben im Alter durch digitale Lösungen. Beschreibung/Zuwendungsantrag für das Projekt, February 2018
5. Clement, J., Kabitzsch, K.: Automatisierte Assistenz in Gefahrensituationen. Technical report, Technische Universität Dresden—Institut für Angewandte Informatik (2013)
6. Müller, K.J.: Gewinnung von Verhaltensprofilen am intelligenten Stromzähler. Datenschutz und Datensicherheit - DuD **34**(6), 359–364 (2010)
7. Löhr, D., Greveler, U., Justus, B.: Identifikation von Videoinhalten über granulare Stromverbrauchsdaten (Date last accessed 23-August-2019)
8. Monacchi, A., Egarter, D., Elmenreich, W., D'Alessandro, S., Tonello, A.M.: Greend: an energy consumption dataset of households in Italy and Austria. In: 2014 IEEE International Conference on Smart Grid Communications (SmartGridComm), pp. 511–516. IEEE (2014)
9. Paganelli, F., Paradiso, F., Turchi, S., Luchetta, A., Castrogiovanni, P., Giuli, D.: Appliance recognition in an OSGi-based home energy management gateway. Int. J. Distrib. Sens. Netw. **11**(2), 937356 (2015)
10. Makonin, S., Popowich, F., Bartram, L., Gill, B., Bajić, I.V.: Ampds: a public dataset for load disaggregation and eco-feedback research. In: 2013 IEEE Electrical Power & Energy Conference, pp. 1–6. IEEE (2013)

11. Filip, A.: Blued: a fully labeled public dataset for event-based nonintrusive load monitoring research. In: 2nd Workshop on Data Mining Applications in Sustainability (SustKDD), p. 2012 (2011)
12. Batra, N., Gulati, M., Singh, A., Srivastava, M.B.: It's different: insights into home energy consumption in India. In: Proceedings of the 5th ACM Workshop on Embedded Systems For Energy-Efficient Buildings, pp. 1–8. ACM (2013)
13. Zico Kolter, J., Johnson, M.J.: Redd: a public data set for energy disaggregation research. In: Workshop on Data Mining Applications in Sustainability (SIGKDD), San Diego, CA, vol. 25, pp. 59–62 (2011)
14. Barker, S., Mishra, A., Irwin, D., Cecchet, E., Shenoy, P., Albrecht, J., et al.: Smart*: an open data set and tools for enabling research in sustainable homes. SustKDD **111**(112), 108 (2012)
15. Reinhardt, A., Baumann, P., Burgstahler, D., Hollick, M., Chonov,H., Werner, M., Steinmetz, R.: On the accuracy of appliance identification based on distributed load metering data. In: 2012 Sustainable Internet and ICT for Sustainability (SustainIT), pp. 1–9. IEEE (2012)
16. Kelly, J., Knottenbelt, W.: The UK-DALE dataset, domestic appliance-level electricity demand and whole-house demand from five UK homes. Sci. Data **2**(1) (2015)
17. Clement, J., Ploennigs, J., Kabitzsch, K.: Detecting activities of daily living with smart meters. In: Ambient Assisted Living, pp. 143–160. Springer, Berlin, Heidelberg (2013)
18. Floeck, M., Litz, L.: Inactivity patterns and alarm generation in senior citizens' houses. In: 2009 European Control Conference (ECC), pp. 3725–3730 (2009)
19. Virone, G., Noury, N., Demongeot, J.: A system for automatic measurement of circadian activity deviations in telemedicine. IEEE Trans. Biomed. Eng. **49**(12), 1463–1469 (2002)
20. Cardinaux, F., Brownsell, S., Hawley, M., Bradley, D.: Modelling of behavioural patterns for abnormality detection in the context of lifestyle reassurance. In: Ruiz-Shulcloper, J., Kropatsch, W.G., (eds.), Progress in Pattern Recognition, Image Analysis and Applications, pp. 243–251. Springer, Berlin, Heidelberg (2008)
21. Elbert, D., Storf, H., Eisenbarth, M., Ünalan, Ö., Schmitt, M.: An approach for detecting deviations in daily routine for long-term behavior analysis. In: 2011 5th International Conference on Pervasive Computing Technologies for Healthcare (PervasiveHealth) and Workshops, pp. 426–433 (2011)
22. Herrero, J.R., Lozano Murciego, Á., Barriuso, A.L., Hernández de la Iglesia, D., González, G.V., Corchado Rodríguez, J.M., Carreira, R.: Non intrusive load monitoring (NILM): a state of the art. In: Advances in Intelligent Systems and Computing, pp. 125–138. Springer International Publishing (2017)
23. Gesetz über den Messstellenbetrieb und die Datenkommunikation in intelligenten Energienetzen (Messstellenbetriebsgesetz—MsbG), November 2019. Messstellenbetriebsgesetz vom 29. August 2016 (BGBl. I S. 2034), das zuletzt durch Artikel 90 des Gesetzes vom 20. November 2019 (BGBl. I S. 1626) geändert worden ist
24. Ethikkommission DGP e.V. Fragen zur ethischen Reflexion (Date last accessed 23-August-2019)
25. General Data Protection Regulation: Regulation (EU) 2016 of the European Parliament and of the Council of on the protection of natural persons with regard to the processing of personal data and on the free movement of such data, and repealing Directive 95/46/EC (2016)

# Performance Analysis of Proposed Database Tamper Detection Technique for MongoDB

Rupali Chopade[1,2(✉)] and Vinod Pachghare[1,2]

[1] College of Engineering Pune, Wellesley Rd, Shivajinagar, Pune, Maharashtra 411005, India
{rmc18.comp,vkp.comp}@coep.ac.in
[2] Savitribai Phule Pune University (SPPU), Pune, India

**Abstract.** Database tamper detection is identifying the change in the old state and new state of database systems. There is a change in the state of database mostly with three kinds of operations namely insert, update and delete. Dropping the database or entire table will also affect the state of the database. Data is very precious to any individual or organization and data tampering will have serious ramifications. Now considering data as an asset of the organization, the protection of this data is supremely important. The tamper detection technique for the MongoDB database is proposed here, which identifies the operation that has altered the part of the database. This paper presents the performance analysis of the proposed technique in terms of computational time.

**Keywords:** MongoDB · Tamper · Delete · Insert · Update · Performance analysis · Database forensics

## 1 Introduction

Database forensics involves the study of database contents and related terms. The internal complexity of the database is the major reason for less research on this topic as compared to digital forensics. Database tamper detection is one of the most important research topics related to databases. The tamper detection from the database pertains to finding the altered state of the database due to data editable commands. These commands include insert, update and delete. All write operations cause the change to the database. To detect the tamper from any database, it is essential to go to the depth of that database. It involves the study of database working style, internal storage system, and similar remnants [1]. Here we have selected MongoDB database, one of the document-based databases from NoSQL. There are many reasons which made this database popular. The major features are security, replication, high performance, scalability and many more [2]. Due to these features, MongoDB sits at the top position among NoSQL databases [3]. The motivation behind the selection of the MongoDB database is its popularity [4] as well as there is not any existing tamper detection technique proposed, yet.


X.-S. Yang et al. (eds.), *Proceedings of Fifth International Congress on Information and Communication Technology*, Advances in Intelligent Systems and Computing 1183, https://doi.org/10.1007/978-981-15-5856-6_39

The tamper in a database can also be detected by comparing the hash values. The varieties of tools like winhex [5], hashcalc [6], etc. are available to calculate and compare the hash. However, the use of hash value in tamper detection will only notify about change in the database state. It is not useful to identify which data has altered and by using which write operation. Accordingly, we have proposed a novel tamper detection technique for the MongoDB database, that will identify which write operation (insert, update, delete) has performed. It will print the result by specifying the name of a write operation, the details related to the operation, database name, table name, and date & time of operation. The operation related details include which data is inserted if it inserts, which data is deleted if it deletes and which data is updated and what is the new data if it is update operation. This technique will also identify if any database or table is dropped.

The remainder of the paper is organized as follows. Section 2 briefs about the related work. The working flow of the proposed technique is given in Sect. 3. The performance analysis of the proposed technique in terms of execution time is presented in Sect. 4. Finally, we conclude the paper.

## 2 Related Work

Very less research work is happened on database tamper detection as per Scopus' results [7] till now. This can be observed from the graph of Fig. 1. The search keyword includes "database tamper detection" in the title of research work.

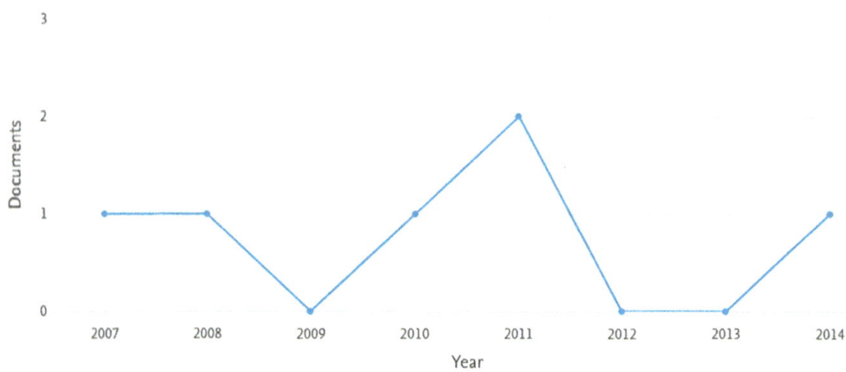


**Fig. 1.** Scopus result for database tamper detection search

Meng-Hsiun et al. proposed a digital watermark based technique [8] to verify database integrity. For the distribution of watermark, support vector regression function is used. This distribution is applicable to numeric attributes. The number of tuples in the table will decide the number of watermark bits. For the embedding phase, a numeric attribute is identified and predicated value is calculated by applying regression function. For watermarking bit value as 1, it will add 1 to predicted attribute value else 1 will be subtracted from it. The trained regression function is used to verify the change in the original and predicted bit value. The change will notify about tamper detection and no change in a bit is no tamper. Extension to the above concept is given by Hsien-Chu et al. [9]. Along with the regression function, the payload information is compressed using the Huffman code. This payload information is useful in the tamper detection phase. Tamper detection from relational databases using fragile watermarking is suggested by Guo [10]. The suggested scheme works with the embedding and verification phase. Initially, entity hash is calculated for every row and attribute. In the embedding phase, the row hash value is calculated with row primary key and grouping key. Similarly, attribute hash is calculated using the attribute name and grouping key. Based on entity hash, rows and attributes are divided into groups. Then embed function is applied to each group. In the verification phase, a secret key is used. The rank of new hash is calculated using a linear permutation ranking algorithm. If there is not any change in entity new hash value will be exactly the same, else it will identify tampering of data. Tamper detection from image database is proposed by V. Prasanna et al. Initially, the thumbnail of an image is generated [11], then base64 string for that thumbnail is computed. In step 3, the generated string along with the watermark is stored in an image database. During verification, the same steps are carried out for the image to be investigated. If there is any change in the value, report the tampered status, otherwise, accept the image as an original image. Tamper detection is required in the medical field also. Nilanjan Dey et al. proposed technique [12] for detecting tamper during transmission of ECG signal. The medical treatment is dependent on the report of the ECG signal, so tampering such kind of signal during transmission may lead to serious effects. The proposed technique is based on a bio-hash and reversible watermarking concept. For secure transmission of medical records web-based intelligent system [13] is proposed by Aniruddha Mukherjee et al. Here the medical record is ECG signal is protected with watermark. The watermark embedded is either a logo of the hospital or electronic patient record and this watermark is considered as a byte array used as a chunk.

## 3 Proposed Technique Working Flow

The proposed technique is based on the concept of the operation log of the MongoDB database, called oplog [14]. Basically, an oplog is a collection available under the local database. In standalone set up of MongoDB database, this collection is not available. The mongod.cfg file should be modified to add the replication information so that the primary replica instance is available on standalone machine [15] and oplog.rs collection will be available. All write operation details will get stored in this collection. The proposed technique will fetch the details from this collection and will store it in the CSV file (Considering only user-defined collections). Whenever any collection is created it will generate a .wt file with the name as a randomly generated number. Using this technique, the hash value of these collection .wt files will be stored in a file. Whenever there is a change in the collection, it's hash value will be changed. By observing the change in the hash value, the details are fetched from the CSV file (which stores the write operation details from oplog.rs collection). This oplog is a capped collection [16] that stores the data until its size limit is reached. The proposed technique will identify the tampered details whenever write operation (insert, update, delete) is performed. The oplog collection does not store any details regarding drop command. One more benefit of this proposed technique is that it will also identify if any database or collection is dropped. For the drop case, it will correctly differentiate whether the table is immediately dropped after creation or any existing table is dropped. In MongoDB, a new collection can be created using insert operation. Our technique will identify this case also, whether the insert is used to create a new collection or it is used for any existing collection. The working flow of this technique is depicted in Fig. 2.

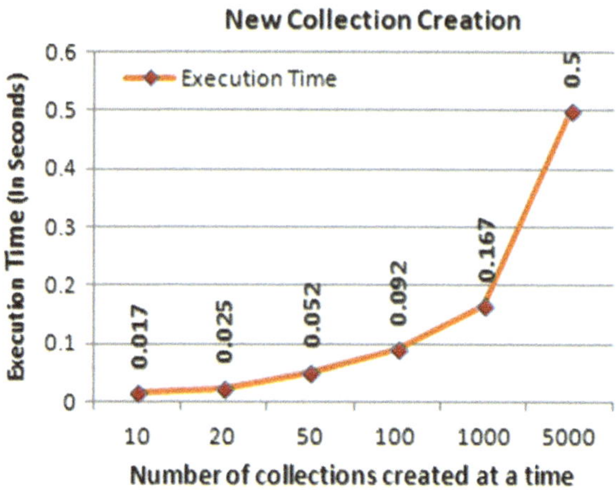


**Fig. 2.** Analysis of new collection creation

## 4 Performance Analysis

The proposed technique is designed using python and experimented on Windows 10 64-bit OS, 32 GB RAM with Intel Xeon 3.60 GHz processor. MongoDB version used is 4.2. This technique will correctly identify cases of write operations performed if any new collection is created if any collection is dropped either existing or newly created. Generally, any tamper detection technique works to identify insert, update and delete operations, but this technique will identify dropped collection/database cases also.

The analysis of the above-mentioned technique in terms of execution time is shown below. The majority of the cases are considered to analyze execution time performance. The database used for performance analysis is a sample database and two datasets are used from GitHub [17].

It can be observed from graphs shown in Figs. 3, 4, 5, and Fig. 6, as the numbers of operations are increased the execution time is also increased. For the proposed technique the execution time depends on the number of operations as well as the size of oplog also. This can be observed from the graph of Fig. 7.

## 5 Conclusion

This proposed technique is useful to keep watch on the MongoDB database. As write transactions are continuously monitored, it is helpful to trace out illegal activity, if any.

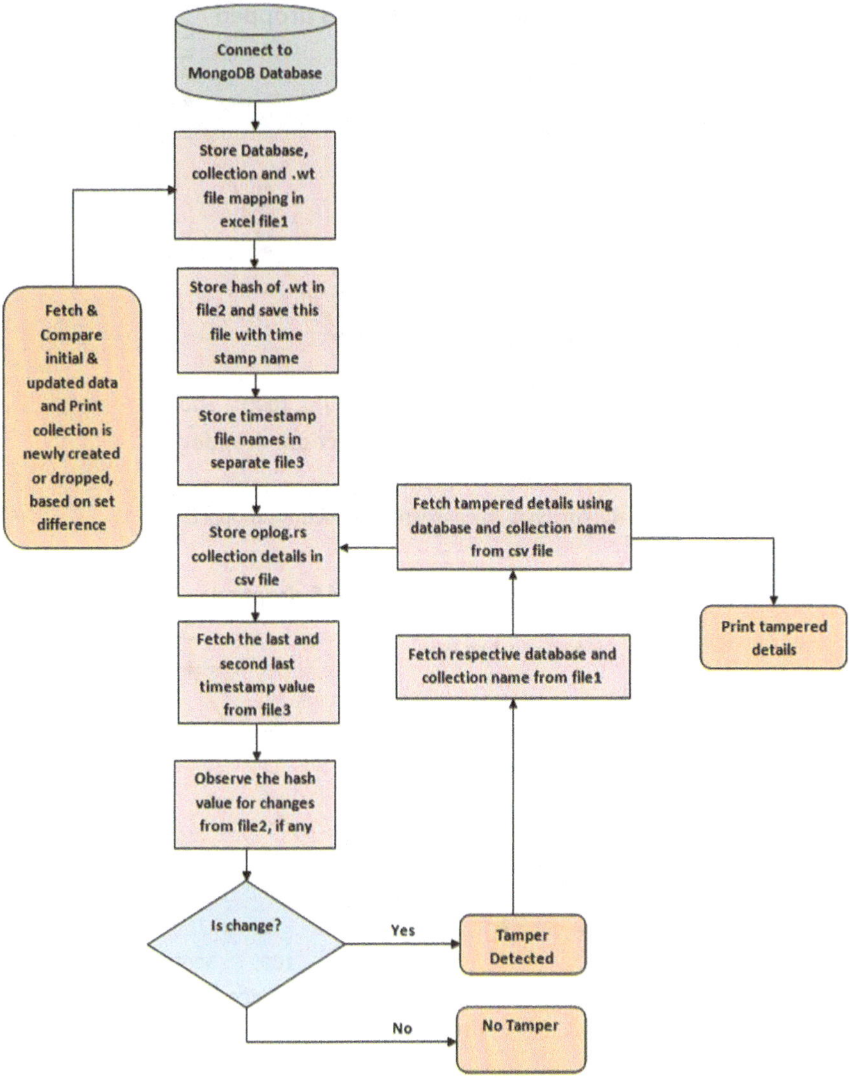


**Fig. 3.** Proposed technique working flow

The technique is based on an oplog. The performance execution of the technique is dependent on oplog size. It can be observed from graphs of Sect. 4 that it performs better when oplog size is minimum. As the oplog size grows, the execution time is more. The overall analysis is also dependent on the number of operations performed at a time. The operations performed in bulk may also affect the execution time performance. The future work of this work is to design tamper detection techniques for other NoSQL databases.

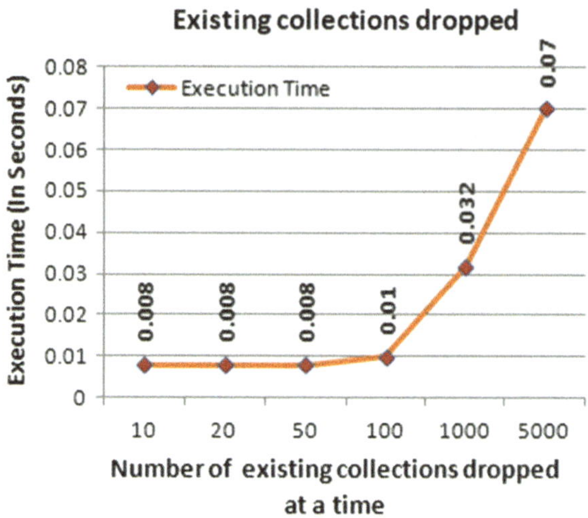


Fig. 4. Analysis when existing collections dropped

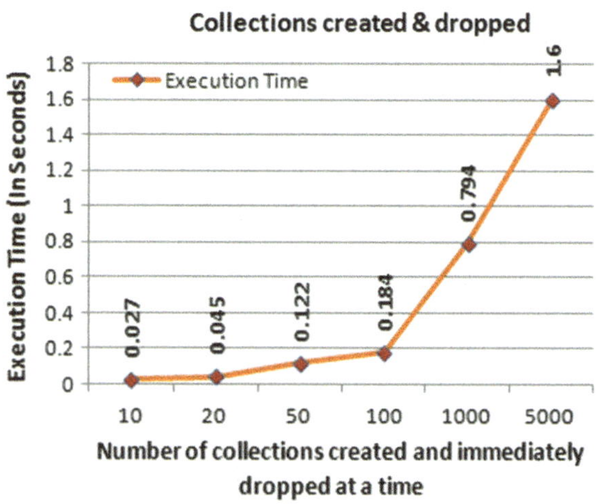


Fig. 5. Analysis when collections are created and immediately dropped

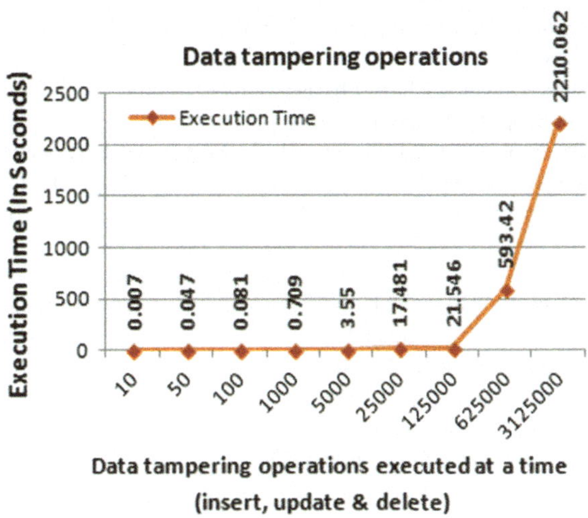


**Fig. 6.** Analysis of data tampering operations

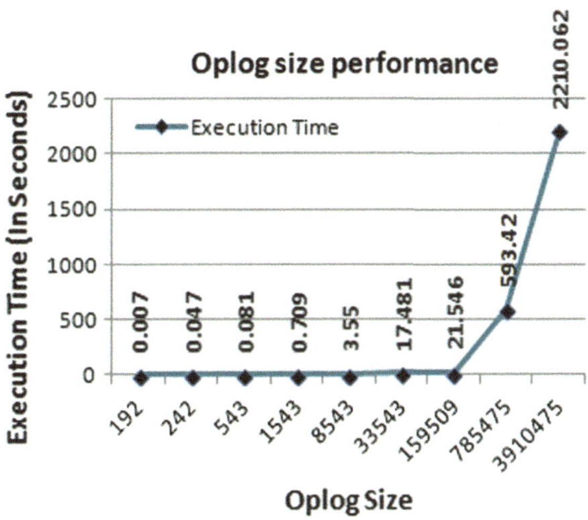


**Fig. 7.** Oplog size performance

## References


1. Chopade, R., Pachghare, V.K.: Ten years of critical review on database forensics research. Digit. Investig. (2019)
2. Architecture Guidel MongoDB. https://www.mongodb.com/collateral/mongodb-architecture-guide
3. DB-Engines Ranking—popularity ranking of database management systems. https://db-engines.com/en/ranking
4. The MongoDB 4.0 Manual—MongoDB Manual. https://docs.mongodb.com/manual/

5. WinHex: Hex Editor & Disk Editor, Computer Forensics & Data Recovery Software. https://www.x-ways.net/winhex/
6. HashCalc 2.02 Free Download. https://hashcalc.soft112.com/
7. Scopus preview—Scopus—Welcome to Scopus
8. Tsai, M.-H., Hsu, F.-Y., Chang, J.-D., Wu, H.-C.: Fragile database watermarking for malicious tamper detection using support vector regression. In: Third International Conference on Intelligent Information Hiding and Multimedia Signal Processing (IIH-MSP 2007), pp 493–496 (2007)
9. Wu, H.-C., Hsu, F.-Y., Chen, H.-Y.: Tamper detection of relational database based on SVR predictive difference. In: 2008 Eighth International Conference on Intelligent Systems Design and Applications, pp. 403–408; Friedman, R., Licher, R.: Hardening Cassandra against byzantine failures (2016). arXiv Prepr arXiv161002885 (2008)
10. Guo, J.: Fragile watermarking scheme for tamper detection of relational database. In: 2011 International Conference on Computer and Management (CAMAN), pp 1–4 (2011)
11. Kumari, V.P., Easwarakumar, K.S.: Tamper detection in multi-format image databases using fragile watermarking. In: 2010 International Conference on Recent Trends in Information, Telecommunication and Computing, pp. 308–310 (2010)
12. Dey, N., Dey, M., Mahata, S.K.: Tamper detection of electrocardiographic signal using watermarked bio-hash code in wireless cardiology. Int. J. Signal Imaging Syst. Eng. **8**, 46–58 (2014). https://doi.org/10.1504/IJSISE.2015.067069
13. Mukherjee, A., Dey, G., Dey, M., Dey, N.: Web-based intelligent EEG signal authentication and tamper detection system for secure telemonitoring. In: Brain-Computer Interfaces, pp. 295–312. Springer, Berlin (2015)
14. MongoDB: Understanding Oplog. http://dbversity.com/mongodb-understanding-oplog/
15. How to setup an oplog on a single MongoDB instance—Cogs and Levers. https://tuttlem.github.io/2014/06/13/how-to-setup-an-oplog-on-a-single-mongodb-instance.html
16. Mongodb The OpLog Explained—Brian Carpio. http://www.briancarpio.com/2012/04/21/mongodb-the-oplog-explain/
17. GitHub—ozlerhakan/mongodb-json-files: A curated list of JSON/BSON datasets from the web in order to practice/use in MongoDB. https://github.com/ozlerhakan/mongodb-json-files

# Machine Learning-Based Classification of Heart Sound Using Hilbert Transform

Pradip Mane(✉), Uttam Chaskar, and Prashant Mulay

College of Engineering Pune, Pune, Maharashtra, India
manepradip18491@gmail.com, umc.instru@coep.ac.in,
mulayprashant@yahoo.in

**Abstract.** Phonocardiogram (PCG) or heart sound signal administers crucial information for the diagnosis of various cardiovascular affliction. The heart sound classification is a confronting task in the field of modern healthcare. This paper confers the heart sound classification using the Hilbert transform envelope technique. The major constituents in the classification stage are preprocessing of PCG signal, features (temporal, spectral, and statistical) extraction, machine learning, and features-based classification of PCG signal. The accuracy and firmness of the proposed method are evaluated using two different datasets with different classes. The heart sound signals are taken from the standard phonocardiogram databases, i.e., PASCAL and PhysioNet/CinC. Evaluation results manifest that the proposed method for PCG signal classification achieves an overall accuracy (A) of 97.7% for the PASCAL dataset and overall accuracy (A) of 98.8% for PhysioNet/CinC dataset. Comparative results manifest that the proposed method is capable of classification of the PCG signal. Further, the method permits extraction of appropriate features for the classification of the PCG signal.

**Keywords:** Evaluation parameter · Heart sound classification · Hilbert transform · Machine learning and MFCC

## 1 Introduction

### 1.1 Background

The phonocardiogram permits the detection of principal heart sounds and other extra sounds and creates a visual recording of these incidents. To obtain the phonocardiogram signal, a microphone is situated on a human's chest and the recorded signal has conspired on a graph. The PCG signal provides decisive clinical information to a specialist for analysis of heart sound abnormalities.

The normal heart sound signal consists of two principal heart sounds which are, namely the first heart sound (S1) and the second heart sound (S2). First heart sound (S1) indicates the start of ventricular systole and second heart sound (S2) indicates the end


X.-S. Yang et al. (eds.), *Proceedings of Fifth International Congress on Information and Communication Technology*, Advances in Intelligent Systems and Computing 1183, https://doi.org/10.1007/978-981-15-5856-6_40

of ventricular systole. In the normal situations, interval S1--S2 (systole) is shorter than interval S2--S1 (diastole) [1, 2]. Figure 1 illustrates the principal heart sound S1, S2, systole, and diastole.

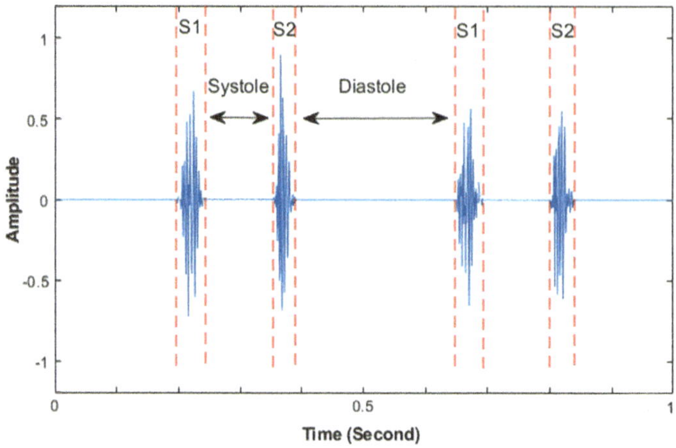


**Fig. 1.** Principal heart sounds S1, S2, systole, and diastole

### 1.2 Literature Review

Many heart sound classification approaches have been presented in previously published works. In the heart sound classification, most of the methods were established on a considerable set of features. Also, classifiers such as k-nearest neighbors (KNN), adaptive neural network (ANN), naive bayes (NB), support vector machine (SVM), neural network (NN), convolutional neural network (CNN), and random forest (RF) have been mostly used. Considering the precise analysis of the heart sound signals is a challenging task in preventive medicine, clinical studies highlight that there is a demand for classification of heart sound signals [2–4]. Despite previous works had promising results, most of the work was developed for normal and abnormal classes. In this paper, the authors used two different datasets with different number of classes.

## 2 Heart Sound Classification

Figure 2 interprets the block diagram of the proposed heart sound classification method, which consists mainly: dataset preparation, preprocessing, feature extraction, machine learning model [5–8], training, testing, and classification. In this section, the authors discuss the heart sound classification in detail.

### 2.1 Dataset Preparation

In this literature, the authors have used two different datasets in which the first dataset subsists of four classes particularly artifact, extrastole, murmur, and normal [9] and

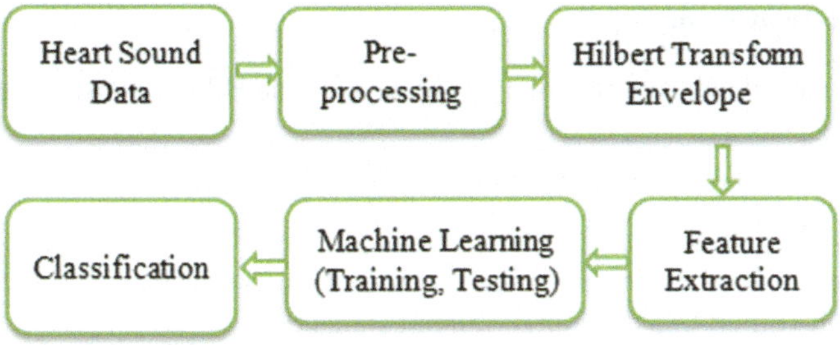


**Fig. 2.** Block diagram of the proposed heart sound classification

the second dataset consists of two classes, namely normal and abnormal [10]. Five-fold cross-validation scheme is applied to analyze the skill of machine learning models. From dataset 1, the authors have eliminated less than 3 s signals. Totally, 216 heart sound signals have been taken for classification. Out of 216 signals, 39 signals are of an artifact class, 45 signals are of extrastole class, 43 signals are of murmur class, and 89 signals are of class. From dataset 2, after eliminating less than 3-s signals and eliminating distorted signals, totally, 1143 signals have been taken for classification. Out of 1143 signals, 557 are of normal signals and 586 are of abnormal signals.

## 2.2 Preprocessing of Heart Sound Signal

In the real time, duration of heart sound signals may be tedious. The tedious data tends to enlarge the processing time, so it is practicable to take 3 s of data for the heart sound analysis. Also, heart sound signals may be corrupted by noise, this noise tends to increase the false alarms. Therefore, the elimination of noise from the heart sound signals is also important [6]. In the preprocessing, step in the first mode, audio data is recouped from wave (.wav) file of heart sound signal. Then a band pass filter is used to eliminate the higher frequencies as the frequency of heart sounds is low in the range between 20 and 250 Hz [11]. Figure 3 demonstrates the signal $p(t)$ which is obtained by preprocessing on heart sound signal (.wav)

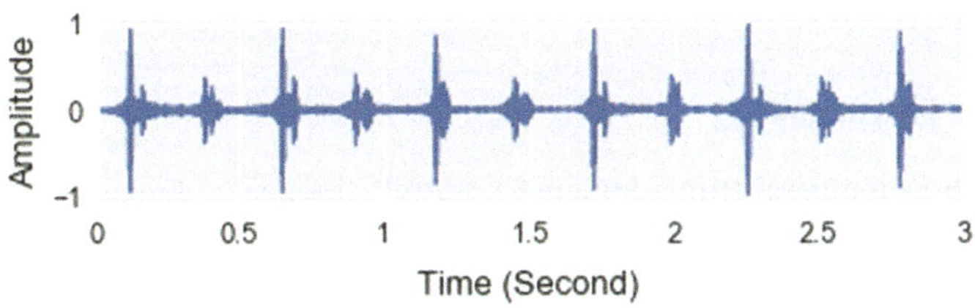


**Fig. 3.** Pre-processed heart sound waveform (.wav)

## 2.3 Hilbert Transform

Let $p(t)$ is a real valued, continuous-time signal, then the Hilbert transform is defined as

$$h(t) = H\{p(t)\} = \frac{1}{\pi} \int_{-\infty}^{\infty} \frac{p(\tau)}{(t - \tau)} \mathrm{d}\tau = p(t) * \frac{1}{\pi t} \tag{1}$$

From convolution of input signal $p(t)$ with the Hilbert transform kernel function $\frac{1}{\pi t}$, we can extract Hilbert transform $h(t)$ of input signal $p(t)$.

The complex analytical signal $g(t)$ which is obtained using input signal $p(t)$ and Hilbert transform $h(t)$ is as given below

$$g(t) = p(t) + jh(t) \tag{2}$$

The signal $g(t)$ has the precursor that all the negative-frequency components of the input signal were excluded by Hilbert transform. The envelope of the input signal waveform $E(t)$ is establish based on complex analytical signal $g(t)$ applying to the absolute function as given below

$$E(t) = |g(t)| = |p(t) + jh(t)| = \sqrt{x^2 + h^2} \tag{3}$$

The advantage of Hilbert transform is that it provides a more consistent and rapid decomposition process [12]. Figure 4 demonstrates the output of the Hilbert transform-based envelope when the PCG signal is normal.

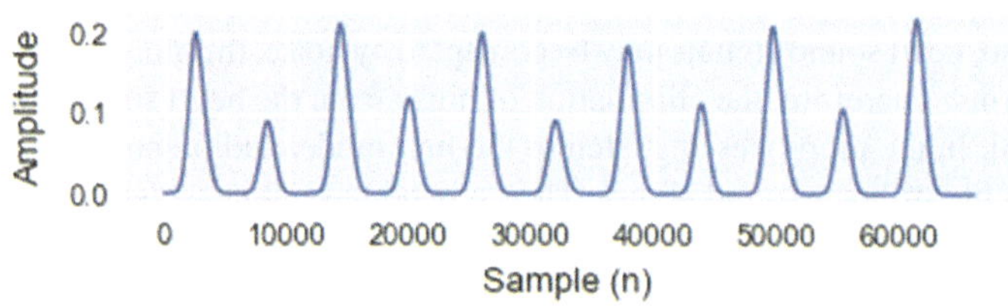


**Fig. 4.** Envelope using Hilbert transform

## 2.4 Feature Extraction

Features such as temporal (time-domain), spectral (frequency-domain), and statistical (statistical-domain) features are extracted from the heart sound signal. It is noted that total 18 features have been used by the authors for the heart sound classification. Table 1 demonstrates the extracted features that are used in this paper. In time-domain features, the authors focused on the number of higher peaks (NHP), the number of lower peaks (NLP), and the zero-crossing rate (ZCR). The authors characterize these parameters by limiting the duration of PCG signal to 3 s. The abnormalities in considered features might produce additional sounds, which in adults indicate pathologies and must be classified as abnormal heart sound.

**Table 1.** Features

| Domain | Features | Total |
|---|---|---|
| Time | Number of higher peaks (NHP)<br>Number of lower peaks (NLP)<br>Zero-crossing rate (ZCR) | 3 |
| Frequency | Mel-frequency cepstral coefficients (MFCCs) | 13 |
| Statistical | Standard deviation of higher peaks (SDHP)<br>Standard deviation of lower peaks (SDLP) | 2 |

In the frequency domain, the authors focused on Mel-frequency cepstral coefficients (MFCCs). As they are enumerating by catching the Fourier transform of the spectrum with logarithmic warping, they contain information about the rate of change in the different spectrum bands [8]. Selecting a large number of Mel-frequency cepstral coefficients may results in more complexity in the models. So, the first 13 cepstral coefficients which are optimal in heart sound analysis have been considered. These 13 MFCCs have been computed from the spectrum of heart sound signal. The mean of each MFCC coefficient have been considered as a feature for heart sound classification. Moreover, some statistical features are considered in this study which are the standard deviation of higher peaks (SDHP) and standard deviation of lower peaks (SDLP).

## 3 Results and Discussion

### 3.1 Classification Results

A machine learning classification model has been implemented in this paper for the classification of the PCG signal. The heart sound classifiers which have been proposed in this study are support vector machine (SVM), k-nearest neighbor (KNN), decision tree (DT), and ensemble classifiers (EC) [5–7]. As the evaluation results of SVM are better than the KNN, DT, and EC. The evaluation results of SVM have been used in this study. The evaluation parameter Precision (P), Recall (R), and Accuracy (A) have been used [5–7] for evaluation of classification. The performance of the support vector machine (SVM) classifier is evaluated through five-fold cross-validation on both datasets. In SVM classifier, the authors considered three submodels which are linear SVM, quadratic SVM, and cubic SVM. Table 2 shows the performance of SVM classifiers for dataset 1 and 2. It is noted that for dataset 1, quadratic SVM has the best accuracy (A), i.e., 97.7% and for dataset 2 also quadratic SVM has best accuracy (A), i.e., 98.8%

### 3.2 Performance Comparisons

In the proposed work, experiments were carried for training and testing the classifier using the PASCAL and PhysioNet/CinC PCG dataset. In this paper, authors have made comparison of performances using results of different methodology used in the previous paper [1–4]. Comparative results have been shown in Table 3. The parameters which

**Table 2.** Performance using support vector machine (SVM)

| Dataset | Classes | Linear SVM | | Quadratic SVM | | Cubic SVM | |
|---|---|---|---|---|---|---|---|
| | | P | R | P | R | P | R |
| 1 | Artifact | 0.92 | 1 | 0.95 | 1 | 0.95 | 1 |
| | Extrastole | 0.96 | 0.88 | 1 | 0.92 | 0.96 | 0.88 |
| | Murmur | 0.95 | 1 | 0.98 | 1 | 0.95 | 1 |
| | Normal | 0.99 | 0.98 | 0.98 | 0.99 | 0.98 | 0.98 |
| | A (%) | 963 | | 977 | | 963 | |
| 2 | Normal | 0.98 | 0.98 | 0.99 | 0.99 | 0.96 | 0.97 |
| | Abnormal | 0.98 | 0.98 | 0.99 | 0.99 | 0.97 | 0.97 |
| | A (%) | 983 | | 988 | | 969 | |

have been used for the comparisons are Accuracy (Acc), Specificity (Sp), Recall (R), and Precision (P). From Table 3 the results obtained by using Hilbert transform method gives better accuracy.

**Table 3.** Performance comparisons with previously published work

| Ref No. | Classification type | Method | Parameters | |
|---|---|---|---|---|
| Varghees et al. [1] | Murmur | EWT, IP | Acc = 95.5<br>R: 98.00<br>P: 97.4 | |
| Wang et al. [3] | N, AS, AR, MS, MR | OMS-WPD, Wavelet Entropy, SVM | Acc = 88.98<br>Sp = 94 | |
| Ahlstrom et al. [4] | N, MR, AS | WT, FD, RQA, SFFS, NN | Acc = 86 | |
| Papadaniil et al. [2] | S1, S2 | EEMD, Kurtosis | Acc = 88.98<br>Sp = 94 | |
| a | N, M, E, A, AN | Hilbert transform | Acc = 98.8% (SVM) | |
| | | | (R)<br>N = 0.99<br>M = 1.00<br>E = 0.88<br>A = 100 | (P)<br>N = 0.99<br>M = 0.95<br>E = 0.96<br>A = 0.92 |

*EWT* empirical wavelet transform; *IP* instantaneous phase; *OMS-WPD* optimum multi-scale wavelet packt decomposition; *NN* neural network; *SVM* support vector machine; *SFFS* sequential forward floating selection; *WT* wavelet transform; *EEMD* ensemble empirical mode decomposition; *FD* fractal dimension; *AR* aortic regurgitation; *MR* mitral regurgitation; *MS* mitral stenosis; *RQA* recurrence quantification analysis; *AS* aortic stenosis; *N* normal; *M* murmur; *E* extrastole; *A* artifact; *AN* Abnormal; [a]denotes results obtained using the proposed method

In this paper, the authors have made another comparison of performances using the results of the dataset 1 and dataset 2 shown in Table 4. The parameters which have been used for the comparisons are Accuracy (Acc), Specificity (Sp), Recall (R), and Precision (P). It shows that results obtain by applying proposed method on dataset 1 and dataset 2 gives better classification results.

**Table 4.** Performance comparisons with previously published work

| Dataset | Ref No. | Accuracy (Acc) | Specificity (Sp) | Recall (R) | Precision (P) |
|---|---|---|---|---|---|
| 1 | Chakir et al. [13] | – | NB = 64.28%<br>KNN = 57.14%<br>SVM = 71.43 | N = 64.28<br>N = 78.57<br>N = 64.28 | N = 78.57%<br>M = 78.57% |
| | Jadhav et al. [14] | ANN = 0.7831 | ANN = 0.7368 | N = 0.7968 | N = 0.9107 |
| | a | SVM = 98.8% | – | N = 0.99<br>M = 1.00<br>E = 0.88<br>A = 100<br>(SVM) | N = 0.99<br>M = 0.95<br>E = 0.96<br>A = 0.92<br>(SVM) |
| 2 | Nassralla et al. [15] | 92% | 98% | 78% | – |
| | Mishra et al. [16] | 97.50% | – | 100% | – |
| | Bobillo et al. [17] | 91.9% | 89.3% | 94.6% | – |
| | Nilanon et al. [18] | – | LR = 0.758<br>SVM = 0.756<br>CNN = 0.813 | LR = 0.726<br>SVM = 0.756<br>CNN = 0.735 | – |
| | a | SVM = 97.7% | – | N = 0.99<br>AN = 0.98<br>(SVM) | N = 0.99<br>AN = 0.98<br>(SVM) |

*NB* Naive Bayes; *KNN* k-nearest neighbors; *ANN* adaptive neural network; *SVM* support vector machine; *LR* logistic regression; *CNN* convolutional neural network; *N* normal; *M* murmur; *E* extrastole; *A* artifact; *AN* abnormal; [a]denotes results obtained using the proposed method

## 4 Conclusion

Accurate classification of principal heart sounds is an integral step in scrutiny of heart sound signals. In this work, the authors have presented the classification of heart sound signal using the Hilbert transform envelope technique. This study has shown that features identified by Hilbert transform envelope technique are promising for the extraction of features for the classification of PCG signal. Also, the machine learning classification models along with extracted features have been used for the performance measurement of classifying human heart sounds. This study provides promising results for the classification of heart sound signal with the use of machine learning classification models on two different datasets. Future work will focus upon the extension of the Hilbert transform envelope technique to the case of identification and extraction of four heart sounds from the raw heart sound signal, facilitating the feature extraction process, and the use of the classified heart sound signal to build a disease recognition tool.

## References


1. Varghees, V.N., Ramachandran, K.I.: Effective heart sound segmentation and murmur classification using empirical wavelet transform and instantaneous phase for electronic stethoscope. IEEE Sensors J. **17**(12), 3861–3872 (2017)
2. Papadaniil, C.D., Hadjileontiadis, L.J.: Efficient heart sound segmentation and extraction using ensemble empirical mode decomposition and kurtosis features. IEEE J. Biomed. Health Inf. **18**(4), 1138–1152 (2014)
3. Wang, Y., Li, W., Zhou, J., Li, X., Pu, Y.: Identification of the normal and abnormal heart sounds using wavelet-time entropy features based on OMS-WPD. Future Generat. Comput. Syst. **37**, 488–495
4. Ahlstrom, C., et al.: Feature extraction for systolic heart murmur classification. Ann. Biomed. Eng. **34**(11), 1666–1677 (2006)
5. Marsland, S.: Machine Learning: An Algorithmic Perspective. Chapman and Hall/CRC (2014)
6. Friedman, J., Hastie, T., Tibshirani, R.: The Elements of Statistical Learning, vol. 1. No. 10. Springer series in statistics, New York (2001)
7. Lim, T.S., Loh, W.Y., Shih, Y.S.: A comparison of prediction accuracy, complexity, and training time of 33 old and new classification algorithms. Mach. Leran. (1999)
8. Zheng, F., Zhang, G., Song, Z.: Comparison of different implementations of MFCC. J. Comput. Sci. Technol. **16**(6), 582–589 (2001)
9. PASCAL Classifying Heart Sounds Challenge. [Online]. Available: http://www.peterjbentley.com/heartchallenge/2016
10. PhysioNet/CinC Challenge 2016. [Online]. Available https://www.physionet.org/challenge/201
11. Phua, K. et al.: Heart sound as a biometric. Pattern Recogn. **41**(3), 906–919 (2008)
12. Downey, A.B.: Think DSP: Digital Signal Processing in Python. O'Reilly Media, Inc. (2016)
13. Chakir, F. et al.: Phonocardiogram signals classification into normal heart sounds and heart murmur sounds. In: 2016 11th International Conference on Intelligent Systems: Theories and Applications (SITA). IEEE, New York (2016)
14. Jadhav, A.R., Ghontale, A.G., Ganesh, A.: Heart sounds segmentation and classification using adaptive learning neural networks. In: 2017 International Conference on Signal Processing and Communication (ICSPC). IEEE, New York (2017)
15. Nassralla, M., El Zein, Z., Hajj, H.: Classification of normal and abnormal heart sounds. In: 2017 Fourth International Conference on Advances in Biomedical Engineering (ICABME). IEEE, New York (2017)

16. Mishra, M. et al.: Classification of normal and abnormal heart sounds for automatic diagnosis. In: 2017 40th International Conference on Telecommunications and Signal Processing (TSP). IEEE, New York (2017)
17. Bobillo, I.J.D.: A tensor approach to heart sound classification. In: 2016 Computing in Cardiology Conference (CinC). IEEE, New York (2016)
18. Nilanon, T. et al.: Normal/abnormal heart sound recordings classification using convolutional neural network. In: 2016 Computing in Cardiology Conference (CinC). IEEE, New York (2016)

# Intelligent Approaches for the Automated Domain Ontology Extraction

Alexander Katyshev[1], Anton Anikin[1,3(✉)], Mikhail Denisov[1], and Tatyana Petrova[2]

[1] Volgograd State Technical University, Volgograd, Russia
Anton@Anikin.name
[2] Volgograd State Socio-Pedagogical University, Volgograd, Russia
[3] Software Engineering School, Volgograd, Russia

**Abstract.** The chapter presents the review of modern approaches for the domain terminology extraction, concept discovery, concept hierarchy derivation, and learning of non-taxonomic relations steps in the ontology learning task. The chapter presents the review of not only various approaches to solving these NLP tasks but also ready-made tools that implement these approaches.

**Keywords:** Semantic link network · Ontology · Ontology learning · Natural language processing · Named entity recognition · Relations extraction · Information extraction

## 1 Introduction

Now in the world over the past 30 years, the volume of various kinds of information tends to constantly increase. To improve the efficiency of solving the problems of accumulation, processing, search, and analysis of information arrays, modern advances in the field of information technology are used. In particular, one of the components of modern promising approaches is the use of ontologies. Ontologies are developed and can be used to solve various problems, including for joint use by people or software agents, for the possibility of accumulation and reuse of knowledge in the specific domain, for the creation of models and programs that operate ontologies, rather than hard-coded data structures, for analysis of knowledge in the specific domain.

The ontology development process usually begins with what is composed of terminological concepts that are used in studies of the properties and characteristics of the terminology presented in it. Then, in a natural language, a list of exact definitions of terms presented in the glossary is created. This method can


X.-S. Yang et al. (eds.), *Proceedings of Fifth International Congress on Information and Communication Technology*, Advances in Intelligent Systems and Computing 1183, https://doi.org/10.1007/978-981-15-5856-6_41

be based on several concepts (class hierarchies). Of the concepts not involved in the compilation of classification trees, class attributes and their possible meanings are distinguished. It is these concepts that establish the basic connections between classes.

The article consists of a description of ontology learning, natural language processing (specifically the description of information extraction task and relation extraction task), and also modern approaches to information and relation extraction from specific domain texts.

## 2 Ontology Learning

### 2.1 Ontology Tasks

The development of a domain ontology consists of the following steps [5]:

- domain terminology extraction;
- concept discovery and hierarchy derivation;
- relations extraction;
- ontology population;

and others.

Concepts, or classes, are considered in a broad sense as the conceptualization of all representatives of an entity or phenomenon. They may contain other concepts, individual instances, or combinations thereof. In ontologies and ontological-type resources, concepts are usually organized into a taxonomy—a hierarchical structure with generic relations. Relations are a type of interaction between concepts. The most common type of relation used in all ontologies is generic relations, also called inclusion, categorization, IS-A, class–subclass, general–private, taxonomic, and hyperhyponymic relations. Relations, like concepts, can themselves form a hierarchy [1].

### 2.2 Ontology Application

In modern search engines, texts are automatically indexed by the set of words that make up these texts.

Such a presentation of texts as a simple set of words ("bag of words")[10] has a large number of obvious drawbacks that make it difficult to find relevant texts, for example:

- redundancy: the word-by-word index uses synonyms that express the same concepts;
- text words are considered independent of each other, which does not correspond to the properties of a connected text;
- polysemy of words: since polysemous words can be considered as a disjunction of two or more concepts expressing different meanings of a polysemantic word, it is unlikely that all the elements of this disjunction are of interest to the user.

These shortcomings are deprived of the so-called conceptual indexing, that is, such indexing when the text is indexed not according to words, but according to the concepts that are discussed in this text. With this technology:

- all synonyms are reduced to the same concept;
- ambiguous words are assigned to different concepts;
- links between concepts and corresponding words are described and can be used in the analysis of the text.

In order to try to implement a scheme of automatic conceptual indexing and conceptual search, you must have a resource that describes the system of concepts of a given specific domain, that is, an ontology in a given specific domain.

# 3 Natural Language Processing

Natural language processing (NLP) is the general direction of artificial intelligence and mathematical linguistics. It studies the problems of computer analysis and synthesis of natural languages. In relation to artificial intelligence, analysis means understanding the language, and synthesis means generating literate text. Solving these problems will mean creating a more convenient form of interaction between a computer and a person. One of the main tasks of NLP is the analysis of texts in natural language. This common task includes several key sub-tasks: text categorization, information extraction, information retrieval, etc.

In this article, we analyze in detail the NLP task of extracting information from the text in a natural language and modern approaches to its solution for automating the construction of domain ontologies.

## 3.1 Information Extraction

The extraction of information is primarily associated with the search for entities and relations. This is one of the key stages of text preprocessing necessary for the implementation of more complex models and programs. Knowledge bases are used to remove homonymy, in word processing, semantic search, question–answer systems, and automatic understanding of the text without a teacher (machine reading). Entities should be categorized. A special place in the extraction of entities is occupied by the problems of identifying named entities and coreference (resolving anaphoric relations). Most of the problems that arise when automatically building a knowledge base on the web are related to the amount and heterogeneity of the data. Today you can find millions of entities, hundreds of thousands of classes, hundreds of types of relations, and hundreds of thousands of facts. The greatest difficulties arise when extracting knowledge from open areas, as well as when processing "temporary" knowledge.

### 3.2 Relation Extraction

There are two key sources for highlighting semantic relations between concepts in a language. First, you can use text boxes—this method was used, in particular, by the authors of Princeton WordNet [7]. Relations are extracted using both rules and machine learning methods on the case with pairs already marked up.

The rules have the form of regular expressions of the type "such (1) as (2), (3), (4)," where a hyperonym falls into slot 1, and hyponyms fall into slots 2–4. Similar heuristics can distinguish relations of other types [3,17].

The second principal source is machine-readable explanatory and encyclopedic dictionaries. Vocabulary definitions are a partially structured text and, as a rule, are constructed according to uniform templates, which facilitates their processing and the creation of general algorithms for identifying relations. The lexical information in the dictionary, in contrast to the plain text, is presented in a concentrated form and provided with labels that can be used as indicators of a relation between the word being defined and part of the definition [12].

## 4 Modern Approaches to Information and Relations Extraction

### 4.1 Using Context-Free Grammar

Grammar features, such as parts of speech, allow you to encode additional information about the language. One of the most effective ways to improve the quality of a model is to introduce grammars and parsers to create lightweight syntactic structures that directly affect dynamic collections of text, which can be of great importance.

To get information about the language in which the sentence is written, we need a set of grammatical rules that determine the components of correctly formed sentences in this language—this is what the grammar gives. Grammar is a set of rules that describe how syntactic units (sentences, phrases, etc.) in a language should be divided into their constituent elements. Using grammars, you can define a variety of rules for assembling phrases or fragments from parts of speech. A context-free grammar (CFG) is a set of rules for combining syntax components into meaningful strings [4].

The task of extracting named entities is currently well studied, and there are many commercial and open solutions for English such as Spacy, Stanford NER, OpenNLP, NLTK [9], MITIE, Google Natural Language API, ParallelDots [16], Aylien, Rosette, and TextRazor.

For example, Stanford NER markup named entities in raw text. The parser marks each token in the sentence, and based on the context of the sentence, the parser can define named entities among the three classes (PERSON, ORGANIZATION, LOCATION) [8].

There are good solutions for Russian too, but they are mostly closed: for example, PullEnti is an SDK for developers of information systems dealing with unstructured data—texts in natural language; functionality: selection of named entities (named entity recognition), morphology, semantics, and various processing procedures; types of entities: persons, organizations, dates, countries, and decrees. All algorithms are rule-based and languages: Russian and Ukrainian.

There are very few open-source solutions for the Russian language. For example, there is a Tomita parser from Yandex. It is designed to extract structured data from natural language text. Facts are extracted using CFG and keyword dictionaries. The parser allows to write custom grammar and add dictionaries for the desired language.

There is one more tool for the Russian language: This is Natasha. Natasha is an analog of Tomita parser for Python and a set of ready-made rules for retrieving names, addresses, dates, amounts of money, and other entities.

Table 1 provides a comparative analysis of several modern tools for extracting facts from the text.

Quality assessment was determined based on the F-measure.

$$F = 2 * \frac{P * R}{P + R}$$

where $P$ is a precision; $R$ is a recall.

**Table 1.** Analysis of modern tools for extracting facts from the text

| Tool | Speed | Settings | Flexibility | Quality |
|---|---|---|---|---|
| PullEnti | Middle | Easy. Input data: raw text and work type | Low. Supports a small set of types of named entities | 0.8672 |
| Tomita parser | Fast | Very hard. User must create some required files including his own CFG and 2–3 setting file | High. Named entities depend on user's CFG | 0.8584 |
| Natasha | Slow | Easy. Python package with ready CFG. Input data: raw text | Low. Supports a small set of types of named entities | 0.6598 |

## 4.2 Neural Networks

Since 2015, most NLP tasks have been solved using vector representations. Each word in the raw text is converted to its corresponding vector. Vectors are built on the basis of the context of use and the mutual occurrence of words in sentences. With these vectors, you can then carry out various mathematical operations. So, for example, the cosine distance between vectors shows how two and several words are similar to each other within the context of this text.

This neural network approach is to solve various NLP tasks such as machine translation, question answering, and others.

With the advent of powerful pre-trained representations, trained using some flavor of a language modeling objective such as ELMO [13], OpenAI [14] GPT, and BERT [6], the de facto technique for NLP has become to take some sort of off-the-shelf model pre-trained on gargantuan amounts of data and fine-tune to your task with some smaller in domain corpus. Indeed, this strategy has successfully achieved tremendous SOTA results on existing NLP benchmarks.

Let us consider the BERT tool in detail. BERT is a neural network from Google, which showed by a large margin state-of-the-art results on a number of tasks [6]. Using BERT, you can create AI programs for processing a natural language: answer questions asked in any form, create chatbots, automatic translators, analyze text, and so on. To submit text to the input of a neural network, you need to present it in the form of numbers somehow. In practice, each word is assigned not one number, but several. For example, it is a vector of 32 numbers. And the distances are measured as the distances between the points that these vectors point to in the space of the corresponding dimension (for a vector 32 digits long, this is a space with 32 dimensions, or with 32 axes). This allows you to compare one word at once with several words that are close in meaning (depending on which axis to count). Moreover, arithmetic operations can be performed with vectors.

This approach is called embeddings [11]. Many packages, such as Python packages, allow the first layer of the neural network to put a special layer of embedding layer, which does this automatically. That is, at the input of the neural network we submit the usual word number in the dictionary, and embedding layer, self-learning, translates each word into a vector of the specified length, say, 32 numbers.

It is much more profitable to pre-train such a vector representation of words on some huge corpus of texts, for example, on the whole Wikipedia, and to use ready-made word vectors in specific neural networks, rather than re-train them every time [2].

In the summer of 2018, OpenAI noticed [15] that if you pre-train a neural network on the transformer [18] architecture on large volumes of text, then it unexpectedly and by a large margin shows excellent results on a variety of different natural language processing tasks. In fact, such a neural network at its output creates vector representations for words and even whole phrases. And by hanging on top of such a language model a small block of a couple of additional layers of neurons, you can train this neural network for any tasks, for example, extracting facts and relations from a text in a natural language.

In addition to BERT, there is another XLNet model [19]. XLNet is an autoregressive language modeling (AR LM). It is trying to predict the next token from the sequence of the previous ones. In classic autoregressive models, this contextual sequence is taken independently from two directions of the original string.

XLNet generalizes this method and forms a context from different places in the source sequence. How it does it? It takes all (in theory) possible permutations of the original sequence and predicts each token in the sequence from the previous ones.

If we draw analogies with the BERT, it turns out that we do not mask the tokens in advance, but rather use different sets of hidden tokens for different permutations. At the same time, the second problem of BERT disappears—the lack of hidden tokens when using the pre-trained model. In the case of XLNet, the entire sequence, without masks, is already input.

Comparison with SOTA results on the test set of RACE, a reading comprehension task shows on the Table 2. * indicates using ensembles. "Middle" and "High" in RACE are two subsets representing middle and high school difficulty levels.

**Table 2.** Comparison with SOTA results on the test set of RACE

| RACE | Accuracy | Middle | High |
|---|---|---|---|
| GPT | 59.0 | 62.9 | 57.4 |
| BERT | 72.0 | 76.6 | 70.1 |
| BERT+OCN* | 73.5 | 78.4 | 71.5 |
| BERT+DCMN* | 74.1 | 79.5 | 71.8 |
| XLNet | 81.7 | 85.4 | 80.2 |

Currently, such models are created for some separate domains. Such models cannot be extended to other domains for solving the problems of extracting concepts and domain relations to automate the construction of domain ontologies, since to solve these problems it is necessary to create and train a separate model for this domain.

## 5 Conclusion

This article has described in detail modern approaches to solving problems such as extracting facts and relationships in a natural language text. In addition, a detailed review and comparison of modern tools and libraries that implement these approaches were carried out. Each of the above tools has its own advantages and disadvantages. But a universal (as well as a multilingual solution) has not been found so far.

**Acknowledgements.** This chapter presents the results of research carried out under the RFBR grants 18-07-00032 "Intelligent support of decision making of knowledge management for learning and scientific research based on the collaborative creation and reuse of the domain information space and ontology knowledge representation model" and 20-07-00764 "Conceptual modeling of the knowledge domain on the comprehension level for intelligent decision-making systems in the learning".

## References

1. Alatrish, E., Tošić, D., Milenkovic, N.: Building ontologies for different natural languages. Comput. Sci. Info. Syst. **11**, 623–644 (2014)
2. Anikin, A., Katyshev, A., Denisov, M., Smirnov, V., Litovkin, D.: Using online update of distributional semantics models for decision-making support for concepts extraction in the domain ontology learning task. In: IOP Conference Series: Materials Science and Engineering, vol. 483, 012073 (2019)
3. Anikin, A., Sychev, O., Gurtovoy, V.: Multi-level modeling of structural elements of natural language texts and its applications. In: Biologically Inspired Cognitive Architectures, pp. 1–8. Springer International Publishing (2018)
4. Bengfort, B., Bilbro, R., Ojeda, T.: Applied Text Analysis with Python. O'Reilly Media, Inc. (Jun 2018)
5. Cimiano, P., Völker, J., Studer, R.: Ontologies on demand? A description of the state-of-the-art, applications, challenges and trends for ontology learning from text, pp. 315–320 (2006)
6. Devlin, J., Chang, M.W., Lee, K., Toutanova, K.: Bert: Pre-training of deep bidirectional transformers for language understanding. In: NAACL-HLT (2019)
7. Fellbaum, C. (ed.): WordNet: an electronic lexical database. In: Language, Speech, and Communication, MIT Press, Cambridge, MA (1998)
8. Finkel, J.R., Grenager, T., Manning, C.D.: Incorporating non-local information into information extraction systems by Gibbs sampling. In: ACL (2005)
9. Loper, E., Bird, S.: Nltk: the natural language toolkit. CoRR cs.CL/0205028 (2002)
10. Ma, S., Sun, X., Wang, Y., Lin, J.: Bag-of-words as target for neural machine translation (2018)
11. Mikolov, T., Chen, K., Corrado, G.S., Dean, J.: Efficient estimation of word representations in vector space. CoRR abs/1301.3781 (2013)
12. Pantel, P., Pennacchiotti, M.: Automatically harvesting and ontologizing semantic relations (2008)
13. Peters, M.E., Neumann, M., Iyyer, M., Gardner, M., Clark, C., Lee, K., Zettlemoyer, L.S.: Deep contextualized word representations. arXiv:1802.05365 (2018)
14. Radford, A.: Improving language understanding by generative pre-training (2018)
15. Radford, A., Narasimhan, K., Salimans, T., Sutskever, I.: Improving Language Understanding with Unsupervised Learning. Tech. rep, Technical report, OpenAI (2018)
16. Singh, P., Varadarajan, S., Singh, A., Srivastava, M.: Multidomain document layout understanding using few shot object detection (2018)
17. Snow, R., Jurafsky, D., Ng, A.: Learning syntactic patterns for automatic hypernym discovery, vol. 17 (2004)
18. Vaswani, A., Shazeer, N., Parmar, N., Uszkoreit, J., Jones, L., Gomez, A.N., Kaiser, L., Polosukhin, I.: Attention is all you need. In: NIPS (2017)
19. Yang, Z., Dai, Z., Yang, Y., Carbonell, J.G., Salakhutdinov, R., Le, Q.V.: Xlnet: generalized autoregressive pretraining for language understanding (2019). arXiv:1906.08237

# S. Park: A Smart Parking Approach

Sneha Singh(✉), Sneha Zacharia, Jibbin P. Mathew, and Hariram K. Chavan

Fr. Conceicao Rodrigues Institute of Technology, Sector 9-A, Vashi, Navi Mumbai 400703, Maharashtra, India
{sneha18121997,jibbinmathew4christ,chavan.hari}@gmail.com

**Abstract.** Being a developing nation, India faces a major issues that is the huge population which is now becoming too large for the spaces available. With increased living standards and affordability, the number of cars owned today has exponentially increased. This has led to the emergence of a major concern of proper management of available spaces while keeping up with the pace of revolutionizing lives by utilizing the SMART way. "SMART PARKING" aims at addressing this plight by providing a mobile-based platform to reserve a parking slot before arriving at desired location. The user is provided with a real-time availability map of the parking space showing occupancy. Depending on the traffic and time required to reach the parking slot, reservation is dynamically altered. It provides the feature of sending alerts based on live tracking to accommodate the unpredictability of reaching the destination due to traffic and adapt the booking timings accordingly by calculation buffer time. In cases where user comes without a booking, slot is provided based on how many vacant parking slots are available. The occupancy of the user's car in the slot is analyzed and utilized to predict the nature of booking to help for future purposes. Monthly subscriptions are provided and user is notified for renewals or updating.

**Keywords:** Pre-booking · Real-time availability · Smart parking · Unpredictability · Buffer time

## 1 Introduction

Cities are meant to play a major role in driving the economic growth as engines of innovation and industry. However, India's cities are choking! Rapid urbanization and lack of planning for urban infrastructure have exerted enormous pressure, which is opening a hazardous portal to parking issues leading to much greater traffic on road, resulting in a frustrated trip for the drivers and most importantly adding to the toxic waste by fuel burning, thus increasing air pollution in concerned area. The number of vehicles per thousand population has


X.-S. Yang et al. (eds.), *Proceedings of Fifth International Congress on Information and Communication Technology*, Advances in Intelligent Systems and Computing 1183, https://doi.org/10.1007/978-981-15-5856-6_42

increased considerably from 317 to 598 during 2005–06 to 2017–18. This triggers to a situation in a way that almost 40% of the roads are occupied by cars searching for a parking slot. Due to lack of open spaces and improper management of the available spaces, very less parking space is available compared to the huge number of cars out there today. It is really inconvenient to not have an option of pre-booking our parking slot prior. So in order to address the same, our project defines a completely automated system for vehicle parking that will overcome the current parking unavailability by letting the user book a parking slot in prior and avoid the hassle of unavailability of a parking slot. As per statistics, Smart Parking can help in achieving reduction in fuel consumption by 220,000 gallons till 2030 and can reach up to 300,000 gallons by 2050, if implemented completely. The project aims at providing a mobile-based platform which allows users to book a parking slot prior to their arrival, adjust spot occupancy based on real-time traffic analysis and provide a completely digitized platform for achieving parking management.

### 1.1 Need for Development

The preeminent reason for developing this project is to provide an optimized parking solution in today's era of technological and smart city development along with the ever-increasing population and thus reducing the chaos in search for a parking slot.

Smart Parking enables users to gain knowledge about the occupancy of parking slots in real time within the concerned geographical region. This information is obtained through installed sensors, which update the current occupancy of the slot to the data repository. The user is presented with the data and directed to a platform for booking a parking slot. It is also enabled with integrating a payment platform. Thus, it aids a user's parking experience by analyzing data and predicting the chances of user being able to find a spot in the concerned region. On deployment, the system as a whole shall aim at reducing car emissions and traffic on the roads by directed routing to desired locations, in turn managing the available spaces in an optimized manner.

### 1.2 System Features

- To provide a user-friendly mobile-based platform enabling reservation of a parking slot before arriving at desired geographic region.
- To provide the user with a real-time availability map of the parking space showing the occupancy and availability of the slot, thus helping the user to decide accordingly.
- To provide an affordable Smart Parking System—The user has to pay only for the time his car occupies the slot, unlike hourly wages used otherwise.
- Providing a systematic management of Car Parking System—Including the feature of sending alerts prior to the users indicating that the booking duration is coming to an end.

- Providing a provision of carpooling to help users traveling to the same destination.
- Providing slots based on how many vacant parking slots are available, if the user has come without a booking.

## 2 System Description

Until recent times, Parking was viewed under models like CLAMP, PARKISM, PARAGENT and MULTILAYERS. With emerging trends, traffic authorities in cities have come up with an implementation model called Parking Guidance and Information (PGI) system for achieving proper management of parking areas. The PGI system gives real-time information about occupancy of parking in an area and routes the users to available slots. The occupancy of slots in a concerned region is viewed in the variable message sign (VMS) board at road signs. This system had several loopholes, hence the idea of smart parking was proposed.

Designing, developing and producing a leading-edge parking technology is called as Smart Parking may be defined as a means of revolutionizing the existing parking management system by extending the latest technological advancements into achieving enhanced parking management. This system extends parking capabilities for users. Using sensors inside the slot, we can get information about occupancy and alert user. Smart Parking system has demonstrated a result-driven accuracy, robustness and cost-efficiency outlook to assure users with the knowledge about vacant slots for parking in the concerned area.

The proposed system allows users to reserve a parking slot for themselves which ensures a way for users to claim their spot, unlike just showing the availability which may not account for dynamism of the real world and result in the spot being occupied when the user reaches the spot.In the Smart Parking system, the slots will be booked for the user, thus ensuring availability and accounting for the dynamic nature of occupancy change. This in turn allows for a greater functional enhancement of the parking management mechanism.

## 3 Literature Survey

There have been several different approaches proposed for the development of intelligent parking system aiming at enhanced management capabilities and user experiences, which poses a major threat to the outlook of Smart Cities having gilt-edge administration of parking spaces as one of its pivotal aspects.

As suggested in [1] approximately 30% of the traffic in a day considering urban landscape can be rooted to vehicles in search for parking spaces. Hence, this opens an urgency for proper utilization of existing spaces and enables fitting administration for parking systems.

In recent times, locating a vacant parking spot poses an enduring difficulty for the drivers, and this problem has aggravated with the exponential increase in car ownership in recent times. Predicaments pertaining to parking and traffic can be viewed under the light of parking management if the users may be presented with prior knowledge about occupancy of the parking system in concerned region.

With the latest advancements in the sensor technology, a lot of urban cities have chosen to deploy IoT-based solutions for gaining information from the environment and act accordingly. According to a study by the International Parking Institute [2], there have been sudden outburst of innovations in the parking industry. Currently, there exist parking systems [3] that deliver real-time occupancy information to the users in concern with parking slots. These utilize sensors installed in the concerned region for evaluating occupancy along with repositories and analysis modules to take decisions of data in practical implementation.

Among the systems proposed, there have been deployments utilizing image processing capabilities [4]. This includes, capturing of the region of the slot through camera and then analyzing its occupancy, and display result. Firstly, the slot image is captured. This picture is segmented resulting in creation of binary images. The image is cleaned to reduce noise and analyze boundaries. Accordingly, the free parking space is allocated.

## 4 Proposed System

The issue of Parking in India is fast becoming a grotesque crisis. Smart Parking is a mobile-based parking management system extending IoT capabilities attempting to digitize the parking industry. The system aims at elucidating the convoluted experience of parking by digitizing the entire process from booking a slot to auditing the nature of occupancy. On the user end, the system yields information about occupancy of parking slots in an area, the rates and directions to reach concerned booked slot. For the parking area managers, it may allow them to digitize the process of tallying bookings, allowing legitimate users and also enable payments from users.

*A. Slot Booking Module* This module implements a means for users to book a parking slot prior to arrival. This not only allows them to ensure availability of slots but also reduces time and unnecessary occupancy of roads in search of slots. This module ensures to assist users in their parking experience. The layout of parking slots in a given region is shown and thus user is allowed to book the desired slot and for desired time. On completion of booking time, an alert is sent to the user to allow them to either extend the booking or check out based on availability and user's response. Based on the booking history, users are prompted for next booking and data is also analyzed to propose new slots in given area of demand.

*B. Live Tracking Module* Live tracking module tracks the user's registered vehicle from the time it starts its commute until it reached the designated parking slot. Based on the current traffic statistics, the buzzer time of arrival of vehicle in the slot is altered, thus providing a dynamic interface ensuring inclusion current road situations.

*C. Payment Module* All payment transactions are made cashless and using "Caros", an application-specific cash. The users are also provided a means to allow monthly subscriptions and memberships.
*D. Carpooling Module* Provision for users to put their cars for carpooling has also been provided. This allows users to earn money and also help reduce number of cars on road and attempt to be an environment friendly alternative.
*E. Blacklisting Module* If a car does not check out at designated time, an alert is sent to prompt user to move their car. If failed to do so, their cars are towed away. Repetition of this makes their name under blacklist and block all their further bookings.

## 5 Implementation

The first step involves user to register on our application. The registration information includes user details along with the registration of the vehicle of commute of the user. This allows us to gain information about what kind of paring slot will be needed by the user. The primitive step involves user pre-booking slot in desired location. This is done after gaining the current availability map of booked and free slots. Once allotted, the user is live tracked from the place of origin until it reaches the parking slot. Based on the current traffic situation, the buzzer time is altered for what time will the estimated time of arrival be manipulated unless the user reaches the slot. On completion of booking time, if slot is available, the user is prompted if he wishes to extend the booking for next slot. Based on availability and user's response, the user either checks out or books the next slot. While checking out, payment is made in cashless approach utilizing "Caros", currency of Smart Parking. Cars may also be made available for carpooling and the user can either provide their car for carpooling or utilize this facility.

## 6 Work Flow Analysis

### 6.1 Use Case Diagram

See Fig. 1.

### 6.2 Flowchart

See Fig. 2.

## 7 Results and Discussions

While implementing "PRE-BOOKING" module, we allow users to book a parking slot prior to their arrival in the parking spot. The users book their preferred parking based on their location as well as the time for which they want to book. On analyzing this booking data, we obtain the following results:

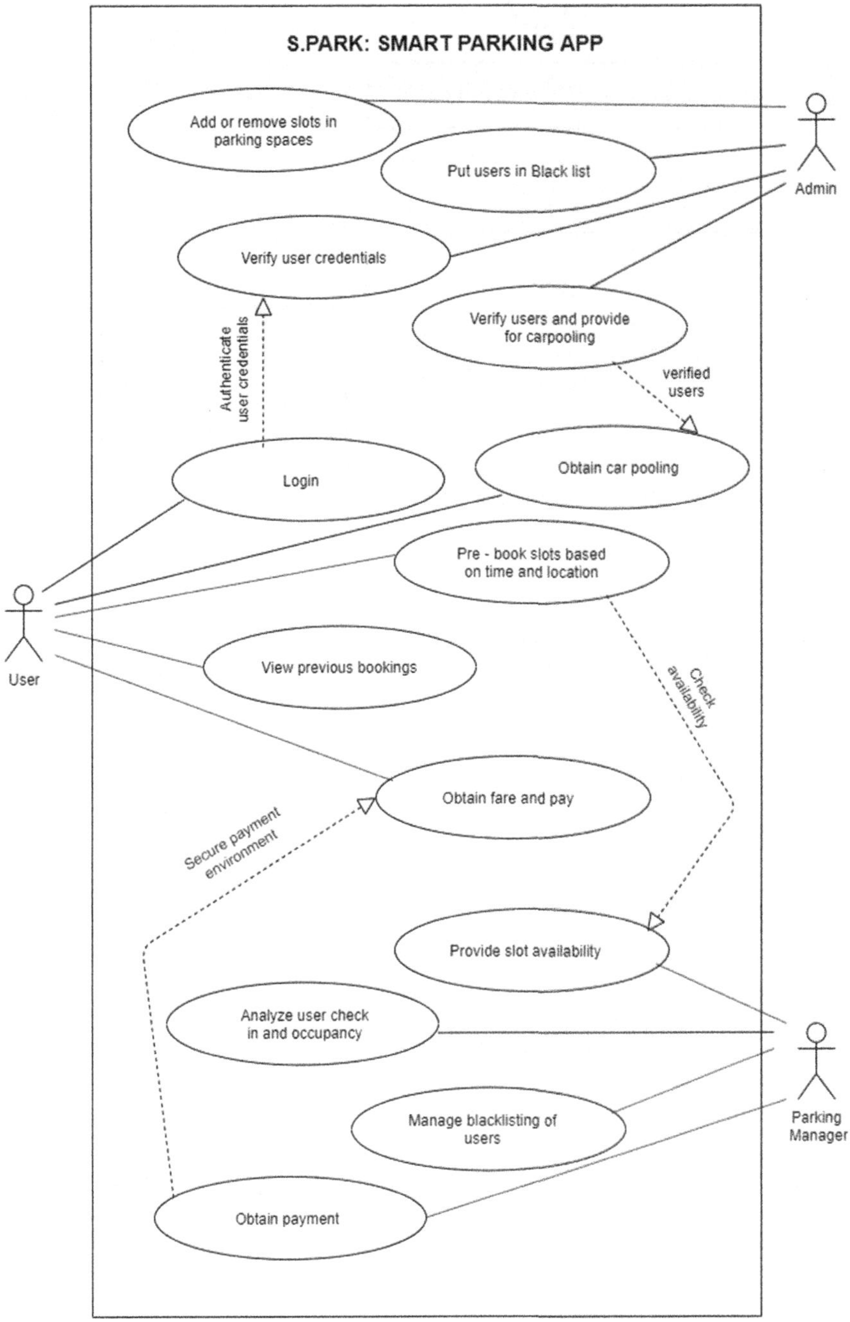


**Fig. 1.** Use case diagram

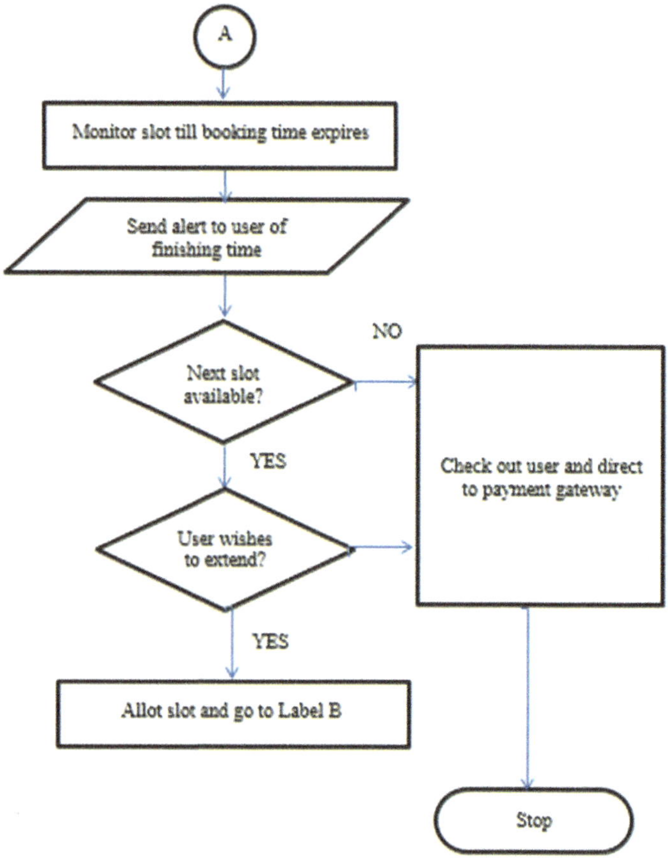


**Fig. 2.** FLowchart for smart parking

- Based on the timing of bookings obtained, we observed there are more parking requests near stations and bus stops in the morning and evening covering the demographics of working professionals.
- Evening time slot is when majority bookings happen near shopping complexes and tourist spots, resulting in the conclusion that preferred time of exploration of local spots is during evening.
- On studying the nature of booking for a user, we analyze which locations are frequently booked and around what time. This information is used to generate recommendations to repeat the bookings,assisting users by not asking for repeatedly filling same details and give a prompt based on regularity in booking.
- The pattern of booking for individual users may also be used to give monthly passes.
- Based on peak hours, we can alter our parking rates to obtain commercial profits.

- By analyzing requests, we can decide what particular location is being requested more and thus expand slot availability with available infrastructure. This may also be used to decide revoking slots in a region based on reduced availability.

On implementing "LIVE TRACKING" module, we track the user from the starting location of the user to the reserved parking slot. On the basis of current traffic scenario, the estimated time of arrival (ETA) is obtained. We analyze the ETA obtained and compare with actual time taken to reach the spot. We obtain the following results:

- Calculate accuracy of ETA (Estimated Time of Arrival) based on obtained ETA and actual time taken to reach the parking spot. when the user starts his travel, an ETA is prompted to determine how long it will take to reach the spot. The vehicle is monitored until it reaches the spot. Calculating difference in actual time of arrival can be used for evaluating accuracy of the previously determined ETA, thus improving accuracy.
- Real-time traffic is analyzed and used alongside ETA to calculate Buffer Time for user occupancy of the slot. This allows to incorporate the dynamism of real world and account for modifying slot time aiding the user experience.

## 8 Conclusion

Smart Parking System is a mobile-based application which aims at aiding a motorist by providing a parking spot that is totally automated and digitally assisted. It focuses on minimizing the chaos by searching a suitable parking slot from the already congested roads by being at par with the growing times of technology. The Slot Booking Module allows the users to book slot on hourly basis which helps in reducing the waiting time put in finding a slot and assuring availability of slot to the user. The Live Tracking Module tracks the user till he reaches the spot and helps in determining the ETA before the car can arrive and occupy the booked slot. The Payment Module allows user to do cashless transactions using Smart Parking cash "Caros". The Carpooling Module allows user to either give their cars for pooling or obtain cars which further reduces the traffic on the roads and also helps the user to earn or save money. The Blacklisting Module allows the application to block users and blacklist them if a user repeatedly fails to adhere to the rules and regulations of the service. Once blacklisted the user will be prevented from making any further bookings. Thus, SMART PARKING aims at providing unified ground to assess and aid the woes of parking by providing a completely automated approach in the best possible way.

The major challenge in implementing this system is the explicit range of hardware and software capabilities which may hamper scalability. The users must be presented with a dependable interface to interact with the system and utilize its capabilities to ease one's parking experience. It must be accessible from mobile applications, web-based services, navigation devices and enhanced built-in services within the user's vehicle. Conclusively, we can state beyond the shadow of doubt that Parking with a Smart approach can revamp and transmogrify the outlook of urbanization, leading to solutions that acknowledge not only a vehicle but also the people, promising to deliver the best solution in parking management.

## References

1. Alfatihi, S., Chihab, S.: Intelligent parking system for car parking guidance and damage notification. http://ieeexplore.ieee.org/document/6498229
2. Khanna, A., Anand, R.: IoT based smart parking system. http://ieeexplore.ieee.org/document/7562735
3. Basavaraju, S.R.: Automatic smart parking system using Internet of Things (IOT). http://www.ijsrp.org/research-paper-1215/ijsrp-p4898.pdf
4. Hanif, M., Hazif, M, Daund, H.: Smart parking reservation system using SMS **17** (2010)

# Scheduling Algorithm for V2X Communications Based on NOMA Access

Emna Bouhamed(✉), Hend Marouane, Ala Trabelsi, and Faouzi Zerai

National School of Electronics and Telecommunications (ENET'COM), NTS'COM Laboratory, Sfax University, Sfax, Tunisia
emnabouhamed94@yahoo.com, marouanehend@gmail.com, trabelsi.ala93@gmail.com, faouzifbz@gmail.com

**Abstract.** 5G network must support a large number of Vehicle-to-Everything (V2X) connections with high throughput and low latency. The existing resource allocation scheme based on orthogonal multiple access (OMA) seems to be unsuitable for dense networks due to the limitation of the spectral bandwidth and the available resources. This work presents a new algorithm named SAVCN (Scheduling Algorithm for V2X Communication based on NOMA), for 5G network. In non-orthogonal multiple access (NOMA) scheme, the same resource can be shared by several transmitters. The proposed algorithm improves network performances in terms of throughput, fairness, and error rate. In fact, SAVCN assigns the available resource blocks (RBs) in order to maximize the throughput by consideration of the minimal distance between transmitters and receivers. It maximizes also the number of served V2X users and minimizes the bit error rate. Simulation results indicate promising performance for SAVCN.

**Keywords:** V2X · NOMA · Scheduling algorithm · Throughput · BER

## 1 Introduction

The rapid evolution of mobile radio technologies, and the increase of the number of connected vehicles of the (V2X) type, makes the task of resource allocation very important for next-generation networks. V2X provides four types of communication: Vehicle-to-Infrastructure (V2I), Vehicle-to-Vehicle (V2V), Vehicle-to-Pedestrian (V2P) and Vehicle to Network (V2N) [1]. These applications require high throughput and some exigencies in terms of quality of service (QoS). V2X is considered as an important alternative of 5G networks. It is impossible to achieve the requirements of 5G, such as low latency, massive connectivity, and high throughput, using orthogonal multiple access (OMA). The orthogonal sharing resources process is not always optimal because it causes serious congestion problems due to the limitation of resources block [2]. In order to reduce access collisions, to support massive, it is very important to apply a new access technique. Alike in future 5G, the resource allocation system needs to be neatly designed to maximize


X.-S. Yang et al. (eds.), *Proceedings of Fifth International Congress on Information and Communication Technology*, Advances in Intelligent Systems and Computing 1183, https://doi.org/10.1007/978-981-15-5856-6_43

the radio resource efficiency, to gratify the dissimilar request of QoS requirements and the target throughput for V2X links corresponding to their supported services.

Non-Orthogonal Multiple Access (NOMA) is developed as an important element of 5G cellular networks. It can broadly be divided into two major categories, the first one is Power-Domain NOMA, the second one is code-domain NOMA [3]. This technology allows us to share more than one user in the same resource block such as a time interval, a sub-carrier, or a code.

To generate the NOMA scheme more useful, a Successive Interference Cancellation (SIC) is applied to the receivers to be decoded to meet with co-channel interference produced by spectrum sharing between different users [4].

Actually, radio resource management (RRM) is becoming a fundamental aspect of design to allow V2X communications. RRM is the central unit in the eNodeB, more precisely at the level of the MAC (Medium Access Control) sub-layer which is responsible for the optimal distribution of the radio resources between the UEs component in the 5G network architecture. RRM makes it possible to increase the use of the efficiency of the radio resources of the system while satisfying the Quality of Services (QoS) according to the different requirements of the target throughput and the type of traffic for V2X links.

In this paper, we suggest a new scheduling algorithm named SAVCN for the next-generation 5G. This algorithm is based on a Block Resource Allocation (RB) policy in a non-orthogonal way. The purpose of the proposed algorithm is to ameliorate the throughput of the system, to maximize the number of users served, to improve equity in the allocation of resources between different users, and to minimize the error rate.

The principal contributions of our work are detailed below:

- We propose a scheduling algorithm based on the NOMA technique for V2X system, which aims to raise the allocation of resources block between all users.
- We choose a set of well-defined users who can share the same RB($K_{\max}$) based on a distance metric.
- We have allocated the same resource block to a set of selected $k_{\max}$ users to take into account not only the distance between Tx and Rx, but again the notion of equity between users.
- To evaluate the performance of our algorithm, we validated the results by simulations taking into account several parameters such as the number of channels, the number of users that can share the same resource block.

We compared the performance of the proposed approach in terms of throughput with the classical OMA process.

The rest of this paper is constructed as follows in Sect. 2, we detail some existing works in literature. In Sect. 3, we present the adopted system model, then we detailed the proposed algorithm SAVCN in Sect. 4. In Sect. 5, We analyze simulation results to evaluate the performance of the proposed algorithm. Section 6 presents a conclusion and some perspectives.

## 2 Related Work

Several articles have dealt with the resource allocation function based on the classical orthogonal multiple access (OMA) technique. The principle of OMA is to serve one single user in each resource block. In [5–7] BER analysis of OMA system has been presented. But, these works do not take into account the throughput criterion.

For code-domain schemas as detailed in [8, 9], the authors examined the fragmented code multiple access methods called non-orthogonal multiple access (SCMA). This technique uses codebooks for multiplexing. Bit error rate (BER) was analyzed in [8] and performance analysis of spectral efficiency (SE) was performed in [9]. However, the criteria of throughput are not considered.

Multi-user shared access (MUSA) [10] is another type of multiple access using the distribution of symbols of each user by a spreading sequence. The purpose of this work is to examine the BER performance of a MUSA system for 5-ary codes and dissimilar sorts of spreading sequences with a short length. But, the criteria of throughput are not considered.

The Power-Domain NOMA (PD-NOMA) was originally proposed as Power Division Multiple Access (PDMA) by assigning users a different power.

Boya et al. in [11] considered a vehicular system with a high density when each vehicle must disseminate its information to its neighborhood. To meet the requirements of different applications such as low latency and high reliability, the authors proposed a scheduling algorithm based on NOMA. When multiple users can transmit their different types of traffic simultaneously on the same channel. Thus, a technique of successively cancelling interference is applied to the receiver to cope with the interference caused by channel sharing. The goal of this algorithm is to minimize V2X collisions and improve spectrum efficiency. However, the error rate and throughput are not addressed in this work.

Chen et al. in [12] proposed a power allocation algorithm to provide better reliability and improve bandwidth efficiency for inter-vehicle (V2V) communication. Indeed, the algorithm designed NOMA-SM is a compromise between two techniques non-orthogonal multiple access (NOMA) and spatial modulation (SM). The proposed algorithm is robust against the harmful effects of V2V environments. The authors analyzed the performance of the algorithm via simulations that take into account the bit error rate (BER) and the spectral efficiency. However, this approach does not consider the throughput criterion. According to the simulation, the distance between the two vehicles is very small.

Rai and Jain in [13] designed a non-orthogonal access method (NOMA) resource allocation algorithm and compared it with the orthogonal multiple base (OMA) method. The proposed algorithm consists of combining Multiple Input Multiple Output (MIMO) schemes with downlink NOMA communication systems. The main objective of this approach is to reduce reliability and improve spectral efficiency. The limitation of this approach does not take into account the bit rate criterion and the bit error rate criterion. Moreover, according to the simulations, the number of V2X users is very limited.

Harshini et al. in [14] studied the concept of NOMA technology for the Spatio temporal Coding MIMO (STC-NOMA) technique. The proposed algorithm aims to improve the reliability of data transfer and to reduce the bit error rate. To reach this goal, the authors used the (STBC) method of transmitting multiple data streams over multiple antennas to exploit different versions of data. In addition, a comparison of the STBC-mono user system with a NOMA-STBC multi-user system is implemented. But, this approach does not consider the throughput criterion.

Xing et al. In [15], have designed a two-user downlink non-orthogonal multiple access technology (NOMA) allocation algorithms in attenuated channels with delay-tolerant data and compared it with another algorithm based on the orthogonal multiple access (OMA) technique. This approach aims to maximize spectral efficiency and to ensure better user equity. However, in this work throughput is not considered.

Table 1 provides a summary of the articles studied.

**Table 1.** Comparison between the algorithms examined

| | Throughput optimization | Error rate minimization | Latency | Fairness | Reliability | S E | S I C | Type of allocation |
|---|---|---|---|---|---|---|---|---|
| [5] | | X | X | | | | | OFDMA |
| [6] | | X | | | | | | SC-FDMA |
| [7] | | X | | | | X | | BDMA |
| [8] | | X | | | | | | SCMA |
| [9] | | | | | | X | | SCMA |
| [10] | | X | | | | | X | MUSA |
| [11] | | | X | | X | X | | NOMA |
| [12] | | X | | | X | X | | NOMA-SM |
| [13] | | | | | X | X | X | NOMA |
| [14] | | X | | | X | | X | NOMA |
| [15] | | | | X | | X | X | NOMA |

The variable parameters used for the system model are given in Table 2.

## 3 System model

In this part, we describe our NOMA-based V2X cell system model, as shown by Fig. 1. The V2X network consists of a single eNodeB and several N vehicles whose transmitters are indicated by Tx and receivers are represented by Rx. The distances between Tx and Rx are variable. Note that $V = \{1, 2, \ldots, N_{\text{Tx}}\}$ represents the set of Tx, while $E = \{1, 2, \ldots, N_{\text{Rx}}\}$ represents the set of Rx. $N_{\text{Tx}}$ and $N_{\text{Rx}}$ represent the number of Tx and Rx, respectively. We assume that the V2X are randomly arranged in the cell.

Each vehicle may send a data packet containing high priority information generally of the safety type around its vicinity. Neighboring users can transmit data in a direct way using D2D mode. The usable bandwidth is split into K sub-channels to ensure data transportation.

**Table 2.** Notations

| Symbol | Description |
|---|---|
| $N$ | Number of V2X |
| $I$ | Set of V2X |
| $N_{\mathrm{Tx}}$ | Number of Tx |
| $V$ | Set of Tx |
| NRx | Number of Rx |
| $E$ | Set of Rx |
| M | Number of available RBs |
| $U$ | Set of RBs |
| $K$ | Number of available channels |
| $K_{\max}$ | Number of users that can share the same RB |
| $W$ | Bandwidth |
| $\eta$ | Spectral density of noise |
| $i$ | Number of time slot |
| $r$ | Distance threshold |
| $N_m = \{1 \leq j \leq N\, d_{j,m} \leq r\}$ | Indicates the set of users verifying the condition on the threshold distance $r$ and subsequently belonging to the receiver communication range Rx($m$) |
| $\gamma_{j,k}$ | It is a binary matrix to denote that the transmitter $j$ can share the sub-channel $k$ for the time slot ($i$) |
| $P_{j,k}^{(i)}$ | It is the transmission power of the transmitter Tx($j$) on the sub-channel $k$ for the time slot ($i$) |
| $H_{j,m,k}^{(i)}$ | Indicate the matrix of the channel formed by a set of sub-channels $K$ between the transmitter Tx($j$) and the receiver Rx($m$) in the time slot ($i$) |
| $s_{j,k}^{(i)}$ | Represents the transmitted symbol of the transmitter Tx($j$) for the time slot ($i$) |
| $\beta$ | Representing a constant defining the shadowing fading |
| $d_{j,m}^{(i)}$ | The distance between two users ($j$) and ($m$) for the time slot ($i$) |
| $\alpha$ | Representing a constant defining the pathloss exponent |

Each Rx receiver can decode the received signals using the SIC (Successive Interference Cancellation) technique. The signal received by Rx for the time slot $i$ on the channel $K$ can be modeled by the following equation:

$$y_{m,k}^{(i)} = \sum_{i \in Nm} \gamma_{j,k}^{(i)} \sqrt{P_{j,k}^{(i)}} H_{j,m,k}^{(i)} s_{j,k}^{(i)} + n_m^{(i)} \quad (1)$$

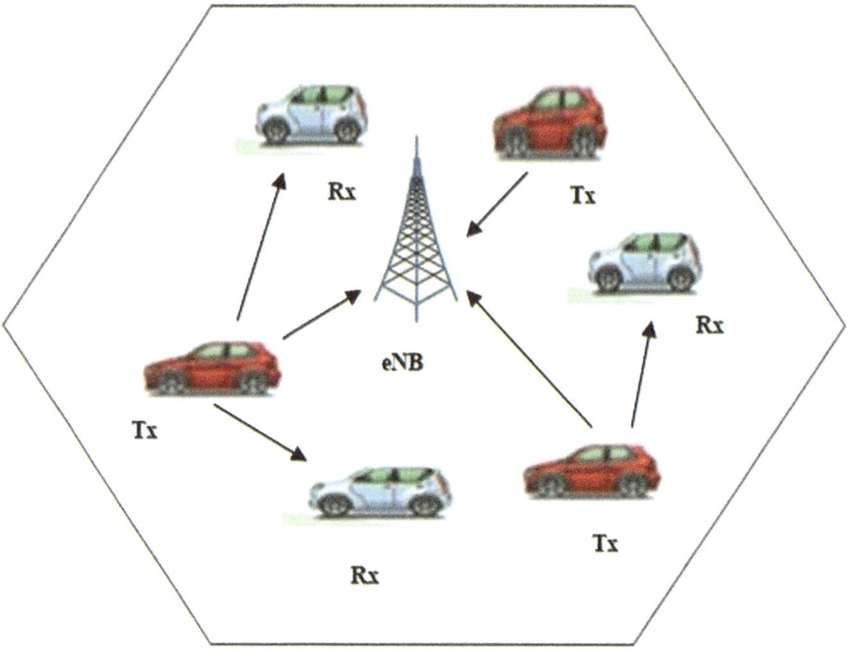


**Fig. 1.** System model

The value of time slot number $(i)$ is equal to one. Equation (2) represents the additive white Gaussian noise for a receiver Rx, where $\sigma_n^2$ is the noise variance

$$n_m^{(i)} \Delta CN(0, \sigma_n^2) \tag{2}$$

The coefficients of the channel $H_{j,m,k}^{(i)}$ can be defined by Eq. (3):

$$H_{j,m,k}^{(i)} = h_{j,m,k}^{(i)} g_{j,m}^{(i)} \tag{3}$$

With,
$h_{j,m,k}^{(i)}$: Complex Gaussian variable (0, 1) showing Rayleigh fading. The gain of the channel is given by Eq. (4):

$$g_{j,m}^{(i)} = \beta(d_{j,m}^{(i)})^{-\alpha} \tag{4}$$

The successive interference cancellation (SIC) is considered an indispensable technical for NOMA technology because it improves network capacity by eliminating inter-channel interference [16]. For NOMA, each receiver Rx($m$) decodes the received signals taking into account a decreasing order of the channel gain. Consider, for example, two transmitters Tx($j$) and Tx($j'$) occupying the sub-channel $k$ in the time slot $(i)$ transmitting to the same user Rx($m$) and gratifying the condition imposed by the Eq. (5):

$$\left|H_{j,m,k}^{(i)}\right|^2 < \left|H_{j',m,k}^{(i)}\right|^2 \tag{5}$$

The receiver Rx($m$) uses the SIC technique to decode and subtract the signal Tx($j$) before decoding the Tx($j'$).

Therefore, the user who has a better SINR (Signal-to-Interference-plus-Noise Ratio) must be decoded. Then subtract the signals of the users who have the lower SINRs. The SINR on the Rx receiver is calculated as follows:

$$\mathrm{SINR}_{(j,m,k)} = \frac{P_j^{(i)} \rho_{j,m,k}^{(i)}}{1 + \sum_{i' \int s_{j,m,k}^{(i)}} P_{j'}^{(i)} \rho_{j',m,k}^{(i)}} \tag{6}$$

With,

$$\rho_{j,m,k}^{(i)} = \left| H_{j,m,k}^{(i)} \right|^2 / \left( n_m^{(i)} \right)^2 \tag{7}$$

$\rho_{j,m,k}^{(i)}$: denotes the SINR of the sub-channel $k$ between the links Tx($j$) and Rx($m$).

$s_{j,m,k}^{(i)}$: represents the set of transmitters Tx($j$) generating interference when the receiver Rx($m$) decodes the signal of the transmitter Tx($j$) on sub-channel $k$.

Following the above computation of the SINR in Eq. (6), the data rate $R$ obtained of Rx($m$) for a time slot ($i$) on the sub-channel $k$ can be formulated by the Eq. (8).

$$R_{j,m,k}^{(i)} = W * (\log_2(1 + \mathrm{SINR}_{(j,m,k)})) \tag{8}$$

## 4 Proposed Algorithm

The main goal of our algorithm is to propose a new opportunistic and efficient scheduling algorithm.

Our scheduling algorithm for V2X communications based on NOMA access (SAVCN) consists of finding a compromise between different performance criteria. The objective of this approach is to maximize the throughput, increase the number of served users, ensure equity between different users, and minimize the bit error rate. The diagram of our proposed algorithm is based on three major steps as illustrated in Fig. 2.

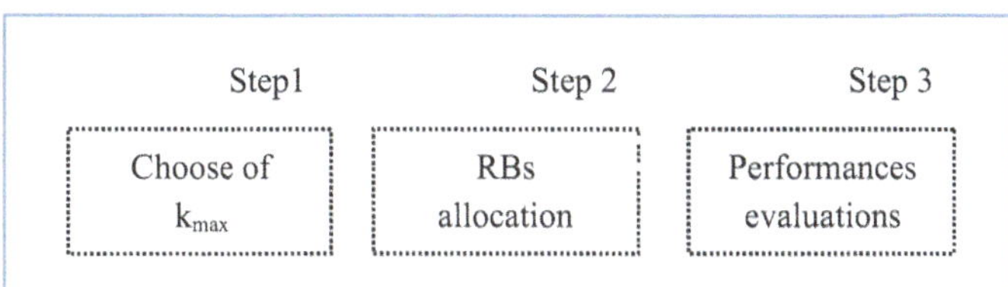


**Fig. 2.** Diagram of the proposed algorithm

For the first step, we try to find the minimum distances between the Tx transmitters and the Rx receivers in the proposed system in order to choose the set of Tx that can share the same RB($k_{\max}$). For this, we insisted on imposing a constraint on the distance between transmitters and receivers. Indeed, the transmitters authorized to send messages to the receivers are those which admit a distance less than or equal to a distance threshold. Of all the transmitters satisfying this condition, we choose only the two possessing the minimum distance as a first approach.

Then, the second step of assigning RB to all selected Tx($k_{max}$) is started. In this context, the principal concept is to allocate resources block when taking into account the minimum distance between the Tx and the Rx, as well as the quality of the Tx channel on each RB expressed by the Signal-to-Interference-plus-Noise Ratio (SINR) value. Indeed, if the scheduler has already assigned a RB to the current time slot ($i$) at the set Tx($k_{max}$), it will have no more RB in the same ($i$) only if all are served. This maximizes the capacity of the system in the cell and ensures better equity between the different users.

The goal of the last step is to evaluate the performance of our algorithm, we validated the results by simulations taking into account several parameters such as the number of RBs, the number of users able to share the same resource block, the number of transmitters and the number of receivers. The performance of the proposed approach was compared with a resource allocation based on the convention OMA algorithm.

The concept of our algorithm is to allocate sub-channels for a set of transmitters. For this, we created a matrix containing the distances between transmitters Tx and receivers Rx. Then, we insisted on putting a constraint on the distance between transmitters and receivers. In fact, the transmitters authorized to send messages to the receivers are those which admit a distance less or equal than a threshold one. The process of allocating resources if only two users can share to same RB is explained by Algorithm 1 (Table 3).

**Table 3.** Parametier setting

| Symbol | Description |
|---|---|
| M1 | Matrix contains distances between Tx and Rx randomly |
| M2 | Matrix contains users satisfying the condition |
| M3 | Matrix contains only users who can share the same RB |

**Algorithm 1**

```
{Inputs}
NTx = x, NRx = y % Number of Tx and Rx
r % distance threshold
Kmax % Number of users that can share the same RB.
{main}
% Choose Kmax
Create matrix M1(x,n) =randi [x',n'] % Matrix contains distances between Tx and Rx
For x=1 : NTx For n=1 :
NRx If M1(x,n) < r
Find (x,n)= M1(x,n) < r % find for distances that are less than the threshold distance
Calculate M2(x,n)=f (M1(x,n)) % matrix contains only users satisfying the desired condition
End if End End For h=1 : Nrx
Tri dist M2 (x,n) % sort users in ascending order
End
// selecting the (Tx,Rx) pair with the min-dist in M2.
// according to value of kmax, selecting the Tx which can be sharing the some RB in M2.
End
```

The main purpose of assigning BRs in our proposed algorithm is to share the same RB with a predefined set of $k_{\max}$ users according to a well-defined criterion. The idea is to find the combination ($\text{Tx}_{i'}$, $\text{RB}_j$) with $1 \leq i' \leq \text{NTx}$ and $1 \leq j \leq M$, with the smallest value of the metric $m_{i',j}$ given by the equation below:

$$m_{i',j} = \min\{M2\} \tag{9}$$

The algorithm below presents the process of allocating radio resources for $k_{\max}$

**Algorithm 2**

**{Inputs}**
M2: Matrix contains users satisfying the imposed condition M3:Matrix contains only users who can share the same RB $K_{max}$ : Number of users that can share the same RB.
**{Main}**
**% Allocation of RBs when $K_{max}$ = 2**
***For*** *k=1 to M do*
*Find (x1, n1) = min M2(x1,n1)*, with (x1,n1) ∈ V x U
% Find the minimum of the n$^{th}$ column (on the same RBn) of the matrix
*Find Tx1 = min(M2)* % find the first user who has a minimum distance in M2
*Copy Tx1 selected in M3*
*Remove the x1$^{th}$ row from M2 V=V \ {x1} % Update V*
*Find Tx2 = min (M2)* % find the second user who has the minimum distance in M2
*Copy Tx2selected in M3*
*Remove the x2 $^{th}$ row from M2 V=V \ {x2} % Update V Assign RB$_k$ à Tx1 and Tx2 U=U \ {n1} % Update U*
**End for**

The performance evaluation of our proposed algorithm is a very important part of our work. This step is explained in detail in the following Sect. 5.

## 5 Simulation Results

The results are generated by evaluating our scheduling on Matlab.

We modify several parameters in the system to evaluate system performance. We consider that the distances between transmitters and receivers are variable. We, therefore, compared our algorithm with another algorithm based on the conventional OMA method [17]. Table 4 presents the parameters used for the simulations.

### 5.1 Throughput Analysis

Throughput growth is one of the main objectives of our project. Indeed, it is defined as the amount of information transmitted on the radio interface in a given time interval [18].

We study the behavior of the throughput according to the number of resources blocks RBs and the number of users who can share the same channel simultaneously $K_{\max}$.

**Throughput depending on the number of RB**

We have studied this section, the throughput behavior as a function of the number of channels.

Figure 3 indicates the performance of SAVCN and the traditional OMA scheduler in sight of throughput depending on the number of resources blocks, with the number of transmitter Tx = 50 and $k_{\max} = 2$.

**Table 4.** Parameters of simulations

| Parameter | Value |
|---|---|
| Bandwidth | 10 MHz |
| Nb of RBs | [30, 50] |
| Nb of Tx | [50, 70] |
| Nb of Rx | [30, 50] |
| Nb of eNodeB | 1 |
| Nb of $K_{max}$ | [2, 5] |
| Duration of $t$ | 1 μs |
| Power of each Tx user on each $K$ sub-channel | [20, 23 db] |
| $r$ | 150 m |

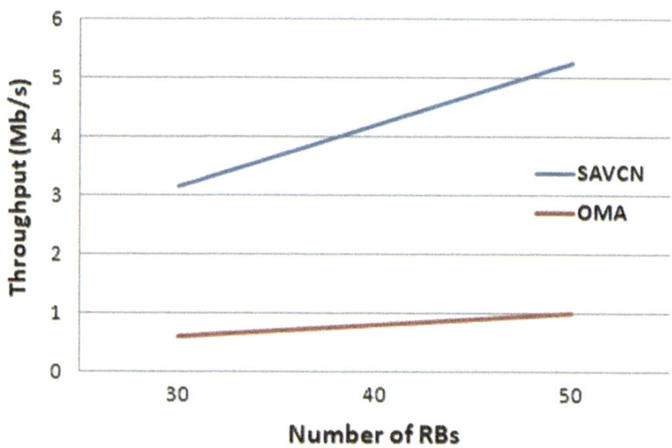


**Fig. 3.** Throughput analysis

When the number of resources blocks increases, the throughput increases. The results indicate that SAVCN reaches an important value compared to the OMA scheduler. Note that the performance of this system is better than the conventional OMA, especially in terms of throughput. Our algorithm is capable to sustain several sizes of messages compliant with the 3GPP standard. Corresponding to [19, 20], the transferred messages possess a size of 50–400 bytes. The results of the simulation demonstrate by Fig. 3 indicates that the rate gate may support inconstant message sizes. Indeed, the proposed algorithm authorizes a message swap of 657.5 bytes when the bit rate is 5.26 Mbps. The results gratify the 3GPP recommendation when the minimum message size is 50 bytes.

**Throughput depending on the number of $K_{max}$**
We study the behavior of the throughput depending on the number of users dividing the identical sub-channel.

Figure 4 shows the throughput depending on the number of users sharing the same sub-channel, with Tx = 70 and Rx = 30.

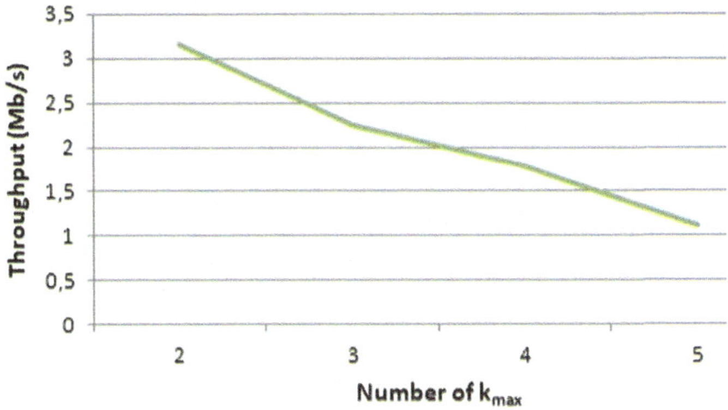


**Fig. 4.** Throughput depending the number of $k_{max}$

When the number of users sharing the same RBs increases, the throughput decreases, because the interference between the users increases, so a complexity at the receiver in the SIC decoding will occur and errors on the detection of the receiver side will be added. So the SINR decreases so the throughput will decrease.

### 5.2 Bit Error Rate

The causes of error are numerous and depend mainly on the nature of the used transmission media.

When propagating signal can be deformed by the mobile radio channel. Therefore, the binary element can be received with a different value from the originally one sent, the entire associated message will be declared as erroneous. Indeed, the error rate depends on several factors such as the number of transmitters, the number of receivers, the number of users sharing the same sub-channel, and the threshold distance.

**BER depending on the number of transmitters and on the receivers**

In this section, we studied the behavior of BER according to the number of transmitters and on the receivers.

Figure 5 presents the variation of the BER depending of the number of Tx, with the number of receivers NRx = 30, M = 50, and $k_{max} = 2$.

With increasing the number of Tx, the BER grows. If we take the case where the V2X network is such that Tx = 70. All transmitters send messages to receivers on different sub-channels, so the probability of error and collision increases.

The number of Rx still influences the BER. Indeed, we have to study the behavior of the BER according to the number of receivers.

Figure 6 represents the variation of the BER as a function of the number of Rx with Tx = 50 and $k_{max} = 2$.

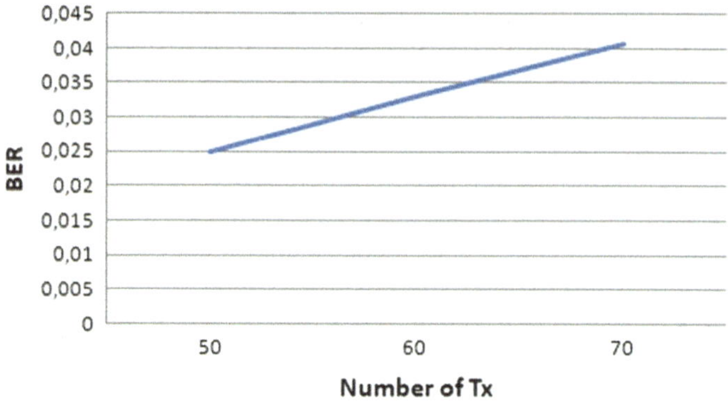


**Fig. 5.** BER analysis

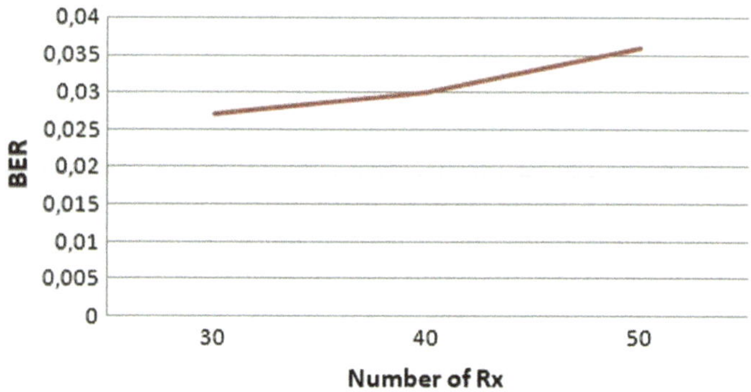


**Fig. 6.** BER depending on the number of Rx

When the number of receivers Rx increases, BER also grows, because the interference also increases, thus a complexity in the decoding will occur SIC, and errors at the detection on the receiving side will be added.

**BER according to $K_{max}$**

In addition, the number of users sharing the same sub-channel ($K_{max}$) influences BER.

Figure 7 shows the variation of the BER based on the number of users sharing the same channel with Tx = 50 and Rx = 30.

When the number of users sharing the same channel increases, the BER grows. As the amount of information elevates, the risk of interference between users also increases, so that the BER increases.

**BER according to the distance threshold**

We studied the behavior of the BER according to the value of $r$.

Figure 8 shows the variation of the BER as a function of the distance threshold between transmitters and receivers with Tx = 60 and $k_{max} = 2$.

We note that as the distance threshold r increases, the error rate also increases. This is normal because the channel disruption becomes stronger, messages sent by a transmitter

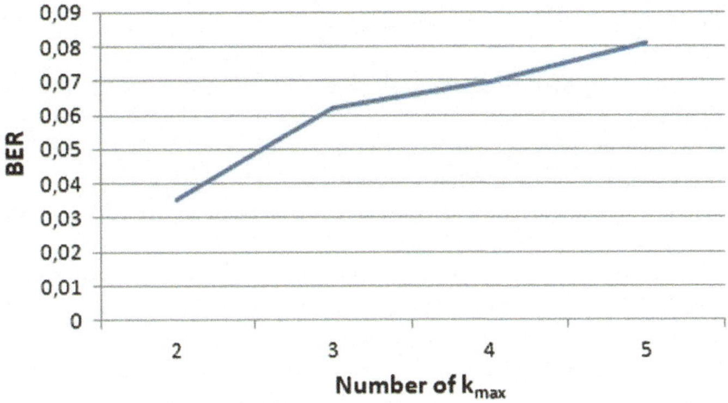


**Fig. 7.** BER depending on the number of $k_{\max}$

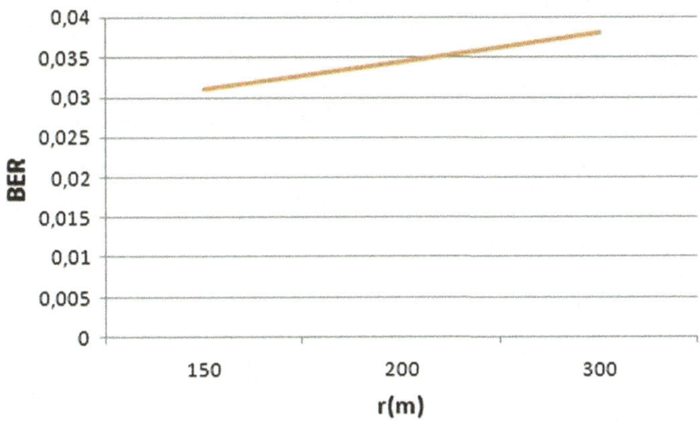


**Fig. 8.** BER depending on the $r$

to a receiver may have significant changes, so errors will appear at the reception. For this, the error rate grows.

## 6 Conclusion

In this paper, we present a new scheduling scheme called SAVCN for 5G networks. This algorithm is based on the optimal allocation of resources block in a non-orthogonal way. Our goal is to maximize the throughput of the system, improve the number of served users, ensure better equity when allocating RBs and minimize the bit error rate. The simulation result shows that the system performance of the proposed strategy is better than the conventional OMA in terms of throughput. In future, we expect to make multitude of enhancements to the SAVCN algorithm, such as adding traffic classifications supported by a V2X network and integrating the power control mechanism and an interference cancellation mechanism.

## References

1. Hegde, A, Festag, A.: Mode switching strategies in cellular-V2X. Int. Federation Autom. Control (2019)
2. Mahmoud, A., Mesut, T., Selahattin, G.: A tutorial on non orthogonal multiple access for 5G and beyond. Wireless Commun. Mob. Comput. (2018)
3. Liu, B., Peng, M.: Joint Resource Block-Power Allocation for NOMA-Enabled Fog Radio Access Networks. arXiv:1903.05808v1[eess SP] (2019)
4. Tran, D.-D., Trany, H.-V., Ha, D.-B., Kaddoum, G.: Cooperation in NOMA networks under limited user-to-user communications: solution and analysis. In: IEEE Wireless Communications and Networking Conference, April 2018
5. Situ, Z., Wang-hei Ho, I., Wang, T., Liew, S.C., Chau, S.C: OFDM Modulated PNC in V2X Communications: An ICI-Aware Approach Against CFOs and Time-Frequency-Selective Channels. IEEE Access (2018)
6. Trivedi, V.K., Kumar, P.: FRFT-SCFDMA scheme for uplink in 5G radio access networks. In: IEEE International Conference on Communications Workshops (ICC Workshops) (2017)
7. Dalela, P.K., Bhave, P., Yadav, P., Yadav, A., Tyagi, V.: Beam Division Multiple Access (BDMA) and modulation formats for 5G: Heir of OFDM? In: International Conference on Information Networking (ICOIN) (2018)
8. Kulhandjian, M., D'Amours, C.: Design of permutation-based sparse code multiple access system. In: IEEE 28th Annual International Symposium on Personal, Indoor, and Mobile Radio Communications (PIMRC) (2017)
9. Liu, L., Sheng, M., Liu, J., Wang, X., Li, J.: Success probability and area spectral efficiency in SCMA wireless networks. IEEE Trans. Veh. Technol. (2018)
10. Eid, E.M., Fouda, M.M., Eldien, A.S.T., Tantawy, M.M.: Performance analysis of MUSA with different spreading codes using ordered SIC methods. In: 12th International Conference on Computer Engineering and Systems (ICCES) (2017)
11. Boya, D., Song, L., Li, Y., Li, G.Y.: NOMA-based low-latency and high-reliable broadcast communications for 5G V2X Services. In: IEEE Global Communications Conference (2017)
12. Chen, Y., Wang, L. et al.: Performance analysis of NOMA-SM in vehicle-to-vehicle massive MIMO channels. IEEE J. Sel. Areas Commun. (2017)
13. Rai, S., Jain, A.: Result analysis of efficient MIMO framework for NOMA downlink and uplink transmission based on signal alignment. Int. J. Comput. Appl. (2018)
14. Harshini, N.M., Kavitha, K.: Performance analysis of STBC-NOMA. Int. J. Pure Appl. Math. (2018)
15. Xing, H., Liu, Y., Nallanathan, A., Ding, Z.: Sum-rate maximization guaranteeing user fairness for NOMA in fading channels. In: IEEE Wireless Communications and Networking Conference (WCNC) (2018)
16. Zhang, N., Ding, T., Kang, G.: Interference balance power control for uplink non-orthogonal multiple access. IEEE Commun. Lett. (2019)
17. Trabelsi, A.D., Marouane, H., Zarai, F., Meddeb-Makhlouf, A.: Dynamic scheduling algorithm based on priority assignment for LTE-V2X vehicular Networks. In: IEEE International Wireless Communications & Mobile Computing Conference (IWCMC) (2019)
18. Nasim, F., Mohamed, O. et al.: Throughput-aware resource allocation for QoS classes in LTE networks. In: International Conference on Computer Science and Computational Intelligence (ICCSCI) (2015)
19. GPP TR 22.885 V14.0.0 Release 14, Technical Specification Group Services and System Aspects; Study on LTE support for Vehicle to Everything (V2X) services (2015)
20. GPP TR 22.186 V15.0.0, Release 15, Service requirements for enhanced V2X scenarios (2017)

# Crime Intelligence from Social Media Using CISMO

Ogerta Elezaj[1](✉), Sule Yildirim Yayilgan[1], Javed Ahmed[1], Edlira Kalemi[2], Brumle Brichfeld[1], and Claudia Haubold[1]

[1] Department of Information Security and Communication Technology, Norwegian University of Science and Technology (NTNU), Trondheim, Norway
{ogerta.elezaj,sule.yildirim,javed.ahmed}@ntnu.no,
{brumle.brichfeldt,caludia.haubold}@stud.ntnu.no
[2] University of Tirana, Tirana, Albania
edlira.kalemi@unit.edu.al

**Abstract.** Nowadays, online social networks (OSNs) are being used as a hosting ground for criminal activities, and the legal enforcement agencies (LEAs) are struggling to process and analyse the huge amount of data coming from these sources. OSNs generate a huge massive volume of unstructured data making it difficult for the LEAs to 'patrol the facts' and to gather intelligence in order to provide it to the legal domain. There is no ontology model, among those found in literature, that allows to exhaustively describe all the aspects of crime investigation targeting data integration, information sharing, collection and preservation of digital evidences by using biometric features, and query answering. To bridge this gap, this paper presents an extended version of our earlier SMONT ontology, called CISMO as a semantic tool suitable for gathering digital evidence from OSNs helping LEAs to develop new investigative systems to counter the threat of different crimes. The new version introduces the core concepts related to crime cases in the police repositories, biometric data and digital evidences collected by OSNs, making it possible for LEAs to classify crimes, investigate hidden crime patterns or predict future crime patterns. CISMO is more concise and has a richer concept knowledge-based compared with the previous version SMONT. We prove the effectiveness of CISMO in a case study covering some general aspects in criminal cases in OSNs, demonstrating how this semantic approach can help LEAs to gather knowledge for crime investigation using natural language processing and machine learning to process messages shared in an online platform and also applying reasoning rules, as semantic inferences.

**Keywords:** Ontology · Crime · Online social networks · Digital evidence · Biometrics · Security · Reasoning


X.-S. Yang et al. (eds.), *Proceedings of Fifth International Congress on Information and Communication Technology*, Advances in Intelligent Systems and Computing 1183, https://doi.org/10.1007/978-981-15-5856-6_44

## 1 Introduction

OSNs have changed the way how people communicate and connect between them. Social media such as Facebook, Twitter, Instagram and Google+ are being used by millions of users every day, and the data which are freely shared in these networks are like a treasure if it is properly processed, and the knowledge is extracted. The data generated in OSNs contain exabytes of information related to the people's day-to-day activities and stand as an important source for big data [1]. What matters most is not the data itself but rather the information and knowledge that can be extracted in order to use it in decision-making in different domains.

Despite the advantages of OSNs in allowing people to stay connected, there is a darker side to these networks, as criminal activities being committed in these platforms are becoming a central problem in every country. In 2014, the FBI's Internet Crime Complaint Center reported that 12% of all logged complaints involved social media equates to 32,330 complaints received during a year.[1] INTERPOL used social media platforms to find out potential witnesses in different terrorism acts as was the case of the London Bridge attack in the UK in 2017.[2] More than 4.7 million counterfeit products were seized in an operation against trafficking of illegal goods, which was carried out by LEAs of 18 countries in collaboration with Europol. In this operation, 16,470 social media accounts and 3400 web sites selling counterfeit products were closed.[3] Of major concern to investigators is the fact that social media facilitates the attraction and the recruitment of new members in extremist groups, becoming a topic of major concern for many legal agencies [2]. It is noted that half of the teenagers have experienced cyberbullying on OSNs, resulting in low self-esteem and consideration of suicide. Moreover, only 1 out of 10 teenagers tells a parent if they have been a victim, which demonstrates that cyberbullying crimes are not being reported, and thus unpunished.[4]

Different types of digital crimes can be collected by OSNs, coming in different forms such as messages, photographs, videos, audios and local-based data. In the crime investigation processes, LEAs have to collect and provide reliable and authenticated evidences to ensure their admissibility in the court. OSNs are one of the key driving factors for the evolution of crime arenas bringing new opportunities and challenges to legal agencies to investigate threats from individuals who may or may not be members of groups. Various data generated by users, known as user-generated content (UGC), can be used to investigate committed crimes or to predict future crime patterns [3]. Digital evidences which contain social media contents have been accepted by different courts to identify suspects, locate witnesses and convict defendants [4].

[1] https://pdf.ic3.gov/2014_IC3Report.pdf.

[2] https://www.interpol.int/en/Crimes/Terrorism/Analysing-social-media.

[3] https://www.europol.europa.eu/newsroom/news/counterfeit-crackdown-hits-two-organised-criminal-groups-more-30-suspects-arrested.

[4] http://www.bullyingstatistics.org/content/cyber-bullying-statistics.html.

However, the characteristics of OSNs data render the existing solutions insufficient to consider the new challenges of LEAs to handle digital evidences in crime investigation and prevention [5]. The exponential growth in the volume, velocity and variability of OSNs data prevents LEAs to efficiently process and manage the large criminal data sets using traditional methods [6]. Due to the heterogeneity, noise and the massive size of unstructured data generated in social media platforms, LEAs have to take real efforts to face the challenges of collection, processing and analyzing the digital evidences in a timely and efficient manner implementing comprehensive solutions.

A well-defined and standardized representation of OSNs data could be achieved using Semantic Web technologies supporting LEAs in structuring and better integrating the crime records and to model formal knowledge in the crime domain. Semantic Web technologies combine a unique addressing mechanism (Uniform Resource Identifiers: URI) with a formal knowledge representation (RDF and OWL) and a common query language (SPARQL). In this research, we use the ontology model as one of the major concepts used in Semantic Web applications, to represent a set of concepts and their relationships within a domain into a machine-made form.

CISMO ontology is used to model the environment of different categories of crime happing in OSNs. With the help of this ontology, data can form an interconnected knowledge base of the different evidence objects extracted from social media [7]. The evidence can also be merged with the reports and evidence streams that exist in LEAs repositories to identify new and implicit knowledge using inference engines. To the best of our knowledge, there is not a generic ontology in the crime domain for OSNs to use it as a knowledge-based tool for data mining applications.

Despite this importance, the existing forensics analyses tools are currently limited to face the identified challenges and very few semantic solutions have been developed to help investigators to cope with new technologies [8]. For this reason, in this paper the existing ontologies used for crime investigation through OSNs content are examined with the objective of better understanding the challenges and gaps unique to crime investigation from OSNs and to provide methods for addressing those challenges and the gaps by developing intelligent systems to sift through massive amounts of online information and to extracts what's useful to the investigators. To the best of our knowledge, there is no ontology model, among those found in the literature that covers each aspect of crime investigation such as data integration, digital evidences, biometric features and data coming from OSNs. The existing ontologies are not generalized and mostly are platform-based, capable to deal with data coming only from particular social media. The main motivation behind the study proposed in this paper is to develop an ontology; as one of the main components of a knowledge-based graph framework introduced by authors to gather intelligence from OSNs in order to assist LEAs to detect and prevent criminal activities. CISMO ontology is based on linked concepts like agents composed of persons or organized criminal networks, institutions involved in crime investigation, digital evidence collection processes and biometric modalities. CISMO is important, firstly, to model relationships among user activities in OSNs and to identify suspects related to different crime categories such as ordinary crime and cybercrime. Secondly, CISMO is conceptualized as the main backbone of an intelligent knowledge-based system used in crime detection and prevention. This framework has the capacity to support

LEAs in crime investigation activities, starting from data collection to the preservation of digital evidence admissible in the court. One of the components of this framework is the ontology used to model and structure the social media content in a well-organized structured way. Once all online activities are stored by considering this structure and integrating them with the data coming from LEAs repositories, investigators can though inference rules and reasoning, infer accurate knowledge related to the crime domain. In order to illustrate on how CISMO can be used as a tool in crime investigation, a case study based on a recently leaked data set collected from Nulled.io,[5] an online forum for distributing cracked software, and trade of leaked and stolen credentials, is presented. Specifically, we first manually classify a subset of the private messages, and we train a machine learning model to classify messages belonging to criminal activities or not. All the classified messages and their related data such as IP of the sender, IP of the receiver and timestamp populate the ontology, and we provide a way for semantically querying the crime ontology.

The remaining part of the paper is organized as follows. Section 2 presents related work on using semantic models and technologies for crime investigation and prevention of OSNs. The methodology used to develop CISMO and its main components is introduced in Sect. 3. Section 4 describes some crime scenarios that can be solved using our ontology. Finally, the conclusions and future work are given in Sect. 5.

## 2 Introduction

In this section, critically we analyze current literature of knowledge preservation in the crime domain. Based on the metanalysis and literature review, a summary of existing ontologies used in the crime domain is provided. The perspective of this review is to find the gaps in the existing semantic solutions in the crime domain. During the review of existing crime ontologies, it has been found out that the existing ontologies in the domain of crime investigation and prevention are sub-domains of various elements that define the crime domain. As the proposed ontology aims to mostly cover all the aspects of the crime investigation in OSNS, including digital evidences by the usage of biometrics features, during this research, we have analyzed different ontologies capable to handle social media data, to collect and represent biometric data and to maintain defensibly the chain of custody of digital evidences. Based on the depth analyses of the existing ontologies, we have considered some re-use of some existing concepts by providing smooth access to these ontologies and at a second stage based on the identified gaps, we extend the existing ontology developed by authors, SMONT [9] by adding new concepts and by providing advanced support in adapting ontologies to crime domain. The existing ontology, SMONT presented by author in previous research, lacks the necessary level of details about the collection and the preservation of digital evidences. The dilemma that we faced was whether to start developing a new ontology from scratch or to examine existing ontologies used to model social media in the crime domain and check if one of them fitted our purposes as it is or in an extended version.

[5] https://archive.org/details/nulled.io_database_dump_06052016.

### 2.1 Ontologies in Social Media

In the literature, few ontologies are proposed for semantic social web, such as Friend Of A Friend (FOAF) [10], Semantically Interlinked Online Communities (SIOC) [11] and Social Semantic Cloud of Tags (SCOT) [12], but none of them are detailed enough to be used for knowledge representation and data integration of crime activities in OSNs. Different researchers have extended these ontologies in order to add new concepts to express the interest domain. In [13], it is proposed an extended version of the FOAF ontology, which is evaluated by W3C Consortium as a good ontology capable to model persons and their relationships [14]. In this work, the FOAF ontology is enriched with new classes and properties related to profilers. Still, this ontology is not a unified semantic model for OSN to describe the content of multiple users from different OSNs such as Facebook, LinkedIn, and Twitter that means that it is platform-dependent and does not offer a generic semantic solution.

SIOC is an open standard ontology developed in 2004 aiming to represent social media content in RDF format and is especially designed for modelling user forums [15], missing many of the OSNs concepts. In 2017, authors in [16] enriched this ontology by presenting a new version exSIOCint, adding new classes and relationships in order to model data coming from web forums and to enhance automatic inherences. However, this is a general-purpose ontology focused on modelling web forums and it is not suitable for the intended problem that CISMO is going to solve, assisting LEAs in daily fighting crimes happening in OSNs. Different ontologies are designed and implemented by different authors to semantically represent the knowledge of OSNs (SocIoS) [17], LODE [18]), but all of them are general-purpose ontologies and their focus is not in modelling formal knowledge for social media forensic.

### 2.2 Ontologies in Forensic Analyses

With the widespread of using ontologies as means for knowledge representation, different ontologies have been proposed and applied in the criminal and legal domain but none of them has a focus on analysing crimes in OSNs. However, we have analyses different semantic solutions in the legal domain in order to get a better understanding of the domain concepts in order to apply them in CISMO.

Osathitporn et al. [19] present an ontology for the criminal legal domain in Thailand. The objective of this ontology is to represent legal elements in the law domain, and its main artefacts are crime and justification. However, the objective of this ontology is totally different from CISMO as it is not used to collect digital evidences from OSNs. Kastrati in [20] presented SEMCON, a criminal ontology, developed to process semantically and contextually Facebook data of different users and to classify these users as suspects or not. This ontology is platform-dependent as it is designed only for Facebook, and its scope is not to model the crime investigation process but only to build up probabilistic predictive models for suspects. The digital evidence collection and preservation are out of the scope of this ontology.

A multilayer semantic framework used to detect crime on OSNs is presented in [8]. This framework is a hybrid solution, and its main component is a global ontology derived by mapping different local ontologies for different OSNs. In this paper, only some parts of these ontologies are presented, and the global on-otology lacks the required level of details of digital evidence gathering. Furthermore, the lack of integration of biometric modalities in digital evidence preservation becomes a serious problem in the crime investigation.

In [21], authors presented a top-level cyber forensics ontology, and its main high-level classes are crime case, criminal, crime type and evidence. This semantic solution does not make an effort to gather digital evidences, but only it suffices with keeping trace of the medium of the digital evidence. Cosic et al. [22, 23] presented DEMF, an ontology to represent semantically the digital chain of custody of digital evidences. The developed ontology is a general solution for forensic investigation, and its scope is not related to the social media. The legal concepts used to represent digital evidences in this ontology are considered in our proposal, serving to us as a method to expand on our ontology related to the digital evidences in OSNs. We have used some of the artefacts such as digital evidence integrity methods, chain of custody based on the possibilities to answer to the six interrogatives of police report writing, known in the literature as 5W+H investigative model (Who, What, When, Where, Why and How) [24]. Moreover, considering the crucial fact that collecting digital evidences from OSNs is a complex task different from classical forensics, new artefacts are added to the digital evidence class of SMONT.

### 2.3 Biometric Ontologies on Crime Investigation

Concentrating on crime investigation in OSNs is very crucial to process multimedia content and to extract biometric features that carry personal information linked with the person's biometric characteristics for all the suspects of a criminal case. CISMO will be used to process the photographs shared in a social media platform, using a facial recognition biometric system to extract biometric characteristics. As biometrics is considered a strong alternative from crime detection, an automatic system should be in place to identify a person on the basis of his physical characteristics such as finger, face, iris and ears. Unfortunately, in the literature there do not exist many attempts focusing on developing biometric ontologies. Authors in [25] proposed a conceptual framework with the core element of human factors ontology for cyber-security, based on socio-cognitive characteristics. In the proposed ontology, there is no evidence of usage of biometric concepts.

The only biometric ontology found is presented in [26], where the authors have developed a biometric ontology and implemented it in a big data environment. This ontology covers a broad range of biometric aspects related to behaviour analyses such as cognitive skills or identification of tacit psychological factors. The ontology is evaluated using the asylum seekers and immigrant identification as a real use case. Some concepts of this ontology can be reused to model biometric characteristics of persons based on the multimedia content shared in that persons' social media profile, as our biometric aspects are not related to behaviour analyses.

From this review, we conclude that no ontology has been developed so far as a complete ontology in order to model all the components of the crime domain capable to gather intelligence from online social networks. Leveraging from existing ontologies (e.g. FOAF, DEMF, SIOC), the objective is to identify the gaps in current semantic solutions and to propose a more generic solution to overcome the identified challenges effectively by representing semantically the crime domain in OSNs.

This research has the following contributions:

1. We proposed a semantical data model for the investigation of crimes in OSNs covering the main aspects of the criminal investigation process such as digital evidence preservation including biometric artefacts.
2. We implemented the data model in an ontology in OWL using Protégé 5.5.0 [27].
3. We populated CISMO ontology with instances from a hacker forum database. The ontology contains 200 classes, 54 object properties and 18 data-type properties.
4. We semantically query the ontology to find suspects and digital evidences for different crimes using the data coming from an online forum. The resulting ontology can be extended for crime prediction and prevention.

In order to achieve our goal, in Sect. 3, we present CISMO; as a unique ontology in the way, it merges important aspects of crime investigation process in OSNs as criminal profile, crime categories, social media content, digital evidence gathering and biometric modalities, enabling a task-driven ontology-developing process.

# 3 CISMO Ontology in OWL

## 3.1 Methodology

Figure 1 presents the CISMO methodology composed by six main steps: (i) domain specification; (ii) consideration of reusing existing artefacts; (iii) conceptualization of key concepts; (iv) implementation in OWL; (v) ontology population; and (vi) ontology evaluation.

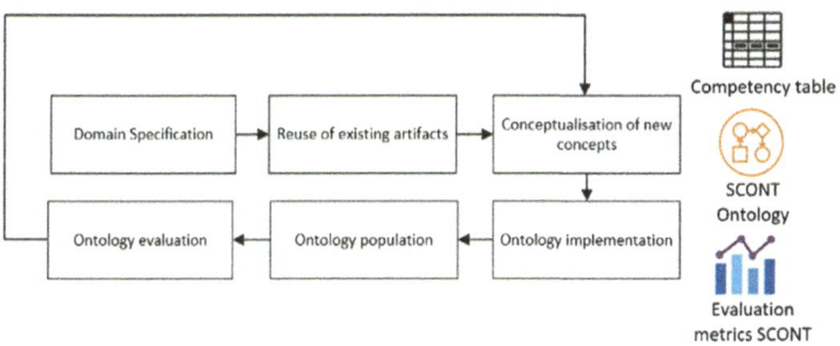


**Fig. 1.** CISMO ontology building methodology

During the first phase, knowledge about crime domain in OSNs is gained, whereas in the conceptualization phase, the main classes and subclasses are defined. Based on the recent literature, the crime domain is analyzed to obtain the required knowledge in order to build CISMO capable to handle digital evidences. Based on the ontology engineering, an integration strategy was applied to import existing knowledge from those fields where concepts are stable and to fit our main goal. The existing ontology, FOAF has been used as a basic to model OSNs artefacts, and it is extended with the new concepts. The output obtained by the first step is the competency table containing competency questions which play an important role as it is considered vital to outline and to constrain the scope of knowledge represented by the ontology [28]. Furthermore, the translation of competency questions into SPARQL queries is used to evaluate the ontology including verification and validation [29].

Specifying competency questions in the specification phase is vital since it allows us to determine the ontology scope. A subset of the CISMO competency questions that cover the main concepts of the crime domain in OSNs is shown in Table 1. Moreover, after identifying the goals of CISMO ontology, concepts and the specific classes to the crime investigation in OSNs domain are defined. From the three main methods found in the literature to construct ontologies, Top-Down, Middle-Out and Bottom-Up [30], we have chosen the Top-Down approach to generate this ontology. In a later stage, the crime domain ontology is developed using an OWL editor tool developed by Stanford University, named Protégé [21], an open-source and free software. At the end, the solution is tested and evaluated using real data of a hacker forum.

**Table 1.** Sample of CISMO competency question

| | Competency questions |
|---|---|
| CQ1 | What crime categories exist in OSNs? |
| CQ2 | Under what conditions should a person be considered as a suspect? |
| CQ3 | Do different profiles in different OSNs belong to the same person? |
| CQ4 | What biometric features are extracted by facial recognition system accessing photographs from social media sites for crime investigation? |
| CQ5 | Can a person of OSNs be considered as a suspect of an *online crime* based on online communication with a victim? |
| CQ6 | Can a person be considered as a suspect of an onsite crime based on the geo-content he shared in a social media? |
| CQ7 | Can a crime be *prevented* based on the person's activities in social media such as following criminal profiles, content of comments or statuses shared and likes to suspect pages or persons? |
| CQ8 | Can digital evidence be collected by using social media artefacts? |
| CQ9 | What are the elements of digital evidences to be admitted in courts? |
| CQ10 | Can the chain of custody of digital evidence be maintained? |

In our proposal, there exists only one global ontology that describing the semantics of each OSNs sources, providing a generic solution for LEAs. Additionally, if new data sources are available or it happens that there are done some changes in the existing data sources, the ontology can be adjusted easily.

## 3.2 Extending SMONT to CISMO

In this section, we present our ontology, developed to store and to arrange all the components of crime evidences in OSNs, providing a common language and a foundation for reasoning. To increase the value of the proposed solution and to generalize its scope, we extended SMONT ontology to a more specific version. The new ontology, named CISMO, is a more detailed ontology aiming at supporting LEAs to collect digital evidences and to increase their integrity keeping their chain of custody to stand up in the court.

Initially, the top hierarchy classes of SMONT included the following classes: person, crimes and crime case Solving. The main extending work lies in the digital evidence collection and keeping its chain of custody in order to prove it in the court. The CISMO ontology generically models key entities relevant for crime investigation and prevention using OSNs and the relations between them. As in Fig. 1, the current ontology, CISMO covers six main concepts, two of which are general concepts like institution and agent, whereas the other four concepts are crime-related concepts (Fig. 2).

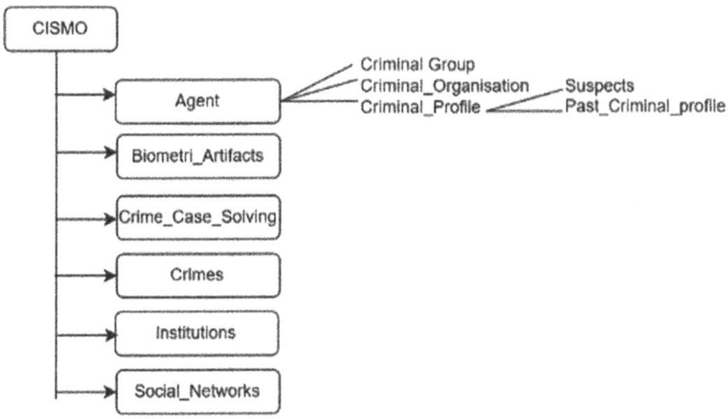


**Fig. 2.** Top classes of CISMO

The agent and institution concepts define the individuals and institutions that benefit from or are subject to the crime case investigation process. The agent class is a superclass of the concepts of person, Criminal_Group, Criminal_Organization and Criminal_Profile. Persons are individual agents; an organization is a group of other organizations or persons which acts 'as one' and a criminal group is a group of persons involved in organized crime. A specific individual involved or suspected for a crime is an instance of the class Criminal_Profile, which has two sub-classes. The sub-class Past_Criminal_Profile stores instances related to previous criminal cases stored in LEA repositories, which might be useful during criminal investigation for checking and linkage of different profiles. Persons, who are declared by the police as main suspects for an open case or individuals that have been considered as suspects after the suspect behaviour on the social networking, will belong to the suspects sub-class. The following fifth subsections briefly describe the CISMO concepts.

**Social Networks**

The class social networks represents information about persons whose data are collected from online social networks providers (classes: Facebook, Google, Instagram, etc.) and from their professional profile. As Facebook is the most popular social networking, in Fig. 3, it illustrates all the sub-classes of the Facebook class.

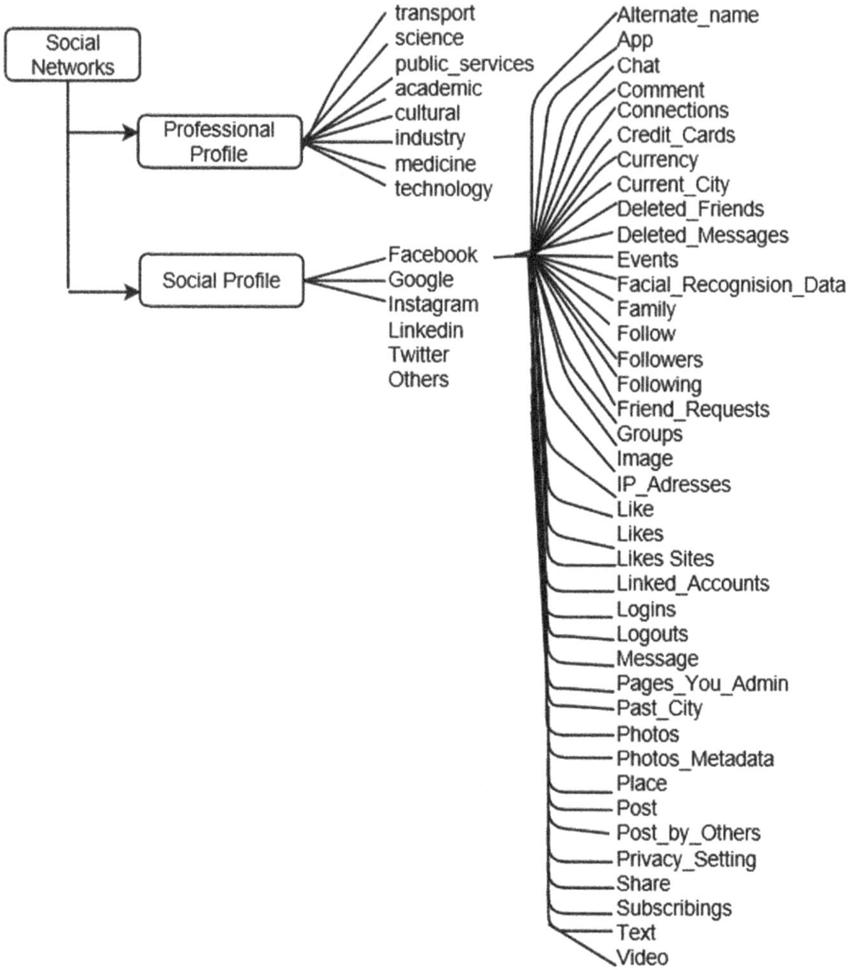


**Fig. 3.** 'Social Networks' concept of CISMO

**Crimes**

As illustrated in Fig. 4, in the crime class we define all the categories of crimes and crime archives. Based on the literature, we tried to identify the taxonomy of crime in two classes which are classical crime ad digital crime. The classical crimes are classified in many categories such as bulgar, corruption, kidnap, sexual crime. The digital crime is classified into three main categories, adult crime, child-crimes and malware. This classification is particularly difficult due to the lack of standardized concepts across countries. Crime archive contains instances of previous crime investigated by LEAS.

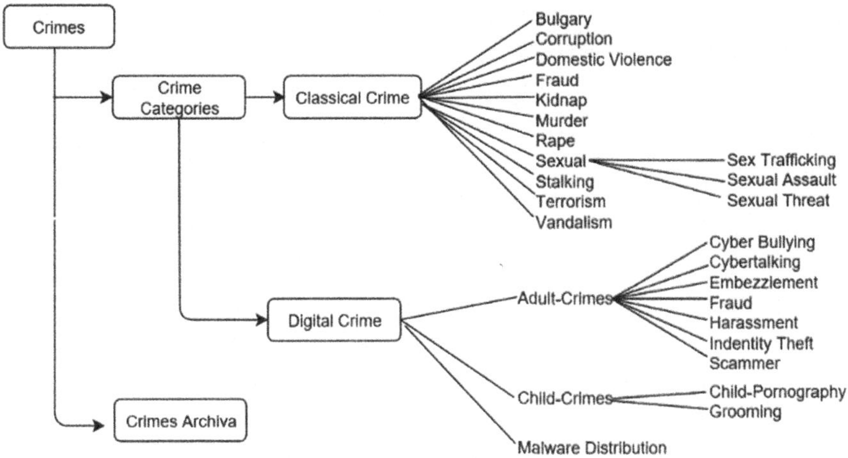


**Fig. 4.** Underlying classes of 'Crimes' concept

**Crime Case Solving**

In the past, reacting to crimes was the main purpose of policing, placing focus on crime investigation, but now the focus has changed to finding new ways for the police officers to shift from crime investigation to prevention using intelligent systems. LEAs need to use modern technologies to access and to manage the data coming from different sources including OSNs to automate insights, to create actionable intelligence in order to be focused on investigating the crime on the front line rather than dealing with time-consuming activities of LEAs. As illustrated in Fig. 5, the concept *Crime_Case_Solving* addresses the wide aspects of the collection and preservation of digital evidences. The module of *Crime_Case_Solving* in CISMO is used to model the activities carried out by LEAs in order to collect and manage the digital evidence of the crimes happening in OSN and to maintain an accurate and complete chain of custody, protecting the integrity of the digital evidence itself. Collecting digital evidences from OSNs is a complex task because the evidences are not saved in the hard drive, and the social media artefacts are stores in different places.

We have taken into account different places of evidences found in OSNs, such as browser history, events, hidden and system files, log files, pictures, images, digital photos, videos, system files and temporary files. In most of the OSNs case investigation, LEAs normally collect the digital evidences in context to the specific case and person and manage them is a tough task, and hence, an intelligent system is required. Many digital forensic tools are developed to collect digital evidences from computers, but less efforts are made to support the collection of digital evidences from different social media platforms, being that OSNs are new field in digital forensics, and there are not commonly accepted standards and guidelines for the forensics investigation based on data coming from OSNs [31].

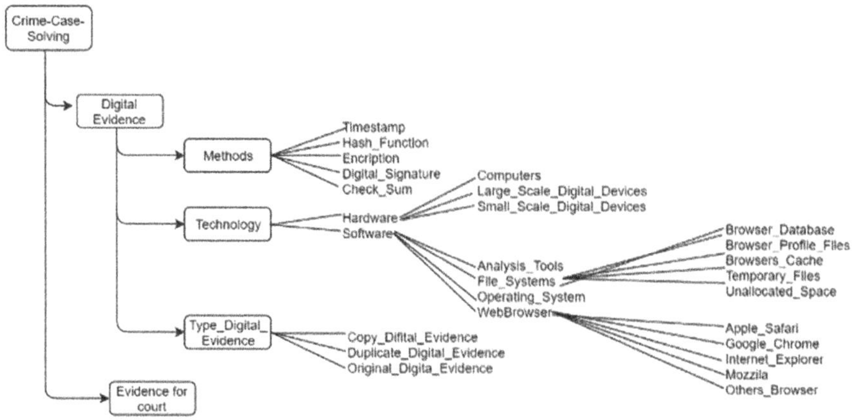


**Fig. 5.** 'Crime_Case_Solving' concept of CISMO

An investigation starts with analysing a specific social networking web page. For example, a victim got a message in a specific date and time with the content 'See you soon' and the icon of a gun, in his account in Facebook. Later, the sender of this message may change it, just deleting the gun icon. An investigator has to analyse the artefacts, and he has to collect immutable elements in order to process with case solving. He has to identify and use different sources for the evidence acquisition, such as suspect's device, victim's device and social network services. All these sources are categorized under the class *Hardware*, belonging to the class *Technology*. The *Hardware* section of CISMO is broken up into three different parts: *Computers, Large-Scale Digital Devices* and *Small-Scale Digital Devices.*

For the purpose of this ontology, small-scale digital devices is one of the most important concepts, broken down into cell phones and PDA, as people use their phones overwhelmingly to text, share and comment via social networks. In the software section of SMONT are presented the tools used to analyse the evidences, the operation system used by suspects and victims, the web browser used to navigate in the social media, and the file system were different forensic tools to try to collect Facebook artefacts of a particular user. The method category focuses on several methods been adapted from the computer science and information security to the domain of digital forensic to prove the integrity of digital evidence, such as timestamp, hash functions, encryption, digital signature and checksum.

In order to examine a system and maintain the chain of custody of a digital evidence defensible, investigators have to freeze it and examine a copy of the original data acquired [32]. So, the digital evidences collected from OSNs using SMONT are categorized in the class Type_Digital_Evidence. Original_Digital_Evidence refers to the physical items and objects related to these items at the time of seizure, Duplicate_Digital_Evidence refers to the duplicate of the evidence on the original physical item, and Copy_Digital_Evidence refers a copy of the original evidence independent from the original physical item. When the integrity of a digital evidence is preserved, it goes under the class Evidence_for_cort.

**Biometric Artefacts**

Using biometric artefacts in crime investigation is becoming an important task for LEAs to narrow down the list of persons accused or suspected of committing a crime and to represent the biometric evidences with strong statistical basis to a court of law. The biometric class is composed of five sub-classes. The Physical_Biometric class includes face, finger, hair, iris, and head. The other category Behaviour_biometrics mainly deals with human behaviours of profiles of social media which are usually extracted by posted videos in their social platforms.

The biometric artefacts of OSNs are collected by routinely seized cameras and mobile devices of suspects. The images, group images and videos retrieved from these devices provide key evidences in the crime investigation. In SMONT, we have defined the common methods used to extract these features, under the class Methods. Biometric processes are evaluated in terms of accuracy, matching score and recognition time, defined under the class Biometric_metrics.

**Object and Data Properties**

CISMO consists of several objects and data properties, and most of them are owned by person and digital evidence. The data properties are used to link the individual of a class to a data value. For instance, a social media post may have date and time recorded, so each message exchanged in a social media platform will have a timestamp associated with it. In order to relate a message of a social media to its timestamp, data-type property has timestamp which was created. On the other hand, object properties are defined to relate individual of two classes.

For example, isSuspected is an object property owned by person and it is used to relate him/her to an instance of crime class. hasbiometric is another property owns by person and is used to relate a person to an instance of biometric artifact class. Figure 6 shows the relationships among the concepts of CISMO, briefly explained above (Fig. 7).

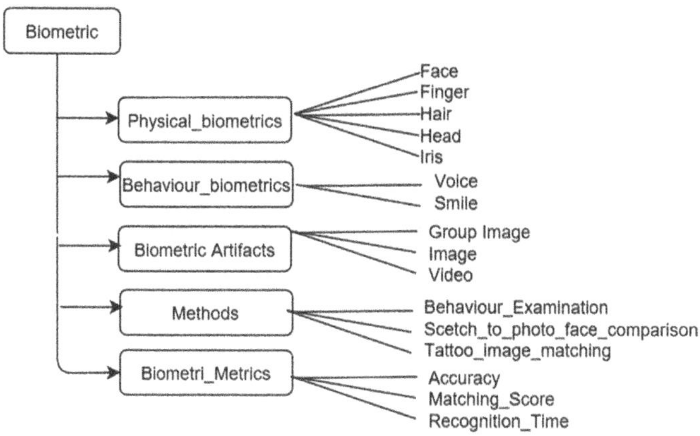


**Fig. 6.** 'Biometric' concept of CISMO

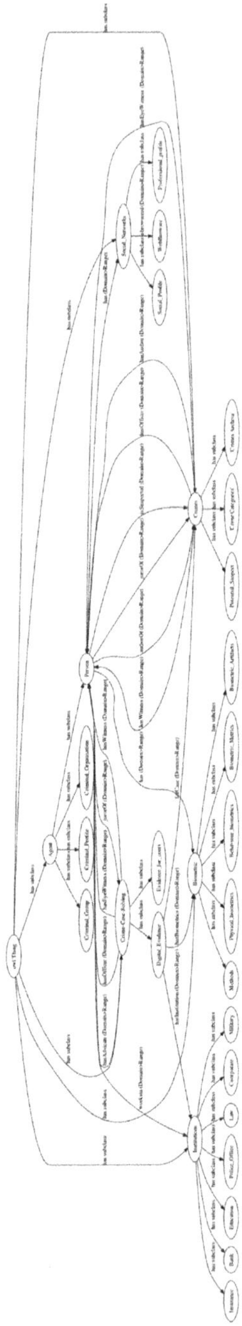

**Fig. 7.** Class dependencies of CISMO

## 4 Case Study

The data generated by users in OSN are proprietary of the social media platforms and LEAs can access their non-public data based on signed agreements. Recently, the USA and the UK have signed a first-of-its-kind agreement to access user-generated data from OSNs companies related to different criminal cases.[6] In this research, we have to evaluate the proposed solution with real data coming from social platforms; but as per confidentiality issue, we could not make use of this data. To evaluate and to show the effectiveness of CISMO, we used a data set from Nulled.io, a forum used by cybercriminals to trade and purchase leaked information, stolen credentials, nulled software, hacking tools and cracks. We do not claim this forum that represents all different categories of crime happening in OSNs, but this data is a treasure trove for LEAs to investigate criminal activities such as illegal sales. During our research, the data which are publicly available are processed preserving the sensitive information in order not to allow direct or indirect identification of members. In this data set, it stored the profile information of 599,085 members and their online activities. In our analyses, we are based on the 800,593 private messages, stored in the table message_posts that have the following fields: message id, message topic id, message date, message post, message post key, message author id and message IP address. This information is relevant for a criminal case to produce digital evidences to be admitted to the court. During the data pre-processing, the welcome messages send by the system or administrators for the new members are deleted. The remaining records are processed to remove HTML tags using Beautiful Soup, an HTML parser. Next, we also use lemmatizers in NLTK to convert nouns and verbs to their lemma. We also removed all the special characters, stop words and punctuation marks from message contents.

As a case study, we randomly selected 500 messages and labelled manually in normal and criminal activity. This data set is used to train different machine learning classifiers, and after the tuning of the parameters of these classifiers, we classified 2000 messages randomly selected from the dump data set. These messages are instances to the CISMO ontology. The method called filtered classifier is used as it allows a filter to be paired up with a classifier. Even though SVM is the most used in the existing literature language processing, we compared its performance to other classifiers such as C4.5 decision tree, artificial neural networks (ANN) and decision tree, and the classifiers performance is presented in Table 2. We applied StringtoWordVector filter to convert string attributes into numeric in and applied the NgramTokenizer that considers the word order in a local context by n-gram features, helping to discover hidden patterns between words which represent a meaningful context. The experiments were conducted using tenfold cross-validation that has been previously explored for text analysis, and it is a recommended approach for small data sets.

The explained experiments are implemented in the Waikato Environment for Knowledge Analysis (WEKA 3.8) [33] and are executed on a PC with Intel® Core i7 processor, 2.1 GHz speed and 8 GB RAM.

---

[6] https://www.cnbc.com/2019/10/04/us-uk-sign-agreement-to-access-data-from-tech-companies-like-facebook.html.

**Table 2.** Classification accuracy of three classifiers

| Group | Filter | Learning methods | Accuracy (%) |
|---|---|---|---|
| Meta | StringtoWordVector GramMaxSize = 3 GramMinSize = 1 | C4.5 | 92.7 |
| | | MLP | 93.4 |
| | | SVM | 95.1 |

The results indicate a better performance of SVM outperforming C4.5 and MLP. What stands out in these tables is that the highest accuracy for the classification of messages was 95.1% with the SVM classifier, whereas the lowest was 92.7% with the C4.5 classifier. Based on this conclusion, we used the SVM classifiers with tuned parameters to classify all 2000 private messages that are used to populate the proposed ontology, as the objective of this research is not to design the best machine learning-based model to classify the messages, but to create a data set to be used for ontology evaluation. The data instances that populated the ontology are presented in Table 3.

**Table 3.** Classification accuracy of the three classifiers

| Message | % | Number of messages |
|---|---|---|
| Non-criminal | 31.2 | 624 |
| Criminal | 68.8 | 1376 |

The classified messages are imported into the ontology using a Protégé plugin called Celfie used to map excel spreadsheets based on predefined rules. Since the aim of this paper is to present the performance of CISMO at detecting suspects from OSNs, a simple case study relying on the data of the mentioned forum is produced to represent a case of online crimes and the need of LEAS to obtain information about online activity of persons suspected for a criminal activity. The CISMO is a .owl file that captures the corpus (2000 RDF triples), that was then loaded in an Apache.

Jena Fuseki server is to process dynamically the queries. This server provides a web service framework to support for querying through SPARQ.

*Reasoning*

The data uploaded in the CISMO ontology present represent facts, and based on these facts, new ones can be defined. Rules can be written for categorizing instances in ontology, as a part of their natural belonging or as a part of other categories. Some reasoning examples in the context of crime detection and prevention could be:

- a person has sent a message to another user in Facebook, that might contain criminal content. This person is a possible source of suspect and digital evidences can be collected based on his online activity. If the message sent by a Social_Profile contains criminal content, this person should be categorized under Criminal_Profile, and especially in the Suspects class.

- a new digital evidence is created if: (i) the message has timestamp and (ii) the message has msgSentBy and (iii) the message has msgSendTo and (iv) the message has content classified as criminal content using NLP. Reasoning rules can be defined to categorize instances in ontology. If the digital evidence has integrity and biometric artefacts, then the digital evidence is categorized under the Evidence_for_court class.

*Online Crime*

We illustrate the potential of using CISMO in detecting online crime by presenting a case study using a small subset of Nulled.io, analysing the content of private messages leading to find out potential suspects engaged with cybercrime. In this scenario, we prove the use of CISMO in analysing the forum data related to online crime and to support LEAs to collect information about person's online activity to better understand the behaviours of offenders and pathways into crime. For the illustration, we show a relevant example (Q1) to the online crime case based on the messages exchanged between users of Nulled.io forum. In the second case, using some fake instances about a crime, we show the effectiveness of our solution followed by the query that processed the interested data and part of the result set.

*Q1: Users who have send messages with criminal content*

```
    PREFIX cismo: <http://www.cismo.org/v1#>
    PREFIX rdf: <http://www.w3.org/...rdf-syntax-ns#> .
    SELECT ?senderProfile WHERE {
    ?msg cismo:msgSentBy ?senderProfile .
    ?msg cismo:msgSentTo ?receiverProfile .
?sender hasAccountIn ?receiverProfile.
    ?msg cismo:hascontent ?content .
VALUES ?content { cismo:Criminal}}
```

This query returns the member IDs of 1376 loaded in the ontology that has criminal content in the messages sent in the forum.

*Q2: Which are the persons that checked in a place where a crime has happened and at the same time that the crime happened?*

```
SELECT (concat(?personName," ",?personSurname) as ?Person) ?checkedIn ?Crime ?cplace
?checkedInDate ?crimeDate
WHERE {
?person cismo:Person.
?person cismo:hasName ?personName.
?person cismo:hasSurname ?personSurname.
?person cismo:owns ?Socia_Profilel.
?person cismo:checkIn ?checkedIn.
?person cismo:checkInOnDate ?checkInDate.
?checkedIn cismo:hasName ?checkInPlace.
?ccase a cismo:Crime.
?ccase cismo:happenedOn ?cplace.
?ccase rdfs:label ?Krimi.
?ccase cismo:happenedOnDate ?crimeDate. BIND(year(xsd:date(?checkedInDate)) as ?checkedInYear).
BIND(month(xsd:date(?checkedInDate)) as ?checkedInMonth).
BIND(day(xsd:date(?checkedInDate)) as ?checkedInDay). BIND(year(xsd:date(?crimeDate)) as ?crimeYear)
BIND(month(xsd:date(?crimeDate)) as ?crimeMonth) BIND(day(xsd:date(?crimeDate)) as ?crimeDay) .
filter(?checkedIn in (?cplace) && ?checkedInDay=?crimeDay && ?checkedInMonth=?crimeMonth &&
?checkedInDay=?crimeDay )}
```

Based on this query, we find out the persons suspected for a crime based on their *check_in* the place where the crime has happened and considering the time when the crime has happened.

## 5 Conclusions

In this paper, a new semantic tool suitable to gather digital evidences from criminal activities happening in OSNs, aiming to guarantee LEAs deeper insights into criminal activities has been introduced. The proposed ontology, called CISMO, is an extended version of a previous ontology developed by authors, with new functionalities to model the core concepts related to crime cases in the police repositories, biometric data and digital evidences collected by OSNs, making it possible for LEAs to classify crimes, investigate hidden crime patterns or predict future crime patterns.

Conducting some experiments with data coming from a hacker forum, we proved the effectiveness of CISMO, showing how this ontology can help LEAs to gather knowledge for crime investigation using natural language processing and machine learning to process messages shared in online platforms and applying reasoning rules, as semantic inferences. The proposed ontology addresses a wide range of aspects related to crime investigation, including concepts digital evidence collection, biometric data and elaboration of data from OSNs simultaneously. CISMO presented in this research should be accepted as a significant attempt to present a criminal ontology for social media. Moreover, it can be improved and extended with new concepts and relationships, and it has to be tested with wider use cases.

Future work will consist in testing this ontology with real use cases obtained from police repositories and real data of OSNs and to evaluate the whole ontology covering a broader range of crimes, which will speed up the crime investigation processes but also made this process more efficient and accurate.

**Acknowledgements.** This work was carried out during the tenure of an ERCIM 'Alain Bensoussan' Fellowship Programme.

## References

1. Yaqoob, I., Hashem, I.A.T., Gani, A., Mokhtar, S., Ahmed, E., Anuar, N.B., Vasilakos, A.V.: Big data: from beginning to future. Int. J. Inf. Manag. **6**(6), 1231–1247 (2016)
2. Sureka, A., Agarwal, S.: Learning to classify hate and extremism promoting tweets. In: Intelligence and Security Informatics Conference (JISIC), 2014 IEEE Joint, pp. 320–320. IEEE, New York, (2014)
3. Mund, B.: Social media searches and the reasonable expectation of privacy. Yale JL Tech. **19**, 238e238 (2017)
4. Mason, S. et al.: Electronic evidence. Reed Elsevier, UK (R.J. Currie, S. Coughlan, Chap. 9: Canada) at p. 293 (2012)
5. Oriwoh, E. et al.: Internet of things forensics: challenges and approaches. In: 9th IEEE International Conference Collaborative Computing: Networking Applications and Worksharing, pp. 608–615 (2013)
6. Oludare, A.I., Jantan, A., Omolara, A.E., Singh, M.M., Anbar, M., Zaaba, Z.F.: Forensic DNA proling for identifying an individual crime. Int. J. Civil Eng. Technol. pp. 755–765 (2018)
7. Loubiri, A., Obaid, A., Sadat, F.: An ontology-based system for social networking for health application support. 2013 International Conference on Advances in Computing, Communications and Informatics (ICACCI), Mysore, 2013, pp. 1152–1157 (2013)
8. Arshad, H., Jantan, A., Hoon, G., Butt, A.: A multilayered semantic framework for integrated forensic acquisition on social media. Digital Invest. **29**, 147–158 (2019)
9. Kalemi, E., Yildirim, S., Domnori, E., Elezaj, O., SMONT: an ontology for crime solving through social media. Int. J. Metadata Semant. Ontol. **12**(2/3) (2017)
10. Goldbeck, J.: The dynamics of web-based social networks: membership, relationships, and change. First Monday **12**(11) (2007)
11. Breslin, J., Bojars, U., Passant, A., Fernandez, S., Decker, S.: SIOC: Content exchange and semantic interoperability between social networks. In: W3C Workshop on the Future of Social Networking, pp. 15–16 2009
12. SCOT Ontology Specification, Rdfs.org, 2016. [Online]. Available http://rdfs.org/scot/spec/. Accessed 27 Jun 2019
13. Kassiri, A. E., Belouadha, F.-Z.: A FOAF ontology 2017 IEEE International Conference on Power, Control, Signals and Instrumentation Engineering (ICPCSI), Chennai, 2017, pp. 3056–3061
14. Golbeck, J., Rothstein, M.M.: Linking social networks on the web with FOAF: a semantic web case study. In: Proceeding AAAI'08 Proceedings of the 23rd National Conference on Artificial Intelligence, vol. 2, pp. 1138–1143 (2008)
15. Breslin, J.G., Harth, A., Bojars, U., Decker, S.: Towards semantically-interlinked online Communities. Semant. Web Res. Appl. p. 500e51 (2005)
16. Kassiri, A.E., F.-Belouadha, Z.: The exSIOCInt Ontology: A SIOC ontology extension. J. Eng. Appl. Sci. **12**, 8160–8166 (2017)

17. Tserpes, K., Papadakis, G., Kardara, M., Papaoikonomou, A., Aisopos, F., Sardis E., Varvarigou, T.: SocIoS: A Social Media Application Ontology, On the Move to Meaningful Internet Systems: OTM 2012 Workshops. Lecture Notes in Computer Science vol. 7567, pp. 574–584 (2012)
18. Shaw, R., Troncy, R., Hardman, L.: LODE: linking open descriptions of events. The Semantic Web, Fourth Asian Conference, ASWC 2009, Shanghai, China, December 6–9, 2009. Proceedings. Lecture Notes in Computer Science, vol. 5926, pp. 153–167. Springer, Berlin (2009)
19. Osathitporn, P., Soonthornphisaj, N., Vatanawood, W.: A scheme of criminal law knowledge acquisition using ontology. In: 2017 18th IEEE/ACIS International Conference on Software Engineering, Artificial Intelligence, Networking and Parallel/Distributed Computing (SNPD)
20. Kastrati, Z., Imran, A. S., Yildirim-Yayilgan, S., Dalipi, F.: Analysis of online social networks posts to investigate suspects using SEMCON. Social Computing and Social Media Lecture Notes in Computer Science, pp. 148–157 (2015)
21. Park, H., Cho, S., Kwon, H.-C.: Cyber forensics ontology for cyber criminal investigation. In: Forensics in Telecommunications, Information and Multimedia, pp. 160–165. Springer, Berlin (2009)
22. Cosic, J., Cosic, Z., Baca, M.: An ontological approach to study and manage digital chain of custody of digital evidence (2011)
23. Ćosić, J., Baća, M.: Leveraging DEMF to ensure and represent 5Ws & 1H in digital forensic domain. Int. J. Comput. Sci. Inf. Secur. **13**(2) (2015)
24. Hart, G.: The five W's: an old tool for the new task of task analysis. Tech. Commun. **43**(2), 139–145 (1996)
25. Oltramari, A., Henshel, D.S., Cains, M., et al.: Towards a human factors ontology for cyber security. In: STIDS, pp. 26–33 (2015)
26. Kanak, A.: Biometric ontology for semantic biometric-as-a-service (BaaS) applications: a border security use case. In: IET Biometrics, vol. 7, no. 6, pp. 510–518 (2018)
27. Protégé 5.5.0. https://protege.stanford.edu/. Last access Sept 2019
28. Noy, N.F., Hafner, C.D.: The state of the art in ontology design: a survey and comparative review. AI Mag. **18**, 53–74 (1997)
29. Zemmouchi-Ghomari, L., Ghomari, A.R.: Translating natural language competency questions into SPARQL queries: a case study. In: First International Conference on Building and Exploring Web Based Environments, pp. 81–86. IARIA (2013)
30. Uschold, M., Gruninger, M.: Ontologies: principles, methods and applications. Knowl. Eng. Rev. **11**(2), 93–155 (1996)
31. Angelopoulou, O., Vidalis, S.: Towards 'crime specific' digital investigation frameworks. In: Proceedings of 3rd International conference on Cybercrime, Security and Digital Forensics, 8–9 June, 2013, University of Cardiff, Cardiff, Wales, UK (2013)
32. Warren, G., Kruse, II, Heiser, J.G.: Computer Forensics: Incident Response Essentials. Pearson Education (2001)
33. Weka 3: Data mining software in java. http://www.cs.waikato.ac.nz/ml/weka/. Accessed 10 May 2019

# Arabic Sexist Comments Detection in Youtube: A Context-Aware Opinion Analysis Approach

Jihad Zahir[1(✉)], Youssef Mehdi Oukaja[2], and Oumayma El Ansari[1]

[1] LISI Laboratory, Cadi Ayyad University, Marrakesh, Morocco
j.zahir@uca.ac.ma
[2] Faculty of Sciences Semlalia, Cadi Ayyad University, Marrakesh, Morocco

**Abstract.** In this chapter, we present an approach to automatize the assessment of attitudes toward violence against women and women's rights, by analyzing Youtube comments, written in Arabic. More specifically, we propose a context-aware approach to opinion analysis in comments by taking into account the polarity of videos to which comments are associated. We build a training set and use it to train a classifier that predicts videos' polarity. The accuracy and precision of the produced video classifier are 98% and 94%, respectively.

**Keywords:** Natural language processing · Opinion analysis · Text mining · Machine learning · Youtube · Violence against women

## 1 Introduction

Violence Against Women (VAW) is one of the most commonly occurring human rights violations in the world, and Arab region is no exception. In fact, at least 37% of Arab women have experienced some form of violence in their lifetime.[1] Forms of violence include physical, psychological, economic and sexual violence. In the Arab region, both attitudes and social norms sustain harmful practices and behavior toward women and hold back progress in eliminating VAW.

A wide variety of direct and indirect, qualitative and quantitative methods are used to gauge people's attitudes toward a subject. However, some of the traditional methods are costly, time consuming and potentially biased. For example, direct methods that use self-reporting tools to evaluate attitudes are heavily affected by social desirability bias: Respondents tend to provide answers that make them appear socially desirable or compliant with social norms.

[1] https://arabstates.unwomen.org/en/what-we-do/ending-violence-against-women/facts-and-figures.


X.-S. Yang et al. (eds.), *Proceedings of Fifth International Congress on Information and Communication Technology*, Advances in Intelligent Systems and Computing 1183, https://doi.org/10.1007/978-981-15-5856-6_45

According to Internet World Stats, Internet penetration by Arabic speakers in 2017 is as high as 41%. The 7th edition of the Arab Social Media Report [6] indicates that social media in the Arab world is *gaining more ground, growing increasingly localized and becoming less youthful*. People in the Arab states, and around the world, use social media to express their thoughts, opinions and sometimes even personal information and details of their daily activities. Natural Language Processing (NLP) techniques can analyze and extract useful information from user-generated content (UGC) on the Web and help measure people's attitudes toward a specific topic of interest.

In this work, we study Youtube comments as a data source to measure Arabic speaking Web users' attitudes toward VAW and women's rights more generally. Actually, Twitter is generally the preferred data source to study public opinions. However, active Twitter users in some Arab states (such as Morocco, Tunisia, Bahrain, Oman, Qatar, Lebanon and Jordan) do not exceed the threshold of 2% [6]. Motivated by this fact, we saw in Youtube comments, written in Arabic, is a good alternative to get a broader representation of more Arab countries.

In summary, the main contributions of this chapter are as follows:

1. We build a training set of videos featuring positive and negative opinions on VAW and women's rights. Such a training set was not available before our work.
2. Based on the training set, we generate multiple video classifiers that automatize the evaluation of Youtube videos addressing a subject related to VAW and women's rights.
3. We use the developed classifiers to provide a context-aware opinion analysis of comments in Youtube.

The remainder of this chapter is organized as follows: In the first section, we present our approach to context-based opinion analysis in Youtube comments. Section 3 describes our methodology for video classification and Sect. 4 provides an overview on the process of polarity scores calculation for comments. The results and performance evaluation are discussed in Sect. 5.

## 2 Our Approach

Many methods have been suggested in the literature to address opinion analysis (i.e., sentiment analysis (SA) and opinion mining (OM)), and they can be roughly grouped into two main categories: (1) machine learning-based and (2) lexicon-based approaches. In the first approach, the OM task is generally modeled as a classification problem, while in the latter, dictionaries of words annotated with their semantic orientation (i.e., polarity) and strength are used to calculate a sentiment score for the evaluated text [7].

The classification-based approach requires an annotated data set to train the classifier. To the best of our knowledge, there is no existing training set presenting examples of comments (or texts) written in Arabic language and annotated depending on the opinion they feature toward VAW. For this reason, a lexicon-based approach seems more practical in our case. However, it would be misleading to rely on the words' polarity only to decide if a comment is positive or not. In fact, the context in which the analyzed comment appears plays a central role in the final decision regarding the overall opinion it is representing. Actually, if we limited ourselves to a lexicon, a comment's polarity would be considered positive if the majority of the words it contains are of positive polarity too. However, in our research, we consider a comment as positive only if it holds a positive opinion toward VAW and women's rights more generally. Therefore, if a comment is associated to a video that promotes VAW, we cannot consider it as positive even if its polarity, according to the used lexicon, was defined as positive.

In this chapter, we present a context-aware approach to opinion analysis that is based on two main phases:

1. **Video Classification**: We build a training set and use it to train a classifier that identifies the polarity of a video, given its title and description.
2. **Comment polarity score calculation**: Using two lexicons, we calculate the polarity score ($Decision_{temp}$) for the comment.

Finally, by combining the results of both phases, we use the decision tree presented in Fig. 1 to obtain the final decision on the overall comment polarity.

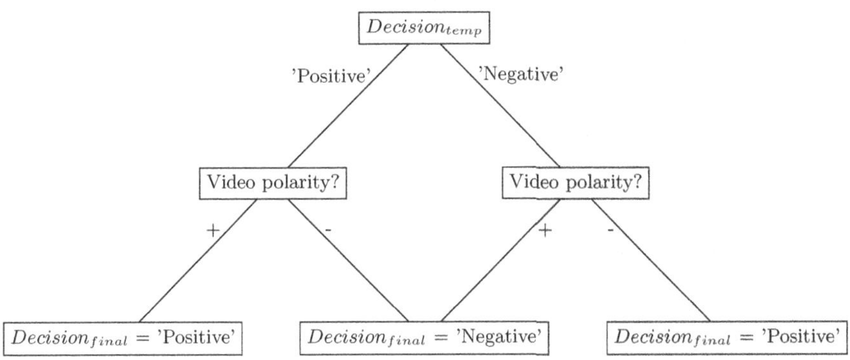


**Fig. 1.** Decision tree for context-based sentiment analysis

## 3 Youtube Video Classification

If we were to watch every single video in order to know how many of them are presenting a positive position or the proportion of those that encourage gender-based violence, it would be fastidious and time consuming. This becomes especially challenging when videos are of a considerable length. Luckily, classification techniques can help us automate the evaluation of the videos. Classification is being a supervised approach in which an algorithm learns how to automatically recognize data classes from annotated data, and we needed to first build our training set.

### 3.1 Building the Training Set

We formulated 12 search queries, written in Arabic, and a dedicated script was developed to collect videos that are relevant to the defined queries. The script is written in Python programming language and uses Youtube API.[2] Figure 2 summarizes different queries with their English translation (QueryEn).

| Query ID | Query (Arabic) | Query (English) |
|---|---|---|
| 1 | عمل المرأة | Women's Work |
| 2 | تعليم المرأة | Women's Education |
| 3 | ضرب المرأة | Beating Women |
| 4 | لاجئات | Refugiees (Feminine) |
| 5 | ختان البنات | Girls Cirumcision (FGM) |
| 6 | التحرش | Sexual Harassment |
| 7 | ولاية المرأة | Women in Power Positions |
| 8 | المرأة و مراكز القرار | Women and Decision Making Positions |
| 9 | المرأة و السياسة | Women and Politics |
| 10 | أهلية المرأة | Women Eligibility (to make decisions) |
| 11 | المرأة و الريادة | Women and Leadership |
| 12 | قيادة المرأة | Women and Leadership/Driving |

**Fig. 2.** Search queries used for data collection from Youtube

We collected 5618 videos. For each video, *Video ID, title, description, number of likes, number of dislikes, number of views* and *publishing channel ID* were collected among other features. To build the training set, we draw a random sample of 600 videos from the set of collected videos. After visualizing each video from the random sample, annotators were invited to indicate the relevant value (from a list of options) for the following attributes:

1. **Type**:
   - *News/TV Show*: Chosen if the video is a re-diffusion or part of a TV show.
   - *Religious*: This option is selected if the video features a religious leader, or a discourse of a religious nature.
   - *Adult*: Assigned to adult content or videos featured by their owners as sexuality education content.
   - *Other*: Selected if none of the options above apply.
2. **Conflict**:
   - *Yes*: If the video presents facts or stories related to women in conflict and humanitarian settings. Otherwise, *No*.
3. **ReportsViolence**:
   - *Yes*: If the video reports on an actual act of violence against women or girls. Otherwise, *No*.

[2] https://developers.google.com/youtube/v3.

4. **isfromcso**
   - *Yes*: If the video is advertising an initiative curated by a civil society organization. Otherwise, *No*.
5. **Ironic**:
   - *Yes*: If the video's title is not consistent with the content and is describing it in an ironic way. Otherwise, *No*.
6. **Polarity**: Provides an evaluation of the position or opinion presented in the video. This attribute takes one of the four following values:
   - *Positive (or Negative)*: Chosen if the content of the video is presenting a positive (or negative) position toward VAW or women's and girls' rights.
   - *Mixed*: If the video presents both negative and positive opinions.
   - *Neutral*: Videos reporting facts without explicitly giving an opinion are marked as neutral. Videos presenting debates are classified as neutral, because different or opposite opinions are generally equally defended. We also consider as neutral, judgments that are given to men and women equally with no differentiation between them, and regardless of their sex. For example, if a video is expressing the opinion that "both women and men should not wear provocative clothes," it is annotated as neutral.

A video was removed from the initial sample if it contained sign language that annotators were unable to interpret, or if it was suspected to report fake or out-of-context facts, or to be owned or published by a movement that is globally recognized as extremist. Duplicates were also removed as some videos appeared in the sample more than once. We provide in Tables 1, 2 and 3 some characteristics of the resulting training set.

**Table 1.** Number of videos reporting on vaw or featuring women in conflict situations

| Attribute | Modality | Video count |
|---|---|---|
| Reports violence | No | 530 |
| | Yes | 44 |
| Conflict | No | 539 |
| | Yes | 35 |

### 3.2 Video Classification

Based on the developed training set, we use multinomial Naive Bayes classifier (for more details see [3]) to build multiple classifiers that identify different categories and features in videos. Before feeding the data into the classifier, we perform some transformations on the initial raw data.

For each video, we join the *description* and the *title* to form one text, we then clean the resulting text by removing stop words, URLs, punctuation, repeating

**Table 2.** Polarity distribution for videos featuring news/TV shows or religious discourse

| Video type | Video polarity | Count |
|---|---|---|
| Religious | Positive | 32 |
| | Negative | 39 |
| | Mixed | 22 |
| | Neutral | 8 |
| News/TV shows | Positive | 111 |
| | Negative | 1 |
| | Mixed | 4 |
| | Neutral | 22 |

**Table 3.** Distribution of videos' polarities

| Video polarity | Count |
|---|---|
| Positive | 239 |
| Negative | 60 |
| Mixed | 38 |
| Neutral | 57 |

letters and diacritics. Next, we transform each text into a vector in a high-dimensional space using TF-IDF vectorizer. The TF-IDF vector is obtained from the product of a *term frequency* (TF) and the *inverse document frequency* (IDF) for each token that appears in the text.

Depending on the target class (i.e., predicted class), adapted samples were drawn from the initial training set to train the classifiers. When the size of the obtained samples was small, based on each original text, we generated new texts by randomly shuffling and joining back the words that constitute the original text.

In the cases where the adapted version of the training set was imbalanced, Synthetic Minority Oversampling Technique (SMOTE) [2] was used to create more examples from the minority class. SMOTE oversampling is done after the vector representation step, while text generation based on words shuffling is performed before vectorization.

## 4 Lexicon-Based Sentiment Analysis in Comments

Opinion mining or sentiment analysis consists in extracting subjectivity and polarity from text. In other words, it is the computational study of people's opinions, sentiments, emotions and attitudes [4]. In our research, we use it to automatically evaluate and capture public opinion toward VAW and, more generally, toward women's rights.

In the field of analysis and automatic extraction of sentiments, the lexicon-based approaches involve calculating polarity for a text from the semantic orientation of words or phrases that constitute it. To get the semantic orientation of words, we refer to lexical resources (dictionaries) containing lists of words annotated with a score representing their semantic orientations or polarities.

In this chapter, we use two publicly available dictionaries to calculate polarity scores for our comments. Both resources are presented in Table 4. For every comment in our database, we extract the words, get their corresponding polarity scores from the dictionary and then aggregate these scores to obtain the overall score of the initial comment.

**Table 4.** Data sources description

| Data source | Description |
|---|---|
| ArSenL[1] | A large-scale standard Arabic sentiment lexicon using a [1] combination of the existing resources. It contains 22,954 positive and 20,331 negative words |
| ArabicSA[2] | Automatically created corpus using a set of 230 arabic words [5] that were manually selected for being highly positive or highly negative. It contains 13,118 positive and 8846 negative words |

[1]http://oma-project.azurewebsites.net/ArSenL/download_intro
[2]https://saifmohammad.com/WebPages/ArabicSA.html

# 5 Results and Discussion

## 5.1 Evaluation Metrics

As explained above, our approach consists in two main phases: (1) video classification, which represents the core of our contribution and (2) lexicon-based sentiment analysis in comments. The latter is a straightforward sentiment scores calculation based on the described lexicons. Therefore, and taking into account the decision tree presented in Fig. 1, the results' quality of our approach depends essentially on the quality of the videos classification.

To evaluate the performance of a classifier, we use two metrics: *Accuracy* and *Precision* calculated using Formulas 1 and 2, respectively.

$$\text{Accuracy} = \frac{\text{True Positive} + \text{True Negative}}{\text{True Positive} + \text{False Positive} + \text{True Negative} + \text{False Negative}} \tag{1}$$

$$\text{Precision} = \frac{\text{True Positive}}{\text{True Positive} + \text{False Positive}} \tag{2}$$

### 5.2 Results

Based on the training set described in Sect. 3.1, we generated multiple binary classifiers, while we only need one classifier that predicts whether a video is featuring positive or negative opinions toward VAW and women's rights, and since the data is available, we decided to train other classifiers that provide other information about the videos. A list of all trained classifiers is provided below:

1. $Model_{pos}$: Predicts whether a video is featuring a positive or a negative opinions toward VAW and women's rights.
2. $Model_{neut}$ and $Model_{mix}$: Verify if an opinion is neutral or mixed, respectively.
3. $Model_{rel}$: Predicts if a video is featuring a religious discourse.
4. $Model_{conf}$: Verifies if the video is presenting women in conflict situations.
5. $Model_{rov}$: Identifies videos reporting on violence cases.

The results show that all models are of very good performance (see Table 5).

**Table 5.** Video classification models: accuracy and average precision

| Model | Predicted classes | Accuracy | Precision |
|---|---|---|---|
| $Model_{pos}$ | "Positive" or "Negative" | 0.9829 | 0.9367 |
| $Model_{neut}$ | "Neutral" or "Not" | 0.9469 | 0.9782 |
| $Model_{mix}$ | "Mixed" or "Not" | 0.9285 | 0.9714 |
| $Model_{rel}$ | "Religious" or "Not Religious" | 0.9647 | 0.9302 |
| $Model_{conf}$ | "Conflict" or "Not Conflict" | 0.9855 | 0.9411 |
| $Model_{rov}$ | "Reports on violence" or "Not" | 0.9651 | 0.9094 |

## 6 Conclusion

In this chapter, we presented a context-based sentiment analysis approach to evaluate sexist opinions in Youtube comments. Our videos classifiers performed very well. In the next phase of the research, we will illustrate the capacity of our approach to inform about attitudes toward VAW and women's rights by a case study that would present the knowledge extracted from collected videos and comments based on the developed classifiers and other text mining techniques.

**Acknowledgements.** This research was funded by Data2X UNF.

## References

1. Badaro, G., Baly, R., Hajj, H., Habash, N., El-Hajj, W.: A large scale Arabic sentiment lexicon for Arabic opinion mining. In: Proceedings of the EMNLP 2014 Workshop on Arabic Natural Language Processing (ANLP), pp. 165–173 (2014)
2. Chawla, N.V., Bowyer, K.W., Hall, L.O., Kegelmeyer, W.P.: SMOTE: synthetic minority over-sampling technique. J. Artif. Intell. Res. **16**, 321–357 (2002)
3. Kim, S.B., Han, K.S., Rim, H.C., Myaeng, S.H.: Some effective techniques for naive bayes text classification. IEEE Trans. Knowl. Data Eng. **18**(11), 1457–1466 (2006)
4. Zhao, J., Liu, K., Xu, L.: Sentiment analysis: mining opinions, sentiments, and emotions (2016)
5. Mohammad, S., Salameh, M., Kiritchenko, S.: Sentiment lexicons for Arabic social media. In: Proceedings of the Tenth International Conference on Language Resources and Evaluation (LREC '16), pp. 33–37 (2016)
6. Salem, F.: Social media and the internet of things towards data-driven policymaking in the Arab world: potential, limits and concerns, p. 7. MBR School of Government, The Arab Social Media Report, Dubai (2017)
7. Taboada, M., Brooke, J., Tofiloski, M., Voll, K., Stede, M.: Lexicon-based methods for sentiment analysis. Comput. Linguist. **37**(2), 267–307 (2011)

# Missing Concept Extraction Using Rough Set Theory

N. F. Nabila[1](✉), Nurlida Basir[1], Nurzi Juana Mohd Zaizi[1], and Mustafa Mat Deris[2]

[1] Universiti Sains Islam Malaysia (USIM), 71800 Nilai, Negeri Sembilan, Malaysia
{fatin,nurlida,njuana}@usim.edu.my
[2] Universiti Tun Hussein Onn Malaysia, Batu Pahat 86400, Johor, Malaysia
mmustafa@uthm.edu.my

**Abstract.** Ontology is used as knowledge representation of a particular domain that consists of the concepts and the two relations, namely taxonomic relation and non-taxonomic relation. In ontology, both relations are needed to give more knowledge about the domain texts, especially the non-taxonomic components that used to describe more about that domain. Most existing extraction methods extract the non-taxonomic relation component that exists in a same sentence with two concepts. However, there is a possibility of missing or unsure concept in a sentence, known as an incomplete sentence. It is difficult to identify the matching concepts in this situation. Therefore, this paper presents a method, namely similarity extraction method (SEM) to identify a missing concept in a non-taxonomic relation by using a rough set theory. The SEM will calculate the similarity precision and suggest as much as similar or relevant concepts to replace the missing or unclear value in an incomplete sentence. Data from the Tourism Corpus has been used for the experiment and the results were then evaluated by the domain experts. It is believed that this work is able to increase the pair extraction and thus enrich the domain texts.

**Keywords:** Concept · Non-taxonomic relation · Ontology · Ontology extraction · Rough set · Similarity precision

## 1 Introduction

There are two types of relations in ontology components such as hierarchical and non-hierarchical relations. The hierarchical relation such as IS-A relation is known as taxonomic relations. While, the non-taxonomic relation is to identify the non-hierarchical relation from texts. In ontology, both components are needed to give more knowledge about the domain texts, especially the non-taxonomic components which are used to describe more about the texts [16]. Most of methods previously for non-taxonomic relation extraction focus on extracting concepts and their relationship in a single sentence. But, those methods unable to represent the whole view of the domain since the texts may have a missing concept of a sentence.


X.-S. Yang et al. (eds.), *Proceedings of Fifth International Congress on Information and Communication Technology*, Advances in Intelligent Systems and Computing 1183, https://doi.org/10.1007/978-981-15-5856-6_46

Thus, this paper proposes a new method based on rough set theory named similarity extraction method (SEM) to suggest the relevance concept for an incomplete sentence. Section 2 provides the literature review of the related topics. Section 3 provides the definition of formula. Section 4 presents a proposed non-taxonomic relation extraction method. Section 5 describes the experiments, which was carried out to evaluate the proposed method and results and Sect. 6 provides the conclusion.

## 2 Literature Review

Works in [1–4] are example studies that focus on extracting non-taxonomic relations component. These works used the predicate that occurs together with concept pair as the potential non-taxonomic relation. The method in [2] has used pre-processing task and MINIPAR parser to identify the relevant concepts and to verify the appearances of relevant concepts. The relation (i.e., verb) that link between two relevant concepts in a sentence then was extracted. Work in [4] also applied Natural Language Processing (NLP) to identify valid concepts of the domain, as similar to [2], and used the technique in data mining techniques to identify the potential relation between valid concepts. Serra et al. [5] used statistical approach to propose a parameterized and extensible method, known as PARNT, which aims to improve the extraction of non-taxonomic relations. Ribeiro [3] proposed a framework for non-taxonomic relation extraction by using three different techniques. The techniques are A Nearly New Information Extraction System (ANNIE) to identify instances of concepts, a Stanford dependency parser is used to identify the dependency relations and concepts, and association technique to determine the most suitable relation (i.e., non-taxonomic relation) for concept pair. This work has tested in tennis sport domain texts.

The hybrid framework in [1] has focused on four features for extracting non-taxonomic relation between treatment and disease entities from biomedical dataset. The datasets are collected from MEDLINE database. The four features such as lexical and semantic based the Unified Medical Language System (UMSL), Natural Language Processing approach and bag of word model. In this work, the treatment and disease entities represent a subject and object of a sentence. Then, verbs that appear together with both entities (i.e., treatment and disease) in a sentence are extracted as non-taxonomic relationships.

Current techniques [1–5] mostly concentrate on obtaining possible non-taxonomic relationships among concepts which appear in a sentence. Nevertheless, the non-taxonomic relationships between concepts may not be extracted whenever there is a sentence without a subject or an object.

Thus, this paper is to propose a technique using rough set theory in ontology extraction to handle the issue of missing concept in domain text in order to enhance the information knowledge.

## 3 Definition of Formula

An information system [6] is a quadruple, $S = (U, A, V, f)$, where $U$ is a non-empty finite set of objects, $A$ is a non-empty finite set of attribute, where $A = C \cup \{d\}$, $C$ is a set of condition attributes and d the decision attribute, such that $f : U \times A \rightarrow V$ for any $a \in A$, where $Va$ is called domain of an attribute a. $S$ is known as incomplete information system if the $U$ has at least one object with unknown value [7, 8]. In incomplete information system, symbol "*" is used to denote the unknown value.

Here, $U$ is a set of sentences in texts. $A$ is a set of terms, where $C$ is a a subject or an object in text and $d$ represents the predicate that occurs together with C in a sentence.

Table 1 shows the information table extracted from several sentences of voting machine texts. The SentenceId (such as $u_1$, $u_2$, ..., $u_{18}$) in Table 1 represents each sentence of domain texts. For each sentence, we identified the subject, the object, and the predicate of a sentence. The missing value of the subject or object is denoted as *.

**Table 1.** Example of dataset in information table

| SentenceID | Subject | Object | Predicate | SentenceID | Subject | Object | Predicate |
|---|---|---|---|---|---|---|---|
| $u_1$ | $S1$ | $O1$ | $P1$ | $u_{10}$ | $S2$ | $O3$ | $P4$ |
| $u_2$ | * | $O2$ | $P2$ | $u_{11}$ | $S3$ | $O2$ | $P5$ |
| $u_3$ | $S2$ | $O3$ | $P3$ | $u_{12}$ | $S4$ | $O6$ | $P7$ |
| $u_4$ | $S3$ | * | $P4$ | $u_{13}$ | $S4$ | $O2$ | $P9$ |
| $u_5$ | * | $O5$ | $P5$ | $u_{14}$ | * | $O2$ | $P3$ |
| $u_6$ | * | $O7$ | $P8$ | $u_{15}$ | $S5$ | $O6$ | $P10$ |
| $u_7$ | $S2$ | $O3$ | $P3$ | $u_{16}$ | $S4$ | * | $P3$ |
| $u_8$ | $S3$ | * | $P5$ | $u_{17}$ | $S4$ | $O4$ | $P3$ |
| $u_9$ | $S2$ | $O4$ | $P3$ | $u_{18}$ | $S2$ | * | $P3$ |

Based on Table 1, most existing non-taxonomic extraction technique is able to extract ten subject--predicate--object (S-P-O) patterns such as in $u_1, u_3, u_7, u_9, u_{10}, u_{11}, u_{12}, u_{13}, u_{15}$ and $u_{17}$. But, the remaining eight sentences (i.e., $u_2, u_4, u_{u5}, u_6, u_8, u_{14}, u_{16}, u_{18}$) of an incomplete pattern of (S-P-O) are not extracted. Hence, the collection of related texts are considered as not properly presented as the potential relations might be missing.

A rough set theory possible can be an alternative to overcome the issue in an incomplete information system [6, 9] and also for categorical data (text in nature). The use of a rough set theory here is to suggest the possible data that could occur and give a better approximation compared to other theory, such as probability and association rules [7, 10, 11]. There are several works that have been used rough set, such as [9] and [10]. In [9], the authors used rough set theory to identify a similar concept in order to generate a concept of ontology. In [10], the authors show that the rough set theory is able to calculate and discover the taxonomic relation. The paper has shown that the rough set theory has better approximation results compared to maximal association rule. Hence, this work study the applicability of a rough set theory to label the most relevant non-taxonomic relation that could occur with the concepts even in incomplete sentence. The paper has shown that the rough set theory has better approximation results compared to maximal association rule.

Therefore, in this paper, the proposed approach will calculate the similarity precision and suggest as much as similar or relevant concepts to replace the missing or unclear value in an incomplete sentence.

## 4 Proposed Method

This section described a similarity extraction method (SEM). The objective of SEM is to improve the extraction of ontology components by giving the most possible relevant value (object or subject) based on the results of similarity value. Figure 1 illustrates the steps of the SEM algorithm.

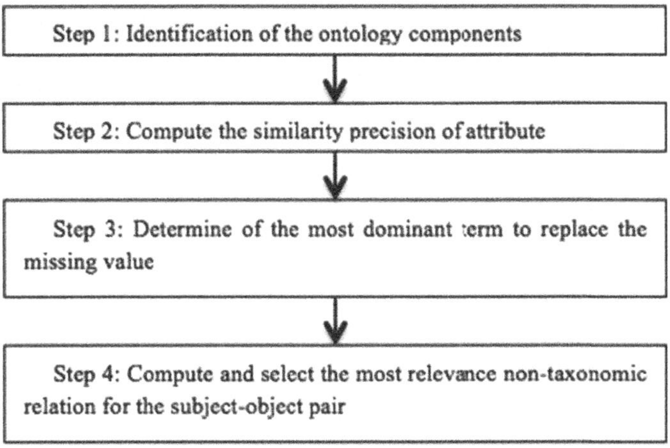


**Fig. 1.** Similarity Extraction Method (SEM) algorithm

### 4.1 Identification of the Ontology Components

This step is to split domain texts into sentences and to extract terms (i.e., concepts and predicates of the sentence) by using preprocessing. First, assign part-of-speech tag to each word. Next, remove the unnecessary words that do not describe the domain, such as a, the, of, with, and, etc. Then, after eliminating the stop word, all the words left are change to its basic word [15]. As an example, the word "visitors" is converted to basic form as "visitor." As a result, a list of terms is produced.

The statistical measurement such as term frequency is applied to the list to determine the relevant terms related to domain texts. Then, the MINIPAR parser [12] is used to identify the appearance (whether as a subject or object in a sentence) of the relevant terms and the parser to identify the grammatical relation between the terms and the predicate. All identified relevant terms with the relations are presented in the information table and we put symbol "*" for missing term.

### 4.2 Compute the Similarity Precision of Attribute

The section explains how to find the similarity attributes among sentences based on a similarity precision formula in [14]. In this work, set of attribute with sim$P$ > =66.7% thresholds will be selected.

**Definition 1** Let $P_B(x) = \{b | b \in B \wedge b(x) \neq *\}$, the similarity precision sim$P$, is defined as

$$\text{sim}P(x, y) = \frac{|P_B(x) \cap P_B(y)|}{|C|}, \tag{1}$$

here, $|C|$ is referred to the cardinality of the set, $B$ refers to the set of attributes. In our work, the synonymous meaning between predicates is identified by using WordNet.

*Example 1* For $u_{18} = \{s2, *, p3\}$, the set of similarity attributes based on Formula (1) is calculated as below:

- $u_7 = \{s2, o3, p3\}, P(u_{18}) \cap P(u_7) = \{2\}$ or $\text{sim}P(u_{18}, u_7) = 2/3 = 0.66$.

In $u_7$, two attributes out of three attributes are similar with attributes in $u_{18}$ such as $s2$ and $p3$.

- $u_9 = \{s2, o4, p3\}, P(u_{18}) \cap P(u_9) = \{2\}$ *or* $\text{sim}P(u_{18}, u_9) = 2/3 = 0.66$
- $u_3 = \{s2, *, p3\}, P(u_{18}) \cap P(u_3) = \{2\}$ or $\text{sim}P(u_{18}, u_3) = 2/3 = 0.66$

### 4.3 Determine of the Most Dominant Term to Replace the Missing Value

This step is to determine the most dominant term to replace the missing value.

**Definition 2** $MDc(x)$ is the most dominant of term subject or object (i.e., $X$) in a set of similar attributes presented as follows:

$$MDC(x) = (|(x)|/|Cs|, x \in C), \tag{2}$$

where (| |) represents the cardinality of the sets, $Cs$ refers to the cardinality of the set of similar attributes.

*Example 2* Based on Example 1, a set of attributes is calculated as follows:

$$\begin{aligned} &u_7 = \{s2, o3, p3\},\ u_9 = \{s2, o4, p3\}, \\ &u_3 = \{s2, o3, p3\},\ \text{ where } Cs = 3, \\ &MD(o3) = 2/3 = 0.67, MD(o4) = 1/3 = 0.33. \end{aligned}$$

Based on Example 2, the $o3$ has higher MD value than $o4$, hence the $o3$ is selected as most dominant term for object in $u_{17}$.

### 4.4 Compute and Select the Most Relevance Relationship for the Subject-Object Pair

The support and the confidence [13] are used to determine the association between the subject-object pair with the relation. Here, $i$ represents the subject--object pair, $j$ refers to predicate. The support value will calculate the number of sentences in a set of domain texts that contain $(i \cup j)$ and is given as below:

$$\text{supp}(i \rightarrow j) = |(i \cup j)| \, / \, |U|$$

The confidence is used to identify the ratio of the number of sentences in a set of domain texts that contain $i \cup j$ that contain $i$, is presented as below:

$$\text{conf}(i \rightarrow j) = \text{supp}(i \cup j)/\text{supp}(i).$$

*Example 3* Based on Table 2, we calculate the confidence for the subject--object pair, i.e., $(s2, o3)$, with predicate as follows:

**Table 2.** A part of tourism dataset

| SentenceID | Subject | Object | Predicate | SentenceID | Subject | Object | Predicate |
|---|---|---|---|---|---|---|---|
| $u_1$ | City | Activity | Conduct | $u_8$ | Hotel | City | Promote |
| $u_2$ | Traveller | Activity | Conduct | $u_9$ | Hotel | Visitor | Accommodate |
| $u_3$ | Tourist | Money | Spend | $u_{10}$ | Visitor | Money | Spend |
| $u_4$ | Tourism | Visit | Involve | $u_{11}$ | Visitor | Money | Spend |
| $u_5$ | Tourism | Tourist | Comprise | $u_{12}$ | * | Money | Spend |
| $u_6$ | Visitor | Money | Spend | $u_{13}$ | Traveler | Activity | * |
| $u_7$ | Traveler | People | Carry | $u_{14}$ | Tourism | Country | Promote |

- $\text{conf}((s2, o3) \rightarrow p3) = 3/4 = 0.75, \text{conf}((s2, o3) \rightarrow p4) = 1/4 = 0.25$

Based on Example 3, predicate $p3$ has the highest confidence value with concept pair (s2, $o3$). Then, the $p3$ is selected as suitable non-taxonomic relation for $s2$ and $o3$.

## 5 Experiments

The goal of SEM is to extract as much as possible non-taxonomic relations from the domain texts to enrich the knowledge of domain texts. For this experiment, a medium collection of a Tourism dataset from Wikipedia websites (http://en.wikipedia.com) and Los Angeles Time website (http://articles.latimes.com) was used. The Tourism dataset consists of 27,876 words. In this experiment, the SEM and other method by [4] are developed and tested using the same domain text. It is due to the capabilities of both methods extract both concepts and relations from the domain text and eventually assign appropriate relations between concept pair. A part of the extracted subject--predicate--object relations from the Tourism corpus is shown in Table 2. In this figure, sentences with missing value are denoted by "*".

Based on Table 2, $u_{12} = \{*, \text{money, spend}\}$ has the missing subject of the sentence. Thus, a set of similarity attributes with sim$P >= 66.7\%$ are calculated and the results are shown as follows:

- $u_3 = \{\text{tourist, money, spend}\}$, Thus, $P(u_{12}) \cap P(u_3) = \{2\}$ or sim$P(u_{12}, u_3) = 2/3 = 0.66$
- $u_6 = \{\text{visitor, money, spend}\}$.Thus, $P(u_{12}) \cap P(u_6) = \{2\}$ or sim$P(u_{12}, u_{10}) = 2/3 = 0.66$
- $u_{10} = \{\text{visitor, money, spend}\}$, $P(u_{12}) \cap P(u_{10}) = \{2\}$ or sim$P(u_{12}, u_{10}) = 2/3 = 0.66$
- $u_{11} = \{\text{visitor, money, spend}\}$, $P(u_{12}) \cap P(u_{11}) = \{2\}$ or sim$P(u_{12}, u_{11}) = 2/3 = 0.66$

Next, we determine the most dominant of the object as below:

- $u_3 = \{\text{tourist, money, spend}\}$, $u_6 = \{\text{tourist, money, spend}\}$,
  $u_{10} = \{\text{visitor, money, spend}\}$, $u_{11} = \{\text{visitor, money, spend}\}$,

MD(visitor) = ¾ = 0.75, MD (tourist) = ¼ = 0.25. Thus, a visitor is selected as most dominant of the object to replace the missing object in $u_{12}$.

While for $u_{13}$, it has missing predicate of the sentence. By using similarity precision formula, $u_2$ has sim$P(u_{13}, u_2)$ greater than 0.5. Therefore, the traveler is preferred to replace the missing predicate in $u_{13}$. For evaluation process, the domain experts are involved to identify all the relevant relations from a given texts as benchmark. Then, the relations produced by the domain experts are compared in against the relations extracted by the SEM, and method [4]. The precision and the recall metrics have been used to evaluate the quality of the knowledge extracted. The quality of correctly extracted relations in domain texts is analyzed by using recall metric and the accuracy of correctly extracted relations is analyzed by using precision metric.

The extraction results for a collection of Tourism domain is shown in Table 3. Based on Table 3, the SEM extract 462 relations with 291 relations are correct relations. While the Serra method [4] extracted 232 correct relations, which is less than the SEM. For this domain, the domain expert has identified 305 correct relations. It shows that the precision value for SEM and Serra method [4] are 62.98% and 53.00%. While, the recall value for SEM is higher than the Serra method [4] with 95.40%. Hence, it shows that SEM able to enrich the domain ontology.

## 6 Conclusions

In conclusion, our work proposed a method that is able to extract potential relations in an incomplete sentence of texts. We used similarity extraction method (SEM) to identify

**Table 3.** Experimental results

| Method | # of extracted triple relations | # of valid triple relations | # of invalid triple relations | Precision (%) | Recall (%) |
|---|---|---|---|---|---|
| Expert | 305 | 305 | – | | |
| SEM | 462 | 291 | 171 | 62.98 | 95.40 |
| Serra method [4] | 232 | 123 | 109 | 53.00 | 40.33 |

a missing concept in a non-taxanomic relation of the domain texts. In conclusion, SEM managed to handle the non-taxonomic relations extraction process and enrich the domain knowledge.

**Acknowledgements.** This work was supported under Fundamental Research Grant Scheme (FRGS) Grant, No: FRGS/1/2018/ICT04/USIM/03/1 and Universiti Sains Islam Malaysia (USIM).

## References

1. Muzaffar, A.W., Azam, F., Qamar, U.: A relation extraction framework for biomedical text using hybrid feature set. Comput. Math. Methods Med. (2015)
2. Punuru, J., Chen, J.: Learning non-taxonomical semantic relations from domain texts. J. Intell. Inf. Syst. pp. 191–207 (2012)
3. Ribeiro, M.: Extraction of non-taxonomic relations from texts to enrich a basic ontology. Unpublished Master Thesis, School of Information Systems and Computer Engineering, University of Lisbon, Portugal (2014)
4. Serra, I., Girardi, R.: A Process for extracting non-taxonomic relationships of ontologies from text. Intell. Inf. Manage. **3**(4), (2011)
5. Serra, I., Girardi, R., Novais, P.: PARNT: a statistic based approach to extract non-taxonomic relationships of ontologies from text. In: 2013 Tenth International Conference Information Technology: New Generations (ITNG), pp. 561–566. IEEE, New York (2013)
6. Pawlak, Z.: Rough sets. Int. J. Comput. Inf. Sci. **11**(5), 341–356 (1982)
7. Kryszkiewicz, M.: Rough set approach to incomplete information systems. Inf. Sci. **112**(1–4), 39–49 (1998)
8. Kryszkiewicz, M.: Rules in incomplete information systems. Inf. Sci. **113**(3–4), 271–292 (1999)
9. Inukai, Y., Gehrmann, A., Nagai, Y., Ishizu, S.: Rough set theory using similarity of objects described by ontology. In: Proceedings of the 51st Annual Meeting of the ISSS-2007, Tokyo, Japan, vol. 51, no. 2 (2007)
10. Bi, Y., Anderson, T., McClean, S.: A rough set model with ontologies for discovering maximal association rules in document collections. Knowl.-Based Syst. pp. 243–251 (2003)
11. Ishizu, S., Gehrmann, A., Nagai, Y., Inukai, Y.: Rough ontology: extension of ontologies by rough sets. In: Symposium on Human Interface and the Management of Information, Springer, Berlin, Heidelberg, pp. 456–462 (2007)
12. Lin, D.: Dependency-based evaluation of MINIPAR. In: Treebanks, Springer, Netherlands, pp. 317–329 (2003)

13. Agrawal, R., Imieliński, T., Swami, A.: Mining association rules between sets of items in large databases. ACM SIGMOD Record **22**(2), 207–216 (1993)
14. Deris, M.M., Abdullah, Z., Mamat, R., Yuan, Y.: A new limited tolerance relation for attribute selection in incomplete information systems. In: 2015 12th International Conference on Fuzzy Systems and Knowledge Discovery (FSKD), pp. 964–970. IEEE, New York (2015)
15. Porter, M.F.: An algorithm for suffix stripping. Program: Electron. Library Inf. Syst. **14**(3), 130–137 (1980)
16. Uschold, M.: Creating, integrating and maintaining local and global ontologies. In: Proceedings of the First Workshop on Ontology Learning (OL-2000) in Conjunction with the 14th European Conference on Artificial Intelligence (ECAI-2000) (2000)

# A Study on Behavioural Agents for StarCraft 2

Ivan Williams and Dustin van der Haar(✉)

University of Johannesburg, Johannesburg, Gauteng, South Africa
201306136@student.uj.ac.za, dvanderhaar@uj.ac.za

**Abstract.** With the recent trend of artificial intelligence, specifically within machine learning, there are some powerful tools that can be utilized to create video game artificial intelligence bots. Bots that can beat professional players or immerse players within the game to the point where enemies are considered intelligent and react to situations similar to how a real human would. However, some of these processes and tasks to create a bot can be an expensive and time-consuming process. In this research paper, we look at two models to building an AI bot and comparing the two, namely a simple reflex model and a recurrent neural network model. From the results, we can see that the recurrent neural network goes further into the tech tree and is able to produce a more complexed set of units as compared to the simple reflex solution. The simple reflex solution, however, is able to reach the win condition by defeating the enemy bot much quicker than the recurrent neural network solution at 5 min and 39 s and costs less in terms of production and complexity. The recurrent neural network solution was also able to get a higher food supply count and spent the most amount of resources in all areas including technology, economy and army supply.

**Keywords:** Artificial intelligence · StarCraft · DeepMind · Neural networks · Reinforced learning · Game agents · Heuristics

## 1 Introduction

StarCraft 2 is a complex real-time strategy game. The objective of the game is to beat your opponent by defeating all their buildings or getting them into a position where they can no longer continue to play the game. To accomplish this, the player must collect minerals and gas scattered around the map, invest these resources into different kinds of troops, each with their own strengths, and attack the enemy to destroy their buildings. In a typical one versus one scenario, the game ends when one of the players surrenders or all the players' buildings are destroyed.

In this article, we discuss two pipelines that were created to play the game of StarCraft 2. We then compare the two pipelines between one another by looking at how long they take to defeat an enemy computer bot, the amount of resources that were collected during the time elapsed and the number of units that were created which are crucial aspects of an RTS game.


X.-S. Yang et al. (eds.), *Proceedings of Fifth International Congress on Information and Communication Technology*, Advances in Intelligent Systems and Computing 1183, https://doi.org/10.1007/978-981-15-5856-6_47

StarCraft 2, however, is just a game, but the tasks that are accomplished by StarCraft 2 agents can be extended to other problem domains. For example, the neural network architecture used within Alpha Star (later discussed in Sect. 4) can be used to model very long sequences of likely actions, with games having an average game time of up to an hour with tens and thousands of moves based on information that could be imperfect. Fundamental problems of creating predictions over some set of data over a long period of time still appear in many real-world problems such as weather predictions, climate modelling and language understanding.

These are just some of the real problems that some of the algorithms used to build a StarCraft 2 agent could address. A StarCraft 2 agent has the potential to make significant advances in these problem domains using the learnings and the developments of the project.

## 2 Problem Background

Real-time strategy (RTS) games have been able to challenge the intelligence of some humans due to its overall complexity. RTS is a great example of a problem that features multiple interacting imperfect decision-making.

The level of AI in the game can play an important role. It can be argued that the immersion of the game can be a primary contributing factor to making a game entertaining [1]. When we look at a game like Grand Theft Auto, for example, the game is modelled around a fictional state of the world in which players interact with. For the game to feel more realistic and immerse the player into the world, the AI bots must be able to simulate typical human behaviour.

Extrapolating this idea of immersion in games into RTS games can be quite a daunting task. Since RTS games require the management of many areas such as base, resource and unit management, the AI bot must be able to play the game whilst managing these different areas and still react to certain situations as a human would. For example, if the AI bot gets attacked, it must be able to defend itself whilst still keeping its uptake in these different areas.

Similar systems for more recent works include StarCraft AIs which compete against one another that is built around a community [2]. Currently, there are three representative StarCraft AI competitions that are run by IEEE, CIG, AIIDE and SSCAIT. Other systems include OpenAI Five's Dota 2 system, IBM's deep blue chess computer and Deep Mind's StarCraft 2 Alpha Star.

### 2.1 Experiment Set-up

The experiment was run in phases, each phase corresponding to a different agent pipeline. Each phase would start by executing a Python start command on the command line interface for Windows 10. Once the command was executed pysc2 would start up its interface to StarCraft 2 and a locally hosted instance of the game would be started up. The corresponding script stated in the start script would then be executed.

The agent was placed into a one versus one scenario against a StarCraft bot where the agent had the goal of defeating the bot. Once the agent was able to defeat the enemy the enemy would surrender and the game would end. The replay of the game would be persistently saved to disc to be later analysed.

In the sections coming up in this research paper, we will be looking at the literature study behind RTS games and why it is a field of interesting domain problems as well as a more in-depth discussion on the model solutions that have been built in Sect. 3. Section 4 discusses the implementation details of the model solutions and how they can solve the problem of creating an agent within StarCraft 2.

## 3 Literature Study

Games, unlike most of the typical problems we encounter, are interesting because they are hard to solve. Games have engaged humans and their faculty of intelligence for years, at an alarming rate. For AI researchers, however, the abstract nature of games makes it an appealing subject of study [3, 4].

Video games are also virtual environments that are predictable to some degree. Simulations within these environments can also be sped up to levels that even the human brain would not be able to compute. For example, an agent might be able to repeat some task 600 times to teach it how to play the game in a few hours, whereas a human might take a month. This makes for an extremely powerful tool when you want to get results in a limited amount of time.

### 3.1 Real-Time Strategy Games

For a very long time, RTS has been able to challenge the intelligence of humans as well as AI alike due to its complexity and the pressure of imperfect decision-making that is placed onto a player [5]. RTS is a great example of a problem that features multiple interacting imperfect decision-makers. They involve many carefully arranged parts or details that are complicated to design, contain partial observability as well as a rapidly diverging action space that is changing frame by frame making rational decision-making elusive.

In a real-time strategy (RTS) game environment, the player and the AI computer must compete for resources distributed across the map. The most common setting for an RTS game is a war-setting, where the player has to engage in war and perform resource management. The resource management process encapsulates obtaining and utilizing these resources to gain the advantage of the other players. Using these resources to build an army and attacking strategic objectives are aspects which make the game a "war game."

### 3.2 StarCraft 2

StarCraft 2 is a balanced game revolving around three different races. Each race has a different play style to the next one. For example, Protoss has units that have shields which can regenerate, whereas the other races are biological. Protoss is focused around the power of its units like the colossus and the carrier. It also makes use of different mechanics to construct units where units will come out in sets of waves depending on how many gateway buildings you have.

### 3.3 The Motivation for Developing AI Agents

The development of a state-of-the-art computer game today requires a substantial amount of investment in terms of money and time. In order to mitigate some of the damages or risks that can happen, it is necessary to evaluate the possible success of a product. Aside from the factors that developers cannot control, such as marketing, one must ensure that a game is attractive to players. For example, if a game is unbalanced or unfair to some users, the game will not be well received [6].

AI research shows that it is possible to use genetic algorithms to improve the performance of AIs in computer games [7, 8]. The purpose of a genetic algorithm (GA) is to identify two partially overlapping subgraphs. These subgraphs form the larger part of another graph with a structure, interrelations and arrangements. The subgraphs could be seen as domains of knowledge in the broader semantic graph.

GAs, however, do have some downfalls, particularly with developing an agent that needs to play StarCraft 2. Since StarCraft 2 is a fast-paced strategy game and requires the agent to make a decision every game step, a GA will tend to struggle with more evaluations that time permits [9, 10]. The information to make a decision within the game is also incomplete which is not good for a GA.

### 3.4 Reinforcement Learning

Designing agents in computer games that rival the skills of the opponent have been a continuously challenging issue in computer games development. Using reinforcement learning, we can automatically control the game level, adapting it to the human skills of the player which balances out the game and makes for a great experience [11].

Q-learning, which is a popular reinforcement learning (RL) algorithm, can be used to address dynamic game balancing. This will ensure a good level of challenge for the game which can avoid the extremes of players getting frustrated. The game may be too hard or simply too easy resulting in the play losing interest in the game.

When trying to get the agent to play the game at a human level, we are faced with two separate problems. One is learning, which entails building an agent that can learn optimal strategies, and the other is adapting which would be providing action selection mechanisms for providing game balance which could mean taking suboptimal actions.

### 3.5 Non-playable Characters

A non-playable character is any character in a game that is not controlled by the player. In the video game, this usually means a character controlled by the computer via algorithmic, predetermined or responsive behaviour. This does not necessarily entail that is pure AI.

Immersion for these NPCs stems off of believability. If the players experience matches what he/she expects, then believability is achieved [12]. The believability of NPCs has spurred research in numerous areas. The belief--desire--intention (BDI) describes agents that set out to achieve a defined set of goals according to a set of beliefs. We also have emotional models that seem to be growing are of research when it comes to NPCs. These models help in increasing the believability of NPC's behaviour or appearance to better mimic human emotion [13].

However, when we look at NPCs in RTS games, the NPC is trying to make decisions that would seem somewhat predictable to the human player. The NPC is trying to react to what the human player is doing to make it seem more realistic. For example, when a human player attacks the NPC, the NPC will react by defending itself. Are its actions defined by making it look more realistic to the human player and thus rendering it incapable of learning from the human player?

This is where an agent and NPC will differ. An NPC is trying to immerse the player by performing actions that are expected by the human player. Above, we mentioned the believability is increased when the human player gets what he/she expects. An agent, however, is mostly involved from learning about what the human player is doing and trying to learn from that to better improve and itself with hopes of defeating the human player. An agent is not involved with making the player more immersed in the game but more about beating the human player and improving its actions.

An agent is a system situated within a part of an environment that senses the environments through its percepts and acts on it over time, in the pursuit of its own agenda as to reflect what it thinks will happen in the future. The agent carries out the best action that will result in the best outcome based on what is has perceived from past experiences through its percepts. An agent is able to perceive its environment through its sensors and act on its environment through its actuators. An agent is able to perceive its own actions but not always.

## 4 Similar Systems

In terms of leading the charge of AI within StarCraft 2, we can see that DeepMind's alpha star has made the most groundbreaking progress [14, 15]. In the fall of 2018, Alpha Star was able to beat one the best StarCraft 2 players by the name of Grzegorz Komincz otherwise known as Mana (5-0) following another successive against teammate Dario also known as TLO. The matches played were standard professional conditions in a 1v1 match scenario.

Alpha star plays the game by interacting directly with the StarCraft 2 engine through a raw interface. This means that it can observe the attributes that belong to itself and that of the enemy's visible units and buildings through the mini-map. This would be the same thing as playing the game from a zoomed-out perspective, model solutions and implementation

### 4.1 Recurrent Neural Network Solution

My recurrent neural network (RNN) solution makes use of vanilla RNN. The neural network is used to predict a build order which will later be executed but the agent as seen in Fig. 1.

The RNN is trained using StarCraft 2 replay data. The replays contain the actions of each player that were made during the game. These actions are then translated to a 1

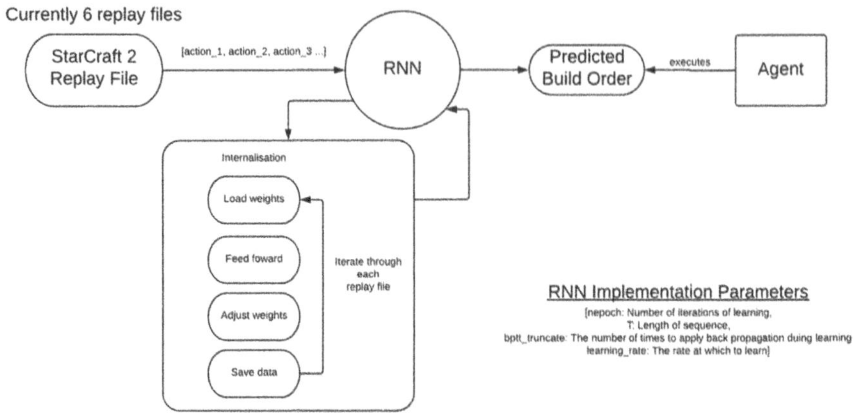


**Fig. 1.** A figure showing the flow of data from the replay data file through the RNN to create a build order which is executed by the agent

by 256 column matrix based on an actions dictionary. These actions are extracted from each replay and processed by the RNN in the form of feed-forward process.

After each replay is processed, the internal RNN weights are adjusted using backward propagation and persistently saved to a text file for the next replay prediction. It then saves the prediction for that replay to a prediction text file. Further details of this process are expressed in Fig. 2.

Once the RNN is done with training, i.e. it has run through all the replays specified, the latest prediction of the network is stored in a text file. The agent must then read this latest prediction (at first the prediction is in matrix form) and transform the actions back into readable actions for the agent to process.

Once all the actions are translated, they are placed into a stack-based list where we pop the action off when the agent executes the agent. The agent will attack with what is has when there are no longer any actions in the stack. The agent will continue to build units and attack until it has won the game by reaching the win condition. Within the context of this environment, the win condition would be for either player to surrender from the game or to have all the players' buildings and units destroyed.

### 4.2 Simple Reflex Solution

The simple reflex agent that was built simply perceives its environment and makes decisions based on what it perceives. This can be seen in Fig. 3.

The main goal of this agent is to accumulate a certain number of military units and attack the enemy player when it hits a certain food count. For e.g. the agent will perceive the food count and if the food count is about to be maximized, the agent will act by constructing a certain building to increase the food supply count. Every game step the agent will check if certain tasks have been completed. In a way, it is a little similar to the RNN solution in the sense that executes somewhat of a build order but performs tasks based on what it perceives from the environment.

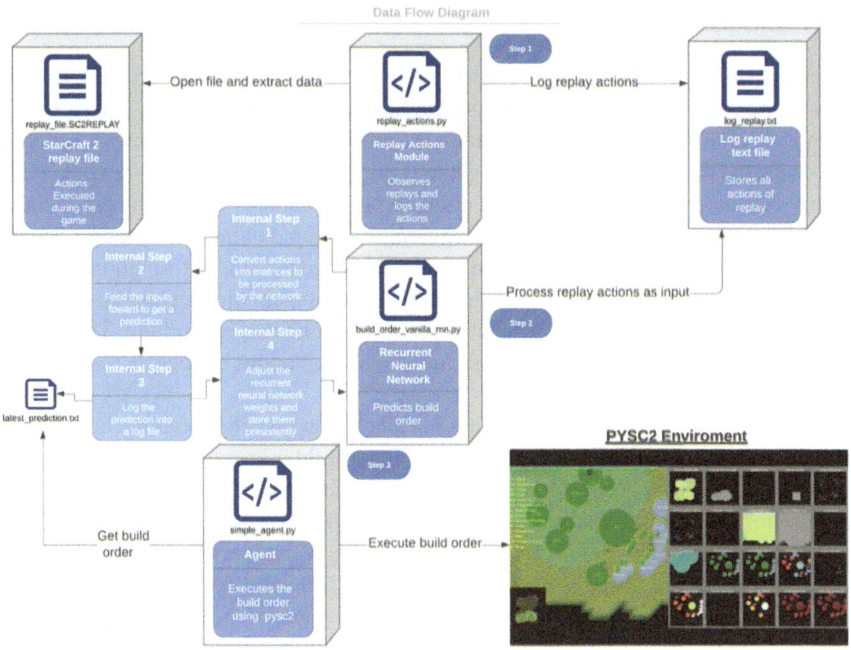


**Fig. 2.** A figure showing the complete data flow on a script level. This figure shows how the scripts persistently save the data and interact with PYSC2

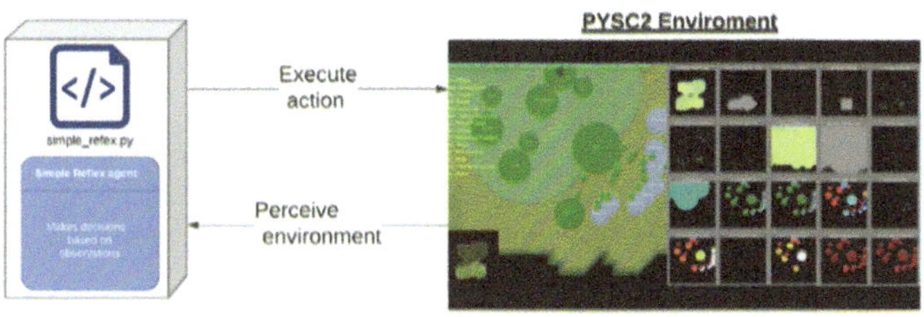


**Fig. 3.** A figure showing the complete data flow on a script level. This figure shows how the scripts persistently save the data and interact with PYSC2

It makes use of a dictionary to store the location of certain buildings. Once a building is placed in a certain location, the key representing that location is flagged so that another building is not placed in that location.

## 5 Results

After both agents were placed into a one versus one scenario against another computer bot, several results were noticed. After each agent played and finished the game, the following results were recorded.

### 5.1 Results Discussion

The results were recorded using blizzards replay interface. Once the game is executed a replay of the game is saved to disc. Using the StarCraft 2 game interface you can extract information about the game like the statistics presented above. The replay also allows you to view the data in real time as the game is played and also allows to rewind and fast forward. Another way to record information about the game is to play the replay in real time and pause the game at certain time segments. The replay interface gives information about the game. Using this information one is able to record the information and log it.

The first metric used was resources spent in terms of technology, economy and army. The second metric used was the amount of resources collected per minute. These are the most important metrics to track because the game is all about collecting resources and spending them the correct way.

The RNN solution spent more resources in all three areas. The economy being worker units and an extra command centre structure used to build more worker units, technology in terms of how far the AI bot was able to move down the tech tree and army units. The simple reflex solution, however, was able to mine more resources per minute but was unable to mine gas. This might be ideal for winning a game in the short term, as this agent was able to beat the opponent faster than the RNN solution, but this will not scale well as the game progresses as players will eventually need to mine gas to move further (Figs. 4, 5 and 6).

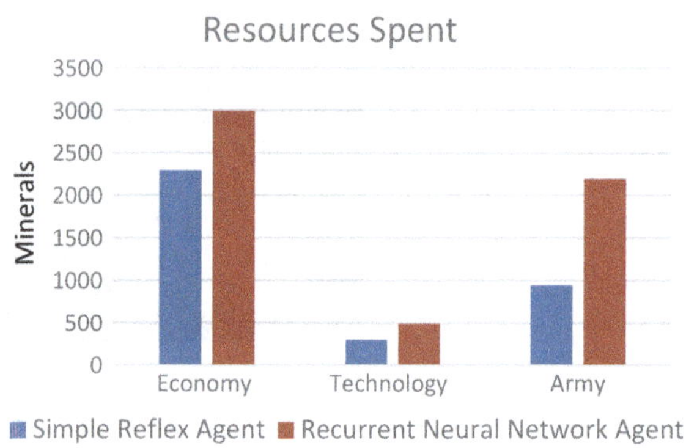


**Fig. 4.** A figure showing the amount resources in terms of minerals spent by each agent on different categories of the game

## 6 Critique and Analysis

In terms of getting winning results, it was observed that the simple reflex agent was the most successful being able to win most of its games. It was able to lose the least amount of resources as compared to the RNN solution. But the RNN-based solution was able to beat the reflex agent in terms of other statistics such as spending more resources and getting a higher food count.

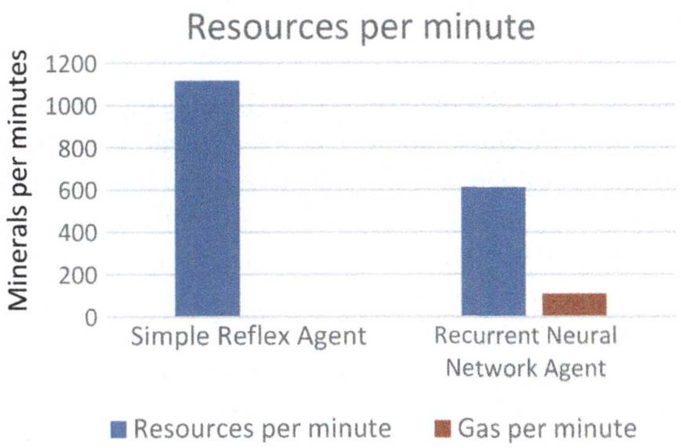


**Fig. 5.** A figure showing the amount resources in terms of minerals and gas spent per minute by each agent

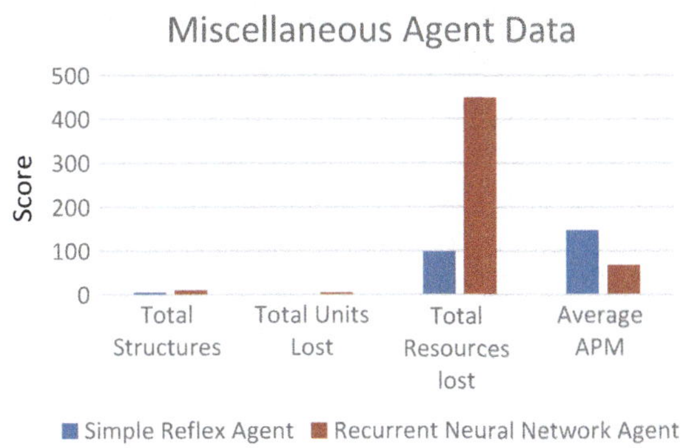


**Fig. 6.** A figure showing the total structures built, the total amount of units lost, the total amount of resources lost and the average actions per minute for each agent

The RNN solution also went further down the tech tree in terms of building units and constructing units that being the hellion which requires gas and an advanced structure called the factory. The simple reflex agent was only able to construct marines. In terms of complexity, it follows the RNN-based solution performed more complicated tasks such as mining gas and constructing advanced structures. The RNN solution also had a more sophisticated build order based on the replay data it was able to analyse, whereas simple reflex solution was only able to make decisions based on what is perceived from its environment.

It can be argued that the RNN solution had a better sense of immersion since it was able to build a wider variety of units and structures. The RNN solution also followed a build order from an actual player which makes it seem more realistic in terms of how a player would play the game.

## 7 Conclusion

When comparing the two pipelines that were created during the project, the RNN solution seemed to be the most powerful solution in terms of creating immersive gameplay as compared to the simple reflex agent. It was able to achieve this going further in tech tree which allowed the agent to build a diverse set of units. Although the simple reflex agent gets the job done faster than the RNN solution by defeating the enemy, it is much more simplistic in the sense that it only constructs one set of units and does not go as far as the RNN solution in the tech tree.

It was found that the RNN solution took 4 min longer than the simple reflex agent to defeat the opponent but was able to get a higher food count supply, mine more minerals including gas in which the simple reflex agent was unable to do and spend more resources in every area including technology, economy and army supply. The RNN solution, however, lost more structures, units and resources during the game.

From the findings, it can be said that the more complexity you add to the agent, the better or more immersive it becomes. The RNN solution was more complexed than the simple reflex agent and it was able to produce more results as well as move further down the tech structure. Due to this fact, it was able to produce a diverse set of units.

## References


1. Algan, Y., Fortin, N.: Computer Gaming and the Gender Math Gap: Cross-Country Evidence Among Teenagers, Emerald Publishing Limited, University of British Columbia and IZA
2. Kim, M., Kim, S., Kim, J., Dey, A.: Evaluation of StarCraft Artificial Intelligence Competition Bots by Experienced Human Players. Games and Playful Interaction, San Jose (2016)
3. Sanchez-Ruiz, A.A..: Predicting the outcome of small battles in StarCraft. In: The Twenty-Third International Conference on Case-Based Reasoning, ICCBR, *Workshop Proceedings,* Frankfurt German (2015)
4. Demasi, P., De, O., Cruz, A.J.: Online coevolution for action games. Int. J. Intell. Games Simul. **2**(2), 80–88 (2003)
5. Yildirim, S., Berg Stene, S.: A survey on the need and use of AI in game agents. In: Society for Computer Simulation International, Proceedings of the 2008 Spring Simulation multiconference (2008)
6. Safadi, F., Fonteneau, R., Ernst, D.: Artificial Intelligence Design for Real-time Strategy Games
7. Christoph, S., Lipski, C., Mahlmaan, T., Mathiak, B.: Using genetically optimized artificial intelligence to improve gameplaying fun for strategical games. In: Proceedings of the 2008 ACM SIGGRAPH symposium in Video Games. ACM, New York (2008)
8. Ponsen, M.J.V., Munoz-Avila, H., Spronck, P., Aha, D.W.: Automatically acquiring domain knowledge for adaptive game UI using evolutionary learning. In: Velso, M.M., Kambhampati, S. (eds.) The MIT Press, Cambridge, pp. 1535–1549 (2005)
9. Tong, C.K., On, C.K., Teo, J., Kiring, A.M.J.: Evolving neural controllers using GA for Warcraft 3-real time strategy. In: IEEE Conferences. Sixth International Conference on Bio-Inspired Computing: Theories and Applications (2011)
10. Goodman, E.D.: Introduction to Genetic Algorithms. Michigan State University, NSF Science and Technology Center (2012)

11. Andrade, R., Ramalho, G., Santana H.: Automatic computer game balancing: a reinforcement learning approach. In: *Proceedings of the Fourth International Joint Conference on Autonomous Agents and Multiagent Systems*, Brazil. ACM, New York (2005)
12. Jennett, C., Cox, A.L., Carins, P., Dhoparee, S., Epps, A., Tijs, T., Walton, A.: Measuring and defining the experience of immersion in games. Int. J. Hum.-Comput. Stud. **66**(9), 641–661 (2008)
13. Mooney, J., Allbeck, M.: Rethinking NPC intelligence—A new reputation system. In: Proceedings of the Seventh International Conference on motion in Games. ACM, New York (2014)
14. Vinyals, O., Ewalds, T., Bartunov, S., Georgiev, P.: StarCraft2: A New Challenge for Reinforement Learning. Blizzard DeepMind (2017)
15. Holocomb, S., Porter, W., Ault, S., Mao, G., Wang, J.: Overview on DeepMind and its AlphaGo Zero AI. In: ICBDE'18: Proceedings of the 2018 of International Conference on Big Data and Education. ACM, New York (2018)

# Analyzing Attention Deviation During Collaterally Proceeding Cognitive Tasks

Yamini Gogna, Rajesh Singla(✉), and Sheela Tiwari

Department of Instrumentation and Control Engineering (ICE), Dr. B.R. Ambedkar NIT Jalandhar, Jalandhar, India
gognayaminiice@gmail.com, {singlar,tiwaris}@nitj.ac.in

**Abstract.** *Background* The brain performs a very significant job in our body by processing the information associated with human critical inclinations, intentions, sensory attention and awareness, execution, and mental state sustenance during a specific task. The attention of every human being gets altered while undergoing two or more cognitive tasks collaterally. *Methods* In this research, subjects were asked to perform two tasks collaterally in which one task was considered as primary task whereas the other task as secondary. The EEG (Electroencephalography) signals of the subjects undergoing those collateral cognitive tasks were recorded using RMS EEG-32 Super Spec machine. *Result* The relative band powers ratio (Theta to Beta band power ratio) helped in tracing the point of time when the attention devoted to the primary task got deviated by the secondary task. An auditory P300 peak generation validated the deviation in this research. *Conclusion* This research track could pave the way for designing a battery that can analyze the subject's performance during multi-tasking. Other than this, such an investigation will help in avoiding the disasters caused by the attention deviation of the operator.

**Keywords:** EEG · Cognitive task · Relative band power · P300

## 1 Introduction

Cognitive study is the branch of neuroscience that explores the problem solving and decision-making skills corresponding to various sensory impulses [1]. To congregate cognitive information from the brain activities, the brain signals are overturned to electrical signals [2].

Whenever the number of tasks to be performed (whether occurring simultaneously or block sequentially) is hiked, due to the limited processing capacity of the brain, the mental effort contribution also gets enhanced. Also, when the task requirements exceed the user's capabilities, the hiked degree of workload on a particular person influences task performance [3]. Not only the performance inhibition but sometimes the attention deviation in such cases can lead to major issues like in case of driving, etc.


X.-S. Yang et al. (eds.), *Proceedings of Fifth International Congress on Information and Communication Technology*, Advances in Intelligent Systems and Computing 1183, https://doi.org/10.1007/978-981-15-5856-6_48

Of all the prevailing neuroimaging modalities like EEG (Electroencephalography), MEG (Magnetoencephalography), ECoG (Electrocorticography), Intracortical Neuron Recording, fMRI (Functional Magnetic Resonance Imaging), NIRS (Near Infrared Spectroscopy), CT scan (Computed Tomography Scan), PET scan (Positron Emission Tomography) and DTI (Diffusion Tensor Imaging), EEG is considered best for not only cognitive studies but also other applications like authentication system, neurological communication establishment, etc. [4]. Features like high sensitivity, noninvasivity, portability, better economical handling, high temporal resolution, and smooth post-processing form the basic reason for the choice of EEG over other modalities [5]. The EEG signals acquired can be further disintegrated into five major frequency domain components as shown in Table 1 [6].

**Table 1.** EEG frequency bands

| S. No. | Frequency band frequency range (Hz) | Salient features |
|---|---|---|
| 1. | Delta band (0.5–4) | Detected in babies and adults in unconsciousness and deep sleep state |
| 2. | Theta band (4–7) | Detected in adults in drowsy, somnolence, sleepy state, reduced consciousness, meditative and cognitive state (working memory) |
| 3. | Alpha Band (8–12) | Provides a link between memory processing and visual processing (Binds the conscious thinking with subconscious mind) |
| 4. | Beta band (12–30) | Indicates active concentration, anxiety, conscious thoughts and logical thinking over frontal and central region |
| 5. | Gamma band (30–100) | Detected in adults during motor functions or perceptions, higher processing tasks, and cognitive functioning |

A brain–computer interface (BCI) is a software and hardware entangled communication system that has the ability to correlate the brain signals with a surrogated output channel for regulating environmental factors, controlling artificial devices, establishing communication for smart tasks, etc. [7–9]. General framework of BCI is shown in Fig. 1.

The BCI research encircling workload analysis is practiced in many crucial fields like aeronautics industry, process plants, defense operations, medical applications, etc. [10]. Also, the arena of non-medical applications of BCI has brought fields like cognitive load detection, cognitive state monitoring, etc. in limelight [11]. Customarily used features of BCI include P300 evoked potential which is the response towards an infrequent visual or auditory stimulus with a latency of approximately 300 ms [12, 13]. The most prominent advantage of P300-based BCI is the absence of training required and its most commonly

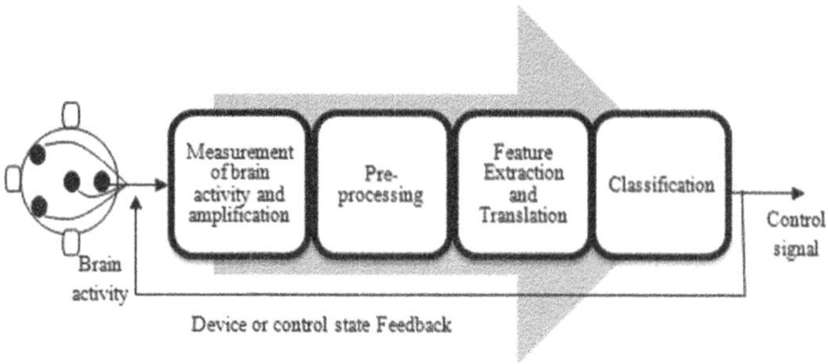


**Fig. 1.** General framework of BCI

practiced paradigm is P300 speller [14]. In research, it was exhibited that P300 peak was obtained whenever difficulty standard was raised during the task performance [15].

A BCI based study presented the fact that cerebral activations could be explored in terms of (i) Relative band powers (ii) Ratio of Theta to Beta band power. Both parameters were studied during the performance of various cognitive tasks in reference to a relaxing span [1]. The performance measurement in consideration of a particular cognitive task is usually done by following two methods: (i) Directly measuring the performance of the primary task using any of the features like band power, reaction time, etc. (ii) Indirectly associating a secondary task with the primary one and study its effect and performance [16]. Power Spectral Density (PSD) analysis, being smoothly processable in case of workload assessment, is considered a much preferred one over other performance measuring parameters [17]. Studies also stated that the introduction of visual attention along with any cognitive task will decrease relative delta and theta band power while increasing the relative alpha and beta band power [18, 19]. Using inferences from such studies, it was concluded that the cognitive load decreases the relative Theta to Beta band power ratio as compared to a resting state [20]. All the studies done until now that have attempted to measure mental workload in real-time environments have limited their analysis to offline mode only. Moreover, the analysis for workload detection with EEG is highly challenging due to factors like high dimensionality of parameter set, establishing controlled laboratory conditions, uncontrollable distractions, high uncertainty about true mental state of the subject, noisy nature of EEG signals.

## 2 Methodology

### 2.1 Participants

12 healthy subjects [6 females and 6 males, aged 18–28 (23.91 $\pm$ 1.90)] willfully participated in this research. All the subjects were right-handed, free from any medication, and had no neurological impairment history. All of the subjects had good hearing and normal or corrected to normal eyesight. The research setup was organized in the Human–Machine Interface (HMI) Laboratory of ICE Department at Dr. B.R. NIT Jalandhar. The subjects were asked not to intake alcohol or caffeine for 12 h before the experiment.

## 2.2 Stimuli

**Description**: The choice of primary cognitive task was made considering the engrossment level of the participants. They were provided with an android platform to play 'Helix Jump' (published by Voodoo) continuously. The game proceeds by guiding the ball through a slowly descending tower. This was done with the rotation of the tower so as to make the ball pass through gaps while avoiding the contact of colored parts. The screenshot of the game is shown in Fig. 2.

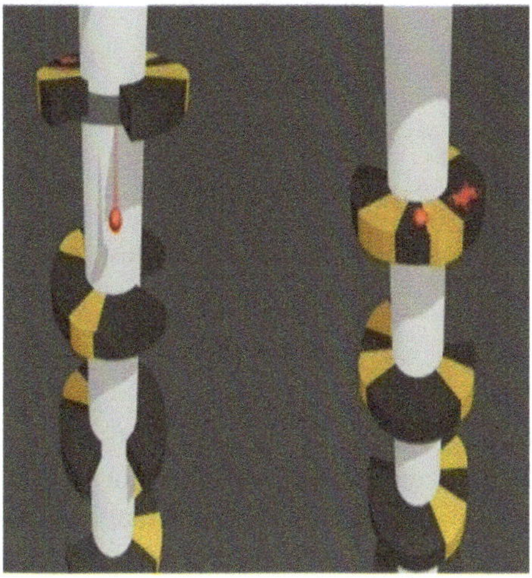

**Fig. 2.** Helix jump (primary Task). *Source* https://play.google.com/store/apps/details?id=com.h8games.helixjump&hl=en_IN

This was followed by the simultaneously occurring auditory task along with the primary task. A recording of 60 monosyllabic English nouns was played in which a target word (specified priorly) was to be counted along with playing Helix Jump. The auditory attention task paradigm is shown in Fig. 3.

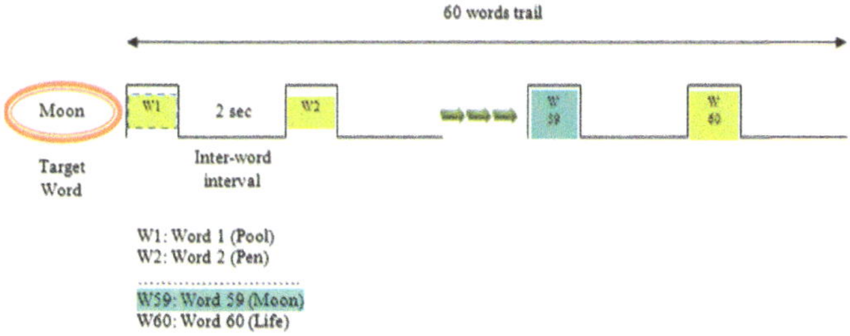


**Fig. 3.** Auditory attention task paradigm (secondary task)

**Protocol**: After a 20 s baseline recording, three gaming trials of 3 min each were performed each with a 30 s break. This was followed by a resting phase of 2 min in which they were asked to keep their eyes closed. The monosyllabic trail contained 60 words, i.e., with an inter-word interval of 2 s. The experimental timeline is shown in Fig. 4. There hence, leading to a whole experiment of 10 min excluding the time consumed in the experimental setup process.

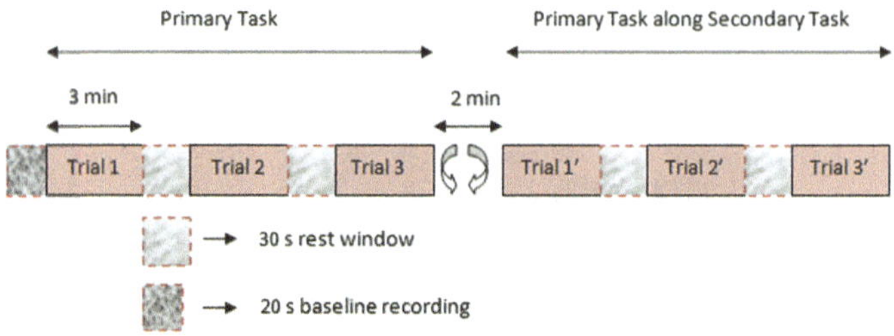


**Fig. 4.** Experimental timeline

## 2.3 Data Acquisition

The Ag/AgCl disc electrodes available with the RMS EEG-32 Super Sec System (Recorders and Medicare Systems, India) were used for recording the data. The electrodes placement was done in accordance with the International Standard 10–20 electrode placement system regulated by American Encephalographic Society [21]. The acquired EEG signals are a difference in potential with respect to time recorded across the electrode on the elicited lobe of the brain and reference electrode (Cz in this case). The sampling rate of the system was set to 1024 Hz/channel.

The participant is asked not to oil the hair on the day of the experiment and scalp of the participant is cleaned with spirit before the electrodes are attached to the scalp. This is done to lower down the scalp-electrode impedance below 5 kΩ for more accurate readings. The EEG paste is used for the placement of the electrodes that further helps in decreasing the impedance. The subject was instructed to refrain from all the possible muscular movements (especially facial ones) to avoid the artifacts. A comfortable sitting position was offered to the subject during the recording (at around 70 cm distance from the simulator screen). The agenda behind the research and the procedure to be followed was explained before the onset of trials. Signals were acquired from three electrodes T7, Cz, and T8.

## 2.4 EEG Preprocessing

The data acquired was preprocessed offline that included using Independent Component Analysis (ICA) for the removal of artifacts (eye blinks, facial movements, etc.), baseline correction, etc. Bandpass filtering (of all the trials) was performed on the data for further feature extraction with the cut-off frequencies: 0.1 and 30 Hz [22]. Also, the same set (second trial sequence) of data (after baseline correction) was windowed down into samples of 1 s so as to obtain the auditory P300 peak.

### 2.5 EEG Feature Extraction

One-second EEG segment was considered for the estimation of power spectral density which can further be extended for real-time applications. Power Spectral Density (PSD) was calculated corresponding to these windowed segments with the help of Welch's method. Theta band power and beta band power analysis was done to study the variation in the visual consciousness as well as active thinking. The theta and beta band power ratio variations pointed to the cognitive load variations during trials 1, 2, and 3 (Primary task alone). While this ratio variation played a significant role in trial 1', 2', and 3' by pointing out whenever the target word was heard (Collaterally occurring primary and secondary tasks).

Figure 5 displays the Theta to Beta band power ratio analysis for a span of one minute, i.e., first 20 s in graph (a), the next 20 s in graph (b), and the last 20 s in the graph (c). All the dips, i.e., decrease in the ratio of Theta to Beta band power imply the introduction of more cognitive load (Auditory Attention Task). This can be incurred from the three graphs of Fig. 5 that the target word was heard six times. In other words, attention from the primary task was deviated six times.

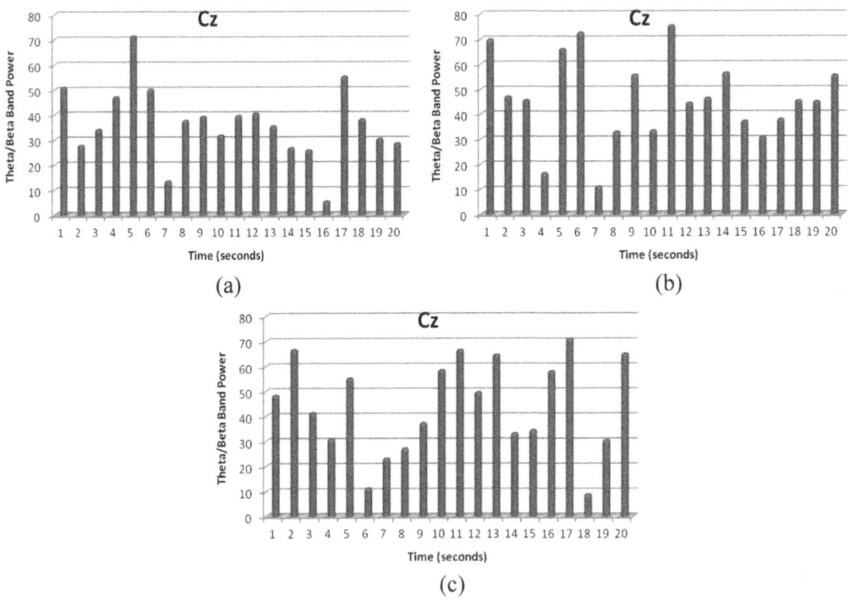


**Fig. 5.** Band power ratio analysis of one minute shown in three parts

## 3 Results

Although the consciousness could be defined from the theta and beta band power analysis separately for the precise calculation of the instant when the attention got shifted, Theta to Beta band power ratio needs to be calculated. As the attention paid by subject is affected by many factors like task tenure, task difficulty level, mental fatigue, ambient environment, mood, etc., the ratio keeps on varying and doesn't remain the same. For every trial (Trial 1', 2' and 3': 3 min each), the attention deviation was studied in slots of 20 s like as shown in Fig. 5.

The auditory attention span EEG recording in time domain was analyzed for the detection of P300 peaks. It was observed that the count of target word tallied by the subject was equal to the number of P300 peaks detected.

## 4 Conclusion

Such investigatory studies can help in alarming at the instants whenever attention deviates from the primary task due to interfering tasks. This is important as attention deviations may lead to disasters in case of air traffic controllers, road traffic controllers, process plants, etc. Such studies shall be converted into cognitive batteries so that it becomes more task-oriented as well as subject-oriented.

In the medical field, such investigations can help in monitoring the improvement pace in patients with vegetative state or minimally conscious state. Responses obtained in such cases can predict the recovery states.

**Acknowledgements.** Authors are grateful to the Voodoo, the developer of 'Helix Jump', an open-source android platform game that was used for this experimentation. The authors' sincere thanks are addressed to Human–Computer Interface Laboratory for the provision of requisite research atmosphere and Ministry of Human Resource Development (MHRD) for granting such explorative opportunities along with support.

## References

1. Sanjeevirayar, B., Ravindran, G., Shenbaga Devi, S.: Analysis of spectral features of EEG during four different cognitive tasks. Int. J. Eng. Technol. **6** (2014)
2. Moretti, D.V., et al.: Individual analysis of EEG frequency and band power in mild Alzheimer's disease. Clin. Neurophysiol. **115**, 299–308 (2004)
3. Loft, S., Sanderson, P., Neal, A., Mooij, M.: Modeling and predicting mental workload in en route air traffic control: critical review and broader implications. Hum. Factors **49**, 376–399 (2007)
4. Cremades, J.G., Barreto, A., Sanchez, D., Adjouadi, M.: Human–computer interfaces with regional lower and upper alpha frequencies as on-line indexes of mental activity. Comput. Human Behav. **20**, 569–579 (2004)
5. Michel, C., Murray, M.: Towards the utilization of EEG as a brain imaging tool. NeuroImage **61** (2011)
6. Ahirwal, M., Londhe, N.: Power spectrum analysis of EEG signals for estimating visual attention. Int. J. Comput. Appl. **42** (2012)
7. Khalid, M.B., Rao, N.I., Rizwan-i-Haque, I., Munir, S., Tahir, F.: Towards a brain computer interface using wavelet transform with averaged and time segmented adapted wavelets. In: 2009 2nd International Conference on Computer, Control and Communication, pp. 1–4 (2009). https://doi.org/10.1109/ic4.2009.4909189

8. Kubler, A.: Towards brain-computer interfacing. An introduction to brain-computer interfacing (2007)
9. Nicolas-Alonso, L.F., Gomez-Gil, J.: Brain computer interfaces, a review. Sensors **12**, 1211–1279 (2012)
10. Hancock, P.A., Verwey, W.B.: Fatigue, workload and adaptive driver systems. Accid. Anal. Prev. **29**, 495–506 (1997)
11. Blankertz, B., et al.: The Berlin brain-computer interface: Non-medical uses of BCI technology. Front. Neurosci. **4**, 198 (2010)
12. Donchin, E., Spencer, K.M., Wijesinghe, R.: The mental prosthesis: assessing the speed of a P300-based brain-computer interface. IEEE Trans. Rehabil. Eng. **8**, 174–179 (2000)
13. Sanei, S., Chambers, J.A.: EEG signal processing (2007)
14. Farwell, L.A., Donchin, E.: Talking off the top of your head: toward a mental prosthesis utilizing event-related brain potentials. Electroencephalogr. Clin. Neurophysiol. **70**, 510–523 (1988)
15. Schultheis, H., Jameson, A.: Assessing cognitive load in adaptive hypermedia systems: physiological and behavioral methods. In: International Conference on Adaptive Hypermedia and Adaptive Web-Based Systems, pp. 225–234. Springer, Berlin (2004)
16. da Silva, F.P.: Mental workload, task demand and driving performance: what relation? Proc. - Soc. Behav. Sci. **162**, 310–319 (2014)
17. Dressler, O., Schneider, G., Stockmanns, G., Kochs, E.F.: Awareness and the EEG power spectrum: analysis of frequencies. BJA Br. J. Anaesth. **93**, 806–809 (2004)
18. Johnstone, S.J., Blackman, R., Bruggemann, J.M.: EEG from a single-channel dry-sensor recording device. Clin. EEG Neurosci. **43**, 112–120 (2012)
19. Corballis, P.M., Funnell, M.G., Gazzaniga, M.S.: Hemispheric asymmetries for simple visual judgments in the split brain. Neuropsychologia **40**, 401–410 (2002)
20. Howells, F.M., Ives-Deliperi, V.L., Horn, N.R., Stein, D.J.: Mindfulness based cognitive therapy improves frontal control in bipolar disorder: a pilot EEG study. BMC Psychiatry **12**, 15 (2012)
21. Jasper, H.H.: The ten-twenty electrode system of the international federation. Electroencephalogr. Clin. Neurophysiol. **10**, 370–375 (1958)
22. van Dinteren, R., Arns, M., Jongsma, M.L.A., Kessels, R.P.C.: Combined frontal and parietal P300 amplitudes indicate compensated cognitive processing across the lifespan. Front. Aging Neurosci. **6**, 294 (2014)

# Melanoma Segmentation Based on Multi-stage Approach Using Fuzzy and Graph-Cuts Methods

Olusoji B. Akinrinade(✉), Pius A. Owolawi, Chunling Du, and Temitope Mapayi

Department of Computer Systems Engineering,
Faculty of Information & Communication Technology,
Tshwane University of Technology, Pretoria, South Africa
sojiakinrinade2014@gmail.com
{owolawipa,duc,mapayit}@tut.ac.za

**Abstract.** Globally, skin cancer is one of the major health problems. While an early diagnosis with the proper management of the disease can successful help in the treatment of the disease, the assessment of the disease by a medical practitioner is time-consuming, subjective, and prone to bias due to variation in the training and experience of dermatologists. Although different automated methods of the disease's diagnosis have been proposed, various problems like image noise due to varying illumination, uneven low contrasts and ambiguities in the non-diseased skin and tumours in the different regions of the clinical image alongside with the edges and boundaries have been highlighted to require accurate discrimination during the use of the automated methods. This is due to the fact that they can lead to inaccurate extraction of the melanoma skin cancer in the medical images as the problems plague the performance of an automated approach of detecting the disease. This study implements a multi-stage image segmentation approach that utilises a fuzzy transformation at the image enhancement stage with graph-cuts technique for a more efficient detection of melanoma skin cancer. This experimental study shows that fuzzy enhancement integrated with graph-cuts technique achieve a very good segmentation performance on the overall image (i.e. foreground and background) with an average accuracy rate of 97.42%. This study also shows that the background segmentation using fuzzy enhancement combined with graph-cuts technique achieved the good background segmentation with an average specificity rate of 99.07%.

**Keywords:** Graph-cuts · Fuzzy · Melanoma · Segmentation · Skin cancer


X.-S. Yang et al. (eds.), *Proceedings of Fifth International Congress on Information and Communication Technology*, Advances in Intelligent Systems and Computing 1183, https://doi.org/10.1007/978-981-15-5856-6_49

# 1 General Introduction

Skin cancer has been identified as a major health problem globally [21]. There are various kinds of skin cancer, namely melanoma, actinic koratosis, atypical moles, basal cell carcinoma, merkel cell carcinoma and squamous cell carcinoma (see Fig. 1).

Melanoma is a skin cancer of the cell called melanocytes, producing melanin in the skin epidermis [35]. Melanoma skin cancer accounts for 75% of all cancer deaths [34]. The global estimation of the mortality rate of melanoma reveals that people older than sixty-five years are more affected by deaths due to the disease than younger people [22].

**Fig. 1.** Different forms of skin cancer [5]

Several risk factors relating to skin cancer have been identified: light skin; light-coloured hair; hazel-coloured, green-coloured or blue-coloured eyes; the susceptibility to burn instead of suntan; past event of grievous burns; as well as a family background chronicle of skin cancer [8]. According to [30], Populations of Caucasoid, in most cases, have a greater possibility of being affected by the cancer of the skin than the populations with dark skin. People with black and brown skin tend to naturally tolerate the exposure of their bodies to a relatively high degree of sunlight without sunburn or risk of developing skin cancer. Alternatively, individuals having pale skin or freckled skin with red-hair and blue eyes are the group with greatest risk of developing cancer of the skin, while the people with dark eyes and hair, who are not often affected by sunburn, are at the medium-level risk of developing skin cancer [2].

With the increasing incidence of melanoma cancer in some sub-groups of the society, the early screening of melanoma is very helpful [3]. The successful treatment of melanoma is related significantly to early diagnosis and management of the disease as the diagnosis is based primarily on visual inspection [28]. Malignant melanoma can be fairly easily excised through a very good prognosis for the patient when it is detected in an early stage that is non-invasive [5]. However, the visual inspection with the interpretation of the clinical images is time-consuming, subjective and even liable to suffer from biasses sometimes for trained dermatologists [33].

A segmentation technique based on the fusion of thresholds was proposed for the detection of lesion borders in [11] after observing that single global thresholding approaches are poor for numerous medical image segmentation. Skin lesions are characterised in [15] using a non-linear random forest (RF) regression model for the extraction of important physiological features from lesions in dermoscopic images. Deep learning technique was proposed by [16] for the segmentation of lesions in clinical images. An ensemble of thresholding approach was implemented in [20] for the automatic delineation of lesion edges in clinical images. An illumination-reflectance correction algorithm based on histogram equalisation and Monte Carlo nonparametric modelling for skin lesion photographs were implemented in [23]. There is, however, a need to further address the correction for local illumination variation. This is because illumination variation can be minimised in an image globally using an image histogram equalisation.

Although the previous studies conducted on the automated detection of melanoma have made some progress, there is a need to improve further on the accurate delineation of melanoma cancer in medical images [24]. This is because non-dermoscopic images are affected by noise caused by illumination variation and light reflection across the images [26]. These noises cause varying low contrasts and ambiguities in regions, intensities, edges, features and boundaries between normal skin and melanoma tumours [7]. If not properly addressed at the pre-processing phase, the image pixels, regions, intensities, edges, features, boundaries and ambiguities will detrimentally affect the segmentation of the disease in the clinical images [25] as well as the performance of the automated diagnosis system of the disease as it is highly dependent on the melanoma segmentation performance. This study proposes a new multi-stage approach utilising fuzzy transformation enhancement and graph-cuts techniques for the efficient delineation of melanoma skin cancer in medical images.

The subsequent sections of this paper are as follow: the methods and techniques such as fuzzy enhancement technique, graph-cuts techniques and the post-processing method utilised in this paper are explored in Sect. 2. The experimental results and discussion of approach investigated in this study with results obtained in the literature are outlined in Sect. 3 while the conclusion is drawn in Sect. 4.

## 2 Methods and Techniques

The section sets out the methods and techniques used in this study for the delineation of melanoma skin cancer in medical images. The multi-stage segmentation approach incorporates an image enhancement stage using fuzzy enhancement, graph-cuts segmentation technique and post-processing stage for the melanoma segmentation.

Complexities such as nonhomogeneous illumination, light reflection across the images nonhomogeneous low contrast result in ambiguities in the image region, intensities, edges, features and boundaries between the healthy skin and the melanoma cancer tumours [26]. These problems require the use of efficient image pre-processing techniques to remove the image pixel ambiguities caused by the image noise and enhance the clinical image to produce good melanoma skin cancer segmented images. A fuzzy enhancement method is used to remove the image pixel ambiguities. It utilises an approach that maps grey-level pixels of the clinical images into membership functions that can improve the clinical image information representation and will be suitably manipulated to enhance an image by minimising the noise due to nonhomogeneous illumination and contrast variation in the clinical image. The steps used are basically fuzzification, inference engine and defuzzification [13]. The use of fuzzy in image enhancement is efficient because the pixels, which are often in the spatial domain, are transformed into fuzzy domain using membership function [27]. The transformation involves three stages and they are: fuzzification process of the image, membership function modification and the process of defuzzification. The image affected non-uniform illumination is converted from in spatial domain to fuzzy domain using the fuzzification process [25].

A fuzzy set ($B$) made up of its finite number of supports $x_1, x_2, \ldots x_n$ is defined as:

$$B = \{(\mu_B(x_i),\ x_i)\} \tag{1}$$

such that the membership function $\mu_B(x_i)$ represents the degree of the membership of an event $x_i$ in '$B$', and it ranges between the interval (0, 1). This function reflects the ambiguity or fuzziness in B. The crossover point in B is equal to $x_i$ when $\mu_B(x_i) = 0.5$. Equation 1 can be rewritten in a union form as:

$$B = \bigcup_i \mu_i/x_i, \quad i = 1, 2, \ldots, n \tag{2}$$

The function p($x_i$), is defined as the degree of possessing property '$p$' over an event $x_i$ and it ranges between the interval (0, 1) [27].

The fuzzy set concept ia applied to the clinical greyscale image $G_{\text{im}}$ which is two dimensional ($M \times N$) of $K$ levels with each image pixel retaining a membership function value that typify the degree of having a certain brightness state that has a relation to some level of brightness $k$, $k = 0, 1, 2, \ldots, K-1$. The fuzzy set of the two dimensional greyscale image $G_{\text{im}}$ is given by [32] as follows:

$$G_{\text{im}} = \bigcup_m \bigcup_n P_{mn}/X_{mn}, \quad m = 1, 2, \ldots, M\ n = 1, 2, \ldots, N \tag{3}$$

where $P_{mn}/X_{mn}$, $0 \leqslant P_{mn} \leqslant 1$ describes the degree of having the property $P_{mn}$ which is the property of brightness level by the pixel $X_{mn}$.

For the purpose of contrast enhancement, a different fuzzy set of $A^{'}$ is generated using the contrast intensification operator (INT) as:

$$A^{'} = \text{INT}(A) \tag{4}$$

The membership function is then computed as:

$$\mu_{A^{|}} = \mu_{\text{INT}(A)}(x)$$

$$= \begin{cases} 2[\mu_A(x)]^2, & 0 \leq \mu_A(x) \leq 0.5 \\ [1 - 2(1 - \mu_A(x))^2] & 0.5 \leq \mu_A(x) \leq 1 \end{cases} \tag{5}$$

The set of values of $\mu_A(x) > 0.5$ are increased and $\mu_A(x) < 0.5$ are decreased when the fuzziness of set A is reduced, thereby enhancing the contrast. Equation (6) can rewritten by applying a transformation function $T_r$ used continuously on $\mu(x)$. In order to enhance the image X, the $P_{mn}$ in equation (4) shall be modified to $P'_{mn}$ and it is further computed as:

$$P'_{mn} = T_r(P_{mn}) = \begin{cases} T'_r(P_{mn}), & 0 \leq P_{mn} \leq 0.5 \\ T''_r(P_{mn}), & 0.5 \leq P_{mn} \leq 1 \\ & r = 1, 2 \ldots. \end{cases} \tag{6}$$

The crossover point $x_c$ is computed as:

$$x_c = l + 0.5 \tag{7}$$

such that all $x_{mn} \gtreqless x_c$ obtainable in the spacial domain shall have the values $P_{mn} \gtreqless 0.5$ obtainable in the property domain. Using $F_{\text{d}}$, which is the denominational fuzzifier, and $F_{\text{e}}$, which is the exponential fuzzifier, the extraction of the fuzzy property $P_{mn}$ of the ($m$ by $n$) pixel of the input greyscale clinical image $x_{mn}$ of $M \star N$ image pixels is computed as:

$$P_{mn} = \begin{cases} \left[ l + \dfrac{x_{max} - x_{mn}}{F_{\text{d}}} \right]^{-F_{\text{e}}} \\ m = 1, 2 \ldots, M; \\ n = 1, 2 \ldots, N; \end{cases} \tag{8}$$

where the maximum grey level is $x_{\max}$ with $P_{mn} = 0$ means dark and $P_{mn} = 1$ means light. Given that $x_{mn} = 0$, the value of $P_{mn}$ is $\alpha$ which finite and positive. The term $\alpha$ is then computed as follows:

$$\alpha = \left( l + \frac{x_{\max}}{F_{\text{d}}} \right)^{-F_{\text{e}}} \tag{9}$$

The $m$th and $n$th pixel of the enhanced image is then computed as:

$$X^1_{mn} = a_0 x_{mn} + a_1 \sum_{Q_1} x_{pq} + a_2 \sum_{Q_2} x_{pq} + \cdots a_k \sum_{Q_k} x_{pq} \tag{10}$$

where $Q_v$, such that ($v = 1, 2, \ldots k$), are within or on the radius $R_1$ that is centred at the image data point $(m, n)$.

A median filter of local window size $k = 6$ is applied on the results obtained after applying the fuzzy transformation-based enhancement to remove different artefacts such as illumination, skin hairs, small stains and lines.

$$P(p, q) = F(m, n) * X^1_{k*k}(m, n) \tag{11}$$

where $X^1(m, n)$ represents fuzzy enhanced image;$P(p, q)$ signifies the different convolved medical images respectively; also median filter is the convolution mask $F(m, n)$ and $k = 6$.

Graph-cuts is an energy minimisation algorithm frequently used for the automated segmentation of images [9,10]. Graph-cuts have been mentioned extensively in the literature for the segmentation of different medical images [14,29,31].

The graph-cuts technique is applied on results obtained from the fuzzy enhancement method investigated in the pre-processing phase to ascertain its effectiveness. A labelled graph denoted as '$G$', and the collection of the graph vertices represented as '$V$' that correspond to the data points of the image denoted by '$p$' of the enhanced clinical image obtained from the pre-processed medical image '$P$' are utilised. The set of the different edges linking the different vertices is specified as '$E$', the sink or background is "$T/0$" while source or foreground is "$S/1$" described as the two different terminal vertices.

The given set '$N$' represents all the different neighbourhood pixel pairs {p, q} of the enhanced medical image indicated as '$P$' and let $L = \{l_1, l_2, \ldots l_i \ldots l_p\}$ be the pixels labelled as '1' & '0' ( $l_i \epsilon \{0, 1\}$) that correspond to melanoma skin cancer and the background of the medical image respectively.

In order to decide whether an image data point is melanoma or the medical image and background, the finest label is then allotted to data point p $\epsilon$ P. Given that the different terminals are alienated, the s/t cut is a collection of boundries that delineates the set of pixels in the medical image into dissimilar classes. The cut-cost denoted by '$C$' is the total weights of all the edges of the melanoma image graph which is portion of the cut. The minimum source/terminal note cut that corresponds to the optimum detection of disease in the medical image amongst the different cuts on the graph are then polynomially computed through a Min-Cut/Max-Flow algorithm as discussed by [10] (see Fig. 2). Given that the factor of relative importance amid the boundary term denoted by B(L) and the regional term denoted by $R(L)$ is $\alpha$, the minimised function that corresponds to the optimum detection of the melanoma skin cancer in the medical image is computed as follows:

$$E(L) = B(L) + \alpha R(L) \tag{12}$$

where $R(L)$ is calculated as:

$$R(L) = \sum_{p \epsilon P} D_p(l_p) \tag{13}$$

and $B(L)$ is calculated as:

$$B(L) = \sum_{p,q \epsilon N} V(l_p, l_q) \tag{14}$$

and $Dp(lp)$, estimates the proportion to which allocating the label fp to image data point denoted by p differs with intensity of the clinical image's data point denoted by $(I_p)$.

$$Dp(lp) = -\ln P(I_p | l_p) \tag{15}$$

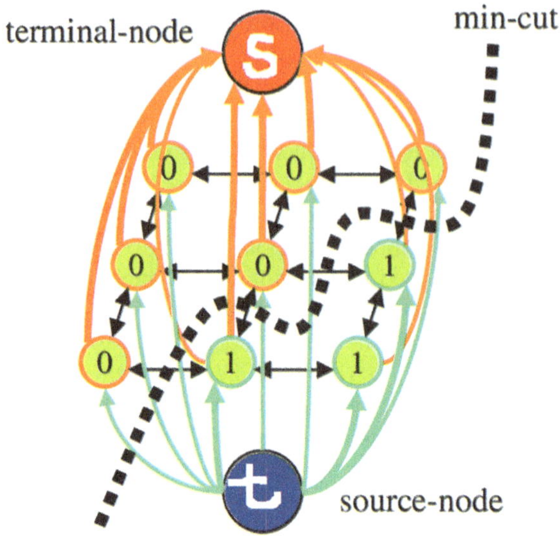


**Fig. 2.** Graph with $S/T$ cut using Min-Cut/Max-Flow

Considering that $(p, q)$ are neighbouring pixels $V(l_p, l_q)$, the penalty is computed as 0 whenever the labels of the pixels are the same. This is indicated as:

$$V(l_p, l_q) = \begin{cases} 1 & \text{If } \; l_p \neq l_q \\ 0 & \text{If } \; l_p = l_q \end{cases} \tag{16}$$

Morphological operation constructed utilising disc-shaped structuring element is used on the resulting image obtained following the application of graph-cuts method. This is implemented for the removal of the remaining noisy pixels in the segmented image in order to enhance the accuracy level of the detected melanoma skin cancer. It is established empirically that the radius 15 effectively removes the noise.

## 3 Experimental Set-up, Results and Discussion

The algorithms for the delineation of melanoma cancer disease in medical images investigated in this study were executed using MATLAB 2014aT M through an Intel(R) Core(TM) i5 3210M CPU (2.50 GHz, 4GB RAM). The datasets used for this study are Dermatology Information System (DermIS) [17] and DermQuest [18] databases. The two databases contain 206 colour images, with their corresponding benchmark images used as ground-truth similarly provided. Out of all the images, 119 are contain melanoma while 87 medical images do not contain melanoma. The medical images containing melanoma in the databases

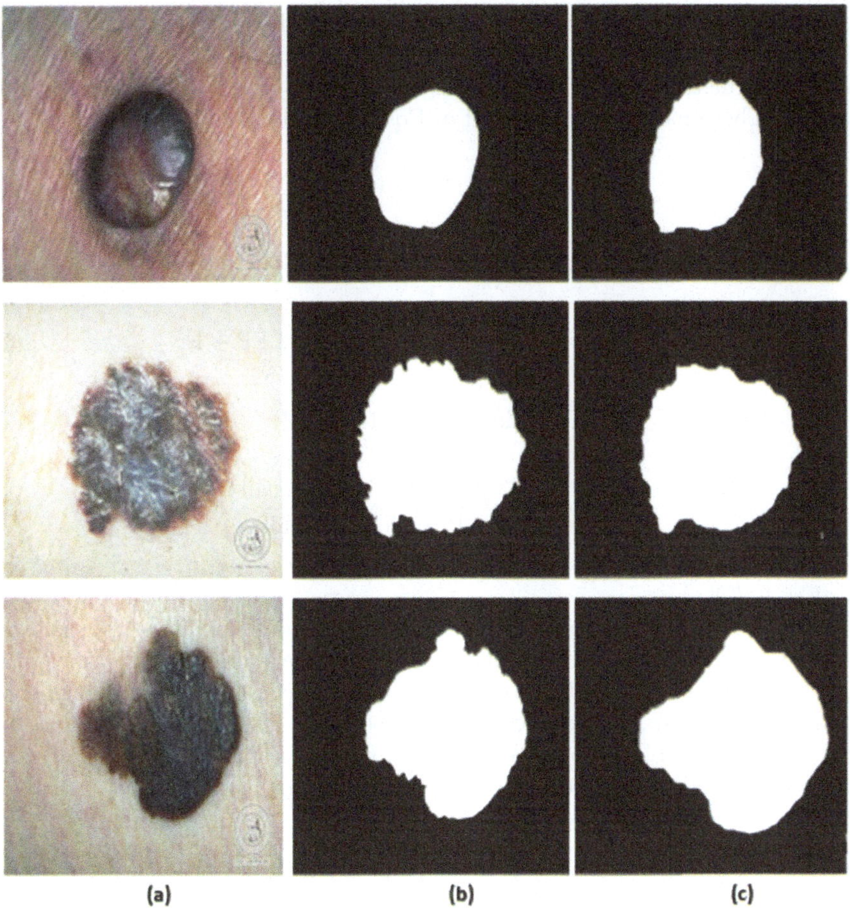


**Fig. 3.** **a** Colour medical images showing melanoma, **b** manually segmented ground-truth, **c** melanoma segmentation results of fuzzy enhancement with graph-cuts technique

are considered for the evaluation of the multi-stage segmentation approach proposed in this study. Figure 3 shows an original coloured medical image containing melanoma, its ground-truth and the visual segmentation results obtained from the method proposed in this paper.

The performance measures used include sensitivity, specificity accuracy, precision and $F$-score rates. The sensitivity metric indicates the ability of the proposed segmentation method to rightly detect the diseased pixels in the medical images, and the specificity metric indicates the ability of the automated segmentation method to rightly detect pixels in the medical image background. The accuracy metric indicates the degree that the detected melanoma disease in the medical image matches the experts benchmark image pixelwise. Precision metric indicates the proportion of rightly classified diseased pixels to the aggregate amount of pixels observed to be melanoma cancer in the medical image. An $F$-score metric calculates harmonic mean value of sensitivity and precision measure. These metrics are outlined in Eqs. (17)–(21) as follow:

$$\text{Sensitivity} = \text{TP}/(\text{TP} + \text{FN}) \tag{17}$$

$$\text{Specificity} = \text{TN}/(\text{TN} + \text{FP}) \tag{18}$$

$$\text{Accuracy} = (\text{TP} + \text{TN})/(\text{TP} + \text{TN} + \text{FP} + \text{FN}) \tag{19}$$

$$\text{Precision} = \text{TP}/(\text{TP} + \text{FP}) \tag{20}$$

$$F\text{-Score} = \frac{2 \times \text{Precision} \times \text{Recall}}{\text{Precision} + \text{Recall}} \tag{21}$$

*where true positive is TP, false positive is FP, true negative is TN and false negative is FN.*

An instance is considered TP when an image data point is appropriately detected as diseased in medical image containing melanoma, and TN is a situation where the non-diseased part of the medical image is appropriately segmented as background. Also, an instance is presumed to be FN if a melanoma diseased pixel is classified to be a non-diseased pixel, while it is FP when a pixel of image background is erroneously detected as the pixel of melanoma.

**Table 1.** Performance comparison of the proposed method with some segmentation methods in the literature

| Method | Average accuracy (%) | Average sensitivity (%) | Average specificity (%) |
|---|---|---|---|
| Dey et al. [19] | 94.81 | 97.74 | 89.71 |
| Celebi & Zornberg [12] | 71.70 | 61.60 | 75.80 |
| Amelard et al. [4] | 86.89 | 91.60 | 89.46 |
| Abbes & Sellami [1] | 76.90 | 97.40 | 48.40 |
| Ashour et al. [6] | 96.24 | 98.25 | 92.46 |
| Proposed method graph-cuts and fuzzy-enhanced | 97.42 | 88.30 | 99.07 |

The results of the methods based on graph-cuts investigated in this research are compared with the results obtained from some methods proposed in the literature (see Table 1). The results obtained for melanoma detection in

[1,4,6,12,19] are obtained from [6]. The graph-cuts-based technique investigated this study achieved higher average specificity value when compared with the results obtained in [1,4,6,12,19]. The graph-cuts-based technique investigated this study achieved higher average accuracy value when compared with the results obtained in [1,4,6,12,19]. The graph-cuts-based technique investigated this study, however, achieved lower mean sensitivity value when compared with the results obtained in [1,4,6,19], but achieved lower mean sensitivity value when compared with the result obtained in [12].

## 4 Conclusion

The study conducted in this paper indicated that while various methods have been proposed for the automatic detection of melanoma in medical images, problems such as noise due to illumination variation, varying low contrast and ambiguities in the normal skin and tumours of medical image region, intensities, edges, features and boundaries. This paper proposed a multi-stage image segmentation approach that utilises a fuzzy transformation-based enhancement to mitigate the problem of non-homogeneous illumination in the clinical images and graph-cuts technique for the automated detection of melanoma skin cancer in the clinical images. This study showed that fuzzy enhancement combined with graph-cuts technique achieved a very good segmentation performance on the overall image (i.e. foreground and background) with an average accuracy rate of 97.42%.

## References

1. Abbes, W., Sellami, D.: Automatic skin lesions classification using ontology-based semantic analysis of optical standard images. Procedia Comput. Sci. **112**, 2096–2105 (2017)
2. Ainger, S.A., Jagirdar, K., Lee, K.J., Soyer, H.P., Sturm, R.A.: Skin pigmentation genetics for the clinic. Dermatology **233**(1), 1–15 (2017)
3. Akinrinade, O., Owolawi, P.A., Du, C., Mapayi, T.: Graph-cuts technique for melanoma segmentation over different color spaces. In: 2018 IEEE International Conference on Intelligent and Innovative Computing Applications (ICONIC 2018), Plaine Magnien, Mauritius, 6th–7th Dec 2018. IEEE (2018)
4. Amelard, R., Glaister, J., Wong, A., Clausi, D.A.: High-level intuitive features (hlifs) for intuitive skin lesion description. IEEE Trans. Biomed. Eng. **62**(3), 820–831 (2014)
5. Amelard, R., Glaister, J., Wong, A., Clausi, D.A.: High-level intuitive features (hlifs) for intuitive skin lesion description. IEEE Trans. Biomed. Eng. **62**(3), 820–831 (2015)
6. Ashour, A.S., Guo, Y., Hawas, A.R., Du, C.: Optimised halftoning and inverse halftoning of dermoscopic images for supporting teledermoscopy system. IET Image Process. **13**(3), 529–536 (2018)
7. Bakheet, S.: An svm framework for malignant melanoma detection based on optimized hog features. Computation **5**(1), 4 (2017)

8. Baxter, A.J., Hughes, M.C., Kvaskoff, M., Siskind, V., Shekar, S., Aitken, J.F., Green, A.C., Duffy, D.L., Hayward, N.K., Martin, N.G., et al.: The queensland study of melanoma: environmental and genetic associations (q-mega); study design, baseline characteristics, and repeatability of phenotype and sun exposure measures. Twin Res. Hum. Genetics **11**(2), 183–196 (2008)
9. Boykov, Y., Funka-Lea, G.: Graph cuts and efficient nd image segmentation. Int. J. Comput. Vis. **70**(2), 109–131 (2006)
10. Boykov, Y., Kolmogorov, V.: An experimental comparison of min-cut/max-flow algorithms for energy minimization in vision. IEEE Trans. Pattern Anal. Mach. Intell. **26**(9), 1124–1137 (2004)
11. Celebi, M.E., Hwang, S., Iyatomi, H., Schaefer, G.: Robust border detection in dermoscopy images using threshold fusion. In: 2010 17th IEEE International Conference on Image Processing (ICIP), pp. 2541–2544. IEEE (2010)
12. Celebi, M.E., Zornberg, A.: Automated quantification of clinically significant colors in dermoscopy images and its application to skin lesion classification. IEEE Syst. J. **8**(3), 980–984 (2014)
13. Chaudhry, H., Rahim, M.S.M., Khalid, A.: Multi scale entropy based adaptive fuzzy contrast image enhancement for crowd images. Multimedia Tools Appl. 1–20 (2018)
14. Chen, X., Udupa, J.K., Bagci, U., Zhuge, Y., Yao, J.: Medical image segmentation by combining graph cuts and oriented active appearance models. IEEE Trans. Image Process. **21**(4), 2035–2046 (2012)
15. Cho, D.S., Haider, S., Amelard, R., Wong, A., Clausi, D.: Physiological characterization of skin lesion using non-linear random forest regression model. In: 2014 36th Annual International Conference of the IEEE Engineering in Medicine and Biology Society, pp. 3349–3352. IEEE (2014)
16. Codella, N.C., Nguyen, Q.B., Pankanti, S., Gutman, D., Helba, B., Halpern, A., Smith, J.R.: Deep learning ensembles for melanoma recognition in dermoscopy images. IBM J. Res. Develop. **61**(4/5), 1–5 (2017)
17. DermIS: Dermatology information system (dermis) database. [online]. Available: http://www.dermis.net. Accessed 25 Mar 2018
18. DermQuest: Dermquest database. [online]. Available: http://www.dermquest.com. Accessed 25 Mar 2018
19. Dey, N., Rajinikanth, V., Ashour, A., Tavares, J.M.: Social group optimization supported segmentation and evaluation of skin melanoma images. Symmetry **10**(2), 51 (2018)
20. Emre Celebi, M., Wen, Q., Hwang, S., Iyatomi, H., Schaefer, G.: Lesion border detection in dermoscopy images using ensembles of thresholding methods. Skin Res. Technol. **19**(1), e252–e258 (2013)
21. Gawkrodger, D.: Occupational skin cancers. Occup. Med. **54**(7), 458–463 (2004)
22. Geller, A.C., Clapp, R.W., Sober, A.J., Gonsalves, L., Mueller, L., Christiansen, C.L., Shaikh, W., Miller, D.R.: Melanoma epidemic: an analysis of six decades of data from the connecticut tumor registry. J. Clin. Oncol. **31**(33), 4172 (2013)
23. Glaister, J., Amelard, R., Wong, A., Clausi, D.A.: Msim: Multistage illumination modeling of dermatological photographs for illumination-corrected skin lesion analysis. IEEE Trans. Biomed. Eng. **60**(7), 1873–1883 (2013)
24. Haider, S., Cho, D., Amelard, R., Wong, A., Clausi, D.A.: Enhanced classification of malignant melanoma lesions via the integration of physiological features from dermatological photographs. In: 2014 36th Annual International Conference of the IEEE Engineering in Medicine and Biology Society, pp. 6455–6458. IEEE (2014)

25. Hasikin, K., Isa, N.A.M.: Fuzzy image enhancement for low contrast and non-uniform illumination images. In: 2013 IEEE International Conference on Signal and Image Processing Applications, pp. 275–280. IEEE (2013)
26. Jaisakthi, S., Chandrabose, A., Mirunalini, P.: Automatic skin lesion segmentation using semi-supervised learning technique. arXiv preprint arXiv:1703.04301 (2017)
27. Kandel, A.: Fuzzy mathematical techniques with applications (1986)
28. Laal, M.: Innovation process in medical imaging. Procedia Social Behav. Sci. **81**, 60–64 (2013)
29. Ming, J.T.C., Noor, N.M., Rijal, O.M., Kassim, R.M., Yunus, A.: Enhanced automatic lung segmentation using graph cut for interstitial lung disease. In: 2014 IEEE Conference on Biomedical Engineering and Sciences (IECBES), pp. 17–21. IEEE (2014)
30. Narayanan, D.L., Saladi, R.N., Fox, J.L.: Ultraviolet radiation and skin cancer. Int. J. Dermatol. **49**(9), 978–986 (2010)
31. Oluyide, O.M., Tapamo, J.R., Viriri, S.: Automatic lung segmentation based on graph cut using a distance-constrained energy. IET Comput. Vis. (2018)
32. Pal, S.K., King, R., et al.: Image enhancement using smoothing with fuzzy sets. IEEE Trans. Sys. Man Cybern. **11**(7), 494–500 (1981)
33. PMM., F.: Contributions to the segmentation of dermoscopic images. Master's thesis (2012)
34. Riker, A.I., Zea, N., Trinh, T.: The epidemiology, prevention, and detection of melanoma. Ochsner J. **10**(2), 56–65 (2010)
35. Yamaguchi, Y., Brenner, M., Hearing, V.J.: The regulation of skin pigmentation. J. Biol. Chem. **282**(38), 27557–27561 (2007)

# Cryptanalysis of Two-Factor Remote User Authentication Scheme for Wireless Sensor Networks

Preeti Chandrakar[1(✉)] and Rifaqat Ali[2]

[1] Department of Computer Science and Engineering, National Institute of Technology, Raipur 492010, India
preet29.chandrakar@gmail.com
[2] Department of Mathematics & Scientific Computing, National Institute of Technology, Hamirpur 177005, India
rifaqatali27@gmail.com

**Abstract.** In Wireless Sensor Networks (WSNs), the real-time data which is highly sensitive, collected by sensor nodes. If anyone acquires this real-time information illegally, then the privacy will be revealed. Therefore, privacy, authentication, and data security are extremely important to access real-time information over an unreliable channel. Authentication is most common security mechanism to protect WSNs data over an untrustworthy network. Recently, Amin-Biswas devised two-factor remote user authentication protocol for WSNs and declared that it is safe from various security threats. However, we identified that their scheme has several security threats like password-guessing, identity-guessing, user-impersonation, session-key temporary information, smart card theft, and lack of forward-secrecy property. The computation cost, smartcard storage cost, and communication cost are also presented in the section of performance comparison.

**Keywords:** Authentication · Cryptanalysis · Security · Wireless sensor networks (wsns)

## 1 Introduction

With the increasing development and popularity of the communication technologies, the wireless sensor networks have been broadly used in different environments for obtaining the specific information [1–6]. WSNs are implemented using sensor nodes which have limited transmission capacity energy and computing resources. The sensor nodes are deployed in various geographical areas for sensing and processing the real-time data. The sensor nodes communicate with


X.-S. Yang et al. (eds.), *Proceedings of Fifth International Congress on Information and Communication Technology*, Advances in Intelligent Systems and Computing 1183, https://doi.org/10.1007/978-981-15-5856-6_50

each other and finally, they transmit the specific data over an insecure channel. On another hand, the open nature of the communication technology makes WSNs more vulnerable to different threats. However, the security and privacy are very sensitive issues in WSNs. Therefore, an authentication scheme is one of the most common and simple security mechanism to facilitate these issues in different network-based applications.

Recently, number of remote user authentication protocols have been devised for WSNs [1,2,5–11]. Das et al. [12] developed a novel dynamic authentication scheme for WSNs and demonstrated that their scheme facilitates appropriate security with efficient computation and communication cost. Later, Wang et al. [13] stated that the protocol [12] has various security threats like privileged insider, server-masterkey disclosure and smartcard security breach. Turkanovic et al. [14] also demonstrate that the protocol [12] is not useful for real-life implementation due to various security weaknesses. Then, Turkanovic et al. [14] devised another authentication scheme to override these-above security weaknesses. Vaidya et al. [6] also devised two-factor mutual-authentication protocol using WSNs and ensured it is resisting security attacks and achieving efficient complexity. After that Kim et al. [15] found that the protocol in [6] has security flaws like gateway node bypassing and user-impersonation. Later, Deebak [16] also identified Vaidya et al. [6] suffers from gateway node bypassing and userkey impersonation attacks. To improve security, Deebak [16] developed an enhanced scheme.

In 2014, a secure authentication protocol for WSNs is designed by Turkanovic et al. [4]. In the protocols [2] and [17] described that the scheme [4] is having various security pitfalls. Later, Amin-Biswas [17] illustrated that the scheme in [4] suffer from password and identity guessing, user-impersonation, sensor node impersonation and smartcard theft attacks. For resolving these issues, Amin and Biswas [17] devised a new scheme for WSNs. In this article, we observed that the scheme [17] does not defend many vulnerabilities like identity and password-guessing, smartcard theft, user-impersonation, revelation of secret parameters, failure to achieve forward-secrecy property and not secure against session-key temporary-information attack.

The remaining of the paper is presented as follows. In Sect. 2, security pitfalls of the scheme [17] are demonstrated. The performance comparison presents in Sect. 3 and finally concluded in Sect. 4.

## 2 Security Pitfalls in Amin-Biswas's Protocol

This section presents security pitfalls in Amin-Biswas's protocol [17] which are explained as follows:

### 2.1 Stolen Verifier Attack

In this attack, attacker theft the information stored in the database of $HGWN$ and later attacker launches various attacks against Amin-Biswas's protocol [17].

The *HGWN* maintains a database that stored the user-specific information such as $\text{TID}_i$ and $\text{DID}_i$, which are fixed for every session. The attacker uses $\text{TID}_i$ and $\text{DID}_i$ to break the Amin-Biswas's protocol, and we describe that their scheme is susceptible to identity and password-guessing and user-impersonation attacks which are described as follows.

**Identity-Guessing Attack:** Usually, $U_i$ chooses low-entropy identity for easy memorization, which is easily guessable by an $\mathcal{A}$. Amin-Biswas proclaimed that their protocol is secure against identity guessing attack. However, we identify that Amin-Biswas's protocol is suffering from off-line identity guessing attack. Suppose attacker $\mathcal{A}$ steals the database of *HGWN* and obtains the all user-specific information $\{\text{TID}_i, \text{DID}_i\}$ from it. Then, $\mathcal{A}$ obtains $U_i$'s smartcard and retrieves all the secret values $\{\text{Reg}_i, A_i, \text{ID}_{\text{GWN}_h}, \text{TID}_i, h(\cdot), r\}$ from it. Now, attacker can successfully launch identity-guessing attack, which is described in **Algorithm 1**.

**Algorithm 1**

1: Attacker ($\mathcal{A}$) extracts $\{Reg_i', A_i', ID_{GWN_h}', TID_i', h(.), r'\}$ from smartcard.
2: $\mathcal{A}$ selects a pair of $\{TID_i, DID_i\}$ from the stolen database of the *HGWN*.
3: **if** $(TID_i' == TID_i)$ **then**
4: $ID_i^*$ is choosen by $\mathcal{A}$ from the dictionary.
5: $\mathcal{A}$ calculates $DID_i^* = h(ID_i^* \parallel r')$
6: **if** $(DID_i^* == DID_i)$ **then**
7: Return $(ID_i^*)$
8: **else**
9: Goto Step 4 and chooses $ID_i^*$ and executes the Steps 4 to 7, until the correct $ID_i$ is not obtained.
10: **end if**
11: **else**
12: Goto Step 2 and selects another pair of $\{TID_i, DID_i\}$ and executes the Steps 3 to 11, until the exact $ID_i$ is not obtained
13: **end if**

**Password-Guessing Attack:** $U_i$ chooses a low entropy password for easy memorization, but low entropy passwords are not secure and more chance to happen password guessing threat. The $\mathcal{A}$ has obtained $\{\text{DID}_i, \text{TID}_i\}$ from the database and extract the smartcard's confidential parameters $\{\text{Reg}_i, A_i, \text{ID}_{\text{GWN}_h}, \text{TID}_i, h(.), r\}$. Then, $\mathcal{A}$ executes **Algorithm 2** to obtain the password.

**User-Impersonation Attack:** The $\mathcal{A}$ obstructs public messages and creates a new modified login message and put forwards to the remote-server. If message is accepted by remote-server, $\mathcal{A}$ is successful to compile this attack. Based on stolen database $\{\text{TID}_i, \text{DID}_i\}$ of the *HGWN* and extracted parameters $\{\text{Reg}_i, A_i, \text{ID}_{\text{GWN}_h}, \text{TID}_i, h(.), r\}$ from smartcard then $\mathcal{A}$ can imitate as user easily.

**Step 1:** First, $\mathcal{A}$ executes **Algorithm 1** and **2** to guess the identity and password of $U_i$.

**Algorithm 2**

1: Attacker $\mathcal{A}$ extracts $\{Reg_i', A_i', ID_{GWN_h}', TID_i', h(.), r'\}$ from smartcard.
2: $\mathcal{A}$ selects a pair of $\{TID_i, DID_i\}$ from the stolen database of the $HGWN$.
3: **if** $(TID_i' == TID_i)$ **then**
4: $PW_i^*$ is choosen by $\mathcal{A}$ from the dictionary.
5: $\mathcal{A}$ calculates $Reg_i^* = h(DID_i \parallel h(PW_i^* \parallel r'))$
6: **if** $(Reg_i^* == Reg_i)$ **then**
7: Return$(PW_i^*)$: $PW_i^*$ is exact password of $U_i$
8: **else**
9: Goto Step 4 and chooses $PW_i^*$ and executes the Steps 4 to 7, until the exact $PW_i$ is not retrieved.
10: **end if**
11: **else**
12: Goto Step 2 and selects another pair of $\{TID_i, DID_i\}$ and executes the Steps 3 to 11, until the exact $PW_i$ is not retrieved
13: **end if**

**Step 2:** After obtaining correct $ID_i$ and $PW_i$ of the user, the attacker calculates $DID_a = h(ID_i^* \parallel r)$ and $PWR_a = h(PW_i^* \parallel r)$.
**Step 3:** After that $\mathcal{A}$ produces $r_a$ and evaluates $B_a = A_i \oplus h(DID_a \oplus PWR_a)$, $C_a = h(ID_{GWN_h} \parallel B_a \parallel r_a \parallel T_a)$ and $D_a = B_a \oplus r_a$, where $T_a$ is current timestamp.
**Step 4:** Finally, $\mathcal{A}$ forwards $\{ID_{GWN_h}, TID_i, ID_{SN_j}, C_a, D_a, T_{a1}\}$ to $HGWN$ over an untrustworthy channel.
**Step 5:** First obtaining the login request message, $HGWN$ checks $T_2 - T_{a1} \leq \triangle T$ and it is confirmed that the $T_{a1}$ would pass the timestamp validity test. Therefore, $HGWN$ proceeds further and extracts $DID_i$ corresponding $TID_i$ from the database. After that $HGWN$ computes $B_a{*} = h(DID_i \parallel TID_i \parallel x_k)$, $r_a^* = D_a \oplus B_a^*$ and $C_a^* = h(ID_{GWN_h} \parallel B_a^* \parallel r_a^* \parallel T_a)$.
**Step 6:** $HGWN$ compares if $C_a^* = C_a$, this condition would obviously true due to the exact values of $ID_i$ and $PW_i$.

## 2.2 Revelation of Secret Parameter

Suppose that $\mathcal{A}$ is wicked user who intercepts all the exchanged message between $\mathrm{SN}_j$, HGWN and FGWN during the session from the unreliable channel. Thus, we assume that $\mathcal{A}$ knows $\{H_j, K_j, T_3\}$, $\{E_i, f_i, G_i, y_k, T_2\}$, $\{\mathrm{TID}_i, \mathrm{AA}_i, \mathrm{BB}_i, \mathrm{TU}_i\}$ and $\{\mathrm{ZS}_i, \mathrm{ID}_{\mathrm{GWN}_f}\}$. We prove that $\mathcal{A}$ can get the secret parameter $P_j$ of sensor-node and $S_{\mathrm{ran}}$ of the gateway-node as follows:

**Step 1:** Initially, malicious $\mathcal{A}$ runs the session as the legitimate user with the $\mathrm{SN}_j$, HGWN and FGWN. $\mathcal{A}$ generates a login message $M_1 = \{\mathrm{ID}_{\mathrm{GWN}_h}, \mathrm{TID}_i, \mathrm{ID}_{\mathrm{SN}_j}, C_i, D_i, T_1\}$, where $D_i = B_i \oplus r_i$, $C_i = h(\mathrm{ID}_{\mathrm{GWN}_h} \parallel B_i \parallel r_i \parallel T_1)$ and $B_i = A_i \oplus h(\mathrm{DID}_i \parallel \mathrm{PWR}_i)$ and sends this message to HGWN.
**Step 2:** From now on $\mathcal{A}$ monitors the session among HGWN and FGWN and captures the exchanged message $\{\mathrm{ZS}_i, \mathrm{ID}_{\mathrm{GWN}_f}\}$ and $\{\mathrm{TID}_i, \mathrm{AA}_i, \mathrm{BB}_i, \mathrm{TU}_i\}$ between HGWN and FGWN, where $\mathrm{ZS}_i = Z_i \oplus S_{\mathrm{ran}}$, $\mathrm{BB}_i = Z_i \oplus r_i$.

**Step 3:** Finally, $\mathcal{A}$ executes the following steps to obtain the secret key $S_{\text{ran}}$ of gateway node and $P_j$ of sensor node.

- $\mathcal{A}$ computes $Z_i = \text{BB}_i \oplus r_i$ and $S_{\text{ran}} = Z_i \oplus \text{ZS}_i$.
- After that $\mathcal{A}$ computes $P_j = \text{PS}_j \oplus S_{\text{ran}}$, where $\text{PS}_j$ is obtained from the public channel.
- Therefore, $\mathcal{A}$ can get to obtain the secret parameters $S_{\text{ran}}$ and $P_j$

## 2.3 Failure to Achieve Forward-Secrecy Property

Forward-secrecy confirms that even if $\mathcal{A}$ got the secret key of the participant entities by some means, the secrecy of the session-key is not disclosed from this disclosure. We demonstrate two instances to prove that the Amin-Biswas's protocol fails to achieve this property.

**Case 1:** Assume that $\mathcal{A}$ is able to reveal the secret key $S_{\text{ran}}$ by some means, he can get $SK$ by performing the Algorithm 3 as given below:

**Algorithm 3**

1: Input: $\langle S_{ran}, ID_{SN_j}, PS_j, f_i, y_k, G_i, K_j \rangle$
2: Output: Session-key $SK$
3: Attacker computes $P_j{=}PS_j \oplus S_{ran}$
4: Attacker computes the random number $r_k{=}P_j \oplus f_i$, $r_i{=}y_k \oplus h(r_k)$ and $r_j{=}K_j \oplus r_k$
5: Attacker computes $DID_i{=}G_i \oplus h(ID_{SN_j} \parallel r_k)$
6: Finally $\mathcal{A}$ calculates $SK = h(DID_i \parallel r_i \parallel r_j \parallel r_k)$

**Case 2:** The attacker got the secret key $S_{\text{ran}}$ by some means and after that he runs the following steps to achieve the session-key $SK$.

**Step 1:** Attacker intercepts the message $\{\text{ID}_{\text{SN}_j}, \text{PS}_j\}$ from the public channel.
**Step 2:** Attacker computes $P_j = \text{PS}_j \oplus S_{\text{ran}}$.
**Step 3:** Attacker intercepts the message $\{\text{TID}_i, \text{ZS}_i, \text{ID}_{\text{GWN}_f}, \text{TID}_i, \text{AA}_i, \text{BB}_i, \text{TU}_i, \text{GG}_i, \text{RR}_F\}$.
**Step 4:** $\mathcal{A}$ computes $Z_i{=}\text{ZS}_i{\oplus}S_{\text{ran}}$, $R_i{=}\text{BB}_i{\oplus}Z_i$, $R_F{=}\text{GG}_i{\oplus}P_j$ and $R_j{=}\text{RR}_F{\oplus}R_F$.
**Step 5:** Finally attacker computes $\text{SK} = h(\text{TID}_i \parallel \text{ID}_{\text{SN}_j} \parallel R_i \parallel R_j \parallel R_F)$.

## 2.4 Known Session-Key Temporary-Information Attack

In this threat, the confidentiality of session-key will be revealed if the ephemeral secret value of the session is disclosed. We provide two instances to prove that Amin-Biswas's protocol suffers from session-key temporary information attack.

**Case 1:** In Amin-Biswas's scheme, SK = $h(\text{DID}_i \parallel r_i \parallel r_j \parallel r_k)$, where $r_i$, $r_j$ and $r_k$ random number generated by the user, sensor node and home gateway node respectively. We supposed that attacker got these ephemeral confidential information by some means and after that he computes $\text{DID}_i = G_i \oplus h(\text{ID}_{\text{SN}_j} \parallel r_k)$, where $G_i$ and $\text{ID}_{\text{SN}_j}$ is easily obtainable through public channel. Finally, $\mathcal{A}$ can easily compute SK = $h(\text{DID}_i \parallel r_i \parallel r_j \parallel r_k)$. Thus, the confidentiality of the session-key is susceptible due to the disclosure of the session-specific temporary-information attack.

**Case 2:** In Amin-Biswas's protocol, the session-key SK = $h(\text{TID}_i \parallel \text{ID}_{\text{SN}_j} \parallel R_i \parallel R_j \parallel R_F)$, where $R_i$, $R_j$ and $R_F$ are random number generated by user, sensor node and foreign gateway node. We suppose that these temporarily generated random number are disclosed to the attacker then he can easily calculate the session-key SK = $h(\text{TID}_i \parallel \text{ID}_{\text{SN}_j} \parallel R_i \parallel R_j \parallel R_F)$, where $\text{TID}_i$ and $\text{ID}_{\text{SN}_j}$ are public parameters. Therefore, the secrecy of the session-key is revealed due to the disclosure of the session-specific temporary information.

## 3 Performance Comparison

In this section, we have discussed the performance comparison of relevant schemes [1,5,8–10,17,18] in terms of several security attributes, smart card storage cost, communication and computation costs.

**Table 1.** The comparison of security features

| Schemes | Security Features Comparison | | | | | | | | | | | | | | |
|---|---|---|---|---|---|---|---|---|---|---|---|---|---|---|---|
| | 1 | 2 | 3 | 4 | 5 | 6 | 7 | 8 | 9 | 10 | 11 | 12 | 13 | 14 | 15 |
| $R^*$ [1] | ✓ | ✓ | ✗ | ✓ | ✓ | ✓ | ✗ | ✗ | ✓ | ✓ | ✓ | ✓ | ✓ | ✓ | ✓ |
| $R^*$ [9] | ✓ | ✓ | ✓ | ✓ | ✓ | ✓ | ✓ | ✓ | ✓ | ✓ | ✓ | ✓ | ✓ | ✓ | ✓ |
| $R^*$ [10] | ✓ | ✓ | ✗ | ✓ | ✓ | ✓ | ✗ | ✓ | ✓ | ✓ | ✓ | ✗ | ✓ | ✗ | ✓ |
| $R^*$ [8] | ✓ | ✓ | ✗ | ✓ | ✓ | ✗ | ✗ | ✗ | ✓ | ✓ | ✓ | ✓ | ✗ | ✗ | ✓ |
| $R^*$ [5] | ✓ | ✓ | ✗ | ✓ | ✓ | ✓ | ✗ | ✗ | ✓ | ✗ | ✓ | ✗ | ✓ | ✗ | ✓ |
| $R^*$ [17] | ✗ | ✗ | ✗ | ✗ | ✓ | ✓ | ✗ | ✓ | ✓ | ✓ | ✓ | ✓ | ✗ | ✗ | ✗ |
| $R^*$ [18] | ✗ | ✗ | ✗ | - | ✓ | ✗ | ✗ | ✓ | ✓ | ✓ | ✓ | ✓ | - | ✗ | ✗ |

*Note* $R^*$: Reference, *1* Prevent password-guessing threat, *2* Prevent identity-guessing threat, *3* Prevent impersonation threat, *4* Prevent smartcard stolen attack, 5 Prevent replay attack, 6 Prevent insider attack, *7* Achieve forward-secrecy property, 8 Exact authentication phase, 9 Prevent sensor-node capture attack, 10 Correct session-key establishment, 11 Facilitate mutual-authentication property, 12 Exact password change phase, *13* Prevent smartcard theft attack, 14 Prevent session-key temporary information attack, 15 Prevent stolen verifier attack.

### 3.1 Security Features

In Table 1, the comparison of security features of relevant schemes is presented. We have shown in table 1 that the protocols [1,5,8,10,17,18] are vulnerable to impersonation attack, session-key temporary-information attack and also not facilitate forward-secrecy property. Next, the protocols [17,18] are suffering from password and identity-guessing attacks, and stolen verifier attack. The protocols [8,18] are not defending the insider attack and the protocol [8] is not providing exact authentication and password change phases. The protocols [8,18] are not resisting insider attack. The protocols [8,17] are suffering from smart card theft attack and the scheme [17] is also suffering smartcard stolen attack. The protocol [5] is also not facilitating exact session-key establishment.

**Table 2.** The cost comparison of communication and smartcard storage (in bits)

| Schemes | | Total | | | | | | SCSC |
|---|---|---|---|---|---|---|---|---|
| | | $U_i$ | | GWN | | $SN_j$ | | |
| | | T | R | T | R | T | R | |
| $R^*$ [17] | Case-1 | 960 | 800 | 1600 | 1440 | 480 | 800 | 800 |
| | Case-2 | 1600 | 1280 | 2720 | 2720 | 640 | 960 | |
| $R^*$ [10] | Case-1 | 1120 | 960 | 1600 | 1600 | 480 | 640 | 800 |
| | Case-2 | 1760 | 1440 | 2880 | 3040 | 480 | 640 | |
| $R^*$ [9] | Case-1 | 800 | 640 | 1600 | 1280 | 480 | 960 | 640 |
| | Case-2 | 1600 | 1280 | 3680 | 3680 | 480 | 800 | |
| $R^*$ [8] | | 672 | 672 | 1344 | 1344 | 672 | 672 | 640 |
| $R^*$ [1] | | 1120 | 480 | 1280 | 1600 | 480 | 800 | 960 |
| $R^*$ [18] | | 960 | 640 | 1120 | 1280 | 320 | 480 | – |
| $R^*$ [5] | | 800 | 480 | 1120 | 1120 | 320 | 640 | 960 |

*Note* $R^*$ Reference, *SCSC* Smartcard Storage Cost; *T* Transmit, *R* Receive

### 3.2 The Communication and Smartcard Storage Costs Comparison

In Table 2, the communication cost of relevant schemes [1,5,8–10,17,18] is discussed. In Table 2, we have elaborated the communication cost for the user, gateway node and sensor node in the context of length of received and transmitted messages. To compute the communication cost, we have presumed that the size of hash function, identity, random number, password, timestamp, encryption/decryption (AES) all are 160 bits. The communication cost for $U_i$, GWN and $SN_j$ in the scheme [17] for **Case-1** are 1760, 3040 and 1280 bits, respectively, and for **Case-2** are 2880, 5440 and 1600 bits, respectively. Next, the communication cost in the scheme [9] for **Case-1** are 1440, 2880, 1440 bits, respectively, and

**Table 3.** The computation cost comparison

| Schemes | | User | Gateway | Sensor node |
|---|---|---|---|---|
| $R^*$ [17] | Case-1 | $7T_H$ | $8T_H$ | $5T_H$ |
| | Case-2 | $8T_H$ | $7T_H$ | $5T_H$ |
| $^*R^*$ [9] | Case-1 | $9T_H$ | $14T_H$ | $7T_H$ |
| | Case-2 | $13T_H$ | $23T_H$ | $6T_H$ |
| $R^*$ [10] | Case-1 | $9T_H$ | $12T_H$ | $4T_H$ |
| | Case-2 | $11T_H$ | $13T_H$ | $4T_H$ |
| $R^*$ [8] | | $7T_H + 2T_S$ | $11T_H + 4T_S$ | $4T_H + 2T_S$ |
| $R^*$ [1] | | $12T_H$ | $16T_H$ | $5T_H$ |
| $R^*$ [18] | | $12T_H$ | $18T_H$ | $6T_H$ |
| $R^*$ [5] | | $10T_H$ | $8T_H$ | $6T_H$ |

for **Case-2** 2880, 7360 and 1280 bits, respectively. The scheme [8] takes 1344, 2688 and 1344 bits for $U_i$, GWN and $\text{SN}_j$. The protocol [1] requires 1660, 2880 and 1280 bits for $U_i$, GWN, and $\text{SN}_j$. The communication cost of the schemes [18] and [5] for $U_i$ (1600 and 1280 bits), $GWN$ (2400 and 2240 bits) and $\text{SN}_j$ (800 and 960 bits). The protocol [10] requires 2080 ($U_i$), 3200 (GWN) 1120 bits ($\text{SN}_j$) for **Case-1** and 3200 ($U_i$), 5920 (GWN) 1120 bits ($\text{SN}_j$) for **Case-2**.

### 3.3 The Computation Cost Comparison

In Table 3, we have delineated the computation cost comparison of related protocols [1,5,8–10,17,18]. This computation cost is described for $U_i$, $SN_j$ and $GWN$ for each authentication cycle separately. The computation cost for $U_i$ of the schemes are [1,5,8–10,17,18] are $7T_H$, $8T_H$, $9T_H$, $13T_H$, $9T_H$, $11T_H$, $7T_H + 2T_S$, $12T_H$, $12T_H$ and $10T_H$, respectively. Where $T_H$ presents the hash function and $T_S$ shows the symmetric key encryption/decryption operation. The computation cost of the protocol [17] for Case-1 are $8T_H$ and $7T_H$ for GWN and $\text{SN}_j$ respectively, whereas for Case-2 both GWN and $\text{SN}_j$ takes $5T_H$. The protocols [9] and [10] for Case-1 and Case-2 take $14T_H$, $23T_H$, $12T_H$ and $13T_H$ for GWN, $7_H$, $6T_H$ and $4T_H$ for $\text{SN}_j$. The computation cost of the protocols [1,5,8,18] for GWN and $\text{SN}_j$ are $11T_H + 4T_S$, $16T_H$, $18T_H$, $8T_H$, $4T_H + 2T_S$, $5T_H$, $6T_H$, $6T_H$ and $6T_H$, respectively.

## 4 Conclusion

Recently, Amin-Biswas developed a secure remote user authentication protocol for WSNs. However, we have found that Amin-Biswas's protocol is not resisting password and identity-guessing, user-impersonation, smartcard theft and known session-key temporary-information attacks. Moreover, it is not providing forward-secrecy property. Apart from it, we have shown the comparison of other existing relevant authentication protocols in terms of costs (computation, smartcard storage, and communication) and security functionalities.

## References


1. Amin, R., Islam, S.H., Biswas, G., Khan, M.K., Leng, L., Kumar, N.: Design of an anonymity-preserving three-factor authenticated key exchange protocol for wireless sensor networks. Comput. Networks **101**, 42–62 (2016)
2. Farash, M.S., Turkanović, M., Kumari, S., Hölbl, M.: An efficient user authentication and key agreement scheme for heterogeneous wireless sensor network tailored for the internet of things environment. Ad Hoc Networks **36**, 152–176 (2016)
3. Das, A.K.: A secure and efficient user anonymity-preserving three-factor authentication protocol for large-scale distributed wireless sensor networks. Wirel. Person. Commun. **82**(3), 1377–1404 (2015)
4. Turkanović, M., Brumen, B., Hölbl, M.: A novel user authentication and key agreement scheme for heterogeneous ad hoc wireless sensor networks, based on the internet of things notion. Ad Hoc Networks **20**, 96–112 (2014)
5. Kumari, S., Om, H.: Authentication protocol for wireless sensor networks applications like safety monitoring in coal mines. Comput. Networks **104**, 137–154 (2016)
6. Vaidya, B., Makrakis, D., Mouftah, H.: Two-factor mutual authentication with key agreement in wireless sensor networks. In: Security and Communication Networks (2012)
7. Das, A.K., Sutrala, A.K., Kumari, S., Odelu, V., Wazid, M., Li, X.: An efficient multi-gateway-based three-factor user authentication and key agreement scheme in hierarchical wireless sensor networks. In: Security and Communication Networks (2016)
8. Lu, Y., Li, L., Peng, H., Yang, Y.: An energy efficient mutual authentication and key agreement scheme preserving anonymity for wireless sensor networks. Sensors **16**(6), 837 (2016)
9. Srinivas, J., Mukhopadhyay, S., Mishra, D.: Secure and efficient user authentication scheme for multi-gateway wireless sensor networks. Ad Hoc Networks **54**, 147–169 (2017)
10. Wu, F., Xu, L., Kumari, S., Li, X., Shen, J., Choo, K.-K.R., Wazid, M., Das, A.K.: An efficient authentication and key agreement scheme for multi-gateway wireless sensor networks in iot deployment. J. Network Comput. Appl. (2016)
11. Jiang, Q., Ma, J., Wei, F., Tian, Y., Shen, J., Yang, Y.: An untraceable temporal-credential-based two-factor authentication scheme using ecc for wireless sensor networks. J. Network Comput. Appl. **76**, 37–48 (2016)
12. Das, A.K., Sharma, P., Chatterjee, S., Sing, J.K.: A dynamic password-based user authentication scheme for hierarchical wireless sensor networks. J. Network Comput. Appl. **35**(5), 1646–1656 (2012)
13. Wang, D., Wang, P.: Understanding security failures of two-factor authentication schemes for real-time applications in hierarchical wireless sensor networks. Ad Hoc Networks **20**, 1–15 (2014)
14. Turkanovic, M., Holbl, M.: An improved dynamic password-based user authentication scheme for hierarchical wireless sensor networks. Elektronika ir Elektrotechnika **19**(6), 109–116 (2013)
15. Kim, J., Lee, D., Jeon, W., Lee, Y., Won, D.: Security analysis and improvements of two-factor mutual authentication with key agreement in wireless sensor networks. Sensors **14**(4), 6443–6462 (2014)

16. Deebak, B.D.: Secure and efficient mutual adaptive user authentication scheme for heterogeneous wireless sensor networks using multimedia client-server systems. Wirel. Person. Commun. **87**(3), 1013–1035 (2016)
17. Amin, R., Biswas, G.: A secure light weight scheme for user authentication and key agreement in multi-gateway based wireless sensor networks. Ad Hoc Networks **36**, 58–80 (2016)
18. Amin, R., Islam, S.H., Biswas, G., Khan, M.K., Kumar, N.: A robust and anonymous patient monitoring system using wireless medical sensor networks. In Future Generation Computer Systems (2016)

# Diving into a Decade of Games for Health Research: A Systematic Review

Paula Alexandra Silva[1(✉)], Sergi Bermúdez i Badia[2], and Mónica S. Cameirão[2]

[1] Department of Informatics Engineering, Center for Informatics and Systems (CISUC), University of Coimbra, Coimbra, Portugal
palexa@gmail.com, paulasilva@dei.uc.pt

[2] Faculty of Exact Sciences and Engineering, Madeira Interactive Technologies Institute, Universidade da Madeira, Funchal, Portugal
{sergi.bermudez,monica.cameirao}@m-iti.org

**Abstract.** Recent years have been characterised by a rising interest in using entertainment computing to monitor, maintain, and improve human health. This is observed in many systems and applications that leverage the benefits of a playful and enjoyable experience to provide a technology-enabled health intervention. This paper reviews one decade of papers (679) published at the intersection of health, entertainment and technology to determine trends, studies' characteristics, type of solutions, domains of application and study purposes. Results show that there is a growing body of research in the area, with the majority of studies providing solutions for rehabilitation and addressing motor conditions related to stroke and/or fitness. Where half of the solutions reported are custom made, the bulk of those studies is performed with the purpose of evaluating the solutions proposed or validating their efficacy. In 80% of the cases, the studies are performed with subjects from the target population with sample sizes that have been steadily increasing over the years.

**Keywords:** Health · Entertainment · Technology · Computing · Review · Games · Simulation · VR · AR

## 1 Introduction

Consumer technologies for gaming and entertainment have become so precise, affordable and pervasive that their application has gone well beyond pure entertainment. With a variety of application domains, from education to business, entertainment computing is increasingly used in the health domain. This gain in popularity has been explained, e.g. with the proliferation of low-cost technologies, the realisation of the potential of entertainment and playful elements in extending motivation, and the widespread access to technology in general [1, 3–5].


X.-S. Yang et al. (eds.), *Proceedings of Fifth International Congress on Information and Communication Technology*, Advances in Intelligent Systems and Computing 1183, https://doi.org/10.1007/978-981-15-5856-6_51

In health, entertainment computing has been used to support the monitoring, detection, treatment, rehabilitation and education of patients and non-patients alike [5]. These technology-enabled solutions have taken the form of simulations, serious games or gamified applications that leverage the potential of virtual/augmented/mixed reality and/or game elements to support a variety of health domains and conditions. Because of the applied research areas at the intersection of health entertainment computing have become so wide, it is important to investigate the research that has been conducted in this area to understand how the field of health entertainment computing has evolved over the years and what specific contributions have been made by that field of research. To our best knowledge, no such analysis has yet been done. This paper engages in a systematic review of the literature of the research published between 2004 and 2014 at the intersection of health, entertainment and technology to develop an understanding of the main contributions and areas of application in the field. Specifically, we investigate the (i) characteristics of the studies, from purpose to participants involved, etc., (ii) types of solutions that have been produced and iii) purposes of the interventions and their main domain areas. A main research question guided the research: How has the field of health entertainment computing evolved over the years? that we broke down into: (i) Which are the most common domains of intervention? (ii) What type of solutions are being produced? and (iii) What are the characteristics of the studies that are being conducted with those solutions?

After presenting our search and data extraction strategy, we present our results and findings, which are based on the full text analysis of 679 articles. We conclude with a discussion and the identification of future avenues of research. This research brings a shed of light into our knowledge and understanding of the multidisciplinary and vast field of health entertainment computing, highlighting strengths and limitations as well as general trends and characteristics.

## 2 Research Approach

The search strategy for this review was as follows: ((health OR rehab*) AND (((serious OR computer OR interactive OR video OR online) AND gam*) OR exergames OR gamification OR ('virtual reality' OR 'augmented reality' OR 'mixed reality'))). To maximise coverage, the authors conducted a systematic search in three databases: IEEE Xplore, PubMed and the ACM Digital Library. Articles included were dated between January 2004 and December 2014; peer-reviewed; in English; involved five or more subjects; were in an application domain related to health; and used entertainment technology. Articles were excluded when: not accessible/possible to locate; in form of book, poster, demo, workshop, keynote, study protocol, review, editorial, letter, commentary, clinical perspective or appraisal.

The above-described searches yielded a total of 6767 articles. Two reviewers screened all titles, abstracts and type of publication for eligibility, according to the inclusion and exclusion criteria. When necessary, the reviewers downloaded the article and skimmed it to determine article eligibility. Articles that were deemed irrelevant were discarded (5147). From the remaining 1485, 124 were not possible to locate/download and 108 were dated outside the date range defined (2004--2014). The full texts of the

1253 articles that remained were downloaded for full text eligibility assessment. Figure 1 shows an overview of the process.

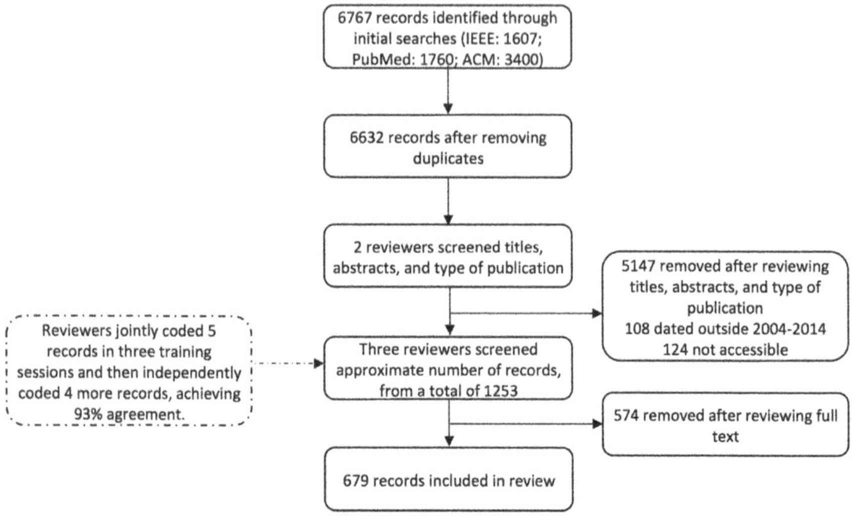


**Fig. 1.** Overview of screening and selection process

Three pre-coding training sessions were carried out involving all reviewers coding five articles to refine goals, themes, concepts and a preliminary data extraction form. Subsequently, a final version of the data extraction form was created that included the following fields: year of publication, type of publication, venue, reason for exclusion, study participants' age/same as target/sex/health condition, health and application domain, stage of the disease, type of intervention, type of technology, purpose of technology, sample size and purpose of the study. Later, reviewers independently coded four articles and an agreement percentage of 93% among reviewers was computed, which was deemed appropriate for subsequent independent coding. Three reviewers divided the remaining articles and examined them independently. Questions and conflicts were resolved by discussion and consensus in regular reviewers' meetings. As a result of this, 574 articles were excluded, and 679 articles were deemed relevant for the research presented in this paper (Fig. 1).

## 3 Results and Findings

Figure 2 shows the yearly trends for the included articles. A tenfold increase of publications can be observed between 2004 and 2014, consisting mainly of conference contributions until 2012, and later matched by journal publications.

Table 1 shows the top five conference and journal publication venues in this area. Whereas there seems to be a large consensus on the most relevant conference venues in the area (top five totalling 171 publications), this is not the case for journal venues (top five totalling 73 publications). Interestingly, a larger number of conference publications

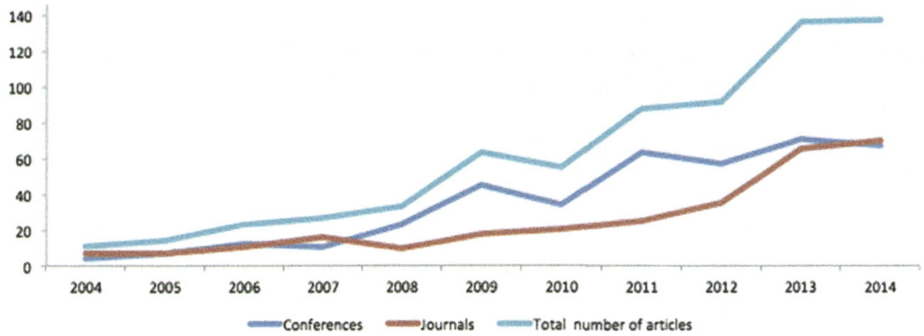


**Fig. 2.** Yearly publication trends

are observed on odd years (Fig. 2). This may be due to two of the most contributing conferences taking place every two years, the International Conference on Virtual Rehabilitation and the IEEE International Conference on Rehabilitation Robotics.

**Table 1.** Publication venues top five

| Conferences---top five | Journals---top five |
|---|---|
| • ACM SIGCHI Conference on Human Factors in Computing Systems (60) | • Journal of NeuroEngineering and Rehabilitation (18) |
| • International Conference on Virtual Rehabilitation (51) | • IEEE Transactions on Neural Systems and Rehabilitation Engineering (17) |
| • IEEE Annual International Conference on Engineering in Medicine and Biology Society (31) | • Studies in Health Technology and Informatics (15) |
| • IEEE International Conference on Rehabilitation Robotics (17) | • Disability and Rehabilitation: Assistive Technology (12) |
| • ACM SIGCHI Conference on Interaction Design and Children (12) | • Annual Review of Cybertherapy and Telemedicine (11) |

The analysis of the studies included in this review presented work at different stages of development and with different purposes. We defined the following categories: conceptualisation, user research (before development), prototyping, evaluation (of the system itself) and validation (of the outcome of the system). Many contributions presented multiple purposes or contributions in different stages. Figure 3 shows that most studies contain a validation study (368) aiming at verifying if the proposed system fulfils its health-related goal. Nonetheless, articles devoted to technical aspects, such as evaluation (293) of the functioning of the system and prototyping (74) are also frequent. There are, however, fewer contributions concerning the conceptualisation and user research phases, which are crucial steps in the development of health entertainment technologies.

The largest sample size identified was 1943 participants, being the most frequent 10, the average 41 and the median 19. Except for a few rare cases, conference publications

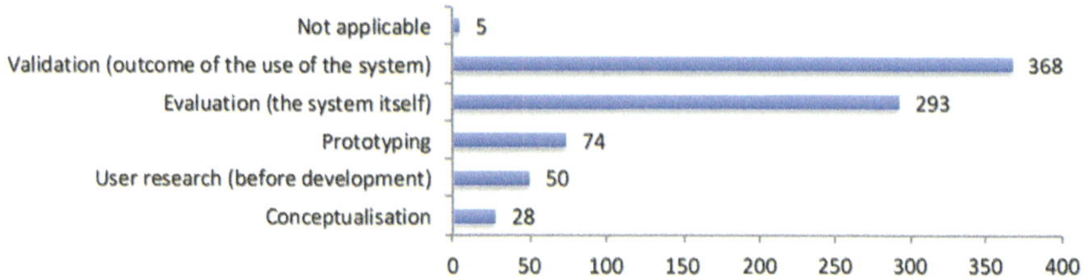


**Fig. 3.** Purpose of the study

tend to present slightly larger sample sizes (mode: 20, mean: 45, median: 23) than journals (mode: 10, mean: 37, median: 16). As for the evolution over time, the sample size has increased during the 2004--2014 decade for both conference and journal publications (Fig. 4), what is consistent with an increase of clinical and field trials and less proof-of-concept studies.

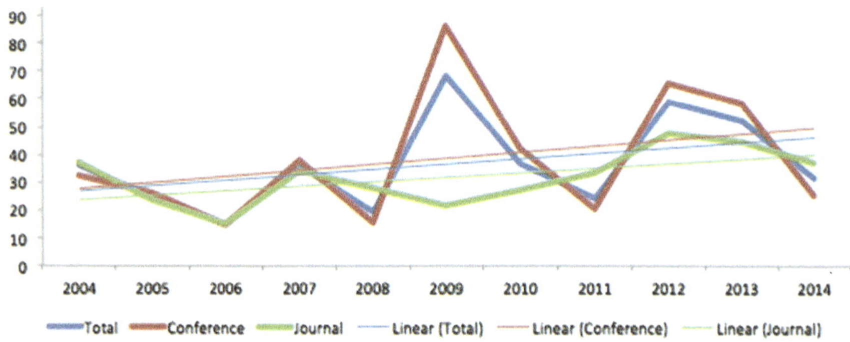


**Fig. 4.** Average sample size across time and corresponding linear regressions

Although sample sizes have increased over time, the information provided on the samples is often insufficient. Most studies do include male and female participants (455), and only a few solely male (24) or female (18) participants. However, a large number of studies (181) do not provide information on the sex of the participants. Similarly, of the included articles, 120 do not report on the age of the participants. The most frequent participants are, in this order, adults (18–49 years old), young senior (50–64 years old) and senior (65–84 years old). Some studies include participants of multiple age groups and hence have been accounted for in the different age ranges. With respect to the condition of participants, 285 studies included patients, 211 included healthy participants, 94 included both, and 73 studies did not indicate the condition of their participants.

To analyse the stage of the disease of the study participants, we used Merril's taxonomy, as summarised by Wattanasoontorn et al. [5], that considers the following stages:

susceptibility, pre-symptomatic, clinical disease, recovery and disability. From the studies included in this review, 160 included participants in susceptibility stage, 24 in pre-symptomatic stage, 249 in clinical disease stage (including transitory or acute conditions), 228 in recovery stage (chronic conditions) and 18 in disability stage, meaning the aim is not to recover from the disease but rather to increase quality of life. Studies considering multiple stages were accounted for in each of the stages. For 63 studies, this information was not available and for 75 this taxonomy was not applicable.

Each article focused on a specific type of digital solution, which we grouped under one of the following categories: virtual reality, augmented reality, mixed reality, and other types of solutions (e.g. games, mobile games, desktop applications, etc.). We found that most studies used virtual reality (384) and few would refer to applications as augmented reality (29) and mixed reality (18). Instead, a large number of studies would refer to other types of technologies (249). It was not possible to determine the type of technology used in 29 of the reviewed articles.

We were also interested in finding out whether the solution in use was (was not) commercial. We found that 50% of the solutions were custom made, while 26% were commercial; for 24% of the articles it was not possible to determine the nature of the solution. While the fact that half of the papers report on custom made solutions is not surprising, especially because this study is sourced on research papers, this also reflects the potential that researchers perceive in embedding entertainment in technological solutions targeting health and also a limited number of commercial alternatives. Hence, this important investment by researchers may foresee an increase of solutions in the market leveraging play, game elements or some sort of entertainment to attain health goals.

To determine the purpose of the technologies in the reviewed papers, we used Wattanasoontorn et al. [5] classification, who created a taxonomy of serious games by target group: patient/non-patient. Targeting patients, Wattanasoontorn et al. [5] considered: health monitoring, detection, treatment or therapy, rehabilitation and education, while for non-patients they considered: health and wellness games, training and simulation games for professional, and training and simulation games for non-professional.

Overall, the results show that studies aiming at patients (664) seem to be more expressive than those targeting non-patients (172). The great majority of articles targeting patients reports on studies which purpose is rehabilitation (315); these are followed by studies aiming at treatment or therapy (205). The smaller number of studies targets health monitoring (50), detection (55) and education (39). When looking into technology for non-patients, the health and wellness category gathers the largest number of studies (156), while training and simulation games for professional (4) and non-professional (12) show residual numbers (Fig. 5). For the correct interpretation of results, it is important to mention that studies sometimes report on more than one single purpose. This is the case of the combination of treatment or therapy plus rehabilitation (63) and of health monitoring and detection together (20). Still, these numbers are less expressive than the ones in categories such as rehabilitation, health and wellness and treatment of therapy alone, which are the three areas that have been receiving the most attention by the research community working in the intersection of health and entertainment technologies.

We also investigated whether interventions were tackling motor, cognitive, social or other domains of intervention. Regarding this, results show that the vast majority of

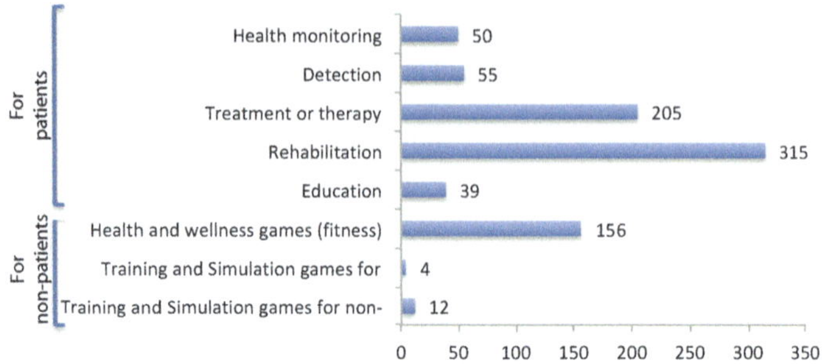


**Fig. 5.** Purpose of technology

studies focused on motor (454) concerns, followed by studies addressing cognitive (132) issues. A smaller number of studies dealt with social (47) or other types of issues (81). In this category, we would include, for example, studies dealing with pain or general health awareness. So far, there has been a tendency to focus on one single domain of intervention, still it is important to note that some studies address more than one domain. This is the case for studies focusing on motor and cognitive (16) or on motor and social (14) domains. While still not expressive, there seems to be a tendency towards holistic approaches in more recent years, where we found the above combinations as well as others combining all three domains (2). It is noteworthy that from a total of 40 studies, 35 are dated 2009 or later.

Each study targeted one or more intervention domains (e.g. balance, stroke, cognitive rehabilitation), and those were labelled as reported by the authors of the articles. This produced a heterogeneous number of labels for each article indicating one or more intervention domains. We computed the co-occurrence of domains by first uniformising the nomenclature for multiple terms referring to the same domain (such as TBI and traumatic brain injury), and second producing a co-occurrence network diagram [2]. The resulting network diagram shows us the most frequent intervention domains as nodes, its size according to the number of occurrences and colour indicating how central the role each domain plays in the network (betweenness centrality). Nodes (domains) are connected through edges according to the frequency of simultaneous appearance. For this analysis, we only considered cases of three or more co-occurrences of domains. This analysis shows that the most central domains were stroke and fitness followed by ageing and Parkinson's. However, the most frequent co-occurrence of terms was stroke, fitness followed by balance, traumatic brain injury and spinal cord injury. The network shows the relations between terms and how they cluster (Fig. 6).

## 4 Concluding Remarks

This paper reviewed health entertainment computing research published between 2004 and 2014. Results show that this area of research has been growing over the years with a sustained increase in conference and journal publications. The largest subset of contributions (~58%) still focuses on the development, prototyping and technical evaluation, with only 54% of the reviewed publications evaluating the outcome of use of

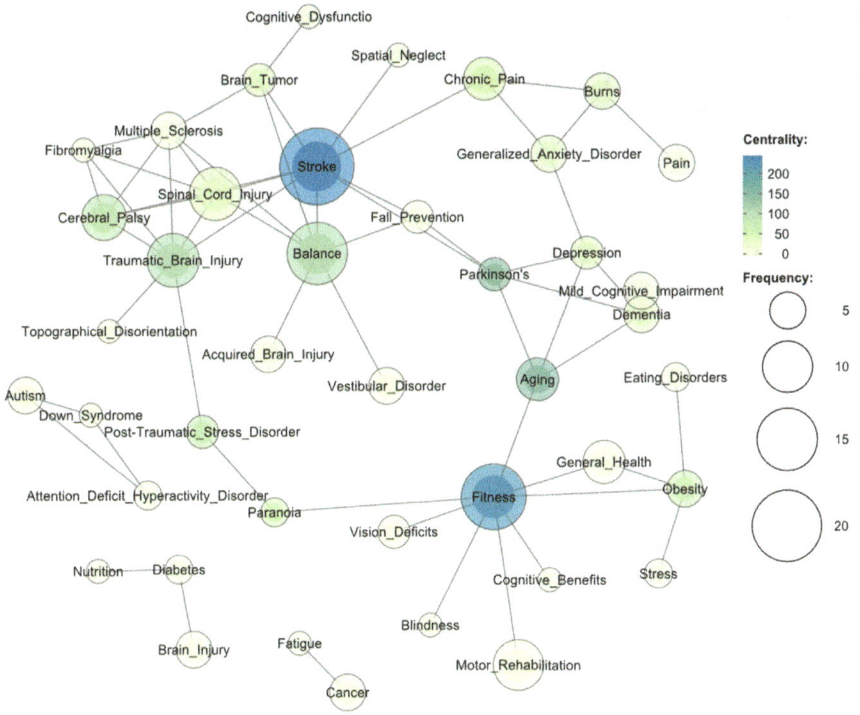


**Fig. 6.** Co-occurrence network diagram of domain of intervention

the proposed tools. This may mean that the impact of some tools is rarely assessed. On a positive note, the sample sizes have grown through the years. More importantly, 80% of the samples included their target population. We did not identify a preference towards a specific type of technology, but data show that most solutions were designed for motor interventions, and for rehabilitation purposes. Finally, the most common intervention domains were stroke and fitness.

This study has some limitations that should be considered while interpreting the results. First, the search criteria and chosen keywords used for identification of the papers may have excluded relevant work. Furthermore, because of the long-time span needed to review the 679 papers, the assessment criteria of the reviewers may have changed over time. To this extends that, unless explicitly stated, information was tagged as 'not available', and that, as in any review process, only the information available was subject to interpretation. Finally, this research is time-bound to 2004--2014.

The authors are currently reviewing the literature up to 2018. Our goal is to expand the results of this study to develop an overall understanding of the field of health entertainment computing and identify strengths, weaknesses and future trends.

**Acknowledgements.** Research supported by BRaNT (PTDC/CCICOM/30990/2017), MACBIOIDI (INTERREG program MAC/1.1.b/098) and NOVA-LINCS (PEest/UID/CEC/04516/2019).

## References

1. Cipresso, P., Giglioli, I.A.C., Raya, M.A., Riva, G.: The past, present, and future of virtual and augmented reality research: a network and cluster analysis of the literature. Front. Psychol. **9** (2018). https://doi.org/10.3389/fpsyg.2018.02086
2. Higuchi, K.: KH Coder 3 Reference Manual. Ritsumeikan University, Kioto, Japan (2016)
3. McCallum, S.: Gamification and serious games for personalized health. Stud. Health Technol. Inform. **177**, 85–96 (2012)
4. Nacke, L.E., Deterding, S.: The maturing of gamification research. Comput. Hum. Behav. **71**, 450–454 (2017). https://doi.org/10.1016/j.chb.2016.11.062
5. Wattanasoontorn, V., Boada, I., García, R., Sbert, M.: Serious games for health. Entertain. Comput. **4**, 231–247 (2013). https://doi.org/10.1016/j.entcom.2013.09.002

# Solutions to Improve the Quality of Higher Education in Vietnam in the Context of Industrial Revolution 4.0

Linh Kieu Phan(✉)

Foreign Languages Department, College of Foreign Economic Relations, Ho Chi Minh City, Vietnam
linhphancofer@gmail.com

**Abstract.** In the digital age, higher education varies widely from the educational environment, the role of teachers and learners to teaching methods. Currently, Vietnam as well as other countries around the world is facing the great challenges of the shortage of a highly skilled workforce. The quality of workforce educated in university has not met the demands of socio-economic development and international integration. Besides, there is a lack of international experienced researchers in higher education institutions. The connection between universities and businesses is not focused, which makes learners not be able to meet job requirements after graduating. Therefore, it is critical to improve the quality of Vietnam's higher education in the context of Industry Revolution 4.0.

**Keywords:** Higher education · Training quality · Lecturer · Industrial revolution 4.0 · Solution · Curriculum · Evaluation

## 1 Introduction

In recent years, Vietnamese higher education has been recognized as a remarkable achievement. Accordingly, there is an improvement in both the quantity and quality of lecturers. By 2017, the number of lecturers was 72,792 which showed an increase of 11% compared to 2015. In particular, the percentage of lecturers with doctoral degrees increased from 15.87% in 2015 to 22.68% in 2017.

University autonomy is considered as a global trend. It is three years since Resolution No. 77/NQ-CP of Government on the pilot renovation of the operation mechanism of public tertiary education institutions during 2014–2017 has been implemented. A good news is that 23 higher education institutions have been successfully adopted. Their transition to the new model brings positive changes in the quality of lecturers and operation, which helps the country's higher education system to move toward university autonomy.

Quality assurance has been also strengthened. During the last two years, there has been a significant increase in the qualified number of higher education institutions with international and domestic accreditation standards. The developments of Industrial Rev-


X.-S. Yang et al. (eds.), *Proceedings of Fifth International Congress on Information and Communication Technology*, Advances in Intelligent Systems and Computing 1183, https://doi.org/10.1007/978-981-15-5856-6_52

olution 4.0 present great challenges and opportunities for continued improvement of higher education in Vietnam.

The position of Vietnamese universities in the Asian rankings has been enhanced. In the school year 2017–2018, there are more Vietnamese higher education institutions listed in different world rankings. In addition, they have made great efforts to gain international rankings (Fig. 1).

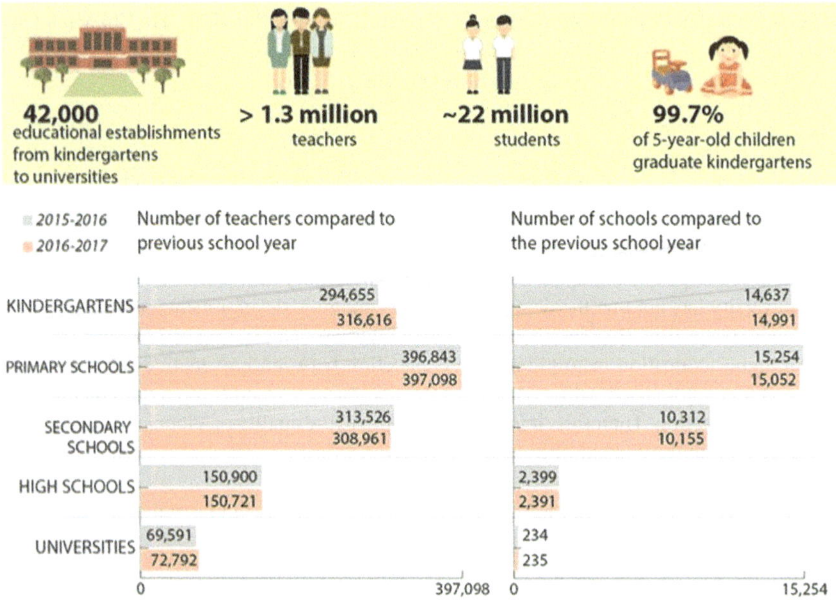


**Fig. 1.** Vietnam has 42,000 educational establishments

Although the Vietnam Ministry of Education and Training also admits that the quality of training as not high as desired, especially postgraduate and connected training, Vietnam has been making a lot of progresses in Education in general and Higher Education in particular. However Research activities, technology transfer, and community service have some limitations. Some universities after operating still do not meet the quality assurance standards under the project of establishment of the university, which results in the low quality of training and enrollment, and the lack of well-qualified resources. Therefore, it is necessary to propose solutions to make full-scale transformation of education and training with the aim of meeting the requirements of industrialization and modernization in socialist-oriented market economy and international integration.

## 2 Objectives

This paper intends to highlight the challenges, opportunities, and direction for improvement of higher education structures and activities in Vietnam as the Industrial Revolution 4.0 manifests its effects on the country, including:

2.1 Fostering fundamental and comprehensive innovation in education from thinking, perspectives to goals, systems, training programs (curriculum, methods, exams, assessments), policies, mechanisms, and conditions for ensuring the quality of education; encouraging innovation at all training levels, in the relationship between the university, family, and society.
2.2 Shifting the training process from equipping only knowledge (people's knowledge improvement) to improving the competence and quality of learners (citizen capacity). Performing objective "Learning with practice, theory with reality, connection education between school, family and society".
2.3 Shifting the education which focuses mainly on quantity objectives to the quality and efficiency objectives. This aims to meet the demand of human resources in society; Implementing standardization, modernization, and democratization of the education system.
2.4 Changing from rigid and incoherent education system to open education system; forming a lifelong learning mechanism associated with a learning society; implementing socialization, democratization, and international integration.
2.5 Focusing training on high-level human resources, fostering talents, the qualities, and competencies for self-study, knowledge self-enrichment, and creativity of learners. Completing the industry structure and training level in line with the human resource development planning of industries and countries and make sure to include a number of training majors at domestic and international levels.

## 3 Higher Education in the Context of Industrial Revolution 4.0

Industrial Revolution 4.0 is based on digital technology to build a hyperlinked world and integrate intelligent technologies to optimize production processes and methods. The most current advanced and innovative technologies are Artificial Intelligence (AI), Big Data, 3D Printing Technology, Self-driving Cars, Robots, Internet of Things (IoT), Nanotechnology, Materials and Sensors Technology and Biotechnology. Achievements of Industrial Revolution 4.0 have been widely applied in production and life such as smart cities, smart houses, transportation systems, hospitals, smart factories, and establishing a connection between devices and machines, even between machines and people.

Industrial Revolution 4.0 provides an opportunity as well as creates an indispensable requirement to build an education system 4.0. Education will accordingly become an ecosystem that everyone can learn at anytime and anywhere with connected devices. Educational institutions become an ecosystem to offer individualized creative products with knowledge, innovation, and creativity of this person.

Education 4.0 is a major change in the goals and ways of training, which creates a shift from transferring knowledge for a group of people to exploit potential and empowering creativity for each individual. Teachers will take new roles as designers, catalysts, mentors to create learning environments. Under the condition of digitalized learning program, learners will have a separate learning path, which can be selected to suit the training objectives. The digital learning system also provides feedbacks on learning performance and suggestions for further learning content.

The range of interaction in education 4.0 is quite vast, the geographical distance, space, and time are blurred. The educational environment does not only take place within the school but also on a global scale. Learners can actively study materials and interact with teachers at any time with a computer or smartphone. The development of online learning methods helps learners save time, effort, and cost. Augmented Reality/Virtual Reality technology (AR/VR) is widely used which helps learners experience and practice skills.

In education 4.0, thanks to the application of AI, Big Data and IoT technologies, university board of management and lecturers can collect data, analyze and assess accurately about learners. They can also monitor the students' learning process at home, check the homework completion, and report learning results to students and their families. Even AI technology can replace teachers in some stages such as taking attendance, marking papers, preparing lessons, and supporting foreign language teaching.

However, Industrial Revolution 4.0 also poses some significant challenges for society in general and higher education in particular. The labor market is at risk of a complete change as robots gradually replace workers. According to statistics from the International Robot Federation, the speed of robot automation in the global industrial sector is reaching a very fast threshold, led by South Korea at the rate of 631 robots per 10,000 employees [1]. The International Labor Organization (ILO) also forecasts that robots put 56% of workers in five Southeast Asian countries, including Vietnam at risk of losing their jobs in the next two decades [2]. Much of Industrial Revolution 3.0 has not been realized and will end up being leapfrogged as Industrial Revolution 4.0 is realized.

Vietnam is becoming one of the countries with the potential of advanced technology development. Clearly, millions of new jobs come from the software sector. However, automation gradually replaces manpower in many areas which make workers have to adapt quickly to the change of production. Therefore, the university should be a place to provide skilled human resources for society. This requires it to change comprehensively both the model, curriculum, and the training method.

In Industrial Revolution 4.0, the interaction between devices and devices, and between devices and people will bring out a new form of production. Some new skills will be required for employees such as problem-solving skills, critical thinking, communication skills, collaboration, creativity, innovation [3]. This is an important feature that not only guides changes in the training programs and creates new majors in universities, but also sets out an essential requirement of "lifelong learning" for workers in Industrial Revolution 4.0.

Under the development of Industrial Revolution 4.0, research and training activities in higher education institutions also face challenges. Many technology corporations have huge scientific, human, and financial potentials which are advantages in the race to turn knowledge into products for life. They have a great deal of practical experience which researchers and lecturers do not have. Moreover, free global trade in higher education creates competition between domestic and foreign universities to attract students.

# 4 Some Solutions to Improve Higher Education in the Industrial Revolution 4.0

## 4.1 Continue to Foster Fundamental and Comprehensive Transformation of Education and Training in the Direction of Focusing on the Competence and Quality of Learners

Develop innovative training programs for undergraduate and postgraduate levels to meet social requirements.

Update learning outcomes of each training majors. Revise and update the learning outcomes to be appropriate with the finalized training programs.

Expand high-quality training programs in the direction of approaching advanced technologies and achievements so that graduates are good at both professional and soft skills, including foreign languages.

Improve students' foreign language proficiency and encourage lecturers to teach specialized subjects in foreign languages.

Open more training majors at postgraduate levels, increase training scale at both undergraduate and postgraduate levels.

## 4.2 Comprehensively Renovate Forms and Methods of Assessing, Testing and Evaluating to Ensure the Truthfulness and Objectivity

Change methods to evaluate learning results.

Develop a project to improve the content, form of assessing, testing, and evaluating training results with a focus on evaluating the understanding and application of knowledge and skills in solving academic and practical problems. Limit the requirement to learn by rote. Assess the improvement of attitude and capacity of learners. Combine in-class evaluation with final-term evaluation, teacher evaluation and student self-evaluation, school reviews, and social assessments.

Separate exam evaluation from teaching evaluation will motivate the renovating the methods of assessing and testing during learning process. While learning must be considered as a continuous process to accumulate knowledge, evaluation should be done during the learning progress and at the final term by using a variety of fair and impartial evaluation methods.

Assess the results of postgraduate education should focus on evaluating the competencies of analysis, criticism, creativity, problem-solving, professional attitudes, and capacity to do and apply scientific and technological research. Practical skills will be in alignment with the training level and the capacity to adapt to the working environment.

## 4.3 Innovate the Evaluation of Training Quality

Study international accreditation systems to apply and evaluate the university's training programs according to international standards when possible.

Complete the quality management system according to ISO standards. Perfect the organizational structure of the Quality Assurance and Training Quality Accreditation Department.

Enhance the participation of the community (Enterprises, faculty members, visiting lecturers, researchers, professors) and students in curriculum development to ensure the sharing and common understanding of the quality of training.

### 4.4 Perform Training Tasks to Make the National Education System to Be Better Towards an Open Education, Lifelong Learning in a Learning Society

Carry out surveys to consider closing training programs that are no longer suitable. Adjust inappropriate programs and develop new programs to meet the needs of production and learners in practice.

Focus on building a number of high-quality training programs such as talented engineers programs to improve the quality of training. Meet the requirements of high-quality human resources. Meet the needs of learners and make a good learning environment and further gradual integration with regional and international higher education.

Focus on high-quality programs to ensure the quality of outcomes but socialize these training programs to improve the quality of training.

Develop training programs in the direction of ensuring the connection among training levels. Focus on building open programs such as part-time training to encourage learners with financial need to be able to work during their studying.

### 4.5 Develop Lecturers and Management Staff to Meet the Requirements of Education and Training Innovation

Develop promotion plans and training plans for teachers and educational administrators to ensure compliance with regulations.

Increase revenue to re-invest in improving the preferential regime for teachers and educational administrators. The recruitment, employment, remuneration, and honor of teachers and educational administrators must be based on the assessment of their professional competence, their performance, and ethics. This will require a cultural change as well as capital reallocation.

Strengthen the implementation of interconnection models, links between domestic and foreign training institutions, especially universities, science and technology organizations, and research institutes to create conditions for international experts and overseas Vietnamese to participate in teaching and research.

Encourage lecturers to improve their professional qualifications by offering financial supports. Facilitate their participation in programs, projects, and courses both at home and abroad.

### 4.6 Accelerate the Process of Digital Transformation and Anticipate the Application of New Technologies

Digital transformation must ensure four elements, including empowering lecturers, interacting with students, optimizing organization, and innovating methods. The university's digital transformation process takes place in all three stages including planning, formulating strategies independently, implementing innovations, and monitoring the impact of technology deployment.

Currently, there are many tools for digital conversation such as Skype video conferencing, GoToMeeting, Blue Jeans, Microsoft Teams conversation. Resource sharing applications include OneNote application, Stream, Reader Analytics reader analysis application, and Look Up Tflat dictionary. Application supports include drawing mindmap (Mindnode, Simplemind), Wolfram Alpha. Many online services support learning such as Power BI tools, Blackboard, WebCT, Desire2Learn, ANGEL, Sakai, and Moodle. Universities need to apply new technology, using versatile tools such as computers, projectors, electronic lectures, smart electronic boards, electronic textbooks, and especially teaching software (E-learning). Accordingly, organizing classes, assigning assignments, limiting time, evaluating, providing documents, receiving feedback, adjusting student activities are all operated on computers.

Build mini studio models using new technology, virtual classrooms, virtual laboratories, virtual equipment, virtual libraries with the support of smart devices. Research and apply AI technology, especially in synthesizing learning information to give useful hints for learners and teachers, enable learners to access standardized curriculum for each individual; in assessing capacity and needs of learners, or overcoming the shortage of teaching staff (such as teaching foreign languages).

## 5 Conclusion

In the current era of scientific and technological revolution and the trend of globalization and development of knowledge economy, higher education plays an important role in the country's education system. However, the current development of Vietnamese higher education does not match with the development of the economy. The issue of reforming the higher education system is always mentioned in all the mass media, educational conferences and seminars and parliamentary forums, etc. It has also attracted the attention of not only lecturers, experts, and educational managers but also the leaders, scholars, organizations, and social groups in country and all around the world. This shows the urgency of the problem and proves the tradition of studiousness and appreciation of talent development in Vietnam as well.

## References

1. The Education Law 2005
2. The Amended Education Law (2009) takes effect on July 1, 2010
3. The Higher Education Law 2012 No. 08/2012/QH13 of the National Assembly

# Access to Smartness: An Intelligent Behaviour Monitoring and Safeguard System for Drivers

Zongliang Wu(✉), Shiwei Wu, Haofan Li, Jichao Zhao, Siyuan Hou, Jingjing Fan, and Yuan Wang

Tianjin University of Science and Technology, 300457 Tianjin, China
652569334@qq.com, wangyuan23@tust.edu.cn

**Abstract.** Advanced Driving Assistant Systems (ADAS) are mature in vehicles. At the same time, automatic driving technology is starting to be applied. But compared with those intelligent cars, old vehicles without any assistance are exposed to the dangers of accidents. And by the fact that, it is unsuitable to install ADAS on old motor vehicles, we are in dilemma to monitor their behaviours of while driving. This paper introduces an intelligent behaviour monitoring and safeguard system with high accuracy, robustness and convenience for drivers in their regular vehicles to guarantee both public's and their own safety. It merges computer vision,and multiple biosensors to collect both drivers behavioural data and health data. Considering the influence of emotion of drivers while driving, their facial expression and driving aggressiveness will be detected as well. For data from biosensors, we use a Statistical-based Bayes data fusion Method to deal with data from multiple sensors and different kinds of data. Using the cloud and Internet of Things (IoT) technology to concentrate those data, the visualized data can be checked and managed on a smartphone. Drivers in unsafe driving states will be warned by a voice assistant. The system can be connected to smart city system in progress, and drivers behaviour and health data is valuable for future data mining as well.

**Keywords:** Driver behaviours · Internet of things · Facial recognition · Advanced driver assistance systems

## 1 Introduction

In recent years, Advanced Driving Assistant Systems (ADAS)and automatic driving technology are in their rapid development, vehicles with smart systems are becoming more and more prevalent. But those technologies concern more about the data from vehicles themselves (OBD data,GPS,etc.) and the surrounding environment (pedestrians, other vehicles, etc.),data from drivers is less con-


X.-S. Yang et al. (eds.), *Proceedings of Fifth International Congress on Information and Communication Technology*, Advances in Intelligent Systems and Computing 1183, https://doi.org/10.1007/978-981-15-5856-6_53

cerned. In fact, most (94% [1]) accidents are estimated to be caused by human factors,and the bad driving behaviour of drivers is the direct cause of many serious traffic accidents,including fatigue driving, distraction, smoking, using mobile phones, not holding the wheel etc. Other special conditions of drivers, such as sudden diseases of drivers,human interference etc., also the cause accidents. However, current driver assistant systems only can warn the driver under the limited circumstances, and cannot detect special situations and potential dangers at all; when an accident occurs, they rely entirely on manual to cry for help, resulting in the delay of rescue time. Besides, there is a great data gap between smart vehicles and unintelligent ones. On the other hand, smart vehicles have little chance to access information from regular ones, bringing inaccurate judgments or difficulties in making decisions.

The purpose of this system is to help urban traffic managers, managers of motorcades and other intelligent systems such as ITS to make decisions, evaluate, supervise drivers, and analyse driving conditions. At the same time, it uses an effective methodology to guarantee the safety of drivers and passengers.

## 2 Related Works

The Method of driver monitoring, which has become one of the key research areas under the umbrella of Intelligent Traffic System (ITS), includes vehicle data monitoring, physiological signal monitoring, and action monitoring.

Vehicle data monitoring is traditional and widely used in modern vehicles. They use sensors to measure the internal status of the vehicle like steering wheel angle, acceleration, rotational velocity etc. The use of mobile smartphone inbuilt sensors to monitor driving behaviour [2] also one of the choice.

For complex states of drivers, such as distraction while driving, physiological signals can be utilized which is the most studied monitoring system. Usually, it is divided into contact method and contactless method. The former one includes the extraction of bio-features such as breathing, electroencephalogram (EEG), electrocardiogram (ECG) etc. In methods of digital signal processing (DSP), the theory of Operational Research is used to assess drivers states [3].

The second one is done by facial detection through cameras. One classic method is using face tracking and PERCLOSE (percentage of eye closure) rate as a real-time measure of fatigue [4]. But, these systems need to be enhanced in weak light conditions and have a low accuracy and robustness.

But the simple fatigue detection cannot meet the satisfaction of monitoring and predicting drivers behaviours. Al Machot et al. put forward a method [5] that uses speech instead of facial detection to recognize emotion. After that, a new type of framework named advanced driver monitoring for assistance systems (ADMAS) [6]is proposed to monitor their emotions and safeguard.

Early prediction of a driver's actions might correct drivers bad behaviours. The article [7] presents an improved deep learning technique to predict a driver's a few seconds early action before the action is performed.

However, less willing is shown in users to install a complex system. Moreover, the efficiency of drivers monitoring systems is questioned, both for their accuracy, of environmental disturbance, convenience,high expense, lacking the automatic cry-for-help function while an accident occurred. In addition, facing the disturbing from humans, which is proved to be a critical cause of serious accidents [8],can not be detected as well.

## 3 System Structure

This section describes the structure of the system and its modules in detail. The system has to be able to monitor and manage the driver's abnormal driving behaviour in real-time and have a safeguard function. The comprehensive framework of the system consists of the following 4 layers:

(a) Physical cabin equipped with IoT terminal and steering wheel with sensors to be connected to the cloud server by Wireless Local Area Networks(WLAN);
(b) Cloud server to store the data of behaviours in a cloud database,and evaluate behaviours of drivers;
(c) An ITS can access the information from cloud database through Application Programming Interface(API);
(d) Client Side where the users (including drivers and manager)can log in and check the visualized data using a smartphone application.

The sketch of the framework is shown in Figure 1. Firstly, the image data of the driver is analysed by the camera, the biological signal is collected by the sensor,thus,the driver's state is detected in real time. After the processing and fusion of data in the vehicle terminal,the data will be transformed into JavaScript Object Notation (JSON) format and sent to the cloud server. The mobile application gets the data from the server and visualizes it. When an abnormal driving behaviour appears, a voice warning will be carried out. When an accident occurs or the driver is injured, an alarm and a GPS position can be automatically sent to police or hospital.

### 3.1 Devices in the Cabin

The equipments inside consist of 2 sections: IoT Terminal and Steering wheel with sensors. Figure 2 depicts an inner layout of the cabin. This physical layer collects data from the driver while driving.

The sensors on the steering wheel are capable to obtain Pulse Oxygen Saturation ($SpO_2$) and Pulse,which are valuable for drivers health monitoring. The most effective scheme is using photoplethysmograph (PPG). Embedded in a transparent outer case made of flexible material,multiple sensors are distributed around the steering wheel. This structure enables the collection of data easy,for driver have to keep touching the wheel while driving. The IoT Terminal equipped in front of the driver serves as the main monitor. The camera collects the real-time image flow. After the facial characteristic analysis, the human body analysis,it obtains the state data and the expression data. The microphone is used

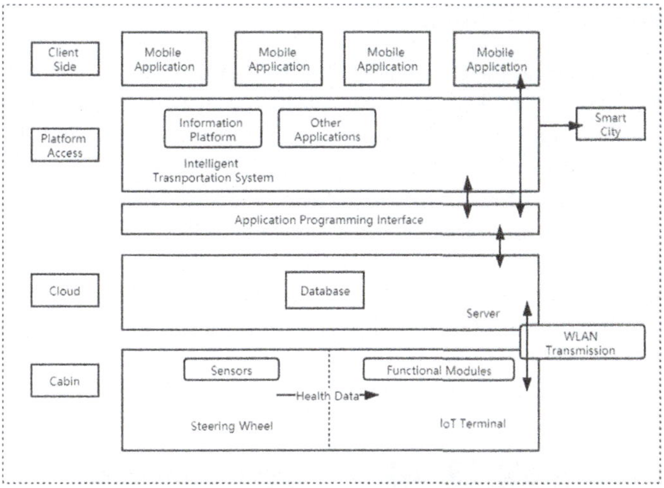


**Fig. 1.** Overall framework of system

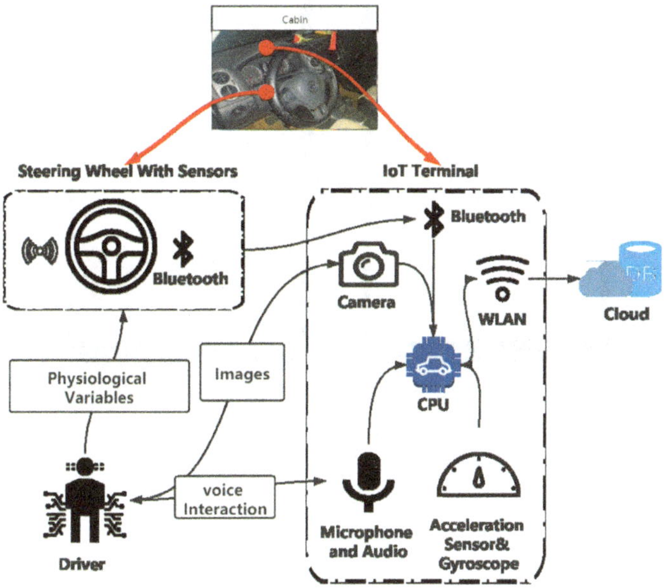


**Fig. 2.** The equipment structure in the cabin

for speech recognition. After being converted to text,it can be used not only for instructions to interact with terminals but also for emotion analysis. Acceleration Transducer and Gyroscope are used to judge whether the driver has aggressive behaviour or not. All those data will be processed by CPU inside and sent to cloud server in a JSON format using WLAN.

### 3.2 Cloud and Application

Cloud Server will contain a Database. The back-end of the server is written in a Java program and has high concurrency, which can respond to a large number of requests at the same time. The client is an Android Application. It contains 2 modules of the individual and the enterprise. As it shows in Fig. 3, the interfaces include the location and route of the vehicle, the real-time states statistics and data management of the driver at a certain time. Abnormal states of the driver recorded according to the time. The emotion of drivers can be shown to improve drivers comfort if necessary. The communication logic and flow among terminal, server and application as shown in Fig. 4, In the enterprise module, a manager of a fleet or a company is able to create a "fleet" after being certificated. The information of each member will be viewed if the one is added into the fleet.

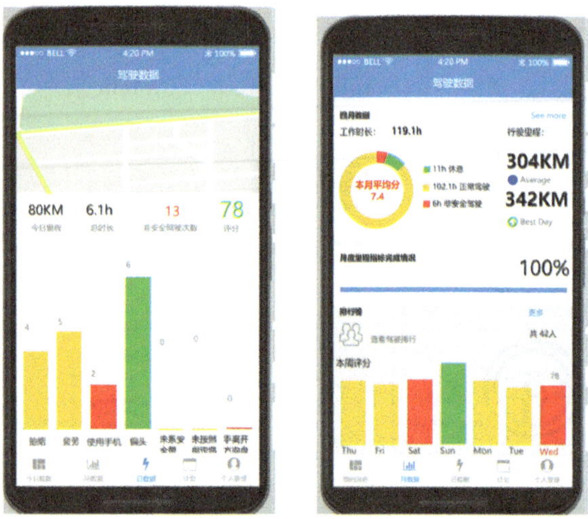


(a) Individual model (b) Enterprise model

**Fig. 3.** A data visualized application indicating drivers' behaviour while driving

## 4 Data Processing and Fusion

### 4.1 Image Analysis

For drivers fatigue and expression on their face, facial recognition will be used. It includes following 5 steps.

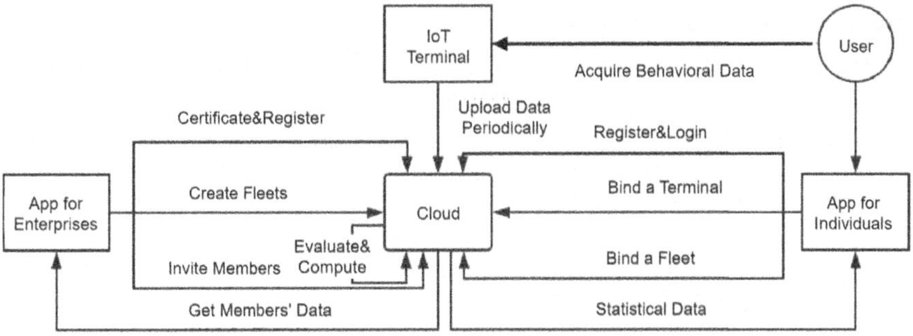


**Fig. 4.** Logic and flow chart of mobile application communication

(1) Face Image Acquisition and Preprocessing.
(2) Facial Landmark Detection: For drivers in their process of driving, disturbances, like trembling and changing of environment, will occur. Thus, the accuracy and robustness of large poses,exaggerated expressions and partial occlusions is acquired. A Fully End-to-End Cascaded Convolutional Neural Network (FEC-CNN) method proposed in [9] significantly improves the accuracy of landmark prediction.
(3) Fatigue and Recognition: We exploit the classic PERCLOSE, which is defined as a parameter where the driver's eyes are closed over a period of time, mentioned in Sect. 2 to recognize fatigue.
(4) Expression Recognition: We make use of a method of recognizing emotions through Facial Micro-Expression (FME) by learning Landmark Features (LMFs) using CNN and Long-Short Term Memory (LSTM)[10].
(5) Driving Behaviours Recognition: Other states including smoking, using phones, not holding the wheel, not facing front,human disturbance and no seat belts are also classified in a trained model in advance.

All those states will be contained in a one-dimensional matrix $S_i$ (7 attributes: fatigue, smoking, using a mobile phone, not facing front, not holding steering Wheel, no seat belts, human disturbance), and expressions are included in a one-dimensional matrix $E_i$ (7 attributes: anger, fear ,disgust, happiness, sadness, surprise and neutral).

### 4.2 Health Data

Heart Rate Variability (HRV) will occur when a driver is of drowsiness. Although PPG signal represents an effective solution in order to measure the HRV, its sampling pipeline shows some noise and distortion even after being filtered. In addition,considering that our PPG sensors are separated from the skin surface by a transparent material, its effects may be affected. Thus, after filtering, the feature level fusion is used, and the heart rate and blood oxygen concentration measurement method based on depth learning is used to learn more features by constructing multiple hidden layer model and a large number of training data, so as to improve the accuracy. The health data will be included in the $H_i$.

### 4.3 Data Fusion

The image information and biological information are fused at the decision level. Data fusion algorithm based on the Bayesian reasoning.

The main idea of designing multi-sensor data fusion method based on Bayesian reasoning is to set $n$ different types of sensors to identify $m$ attributes of the same target. Firstly, the data observed by $n$ sensors for $m$ hypothetical $A_j$ $(i = 1, 2, \ldots, m)$ are classified. Get a set of object declaration for different attributes: $B_i$ $(i = 1, 2, \ldots, n)$; Second, calculate the conditional probability (likelihood function) of each description in the case where each hypothesis$A_j$ $(i = 1, 2, \ldots, m)$ holds.

Considering that each sensor is observed independently, so each target description $B_i$ is also independent of each other, then it is necessary to calculate the joint likelihood function of $n$ target descriptions under m conditions.

Finally, the posterior density of all kinds of assumptions $A_j$ in the case of $n$ target descriptions is calculated by using Bayesian formula. And then the maximum posterior density can be obtained. Use it as a basis for deciding whether to accept assumption $A_j$. abandoning redundant information and completing data fusion.

## 5 Application Scenarios

This monitoring and safeguard system combines various technologies including facial expression recognition, biosensors, IoT and data fusion to enable regular vehicles to be smarter to some extent. Regular vehicles and their drivers will be linked to an ITS or a Vehicle Networking. A city ITS platform can easily access our driver behavioural database. For users of diverse demand, we designed the following 2 scenarios:

(1) For individual users, in addition to getting feedback on their driving behaviour from the mobile application, they can also create car fleet with relatives or friends to view their driving data to protect the personal safety of family and friends. Their injuries in accidents will be informed at once.
(2) For enterprise users, they can manage a group of vehicles and their drivers. By supervising the behaviours of employees, enterprises can solve the problems of performance evaluation on the road at regular periods. All the unsafe actions will be quantified and visualized. If expression data is allowed to be accessed, it will be utilized to improve the interior condition or service by enterprises like automobile service companies. Also, the monitoring of drivers behaviour can be exploited in the training and assessment of intern drivers.

## 6 Conclusion

This work aimed to overcome some defects of existing systems and provide smart access to regular vehicles. The system combining the facial expression and speech recognition,biosensors,data fusion and IoT equipment increases the accuracy,robustness and efficiency of management,and is easy to implement and use by both individuals and enterprises. The design of the system is valuable to the construction of intelligent traffic system network, which can use the collected data from drivers to solve many driving problems and to provide a more convenient and safe road transportation environment.

Future works foresee the validation of the monitoring and safeguard system in the improvement of safe driving. Besides,more value of those data from drivers is still waiting for mining. Considering the lack of support of large scale data,analysis and prediction of unsafe behaviour still remain to be done.

**Acknowledgements.** This study is part of Intelligent Driving Safeguard System Based on Facial Micro-expressions and Head Features Project, supported by National Students' Innovation and Entrepreneurship Training Program, and College of Artificial Intelligence, Tianjin Univ. of Sci. & Tech.

## References

1. Barnhizer, D.: Lawyers and self-driving cars: an example of job loss from ai/robotics. In: Artificial Intelligence/Robotics and the Death of Work and Democracy. http://works.bepress.com/davidbarnhizer/119/ (2016)
2. Chhabra, R., Verma, S., Rama Krishna, C.: Detecting aggressive driving behavior using mobile smartphone. In: Krishna CR, Dutta M, Kumar R (eds.) Proceedings of 2nd International Conference on Communication, Computing and Networking, pp. 513–521. Springer, Singapore (2019)
3. Taherisadr, M., Dehzangi, O.: Eeg-based driver distraction detection via game-theoretic-based channel selection. In: Fortino, G., Wang, Z. (eds.) Advances in Body Area Networks I, pp. 93–105. Springer International Publishing, Cham (2019)
4. Wang, H., Zhou, L.B., Ying, Y.: A novel approach for real time eye state detection in fatigue awareness system. In: 2010 IEEE Conference on Robotics, Automation and Mechatronics, pp. 528–532 (2010). https://doi.org/10.1109/RAMECH.2010.5513139
5. Machot, F.A., Mosa, A.H., Dabbour, K., Fasih, A., Schwarzlmüller, C., Ali, M., Kyamakya, K.: A novel real-time emotion detection system from audio streams based on bayesian quadratic discriminate classifier for adas. In: Proceedings of the Joint INDS'11 ISTET'11, pp. 1–5 (2011). https://doi.org/10.1109/INDS.2011.6024783
6. Izquierdo-Reyes, J., Navarro Tuch, S., Ramirez-Mendoza, R.A., Bustamante-Bello, R., Avila-Vázquez, R.: Advanced driver monitoring for assistance system (ADMAS) based on emotions (2016)
7. Gite, S., Agrawal, H.: Early prediction of driver's action using deep neural networks. Int. J. Inform. Retriev. Res. **9**, 11–27 (2019). https://doi.org/10.4018/IJIRR.2019040102

8. Zhawei, N. Ascertaining nature of snatching steering wheels of moving buses and leading to vehicle damage and body injury, pp. 7–9. People's Judicature (2019). https://doi.org/10.19684/j.cnki.1002-4603.2019.08.003 (in Chinese)
9. He, Z., Kan, M., Zhang, J., Chen, X., Shan, S.: A fully end-to-end cascaded cnn for facial landmark detection. In: 2017 12th IEEE International Conference on Automatic Face Gesture Recognition (FG 2017), pp. 200–207 (2017). https://doi.org/10.1109/FG.2017.33
10. Choi, D.Y., Kim, D.H., Song, B.C.: Recognizing fine facial micro-expressions using two-dimensional landmark feature. In: 2018 25th IEEE International Conference on Image Processing (ICIP), pp. 1962–1966 (2018). https://doi.org/10.1109/ICIP.2018.8451359

# Regulatory Modeling for the Enhancement of Democratic Processes in Smart Cities: A Study Based on Crowdlaw—Online Public Participation in Lawmaking

Marciele Berger Bernardes[1](✉), Francisco Pacheco de Andrade[1], and Paulo Novais[2]

[1] Escola de Direito, Universidade do Minho, Braga, Portugal
marcieleberger@gmail.com, franc.andrade@gmail.com
[2] Escola de Engenharia, Universidade do Minho, Braga, Portugal
pjon@di.uminho.pt

**Abstract.** The advent of Information and Communication Technologies (ICTs) brought a fast development of urban centers, and a debate emerges on how to use ICTs to enhance the development and quality of life in cities and how to make these more efficient. The object of this paper has as the analyze whether or not the traditional regulatory model, based on a system of sanctioning behavior divergent from the normative prescription by the postulates of digital governance. For the elaboration of this paper, we use the method of deductive approach. This way, along with the prominent literature and the experience of good international practices, we must recognize the need for an "intelligent" regulatory modeling thus being, we presented a contribution to building a new legal paradigm toward the enhancement of democratic processes in smart cities, structured on the postulates of Crowdlaw (collective production of the legislative process). Last, we believe that the contributions arising out of this work may fill some of the gaps existing in terms of legal theory production on the regulatory modeling for participative governance.

**Keywords:** Regulatory modeling · Participatory governance · Smart cities · Crowdlaw

## 1 Introduction

The quick introduction of new technologies, associated with the acceleration of urban development, led to the creation of smart cities. In the legal sphere, there was a normative profusion on the subject of technological development and administrative modernization. However, the practice of the majority of administration shows that the reproduction of hierarchical normative production's structure (top down) is not enough and is inefficient in terms of taking advantage of the creative potential of people. In this sense, this work justifies the need to incorporate the citizen in the axis of this transformation.


X.-S. Yang et al. (eds.), *Proceedings of Fifth International Congress on Information and Communication Technology*, Advances in Intelligent Systems and Computing 1183, https://doi.org/10.1007/978-981-15-5856-6_54

In the face of this, the traditional normative structure becomes questioned and even (why not say it?) challenged to (re)design its models, well aware of the current social values, not limited to the "cold letters of the law." Thus being, this study approaches the "new" form of normative production in the light of directives that embody what Canotilho [1] called "Modern Statehood." This aims at adapting and widening the constitutional principles of the State of Law to the new demands coming from society.

Thus, diverse forms of integration of organs and entities in the performance/development of public policies are connected toward broader participation and democratic openness. In face of this, the write contribution and novelty are part of the outstanding literature and the experience of good international practices on the need for a "smart" regulatory modality (in which governments—in collaboration with companies and citizens—are able to adopt experimentation policies such as enabling or dialogue, learning and adapting, in favor of a more efficient regulatory ecosystem).

From this side, and with the aim to reach the proposed targets, this study was divided in four items. The first one is destined to discourse about the contemporary settings; next, the study is directed to presentation the regulatory modeling for the enchantment of democratic processes in smart cities; in the third item it is made an analysis of the Crowdlaw legal system. Finally, upon this, it is intended to show the ways to be explored for the enhancement of democratic processes in smart cities.

## 2 Contemporary Settings

Following the teachings of Castells [2], contemporary societies experience a structural process of transformation, in different axes, with reflections mainly in the economic, cultural, and political dimensions. Thus being, such changes led humanity to a new age, the informational age.

Concerning the legal sphere, Canotilho [1] points out that the main traditional principles of State of Law go through a process of widening (as it happens with "geological layers") and, from that, they sustain the material and processual bases of what he called "Modern Statehood." This form of Statehood would be characterized, in general terms, for the capacity of institutionalizing new channels of intercommunication between State and society, seen from the understanding that knowledge must not be centralized.

The revision of the State model thus occurs within the emergence of a new model of economic, social, and technological development—the informational model. This new paradigm is allied to urban development, understood through the phenomenon of smart cities.

It is now clear that the configuration of urban contexts in a network has affected organizational and institutional tools of governance, which is becoming more and more insufficient and inadequate in the current situation. What corroborates the increase of a democratic deficit, mainly due to a lack of legitimacy of institutions with relation to social demands.

That means that smart cities predicate plurality, complexity, and dynamics and that these requirements are in counter-cycle with the traditional hierarchical, centralized normative production ways, operating on the logic of a binary system with assigning conducts and sanctions.

We must register that it is not enough a whole principled construction around this Modern Statehood if there is no concrete mechanism for its effectiveness. Moreover, it is to note that specific regulatory modeling to deal with the specificities and challenges of the new state paradigm is just one of these instruments. However, if well used, it may well be fundamental in order to minimize the governance deficits toward renewed processes of individual and collective autonomy, either in the context of the State or in the relationship State/Market/Society (Citizen). In this perspective, the theories on regulatory modeling are adequate for the understanding of this thematic. From the decade of the 80s, the issue of governance assumes that each regulated environment has a logic of its own.

## 3 Regulatory Modeling for the Enhancement of Democratic Processes in Smart Cities

As we have seen, the regulation system cannot stay (anymore) restricted to the figure of the State that centralizes power and regulates the market and society through static criteria. Such a system must open to including in it the agents to be ruled.

Thus being, the proposals for analysis of a reformulation of the regulatory modeling, in the current situation, are analyzed from three fundamental theoretical sides: Responsive Regulation, Smart Regulation, and Really Responsive Regulation.

### 3.1 Theory of Responsive Regulation

The theory of Responsive Regulation was conceived, in the decade of the 80s, by Ayres and Braithwaite [3]. The scholars have departed from the assumptions of the Theory of the Behavioral Law and Economics (BLE), confronting it with the classical theory of regulation.

So, from an empirical study, the authors concluded that the classical theory of regulation, anchored on the utilitarian paradigm (that a norm will be effective just if it is advantageous), could not be applied to all the regulated agents consistently. From the verification that state entities should be sensible and responsive to how each regulated agent is organized, Ayres and Braithwaite [3] proposed a methodology of structured dialogue, involving the regulatory entity and the regulated agents in a perspective of supervising and collaboration.

This way, they depart from the assumption that the hierarchical regulatory model (top down) and imposed (assigning negative sanctions) is important for the maintenance of the regulatory system, but it is not enough to ensure the effectiveness of the law (compliance). The authors then propose multiple strategies, with different levels of intervention and coercion, translated to the figure of a responsive regulatory pyramid.

We must refer that the theory of Responsive Regulation keeps in the State the role of regulation, seen as a factor of normative effectiveness, but also of maintenance of the state figure as an entity concentrating the regulatory function. Moreover, there remains the primary focus falling upon the theory of the Responsive Regulation, considering that the State must exercise the regulatory function. Nevertheless, in the same measure, it must combine regulatory measures that assign to the market itself the power of solving problems and auto-regulating itself, and these were some reasons that justified the development of the theory of Smart Regulation, as follows.

### 3.2 Theory of Smart Regulation

As we have seen, responsive regulation was presented as an alternative in order to overcome the gaps in the classic model of state regulation. Regardless of its practical application, the final aim of auto-regulation started to be questioned in what concerns the capacity to give answers to demands and to highly complex social problems. For instance, environmental issues and the problems of collusion between different normative instruments regulating these issues must consider that the divergences turn unfeasible an effective compliance with the norms.

Within this debate, it becomes clear the doctrine of Gunningham et al. [4] characterized mainly by the "regulatory pluralism" aiming at enhancing the performance of responsive regulation. The authors propose that this must be encompassed in the self-regulation or co-regulation mechanisms. In this sense, the theorists of Smart Regulation propose that other agents, besides the state regulator, should participate actively in the regulation process.

It is, we may say, a regulatory technology based on the pyramidal model of sanctions (possibility of echeloning of sanctions). Meanwhile, the regulatory pluralism is built from the interlocution between State (mechanisms of command and control), market (auto-regulatory mechanisms), and society. It is thus a tridimensional structure.

Such strategies were not exempt from critics, based as Moreno states on the following arguments. First, theories that are sustained by normative effectiveness alone are not enough to comply with all the requirements of the legal order (considering the difficulty of establishing objective parameters in order to assess whether the regulated agents comply with the norm). Following the regulatory pluricentric of Smart Regulation, we may face difficulties in operationalization, namely from the lack of clarity concerning the responsibility of the regulated agent. Regulating agents may adopt actions and behaviors not conforming with the scope of the norm; the process of construction/adaptation of the regulatory system, involving a multiplicity of actors and of interests, will tend to be time-consuming and costly.

For all those reasons, Moreno [5] postulates that such strategies must incorporate, in the step of regulatory production, communication system/debate, these and other measures are postulated by the theorists of the Really Responsive Regulation.

### 3.3 Theory of Really Responsive Regulation

The Really Responsive Regulation considers the complexity of social-technical interactions (with effect on the current regulatory paradigm). It departs from premise according which the State must be responsive to the specificities of the market (Responsive Regulation) and promote the dialogue between multiple regulated agents—State, market and society (*Smart Regulation*); Baldwin and Black [6] postulate that regulation must also be responsive to the regulated system (to its own effects), meaning that there must be a monitoring and periodic review of the effectiveness of the regulatory strategy.

The main idea of the proponents of RRR is that a regulatory theory is only important if it does help to overcome the challenges with which the regulatory agents are faced in practice, justifying the need to adopt strategies previously to the creation of the norm (ex ante), of monitoring (pari passu), and supervise and enforce the norm (ex post). With this, the authors point out that really smart standards should explore and deepen the debate on "transparency, evaluation and monitoring" [6].

For this, the theoretical supporters of RRR sustain that the design of the regulatory modeling must consider five circumstances: (1) Cognitive and Operational Structures of the regulated agents. "Configurations of attitude": The social signs that the regulated agents send to the regulator for communicating the level of acceptance of the regulating agenda and of the regulator itself in the performance of its functions. (2) Wider institutional regulatory environment: to recognize and to answer to restrictions and opportunities presented by the institutional environments. (3) Different logics and strategies of regulatory tools: Each regulatory modeling may have its own logic, such as, for instance, punishment, rehabilitation, and restauring. An intelligent regulation must offer strategies to balance the tensions[1] between the norms. (4) The own design of the regulatory regime: By the application of the regulatory model the regulator must be capable of measuring whether or not the tools are efficient in order to reach the desired goals. (5) The changes eventually may occur in each one of these circumstances: The regulatory strategies need to adapt to changes and new priorities, either arising out of internal or of external factors [6]

That said, RRR is based on a new form of regulatory modeling, more flexible. With a holistic vision, involving the market and also society. In a perspective of collaboration, through the participation in public policies. And, besides that, the construction of the regulated system in your different phases (diagnosis, modification, and evaluation of the normative scenario).

From the synthesis of the models and respective critics, it may be said that the theory of Really Responsive Regulation was the one that better synthesized the critics directed both to the model of hetero-regulation and the models of auto-regulation and co-regulation. Moreover, it has evolved in a way to encompass the complex social demands,

[1] The tension among the norms are referred to as responsiveness between regulation and the legal system in general. In this sense, Saddy and Greco [7] illustrate this scenario on the study "Terms of Adjustment of Conducts in sanctionative procedures in the Brazilian State)." In which the authors discourse on the different logics of legislations that regulate the economic activity, foreseeing divers normative species (fines, warnings) on the Federal Law of Administrative Procedure in this country. Such sanctions may be commuted, meaning that it is up to the legislator to apply the sanctions or to dissuade the infractor from celebrating agreements (replacement/suspension) through the Agreement of Adjustment of Conduct. Greco illustrates this scenario in the study "Terms of Adjustment of Conducts-TAC in sanctionative procedures in the Brazilian State," according to the best public interest. Meanwhile, these authors punctuate that not always Brazilian laws allow the use of TAC, through regulatory agencies, identifying three possible situations. "(I) cases in which sectorial legislation expressly establishes the possibility of TAC by regulatory agencies. (II) cases in which sectorial legislation does not establish such possibility, but regulatory norms authorize negotiation, case by case, to overcome infractions established at an infra-legal plan. Furthermore, (III) cases in which there is no provision in legislation, nor normative resolution to the case." It is thus suggested that in complex scenarios, it is up to the administrations to adopt a vanguard position, electing consensual strategies of regulation, also called soft-regulation. The current position of the administration must be tied "not only with the act of decision, as it was the administrative tradition, but well beyond that, also with the result of what was decided." Saddy and Greco [7].

thus being elected as this study's paradigm in the construction and of regulatory modeling for the enhancement of democratic processes in smart cities.

Next, it will be presented some international cases that are establishing the connection between these new paradigms (responsive laws, incorporation of technologies in its routine and management of the problems of administration in a shared way), toward intelligent and participative governance.

## 4 Crowdlaw: Legal System Qualified by "New"[2] Agents

In front of the scenario mentioned above, it is proved the thesis according to which smart cities are nor just technology: It is not enough the presence of the municipal administration in a government portal. Smart cities are those that developed its planning by integrating leadership and political commitment and placing social factors at the core of its approach. This issue is confirmed by the postulates of the theory of Really Responsive Regulation, when it is supported that, in intelligent environments, the citizen assumes a central role—and "active" one. However, it must also be "responsive" to the conditions of the system.

The "active" role of citizens echoes in legislative processes and gets strength with the movement "Crowdlaw,[3]" a strategy of collective production of the legislative process, "creating opportunities for the persons to provide knowledge, information and opinions in search of the law production, of better informed regulation, towards enhancing the quality of legislative results and efficiency of government" GovLab [9], as it is shown in Fig. 1.

As illustrated, the process of citizen participation may operate in each of the five phases of the legislative process. The first phase "formation of the agenda"; the second phase "Insertion in the political agenda"; third phase integrates the "formulation of public policy"; the fourth phase "implementation" at last, the fifth phase "policy's evaluation," ends the cycle of participation with supervision and monitoring of the results of the adopted legislation by the citizens.

Still, according to GovLab researches, the main trend of Crowdlaw in present days is the use of tools and channels to allow various types of digital interactions with citizens.

---

[2] Although the figure of the agents is not a new one, (citizens, private agencies, universities have always existed), they assume an important role in the intelligent urban context. After all, they are "new," considering the function they assume, which was before exclusively assumed by the public power regulator.

[3] Crowdlaw is an emerging practical field developed by GovLab, a research and action center based on Tandon School of Engineering of New York University. Beth Simone Noveck and Stefaan Verhulst founded the referred Center, in 2012, to promote the design of more open, more efficient, and networked governmental institutions using data, technology, and crowdsourcing. In the year of 2014, GOVLAB adopted the term Crowdlaw to refer to every possibility of citizens becoming collaborators and co-creators of all types of legislative activities, including legislation, regulation, and constitution. Available information in http://www.thegovlab.org/project-crowdlaw.html. Access in: August 2019. GOVLAB [8].

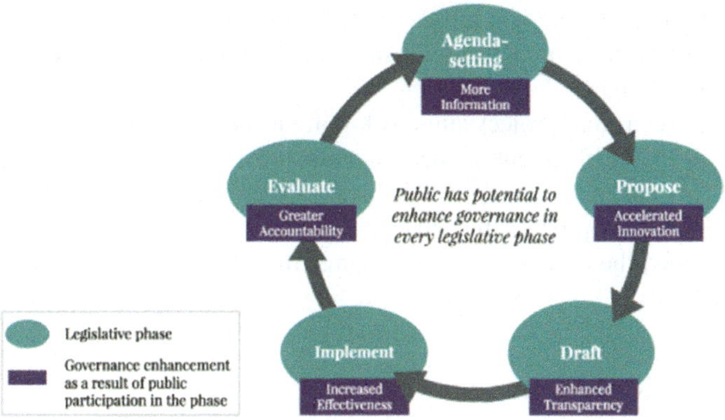


**Fig. 1.** Phases of the legislative process and the aspects of governance that the public can help to enhance each GovLab [8]

For that, it may be referred to the experience[4] about the Government of Taiwan (Ministry of the Digital Government of Taiwan), which has been "developing solutions for intelligent interactions of social commitment, using tools developed with open-source software."

So, it must be identified which actions the municipal administration must adopt in order to ensure this virtuous circle. From this perspective, GovLab presents a script with the main elements of a legal system turned to popular participation, which is: (1) Creating a Laboratory of Open Assemblies: placed within the municipal Assembly, with powers to bind the actions of this one, and with different responsibilities (to moderate and to facilitate the commitment and to undertake training for the active popular participation); (2) to declare popular participation as a right: fomenting co-responsibility and recognizing that every citizen has the right to participate; (3) to articulate the purpose of participation: to include individuals, civil society organizations, corporations, and public servants in all the steps of the legislative process; (4) to promote active participation: adopting measures to promote an active posture of the citizens in the development and implementation of the law; (5) transparency and responsibility: Active participation assumes that persons have the right to know how legislation is promulgated, what is promulgated, by whom, and with what purpose; (6) incorporating commitment in all phases of the legislative process: Participation must be a praxis of the legislative pattern;

[4] Many other interesting projects came to our knowledge. Just to mention a few more, we highlight a project from Iceland, a country that is considered by many as a "Better Reykjavik," and popular participation in the different steps of the legislative process, see the case of Madrid, Law 39/2015, article 133. "Participation of citizens in the procedure of elaboration of norms classified as laws and regulations. 1. Before the elaboration of the project of law or the law, a public consultation must be undertaken through the portal of the respective Administration, in order to get the opinion of individuals and more representative organizations that might be affected by the future norm." Moreover, still, the case of Brazil, Resolution Project n° 217, of 2017, whose aim is to modify the Regulation of the Chamber of Deputies, in a way to include dispositions that turn the public consultation (of the phase of the report of projects law) into a phase of the legislative process. Ferri [10].

(7) evaluation and test: The law must require continuous researches with participants in order to evaluate the level of consciousness in the community on the law of popular participation GovLab [8].

As may be seen, this project aims to highlight the possibilities of qualifying the legal system by the involvement of "new" agents. It is important to stress the role of the centrality of the citizen, as an "active" subject of the process of formulation of public policies (the legislative process in itself). And, besides that, as a "responsive" subject in front of the policy, the citizen must thus change his attitude for the prosecution of the normative finalities.

## 5 Final Considerations

The scenario presented here has made possible to launch the bases for the construction of a new legal paradigm, such as presented in the third item of this paper. Following this approach, we had a great contribution of the theories of Responsive Regulation, Smart Regulations, and Really Responsive Regulation. Moreover, we got the perception that (in a world in which public functions migrate to the digital environment), governments and regulators must "find ways to adapt continuously to a new fast-changing environment. They must reinvent themselves in order to understand better what it is being regulated," Schwab [11].

From this logic, and to ensure passage from a vicious circle to a virtuous circle, it is recommended that the municipal public sector adopt the following strategies:

- Regulating the right of popular participation, considering the technological potentialities and, mainly, ensuring the participation in all these phases of the legislative process, in order to legitimate the decisions taken by regulators;
- Foreseeing the creation of a permanent institution, the Urban Innovation Laboratory (v.g "Open Assemblies Laboratory"[5]), with the responsibility of implementing all the cycle of popular participation strategies;
- Ensuring the participation of diverse actors involved in the "innovation ecosystem" is possible (and even desirable) the inclusion of government, citizens, academy, and private sector.

Such regulatory acting and design of popular participation—in all the phases of the legislative process (not just in the physical world but even in the digital environments v.g. government portals)—would thus serve as a minimum warranty of leverage of participative governance. Entirely responsively committing the involved parties.

[5] By way of illustration, it may be referred the case "City Observatory," adopted by the City of Madrid/ Spain, characterized by a permanent organ of citizen's participation competent for debating the cities issues, to propose the convocation of widespread consultation and to make recommendations to the municipal organs. The referred Observatory was regulated by "Acuerdo del Pleno," of the 20 January 2020. Available in: https://www.madrid.es/UnidadesDescentralizadas/UDCMedios/noticias/2019/01Enero/29Martes/Notasprensa/Reglamento%20Observatorio%20de%20la%20Ciudad/ficheros/Acuerdo%20de%20Pleno%20Rto%20Observatorio%20original.pdf. Access in: Jan 2020.

**Acknowledgements.** This work has been supported by the CAPES Foundation, Ministry of Education of Brazil (by CAPES under Grant nr. BEX—1788/15-9), and also to the JusGov-Research Centre for Justice and Governance, and to the FCT Fundação para a Ciência e Tecnologia within the Project Scope: UID/CEC/00319/2019.

## References

1. Canotilho, J.J.G.: Constitucionalismo e a geologia da Good Governance. In: Brancosos e a Interconstitucionalidade Idearios dos discursos sobre a historicidade constitucional, pp. 325–334. Almedina, Coimbra (2008)
2. Castells: Manuel. A sociedade em rede. 2. ed. Traduzido por Roneide Venâncio Majer. São Paulo: Paz e Terra, 1999. v. 1
3. Ayres, I., Braithwaite, J.: Responsive Regulation. Transcending the Deregulation Debate. Oxford University Press, New York (1992)
4. Gunningham, N., Grabosky, P., Sinclair, D.: Smart Regulation: Designing Environmental Policy. Oxford University Press, Oxford (1998)
5. Moreno, N.A.: Tecnologias regulatórias piramidais. Responsive regulation e smart regulation. Revista de Direito Público da economia. Ano 13, n. 49. Jan/mar 2015. Belo Horizonte: Editora Forum, p. 125–158, p. 126
6. Baldwin, R., Black, J.: Really responsive regulation. Mod. Law Rev. **71**(1), 59–94 (2008)
7. Saddy, A., Greco, R.A.: Termo de Ajustamento de Conduta em procedimentos sancionatórios regulatórios. Revista de Informação Legislativa. Ano 52 Número 206 abr./jun. 2015, p. 167
8. GOVLAB: #2. CrowdLaw-introduction (2017). Available in https://crowd.law/crowdlaw-introduction-baef458325ae. Access in Aug 2019
9. GOVLAB:#4. CrowdLaw, framework for institutionalizing public engagement in lawmaking (2017). Available in <https://crowd.law/crowdlaw-model-legislation-annotated-30f864593e2. Access in Aug 2019
10. Ferri, C.: CrowdLaw: when the intelligence of crowds knocks at the State's door. Disponível em (2018). http://thegovlab.org/crowdlaw-when-the-intelligence-of-crowds-knocks-at-the-states-door/. Access in Aug 2019
11. Schwab, K.: A quarta revolução industrial, p. 66. Levoir, Portugal (2017)

# Measuring Physical Fitness for Military Cadets Officers

Mohar Kassim and Shahrulfadly Rustam(✉)

Defence Fitness Academy, National Defense University of Malaysia, Aras 2, Bangunan PTMK, Kem Sungai Besi, 57000 Kuala Lumpur, Malaysia
shahrulfadly@upnm.edu.my

**Abstract.** The research objective measures the muscle lower body of level of fitness included in armed forces cadet commissioner in Malaysia. This research uses standing broad jump as test battery to measure muscle lower body fitness and to gauge the sample power of fitness. 212 male respondents ($N = 212$) comprising of military cadet commissioner of the NDUM were chosen into research. The standing broad jump test was utilized as an instrument for this research. Quantitative research as a quasi-experiment is the technique embraced for this research. The quasi-experiment technique is utilized to gauge and assess the degree of physical fitness of muscle lower body especially for leg power fitness among the military cadet officers. The research plan is a quasi-experiment study with pre- and post-test where data were gotten through down to testing at the outdoor. The information broke down utilizing the SPSS programming adaptation 20.0 to figure standard deviation, mean and paired sample t-test in measure the lower body of physical fitness of the military cadet officers. The result appears that the standard deviation and mean for measure the muscle lower body fitness is ($m$ = 204.01), (SD = 23.197) for pre-test and ($m$ = 222.21) (SD = 20.716) for post-test. The paired-sample $t$-test for assessing the level of fitness in muscle lower body for pre- and post-test is significantly different ($p < 0.05$). The ramifications of research the military cadet officers can recognize the degree of physical fitness of their muscle lower body especially for leg power fitness.

**Keywords:** Physical fitness · Armed forces cadet officers · Standing broad jump

## 1 Introduction

Throughout different exercises or day-by-day schedules, respect lacking of physical activity performance or entertainment, direct or substantial physical activity, we won't have the option to actualize it without ideal level of fitness. This is the motivation behind why level of fitness is significant in individual being. Level of fitness is likewise a situation that somebody can do his day-by-day work without torpidity, utilizing negligible vitality while having additional vitality that empowers him to perform all the more testing exercises [1]. The entire task involves in our day by day in life, we need well-being that


X.-S. Yang et al. (eds.), *Proceedings of Fifth International Congress on Information and Communication Technology*, Advances in Intelligent Systems and Computing 1183, https://doi.org/10.1007/978-981-15-5856-6_55

incorporates body, psychological, physiological and community viewpoints. This is all inclusive idea that enables an individual to live superbly and vivaciously [2].

Level of fitness is the capacity to contend in a decent and excellent living. What's more, fitness additionally incorporates different viewpoints, for example, social, mental and physiology fitness, which are dictated by the person's condition, heredity, qualities and way of life. Physical fitness exercises are the epitome of learning characteristics and qualities contained in the instructive projects.

It is commonly understood that military workforce requires a particular level fitness to fulfill the physical need of war. Officers are required to bounce, crawl, hop, climb, push, run from spread to cover, and convey overpowering burdens over extensive stretches in their missions. These are the necessary endeavors put upon the military faculty. Thus, from the military point of view, the key quantifiable fitness segments incorporate versatility, quality, perseverance and adaptability [3].

The individual component and well-being in army workforce is the benchmark and are remain as the noticeable attributes of officers the world over, even to the degree of being better than weapons upgrades and development [4, 5]. In view of occasions ever, physical planning is of most extreme significance for the two troopers and officials in guaranteeing a considerable order and authority.

Well-being is a significant guide on the security workforces and military health [6]. Security powers faculty ought to have an elevated level of physical fitness since they are constantly occupied with physically requesting exercises particularly in military exercise and during war [7]. Observing of the physical fitness level into security forces and military vigor is significant from the exhibition perspective, yet in addition to survey their battle capacity.

Power as the capacity to express high paces of force improvement, which is frequently, identified with a competitor's general quality levels and capacity to express high power yields [8, 12]. Furthermore, he likewise recommended capacity to convey big paces of power advancement and big force yields are basic execution attribute key to the accomplishment in most games [8]. This announcement is upheld by [9], who expressed that these capacities are viewed as among the most significant games execution qualities, particularly in exercises that depend on bouncing, altering of course, and additionally run execution. Power is the aftereffect of strong solid quality as a flitting motivator in the midst of a given improvement. The last referenced, routinely implied as peak power (PP), is regularly associated with touchy developments, for example, run and hopping, can be an essential variable identified with achievement in a given preparing.

## 2 Research Statement

Well-being measure is the basis to utilized aid to translate volume of fitness individual in an athletic ability [10]. Well-being measure will assist physical trainer with checking the volume of fitness level of armed force officials occupied with physical exercises. Moreover, through fitness measure, armed force physical exercise can recognize of qualities and shortcomings of armed forces staff in the Malaysian Army.

Information on the volume of fitness among army staff can assists armed force vigor coaches give suitable army preparing to build the degree of well-being [3]. Creating physical fitness measure for military officials in Malaysia is critical in order to evaluate and keep up presentation volume of fitness of Malaysian military officials [3].

The nonattendance of a physical fitness measure brings about the act of utilizing tests that do not fit in with standard technique for estimating the physical fitness level [5]. This can mess major up to Physical Trainers. Up until this point, no physical fitness measure is received in deciding the physical fitness level, particularly relating to intensity of armed force officials in Malaysia [5]. As of now, there are no test techniques and standard physical fitness measure being presented and actualized in the Malaysia Armed Forces.

Thus, military officials who experience physical preparing or physical fitness session could not realize their physical fitness level in light of the fact that their exhibition was not estimated and evaluated methodically. Since there is no standard physical fitness measure, the military physical trainer needs to depend on conflicting estimation tests, in this manner the test scores acquired do not exhibit the real degree of physical wellness of the real military officials.

Thus, the researcher intends to research the volume of fitness among the armed forces Cadet Officer in Malaysia along of improvement of a standard physical fitness measure among Military Cadet Officers.

## 3 Research Objective

This study is conducted with the aim of measuring a physical fitness level for muscle lower body and to evaluate the muscle power level among the Malaysian Military Cadet Officers.

## 4 Method of Research

The design of this research is a quasi-experiment of quantitative research. This technique is utilized gauge, assess the degree of well-being, and create level of fitness measure. This trial perception research utilizes a power component of physical fitness test battery.

The structure of this research is quasi-experimental study that form pre- and post-test. In a quasi-experimental research structure, information is gotten through the useful test in the field [11]. This research including 20–22-year-old Military Cadet Officers from the National Defence University of Malaysia partook in this study. A sum of 212 male Military Cadet Officers were chosen as a sample. The motivation behind the examination is to measure vigor of muscle lower body utilizing the standing broad jump test battery (Tables 1 and 2).

**Table 1.** Fitness test battery

| Physical fitness component | Test battery |
|---|---|
| Power | Standing broad jump |

**Table 2.** Analysis of data

| Part A | Analysis type | Measurement tool |
|---|---|---|
| A | Sample background | Frequency, percentage |
| B | Pre- and Post-test fitness level for muscle lower body | Mean, standard deviation<br>Paired sample $T$-test |

## 5 Finding Result

Every single crude information including outcome of chosen vigor test battery were gathered and dissected utilizing SPSS adaptation 20.0 to get the outcome of the standard deviation and mean to be utilized in measure the physical fitness level. Paired $t$-test is utilized to evaluate pre- and post-test muscle lower body power level for the Military Cadets Officers. This examination utilized the standing broad jump test to find out the leg explosive force.

This regularizing research is utilized to build up a physical fitness measure utilizing a chose testing equipment and factual derivation is utilized to examine degree of vigor as indicated by lifetime and sexual orientation. Paired T-test sample were utilized watch distinction allying pre and post-test vigor volume of the subjects. Physical fitness volume of sample was examined with taking a gander of the outcome during the presentation of the battery tests. The data analysis process included a few stages, for example, putting away, uncording, disconnecting, and computing the information. Every one of the estimations of the test battery was gathered and recorded. Corrections were made to fitness tests information to guarantee the information is finished and as per built up techniques. For this examination, the researcher set three degrees of estimation. First, the pre-estimation level. Preceding the estimations, the researcher informed the respondents on the motivation behind the estimation.

Second, is the level estimation, where the subjects were separated into three gatherings dependent on their age (20 years, 21 years and 22 years). In the wake of separating the subjects into three gatherings, the subjects were required to fill the example score structure.

Third, is the process of study where each test battery was estimated by the fitness segments. Information taken from estimations made during the pre- and post-test was then investigated to decide the fitness level of the subjects.

### 5.1 Sample Background

See Table 3.

### 5.2 Data for Measure for Power Fitness Level

Table 5 appears level of fitness measure create to the standing broad jump test results where the mean and standard deviation is 204.01 and 23.197, respectively. There are 5 categories used to classify the grade in the physical fitness measure: Excellent (5),

**Table 3.** Age descriptive data

| Subject | *N* | Age | Frequency | Percentages | Valid percent | Cumulative percent |
|---|---|---|---|---|---|---|
| Male | 212 | 20 | 43 | 20.3 | 20.3 | 44.3 |
| | | 21 | 94 | 44.3 | 44.3 | 79.7 |
| | | 22 | 75 | 35.4 | 35.4 | 100.00 |

**Table 4.** Analysis descriptive on sample standing broad jump

| Descriptive statistic | Test score |
|---|---|
| Minimum | 150 |
| Maximum | 285 |
| Mean | 204.01 |
| Standard deviation | 23.197 |
| Skewness | 0.167 |
| *N* | 212 |

Good (4), Fair (3), Poor (2), and Very Poor (1). Evaluating of physical fitness empowers the various degrees of potential in a specific gathering as per the capacities of every person to be distinguished. Scores from 5 (most elevated) to 1 (least) are given to show the accomplishment level. From Table 5, the good performance is 'Excellent' having outcome of 5 starts from 241 cm, 'Good' with the outcome of 4 is from 216 to 240 cm, 'Fair' with the outcome of 3 is from 192 to 215 cm while 'Poor' with the outcome of 2 is from 168 to 191 cm and lastly, 'Very Poor' with the outcome of 1 is from 167 cm and below (Table 4).

**Table 5.** Physical fitness measure for muscle lower body of power

| Category | Distance (cm) | Score |
|---|---|---|
| Excellent | 241< | 5 |
| Good | 216–240 | 4 |
| Fair | 192–215 | 3 |
| Poor | 168–191 | 2 |
| Very poor | <167 cm | 1 |

*Note* cm—centimeter

### 5.3 Data for Muscle Lower Body Power Level of Fitness

Paired sample $T$-test perusal utilized to analyze the score of mean of pre-test muscle lower body of power and post-test muscle lower body of power. The discoveries demonstrated the mean score of post-test muscle lower body of power ($m = 222.21$, sd $= 20.716$) is more greater contrasted with the score of mean of pre-test ($m = 204.01$, sd $= 23.197$), $t\ (211) = -11.306$, $p < 0.05$. From the finding, it was found that the level of muscular fitness of the lower body of the Malaysian Military Cadet Officer was in good stead with the increase from pre- to post-test. This indicates that military cadet officers have high leg muscle power in performing tasks as security forces (Tables 6, 7 and 8).

**Table 6.** Frequency and percentages for level of fitness level for power amid Military Cadet Officers NDUM

| | Frequency | Percent | Valid percent | Cumulative percent |
|---|---|---|---|---|
| Excellent | 38 | 17.9 | 17.9 | 18.9 |
| Good | 100 | 47.2 | 47.2 | 65.1 |
| Fair | 61 | 28.8 | 28.8 | 93.9 |
| Poor | 13 | 6.1 | 6.1 | |
| Very poor | 0 | 0.0 | 0.0 | |
| Total (*N*) | 212 | 100.0 | 100.0 | |

**Table 7.** Analysis of Paired Samples for pre- and post-test level of fitness for muscle lower body power

| Pair 1 | Mean | *N* | Std. deviation | Std. error mean |
|---|---|---|---|---|
| Pre-test explosive power | 204.01 | 212 | 23.197 | 1.593 |
| Post-test explosive power | 222.21 | 212 | 20.716 | 1.423 |

## 6 Conclusion

Physical fitness for the most part incorporates psychological, body, physiological, scholarly, profound and social perspectives. Level of fitness is a complete idea to enables people to reside a superior, invigorated and sure life. Along these lines, in this cutting edge and propelled age, physical fitness is significant for an individual's well-being. Disintegration of the physical state of an individual will affect the person's social mental quality. As it were, a person's psychological capacity and capacity to respond might be influenced in the event that he is physically less dynamic.

**Table 8.** Paired samples test for pre- and post-test level of fitness for muscle lower body power

| | Paired differences | | | | | *t* | *df* | Sig. (2-tailed) |
|---|---|---|---|---|---|---|---|---|
| | Mean | Std. deviation | Std. error mean | 95% confidence interval of the difference | | | | |
| | | | | Lower | Upper | | | |
| Pair 1 Pre-test power–Post-test power | −18.203 | 23.443 | 1.610 | −21.377 | −15.029 | −11.306 | 211 | 0.000 |

By distinguishing the physical fitness level, one can pick the sort of sports and profession intrigue fittingly. The aftereffects on trial and estimation permit mentors and specialists of sport science to operate level of fitness execution data to be the reason for putting gatherings of people dependent on their genuine physical fitness level during instructional courses. Information on the person's level of fitness brings mentors and sports science professionals plan a techniques and plan of action that are successful for the competitors or players.

This research was led to build up the physical fitness measure for the Military Cadet Officers dependent on Malaysia's segment factors. This investigation additionally helps the exercise physical of army evaluate a degree level of fitness between the Malaysian Military Cadet Officers particularly at National Defense Universities of Malaysia. The foundation of a standard level of fitness measure may upgrade the nature level of fitness between armed force officials when all is said in done. In addition, the physical fitness measure can be utilized by physical coaches to design and recognize proper sorts of preparing for Military Cadet Officers.

**Acknowledgements.** The authors would like to thank the Defense Fitness Academy and the Research Management Center of the National Defense University of Malaysia for the chance to managing and produce this research under National Defense University of Malaysia research grant UPNM/2016/GPJP/4/SS/2. Rewards go to the National Defense University of Malaysia Military Cadet Officer Unit, for participating in this research.

## References

1. Kassim, M., Norfazilawati, O., Rustam, S., Edawaty, U.: Physical fitness status among Female Reserve Officers Training Unit (ROTU). The pre-test. Int. J. Eng. Technol. **7**(3.7), 192–196 (2018)
2. Kassim, M, Sheikh Ahmad, S.K., Muda, B.B.: Konsep Latihan Dalam Sukan. Pekan Ilmu Publication Sdn Bhd, Ampang, Selangor (2016)
3. Rustam, S., Kassim, M.: Physical fitness index for assess fitness speed among Army Reserve Officer Training Unit Cadet in Malaysia. J. Phys. Conf. Ser. (2018). https://doi.org/10.1088/1742-6596/1020/1/012009
4. Roy, T.C, Springer, B.A, McNulty, V., Butler, N.L.: Physical fitness. Military Med. **175**(8), 14–20 (2010)

5. Rustam, S., Kassim, M.: Developing a physical fitness index to evaluate explosive power. Eur. Proc. Soci. Behav. Sci. (2018). https://doi.org/10.15405/epsbs.2018.05.86
6. Erikssen, G.: Physical fitness and changes in mortality: the survival of the fittes. Sport Medicine. **31**(2), 121–127 (2001)
7. Knapik, J., Daniels, W., Murphy, M., Fitzgerald, P., Drews, F., Vogel, J.: Physiological factors in infantry operations. Eur. J. Appl. Physio. Occup. Physiol. **60**(3), 223–238 (1990)
8. Stone, M.H., Moir, G., Glaister, M., Sanders, R.: How much strength is necessary? Phys. Theraphy Sport **3**, 88–96 (2002)
9. Haff, G.G., Carlock, J.M., Hartman, M.J., Kilgore, J.L., Kawamori, N., Jackson, J.R., Morris, R.T., Sands, W.A., Stone, M.H.: Force-time curve characteristics of dynamic and isometric muscle actions of elite women olympic weightlifters. J. Strength Condition. Res. **19**, 741–748 (2005)
10. Ahmad, H.: Pengujian pengukuran dan penilaian Pendidikan Jasmani, 1st. Dubook Press Sdn Bhd, Selangor (2015)
11. Ahmad, H., Gunathevan.: 900 push-up test norms for sport science students Sultan Idris Education University. Int. J. Develop. Emerg. Econ. **3**:1–9 (2015)
12. Kassim, M., Rustam, S.: Developing norms for selected physical fitness battery as gauge for fitness assessment among Army Reserve Officer Training Unit Cadet. Adv. Sci. Lett. **24**(7), 5136–5138 (2018)

# System Architecture of a Smart Fall Detection System

Ramachandran Anita[1](✉) and Karuppiah Anupama[2]

[1] BITS Pilani, Bengaluru, India
anita.ramachandran@pilani.bits-pilani.ac.in
[2] BITS Pilani, KK Birla Goa Campus, Goa, India
anupkr@goa.bits-pilani.ac.in

**Abstract.** With advances in medicine and healthcare systems, the average life expectancy of human beings has increased to more than 80 years. As a result, the demographic old-age dependency ratio (people aged 65 or above relative to those aged 15–64) is expected to increase, by 2060, from ~28 to ~50% in the European Union and from ~33 to ~45% in Asia [1]. Therefore, the percentage of people who need additional care is also expected to increase. Geriatric health care, which pertains to care for the elderly, has gained a lot of prominence in the recent years, with specific focus on fall and sleep apnea detection systems because of their impact on public lives. In the recent years, there has been widespread application of Internet of things (IoT) and machine learning in the geriatric healthcare domain because of the potential cost reduction such technologies can bring in. In this paper, we present the architecture and design of an end-to-end geriatric healthcare system, with focus on wearable device based fall detection using machine learning. We explain the major components of the system architecture, under a certain deployment scenario, and present the communication protocol between these system components. We also present the salient aspects of the multi-channel variable time-division multiple access (multi-channel V-TDMA) MAC protocol designed to suit the requirements of such a system. This protocol combines the strengths of both standard time-division multiple access (TDMA) that is modified to support flexibility and frequency-division multiple access (FDMA).

**Keywords:** Health care · Machine learning · Fall detection · MAC protocol · Multi-channel V-TDMA

## 1 Introduction

In this paper, we explain the design of an end-to-end geriatric healthcare system that can be deployed in elderly care homes. The design presented here focuses on the application of machine learning (ML) in fall detection and describes the system components that enable this, and the communication mechanism required between these components.


X.-S. Yang et al. (eds.), *Proceedings of Fifth International Congress on Information and Communication Technology*, Advances in Intelligent Systems and Computing 1183, https://doi.org/10.1007/978-981-15-5856-6_56

While this paper explains the design and the work done from the perspective of fall detection, the same architecture can be extended to include other functionalities such as sleep apnea detection and long-term health monitoring.

There has been substantial research into the design and development of end-to-end healthcare systems. For example, a system architecture for smart health care for continuous monitoring of assisted-living residents is proposed in [2]. The system comprises of sensors for biophysical and environmental monitoring, backend databases and end devices for visualization, and all components are connected by a backbone network wirelessly or overlaid on to an existing wired infrastructure. Other architectures for remote healthcare monitoring include [3, 4].

The application of machine learning in fall detection, using data generated by various means such as wearable devices, environment sensors, and vision based systems, is also being actively researched. The main challenge is to create a model that detects falls accurately, while keeping the design of the fall detection system minimal and non-intrusive. Wearable devices equipped with kinematic sensors and vital signs sensors are commonly used to enable analysis around performance of machine learning models. Hence, in our system, the device that enables monitoring of vitals is a wearable device, worn by the elderly in an elderly care home. Many papers have been published which use one or more of the kinematic sensor parameters with threshold-based or ML-based techniques. For example, Li et al. [5] proposed a three-step algorithm based on activity intensity analysis, posture analysis, and transition analysis, with signals reported by accelerometer and gyroscope, in order to separate falls from activities of daily life (ADLs) and near-fall conditions. In [6], the data set used was generated from accelerometer and gyroscope, placed at the waist level. After feature extraction and feature selection, classification was performed using Naïve Bayes, least squares method (LSM), artificial neural network (ANN), support vector machine (SVM), and k-nearest neighbor (kNN) algorithms. Hossain et al. [7] attempt to distinguish falls from ADLs and compares SVM, kNN, and complex tree algorithms applied on data generated by accelerometers. The paper compared the performance of these algorithms with respect to accuracy, precision, and recall, on ADLs and four types of falls (forward, backward, right, and left). It was observed that the accuracy and precision of SVM were the highest, while complex tree performed better in terms of recall analysis. In [8], the authors performed experiments using an Android application deployed on a smartphone, to detect falls. The parameters monitored included pulse rate and oxygen saturation in addition to accelerometer readings. Sensor fusion techniques have been experimented in some cases, where more than one type of sensor is used, such as in [9].

In our system, a wearable device collects various parameters of the person wearing the device at periodic intervals. The deployment scenario we assume is an elderly care home comprising multiple rooms for the elderly. The network topology of the system consists of three levels of components hierarchy---the first comprising individual wearable devices, the second comprising data collection points per room, and the third comprising a data collection and analysis point for the whole elderly care home. Since the system is time critical, it is important to have a network protocol stack that offers low latency and reliable communication between the entities of the system. The shortcomings of the traditional time-division multiple access (TDMA) and carrier sense multiple access (CSMA) schemes render them inefficient for the network design for this system. Many MAC protocols have been developed which addresses the shortcomings of the TDMA and CSMA schemes, along with optimizing energy consumption. Some of these protocols include B-MAC [10], S-MAC [11], WiseMAC [12], T-MAC [13], E-MACS [14], and L-MAC [15]. Flexible TDMA (FTDMA) [16] is developed for mobile sensor networks. In this, the leader broadcasts the number of available slots and fixed slots to the mobile nodes. The mobile nodes send a request message in any one of the free slots. The leader acknowledges the request, allocates slot to the mobile node, and broadcasts the updated schedule in the next cycle. However, the mobile nodes in the communication range of two or more leaders will suffer interferences and collisions.

The scope of this paper is to explain the architecture of the healthcare system, present results from our analysis of application of machine learning for fall detection, describe the design of the network protocol stack, and results from simulation of the same.

## 2 Architecture

The healthcare system we are working on is based on wearable devices. The wearable sensor node measures various parameters of the person wearing the node, and it consists of the following sensors: three-axis accelerometer, gyroscope, magnetometer, and optical heart rate sensor. Since these nodes are mobile because they are worn by the residents, we refer to them as mobile nodes (MN). Each room is equipped with a room coordinator (RC) that collects the data reported by the wearable device. The RCs relay the collected data to a central coordinator (CC), which is a single, powerful node deployed for the entire elderly care home. The CC analyzes the data from each wearable device and applies machine learning classifiers to detect abnormalities such as a fall.

A tree-based network is formed where CC is the root of the tree. Several RCs are connected to CC directly or through a multi-hop path formed via intermediate RCs and multiple MNs are connected to individual RCs. Figure 1 represents the network topology of the healthcare system.

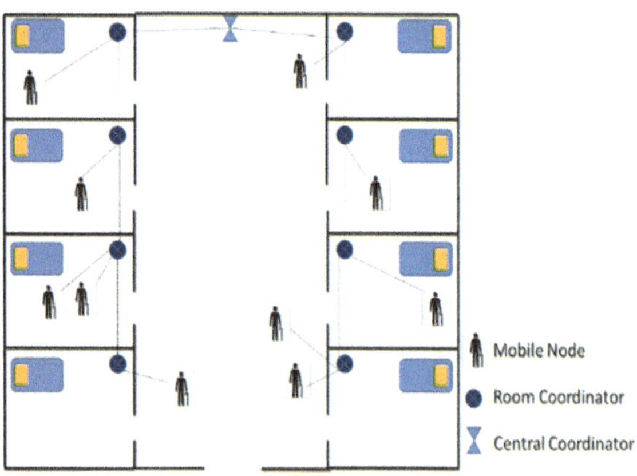


**Fig. 1.** Network topology diagram

# 3 Methodology

## 3.1 Machine Learning for Fall Detection

In order to train machine learning classifiers, we generated our own data set, from experiments performed by 10 different volunteers, in controlled environment. We used the TicWatch S smartwatch, which included a three-axis accelerometer, three-axis magnetometer, three-axis gyroscope, linear acceleration, and optical heart rate sensor. The volunteers were aged between 20 and 25 years and were of good health condition. The height ranged from $5'1''$ to $5'8''$, and weight from 40 to 75 kg. The activities simulated by the volunteers included different types of falls and ADLs. A total of 14 ADLs and 6 fall activities were simulated, on two separate rounds by each volunteer, resulting in a data set of about 110,000 lines. A software application was developed on the TicWatch S device to collate the collected data into a csv file.

We applied five classifiers---ANN, kNN, Naïve Bayes, XGBoost, and random forest---for fall detection using this collected data set and observed their accuracy for fall detection. In order to improve the performance of the classifiers, we performed feature engineering on the data set, by using a rolling window technique. This was suited to the time series data that we are analyzing, and the window size we applied was 20. For every window, features based on statistical models were derived, for the values reported by the magnetometer, gyroscope, accelerometer, and heart rate sensors, within that window. The features thus derived included mean, median, standard deviation, variance, skew, and kurtosis, of the attribute values within a window. The features that were the most important in making the classifiers perform best were noted.

## 3.2 Design of Network Protocol Stack

We designed a multi-channel V-TDMA protocol, in which there are two levels of hierarchy in the network: one level between RC and MN and the second level between RC and CC. Data collected by the MNs includes kinematic and biological parameters of the elderly person wearing the device. This data is relayed by the MNs to the RC via a single-hop communication path. The RCs collect data transmitted from all the MNs within its range, and relays the same to the CC, where the data is analyzed. This is enabled by a multi-hop path between RCs and CC. The major design considerations of the multi-channel V-TDMA protocol are with respect to network discovery, timeslot allocation map advertisement, data transmission and forwarding, and frequency switching.

(i) **Network discovery**: Network discovery is performed using a broadcast message initiated by the CC, which is received by the RCs closest to the CC. This message is rebroadcast by those RCs, to the RCs at the next level, and subsequently by them further away, in a hierarchical manner. This flooding of the broadcast messages with the broadcasting node's ID and location coordinates helps forming the network topology.

(ii) **Timeslot allocation map advertisement**: Once the network formation is complete, each RC broadcasts a timeslot allocation map which publishes the total number of slots, occupied slots, the frequency of communication (channel) for

MN-RC data transmission and IDs of MNs to which the slots are allotted (MN1 to MNm). This message will be received by all MNs which are present in the communication radius of the RC. This message enables the MNs to know which time slots they can use to transmit data to the RC.

(iii) **Data transmission**: Each MN scans through the list of fixed slots in the message broadcast by the RC, to look for its time slot in the timeslot allocation map. If the MN finds its ID in the list of occupied slots, it sends the data in that slot; else it picks up random slot from the free list of slots and sends a time slot request message to RC. RC upon receiving the timeslot request message marks that time slot against the MN.

(iv) **Data forwarding**: The process of data forwarding from RC to CC happens via the multi-hop path between RC and CC. RC at Level $(i + 1)$ transmits data collected from its MNs to its neighboring RC at Level $(i)$, during its RC-CC data transmission slot. This cycle is repeated every fixed time interval, after the RC-CC communication.

(v) **Frequency switching in multi-channel V-TDMA**: In order to ensure minimal packet loss in the network, each individual RC-MN subnetwork uses a unique frequency for data transmission. All other communications (RC-RC, CC-RC) occur at a different frequency. An appropriate frequency switching mechanism for RCs and MNs is implemented to support this.

## 4 Simulation and Results

### 4.1 Machine Learning for Fall Detection

We analyzed the performance of five different classifiers (ANN, kNN, Naïve Bayes, XGBoost, and random forest) on the raw data set, after minimal preprocessing. Subsequently, in order to improve the accuracy of the classifiers, we applied feature engineering using statistical techniques and feature extraction, on the raw data set. We observe that this improved the accuracy of XGBoost, Naïve Bayes, and random forest classifiers. The performance of the classifiers were compared before and after feature extraction. Figures 2 and 3 show the results. We also note that among the five classifiers we analyzed, random forest gave us the best accuracy across all scenarios we experimented with. This was in the case where the feature set consisted of the mean and median of parameters reported by accelerometer, gyroscope, magnetometer, and optical heart rate sensors, while using a size of 20 for the rolling windows.

### 4.2 Multi-channel V-TDMA

The design for multi-channel V-TDMA was simulated on Cooja simulator for a network topology containing 1 CC and 6 RCs. The RCs were scattered across three different levels, with 2 RCs per level. The following densities of MNs per RC were simulated: 5, 10, 15, and 20, corresponding to different patterns of mobility and topology changes. The overall frame size considered was 1000 ms. The network was simulated for 15 min, and the results were averaged across 5 runs of the simulation. Mobility of MNs was

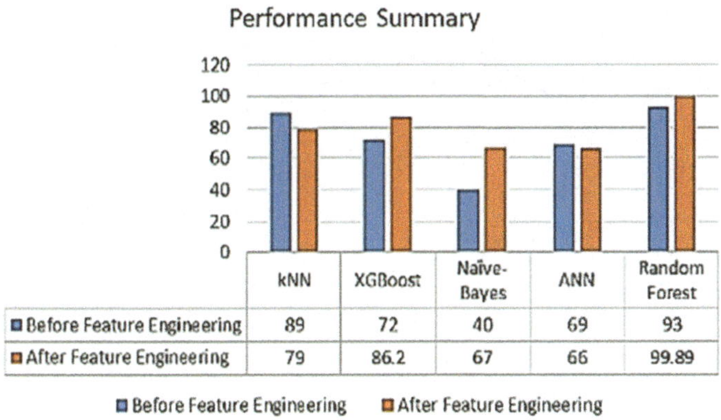


**Fig. 2.** Performance Summary of ML classifiers before and after feature engineering

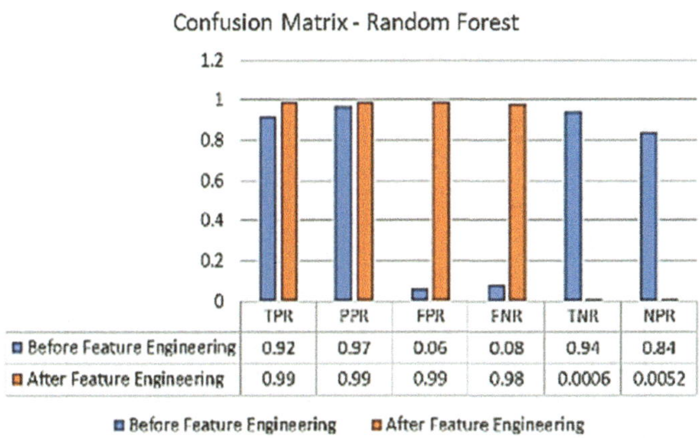


**Fig. 3.** Confusion matrix for random forest before and after feature engineering

induced every one minute to simulate changes in network topology. Figures 4 and 5 show the screenshot of the simulated network.

The performance parameters collected were average delay, maximum delay (where delay is the time taken for a data packet originated by an MN, to reach the CC), and packet loss ratio. The performance was then compared with the standard TDMA and FTDMA [1]. The results are shown in Figs. 6, 7 and 8.

It was observed that multi-channel V-TDMA performs better than standard TDMA and FTDMA with respect to average delay, maximum delay, and packet loss ratio, for the two-level hierarchical network.

## 5 Conclusions

In this paper, we presented the system architecture for an end-to-end geriatric healthcare system, for a given deployment scenario. Machine learning techniques and the impact of

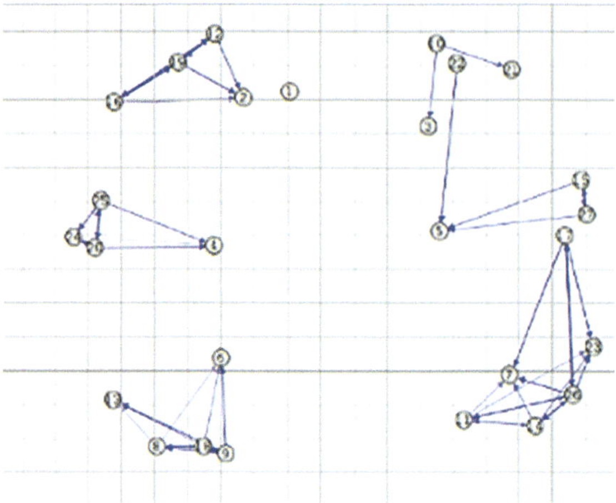

**Fig. 4.** MN-RC communication

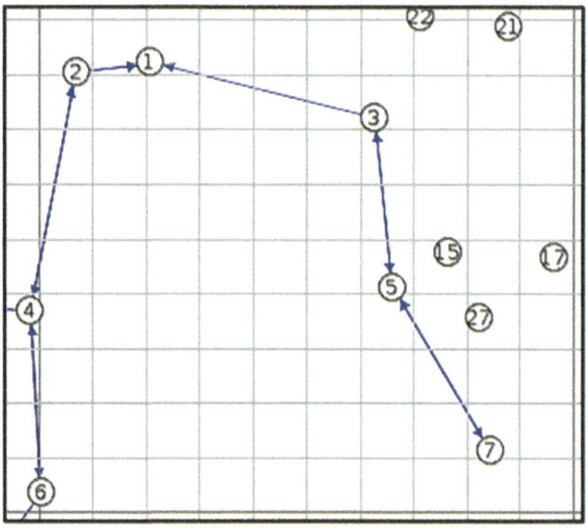


**Fig. 5.** RC-CC communication

feature engineering were evaluated for fall detection, on an experimental data set that was collected using wearable devices. The communication between the components of the hierarchical system is enabled via the proposed multi-channel V-TDMA MAC protocol. We simulated this protocol on Cooja and compared the performance of multi-channel V-TDMA with basic TDMA and FTDMA. The results show that multi-channel V-TDMA performs better than basic TDMA and FTDMA with respect to average delay, maximum delay, and packet loss ratio. As future work, we plan to develop the prototype of a wearable device with STM32 interfaced with appropriate sensors for data collection. We plan to implement the multi-channel V-TDMA on real-time test bed for further analysis, with appropriate hardware boards for room coordinators and central coordinator. The

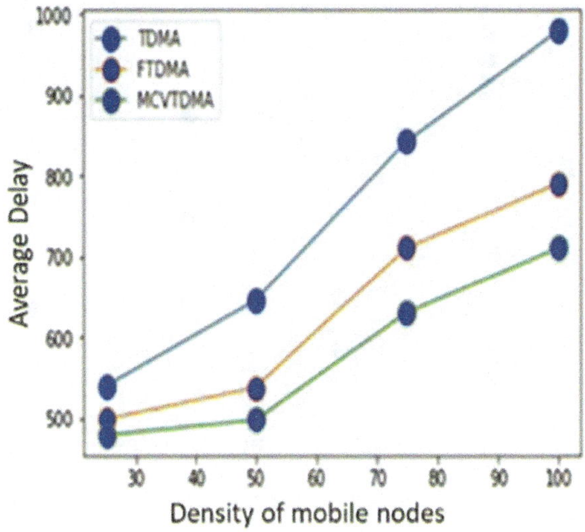


**Fig. 6.** Average delay

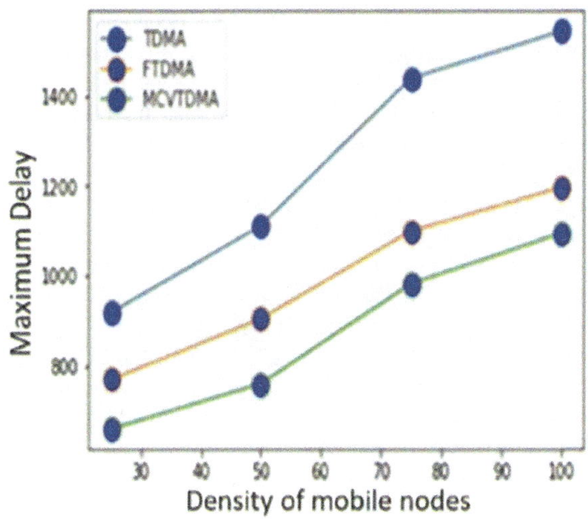


**Fig. 7.** Maximum delay

trained machine learning classifier will be deployed on the central coordinator to analyze the data for detection of falls.

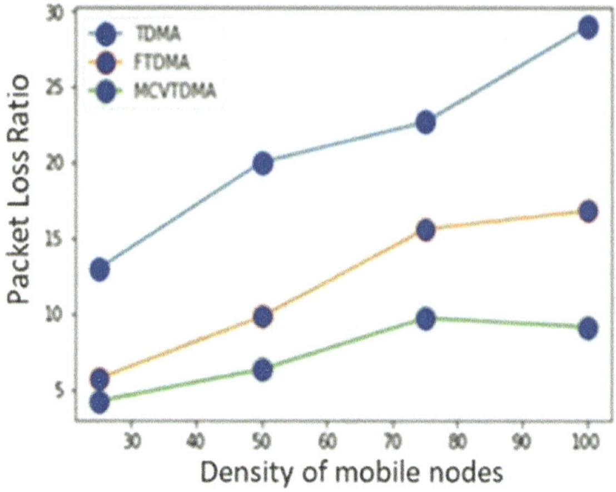


**Fig. 8.** Packet loss ratio

## References


1. The 2015 Ageing Report EUROPEAN ECONOMY—Economic and budgetary projections for the 28 EU Member States (2013–2060), ISSN 1725-3217 (online), ISSN 0379-0991 (print)
2. Virone, G., Wood, A., Selavo, L., Cao, Q., Fang, L., Doan, T., He, Z., Stankovic, J.A.: An advanced wireless sensor network for health monitoring. In: Transdisciplinary Conference on Distributed Diagnosis and Home Healthcare (D2H2), pp. 95–100 (2006)
3. Shen, X., Zeng, J., Liu, T., Hu, C.: Remote healthcare monitor system. In: 3rd International Conference on Biomedical Engineering and Informatics. pp. 1901–1905. Yantai (2010). https://doi.org/10.1109/bmei.2010.5639530
4. Chauhan, J., Bojewar, S.: Sensor networks based healthcare monitoring system. In: International Conference on Inventive Computation Technologies (ICICT), pp. 1–6. Coimbatore (2016). https://doi.org/10.1109/inventive.2016.7824814
5. Li, Q., Stankovic, J.A., Hanson, M.A., Barth, A.T., Lach, J., Zhou, G.: Accurate, fast fall detection using gyroscopes and accelerometer-derived posture information. In: Sixth International Workshop on Wearable and Implantable Body Sensor Networks, pp. 138–143. Berkeley, CA (2009). https://doi.org/10.1109/bsn.2009.46
6. Vallabh P., Malekian R., Ye N., Bogatinoska D. C.: Fall detection using machine learning algorithms. In: International Conference on Software, Telecommunications and Computer Networks (SoftCOM), pp. 1–9. Split (2016). https://doi.org/10.1109/softcom.2016.7772142
7. Hossain, F., Ali, M.L., Islam, M.Z., Mustafa, H.: A direction-sensitive fall detection system using single 3D accelerometer and learning classifier. In: Proceedings of the 2016 International Conference on Medical Engineering, Health Informatics and Technology (MediTec), pp. 1–6. Bangladesh (2016). https://doi.org/10.1109/meditec.2016.7835372
8. Koshmak, G.A., Linden, M., Loutfi, A.: Evaluation of the android-based fall detection system with physiological data monitoring. In: Annual International Conference of the IEEE Engineering in Medicine and Biology Society (EMBC), pp. 1164–1168. Osaka (2013). https://doi.org/10.1109/embc.2013.6609713
9. Shahiduzzaman, Md.: Fall detection by accelerometer and heart rate variability measurement. Global J. Comput. Sci. Technol. **15**(3) (2015)

10. Polastre, J., Hill, J., Culler, D.: Versatile low power media access for wireless sensor networks. In: Proceedings of the 2nd International Conference on Embedded Networked Sensor Systems, New York, USA (2004), https://doi.org/10.1145/1031495.1031508
11. Ye, W., Heidemann, J., Estrin, D.: An energy-efficient MAC protocol for wireless sensor networks. In: Proceedings of the Twenty-First Annual Joint Conference of the IEEE Computer and Communications Societies, vol. 3, pp. 1567–1576. New York, USA (2002). https://doi.org/10.1109/infcom.2002.1019408
12. El-Hoiydi, A., Decotignie, J.D.: WiseMAC: An ultra low power MAC protocol for multi-hop wireless sensor networks. In: Nikoletseas, S.E., Rolim, J.D.P. (eds.) Algorithmic Aspects of Wireless Sensor Networks (ALGOSENSORS 2004). Lecture Notes in Computer Science, vol. 3121. Springer, Berlin (2004)
13. van Dam, T., Langendoen, K.: An adaptive energy-efficient MAC protocol for wireless sensor networks. In: Proceedings of the 1st International Conference on Embedded Networked Sensor Systems (SenSys'03), pp 171–180. New York, USA (2003). https://doi.org/10.1145/958491.958512
14. Nieberg, T., Dulman, S., Havinga, P., van Hoesel, L., Wu, J.: Collaborative algorithms for communication in wireless sensor networks. In: Ambient Intelligence: Impact on Embedded System Design, pp 271–294. Kluwer Academic Publishers, New York (2003)
15. van Hoesel, L.F.W., Havinga, P.J.M.: A lightweight medium access protocol (LMAC) for wireless sensor networks: reducing preamble transmissions and transceiver state switches. In: 1st International Workshop on Networked Sensing Systems, pp. 205–208. Tokio (2004)
16. Abdullah-Al-Wadud, M.: A TDMA scheme for mobile sensor networks. Int. J. Distrib. Sensor Networks (2013). https://doi.org/10.1155/2013/907583

# Determining Optimal Parallel Schedules in Tree-Based WSNs Using a Realistic Interference Model

Aleksandar Ilić[1(✉)], Peter Langendörfer[1], Stefan Weidling[1], and Mario Schölzel[1,2]

[1] IHP – Leibniz-Institut für innovative Mikroelektronik, Im Technologiepark 25, 15236 Frankfurt (Oder), Germany
ilic@ihp-microelectronics.com

[2] Hochschule Nordhausen,Weinberghof 4, 99734 Nordhausen, Germany

**Abstract.** Schedule length minimization in TDMA protocols for single-channel raw data aggregation convergecast is a topic that is often studied. Many algorithms that solve this problem have been proposed. However, the impact of interference introduced with parallel transmissions in such networks was not studied in detail so far. Most of the protocols use a simple 2-hop interference model, which considers the interference range to be equal to the transmission range. We show that this model does not provide satisfactory results, and we propose an adaptable interference model, derived from real-world measurements. Using extensive simulations, we prove that the proposed model can significantly increase network throughput when used to calculate conflict graphs for any of the existing parallel scheduling algorithms.

**Keywords:** Wireless sensor networks · Medium access protocols · Time division multiple access · Graph coloring algorithms · Convergecast · Interference

## 1 Introduction

A very common data collection scenario is the raw-data convergecast, a scenario in which one node, the sink, collects the data from a multi-hop network. For routing packets to the sink in such a network, a tree-shaped routing topology is used. When data generation is high, time division multiple access (TDMA) medium access protocols (MAC) are used for resolving medium access. These protocols divide time into slots and assign a certain number of time slots to each node in the network. When a node has data to transmit, it may only transmit it during the assigned time slot. Such a slot assignment is called the schedule. Assigned time slots form a cycle, and cycles are repeated until the protocol is terminated.


X.-S. Yang et al. (eds.), *Proceedings of Fifth International Congress on Information and Communication Technology*, Advances in Intelligent Systems and Computing 1183, https://doi.org/10.1007/978-981-15-5856-6_57

Some examples of this application scenario are traffic monitoring [10], environmental monitoring [15], and industrial process monitoring and control [11]. All those applications require the measured data to be transferred to the network sink quickly and reliably. If a random access MAC protocol is used, a large number of collisions are caused by many simultaneous transmission attempts; and this in turn results with low packet delivery ratios and high latency. To solve this issue, TDMA protocols that allow high data traffic without collisions are used. To be efficient, a TDMA protocol must allocate an appropriate number of slots to each node, consider traffic flow, and minimize the schedule length. Schedule length reduction is achieved by the utilization of parallel transmissions. Whether two transmissions can be scheduled at the same time or not is decided by considering an interference model. If the interference model is not accurate, parallel transmissions might interfere with each other, potentially causing a significant packet loss.

Though finding the minimal schedule for convergecast is proven NP-complete by Choi et al. [2], many scheduling algorithms that find approximate solutions were proposed [6,13,14]. These algorithms find a schedule with a length close to the minimal, or even equal to it when the network topology meets certain requirements. However, these solutions fail to consider the real interference conditions in the network. Instead, they use the 2-hop interference model that assumes the interference range to be equal to the communication range. Some studies, however, suggest that the interference range can be up to three times larger than the communication range [19], depending on the physical environment. On the other hand, the 2-hop model allows simple and distributed protocol implementation, resulting in short network setup time, lower implementation costs, and higher reliability. In this paper, we propose an adaptable interference model, which keeps all these benefits, while effectively eliminating the interference in the network.

The proposed model was derived from a wireless sensor network for gathering traffic data from sensors located alongside the highway. We had access to an existing network on the highway that consists of a number of measuring stations, each of which is provided with a solar panel for energy supply and a GSM module for data transmission. With the intention of replacing the GSM modules with 868 MHz radio modules, we have already used this infrastructure to measure radio range and packet reception error rates [17]. Based on these results, we have obtained a realistic path loss model and bit error rate (BER) model for our radios. Since interference model evaluvation requires measurements from many different networks with different topology, we have performed evaluation in the simulator, using this radio model.

The main contributions of this paper are the evaluation of the 2-hop interference model and the proposed adaptable interference model. Using extensive simulations, we show that the 2-hop interference model does not eliminate interference effectively. We then study how the chosen interference range affects network performance. Based on the theoretical analysis and the simulation results, we propose an adaptable interference model that selects the interference range size in hops based on the application scenario. The comparison of our measurements with the simulations we did clearly shows that our adaptable model fits well with real-world conditions.

## 2 Related Work

The problems of schedule minimization and interference elimination in convergecast networks are most commonly studied together. Because real interference is hard to determine, some authors suggest dividing the network into clusters and then using different radio channel for every cluster [7,18]. Schedule for each cluster is then calculated without the utilization of parallel transmissions. Such an interference-free schedule provides good results for small network sizes, but its performance decreases drastically with the network size. In [9], the authors assume that interference in the network is already known, and then propose the graph coloring algorithm that adjusts the number of colors assigned to a node to its traffic demand. Ergen and Varaiya start from the same assumption and then improve the algorithm to consider data flow in the network [3].

A large number of protocols use the 2-hop interference model [1,6,13,14,16]. Some of them use this model to calculate the interference and create conflict graphs, while the others combine this model with minimum spanning tree routing topology to eliminate the need for constructing conflict graphs. In such topology, the minimum distance between two nodes in the routing tree is the same as the minimum distance between them in the network graph, allowing each branch to be scheduled independently. TreeMAC uses this property to produce schedule length of $3N$, where $N$ is the number of total nodes in the network excluding the sink. Protocol designs based on the 2-hop interference model provide important theoretical conclusions. However, they fail to consider real interference ranges in the network leading to reduced performance.

Apart from this, several other interference studies have been conducted. Interference models can be node-based or link-based. A node-based model defines interference neighborhood for each node in the network. If two nodes cannot be scheduled in parallel, they are in each other's interference neighborhoods. For link-based models, a set of interfering links is defined in a similar way for each link in the network. A summary of different interference models in use is given in the following subsections.

### 2.1 The 2-Hop Interference Model

The 2-hop interference model is a node-based model built on the assumption that the interference range is equal to the communication range [16]. According to this model, interference neighborhood of a node is comprised from its 2-hop neighbors, as depicted in Fig. 1. Since this is a node-based model, two nodes are in the interference range if any of the transmissions from one node affects any of the transmissions from the other. This means that some of the possible parallel transmissions are not allowed. For example, when node $T$ in Fig. 1 is transmitting, node $N2$ could transmit to node $N3$ in parallel; but this transmission cannot be scheduled when this model is used. This problem can be solved by utilization of link-based scheduling [2,8].

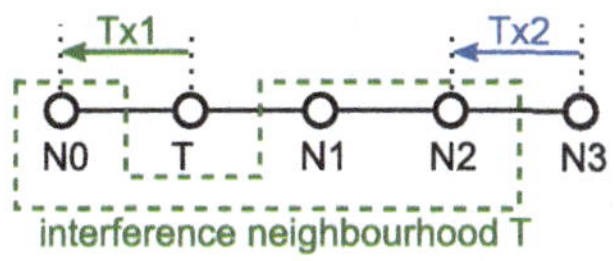


**Fig. 1.** Interference neighborhood obtained using the 2-hop interference model

### 2.2 Protocol Interference Model

This model defines interference neighborhood based on physical distances between the transmitter, the receiver, and the source of interference [4]. Let $v_{t1}$ and $v_{t2}$ be two nodes transmitting in parallel, and $v_{r1}$ the receiver for the transmission from node $v_{t1}$. The transmission is considered interference-free if the following is satisfied $\|v_{t1} - v_{r1}\| \leq \beta \cdot \|v_{t2} - v_{r1}\|$, i.e., the distance from the transmitter to the receiver is at least $\beta$ times smaller than the distance to the interference source. This model can only be used if all nodes have a positioning system, or if node locations are programmed before deployment.

## 3 The Adaptable Interference Model

To solve the aforementioned issues, we propose the interference model based on the distance between nodes in hops. The proposed model efficiently eliminates the interference in the network while keeping all the benefits provided by the simplicity of the 2-hop model. Since interference range depends heavily on the radio medium characteristics, we propose that the number of hops is adjusted to the application scenario and give the guidelines on how can this be achieved.

### 3.1 Path Loss Modeling

When the network is located outdoors, the two-ray ground-reflection model can be used to model path loss. This model calculates path loss as:

$$L = G_{\mathrm{t}}\, G_{\mathrm{r}} \frac{h_{\mathrm{t}}^2 h_{\mathrm{r}}^2}{d^k} \tag{1}$$

where $G_{\mathrm{t}}$ is transmitter antenna gain, $G_{\mathrm{r}}$ receiver antenna gain, $h_{\mathrm{t}}$ transmitter height and $h_{\mathrm{r}}$ receiver height. Term $d$ is the distance between the nodes, and $k$ is the path loss coefficient that depends on the distance. In the far-field region, when $d \gg \frac{4\pi h_t h_r}{\lambda}$, the path loss factor is equal to four. In the near field region, it is equal to two, reducing the path loss model to the free-space path loss model. In indoor networks, the characteristics of the physical environment have a big impact on path loss. Nonetheless, in [12], it is shown that a satisfactory path loss model for the indoor environment can be obtained by adjusting the coefficient $k$. For large open buildings (like factory halls) it is equal to two, whereas for buildings with many rooms it is between three and four.

### 3.2 Selecting the Interference Range

Since only distances between nodes in hops are known and not physical distances between them, the size of the interference range in hops should be chosen in such a way to ensure that a particularly unfavorable node arrangement cannot create significant interference. We identify such arrangements for different interference ranges and chose the lowest that allows interference-free transmitting. Since interference range depends on the path loss, we distinguish two scenarios; the outdoor network with two-ray ground reflection path loss, and the indoor network with free-space path loss. We assume that there is no interference if SIR value is larger then 10 dB. In the outdoor case, this is satisfied when the distance from the receiver to the transmitter is two times smaller than the distance from the receiver to the interference source. This was obtained in the same way as in [19], using (1).

The worst-case for the 2-hop range is shown in Fig. 2a. Here, node $N4$ is the receiver, and node $N2$ is the interferer. In this case, the distances are almost equal, meaning substantial packet loss due to interference. The required distance ratio larger than two can only be guaranteed when four hops are used as interference range as it can be observed from Fig. 2b. Therefore, we propose the 4-hop interference range for outdoor networks. For indoor networks, because of the lower path loss, the distance ratio larger than three is required. For that case, using similar analysis, we find the 5-hop interference range to be sufficient.

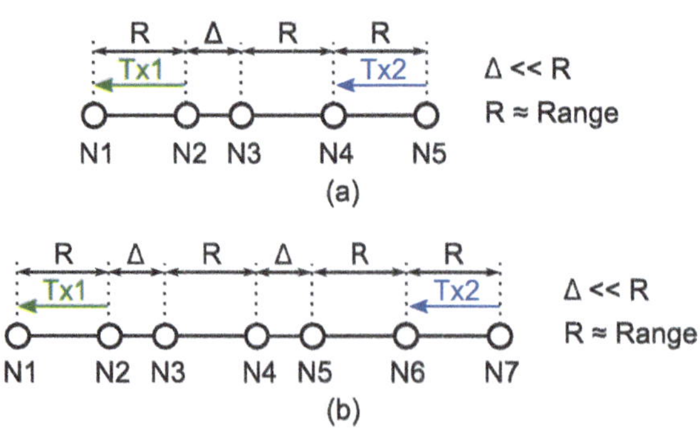


**Fig. 2.** Worst-case scenario when: **a** 2-hop interference range is used; **b** 4-hop interference range is used

# 4 Evaluation

The correct evaluation of the interference model requires that the model is tested on many networks with different topology. We first evaluate the proposed model on a large number of randomly generated linear networks, assuming that the same will hold for tree networks. The schedules used for the evaluation are generated using the TreeMAC protocol, but they hold for any other protocol.

## 4.1 Radio Model

To obtain relevant simulation results, we have fitted the radio model using the measurement results from our previous work [17], in which the dependency of the packet loss rate (PER) and RSSI on the distance for 868 MHz radios was measured. The two-ray ground reflection model was used to model the path loss, and the simulated results matched the measurements, as it can be seen on upper graph in Fig. 3. The bit error rate (BER) model of the radio was then adjusted to match the measurements. The comparison of the measured and simulated dependency of PER on the distance is shown in lower graph in Fig. 3.

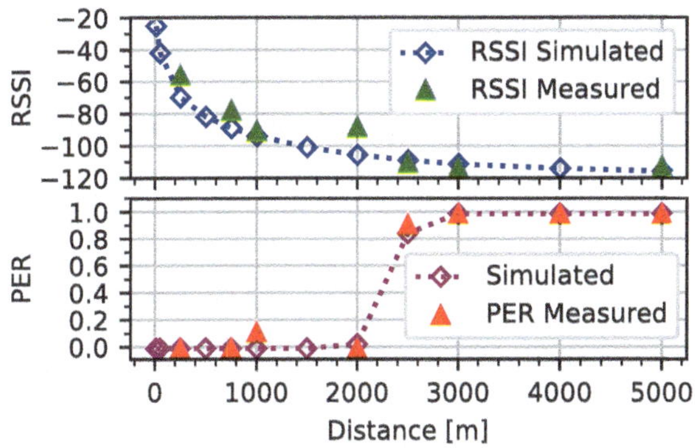


**Fig. 3.** Comparison of the radio model used for simulations with measurement results. The $x$ axis is shared by both graphs

## 4.2 Performance Parameters

To evaluate the results, two metrics are used, data delivery fairness, and maximal throughput. To calculate the fairness, we adopt the model from [5], which defines it as:

$$\phi = \frac{(\sum_{i=1}^{N} n_i)^2}{N \cdot \sum_{i=1}^{N} n_i^2} \quad (2)$$

where $\phi$ is fairness, $N$ is the network size without the sink, and $n_i$ is the number of packets the sink received from the node with index $i$. The value of fairness defined in this way ranges from 0 to 1. Fairness shows if data is equally collected from all nodes in the network. If interference between parallel transmissions is not successfully eliminated, fairness will be reduced when throughput increases. When interference is high in tree networks, throughput will still continue to rise because sink's children can send their own packet instead of forwarding lost packets from other parts of the network. This means that when lower interference range is used, maximum achievable throughput will be higher, but fairness will be much lower. As we are interested in maximizing throughput while keeping fairness high, we define the maximum acceptable throughput as the throughout at which fairness drops to 0.95.

### 4.3 Outdoor Network Case

To test the proposed interference model, ten linear random networks comprised of 24 nodes and one sink were generated. The networks were generated in such a way that every node can communicate with only two other nodes, it's parent and child. In such a network, the maximum number of hops a packet needs to travel to the sink is 24. This is a large value, and successful data gathering can be achieved only under minimal interference. To conduct the simulations, the previously introduced radio model was used, with the data rate of 100 kbps. Packets were generated randomly over a period of 100 s and total packet length including the headers was 20 bytes and TDMA slot size including guards was set to 3 ms. The simulations were repeated for different traffic generation rates and different values of interference range for each of the generated networks. The resulting values were averaged to obtain results shown in Figs. 4 and 5.

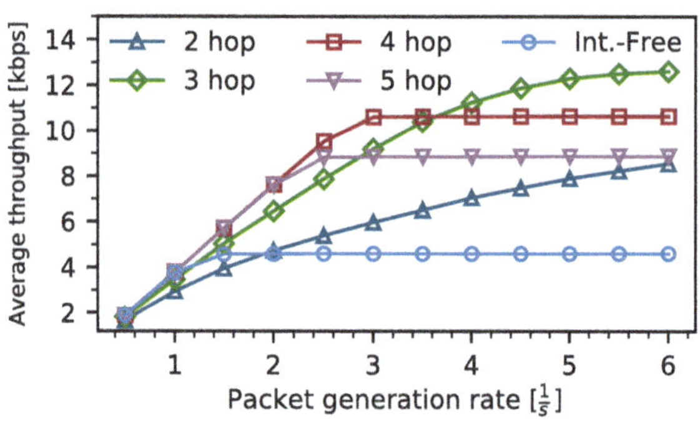


**Fig. 4.** Average fairness of linear network versus packet generation rate for different interference ranges and for schedule without parallel transmissions (interference-free)

The results confirm the assumption that the 2-hop criterion is not sufficient for parallel scheduling. Figure 4 shows that when this model is used, significantly lower network throughput can be achieved. Larger interference ranges provide better results despite longer schedules because fewer number of packets are lost

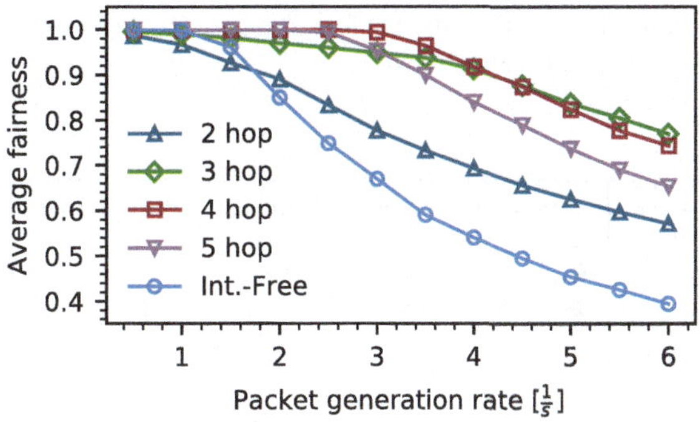


**Fig. 5.** Average fairness of linear network versus packet generation rate for different interference ranges and for schedule without parallel transmissions (interference-free)

due to interference. As packet generation rate increases, the throughput obtained using interference range of 3-hops is initially low, but it increases afterward, surpassing all the other ranges at the end. However, this does not mean that this range performs better, because the fairness starts dropping much sooner than for higher ranges. The throughput value becomes larger than in the 4-hop case only when fairness drops significantly. This occurs because nodes close to the sink have many time slots assigned for forwarding data and if this data is lost, they can send their own packets in these slots (when available). When generation rate is very high, nodes close to the sink will always have their own packet to send, thus resulting in low fairness and high throughput. We can conclude that the 4-hop range provides the best results as assumed. In the lower part of the range, it provides the highest values of both interference and throughput. The 3-hop range provides higher throughput in the upper part of the range, but significantly lower fairness.

To determine how protocol performance scales with the increased number of hops, the simulations were performed for different network sizes. For every network size, ten different networks were generated and tested with random traffic. For each network size, the maximum throughput was calculated as throughput value when fairness drops to 0.95. The simulated dependency of maximum throughput on the network size for different interference models is shown in Fig. 6. The results show that for networks with size of up to 12 nodes, protocols without parallel transmissions (interference-free) achieve good results. This is because for small networks only a few nodes can be scheduled in parallel. The 3-hop range also performs better for smaller network sizes. The reason for this is that in a network of small size it is less probable for a significantly inconvenient network topology to occur. An important observation is that for the 4-hop interference range, the network throughput is nearly constant. This is in accordance with theoretical results, as it delivers one packet to the sink in every frame (one frame equals 5 slots when 4-hop range is used), when there is no loss due to interference. Hence, the results show that 4-hop interference range eliminates interference due to parallel transmissions.

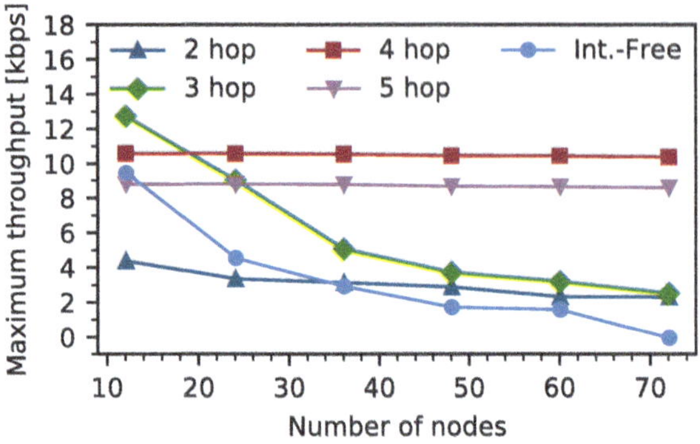


**Fig. 6.** Maximum throughput versus networks size for different interference ranges

### 4.4 Indoor Network Case

Since path loss depends heavily on the physical medium characteristics, the interference range needs to be adjusted to the application scenario of the network. Indoor networks are the most challenging because of very irregular signal propagation due to the presence of various obstacles. As mentioned in [12], the path loss can be roughly approximated by varying the path loss coefficient $k$ in (1). The proposed 5-hop interference range for this case was based on the worst-case scenario in terms of both topology and the value of $k$. Therefore, it represents a safe choice, and it should be adjusted to the particular network if possible. To verify the assumed values, the same simulations as in the outdoor case were performed. The dependency of maximum throughput on the network size is shown in Fig. 7. In this case, the results show a higher dependency on the network size. For large networks (with large tree height) the interference range has to be increased to six, while for smaller ones five hops are sufficient.

## 5 Conclusion

In this paper, we discussed different single-channel raw data convergecast scheduling algorithms. We identified imprecise interference models as the main issue in the existing protocols of this type. We then proposed an interference model that can efficiently eliminate interference without introducing additional complexity. We have performed extensive simulations to prove the effectiveness of the proposed model. The results show that the proposed model offers significantly improved performance in terms of both throughput and fairness compared

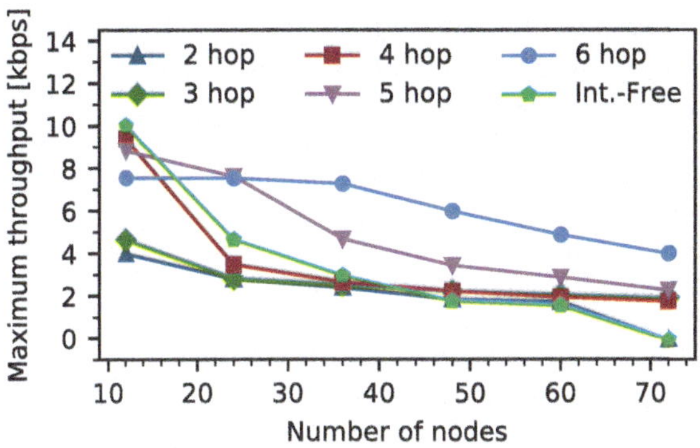


**Fig. 7.** Maximum throughput versus networks size for different interference ranges

to the commonly used 2-hop interference model. We varied the interference range size in hops to confirm that the proposed ranges offer the best performance. The results show that there is a trade-off between fairness and throughput. The proposed ranges provide high fairness. If shorter ranges are chosen, the throughput might increase, but fairness will decrease because interference will be high in some parts of the network.

## References


1. Alicherry, M., Bhatia, R., Li, L.E.: Joint channel assignment and routing for throughput optimization in multiradio wireless mesh networks. IEEE J. Sel. Areas Commun. **24**(11), 1960–1971 (2006)
2. Choi, H., Wang, J., Hughes, E.: Scheduling for information gathering on sensor network. Wireless Netw. **15**, 127–140 (2007)
3. Ergen, S.C., Varaiya, P.: TDMA scheduling algorithms for wireless sensor networks. Wirel. Networks **16**(4), 985–997 (2010)
4. Gupta, P., Kumar, P.R.: The capacity of wireless networks. IEEE Trans. Inf. Theory **46**(2), 388–404 (2000)
5. Hull, B., Jamieson, K., Balakrishnan, H.: Mitigating congestion in wireless sensor networks. In: SenSys'04—Proceedings of the Second International Conference on Embedded Networked Sensor Systems, pp. 134–147 (2004)
6. Incel, Ö.D., Ghosh, A., Krishnamachari, B., Chintalapudi, K.: Fast data collection in tree-based wireless sensor networks. IEEE Trans. Mob. Comput. **11**(1), 86–99 (2012)
7. Jolly, G., Younis, M.: An energy-efficient, scalable and collision-free MAC layer protocol for wireless sensor networks. Wirel. Commun. Mob. Comput. **5**(3), 285–304 (2015)
8. Kang, H., Zhao, Y.n., Mei, F.: A graph coloring based TDMA scheduling algorithm for wireless sensor networks. Wirel. Pers. Commun. **72**(2), 1005–1022 (2013)
9. Lai, N.L., King, C.T., Lin, C.H.: On maximizing the throughput of convergecast in wireless sensor networks. In: Advances in Grid and Pervasive Computing, pp. 396–408 (2008)

10. Mounib, K., Mouchine, G., Hussein,T.M.: WSN architectures for intelligent transportation systems. In: 2009 3rd International Conference on New Technologies, Mobility and Security (2009)
11. Pöttner, W.B., Seidel, H., Brown, J., Roedig, U., Wolf, L.: Construcitng schedules for time-critical data delivery in wireless sensor networks. ACM Trans. Sen. Netw. **10**(3) (2014)
12. Seidel, S.Y., Rappaport, T.S.: 914 MHz path loss prediction models for indoor wireless communications in multifloored buildings. IEEE Trans. Antennas Propag. **40**(2), 207–217 (1992)
13. Shashidhar, G., Ying, Z., Qingfeng, H.: Distributed time-optimal scheduling for convergecast in wireless sensor networks. Comput. Netw. **52**(3), 610–629 (2008)
14. Song, W., Huang, R., Shirazi, B., LaHusen, R.: TreeMAC: localized TDMA MAC protocol for real-time high-data-rate sensor networks. In: IEEE International Conference on Pervasive Computing and Communications, pp. 1–10 (2009)
15. Talzi, I., Hasler, A., Gruber, S., F. Tschudin, C.: PermaSense: investigating permafrost with a WSN in the Swiss Alps. In: Proceedings of the 4th Workshop on Embedded Networked Sensors (EmNets 2007), pp. 8–12 (2007)
16. Wang, W., Wang, Y., Li, X.Y., Song, W.Z., Frieder, O.: Efficient interference-aware TDMA link scheduling for static wireless networks. In: IEEE GLOBECOM, pp. 262–273 (2006)
17. Weidling, S., Todtenberg, N., Basmer, T., Schölzel, M., Masser, M., Taubert, M.: Outdoor range measurements in sub-GHz license-free radio bands under realistic conditions. In: MobiWac'17 Proceedings of the 15th ACM International Symposium on Mobility Management and Wireless Access, pp. 71–74 (2017)
18. Xiang, X., Liu, W., Wang, T., Xie, M., Li, X., Song, H., Liu, A., Zhang, G.: Delay and energy-efficient data collection scheme-based matrix filling theory for dynamic traffic IoT. EURASIP J. Wirel. Commun. Network. (2019)
19. Xu, K., Gerla, M., Bae, S.: Effectiveness of RTS/CTS handshake in IEEE 802.11 based ad hoc networks. Ad Hoc Netw. **1**(1), 107–123 (2003)

# Using Mathematical Models for Analysis and Prediction of Payment Systems Behavior

Victor Dostov[1,2](✉), Pavel Shust[1,2], and Svetlana Krivoruchko[3]

[1] Saint-Petersburg State University, St. Petersburg, Russia
leonova.v.anna@gmail.com, shoust@npaed.ru
[2] Russian Electronic Money and Remittance Association, Moscow, Russia
[3] Financial University Under the Government of the Russian Federation, Moscow, Russia
skrivoruchko@fa.ru

**Abstract.** This research focuses on the behavior of isolated payment systems. Our previous analysis of the isolated 'consumer-to-business' (C2B) and 'peer-to-peer' (P2P) payment systems has shown that their behavior can be analyzed using modifications of the Bass equations. In this paper, this approach is extended to the analysis to 'hybrid' payment systems with both C2B and P2P functionality using Ricatti equations. We derived universal solution which contains previously obtained C2B and P2P solutions as particular cases. The results are illustrated by analytical solutions for different system parameters and practical examples.

**Keywords:** P2P · B2C · Payment system · The dynamics of auditorium spread · Bass equation · Ricatti equation

European Central Bank in [1] defines a payment system as a complete set of instruments, intermediaries, rules, procedures, processes, and interbank funds transfer systems which facilitate the circulation of money in a country or currency area. Particular case of payment system is retail payment system where payer is an individual and payee is an individual (so-called peer-to-peer or P2P payment systems) or legal entity, for example, an online shop (so-called C2B payment systems). According to Laffont [2], the principal difference between these two types is existence of cooperation of both sides of the payment process. In C2B (according to Laffont—'one-sided') systems, customers make payments to the stores with infinite capacity. In such situation, payer acts independently from the payee.

P2P payments (considered by Laffont [2] as 'two-sided system') are made between individuals within limited audience and payer and payee mutually defines each other's behavior. International remittances between individuals are pure example of P2P payments. Local card-to-card transfers between individuals implemented by Sberbank in Russia and other players [3] are also a good example.


X.-S. Yang et al. (eds.), *Proceedings of Fifth International Congress on Information and Communication Technology*, Advances in Intelligent Systems and Computing 1183, https://doi.org/10.1007/978-981-15-5856-6_58

Retail payments are developing in modern society very fast. According to BCG 2019 report [4], the volume of non-cash payments is now growing by 5,9% per year average worldwide, but for young economies, e.g., for Russia and Eastern European countries, the annual growth may be around 10%. This growth follows fast functional and technological transformation, completely new systems, like cryptocurrencies, arise. Internally, such systems may be very complicated, but, as it was demonstrated by different authors, for example, by Dostov and Shust [5] that from customer's point of view their main function is reduced to fund transfers from a payer to a payee: the process often called 'commodization.' As we show below, sometimes only three parameters are enough to accurately describe payment system real-life behavior.

In our previous papers [6, 7], we studied a general evolution of payment system audiences, both classical and cryptocurrencies using simple Baas-like equations. By solving these equations, we derived expressions which demonstrated good quantitative approximation to real data in different geographies and payment system types. Also, by analyzing very basic mathematical properties of such equations we explained few quantitative phenomena, such as inability to predict long-term payment system behavior of payment systems based on early stage (startup) behavior data and extremely strong competition in remittance systems. In latest decades, the idea of using simple mathematical models for description of social--economy behavior became quite popular, e.g., for explaining of uneven distribution of wealth in society [8, 9], just to mention an example. Inspired by this trend, we decided to generalize models used in [6, 7] for more complicated types of payment systems.

In [6, 7], we considered idealized payment systems with following parameters:

1. current number of users $x$;
2. the maximum number of users, for example, the entire audience of a given country, $N$;
3. audience capture rate, which reflects the probability that a given user will start using the service: $a > 0$ within a given period.
4. audience fatigue rate, which reflects the probability that a given user will stop using the service: $b > 0$ within a given period.

Parameters $a$ and $b$ may also as an average reverse time of corresponding user decision start or stop to use the payment system.

Using the above parameters and assumptions, by mathematical and logical analysis, we were able to draw equations that describe well enough the dynamics of growth and development of payments in pure B2C and P2P markets. For pure C2B markets, we derived differential equation:

$$\frac{dx}{dt} = a(N - x) - bx \tag{1}$$

The solution of (1) yields:

$$x = x_{\infty}\left(1 - \left(1 - \frac{x_0}{x_{\infty}}\right)e^{-\frac{t}{T}}\right) \tag{2}$$

$$x_{\infty} = \left(\frac{a}{a+b}\right)N \tag{3}$$

$$T = \frac{1}{a+b}; \tag{4}$$

where $x_{\infty}$ is an asymptotic value of $x$ for $t \to \infty$. For P2P markets, the same approach leads to the equation, known in mathematics as Ferhulst equation

$$\frac{\mathrm{d}x}{\mathrm{d}t} = a(N - x)x - bx \tag{5}$$

which is well-studied in relation to biological models and innovation diffusion processes [10]. It has an analytical solution which can be written as:

$$x = \frac{x_{\infty} x_0 \mathrm{e}^{rt}}{x_{\infty} - x_0 + x_0 \mathrm{e}^{rt}} \tag{6}$$

where $x_0$ is initial user auditorium at $t = 0$ and

$$x_{\infty} = \frac{(aN - b)}{a} \tag{7}$$

is an asymptotic stationary solution (6), and reverse system evolution time

$$r = aN - b. \tag{8}$$

In current practice, most payment systems allow both C2B and P2P transactions. For example, the biggest e-money system, PayPal, from very beginning of its operation allowed both wallet-to-wallet transactions as well as payments from wallets to online shops. Recently, the trend toward this 'universalization' became very common. Card systems, such as VISA or Mastercard, originally focused on C2B transaction, later introduced card-to-card (essentially, P2P) transfer functionality. On the other hand, remittance systems that specialized in P2P transfers now allow payments to online shops. Such changes may yet not have drastic effects on their business but will definitely reshape their business models in future. For example, the largest Russian commercial bank Sberbank saw the value of card-based P2P transactions exceed card payments in stores in 2018 [11].

To provide empirical description of such hybrid payment system, we combine (1) and (6) introducing two capture rates $a_p$ and $a_b$ corresponding to P2P and C2B payments obtaining the following equation:

$$\frac{\mathrm{d}x}{\mathrm{d}t} = a_p(N - x)x + a_b(N - x) - bx \tag{9}$$

Or, combining powers of $x$ we obtain well-known Ricatti equation [12] with constant coefficients

$$\frac{\mathrm{d}x}{\mathrm{d}t} = -a_p x^2 + \left(a_p N - a_b - b\right)x + a_b N \tag{10}$$

Its stable stationary solution $x_\infty$ is defined by equation

$$0 = a_p x^2 - \left(a_p N - a_b - b\right)x - a_b N \tag{11}$$

as following

$$x_\infty = \frac{\left(a_p N - a_b - b\right) + \sqrt{\left(a_p N - a_b - b\right)^2 + 4a_b a_p N}}{2a_p} \tag{12}$$

Or, introducing $D$ as

$$D^2 = \left(a_p N - a_b - b\right)^2 + 4a_b a_p N \tag{13}$$

we can rewrite (12) in more compact form as

$$x_\infty = \frac{\left(a_p N - a_b - b\right) + D}{2a_p} \tag{14}$$

with corresponding choice of sign of $D$.

With $a_b \to 0$

$$x_\infty = \frac{a_b N}{(a_b + b)} \tag{15}$$

With $a_p \to 0$

$$x_\infty = \left(a_p N - b\right)/a_p \tag{16}$$

These expressions coincide evidently with early obtained (3) and (7).

For arbitrary parameters Eq. (10) allows analytical solution

$$x = \frac{1}{2a_p}\left(-D\tanh\left(\left(C - \frac{1}{2}tD\right)\right) + \left(a_p N - a_b - b\right)\right) \tag{17}$$

where $C$ is an arbitrary constant. It is interesting that audience growth rate curve shape depends only on limited number of parameters, constant $C$ and reverse time of system evolution $D$

$$\frac{\mathrm{d}x}{\mathrm{d}t} == \frac{D^2}{4a_p \cosh\left(C - \frac{1}{2}tD\right)} \tag{18}$$

A maximum of audience growth rate, if being achieved, is reached at

$$t\mathrm{max} = \frac{2C}{D}. \tag{19}$$

If the initial condition is $x(0) = 0$, then arbitrary constant $C$ is

$$C = arctanh\left(\left(a_p N - a_b - b\right)/D\right) \tag{20}$$

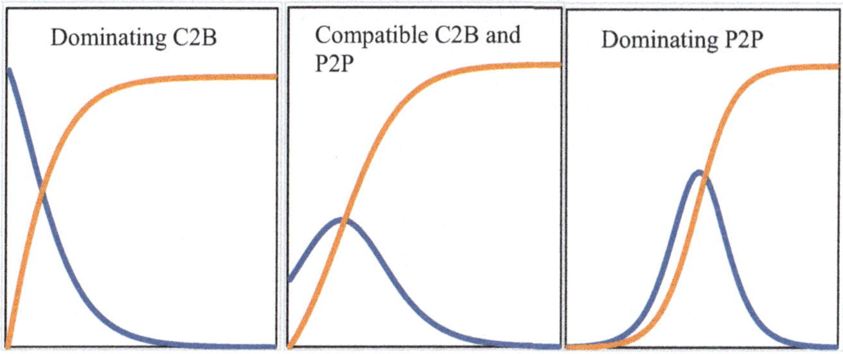


**Fig. 1.** Transition from dominating C2B type, $a_b \gg a_pN$ to dominating P2P type, $a_b \ll a_pN$ for number of users (red) and auditorium growth rate (blue)

In limited cases of $a_p \to 0$ or $a_b \to 0$, (17) reduces to previously obtained pure B2C and P2P solutions. In Fig. 1, we can see a transition from C2B-like to P2P-like solution for different relation between $a_b$ and $a_pN$, depending on which type of transactions is overcoming.

Of course, due to empirical character of the model, this development should be rather considered for qualitative description of payment systems rather than as deep quantitative analysis. However, as previously demonstrated in [6], this is a good basis for quantitative description and further improvement of formulae may strengthen interpolation quality. As an advantage to previous approach, we do not need to make assumption about which type of transactions dominate, as we can just apply (17) or (18) to real data. As an example, we applied (18) to Webmoney e-money wallets statistics [13] with quite good coincidence as shown in Fig. 2. Calculated values of $C = 2.2$ and $D = 0.33$ yield $C/D \gg 1$ which means dominance of P2P transactions, as previously proven in [6].

We also applied our calculations to Russian Central Bank data on prepaid payment cards [14]. We used (17) and obtained with $C = 2$ and $D = 0.19$ quite good coincidence shown in Fig. 3.

One should not overestimate detalization of model and always should take its empirical origin into account. However, good agreement with real data and qualitative trends allows to use the approach as a basis for more detailed investigations. In future, we plan to extend this approach to describe competition of payment systems which we started in [6] by introducing systems of differential equations similar to (1) for B2C segment and equations similar to (5) for P2P segment. More general approach will also allow to evaluate the stability of solutions. We also expect that proposed generalization of Baas-type equations might be useful for drawing more accurate models for explaining dynamics of customer acquisition in much wider class of models beyond payment system segment.

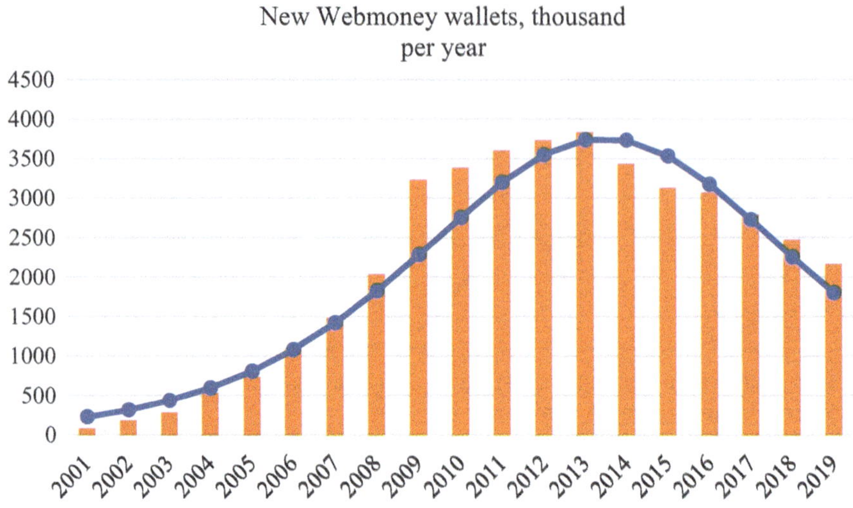


**Fig. 2.** Webmoney eWallets growth rate statistics (histogram) and calculation (curve)

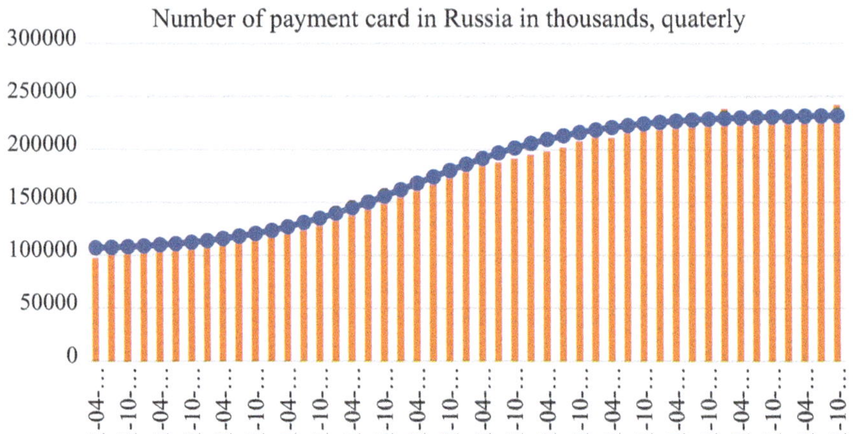


**Fig. 3.** Number of issued payment card in Russia, quarterly. Real data (histogram) and model-based calculation (curve)

## References

1. European Central Bank: The payment system. In: Kokkola, T. (ed.) (2010). https://www.ecb.europa.eu/pub/pdf/other/paymentsystem201009en.pdf
2. Laffont, J.-J.: Regulation and Development. p. 440. Cambridge University Press, NY (2005)
3. Sberbank: Tinkoff Bank launch money transfers using mobile phone number. In: AK&M. 13 July 2018. http://www.akm.ru/eng/news/2018/july/16/ns6071431.htm
4. BCG Global Payments Report: Tapping into Pockets of Growth (2019). https://www.bcg.com/en-us/publications/2019/global-payments-tapping-into-pockets-growth.aspx

5. Dostov, V.L., Shust, P.M.: Evolution of the electronic payments industry: problems of quality transition. In: Russian Academy of National Economy and Public Administration under the President of the Russian Federation. Moscow (2017). (in Russian). https://doi.org/10.2139/ssrn.2970963
6. Dostov, V., Shoust, P., Popova, E.: Using mathematical models to describe the dynamics of the spread of traditional and cryptocurrency payment systems. In: Computational Science and Its Applications—ICCSA 2019. Springer, 19th International Conference Saint Petersburg, Russia, July 1–4, 2019 Proceedings, Part II, pp. 457–471 (2019)
7. Dostov, V., Shust, P., Popova, E.: Using mathematical models for analysis and prediction of payment systems behavior. In: 34th International Business Information Management Association Conference (IBIMA), Madrid, Spain, 13–14 Nov 2019, ISBN 978-0-9998551-3-3, IBIMA Publishing, King of Prussia, PA, forthcoming
8. Zitikis R.: Analysis of indices of economic inequality from a mathematical point of view. In: RePAd Working Paper Series lrsp-TRS366, Département des sciences administratives, UQO (2002)
9. Ispolatov, S., Krapivsky, P. Redner, S.: Wealth Distributions in Models of Capital Exchange. Quantitative Finance Papers (1997). https://doi.org/10.1007/s100510050249
10. The Bass Model: Bass's Basement Research Institute. http://bassbasement.org/BassModel/Default.aspx
11. Eremina A., Yumabaev D.: Russians make more C2C transactions that card payments in stores. Vedomosti, 16 Mar 2018
12. Simmons, G., Robertson, J.: Differential equations with applications and historical notes, 2nd edn. In: International Series in Pure and Applied Mathematics, p. 640. McGraw-Hill (1991)
13. WebMoney Statistics and company information. https://www.webmoney.ru/eng/information/statistic/
14. Central Bank of Russia Statistics: The number of payment cards issued by credit institutions and the Bank of Russia. https://cbr.ru/statistics/psrf/sheet013

# A Pedagogical Game Design Document (Ped-GDD) to Promote Teachers' Engagement in the Kingdom of Saudi Arabia

Alaa Saggah[1,2(✉)], Anthony Atkins[1], and Russell Campion[1]

[1] Staffordshire University, Stoke on Trent ST4 2DE, UK
saggah.a@gmail.com, {a.s.atkins,r.j.campion}@staffs.ac.uk
[2] Umm Al-Qura University, Makkah, Kingdom of Saudi Arabia

**Abstract.** Current research suggests that using gamification in a pedagogical context can provide a positive learning experience to students. Integrating teachers' pedagogical input into the early stage of game design is an understudied area, and although the teachers' role is acknowledged, the way of communicating teachers' input is still undetermined. The current literature suggests the practice of using a Gamification Design Document (GDD) to illustrate a game requirements plan is a useful approach. In this paper, we discuss the development of a Pedagogical Game Design Document (Ped-GDD) using an Agile Holistic Framework to Support Teachers in Pedagogical Gamification Design. The paper also suggests the benefits of extending the research of the Ped-GDD further to be part of an e-Government scheme in the Kingdom of Saudi Arabia (KSA) to promote resources collaboration among teachers in educational software tool developments.

**Keywords:** Gamification · Pedagogical game design document · Educational games · E-government · Kingdom of Saudi Arabia

## 1 Introduction

Current literature suggests that gamification can promote students' engagement and influence their experience positively [1–3]. This paper discusses the development of a Pedagogical Game Design Document (Ped-GDD) using an Agile Holistic Framework to Support Teachers in Pedagogical Gamification Design (AH-PGD) which is illustrated in Fig. 1. The framework is composed of four distinct stages: Stage 1 Requirements; Stage 2 Gamification design; Stage 3 Implementation; and Stage 4 Testing and evaluation. The focus of this research is the gamification design (which is shown as the second stage in Fig. 1). The process of developing the framework elements and sub-elements has been documented in earlier research [4–6]. The aim of Ped-GDD is to facilitate communication among teachers and game developers.


X.-S. Yang et al. (eds.), *Proceedings of Fifth International Congress on Information and Communication Technology*, Advances in Intelligent Systems and Computing 1183, https://doi.org/10.1007/978-981-15-5856-6_59

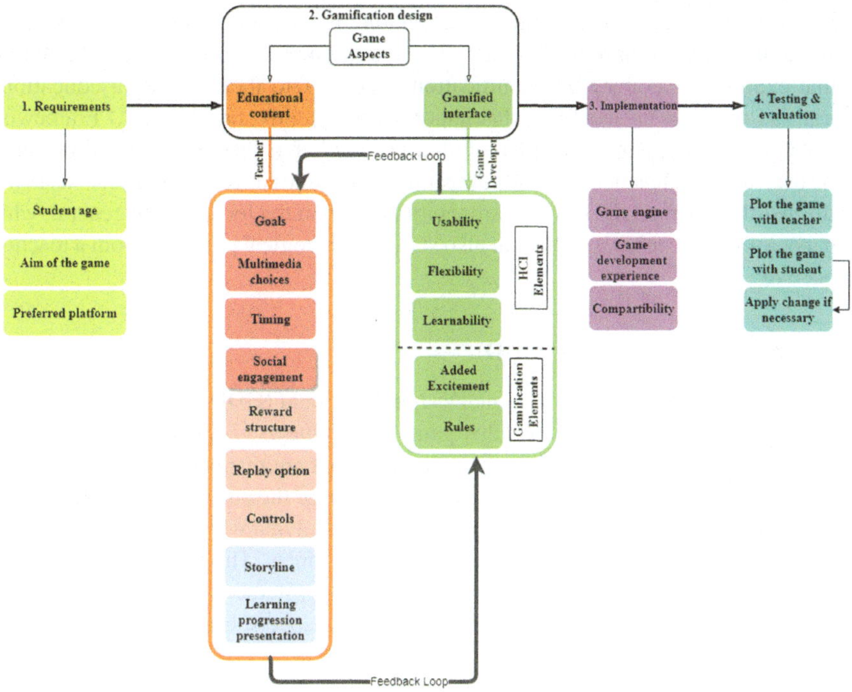


**Fig. 1.** An agile holistic framework of pedagogical gamification design (AH-PGD)

For future work, it is proposed to develop this concept as part of e-Government implementation for educational software tools development to promote teaching collaboration on a national level in the Kingdom of Saudi Arabia (KSA) which will contribute to a dramatic improvement in resourcing and help improve educational support to teachers and educational institutions in KSA.

This paper is organized as follows: Sect. 2 discusses the previous related work and provides a literature review. Section 3 outlines the Pedagogical Games Design Document (Ped-GDD) and gives an example of its application for teaching English in a school environment. Finally, the conclusion and proposed future work are presented in Sect. 4.

## 2 Related Work/Literature Review

### 2.1 Pedagogical Gamification

Previous work suggests [4] a collaborative approach to facilitate teacher engagement in the design process of gamification and enable a better fit for pedagogic requirements. Teachers recognize the importance of collaboration with game developers in turning the conceptual design into an electronic game for teaching purposes [6]. However, the form of communication between teachers and game developers was yet to be determined. The paper proposes a template that can be used to communicate the pedagogical input of teachers to the game developers to help facilitate collaboration to achieve a pedagogical gamification approach. The template will contain an example of an educational game requirements list for a particular lesson. The intention is for the teacher to be able to use that example to convey the gamification concept into their choice of educational material, as outlined in Sect. 3 of this paper. Consequently, the AH-PGD Framework Stage 2 Gamification design template with associated elements as illustrated in Fig. 1, will be used to assist teachers in communicating with games developers. The next section will discuss the current GDDs which are templates generated for games in general, while this paper aims to develop a GDD that also serves pedagogical purposes from a teaching perspective.

## 2.2 Game Design Document (GDD)

The current literature identifies a Game Design Document (GDD) as a template used in the game development industry to gather game requirements [7–9]. The developed AH-PGD Framework has eleven gamification elements that are mapped to the current GDD components outlined in Table 1. One of the challenges of the current GDD to teachers is the use of technical terms, e.g., mechanics, gameplay, dynamics, without further illustration, which might be challenging for teachers to address. The Ped-GDD is designed to accommodate different levels of teachers' computer literacy. Another challenge for teachers is the inclusion of commercial and business aspects, e.g., marketing, which is applicable in [10].

**Table 1.** Mapping the proposed Ped-GDD to the existing examples of GDD from the literature

| Gamification elements (Framework components) | GDD [7] | GDD Ten-Pager [8] | GDD template [9] | GDD community [10] | GDD [11] |
|---|---|---|---|---|---|
| Goals | objectives, levels, challenges | | Milestone | Goals, Progression & challenge | |
| Multimedia choice | Aesthetic what players will hear and see | | 2D, 3D, sound, animation | Music& sound, Art style | |
| Timing | | | Time scale | | |
| Social engagement | enemy | Enemies, Multiplayer | Team size | | |
| Reward structure | Rewards | Collectibles, Monetization | | Losing (as in score mechanics) | |
| Replay option | | | | | |
| Controls | | | | User skills | |
| Storyline | player avatar | Story | Theme, story | Characters, Story, theme, Story Progression | Core concept |
| Learning progression representation | | Progression | | Progression & challenge | |
| Rules | rules of how elements can interact with others | A mechanic is an item or element that **players interact with to create** or aid with gameplay | | | |
| Added excitement | | Power-ups, bonus material | | Items & powerups | |
| Number of matching | 6/11 | 6/11 | 5/11 | 7/11 | 1/11 |

There are similar elements among the Ped-GDD and the current GDDs with differences in the wording with the meaning remained the same. For example, Goals in this Ped-GDD was referred to as objectives by [7], as a milestone by [11], challenge by [10], as illustrated in Table 1. Another example of social engagement in Ped-GDD is referred to as enemies by [7, 8], team size by [11], as illustrated in Table 1.

The overall number of gamification elements studied in Ped-GDD is eleven. The number of elements that were mapped to the existing GDD and the equivalent ranged from one to seven, as illustrated in Table 1. Ped-GDD has eleven gamification elements that include some sub-elements built into the framework from pedagogical consideration, which supports the holistic aspect of the framework.

## 3 Pedagogical Game Design Document (Ped-GDD)

The Ped-GDD is a document that has two major collaborators, teachers and game developers. Teachers' task encompasses nine elements which are as follows: Goals, multimedia choice, timing, social engagement, reward structure, replay option, controls, storyline, and learning progression presentation. Each of these nine elements has at least one sub-element, as illustrated in Table 2. The Ped-GDD provided in Table 2 includes an example of a game to teach English letters.

## 4 Conclusion

There is a need to develop a Pedagogical Game Design Document (Ped-GDD) template that facilitates collaboration between teachers and game developers by including pedagogical gamified requirements. A flexible template is required that allows teachers to use the elements collectively or separately while giving game developers a chance to modify and or add the supporting requirements. The Ped-GDD template provides illustrations and examples to simplify the process for the teachers and assure consistency with elements' and sub-elements' meanings.

As research suggests that gamification is 'a modern educational approach' [3], this paper suggests future work to combine the benefits of pedagogical gamification with e-Government educational initiatives to enable more extensive collaboration among teachers nationwide in educational software tools development. The intention is that e-Government educational initiative from Vision 2030 of Saudi Arabia will also help to build a national repository of Ped-GDD examples of lessons to minimize the effort and maximize the impact and improve resources. The repository will hold the examples of gamification applied to lessons that can be shared and modified without the need to duplicate the effort. Abu-Shanab and Al-Sayed [12] suggested using gamification mechanics such as points or coupons to promote the use of e-Governments among citizens. The adoption of technology has been an area of study on emerging technologies, which also applies to e-Government. Alloghani et al. [13] suggest that applying the gamification concept into the e-Government application would help to overcome the technology adoption barrier. Furthermore, [13] mentioned the positive experience of gamification in education on a governmental level, such as providing feedback and allowing better planning for improvements.

**Table 2.** Ped-GDD illustrative example of a game to teach english letters

| This lesson scenario is to enable students to learn English alphabets from A-Z (26 letters). The example scenarios can be applied to all the letters. The target audience for these English alphabets is early-stage learners. | | | |
|---|---|---|---|
| Gamification Elements | Example | 2- Multimedia choice<br>• Suggestion list of multimedia | 3- Timing for individual tasks<br>• Below average (A-)<br>• Average (A)<br>• Above average (A+) |
| 1- Goals<br>• Learning milestone<br>• Suggestion of measurement<br>• Difficulty level | To be able to identify the letter *(n), e.g., A* shape<br>Choosing the image from different options, where only one of them is correct. | A picture to illustrate the letter<br> | (A-) 2 minutes<br>(A) 1.5 minutes<br>(A+) 1 minute |
| | To be able to identify the letter *(n), e.g., A* sound<br>Choosing the sound from different options, where only one of them is correct. | Audio files to play the sound<br> | (A-) 3 minutes<br>(A) 2 minutes<br>(A+) 1.5 minute |
| | To be able to how to write the letter *(n), e.g., A*<br>Tracing a dotted image of the letter using the correct start and finish point in writing the letter, also, how closely the line followed the letter dotted image. | A dotted image illustrates the letter *(n), e.g., A,* and indicates the start and finish point of writing the letter.<br> | (A-) 2 minutes<br>(A) 1.5 minutes<br>(A+) 1 minute |
| | To be able to identify the letter *(n), e.g., A* in words<br>Choosing the word(s) from different options, where only a few have the letter in the word. | List of words; some of them have the letter *(n), e.g., A* while others have not.<br>Cat Fun Eat Dog Air | (A-) 3 minutes<br>(A) 2 minutes<br>(A+) 1.5 minutes |
| 4- Social engagement<br>• Individual play<br>• Collaborative play<br>• Cooperative play<br>• Competitive play<br>• Number of players | An individual play. Every child would be expected to complete the game individually as an in-class practice. | | |
| 5- Reward structure<br>• Supporting list of E-rewards | Points collecting scheme will be incorporated into the game for choosing the right audio file that represents the letter *(n), e.g., A* sound. | | |
| 6- Replay option<br>• Not allowed<br>• Number of allowed attempts | Allowed, unlimited<br>The intention is to allow an unlimited number of attempts to build and reinforce the learning milestones based on an individual's need. Also, the game has no competition plan; therefore, the replay will not affect the group. | | |
| 7- Controls<br>• Supporting list of control options | Touchscreen, mouse depends on the preferred and available platform | | |
| 8- Storyline<br>• Game duration<br>• Game span<br>• Game standard (Comprehensive, Facile, None) | Overall the game duration between 7 and 10 minutes for each letter<br>3 days span for each letter, and the game will be going on for all the letters, which is a semester.<br>The story standard applied is a facile story that includes a character and a visual representation of the full letters to emphasize the attachment. However, there will be no plot planed. | | |
| 9- Learning progression representation<br>• Supporting list | Progress bar as visual scale<br> | | |

## References


1. ter Vrugte, J., de Jong, T., Vandercruysse, S., et al.: How competition and heterogeneous collaboration interact in prevocational game-based mathematics education. Comput. Educ. **89**, 42–52 (2015). https://doi.org/10.1016/j.compedu.2015.08.010
2. Morillas Barrio, C., Munoz-Organero, M., Sanchez Soriano, J.: Can gamification improve the benefits of student response systems in learning? An experimental study. IEEE Trans. Emerg. Top Comput. **4**, 429–438 (2016). https://doi.org/10.1109/TETC.2015.2497459
3. Skalka, J., Drlik, M.: Priscilla-proposal of system architecture for programming learning and teaching environment. In: IEEE 12th International Conference on Application of Information and Communication Technologies, AICT 2018—Proceedings. Institute of Electrical and Electronics Engineers Inc. (2018)
4. Saggah, A., Campion, R., Stanier, C.: An Investigation of the Role of the Teacher in Gamified Learning in Primary Schools. EDULEARN18 Proceedings, pp. 7700–7708. IATED Academy, Palma, Mallorca (2018)
5. Saggah, A., Campion, R., Stanier, C.: A Collaborative Gamification Design Framework in an Educational Context. ICERI2018 Proceedings, pp. 2410–2414. IATED, Seville (2018)
6. Saggah, A., Campion, R., Atkins, A.: An agile holistic framework to support teachers in pedagogical gamification design. In: 2019 International Conference on Advances in the Emerging Computing Technologies (AECT). IEEE, Madinah (2020) (in Press)
7. Salazar, M.G., Mitre, H.A., Olalde, C.L., Sanchez, J.L.G.: Proposal of game design document from software engineering requirements perspective. In: 2012 17th International Conference on Computer Games (CGAMES), pp. 81–85. IEEE (2012)
8. Rogers, S.: Level up!: The Guide to Great Video Game Design. John Wiley & Sons, Chichester (2014)
9. gamedesigning.org: How to Create a Game Design Document. In-Depth Guide (2018). https://www.gamedesigning.org/learn/game-design-document/. Accessed 20 Mar 2019
10. Gonzalez, L.: Gamasutra: Leandro Gonzalez's Blog—How to Write a Game Design Document. In: Gamasutra.com (2016). https://www.gamasutra.com/blogs/LeandroGonzalez/20160726/277928/How_to_Write_a_Game_Design_Document.php. Accessed 20 Mar 2019
11. Stanley, B.: Game Design Document Template—Unity Forum. In: Unity Forums (2014). https://forum.unity.com/threads/game-design-document-template.240038/. Accessed 20 Mar 2019
12. Abu-Shanab, E.A., Al-Sayed, M.R.: Can gamification concepts work with e-government? J. Inf. Technol. Res. **12**, 44–59 (2019). https://doi.org/10.4018/JITR.2019070103
13. Alloghani, M., Hussain, A., Al-Jumeily, D. et al.: Gamification in e-governance: development of an online gamified system to enhance government entities services delivery and promote public's awareness. In: ACM International Conference Proceeding Series. Association for Computing Machinery, pp 176–181 (2017)

# Limitations in Thesis Development of Systems Engineering: Knowledge and Skills

Sussy Bayona-Oré(✉)

Universidad Autónoma del Perú, Dirección de Investigación, Panamericana Sur Km. 16.3, Villa EL Salvador, Peru
sbayonao@hotmail.com

**Abstract.** The thesis development is a combined process that requires the student's knowledge, skills, and self-determination to complete the thesis and reach the desired degree. Despite the efforts made by students, research supervisor and the abundance of information that exists on research methodologies, a good percentage of the students complete the studies but do not complete the research. This research shows the results of a study directed at sixty-six undergraduate students with advanced abilities and knowledge to develop their thesis. The conclusions of the research show the necessity to supplement the knowledge and need for the reinforcement of a research supervisor.

**Keywords:** Limitations · Thesis · All but thesis · Supervisor

## 1 Introduction

The thesis is the last formal step to complete a professional career, but few students finish their research. "A considerable number of students struggle with the thesis elaboration process, resulting in delays, interruptions, and non-completion of their degrees" [1]. This situation suggests paying more attention and proactive planning on the part of the study centers [2]. The development of the thesis has been called by various authors as a combined process and requires a series of research skills, knowledge, and self-determination on the part of the student to complete it successfully. An academic degree is obtained, after successful completion of the studies [3]. Non-tangible indicators can measure success, such as "the development of research skills, the growth of personal development, and the contribution to your country" [4]. However, not all students manage to complete the thesis in time, a situation that is known as all but thesis (ABT).

In spite of the abundant bibliography of research methodology that exists, the students do not present the thesis in the expected time, and the thesis development is difficult [5]. Also, several authors acknowledge that in this process, the assistance of a supervisor is necessary. In this regard, several studies agree to consider thesis supervisor as a factor to complete the thesis successfully [6–8]. These studies highlight the relationship between the thesis and the supervisors as a critical factor to complete their candidacy.


X.-S. Yang et al. (eds.), *Proceedings of Fifth International Congress on Information and Communication Technology*, Advances in Intelligent Systems and Computing 1183, https://doi.org/10.1007/978-981-15-5856-6_60

With the promulgation of the University Law, 30220 [9], students must carry out research work to complete university studies and establish obtaining of professional degree will only be achieved with the preparation, approval, and support of an investigation. To achieve this goal, the different specialties of the universities teach courses aimed at supporting the student in the elaboration of the research work or thesis.

In systems engineering, the undergraduate research is so important because it allows the student to apply the scientific method and develop research skills, the development of new knowledge, and critical thinking skills to solve problems of different sectors through the incorporation of Information and communication technologies (ICT) and improve user experience. Besides, the thesis is the instrument that evidences the student's learning, since in this process, the student applies knowledge or acquires new and develops research skills.

The purpose of this study is to find the challenges that students face when they prepare the thesis. Knowing them is essential to help reduce the ratios of students who have finished their studies, but have not obtained the degree. The motivation to carry out this research is because there are few works on the difficulties faced by students to complete the thesis in Peruvian universities. This study focuses on the activities in the development of the research, level of knowledge, and skills achieved by undergraduate students.

This article has been structured into five sections. Section 2 describes the theoretical framework. In Sect. 3, the research methodology is presented. Section 4 shows the results. Finally, Sect. 5 presents conclusions and future work.

## 2 Related Work

### 2.1 Thesis Development Process

The thesis is the result of a rigorous and systematic process of scientific analysis to obtain an academic degree and contribute to a field of research. In the elaboration of the thesis, there are other actors with different weights beside the student [8, 10]. This process is in four aspects: informative, methodological, technological, and communicative. The informative is related to state of art and critical analysis of information. The methodology is relevant to knowledge and application of the process to develop the research, the technological one with the use of technology as processors of text, spreadsheets, tools for data exploitation, and communicative is related to competences to establish structure and content, organization of the thesis, and grammar [8].

The writing of a thesis to be taken as a piece of research to be completed as part of a degree [11]. It requires accuracy and order, so that students' knowledge, skills, attitudes, and values are evident in the thesis work [12]. Thus, the thesis is an effort that requires skills, abilities, determination, and aptitude for successful and timely completion [13]. Writing the thesis in the form of research work is considered one of the challenges of a researcher's life [12], and this process is termed as complex [14]. This process followed by the dissemination of thesis results to the scientific community and peer-reviewed and indexed journals; however, not all these are disseminated.

### 2.2 Limitations in the Thesis Preparation Process

Ensuring that students enrolled in undergraduate programs can obtain their degrees promptly are of interest to students, institutions of higher education, and society. Among the variables that influence student performance in the development of the thesis, we can mention (1) relationship between the candidate and the thesis supervisor [1], (2) attitude and motivation of the student, (3) the average entrance of students [15], and (4) language skills and communication [16] among others. Among the variables of the supervisor, it has been shown that a person with experience on supervision, research experience, and workload constitute success factors of the thesis [17].

Other aspects that have been discussed in the research, in order not to complete the thesis, are aspects of health (illness, pregnancy), psychological, related to the preferences established by the student or difficulty of balancing the activities of personal life and professional activities. In events of working scholars, they may choose not to want to balance and not devote significant sectors of their personal life. The decision will be intentional and conscious but that affects them because there are external influences, for example, mandatory legal frames. Studies on determinants to complete the research using the qualitative approach and few contemporary studies analyze the factors of success and failure related to the thesis works. Existing studies focus on completion variables and not non-completion variables [18].

### 2.3 Thesis Development Supervisor

The supervisor is the person who helps the student in his research and maintains a teaching and learning relationship with one or more of his peers. He is the one who evaluates the research critically and constructively. It is not who does the exposition to the student, but it does provide time, experience, and support to the person preparing the thesis.

According to the literature, the failure of the supervisor is related to manage the negative feelings of the supervised. That is the reason why they stay in the middle of the process due to solo work, and lack of self-motivation, organization, and independence. Self-motivation, being organized, and independent are factors of success to complete the thesis.

The supervisor must deal with issues such as (1) the average admission to grade [16] reported as a factor to complete the thesis, in addition to establishing strategies to achieve the social and academic integration of the student that has been considered another factor by [19], (2) with communication skills and language proficiency skills [16] considered as limiting in the thesis development process, (3) provide constructive and timely feedback, as well as encourage in these cases with the phenomenon of blocking the writer, and (4) supporting the thesis supervisor through all stages of the thesis development process is an indispensable factor [16].

### 2.4 Research Methodology and Thesis

The research methodology is an important matter because the understanding of research techniques contributes to the development of an academic discipline that integrates knowledge, concepts, theories, and tools related to how social and natural systems and entities can be studied [20]. The teaching of research methodology not only requires theoretical knowledge but also to participate in field research and contribute to its growth and progress. Its importance lies in the fact that this subject provides the student with the knowledge and skills necessary to solve various problems in society [21, 22].

Several authors agree that the learning outcomes associated with the courses of research methods are lower [23–25]. This situation may occur because some students may question the usefulness and benefits of the courses at the end of the degree. How these courses are designed and taught is diverse, and there is a tendency to define the classes to qualitative and quantitative approaches. Taken into account that the elaboration of the thesis in many cases is the student's first contact with scientific research.

Another aspect to mention is that research methodology teachers have a wide range of students in different disciplines. Some teachers are specialists both in the area of knowledge related to the thesis and research methodology. However, in their vast majority, they have not acquired a formal degree in the field, also, that few are also considered experts in research methodology [20].

Another study [12] shows the difficulties that students have in selecting the most appropriate research methodology to conduct their research, present problems with data collection, in ensuring the validity of the questionnaire, perform the data analysis or determine the sample. Define the problem, review the literature, build the theoretical framework, design a strategy, and verify the hypothesis mentioned as highly complex processes [26]. That is, as part of the research methodology, students are expected to use systematic proposals and procedures to investigate well-defined problems [23].

## 3 Research Methodology

The research carried out is of the quantitative type. It is transversal because the data is collected and analyzed at a given time. The population consisted of undergraduate students of the systems engineering, offered by a private university. The students participating in the research were in the last year of their professional career and were studying research subjects. The research courses are designed to provide the knowledge and skills for research and the gradual development of the undergraduate thesis. Also, students have a statistics course whose content includes: descriptive statistics, quantitative and qualitative methods, and study design. The sample is non-probabilistic and consists of 66 students. Two instruments were applied: one linked to the adviser and the other related to the knowledge and skills developed for the thesis development by the student. In this article, we will focus on the second instrument.

The questionnaire consists of two parts: the first part oriented to sociodemographic data and the second part to the thesis development process. This article studies the thesis development process in the phases of literature review, problem definition, and methodology. The collected data were processed using SPSS statistical software. A descriptive analysis of the study variables, through frequency tables for the qualitative variables and measures of central tendency for the quantitative variables, were obtained.

# 4 Results

This section presents the results of the research carried out to know the knowledge and skills developed by the students who develop theses. The results correspond to 66 students of which, 17% are female, and 83% are male, 62% between 21 and 22 years, 8% 23 years, and 10% 24 years. Figure 1 shows the distribution of students by gender, who participated in the research.

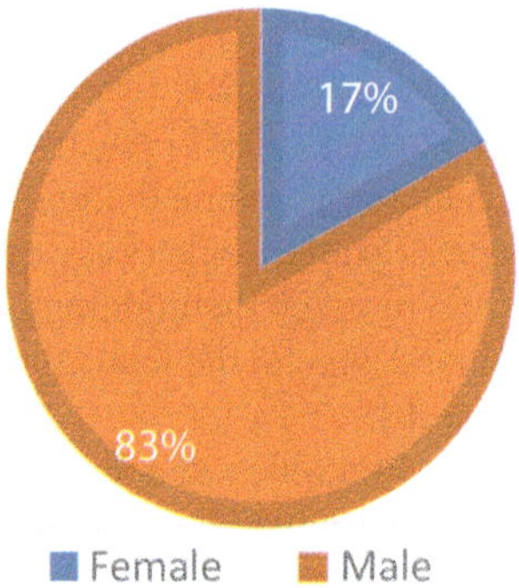


**Fig. 1.** Distribution of students according to gender

In examining the results, students anticipated developing abilities and necessary knowledge of techniques and tools for the development of the thesis. In other words, the ideal percentage expected for each activity is 100% on the Likert scale 4 (totally agree).

## 4.1 Literature Review/Theoretical Framework

Once the idea of the research problem is conceptualized, the state of art of the topic is reviewed. The results related to the literature review activities show that the percentages on the search for information relevant to the research to be carried out do not reach the ideal value. There is a need for students to improve their research skills to extract relevant information from the databases by using extraction sheets. Having relevant data extraction sheets is a good practice used in the systematic review of the literature. Likewise, the knowledge and skills of teachers and students improved in the use of tools to make automatic references in the text. Table 1 shows the results for each of the activities by scale expressed as a percentage.

## 4.2 Problem Definition

Students must have the skills to examine the information from the literature review and develop critical thinking to contrast the positions and approaches of different authors. Likewise, redefine the problem to be solved and the variables that intervene. In practice in thesis supervision, students show challenges in the final approach to the problem and in writing the research objectives. In the supervision process, these aspects are reinforced so that students reach the desired level. Table 2 shows the results for each of the activities by scale expressed as a percentage.

**Table 1.** Activities related to literature review

| Literature review/theoretical framework | 1 | 2 | 3 | 4 |
|---|---|---|---|---|
| Search for relevant information in electronic journals | 1.5 | 6.1 | 56.1 | 36.4 |
| Create data extraction sheets | 0.0 | 24.2 | 48.5 | 27.3 |
| Use a referral system | 0.0 | 24.2 | 45.5 | 30.3 |
| Use tools to make automatic references | 9.1 | 25.8 | 36.4 | 28.8 |

*1* totally disagree, *2* disagree, *3* agree, *4* totally agree

**Table 2.** Activities related to problem definition

| Problem definition | 1 | 2 | 3 | 4 |
|---|---|---|---|---|
| Contrasting approaches and positions of various authors | 0.0 | 21.2 | 50.0 | 28.8 |
| Make a critical evaluation of the different positions | 1.5 | 21.2 | 53.0 | 24.3 |
| Pose the research problem to be solved | 0.0 | 3.0 | 74.2 | 22.8 |
| Write the research objectives | 0.0 | 3.0 | 66.7 | 28.8 |

*1* totally disagree, *2* disagree, *3* agree, *4* totally agree

### 4.3 Methodology

The methodology focuses on research questions erected, how essential data is collected, processed, analyzed, and interpreted [20]. Knowing the types of research allows you to respond appropriately to research questions. Defining the population and defining the sample are important aspects that students should know. In this case, teachers are knowledgeable about the subject related to the thesis but have not acquired a formal degree of supervisors. Likewise, these results reveal the importance of an interdisciplinary work in which the specialist in the subject, a specialist in methodology, and the statistical professional participates. Table 3 shows the results for each of the activities by scale expressed as a percentage.

**Table 3.** Activities related to research methodology

| Methodology | 1 | 2 | 3 | 4 |
|---|---|---|---|---|
| Choose a type of research to answer the question posed | 0.0 | 10.6 | 54.5 | 34.8 |
| Define the variables to study | 0.0 | 7.6 | 60.6 | 31.8 |
| Make sufficient delimitation of the population | 0.0 | 13.6 | 60.6 | 25.8 |
| Make an appropriate selection of the sample to be performed | 0.0 | 16.7 | 48.5 | 34.8 |

*1* totally disagree, *2* disagree, *3* agree, *4* totally agree

### 4.4 Thesis Supervisor

Different authors agree on the importance of the supervisor-supervised relationship. Thesis supervisors should not only know but also proven research experience. The behavior of the supervisor is crucial in the stages of the thesis work, in which the student must collaborate with the fulfillment of the tasks. The experience and research production of the supervisor is a factor that positively affects the culmination of the thesis. The results show that teachers know the rules and regulations of thesis supervision (mean = 3.4), demonstrate knowledge in research methodology (mean = 3.3), and demonstrate confidence in their ability as supervisor and researcher in an average vaster than 3.3 of an ideal value equal to 4.

## 5 Conclusions

Various studies conducted on factors that condition that a student does not culminate from an emotional, family, organizational, supervision, and among others. An important aspect is the knowledge and development students' research skills for thesis development. In this article, the results of quantitative research presented to show the situation of knowledge and skills developed by undergraduate students on a Likert scale from 1 to 4 (4 the ideal value), in three aspects: literature review, problem definition, and methodology. The results of the study show the need for students to improve their knowledge and research skills to prepare the thesis. The supervisors should not only know about the subject but also about research methodology coupled with the experience in research and statistics. As future work, studies continued on the limitations that students have in the thesis development process, in each of its stages, as well as identifying the technical, organizational, personal, and social factors that impact the thesis development process.

## References

1. Rauf, F.A.: Challenges of thesis work: towards minimizing the non-completion rate in the postgraduate degree program. Eur. J. Bus. Manag. **8**(7), 113–124 (2016)
2. Nouri, J., Larsson, K., Saqr, M.: Identifying factors for master thesis completion and non-completion through learning analytics and machine learning. In: Scheffel, M., Broisiisin, J., Pammer-Schindler, V., Ioannou, A., Schneider, J. (eds.) Transforming Learning with Meaningful Technologies. EC-TEL 2019. Lecture Notes in Computer Science, vol. 11722 (2019)
3. Woodrow, L.: Academic success of international postgraduate education students and the role of English proficiency. Univ. Syd. Pap. TESOL **1**, 51–70 (2006)
4. Singh, J.: Human capital or talent development? A paradigm shift by Malaysian higher education. In: de Wit, H., Gacel-Ávila, J., Jones, E., Jooste, N. (eds.) The Globalization of Internationalization: Emerging Voices and Perspectives, pp. 153–165. Routledge, London (2017)
5. Gallo, J., López, J.: El tutor de tesis: sugerencias para su desempeño. Pedag. Prof. **12**(1) (2014)
6. Roach, A., Christensen, B.K., Rieger, E.: The essential ingredients of research supervision: a discrete-choice experiment. J. Educ. Psychol. **111**(7), 1243–1260 (2019)

7. Vera, L., Vera, C.: Desempeño del tutor en el proceso de acompañamiento en la producci6n científica. TELOS. Rev. Estud. Interdiscip. Cienc. Soc. **17**(1), 58–74 (2015)
8. Bayona, S.: Supervisors performance and thesis development on information technology: graduate students perceptions. In: Proceedings of the 13th Iberian Conference on Information Systems and Technologies (CISTI), Caceres, pp. 1–5 (2018). https://doi.org/10.23919/cisti.2018.8399224
9. Ley 30220: Ley Universitaria, Diario Oficial El Peruano, 9 julio, pp. 52713–527233 (2014). http://www.leyes.congreso.gob.pe/Documentos/Leyes/30220.pdf. Accessed 2018/12/12
10. Rosas, A., Flores, D., Valarino, E.: Rol del tutor de tesis: competencias, condiciones personales y funciones. Investig. postgrado **21**(1), 153–185 (2006)
11. Hart, C.: Doing Your Masters Dissertation. Sage, London (2005)
12. Prasad, H.: Writing thesis in English education: challenges faced by students. J. NELTA Gandaki (JoNG) **1**, 44–52 (2019)
13. Agu, N., Oluwatayo, G.: Variables attributed to delay in thesis completion by postgraduate students. J. Emerg. Trends Educ. Res. Policy Stud. **5**(4), 435–443 (2014)
14. Hodza, F.: Managing the student-supervisor relationship for successful postgraduate supervision: a sociological perspective. S. Afr. J. High. Educ. **21**(8), 1155–1165 (2007)
15. House, J., Johnson, J.: Predictive validity of graduate record examination scores and undergraduate grades for length of time to completion of degree. Psychol. Rep. **71**(3), 1019–1022 (1992)
16. Jiranek, V.: Potential predictors of timely completion among dissertation research students at an Australian faculty of sciences. Int. J. Doctoral Stud. **5**(1), 1–13 (2010)
17. Pitchforth, J., et al.: Factors affecting timely completion of a PhD: a complex systems approach. J. Scholarsh. Teach. Learn. **12**(4), 124–135 (2012)
18. Ishitani, T.T.: Studying attrition and degree completion behavior among first-generation college students in the United States. J. High. Educ. **77**(5), 861–885 (2006)
19. Pascarella, E.T., Terenzini, P.T.: Predicting voluntary freshman year persistence/withdrawal behavior in a residential university: a path analytic validation of Tinto's model. J. Educ. Psychol. **75**(2), 215–226 (1983)
20. Daniel, B.K.: Reimaging research methodology as data science. Big Data Cogn. Comput. **2**(1), 4 (2018)
21. Kilburn, D., Nind, M., Wiles, R.: Learning as researchers and teachers: the development of a pedagogical culture for social science research methods? Br. J. Educ. Stud. **62**(2), 191–207 (2014)
22. Nind, M., Kilburn, D., Luff, R.: The teaching and learning of social research methods: developments in pedagogical knowledge. Int. J. Soc. Res. Methodol. **18**, 455–461 (2015)
23. Daniel, B.K.: Contestable professional academic identity of those who teach research methodology. Int. J. Res. Method Educ. **41**(5), 548–561 (2017)
24. Earley, M.A.: Lessons learned from students' research experiences. Available online: http://jrp.icaap.org/index.php/jrp/article/view/93/75. Last accessed 2019/01/11 (2007)
25. Murtonen, M.: University students' understanding of the concepts empirical, theoretical, qualitative and quantitative research. Teach. High. Educ. **20**, 684–698 (2015)
26. Ochoa, L., Cueva, A.: Tesis y deserci6n: entre el compromiso y el obstáculo. Universidad, Bogotá (2012)

# Mapping the Relationship Between Hedonic Capacity and Online Shopping

John Magnus Roos[1,2,3,4](✉)

[1] School of Business, Economics and Law, University of Gothenburg, Gothenburg, Sweden
magnus.roos@chalmers.se
[2] School of Health Sciences, University of Skövde, Skövde, Sweden
[3] Faculty of Textiles, Engineering and Business, University of Borås, Borås, Sweden
[4] Department of Space, Earth and Environment, Chalmers University of Technology, Gothenburg, Sweden

**Abstract.** In the present study, the relationship between hedonic capacity and online shopping is explored through a Swedish nationally representative sample. A survey was distributed to 3000 citizens. The number of respondents was 1591 (response rate: 53%). Ordinal regression analyses were conducted in order to test the association between hedonic capacity and online shopping. The dependent variable was online shopping frequencies. Gender, age, and individual income were control variables. Our findings indicated that hedonic capacity was positively associated with online shopping ($p < 0.001$). The findings propose that online shopping primarily is triggered by emotions and affect rather than reasoning and cognition. Such insights can be used in strategical marketing and technological decisions by academy and industry, as well as in Web site design and communication.

**Keywords:** Online shopping · E-shopping · Hedonic capacity · Hedonic shopping

## 1 Introduction

The number of citizens who shop online, the number of online purchases per citizen, and the total amount of money that citizens spend on purchases via the Internet has increased in Sweden as well as in other developing countries [6, 7, 11, 17, 21]. In 2018, the total Swedish retail sales was SEK 787 billion, while the total retail sales on the Internet was SEK 77 billion. This means that retail sales on the Internet represented 9.8% of total retail sales in Sweden [20]. According to the Swedish Trade Federation, the Swedish retail sales on the Internet will grow rapidly the coming years and correspond to one third of the total retail sales already in 2025 [25]. The number of occasions that Swedish citizens, on average, shop something online has increased noteworthy in the last decade, from 9 occasions in 2008 to 24 occasions in 2018 [20].


X.-S. Yang et al. (eds.), *Proceedings of Fifth International Congress on Information and Communication Technology*, Advances in Intelligent Systems and Computing 1183, https://doi.org/10.1007/978-981-15-5856-6_61

Clearly understanding online shoppers are advantageously in ICT design, marketing communication, and overall strategical decisions [28]. Traditionally, online shopping has been perceived as utilitarian rather than hedonic [28]. People in general differ in their hedonic capacity, which is the capability for experiencing pleasure, excitement, enthusiasm, and enjoyment in everyday life [15]. The aim of the present study is to explore the relationship between hedonic capacity and online shopping. Since online purchases primarily seems to be associated with utilitarian shopping and not hedonic shopping [28], people with a low degree of hedonic capacity are expected to shop more online than people with a high degree of hedonic capacity. Therefore, the following hypothesis has been formulated:

Hedonic capacity is negatively related to online shopping.

## 2 Theoretical Framework

### 2.1 Utilitarian- and Hedonic Shopping

In the off-line environment, consumer researchers have for a long period of time recognized that consumers shop differently depending on whether they shop for fun, so-called hedonic or experiential shopping, or for efficiency, so-called utilitarian or goal-directed shopping [1]. In utilitarian shopping, consumers are looking for functional, instrumental, and practical benefits. In general, utilitarian shopping begins with a problem that the consumer needs to solve (e.g., something has broken). Utilitarian shopping is normally linked to necessary requirements that must be fulfilled [5, 14]. In the marketing literature, utilitarian shopping has been described as efficient, rational, and task oriented [13]. Utilitarian consumers are transaction oriented and desire to purchase what they want fast and without distraction.

In hedonic shopping, consumers are looking for esthetics, joyful, and pleasurable benefits [5]. Hedonic benefits are related to expectations that consumers aspire to meet and are perceived as luxuries and pleasure in the consumption experience [4, 5, 14].

### 2.2 Online Shopping as Utilitarian Shopping

The traditional view of online shopping is that it has been more related to utilitarian shopping than to hedonic shopping. Online shopping is traditionally perceived as "work" and evaluated in terms of work-related performance (e.g., success and accomplishment) [26]. One reason for this is that pioneer users of the Internet tended to be task and goal oriented, with a strong internal locus of control, rather than spontaneous with a strong external locus of control [12]. There are at least four reasons to why online shopping is more utilitarian than hedonic [28]:

1. Convenience. Online shopping saves time and effort.
2. Selection. Online shopping offers an enormous supply, and thereby a better opportunity to optimize the utility of the purchase.
3. Information. Online shopping facilitates search for information such as product specifications and prices.

4. Lack of sociality. Online shopping offers shopping absent of influences and distractions from others, such as sale people, spouses, crowd, and lines.

Online shopping has primarily been related to cognitive consumer decision models, while affective consumer decision models are more commonly related to shopping in brick-and-mortar stores [28].

# 3 Methodology

## 3.1 Sample and Procedure

The data were collected by the SOM Institute, a survey research organization at the University of Gothenburg. For more than 30 years, the institute has gathered data that is representative of the Swedish population. The data covers several areas, such as political views, media habits, and consumer behaviors. The SOM Institute use systematic probability samples based on registration addresses [24].

The present study consists of data from the cross-sectional National SOM survey of 2012 [27]. The survey included measures of hedonic capacity, online shopping, individual income, age, and gender. The survey was distributed to a nationally representative sample of 3000 selected individuals in Sweden, aged 16–85. The number of respondents was 1591 (response rate: 53%). The average age was 50.7 years. Of the 1591 respondents, 819 were women and 772 were men. The surveys were sent out on September 14, 2012, followed by a series of reminders distributed via post and telephone, ended on February 21, 2013 [27].

The National SOM survey of 2012 is in accordance with the Swedish law on research ethics and approved by the regional ethical review board in Gothenburg [27].

## 3.2 Measures

Online shopping was measured through a one-item question: "How often have you purchased an article or a service on the Internet the last twelve months?". The respondents were asked to indicate purchase rate on a seven-point scale, from "Never" (1) to "Every day" (7). In total, 1567 respondent answered the question related to online shopping. Among them, 30.3% had not bought anything online during the last 12 months; 14.2% had bought something online one time during the last 12 months; 25.1% had bought something one time per 6 months; 25.7% had bought something every month, 3.8% had bought something every week, 0.4% had bought something several times a week, and 0.4% had bought something online every day.

Hedonic capacity was measured by a three-items scale [9, 10]. A four-point Likert scale ranging from 1 ("strongly disagree") to 4 ("strongly agree") was used for the three items: "I often feel exhilarated," "I often enjoy life," and "I am in good mood when socialize." The hedonic capacity measure was constructed by averaging the responses to the three items. Only respondents who had answered all three items received a mean score and were thereby included in the analyses ($N = 1482$). The measure of Hedonic capacity shows a satisfactory level of validity in terms of construct, convergent, and discriminant validity. The reliability in terms of test-retest reliability is also satisfactoriliy high [9, 10]. Previous researchers [10] reported Cronbach alpha coefficients of 0.54 and 0.65 in two different samples of the Hedonic capacity scale, while the Cronbach alpha coefficients in the present study was 0.62. ($\alpha = 0.62$, $p < 0.01$, $M = 3.14$, SD = 0.50, $N = 1482$).

Prior research has found that gender and age influence Internet usage; men use the Internet more than women and younger people use the Internet more than older [12, 22]. Previous research has also shown that individual income is positively related to overall Internet usage [16, 18, 19], and especially to online shopping [8, 21]. Therefore, the respondent's age, gender, and individual income were used as control variables in the present study. Gender was coded as a dummy variable (male = 0; female = 1). Age was measured through an open question. Individual income was measured through a seven-point Likert scale, ranging from 1 (< SEK 10,000 per month) to 7 (> SEK 60,000 month).

## 4 Findings

Two ordinal regression models were conducted in order to explore the relationship between hedonic capacity and online shopping. In the first model, hedonic capacity was used as a single predictor, reasoning that hedonic capacity is in focus for the present study. In the second model, gender, age, and individual income were added as control variables, in order to investigate potential influences on the relationship between hedonic capacity and online shopping.

Hedonic capacity was positively associated with increasing odds ratios (OR) for online shopping, both in the univariate model 1 (OR, 1.62; 95% confidence interval [CI], 0.30–0.67; $p < 0.001$) and in the multivariate model 2 (OR, 1.43; 95% CI, 0.17–0.56; $p < 0.001$) (Table 1).

Higher individual income (e.g., an increase of 10,000 SEK per month) was associated with increasing OR for online shopping (OR, 1.49; 95% CI, 0.32–0.48; $p < 0.001$). Aging (in the range of 16–85) was associated with decreasing OR for online shopping (OR, 1.06; 95% CI, −0.05 to −0.07; $p < 0.001$). We did not find any significant ($p < 0.05$) association between gender and online shopping (Table 1).

## 5 Discussion

### 5.1 Discussion of the Findings

Hedonic capacity was positively associated with online shopping. This observed association between high levels of hedonic capacity and more frequent online shopping went in the opposite direction to our hypothesis.

A tentative explanation for the positive relationship between hedonic capacity and online shopping is that people with a higher hedonic capacity to a larger extend can utilize the advantages from online shopping (e.g., convenience, selection, and information) in order to make their everyday life even more pleasurable and excitement. Several online shopping activities seem especially suited for people with a high degree of hedonic capacity, such as auctions, involvement with a hobby, and bargain hunting [28]. The

**Table 1.** Ordinal regression analysis to predict online shopping

| Predictors | Estimate | SE | *p* value | OR | 95% CI |
|---|---|---|---|---|---|
| *Model 1* | | | | | |
| Hedonic capacity | 0.48 | 0.09 | **0.000** | 1.62 | [0.30, 0.67] |
| *Model 2* | | | | | |
| Hedonic capacity | 0.36 | 0.10 | **0.000** | 1.43 | [0.17, 0.56] |
| Individual income | 0.40 | 0.04 | **0.000** | 1.49 | [0.32, 0.48] |
| Age | −0.06 | 0.00 | **0.000** | 1.06 | [−0.07, −0.05] |
| Gender (male) | −0.02 | 0.10 | 0.880 | 0.99 | [−0.21, 0.18] |

*Note* Bolded values entries statistically significant effects ($p < 0.001$). SE = standard error; OR = odds ratio; CI = confidence interval. The scales are as following: online shopping, 1–7; hedonic capacity, 1–4; individual income, 1–7; age, 16–85; gender, 0 or 1

hypothesis is based on a bygone era, 1990–2001 [13, 26, 28]. Since then, the Internet landscape has changed drastically, especially with regards to electronic commerce (e.g., supply of more luxury goods, marketing, Web design, and Big Data consumer tracking). It is likely that online shopping from the beginning was particularly suitable for people with a low degree of hedonic capacity and that online shopping were perceived as a work-related task, rather than a leisure activity [13, 26].

We suggest further research in order to understand why people with a high degree of hedonic capacity shop online more frequently? Perhaps, hedonic capacity is positively related to high frequent shopping in general, both online and in brick-and-mortar stores. Perhaps, people low on hedonic capacity are doing a larger amount of their shopping online, compared to people high on hedonic capacity. The design of the present study is insufficient for answering these questions, but we suggest that further research focuses more on a comparative analysis of online shopping and physical store shopping. Such research might also gain from including a personality trait related to utilitarian shopping, such as need for cognition (the degree to which a person likes to think about things and process information) [23]. Furthermore, we do not know if people high on hedonic capacity primarily are doing utilitarian shopping online and hedonic shopping in physical stores. In other words, it might be that online shopping primarily is related to utilitarian shopping even if it is positively related to the consumers' hedonic capacity. There are still many questions without answers and therefore good reasons for further research.

### 5.2 Discussion of the Methodology

We would like to discuss some methodological limitations with the present study. First, the cross-sectional sample only allows us exploring relationships between hedonic capacity and online shopping, and not causal effects. However, it might be argued that hedonic capacity (as other personality traits) causes online shopping (as other behaviors) and not vice versa [3].

Second, online shopping is measured by a one-item question: "How often have you purchased an article or a service on the Internet during the last twelve months?". We recommend future researchers to use an index or query battery instead, which would also allow for comparative analyses between hedonic and utilitarian shopping, and across consumption categories such as clothes [6], consumer electronics [7], and groceries [11, 17].

Third, online shopping is measured by self-reported purchase frequencies and not actual purchase frequencies. Estimating actual behavior through self-estimation causes a well-known problem in social psychology [2].

Fourth, although the goal was to represent the Swedish population in the 16–85 age range, the sample includes slightly fewer younger citizens. Furthermore, married people are slightly over-represented, while foreign citizens are slightly underrepresented [27].

Finally, caution should also be taken regarding the generalization of the study's results to other places and times. Online shopping varies between different cultures, among other things, depending on economic and technological developments [20, 21, 23]. We suggest more up-dated nationally representative studies on hedonic capacity and online shopping in Sweden and elsewhere.

## 6 Conclusion

The key finding of the present study is that high-frequent online shoppers are hedonic in their characters. If academy and industry are perceiving online shoppers as primary utilitarian, they might miss their most important target group – people with a high degree of hedonic capacity. Furthermore, they might develop and use inappropriate business models, which overemphasize the role of cognitive aspects in consumers' online decisions and underemphasize the role of affects and emotions. Insights on the relationship between hedonic capacity and online shopping can and should be used in strategical marketing and technological decisions, as well as in Web design and communication [28].

**Acknowledgment.** Thanks are due to the SOM Institute at the University of Gothenburg and the Swedish National Data Service for providing access to the data. Thanks, are also due to the Centre for Retailing, CFR (School of Business, Economics and Law, University of Gothenburg), for funding participation in the fifth International Congress on Information and Communication Technology (ICICT).

## References


1. Barbin, B.J., Darden, W.R., Giffen, M.: Work and/or fun: measuring hedonic and utilitarian shopping value. J. Consum. Res. **20**, 644–656 (1994)
2. Baumeister, R.F., Vohs, K.D., Funder, D.C.: Psychology as the science of self-reports and finger movements: whatever happened to actual behavior? Perspect. Psychol. Sci. **2**(4), 396–403 (2007)
3. Chamorro-Premuzic, T.: Personality and Individual Differences. BPS Blackwell, Glasgow (2012)
4. Chitturi, R.: Emotions by design: a consumer perspective. Int. J. Des. **3**, 7–17 (2009)
5. Chitturi, R., Raghunathan, R., Mahajan, V.: Delight by design: the role of hedonic versus utilitarian benefits. J. Mark. **72**, 48–63 (2008)
6. Cullinane, S., Browne, M., Karlsson, E., Wang, Y.: Retail clothing returns: a review of key issues. In: Wells, P. (ed.) Contemporary Operations and Logistics. Achieving Excellence in Turbulent Times, pp. 301–322. Palgrave Macmillan, London (2019)
7. Duch-Brown, N., Grzybowski, L., Romahn, A., Verboven, F.: The impact of online sales on consumers and firms. Evidence from consumer electronics. Int. J. Ind. Organ. **52**, 30–62 (2017)
8. Farag, S., Krizek, K.J., Dijst, M.: E-shopping and its relationship with in-store shopping: empirical evidence from the Netherlands and the USA. Transp. Rev. **26**(1), 43–61 (2006)
9. Gunnarsson, M., Gustavsson, P., Holmberg, S., Weibull, L.: Statistical evaluation of six short five factor model personality measures aiming for potential inclusion in the SOM Institute's national surveys. SOM-rapport nr 2015:27. University of Gothenburg, The SOM-Institute, Gothenburg, Sweden (2015)
10. Gustavsson, J.P., Jönsson, E.G., Linder, J., Weinryb, R.M.: The HP5 inventory: definition and assessment of five health-relevant personality traits from a five-factor model perspective. Pers. Individ. Dif. **35**, 69–89 (2013)
11. Hagberg, J., Holmberg, U.: Travel modes in grocery shopping. Int. J. Retail Distrib. Manag. **45**(9), 991–1010 (2017)
12. Hills, P., Argyle, M.: Uses of the Internet and their relationships with individual differences in personality. Comput. Hum. Behav. **19**(1), 59–70 (2003)
13. Hoffman, D.L., Novak, T.P., Schosser, A.: Consumer Control in Online Environment. Working Paper. eLab, Owen Graduate School of Management, Vanderbilt University (2000)
14. Kievetz, R., Yang, H.-W.: Passion for online shopping: the influence of personality and compulsive buying. Soc. Behav. Pers. **36**(5), 693–706 (2008)
15. Meehl, P.E.: Hedonic capacity: some conjectures. Bull. Menninger Clin. **39**, 295–307 (1975)
16. Mocnik, D., Sirec, K.: The determinants of Internet use controlling for income level: cross-country empirical evidence. Inf. Econ. Policy **22**(3), 243–256 (2010)
17. Nilsson, E., Gärling, T., Marell, A., Nordvall, A.: Who shops groceries where and how? The relationship between choice of store format and type of grocery shopping. Int. Rev. Retail Distrib. Consum. Res. **25**(1), 1–19 (2015)
18. NTIA: Falling through the net: toward digital inclusion. A report on Americans' access to technology tools. National Telecommunication and Information Administration (NTIA), Economic and Statistics Administration, U.S. Department of Commerce. The Secretary of Commerce, Washington, DC (2000)
19. Roos, J.M.: The winner takes IT all: Swedish digital divides in global internet usage. In: Alexandrov, D., Boukhanovsky, A., Chugunov, A., Kabanov, Y., Koltsova, O. (eds.) Digital Transformation and Global Society. DTGS 2018. Communications in Computer and Information Science, vol. 859, pp. 3–18. Springer, Cham (2018)

20. Roos, J.M.: Konsumtionsrapporten: Orosmoln. University of Gothenburg, Centre for Consumer Research, Gothenburg (2019)
21. Roos, J.M.: Personality and e-shopping: insights from a nationally representative study. In: Alexandrov, D., Boukhanovsky, A., Chugunov, A., Kabanov, Y., Koltsova, O., Musabirov, I. (eds.) Digital Transformation and Global Society. DTGS 2019. Communications in Computer and Information Science, vol. 1038, pp. 257–267. Springer, Cham (2019)
22. Roos, J.M., Kazemi, A.: Personality traits and Internet usage across generation cohorts: insights from a nationally representative study. Curr. Psychol. 1–11 (2018). https://doi.org/10.1007/s12144-018-0033-2
23. Solomonson, M.R., Bamossy, G.J., Askegaard, S.T., Hogg, M.K.: Consumer Behaviour: A European Perspective. Pearson, Edinburgh (2013)
24. SOM Institute Homepage: https://www.som.gu.se. Last accessed 27 Jan 2020
25. The Swedish Trade Federation (Svensk Handel): Det stora detaljhandelsskiftet (2018)
26. Thompson, C.J., Locander, W.B., Pollio, H.R.: The lived meaning of free choice: an existential-phenomenological description of everyday consumer experiences of contemporary married women. J. Consum. Res. **17**(3), 346–361 (1990)
27. Vernersdotter, F.: Den nationella SOM-undersökningen. In: Weibull, L., Oscarsson, H., Bergström, A. (eds.) Vägskäl, pp. 651–686. University of Gothenburg, The SOM Institute, Gothenburg (2013)
28. Wolfinbarger, M., Gilly, M.C.: Shopping online for freedom, control and fun. Calif. Manag. Rev. **43**(2), 34–55 (2001)

# Prototyping PLCs and IoT Devices in an HVAC Virtual Testbed to Study Impacts of Cyberattacks

Aaron W. Werth(✉) and Thomas H. Morris

University of Alabama in Huntsville, Huntsville, AL 35899, USA
{aww0001,tommy.morris}@uah.edu

**Abstract.** This work describes a virtual testbed of a heating, ventilation, and air conditioning (HVAC) system that has been developed. The testbed incorporates a programmable logic controller (PLC) and is applicable to Internet of Things (IoT) devices. The PLC's ladder logic program uses hysteresis control and multiple modes of operation. This form of control was selected because of its common use in industry and even in residential applications. The purpose of this work is to demonstrate that by using modern tools and platforms, such as OpenPLC, a controller for an HVAC system can be prototyped and used to research and explore possible cyberattacks and their effects in various cyberphysical systems. In particular, possible and plausible implementations of the aforementioned hysteresis controller in ladder logic are studied and described to the reader. Also, in the experimental results, this work explores how these implementations can be compromised by an injection attack to change settings and by a malicious ladder logic upload in the virtual testbed. The impacts of cyberattacks in terms of safety and cost are discussed also.

**Keywords:** Cybersecurity · Virtual testbed · OpenPLC · HVAC · IoT · PLC

## 1 Introduction

In recent times, cyberphysical systems—a generic term encompassing industrial control systems (ICS) and Internet of Things (IoT) devices—have been compromised by cyberattacks. There have been many examples of cyberattacks targeting ICS and IoT according to the research literature [1] and even the news [2]. Despite not being a case of a cyberattack that directly involved a cyberphysical system, the attack on the main network of Target stores is also noteworthy. Hackers were believed to have compromised the company servicing the HVAC system who had access to the network for billing purposes. Malicious software could then be installed on Point of Sale (PoS) machines involved with debit/credit transactions at target [3]. A cybersecurity expert stated that heating, ventilation, and air conditioning (HVAC) systems are sometimes connected as


X.-S. Yang et al. (eds.), *Proceedings of Fifth International Congress on Information and Communication Technology*, Advances in Intelligent Systems and Computing 1183, https://doi.org/10.1007/978-981-15-5856-6_62

computerized devices to the main network of businesses to allow information concerning the HVAC systems, such as energy usage and cost, to be collected. The expert also speculated that the HVAC company serviced the HVAC system through its network credentials [4]. Although Target was likely not set up this way, this description allows for interesting possible scenarios to study.

The idea and desire behind IoT is to have devices and appliances used in everyday life more connected through computer networks and the Internet to facilitate ease of use. IoT devices have similarities to ICS in that they are embedded devices and may interface with the physical world. With this development, however, comes risks due to vulnerabilities to cyberthreats. Although the above example of an HVAC does not include the case where the adversary is able to hack the HVAC system to affect the physical world, the example does illustrate that there is a concern for such cyberphysical systems if the HVAC is linked to other devices through networks. This current work will explore the following situation: A malicious entity has access to the HVAC device by means of a network and therefore can maliciously alter the HVAC's settings and programming.

The study of cyberphysical systems, including cybersecurity of these systems in particular, is an interdisciplinary pursuit. Concepts of control systems and computer science as well as those of many engineering fields have become intertwined. Malicious changes to software or to settings in programmable logic controllers (PLCs) or other devices and computer systems associated with the PLC have implications in the physical world that fall under the domains of electrical, mechanical, civil, chemical, and control systems engineering as well as others.

A literature search of related works was conducted. To better understand the implications of computerized control systems and their impacts in various physical domains, Hahn et al. describe a useful framework consisting of a model of a cyberphysical system and a kill chain [5]. The model consists of three main layers: (1) cyberlayer, (2) control layer, (3) and the physical layer. When an electrical engineer or specifically a control system engineer designs a control system for a given cyberphysical system, the engineer uses mathematical abstractions or creates a block diagram in special design software. The designs must be directly compiled or written in some common programming language which will then be compiled for the actual hardware of the PLC that will control the physical process. The control layer is thus mapped to the cyberlayer, and it exists in actuality as a compiled program in binary machine code form. It resides in the memory of the PLC's hardware and is processed by the CPU of the hardware. The hardware itself is connected to actuators that influence a physical process that can be described by an appropriate domain expert. Sensors allow the PLC to have information about the physical process. The kill chain described in Hahn's work can be used to understand how violations of one or more of the components of cybersecurity (availability, integrity, and confidentiality) can lead to violations of properties of the physical process, namely stability, safety, and efficiency. Also, the paper by Huang et al. [6] is a seminal work from the past that investigates the physical and economic consequences of cyberattacks against integrity and availability in control systems. Other recent works focus on cybersecurity experiments with HVAC systems. The work [7] discussed an actual testbed of an industrial blower representing an HVAC system. This is in contrast to this work, which uses a virtual testbed as an inexpensive means of prototyping. Also, the work

[8] discusses simulations of smart HVAC devices and how they optimize energy usage based on cost and how they can be compromised by misleading reported prices due to cyberattacks. This current work, on the other hand, focuses on more direct attacks, namely malicious changes in settings and uploaded programs.

The remainder of this paper is organized as follows. Section 2 covers background material on pertinent theory. Section 3 covers modeling the HVAC system and its implementation. Section 4 consists of the experimental results, and Sect. 5 is the conclusion.

## 2 Background

### 2.1 Virtual SCADA Testbeds

To experiment with various scenarios of ICS and IoT, virtual testbeds may be developed and used. This work employs the virtualized framework of a supervisory control and data acquisition (SCADA) system described in the work by Alves et al. [9]. This framework consists of five components with an emphasis on OpenPLC to serve as the platform to emulate PLCs on a virtual network. These five components are the (1) physical system, (2) cyberphysical link, (3) PLC, (4) network, and (5) human-machine interface (HMI) or centralized computer system for remote monitoring and control. For a virtual testbed, the PLC and HMI exist as virtual machines on a virtual network, and the physical system is represented by a simulator, such as MATLAB Simulink. These are illustrated in Fig. 1 along with important flows of information.

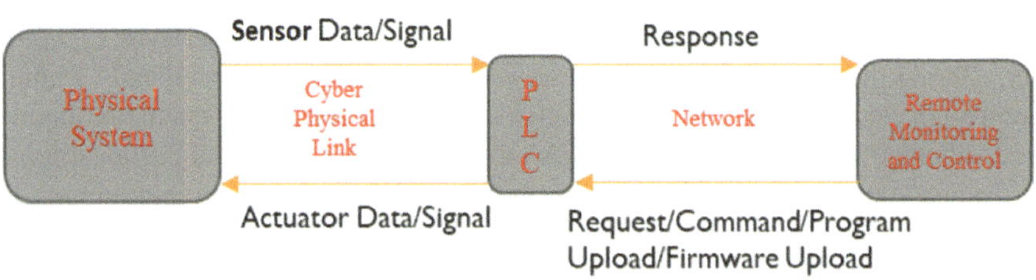


**Fig. 1.** SCADA system decomposed into major components

### 2.2 PLCs and Their Function

One of the main focuses of this work is the PLC (or IoT device) itself, which has the local ladder logic program responsible for controlling the physical system or plant through sensors and actuators. It is important to understand the PLC's role and function in the SCADA system and how the PLC is programmed and designed to control the physical system. A PLC typically has a set of registers and coils that make up its memory. These may be used for input and output and for general information and settings used by the internal program running on the PLC. OpenPLC, the tool used in this work, emulates this setup [10]. As far as the programs are concerned running on the PLCs, PLCs in various engineering domains (i.e., electrical, chemical, robotics, etc.) are typically programmed with a form of feedback control [11]. In this scheme, the sensor information that is indicative of the current state of the physical process under control is relayed to the PLC, so that the PLC can act through its actuators to influence the behavior of the physical system according to a desired objective. The state of the physical system may include such physical quantities as temperature, voltage, and pressure depending on the application. This state is typically referred to as the process variable (PV). Generally, the control algorithm or program in the PLC has a certain goal or set point to achieve that is determined by an engineer or user. The set point (SP) is the desired value for the process variable. The controller must manipulate the actuators, which directly act on the controlled variable (CV) to influence the process variable to reach the set point or in some cases to be maintained between two set points. PLCs are connected to actuators and sensors through wires or simple network protocols. The programs designed for PLCs are typically made in ladder logic or at least one of the five languages for PLCs according the IEC 61131-3 standard, which OpenPLC is compliant with [10]. A diagram that illustrates these feedback control systems can be seen in Fig. 2.

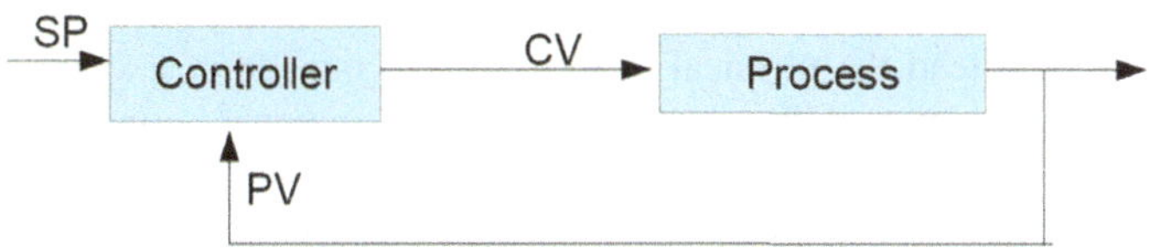


**Fig. 2.** Feedback control system

### 2.3 The PLC/Ladder Logic Program

The hysteresis controller, also known as an on-off controller, is a common type of controller implemented as a program for the PLC. This controller uses two main discrete states for the actuator, rather than a range of values that the actuator may take on. The actuator is said to be either "on" or "off" in other words. A well-known example is a thermostat, which can either turn on or off the heat according to its programming. With a thermostat, as well as other on-off controllers, there is typically a main set point determined by a user and two set points—high and low set points—that are at offsets slightly above and below the main set point. With the two states of the actuator, this type of controller can cause the process variable to either increase or decrease to keep the process between the two set points. If the PV reaches or goes below the low set point ($SP_{Low}$). Then, the controller must choose the actuator state that causes the PV to increase ($CV_{inc}$). On the other hand, if the PV reaches the high set point ($SP_{High}$) or is above it, the control must choose the actuator state that causes it to decrease ($CV_{dec}$). The following piecewise Formula (1) defines the behavior that the controller must have for its actuator, which affects the control variable:

$$CV(t) = \begin{cases} CV_{inc}, & PV \leq SP_{Low} \\ CV_{dec}, & PV \geq SP_{High} \\ \text{unchanged}, & SP_{Low} < PV < SP_{High} \end{cases} \quad (1)$$

**Modes of Operation in PLCs** The type of controller described above may be considered automatic since the controller can regulate a physical process without direct intervention except for changes in settings done at a supervisory level. Examples of settings are the high and low set points. The ladder logic implemented in a PLC may consist of multiple modes that can be selected by the user. One mode is manual where an operator who is communicating with the PLC via an HMI over a network has more direct control of the actuators. The automatic modes could be, for example, hysteresis control as in the case of this work. The current mode may be stored as an integer in a register of the PLC. The value "1" for example may represent manual mode and a "2" may represent automatic mode, etc.

**Safety and Cost Concerns** For a given plant, there are general ranges of values for the process variable that are acceptable from a safety consideration. In the context of an HVAC system, this may be considered more of a matter of comfort than avoiding danger. However, there is the case when an HVAC system is used to regulate the temperature of a server room or supercomputer, where it is paramount to keep the temperature within a certain range. Otherwise, equipment may be damaged. There are some ways to choose parameters in the hysteresis controller that would be unacceptable since such values of parameters could lead the physical process to veer outside of a safety region. For example, a high set point could be chosen that is above an upper bound of a safe region of behavior for the process variable. If the PLC were to drive the PV outside of these safety conditions based on its program or settings, this would be unacceptable. Also, there may be some control schemes that are more optimal in terms of cost. The reason that there is a cost associated with the control system is that actuators generally have a cost associated with them when they are active or switched on. In that case, the actuators are consuming fuel or electricity, which cost money.

## 2.4 Cyberattacks

If the adversary desires to take advantage of the controller by changing settings to make the controller unstable, the attacker will have to have gained access to the network and proceed to inject command packets into the PLC to change the values of various registers. These registers may represent the various parameters of the on/off or proportional/integral/derivative (PID) controllers. However, it is possible that the designer of the software does not have those parameters as register values but simply has them stored as constants in the program. To change these constants, the hacker would need the ability to change the memory that the code resides on or to modify the code before it is uploaded to the PLC or to spoof the computer running the engineering software and send the code to the PLC [12]. Other methods include performing Man-in-the-Middle (MitM) attacks to change parts of the code when it is in transit to the PLC. Many of these types of attacks are described in [13], where experiments were conducted using an Alan Bradley or a Siemens PLC. What differentiates this current work from that one is that this work uses a virtual testbed environment and OpenPLC to emulate the PLC. Also those previous works do not examine the many ways the integrity attack can influence the PLC that this work does in terms of impact to the physical system. The major attacks selected in this work are based on the threat model described in [14] where the attacker seeks to cause harm by choosing values for parameters used by the ladder logic program to lead to instability or a violation of a safety condition. These parameters may exist as registers in the PLC or as constant values embedded in the ladder logic. Works [15, 16] also include similar attacks. Note that other relevant attacks for this work may not just simply change parameters, but also the control logic.

## 3 Modeling an HVAC System

The focus of this work is an HVAC system because of the importance of HVAC systems to society. Many businesses and residential areas use HVAC systems to keep the temperature of the building at a level reasonable to occupants. A "smart home" would most likely have the HVAC controller networked with a centralized computer system or server. The HVAC systems are generally controlled using an on-off controller to maintain the temperature near a set point chosen by the user. From this set point, a high set point and a low set point are determined, where the high set point is the user-defined set point added to some constant value. Similarly, the low set point is a constant value subtracted from a user-defined set point. The physical system under control of the HVAC is governed by Newton's laws of heating and cooling. In this work, a model of the physical system is developed in MATLAB based on a paper by Balan et al. [17] and an example in MATLAB Simulink [18], which rely on these laws. The model of this work uses heating and cooling systems to influence the temperature inside of the building. When hot air from the heating system flows into the surrounding air, the rate of heat transferred is proportional to the difference between the temperature of the hot air and the temperature of the room or the surrounding air. The document [18] describes the equations that govern this behavior, which are implemented in a Simulink model. They are also implemented in the Simulink model of this work (Fig. 3) in the red block (The heating system) in Fig. 3, which is similar to the original Simulink model except for the color. In this current work, an additional block was added nearly equivalent to the heating system. This block shown in blue is the cooling system. There is only one main difference between the two blocks. The heating system involves blowing air at a temperature of 50 °C. However, for the cooler, the value is changed to 0 °C to simulate cooled air flowing from the fan. Also, the Simulink model used in this work, which is based on the example provided in MATLAB takes into account heat loss to the external environment of the building. A key difference between the model used in this work and the example is that the original block for hysteresis control was replaced with a block that allows for UDP communication with an OpenPLC virtual machine as described in [9]. Two digital actuators are used for turning on the heater and the cooler. These directly manipulate the controlled variables. One analog sensor is used for measuring the temperature, which is the process variable.

For this work, instead of the attack originating from the HVAC system to affect the payment system, where both systems are on the same network, the focus of this work is to affect the physical world. Potential attacks against the HVAC would be those that change settings in adverse ways such as merely changing the set point to a high value or low value, where the attacker has the goal of either causing discomfort or to increase the energy bill. Another potential attack would be for the attacker to reprogram the device

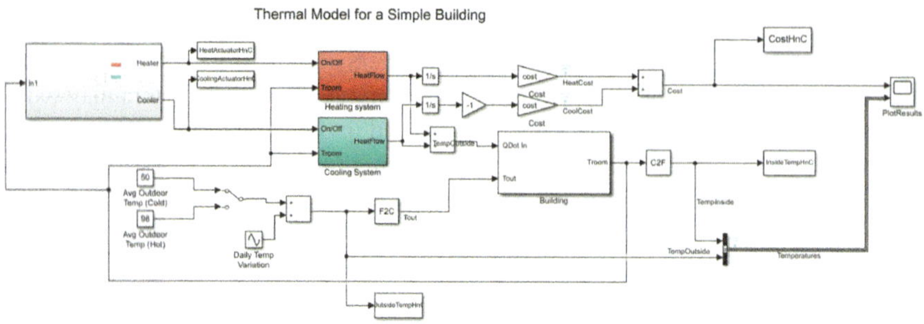


**Fig. 3.** MATLAB Simulink model of thermal system

with new ladder logic. The following is the MATLAB diagram (Fig. 3) for this type of physical system. For this work, a virtual testbed is used, where the PLC is emulated with OpenPLC running the ladder logic, and the physical system is implemented and simulated in MATLAB.

In this work, the PLCOpen editor [19] is used to create the ladder logic. An explanation of the ladder logic of the HVAC system is shown in Fig. 4: To aid in this discussion of the ladder logic for this testbed, the ladder logic is divided into several sections as indicated by the comments within the ladder logic. These sections are (1) Off Mode, (2) Manual Mode, (3) Auto Mode—Heat, (4) Auto Mode—Cool, (5) Output.

For the first section, which corresponds to "off mode," when the register mode is set to 0, the output of the "EQ" block in the rung of that section is set to one or "High." The "EQ" works by comparing both of its inputs to determine if they are equal. If they are, the output is "High." Otherwise, the output of the "EQ" block is set to "Low." A "high" value for output of the "EQ" block causes the "heat" and "cool" coils to be reset meaning both coils are set low. These specific coils are associated with the states of the actuators for the heating and cooling systems.

For the second section (Manual Mode), there are two rungs. For the first rung, when the heat register is set to 1, the output of the "EQ" block in the rungs becomes "High." Thus, the coil ("Heat_man") is set high. Likewise, when the cool register is set to 1, the coil ("Cool_man") is set high. These coils serve as flags for the output section of ladder logic. The two registers (heat and cool registers) are considered settings that may be set by user at an HMI that can send command packets to the PLC to set these registers.

For the third section (Auto Mode—Heat), two rungs are used to implement hysteresis control. If the temperature as indicated by the temp register is equal to or greater than the high set point, then the "heat_auto" coil is reset to "Low" or false. The "GE" block is used to make this comparison. If, on the other hand, the temp register is equal to or less than the low set point, then the "head_auto" coil is set to "High" or true. The "LE" block is used to achieve this.

For the fourth section (Auto Mode—Cool), two rungs are used to implement hysteresis control like the previous section, but the set and reset operation is performed in reverse for the "cool_auto" coil. Using this mode assumes that the outside temperature is higher than the inside temperature, and there is a tendency for the inside to become

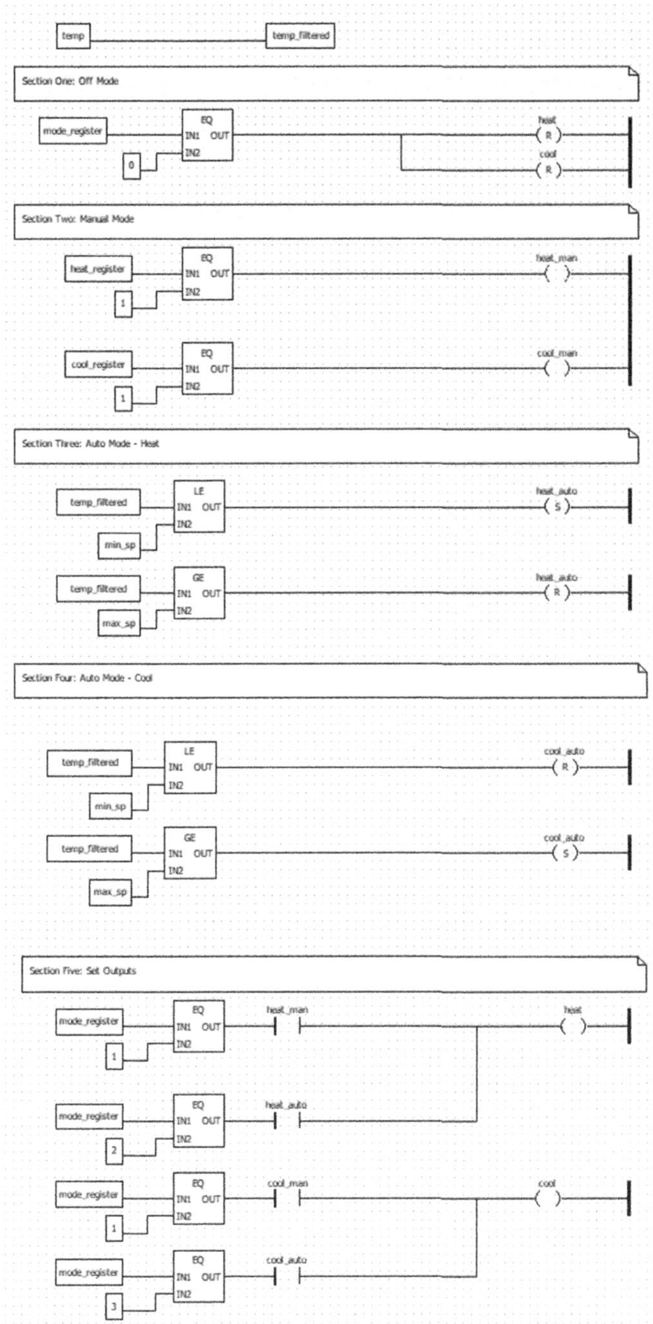


**Fig. 4.** Ladder logic for hysteresis control of HVAC with multiple modes of operation

warmer until an equilibrium is reached. Therefore, the PLC must maintain the desired temperature by cooling the building.

The remaining fifth section of the ladder logic is the portion that deals with the output. The logic of this section is such that the following operations are performed: If the mode register is equal to 2 ("Auto Mode—Heat") and the "heat_auto" flag which is determined by section 3 is set "High," then "heat" coil is set "High." If the mode register is equal to 3 ("Auto Mode—Cool") and the "cool_auto" flag which is determined by section 3 is set "high," then "heat" coil is set "High." Also when the mode register is set to 1 for manual mode, the flags "heat_man" and "cool_man" determine the states of the "heat" and "cool" coils, which direct the actuator states.

# 4 Experimental Results

## 4.1 Setup

Several experiments in this work were performed for the HVAC system to illustrate various scenarios where a cyberattack can induce harm physically. The experiments were performed on a virtual testbed that included the Simulink model of the HVAC system as well as virtual machines for the PLC and a malicious rogue device on the network. The attacks were designed to be launched from the rogue device, which runs the Linux operating system and has Python-based tools to launch the cyberattacks. One of the tools uses the pymodbus library to be able to inject Modbus write command packets to registers in a given PLC target. The scenarios were (1) normal conditions, (2) malicious ladder logic upload, and (3) injection attack to change settings.

## 4.2 Discussion

In normal conditions, the HVAC system is running in automatic heat mode, which is a hysteresis mode with a user-defined setting of 22 °C. The program can use this value to compute high and low set points at offsets above and below the user setting. The goal of the ladder logic program in that mode is to maintain the temperature between those two set points. An illustration of this is seen in Fig. 5 on the left. What can be observed is that the temperature rises when the heater is turned on as expected until it is slightly above the user-defined set point. At which point the heater is turned off allowing the room temperature to decrease until it reaches a point slightly lower than the user-defined set point. The decrease in temperature is expected because of the heat transfer from the inside of the building to the outside, which is colder. This pattern repeats itself. In the case of a malicious upload, a program is sent with some changes to the rungs in section 3 of the ladder logic that causes the heat to always be turned on. At the same time, the cooling system is activated to regulate the temperature. If the temperature is greater than or equal to the high set point, the cooling system is turned on to bring the temperature down. If the temperature falls to a level equal to or less than the low set point, the cooling system turns off to allow the heat of the other system to bring the temperature up. This can be seen in Fig. 5 (center). In this case, the goal of the attack is to regulate the temperature and deceptively fool the user that everything is normal since the HVAC system maintains the desired temperature. However, the cost of running HVAC becomes much greater because the heating and cooling systems are used to a greater extent. For a scenario involving a cyberattack, a change in a register or multiple registers can influence the behavior of the system adversely. In this case, a rouge device on the network can send a command to the PLC to switch to manual mode and then to turn on the heat, which can lead to unfavorable conditions (Fig. 5 on right), where the temperature becomes high. Operating costs associated with the three scenarios over time can be seen in Fig. 6.

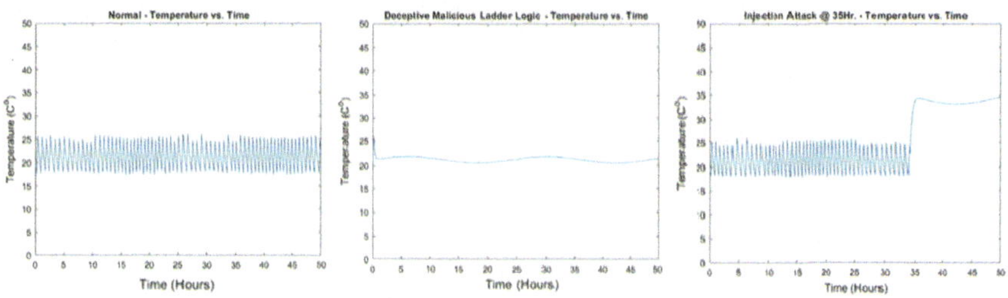


**Fig. 5.** Scenarios

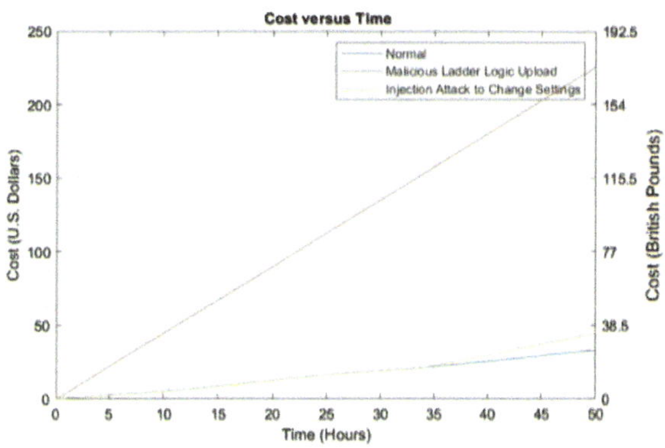


**Fig. 6.** Operating cost of running HVAC over time

## 5 Conclusion

This work demonstrates how OpenPLC and an overall virtual testbed may be used as a prototyping tool for a controller and its ladder logic. In particular, this work showed that a testbed can be constructed for hysteresis control used for an HVAC system with multiple operating modes. This virtual testbed can allow for a researcher to gain insight into the impacts of cyberattacks that include harm in terms of safety or cost as in the case of this work. Future work will be to investigate more advanced and extensive attacks that can include MitM attacks to deceive HMIs which were used in [9, 20].

For example, an attack can perform an MitM attack, so that the packets that contain information regarding the current temperature or energy usage as detected by the PLC are modified in transit before traveling to the HMI. This is done to deceive the HMI concerning the true state of process variables. Possible mitigations or security controls to prevent or deter cyberattacks have also been investigated in the literature that include encryption, authentication, and intrusion detection systems [21]. These may be applied to the testbed to study the effectiveness of these mitigations. Other control systems, such as PID controllers, may be studied to have a greater understanding of how a cyberattack can manipulate gains, which may be stored in registers in the ladder logic or as constants embedded in the code of the PLC. Some of the methods of attack described in this work may also be applied to the PLC with PID control to study impacts to the cyberphysical system.

## References


1. Miller, B., Rowe, D.: A survey SCADA of and critical infrastructure incidents. In: Proceedings of the 1st Annual Conference on Research in Information Technology, pp. 51–56 (2012)
2. The biggest Internet of Things, smart home hacks of 2019. https://www.zdnet.com/pictures/the-biggest-internet-of-things-smart-home-hacks-over-2019/11/ (2019)
3. Target Data Breach: A "kill chain" analysis of the 2013 target data breach (2014)
4. Krebs, B.: Target Hackers Broke in via HVAC Company. https://krebsonsecurity.com/2014/02/target-hackers-broke-in-via-hvac-company/ (2014)
5. Hahn, A., Thomas, R.K., Lozano, I., Cardenas, A.: A multi-layered and kill-chain based security analysis framework for cyber-physical systems. Int. J. Crit. Infrastruct. Prot. **11**, 39–50 (2015)
6. Huang, Y.-L., Cárdenas, A.A., Amin, S., Lin, Z.-S., Tsai, H.-Y., Sastry, S.: Understanding the physical and economic consequences of attacks on control systems. Int. J. Crit. Infrastruct. Prot. **2**(3), 73–83 (2009)
7. Morris, T., Vaughn, R., Dandass, Y.S.: A testbed for SCADA control system cybersecurity research and pedagogy. In: Proceedings of the Seventh Annual Workshop on Cyber Security and Information Intelligence Research, p. 1 (2011)
8. Zhang, Y., Eisele, S., Dubey, A., Laszka, A., Srivastava, A.K.: Cyber-physical simulation platform for security assessment of transactive energy systems. arXiv preprint arXiv:1903.01520 (2019)
9. Alves, T., Das, R., Werth, A., Morris, T.: Virtualization of SCADA testbeds for cybersecurity research: a modular approach. Comput. Secur. **77**, 531–546 (2018)
10. Alves, T., Morris, T.: OpenPLC: an IEC 61131-3 compliant open source industrial controller for cyber security research. Comput. Secur. **78**, 364–379 (2018)
11. Erickson, K.T.: Programmable Logic Controllers: An Emphasis on Design and Application. Dogwood Valley Press (2005)
12. Govil, N., Agrawal, A., Tippenhauer, N.O.: On ladder logic bombs in industrial control systems. In: Computer Security, pp. 110–126. Springer, Cham (2017)
13. Yoo, H., Ahmed, I.: Control logic injection attacks on industrial control systems. In: IFIP International Conference on ICT Systems Security and Privacy Protection, pp. 33–48. Springer, Cham (2019)
14. Werth, A., Morris, T.H.: A specification-based intrusion prevention system for malicious payloads. In: National Cyber Summit, pp. 153–168. Springer, Cham (2019)
15. Gao, W., Morris, T.H.: On cyber attacks and signature based intrusion detection for modbus based industrial control systems. J. Digit. Forensics Secur. Law **9**(1), 3 (2014)

16. Morris, T.H., Gao, W.: Industrial control system cyber attacks. In: Proceedings of the 1st International Symposium on ICS & SCADA Cyber Security Research, pp. 22–29 (2013)
17. Balan, R., Donca, R., Balan, A., Pleşa, A., Pacurar, L., Muresan, V.: Thermal modelling and temperature control of a house. Roman. Rev. Precis. Mech. Opt. Mechatron. **39**, 59–62 (2011)
18. Thermal Modal of a House: https://www.mathworks.com/help/simulink/slref/thermal-model-of-a-house.html (2012)
19. Plcopen-editor: https://www.openplcproject.com/plcopen-editor (2019)
20. Alves, T., Das, R., Morris, T.: Virtualization of industrial control system testbeds for cybersecurity. In: Proceedings of the 2nd Annual Industrial Control System Security Workshop, pp. 10–14. ACM (2016)
21. Giraldo, J., Sarkar, E., Cardenas, A.A., Maniatakos, M., Kantarcioglu, M.: Security and privacy in cyber-physical systems: a survey of surveys. IEEE Des. Test **34**(4), 7–17 (2017)

# Making High Density Interval from Fuzzy $\alpha$-Cut

## A Deterministic and an Interval-Based Possibility-to-Probability Transformation

Tanarat Rattanadamrongaksorn(✉)

International College of Digital Innovation, Chiang Mai University, Chiang Mai, Thailand
tanarat.r@cmu.ac.th, http://icdi.cmu.ac.th

**Abstract.** A probability distribution is constructed using a deterministic method by gradually slicing a fuzzy number and incrementally transforming a set of $\alpha$-cuts into high density intervals. The test of the proposed method was conducted by artificial examples to make a comparison arbitrarily with other commonly used methods and measured by common statistics. The result showed a good, however not the best, possibility-to-probability transformation.

**Keywords:** Artificial intelligence · Expert system · Knowledge representation · Confidence interval · Credible interval · Aleatory · Epistemic · Uncertainty · Prior probability · High density interval

## 1 Introduction

Representing knowledge from an expert is frequently, conveniently, and efficiently achieved by a fuzzy set. A lack of statistical data leads to a reliance on subjective evaluation in representing unknown quantity in a problem. This type of data is mainly in a form of expert opinion that could be decoded from a linguistic term into mathematical quantity by fuzzy set theory. Essentially, the use of fuzzy number, $A$, is to map each variable, $\theta$, in a universe of discourse, $\Omega$, to the corresponding possibility, $\pi(\theta)$, due to the uncertainty inherited from expert's incomplete knowledge, $A = \{\theta \in \Omega | \pi(\theta) \in [0, 1]\}$, where $\pi = 0$ indicates the impossible and $\pi = 1$ indicates the most possible. The common shapes of fuzzy number are the triangle and trapezoid and can be written in the general form of fuzzy number, $A = (a, b, c, d)$, as, for instance, $A = (0, 5, 10)$ and $A = (0, 3, 5, 10)$, respectively.

However, there exist some inconveniences in processing the possibility data, especially on the application side of tools for fuzzy data. Thus, the transformation from possibility to probability emerges to bridge such


X.-S. Yang et al. (eds.), *Proceedings of Fifth International Congress on Information and Communication Technology*, Advances in Intelligent Systems and Computing 1183, https://doi.org/10.1007/978-981-15-5856-6_63

deficiencies. The obvious benefits are the more standardized and the more widely accepted in applying probability over fuzzy set. One of our applied areas is data fusion by Bayes theorem. In this case, instead of assuming a prior probability in a traditional way, we present the informative prior translated from linguistic terms.

The following subsections contain discussions on possibility-to-probability transformations (Sect. 3), examples (Sect. 4), and conclusion (Sect. 5).

## 2 Possibility-to-Probability Transformation

In this research, we considered only one-way transformation from possibility to probability due to our specific requirement. Three methods of the transformation are included as companions. Let $P$ be a probability and $\pi$ be a possibility of parameter, $\theta$. First, the rule of consistency implying the more possibility, the more probability [5] suggests a heuristic with the transformation by

$$P_i(\theta) = \frac{\pi_i(\theta)}{\sum_{j=1}^{N} \pi_j(\theta)}; \quad i, j = 1, \ldots, N; \tag{1}$$

where $N$ is number of parameters with indicators, $i$ and $j$. Another widely adopted method [1], that is also used as a major baseline in this work, relies on a stochastic experiment that is governed by three rules of (1) probability–possibility consistency that increases or, at least, keeps information after transformation or $P(\theta) \leq \pi(\theta)$; (2) preference preservation that maintains the order between possibility and probability, i.e., $\pi(\theta) > \pi(\theta') \Leftrightarrow P(\theta) > P(\theta')$; and (3) least commitment stating that the transformation is as uncertain as possible, $P(.) \sim Unif(0,1)$ that can be seen as [1]:

$$P_i(\theta) = \sum_{k=i}^{T-1} \left( \frac{\pi_k(\theta) - \pi_{k+1}(\theta)}{k} \right); \quad i = 1, \ldots, T-1; \tag{2}$$

where $k$ and $T$ are the transforming order and number of loops required for the simulation, respectively. May the abovementioned transformations be referred later as Zadeh's [1], Dubois' [2], and Klir's [3], respectively. On the other hand, an interesting method that maintains the level of information before and after transformation is written by the equation [3]

$$P_i(\theta) = \frac{\pi_i^{1/\delta}(\theta)}{\sum_{j=1}^{N} \pi_{j+1}^{1/\delta}(\theta)}; \quad i, j = 1, \ldots, N; \tag{3}$$

where $\delta$ is between the range of (0,1) and used in controlling the degree of change probabilities or shape of distribution.

It must be noted that the transformation might be misleading. The quantities of the two are known standing on different concepts. While the possibility reflects incomplete knowledge about a problem, the probability represents the outcome of stochastic experiment, i.e., epistemic and aleatory uncertainties, respectively. As a result, the domains of application are restricted in reality and must be utilized carefully.

## 3 Proposed Transformation

Our fundamental concept has been evolved from the idea of interpreting confidence interval as corresponding alpha-cut [2], however implementing such an idea is fairly difficult in terms of computation complexity. Several interval-based statistics were experimented. The best alternative appears to be the simplest in computation, formula, and interpretation.

Let $P(\theta)$ be a probability function of a random variable, $\theta$. High density interval (HDI) is a set of variables whose probabilities are greater than or equal to a threshold probability, ($\gamma \in [0, 1]$), as indicated by the following expression:

$$^{\gamma}H(P_{\gamma}) = \{\theta \in \Theta | P(\theta) \geq P_{\gamma}\}, \tag{4}$$

where $P_{\gamma}$ is the largest probability such that

$$P(^{\gamma}H(P_{\gamma})) \geq 1 - \gamma.$$

On the other hand, alpha-cut ($\alpha$-cut) is a set of all variables whose possibilities are greater than or equal to the specified constant ($\alpha \in [0, 1]$) as defined by the following equation:

$$^{\alpha}A(\alpha) = \{\theta \in \Theta | \pi(\theta) \geq \alpha\}, \tag{5}$$

where $\alpha$ is a constant of possibility level.

Based on the pattern observed from Eqs. 4 and 5, the resemblance between the two concepts is noticed and leads to our proposal that *"A corresponding probability of high density interval would be equal to and could be derived from a particular ($\alpha$-cut) level of possibility"*, given the expert's judgment is correct. The proposed transformation is based on such a notion and expressed mathematically as follows:

$$^{1-\alpha}H(P_{1-\alpha}) = {}^{\alpha}A(\alpha), \tag{6}$$

where $\gamma = 1 - \alpha$.

For a convincing example, the estimation of a financial asset of 100–120 dollars with 0.95 or high possibility ($^{.95}A$) should consequently result, for a number of time, in the same price bracket with top-5-percent probability ($^{.05}H$). Thus, the 5-percent HDI could be built by the values of 100–120 dollars at 0.95 possibility level. We, therefore, propose the construction of probability distribution by gradually slicing fuzzy number and incrementally transforming $\alpha$-cut into HDI as shown in the following steps:

(a) Divide a unit possibility into a set of alphas ($\alpha_i \in [0, 1]; i = 0, 1, 2, \ldots, N$. The number of increments, $N$, is sufficiently large in order to accommodate the linear operation. $N = 100$ produces the appropriate result as graphically shown in the examples (Sect. 4));
(b) Initialize $i = 0$ at $\alpha = 1$, calculate the $\alpha$-cut ($^{\alpha}A_i$);
(c) Interpret, via Eq. 6, the range of variables obtained in the previous step as an HDI for the new probability distribution ($P_i(\theta)$);
(d) Move to the next $i$ and repeat step (b) to obtain the $\alpha$-cut ($^{\alpha}A_i$);

(e) Compute the relative position of probability (the initial probability recommended is $p_0 = 1$) of the incremental HDI by a simple geometric calculation:

$$\Delta P_i = \frac{P\left({}^{1-\alpha}H_i(P_{1-\alpha}))\right) - P\left({}^{1-\alpha}H_{i-1}(P_{1-\alpha}))\right)}{.5\left({}^{1-\alpha}H_i^{\max} - {}^{1-\alpha}H_i^{\min} - {}^{1-\alpha}H_{i-1}^{\max} + {}^{1-\alpha}H_{i-1}^{\min}\right)}, \tag{7}$$

where $\Delta.p_i$ is the increment in probability of HDI from $i-1$ to $i$. The ${}^{1-\alpha}H_i^{\max}$ is the short form of the maximum HDI at a particular probability, $max({}^{(1-\alpha)}H_i(P_{1-\alpha}))$, and so on;

(f) Calculate

$$P_i = P_{i-1} + \Delta P_i, \tag{8}$$

which is a cumulative HDI at order $i$;

(g) Repeat the above steps from (d) to (e) for every $i$ until $\alpha = 0$; and

(h) Reset the probability axis to the origin by

$$P_i^*(\theta) = P_i(\theta) - \min(P_i(\theta)){:}\forall i, \tag{9}$$

where $P_i^*(\theta)$ is the transformed probability, thus the new or transformed probability distribution is obtained.

Although this work is developed under the framework of Bayesian statistics, the same technique can be applied to the frequency-based statistics as well.

## 4 Examples

We continued our ongoing investigation in the domain of prior probability distribution for the applications in Bayesian statistics [4] by transforming fuzzy number into prior probability distribution. In this research, we experimented the transformations with two commonly used shapes of fuzzy number, i.e., triangle and trapezoid. Fuzzy data demonstrated in the computation were artificial, however, were sufficient in representing most common cases in actual priors such as expert estimation of financial data of stock, real estate, and trading. The data and results are shown in Table 1.

Moreover, in order to compare the results of different methods, some operations were set in place for a purpose of comparison by

- Instead of separately sampling from the each support of transformation, we selected a fixed set of samples for every experiment.
- The transformed probabilities were normalized for a better visualization and had no effect in our usage of prior in the Bayesian framework.
- Currently, to our best knowledge, there is no standard technique of the transformation; therefore, the examination was not on any baseline but, rather, to find out the characteristics of each method.

Each sample was run on 10, 100, and 1000 transformations on each test. The examples were observed in three criteria, i.e., shapes of the transformed probability distribution, point estimations (mean/median/mode), and variation or interval estimation (variance).

**Table 1.** Data and statistical results of the transformed probabilities

| No | Shape | Fuzzy data | Method | Size | Mean | Variance | Median | Mode |
|---|---|---|---|---|---|---|---|---|
| i(L) | Triangular | (0,5,10) | Zadeh | 10 | 5.00 | 4.12 | 4.42 | 4.87 |
| | | | Dubois | 10 | 4.53 | 2.89 | 3.33 | 4.87 |
| | | | Klir | 10 | 5.00 | 2.55 | 4.42 | 4.87 |
| | | | Proposed | 10 | 5.00 | 2.79 | 4.42 | 4.87 |
| i(R) | Trapezoidal | (0,3,5,10) | Zadeh | 10 | 4.60 | 4.30 | 2.65 | 2.92 |
| | | | Dubois | 10 | 4.05 | 2.48 | 1.34 | 2.92 |
| | | | Klir | 10 | 4.42 | 2.78 | 2.65 | 2.92 |
| | | | Proposed | 10 | 4.47 | 3.20 | 2.65 | 2.92 |
| ii(L) | Triangle | (0,5,10) | Zadeh | 100 | 5.00 | 4.17 | 5.00 | 5.00 |
| | | | Dubois | 100 | 4.99 | 2.68 | 5.00 | 5.00 |
| | | | Klir | 100 | 5.00 | 2.50 | 5.00 | 5.00 |
| | | | Proposed | 100 | 5.00 | 2.78 | 5.00 | 5.00 |
| ii(R) | Trapezoidal | (0,3,5,10) | Zadeh | 100 | 4.59 | 4.30 | 3.00 | 3.30 |
| | | | Dubois | 100 | 4.43 | 0.36 | 2.77 | 3.23 |
| | | | Klir | 100 | 4.39 | 2.66 | 3.00 | 3.30 |
| | | | Proposed | 100 | 4.46 | 3.28 | 3.00 | 3.30 |
| iii(L) | Triangle | (0,5,10) | Zadeh | 1000 | 5.00 | 4.17 | 4.99 | 5.00 |
| | | | Dubois | 1000 | 4.97 | 2.80 | 4.99 | 5.00 |
| | | | Klir | 1000 | 5.00 | 2.50 | 4.99 | 5.00 |
| | | | Proposed | 1000 | 5.00 | 2.78 | 4.99 | 5.00 |
| iii(R) | Trapezoidal | (0,3,5,10) | Zadeh | 1000 | 4.59 | 4.30 | 2.99 | 3.29 |
| | | | Dubois | 1000 | 4.08 | 0.42 | 2.77 | 3.29 |
| | | | Klir | 1000 | 4.40 | 2.66 | 2.99 | 3.29 |
| | | | Proposed | 1000 | 4.46 | 3.28 | 2.99 | 3.29 |

*Remark* The (L) and (R) indicate Left and Right columns shown in Fig. 1

First, the visual inspection primarily captured the patterns of changes in the transformations of triangles and trapezoids resulting in the transformed probability distributions as demonstrated in Fig. 1 in the left and right columns, respectively. There was no much to say about the results from the Zadeh's approach (dotted-dash) since the transformation of then fuzzy numbers gave the exact shapes of probability distributions. By such an indifference, this method would not be included in our detailed consideration. The Dubois' (dashed) gave coarse edges of distributions in smaller number of transformations but became smoother once the number grew. In the case of Klir's (dotted) and proposed transformation (solid), both yielded robust outcome that was not varied on different number of transformation across the experiments. It may be able to claim a consistent property from the proposed methodology for the fact of deterministic type of computation.

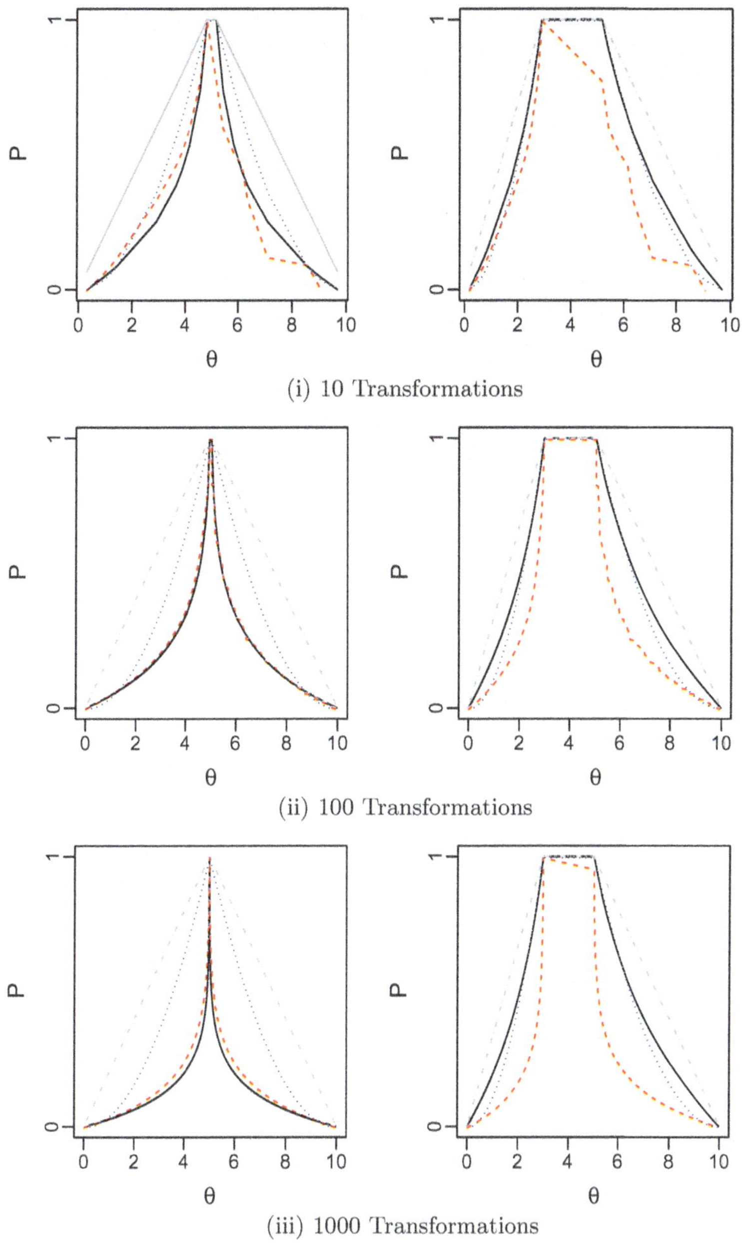


**Fig. 1.** Comparison between the results of transformed probabilities from triangular (left) and trapezoidal (right) fuzzy numbers by the Zadeh's (dotted-dash/gray), Dubois's (dash/red), Klir's (dotted/blue), and proposed (solid/black) methods on different number of transformations

Second, most of the transformations had an agreement on the mean (e.g., 5.00 on i(L)). However, the method of Dubois' resulted in a different output (4.53 on i(L)). This would be more different in the actual work due to more randomness added from the stochastic steps. In contrast, there was not much difference in the other cases of the higher number of transformation (4.99 and 4.97 in ii(L) and iii(L), respectively, versus 5.00 by other methods).

Third, the comparison was also made numerically in Table 1. However, some biases may be noticed in the medians and modes due to the values sampled inclined by the intended asymmetrical shapes of the samples. Although we see some disadvantages of the Dubois' earlier, the consideration on the variation was advantageous. The variations were quite different as can be seen in the case of trapezoids (as lowest in a group as 0.42 in iii(R) and 0.36 in ii(R)). Even though the transformation of the Dubois' required more work and, therefore, was slower than other methodologies, the variances were quite lower comparing to other methods. Additionally, the HDI transformation performs equally well in the triangular samples.

## 5 Conclusion

The deterministic and interval-based type of possibility-to-probability transformation was proposed as an alternative prior in another ongoing research. Comparing to the three commonly used methods on the criteria of processing time, robustness, and consistency, it was found that there was still no best transformation available for every case. The interval-based transformation showed, however, a good solution in most cases. In other words, through the experiments in different iterations, the proposed method kept the satisfied, not the best, performance. However, the accuracy between the new and conventional method has not been confirmed due to various standards of transformation at the moment which, therefore, needs more elaborated examinations in future work. At the current stage of implementation, however, the transformation is sufficient for our current research interest.

## References

1. Dubois, D., Prade, H., Sandri, S.: On possibility/probability transformations. In: Lowen, R., Roubens, M. (eds.) Fuzzy Logic: State of the Art, pp. 103–112. Springer, Dordrecht (1993)
2. Goodman, I.R., Mahler, R.P., Nguyen, H.T.: Mathematics of Data Fusion, vol. 37. Springer Science & Business Media, Berlin (2013)
3. Klir, G.J., Geer, J.F.: Information-preserving probability-possibility transformations. In: Fuzzy Logic, pp. 417–428. Springer, Berlin (1993)
4. Rattanadamrongaksorn, T., Sirikanchanarak, D., Sirisrisakulchai, J., Sriboonchitta, S.: Adjusting beliefs via transformed fuzzy priors. J. Phys. ConferenceSeries **976**(1), 012005 (2018)
5. Zadeh, L.A.: Fuzzy sets as a basis for a theory of possibility. Fuzzy Sets Syst. **1**(1), 3–28 (1978)

# Identification of Construction Era for Indian Subcontinent Ancient and Heritage Buildings by Using Deep Learning

Md. Samaun Hasan[1,2(✉)], S. Rayhan Kabir[3,4], Md. Akhtaruzzaman[3], Muhammad Jafar Sadeq[3], Mirza Mohtashim Alam[1], Shaikh Muhammad Allayear[1,4], Md. Salah Uddin[1], Mizanur Rahman[1], Rokeya Forhat[4], Rafita Haque[3], Hosne Ara Arju[5], and Mohammad Ali[6]

[1] Department of MCT, Daffodil International University, Dhaka, Bangladesh
{hasan.mct,mirza.mct,drallayear.swe,salah.mct, mizan.mc}@diu.edu.bd

[2] Department of Archaeology, Jahangirnagar University, Savar, Dhaka, Bangladesh

[3] Department of Computer Science and Engineering, Asian University of Bangladesh, Dhaka, Bangladesh
{rayhan923,azaman01,jafar,rafitahaque}@aub.edu.bd, rayhan561@diu.edu.bd

[4] Department of Software Engineering, Daffodil International University, Dhaka, Bangladesh
rokeya35-1028@diu.edu.bd

[5] Department of Sanskrit, Rajshahi University, Rajshahi, Bangladesh
arju.ru.1@gmail.com

[6] Department of Graphic Design, Crafts and History of Art, Rajshahi University, Rajshahi, Bangladesh
alifineartsru@gmail.com

**Abstract.** The Indian subcontinent is a south geographic part of Asia continent which consists of India, Bangladesh, Pakistan, Sri Lanka, Bhutan, Nepal, and Maldives. Different rulers or the empire of different periods have built various buildings and structures in these territories like Taj Mahal (Mughal Period), Sixty Dome Mosque (Sultanate Period), etc. From archaeological perspectives, a computational approach is very essential for identifying the construction period of the old or ancient buildings. This paper represents the construction era or period identification approach for Indian subcontinent old heritage buildings by using deep learning. In this study, it has been focused on the constructional features of British (1858–1947), Sultanate (1206–1526), and Mughal (1526–1540, 1555–1857) periods' old buildings. Four different feature detection methods (Canny Edge Detector, Hough Line Transform, Find Contours, and Harris Corner Detector) have been used for classifying three types of architectural features of old buildings, such as Minaret, Dome and Front. The different periods' old buildings contain different characteristics of the above-mentioned three architectural features. Finally, a custom Deep Neural Network (DNN) has been developed to apply in Convolutional Neural Network (CNN) for identifying the construction era of above-mentioned old periods.


X.-S. Yang et al. (eds.), *Proceedings of Fifth International Congress on Information and Communication Technology*, Advances in Intelligent Systems and Computing 1183, https://doi.org/10.1007/978-981-15-5856-6_64

**Keywords:** Computational archaeology · Digital cultural heritage · Computer vision · Feature detection · Deep neural network (DNN) · Convolutional neural network (CNN) · Deep learning · Machine learning

## 1 Introduction

Building detection and feature detection are vital research areas in the study of computer vision. There are numerous old and ancient building sites in the Indian subcontinent region, such as Taj Mahal (Mughal era), Sixty Dome Mosque (Sultanate era), etc. Generally, the archaeologists can identify the construction period of old building by using its architectural characteristics or features. In this point of view, this research has established a computational technique for recognizing the construction period of old architectures by differentiating the building's architecture.

In previous years, some researches have been published, where computer vision is used in archaeology sections [1, 3]. An artificial neural network based feature recognition technology is used to identify the features of the ancient structure [4]. A CNN method focuses on visualization for primitive Maya hieroglyph [5]. The deep learning method is being utilized for recognizing the ancient Roman coin [6]. China's ancient warrior terracotta has been visualized by computer vision [7] and it is effective for 3D modeling [8]. Photogrammetric method has been enabled the image analysis of the Turkish ancient heritage site [9]. Moreover, some researches have been revealed where machine learning is also used in period identification [10, 11].

Furthermore, any technique for recognizing the building period of old architectural structures like the old building, mosque, and temple is not available. That's why this research has committed a technique that helps the archaeologists for recognizing the construction period by detecting the old spectacular architecture.

For establishing the CNN, a deep learning model has been developed where four features detection methods are applied. These are Canny Edge Detector [12], Hough Line Transform [13], Find Contours [14], and Harris Corner Detector [15]. After utilizing these methods, three diverse architectural features have been classified which are Dome, Minaret, and Front because different old structures contain different forms of these three features. A deep feed-forward neural network [16] model has been developed where three features have been used for identifying the era. Moreover, this research has identified three ruler periods, such as the Mughal period (1526–1857), Sultanate period (1206–1526), and British period (1858–1947).

Recently a deep learning model has been expressed [17] for identifying the old era for ancient buildings. Here, only Canny Edge Detector method and two eras' (Mughal and Sultanate) datasets have been used. The updated research has developed a more custom neural network where the remaining methods (Hough Line Transform, Find Contours, and Harris Corner Detector) has been utilized. Moreover, the British dataset is used here in addition to Mughal and Sultanate datasets.

The contributions of this manuscript are in three areas: (1) Identifying construction era based on Dome, Minaret and Front features of Mughal (1526–1857), Sultanate (1206–1526) and British (1858–1947) eras' buildings; (2) Edge, Line, Contour and

Corner elements are raised for identifying the different features (Dome, Minaret and Front) of different heritage buildings; (3) A Deep Neural Network (DNN) has been developed and applied in CNN where three features (Dome, Minaret and Front) of four different methods (Canny Edge Detector, Hough Line Transform, Find Contours, Harris Corner Detector) have been used for classifying old periods.

## 2 Era Identification Process

This research has illustrated a computational archaeological model that has described how a program can identify the construction era of an old building. At first, a photo was sent to Canny Edge Detector, Hough Line Transform, Find Contours, and Harris algorithm functions. These techniques have been used for collecting the features of Dome, Minaret, and Front from the old building image. The architecture and process of the era identification model have been illustrated in Fig. 1.

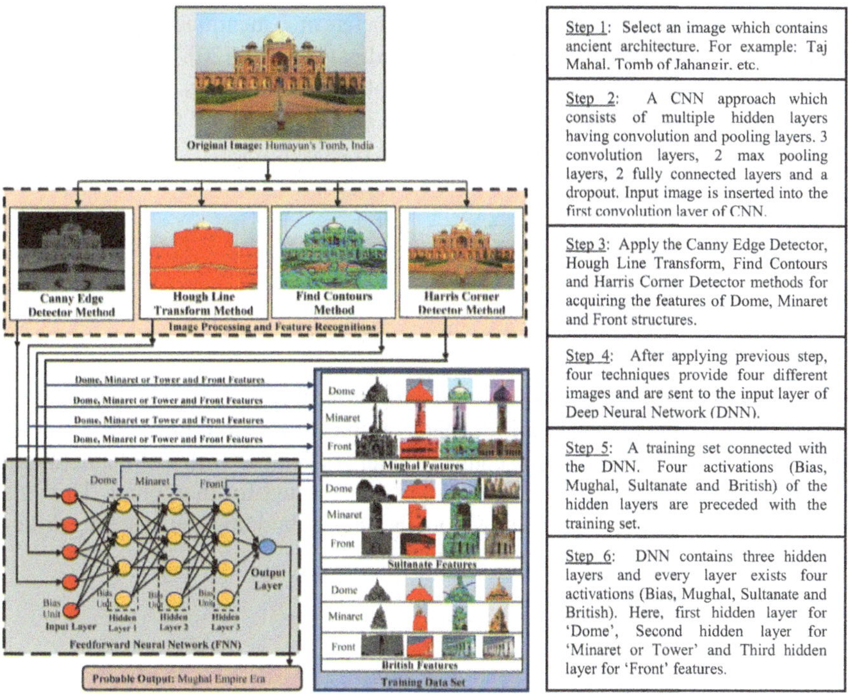


**Fig. 1.** Process and steps of the era identification for Indian subcontinent old buildings

## 3 Experimental Methods

### 3.1 Canny Edge Detection

Edge recognition covers a diversity of mathematical processes that goals at identifying the points in an image. In this experiment, Canny edge detection method has been utilized for acknowledging the edges from a photo. At first, vertical direction ($G_y$) and horizontal direction ($G_x$) were filtered by finding the gradient intensity of an image. After applying the Canny algorithm, gradient was constantly perpendicular to edges and it was rounded to the angles for illustrating vertical, horizontal, and diagonal directions. The direction and edge gradient [18] for each pixel were found as follows:

$$\text{Edge\_Gradient}\,(G) = \sqrt{G_x^2 + G_y^2} \tag{1}$$

$$\text{Angle}\,(\theta) = \tan^{-1}\left(\frac{G_y}{G_x}\right) \tag{2}$$

### 3.2 Hough Line Transform

Hough line transform is a feature extraction technique. It was related to the line identification on the picture. In this technique, the parameters $m$, $b$ mentioned [19] for Cartesian coordination and parameters $r$, $\theta$ for Polar coordinate system [20]. These coordination approaches were used for identifying the line of ancient buildings. In this research, a line had been represented as $y$ where, $y = mx + b$ or in parametric form, as $r = x \cos \theta + y \sin \theta$. Hence, the line equation for an image is as follows:

$$y = \left(-\frac{\cos\theta}{\sin\theta}\right)x + \left(\frac{r}{\sin\theta}\right) \tag{3}$$

### 3.3 Find Contours

Contours can be narrated entirely as a curve or turn joining all the continuous points' boundary and it is an adjuvant tool for shape or object detection. Image Moment technique has been used for finding the counters of an image. The spatial structure moment of an image was declared as $m_{ij}$ where $i$ and $j$ are nested for loop order. This image moment had detected different features from the ancient buildings by matching different shapes. The image moment [21] was computed as:

$$m_{ij} = \sum_{x,y} \left(\text{array}(x, y) \cdot x^j \cdot y^i\right) \tag{4}$$

### 3.4 Harris Corner Detection

Harris corner detection method extracts the corners and concludes the features of an image. It generally searches the corners in image intensity for a prolapsed of $(u, v)$. In this method, there is a Window function that is Gaussian Window and gives weights to the image pixels down. In Eq. (5) [22], $E$ is the distinction between the original and the moved window. Here, $I$ parameter is the image intensity. The window's translocation in the direction $x$ is $u$ and the direction $y$ is $v$. Window $w(x, y)$ is at $(x, y)$ position. The

$I(x+u, y+v)$ portion is moved window's intensity. Last portion $I(x, y)$ is the original intensity. The window function $w(x, y)$ is a Gaussian function.

$$E(u, v) = \sum_{x,y} w(x, y)[I(x+u, y+v) - I(x, y)]^2 \tag{5}$$

### 3.5 Training Dataset and Classification

For recognizing the construction period or era three classifications had been created for Sultanate, Mughal, and British periods. After using the feature detection techniques, a Decision Tree [23] had been created based on output of above techniques. Decision tree creates classification in the form of a tree formation. It improves an "if-then" ruleset which is reciprocally exclusive. These rules are learned orderly using the training data one at a time. Table 1 showed the types of data classification of the training dataset. Here, the data were classified by three periods (Mughal era, Sultanate era, and British era). Every era contains three different features (Dome, Minaret, and Front) of four different methods (Canny Edge Detector, Hough Line Transform, Find Contours, and Harris Corner Detector).

## 4 Deep Neural Network (DNN) Model

In this experiment, a DNN approach has been developed. In the input layer of DNN, there are five nodes ($x_1, x_2, x_3, x_4$, and bias unit). The inputs of the input layer have been displayed in Table 2, Figs. 1 and 2. The mathematical structure [24, 25] of the node at neural network in this experiment has been illustrated in Fig. 2. Here, $a$ is Activation, $b$ is Bias and $W$ is the 'Weight' of input layer. A bias unit allows changing the activation to the left or right, which is used for successful learning.

From Fig. 2, the equation for each activation node ($a$) is as follows:

$$\text{For hidden layer 1:} \quad a_i^{(L)} = W_i^{L-1} x_i + b_i^{L-1} \tag{6}$$

$$\text{After hidden layer 1:} \quad a_i^{(L)} = W_i^{L-1} a_i^{(L-1)} + b_i^{L-1} \tag{7}$$

Here, Index = $i$; Activation = $a$; Current Layer = $L$; Previous Layer = $L - 1$; Input node = $x$; Bias Unit = $b$. The computational algorithm of the developed DNN is represented as follows:

**Layer, $L = 2$ (Hidden Layer 1):**

$$a_1^{(2)} = f(W_1^{(1)} x_1 + W_4^{(1)} x_2 + W_7^{(1)} x_3 + W_{10}^{(1)} x_4 + b_1^{(1)}) \tag{8}$$

$$a_2^{(2)} = f(W_2^{(1)} x_1 + W_5^{(1)} x_2 + W_8^{(1)} x_3 + W_{11}^{(1)} x_4 + b_2^{(1)}) \tag{9}$$

$$a_3^{(2)} = f(W_3^{(1)} x_1 + W_6^{(1)} x_2 + W_9^{(1)} x_3 + W_{12}^{(1)} x_4 + b_3^{(1)}) \tag{10}$$

**Table 1.** Training dataset and classification of Mughal, Sultanate and British eras

| Features | Canny edge detection | Hough line transform | Find contours technique | Harris corner detector |
|---|---|---|---|---|
| *Mughal era (1526–1540, 1555–1857)* | | | | |
| Dome |  |  |  |  |
| Minaret |  |  |  |  |
| Front | 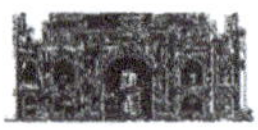 |  | 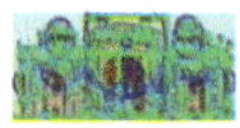 |  |
| *Sultanate era (1206–1526)* | | | | |
| Dome |  |  |  |  |
| Minaret |  |  |  |  |
| Front |  |  |  |  |
| *British era (1858–1947)* | | | | |
| Dome |  |  |  |  |
| Minaret |  |  |  |  |
| Front | 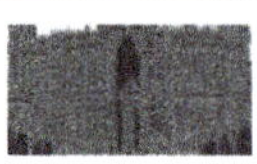 |  |  |  |

**Layer, $L = 3$ (Hidden Layer 2):**

$$a_1^{(3)} = f(W_1^{(2)} a_1^{(2)} + W_4^{(2)} a_2^{(2)} + W_7^{(2)} a_3^{(2)} + b_1^{(2)}) \tag{11}$$

$$a_2^{(3)} = f(W_2^{(2)} a_1^{(2)} + W_5^{(2)} a_2^{(2)} + W_8^{(2)} a_3^{(2)} + b_2^{(2)}) \tag{12}$$

$$a_3^{(3)} = f(W_3^{(2)} a_1^{(2)} + W_6^{(2)} a_2^{(2)} + W_9^{(2)} a_3^{(2)} + b_3^{(2)}) \tag{13}$$

**Table 2.** Input and inputs of DNN

| Input layer | Input image |
|---|---|
| $x_1$ | Canny edge detection image |
| $x_2$ | Hough line transform image |
| $x_3$ | Find counter image |
| $x_4$ | Harris corner detection image |
| Bias unit ($b$) | +1 |

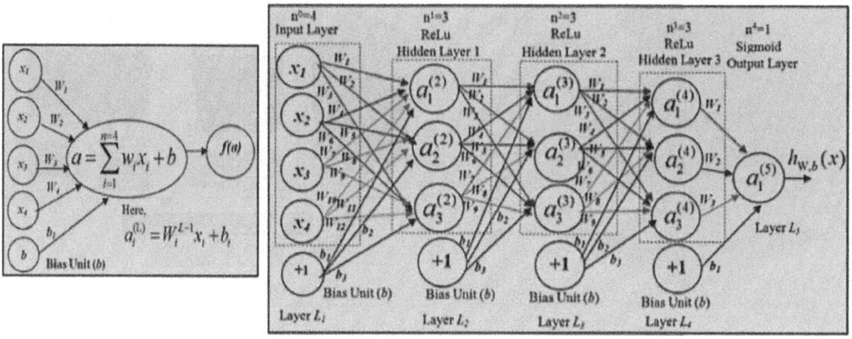


**Fig. 2.** Deep neural network (DNN) for construction era identification

**Layer, $L = 4$ (Hidden Layer 3):**

$$a_1^{(4)} = f(W_1^{(3)}a_1^{(3)} + W_4^{(3)}a_2^{(3)} + W_7^{(3)}a_3^{(3)} + b_1^{(3)}) \tag{14}$$

$$a_2^{(4)} = f(W_2^{(3)}a_1^{(3)} + W_5^{(3)}a_2^{(3)} + W_8^{(3)}a_3^{(3)} + b_2^{(3)}) \tag{15}$$

$$a_3^{(4)} = f(W_3^{(3)}a_1^{(3)} + W_6^{(3)}a_2^{(3)} + W_9^{(3)}a_3^{(3)} + b_3^{(3)}) \tag{16}$$

**Layer, $L = 5$ (Output Layer):**

$$h_{W,b}(x) = a_1^{(5)} = f(W_1^{(4)}a_1^{(4)} + W_2^{(4)}a_2^{(4)} + W_3^{(4)}a_3^{(4)} + b_1^{(4)}) \tag{17}$$

In Fig. 2, we have applied node to also denote the inputs to the network. The nodes labeled "+1" are called bias units corresponding to the intercept. We denoted $n^i$, the number nodes (without bias unit) in neural network. Weight $W_i^{(L-1)}$ denoted the parameter which connected with the link between $i$ unit in layer $L$ and this weight comes from previous layer $L - 1$. The bias units don't have inputs and links going into them. The bias units always output the value +1. Here, we have denoted the activation $a_i^{(L)}$ of unit $i$ in layer $L$. For $L = 1$, we declared $a_i^{(L)} = x_i$ to denote the $i$th input. The parameters $W$, $b$ defines the hypothesis $h_{w,b}^{(x)}$ that outputs a real number.

## 5 Convolution Neural Network (CNN) Model

A CNN has been created which is based on the developed DNN model. Generally, the CNN consists of multiple hidden layers having convolution and pooling layers. Here, CNN has been developed on three convolution layers, two max-pooling layers, two fully connected layers, and a dropout (Fig. 3). After using the feature detection methods we had got two datasets: training set and test set. Then the neural network model provides the prediction result of the old period.

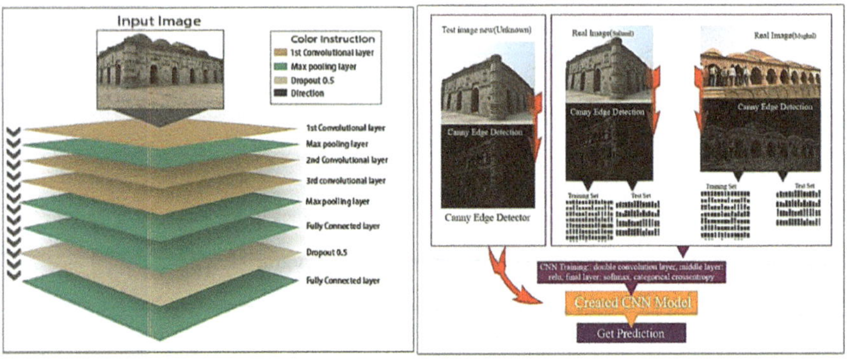


**Fig. 3.** CNN model for era identification of the old or ancient building

## 6 Results and Analysis

The outputs of the developed model consist of identifying the construction era, where a program provides a probable output by learning the ancient buildings' features. This work has indicated how a computer program learns several old buildings' features such as Dome, Minaret, and Front. For evaluating the performance of such systems, the data in the matrix has been used. The CNN model has been trained with the modified dataset and calculated the accuracy. Figure 4 has shown the composition of the CNN model, where the process successfully predicted the period from the picture of the ancient or old heritage building. Accuracy is also used as a statistical grade of the test calculations. The law for calculating accuracy is as follows:

$$\text{Accuracy} = \frac{(\text{TP} + \text{TN})}{(\text{TP} + \text{TN} + \text{FP} + \text{FN})} \times 100\% \qquad (18)$$

where, TP = True Positive; FP = False Positive; TN = True Negative; FN = False Negative.

In this research, total tested 500 images data have been used. The Sultanate era contains 270 data, Mughal era 130, and British era 100 data. We get TP = 254, TN = 227, FP = 3, FN = 16. Following the above Eq. (17) for the raw data, 96.20% accuracy achieved from this research.

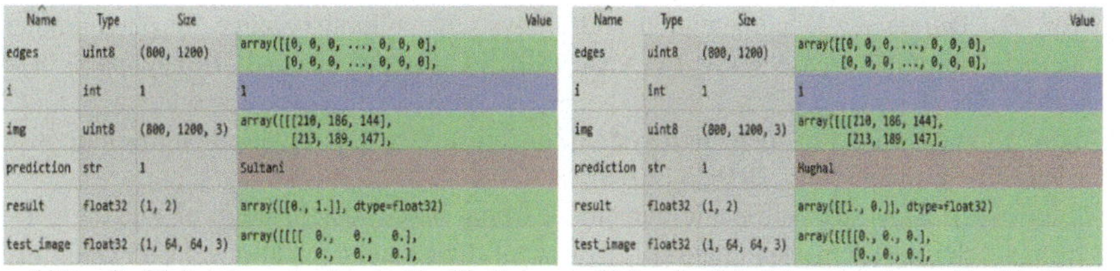


(a) Result of Sultanate construction era identification (b) Result of Mughal construction era identification

**Fig. 4.** Results of construction era identification of old buildings by using CNN model

## 7 Conclusion

This study has represented a model that demonstrates how an intelligent program can identify the construction era from an ancient or old heritage building. This research is mainly focused on the construction period and features of the heritage building by using artificial neural network and feature detection techniques. This research achieved much better accuracy over the previous method by using three periods (Mughal, Sultanate, and British eras) and four feature detection methods (Canny Edge Detector, Hough Line Transform, Find Contours, and Harris Corner Detector).

Still there are some limitations to this study. There are further issues to be resolved. Furthermore, if the model is tested on the low pixel picture, it cannot determine the target result. This drawback would also lead this research to the future work to make the raised model more robust and more significant to recognize precise objects from the image. These issues will be looked forward to solve in the future experiment.

## References


1. Tang, T., Chen, B., Hu, R.: Combined with DCT, SPIHT and ResNet to identify ancient building cracks from aerial remote sensing images. In: Liang, Q., et al. (eds.) Artificial Intelligence in China. Lecture Notes in Electrical Engineering, vol. 572, pp. 313–318. Springer, Singapore (2020)
2. Kabir, S.R., et al.: Performance analysis of different feature detection techniques for modern and old buildings. CEUR Workshop Proc. **2280**, 120–127 (2018)
3. Barceló, J.A.: Computational Intelligence in Archaeology. Universidad Autonoma de Barcelona, Spain (2008)
4. Zou, Z., et al.: Feature recognition and detection for ancient architecture based on machine vision. In: Proceedings of SPIE 10602, Smart Structures and NDE for Industry 4.0, p. 1060209 (2018)
5. Can, G., et al.: How to tell ancient signs apart? Recognizing and visualizing Maya Glyphs with CNNs. ACM J. Comput. Cult. Herit. **11**(4), Article no. 20 (2018)
6. Schlag, I., Arandjelovic, O.: Ancient Roman coin recognition in the wild using deep learning based recognition of artistically depicted face profiles. In: 2017 IEEE International Conference on Computer Vision Workshops, Venice, Italy, pp. 2898–2906 (2017)
7. Bevan, A., et al.: Computer vision, archaeological classification and China's terracotta warriors. J. Archaeol. Sci. **49**, 249–254 (2014)
8. Brutto, M.L., Meli, P.: Computer vision tools for 3D modeling in archaeology. Int. J. Herit. Digit. Era **1**(1), 1–6 (2012)

9. Toz, G., Duran, Z.: Documentation and analysis of cultural heritage by photogrametric methods and GIS: a case study. In: XXth ISPRS Congress, Istanbul, Turkey, pp. 438–441 (2004)
10. Min, Y., et al.: Real time detection system for rail surface defects based on machine vision. EURASIP J. Image Vide. **2018**, 3 (2018)
11. Zdravevski, E., et al.: Automatic machine-learning based identification of jogging periods from accelerometer measurements of adolescents under field conditions. PLoS ONE **12**(9), e0184216, 1–28 (2017)
12. Ramnarayan, Saklani, N., Verma, V.: A review on edge detection technique "Canny edge detection". Int. J. Comput. Appl. **178**(10), 28–30 (2019)
13. Tatsubori, M., et al.: A probabilistic Hough transform for opportunistic crowd-sensing of moving traffic obstacles. In: 2018 SIAM International Conference on Data Mining, California, USA, pp. 217–215 (2018)
14. Soomro, S., Munir, A., Choi, K.N.: Hybrid two-stage active contour method with region and edge information for intensity inhomogeneous image segmentation. PLoS ONE **13**(1), Article: e0191827 (2018)
15. Sun, Y., Ientilucci, E., Voisin, S.: Improvement of the Harris corner detector using an entropy-block-based strategy. In: SPIE 10644, Algorithms and Technologies for Multispectral, Hyperspectral, and Ultraspectral Imagery XXIV, 1064414, Florida, United States (2018)
16. Zheng, W., Wang, H.B., et al.: Multi-layer feed-forward neural network deep learning control with hybrid position and virtual-force algorithm for mobile robot obstacle avoidance. Int. J. Control Autom. Syst. **17**(4), 1007–1018 (2019)
17. Hasan, M.S., et al.: Heritage building era detection using CNN. IOP Conf. Ser. Mater. Sci. Eng. **617**(1), Article: 012016 (2019)
18. Mordvintsev, A., Revision, A.K.: Canny Edge Detection. OpenCV-Python Tutorials (2013)
19. Hough Line Transform: OpenCV (2017)
20. Xu, G., Zheng, A., Li, X., Su, J.: A method to calibrate a camera using perpendicularity of 2D lines in the target observations. Sci. Rep. 6, Article number: 34951 (2016)
21. Structural Analysis and Shape Descriptors: OpenCV (2014)
22. Nelli, F.: OpenCV & Python—Harris Corner Detection—A Method to Detect Corners in an Image. Meccanismo Complesso (2017)
23. Mesarić, J., Šebalj, D.: Decision trees for predicting the academic success of students. Croat. Oper. Res. Rev. **7**(2), 367–388 (2016)
24. Higham, C.F., Higham, D.J.: Deep learning: an introduction for applied mathematicians. SIAM Rev. **61**(4), 860–891 (2019)
25. Durstewitz, D., Koppe, G., Meyer-Lindenberg, A.: Deep neural networks in psychiatry. Mol. Psychiatry **24**, 1583–1598 (2019)

# Lock-Free Parallel Computing Using Theatre

Christian Nigro[1] and Libero Nigro[2(✉)]

[1] Independent Computing Professional, Rende, Italy
christian.nigro21@gmail.com
[2] DIMES, University of Calabria, Quattromiglia, Italy
l.nigro@unical.it

**Abstract.** Theatre is a control-based actor system currently developed in Java, whose design specifically addresses the development of predictable, time-constrained distributed systems. Theatre, though, can also be used for untimed concurrent applications. The control structure regulating message scheduling and dispatching can be customized by programming. This paper describes a novel implementation pTheatre (Parallel Theatre), whose control structure can exploit the potential of parallel computing offered by nowadays multi-core machines. With respect to the distributed implementation of Theatre, pTheatre is more lightweight because it avoids the use of Java serialization during actor migration, and when transmitting messages from a computing node (theatre/thread) to another one. In addition, no locking mechanism is used both in high-level actor programs and in the underlying runtime support. This way, common pitfalls related to classic multi-threaded programming are naturally avoided, and the possibility of enabling high-performance computing is opened. The paper demonstrates the potential of the achieved realization through a parallel matrix multiplication example.

**Keywords:** Actors · Reflective control structure · Theatre · Java · Lock-free data structures · Multi-core parallel computing · Dense matrix parallel multiplication

## 1 Introduction

Actors [1] represent a well-established formal computational model for the development of general concurrent distributed systems. The model is founded on the concept of an *actor*, which is a basic, modular, and isolated software entity that shares no data and communicates with its peers solely by *asynchronous message passing*. In its basic formulation, the actor model features one thread per actor and admits an input mailbox where all the incoming messages get stored. It is the responsibility of the thread that of extracting, if there any, one message at a time from the mailbox and processing it by updating the local data status and possibly creating new actors and sending messages to known actors (*acquaintances*) including itself. Message processing is *atomic* and follows the *macro-step semantics* [2]: a new message can only be processed when current message processing is finished. In the last years, the actor model emerged as a valid


X.-S. Yang et al. (eds.), *Proceedings of Fifth International Congress on Information and Communication Technology*, Advances in Intelligent Systems and Computing 1183, https://doi.org/10.1007/978-981-15-5856-6_65

alternative [3] to classical multi-threaded programming with data sharing and the use of *locks*, which are notoriously prone to subtle synchronization errors and deadlocks [4].

Actors are currently supported by such frameworks as Salsa, ActorFoundry, Scala/Akka, Erlang, CAF, Rebeca, etc. some of which have been experimented in the construction of large web-based software systems.

Classical actors, though, are more suited to untimed systems where non-determinism regulates message delivery. Extensions to actors have been defined to allow modelling, analysis and implementation of distributed time-constrained systems [5–8].

Theatre [6–8] distinguishes from similar actor-based frameworks by its control-centric character, which in turn favours time predictability and model transformation in development, thus facilitating a smooth transition from analysis down to design, prototyping and final, model compliant, implementation of a system. A general, distributed implementation of Theatre was achieved in pure Java and it is described in [9]. The possibility of supporting hybrid actors during modelling and analysis of cyber-physical systems has recently been shown in [10, 11].

The work described in this paper argues that Theatre can effectively also be used for parallel untimed applications. The paper specifically proposes Parallel Theatre (pTheatre), an original realization of Theatre which optimizes the implementation in [9] so as to enable high-performance computing on today's multi-core machines.

The rest of this paper is structured as follows. In Sect. 2, the basic concepts of Theatre are briefly presented. Then the design and lock-free implementation of pTheatre are discussed in Sect. 3. Section 4 proposes a scalable application of pTheatre to parallel matrix multiplication. Section 5 discusses the performance issues of the case study. Finally, Sect. 6 concludes the paper with an indication of future work.

## 2 The Theatre Actor System

### 2.1 Programming Concepts

Theatre actors have no internal thread and no local mailbox. Rather, actors execute on computing nodes said theatres. In a theatre, a *reflective control layer* (control machine) is used which, transparently, captures sent messages, buffers them and regulates message delivery according to a strategy that ultimately depends upon the application requirements. The control machine can reason on time (simulated time or real-time). A library of control machines was developed (see [9]). An Actor is at rest until a message arrives. Message processing is atomic and cannot be pre-empted nor suspended. In Java, a Theatre actor is programmed by a normal class that derives from the Actor base class. Selected methods (said *message servers -msgsrv-* as in [5]) specify how corresponding messages will be responded. For initialization purposes (see also Fig. 1), instead of relying on class constructors, an explicit message like init() can be used. Differently from normal Java methods, which have a synchronous invocation semantics, message servers can only be invoked by a non-blocking *send* operation, which can carry timing information [8, 9]. Therefore, message servers have an *asynchronous invocation semantics*. Message servers can have arguments but have no return type. To return a result, a message server must send back to its requestor an explicit message with the result as an argument. As a basic assumption, a message server is expected, normally, to have a negligible time duration. The control machine of a theatre provides the execution context (stack frame) to local actors. Therefore, *concurrency*, within a theatre, is *cooperative* not pre-emptive, and naturally emerges from *message interleaving*. The control machine repeats a *basic control loop*. At each iteration, a pending message is selected and dispatched to its target actor by invoking a corresponding message server. At message server termination, the control loop is re-entered, any sent messages scheduled, then the next iteration is started. The control loop is actually continued until a final condition eventually holds: for example, in a concurrent application, message exchanges are exhausted; in a simulation application, the assigned simulation time limit is reached. A real-time application, instead, typically executes in a non-stop way. A time-sensitive control machine ensures timestamped sent messages are delivered in timestamp order. When messages have the same timestamp, a Lamport's logical clock is used (added meta-data in messages) to deliver messages according to their logical clock (*generation time*). A concurrent control machine ultimately delivers messages according to their sending order. Controlling the message delivery order contributes to the determinism of a Theatre application.

## 2.2 System-Level Concepts

A distributed Theatre system is a federation of theatres (logical processes, e.g. JVM instances or address spaces) currently implemented in Java [9]. Each theatre hosts a *transport layer*, a *control layer* (control machine) and a *collection of local* application *actors*. Theatre is based on *global time*. A *time server* component can be used (attached to a given theatre) to ensure global time is kept updated. Theatres (their control machines) coordinate each other with the time server, to preserve the notion of global time. The transport layer can be based, in a case, on the TCP transport, thus ensuring sent messages from a theatre and directed to the same destination theatre are received in first-in-first-out order. Whereas concurrency in the same theatre is cooperative, actors (message servers) belonging to distinct theatres, allocated, e.g. to distinct physical processors of a multi-computer, can be executed in truly *physical parallelism*. At initial configuration time, a *socket network* is established among theatres which, in a case, can be a complete mesh. Required socket connections are specified in an xml configuration file, which also declares the control machine type to instantiate on each theatre. A master theatre is elected which actually supervises the creation of the socket network and bootstraps system execution by creating and initializing first actors. In a Theatre system, actors are assumed to have *universal names*, that is unique system-level identifiers (strings). Actors are created in a theatre, then they can be migrated (*move*() operation) on a different theatre for load-balancing issues. Inter-theatre messages and migrated actors use the Java serialization mechanism. Intra-theatre messages, instead, are normal Java objects. When an actor is migrated, a *proxy* version of itself is kept in the originating theatre, which stores the last known address (theatre URL) of the migrated actor, and acts as a message *forwarder*. Dispatching a message to a proxy actor, causes an external message to be generated and routed to its destination through the transport layer. A *local actor table* (LAT) is maintained into each theatre to store information about proxy/actual actors. When a migrated actor comes back to a theatre where a proxy version of it exists, the proxy gets replaced by the actual actor along with its full data status. Messages exchanged by theatres can be *control messages* or *application messages*. Application messages refer to the core business of the system. They are dispatched and processed by actors. Control messages, instead, are transparent to actors. They are intended to be received and processed by the control machine themselves. Control messages are used to start/stop a federation system, to interact (according to a protocol) with the time server to keep global time aligned, etc.

## 3 Implementing pTheatre in Java

pTheatre (Parallel Theatre) represents an optimization of basic distributed Theatre, aimed at enabling high-performance parallel computing on multi-core machines. The following are some main points about the design and implementation in Java of pTheatre.

1. Theatres are mapped onto Java threads (one thread per theatre).
2. A Theatre system coincides with one JVM instance. Therefore, all theatres of a parallel application share a common address space.
3. Actor universal naming reduces to the use of Java actor references. An actor's reference persists when the actor is moved from a theatre to another.
4. The message communication mechanism is lightweight and highly efficient because transferring a message (or migrating an actor) reduces to transferring a message object reference. No object serialization is needed.
5. No local actor table (LAT) per theatre is now required. Actors hold in an internal field the identity of the execution theatre. The move() operation just changes the running theatre reference in the moved actor.
6. The transport layer of a theatre no longer uses socket connections. It only contains a lock-free message queue (inputBuffer) managed by a couple of send/receive operations. The inputBuffer is used by external theatres for sending inter-theatre control or application messages. The inputBuffer is emptied, one message at a time, if there are any, at each iteration of the control machine loop. Control messages are directly managed by the control machine. Application messages get scheduled on the local message queue of the control machine.
7. Being untimed, a pTheatre application does not need a time server component. However, a "time server", implemented as a passive object, is still used as a global detector of the termination condition of a whole pTheatre application. When a control machine finds it has no messages in its local message queue, a stop control message is sent to the theatre hosting the time server. No further stop messages are sent, provided no new external messages are received. Would an external message arrive, a subsequent stop message will be re-issued when the message queue empties again. The stop message carries the control machine counters of sent/received external messages. The time server holds in two maps the identity of the stop requesting theatre and the values of its message counters. When all the theatres of an application have requested a stop and there are no in-transit messages among theatres (the total number of sent messages is found equal to the total number of received messages), the time server broadcasts a terminate control message to theatres which then break the control machine event loop.

8. Newly developed classes include Theatre, PConcurrent, PCTimeServer, PTransportLayer. Theatre extends Thread. Its default run() method just activates the control machine event loop (the controller() method). The run() method is redefined "on-the-fly" when the Theatre instance acts as the master theatre which configures and launches a parallel application (see Fig. 1). PConcurrent is a control machine which processes messages in the sending order. PCTimeServer implements the protocol for system termination.
9. Considering that actors have no built-in thread, a whole pTheatre system runs without any use of locks. All of this simplifies and contributes to safe concurrent programming, and it has the potential to deliver a high execution performance (see the example in the next section).

## 4 An Example Using Parallel Matrix Multiplication

The goal of the following case study is twofold: to illustrate the actual programming style of pTheatre in Java, and to show the achievable execution performance. The example is concerned with the Geoffrey Fox dense parallel matrix multiplication algorithm [12, 13], which is one of the most memory efficient algorithms [14]. Two squared matrixes $N \times N$ of doubles $a$ and $b$ are considered, and their product matrix c calculated. Toward this, matrixes are decomposed into squared sub-blocks of $\frac{N}{\sqrt{P}} * \frac{N}{\sqrt{P}}$ elements, where $P$ is the number of available processors and $\sqrt{P}$ is assumed to be an integer number. Processors are organized into a $\sqrt{P} * \sqrt{P}$ toroidal grid. Each processor $P_{ij}$ is initially assigned the sub-blocks $a_{ij}$, $b_{ij}$ and it is responsible of computing the sub-block $c_{ij}$ of the product matrix. In the practical experiment, $P = 4$ processors (cores) are assumed. The Java model is split into three classes: Configurer, Collector and Processor. In the main() method of the Configurer (see Fig. 1) four theatres are created. Theatre 0 is the master theatre (it redefines the run() method) and creates actors, initializes them and activates the parallel system. The Collector actor (not shown for brevity) receives from the processor actors the computed sub-blocks $c_{ij}$ and composes them into the final result matrix $c$. The Processor actor class realizes the Fox algorithm. The configurer, as well as the time server, are allocated to theatre 0. Each separate processor is allocated to a distinct theatre (core) from 0 to 3.

The processor class is shown in Fig. 2. It is realized as a finite state machine. Each processor computes its sub-block $c_{ij}$ by iterating $p$ times, where $p = \sqrt{P}$, three basic phases expressed as internal states: *broadcast*, *computing* and *rolling-up*. During broadcast, only one processor in a row, starting from the diagonal position, broadcasts (by right piping and toroidally) its sub-block $a_{ij}$ to all the remaining processors in the row. In the BROADCAST state, either a processor is sending its $a_{ij}$ sub-block or it is waiting a sub-block from the left (toroidally) processor in the row. Let $t$ be the received sub-block during a broadcast phase. In the computing phase, each processor updates its $c_{ij}$ sub-block as: $c_{ij} = c_{ij} + t * b_{ij}$. In the subsequent rolling-up phase, each processor rolls up in the grid (upper in the column and always toroidally) its $b_{ij}$ sub-block. As the iterations end, the processor sends its $c_{ij}$ sub-block to the collector actor through a "report" message. Due to the asynchronous nature of the algorithm in Fig. 2, in the case

```
public class Configurer{
   static double[][] block( double[][] m, int d, int i, int j ){...}
   public static void main( String[] args ){
     new Theatre( 3, 4, new PTransportLayer(), new PConcurrent() );
     new Theatre( 2, 4, new PTransportLayer(), new PConcurrent() );
     new Theatre( 1, 4, new PTransportLayer(), new PConcurrent() );
     new Theatre( 0, 4, new PTransportLayer( new PCTimeServer() ), new PConcurrent() ){
        public void run() {
           final int N=...;
           double[][] a=new double[N][N];
           double[][] b=new double[N][N];
           //fill values to a and b
           int P=4; //nr of processors
           int dim=N/(int)Math.sqrt(P); //sub-block dimension
           long start=System.currentTimeMillis();
           Collector c=new Collector();
           c.send( "init",N,P ); c.move(0);
           int p=(int)Math.sqrt(P); //nr of processors in a row/column
           Processor[][] grid=new Processor[p][p];
           for( int i=0; i<p; ++i )
              for( int j=0; j<p; ++j )
                 grid[i][j]=new Processor();
           for( int i=0; i<p; ++i )
              for( int j=0; j<p; ++j ){
                 grid[i][j].send( "init",i,j,p,block(a,dim,i,j),block(b,dim,i,j),
                    grid[i][(j+1)%p] /*right*/, grid[(i-1+p)%p][j] /*above*/, c );
                 grid[i][j].move(i*p+j);
              }
           for( int t=0; t<P; ++t ) Theatre.getTheatre(t).activate();
           Theatre.getTheatre( Thread.currentThread().getName() ).
              getControlMachine().controller();
           System.out.println("Parallel WCT="+(System.currentTimeMillis()-start)+" msec");
        }
     };
   }//main
}//Configurer
```

**Fig. 1.** The configurer class for the pTheatre implementation of the Fox algorithm

a processor receives a message which is not expected in the current state, the message is later postponed by re-sending it to itself.

## 5 Experimental Analysis

The pTheatre implementation of Fox's algorithm for parallel matrix multiplication was executed on a Win 10, Intel Core i7-7700 CPU@3.60 GHz, 32 GB of memory, by varying the matrix size *N* from 100 to 3000. The emerged execution time (parallel wall clock time P_WCT) was then compared with that (sequential wall clock time S_WCT) of the classical sequential implementation of the matrix multiplication based on three nested for loops. Each experiment was repeated five times. The

```
public class Processor extends Actor{
  private int i, j, p, it=0, broadcastIteration;
  double[][] a, b, t, c;
  private Processor right, above;
  private Collector collector;
  private static byte CREATED=0,
    BROADCAST=1, COMPUTE=2,
    ROLLING_UP=3, STOP=4;
  private byte cs=CREATED; //current status

  @Msgsrv
  public void init( Integer i, Integer j,
           Integer p,
           double[][] a, double[][] b,
           Processor right, Processor above,
           Collector collector){
    this.i=i; this.j=j; this.p=p; this.a=a;
    this.b=b; this.right=right;
    this.above=above;
    this.collector=collector;
    this.c=new double[a.length][a.length];
    broadcastIteration=(i-j+p)%p;
    if( it==broadcastIteration )
      send( "broadcast", a, i, j );
    cs=BROADCAST;
  }//init

  @Msgsrv
  public void broadcast( double[][] a,
    Integer i, Integer j ){
    if( cs!=BROADCAST )
      send( "broadcast", a, i, j );
    else{
      this.t=a;
      if( j!=(this.j+1)%p )
        right.send( "broadcast", a, i, j );
      send( "compute" ); cs=COMPUTE;
    }
  }//broadcast

  @Msgsrv
  public void compute(){
    if( cs!=COMPUTE ) send( "compute" );
    else{
      update(); it++;
      above.send( "rollingUp", b );
      cs=ROLLING_UP;
    }
  }//compute

  @Msgsrv
  public void rollingUp( double[][] b ){
    if( cs!=ROLLING_UP )
      send( "rollingUp", b );
    else {
      this.b=b;
      if( it==p ){ send( "stop" ); cs=STOP; }
      else{
        if( it==broadcastIteration )
          send( "broadcast", a, i, j );
        cs=BROADCAST;
      }
    }
  }//rollingUp
  @Msgsrv
  public void stop(){
    collector.send( "report", c, i, j );
  }//stop
  private void update(){
    //c=c+t*b
    for( int row=0; row<t.length; ++row )
      for( int col=0; col<t.length; ++col ){
        double tmp=0.0D;
        for( int k=0; k<t.length; ++k )
          tmp+=t[row][k]*b[k][col];
        c[row][col]+=tmp;
      }
  }//update
}//Processor
```

**Fig. 2.** The processor actor implementing Fox's algorithm

speedup of the parallel implementation was then calculated, for each experiment, as best-case(S_WCT)/worst_case(P_WCT), to smooth out Operating System dependencies. Experimental results are collected in Table 1.

As one can see from Table 1, for low values of the matrix size, the speedup is low because the sequential implementation outperforms the more complex parallel implementation. Some super speedups emerge when a matrix size of 900–1500 is adopted. A speedup value close to the ideal value of 4 (according to the Amdahl law, when 4 cores are used) occurs for $N = 2000$. Obviously, such values are the result of complex combinations of hardware features (cache contents) with the efficient implementation of

**Table 1.** Speedup of parallel versus sequential program (4 cores)

| Matrix size $N$ | S_WCT (ms) | P_WCT (ms) | Speedup |
|---|---|---|---|
| 100 | 3 | 33 | 0.09 |
| 200 | 10 | 40 | 0.25 |
| 300 | 36 | 63 | 0.57 |
| 400 | 88 | 81 | 1.1 |
| 500 | 170 | 98 | 1.73 |
| 600 | 296 | 152 | 1.95 |
| 700 | 506 | 213 | 2.38 |
| 800 | 858 | 252 | 3.4 |
| 900 | 1827 | 448 | 4.07 |
| 1000 | 5364 | 812 | 6.61 |
| 1500 | 30,445 | 6352 | 4.79 |
| 2000 | 81,972 | 20,907 | 3.9 |
| 2500 | 205,064 | 55,237 | 3.71 |
| 3000 | 350,438 | 106,753 | 3.28 |

pTheatre. On the other hand, similar super speedups were reported, for the Fox algorithm, also in [14].

## 6 Conclusions

This paper proposes pTheatre, an original, efficient, and totally lock-free Java implementation of the Theatre actor system [6–9]. pTheatre mission is that of supporting high-performance computing on modern multi-core machines. The potential of pTheatre was demonstrated in the paper by an actor-based, asynchronous implementation of the Geoffrey Fox parallel matrix multiplication algorithm [12, 13].

Prosecution of the research is directed to the following:

- Improving the implementation of pTheatre and applying it to complex parallel algorithms.
- Completing a specialization of pTheatre for high-performance agent-based modelling and simulations over multi-core machines, according to a conservative synchronization strategy.
- Porting pTheatre over the GPU, for supporting massive, data-parallel applications.

## References

1. Agha, G.: Actors: a model of concurrent computation in distributed systems. PhD Thesis, MIT Artificial Intelligence Laboratory (1986)

2. Karmani, R.K., Agha, G.: Actors, pp. 1–11. Springer US, Boston, MA (2011)
3. Agha, G., Palmskog, K.: Transforming threads into actors: learning concurrency structure from execution traces. In: Principles of Modelling, pp. 16–37. Springer, Cham (2018)
4. Lee, E.A.: The problem with threads. Computer **39**, 33–42 (2006). https://doi.org/10.1109/MC.2006.180
5. Jafari, A., Khamespanah, E., Sirjani, M., Hermanns, H., Cimini, M.: PTRebeca: modelling and analysis of distributed and asynchronous systems. Sci. Comput. Program. **128**, 22–50 (2016)
6. Cicirelli, F., Nigro, L.: Control centric framework for model continuity in time-dependent multi-agent systems. In: Concurrency and Computation: Practice and Experience, vol. 28, no. 12, pp. 3333–3356. Wiley (2016)
7. Cicirelli, F., Nigro, L., Sciammarella, P.F.: Model continuity in cyber-physical systems: a control centered methodology based on agents. Simul. Model. Pract. Theory **83**, 93–107 (2018). https://doi.org/10.1016/j.simpat.2017.12.008
8. Nigro, L., Sciammarella, P.F.: Qualitative and quantitative model checking of distributed probabilistic timed actors. Simul. Model. Pract. Theory **87**, 343–368 (2018)
9. Cicirelli, F., Nigro, L., Sciammarella, P.F.: Seamless development in Java of distributed real-time systems using actors. Int. J. Simul. Process Model. **15**, Nos 1/2, 13–29 (2020)
10. Nigro, L., Sciammarella, P.F.: Verification of a smart power control systems using hybrid actors. In: IEEE WorldS4, London, 30–31 July 2019
11. Cicirelli, F., Nigro, L.: Home energy management using theatre with hybrid actors. In: IEEE/ACM 23rd International Symposium on Distributed Simulation and Real Time Application (DS-RT 2019), Cosenza, Italy, 7–9 Oct 2019
12. Fox, G.C.: Solving problems on concurrent processors. In: General Techniques and Regular Problems, vol. 1. Prentice-Hall (1988)
13. Fox, G.C., Otto, S.W., Hey, A.J.G.: Matrix algorithms on a hypercube I: matrix multiplication. Parallel Comput. **4**, 17–31 (1987)
14. Gergel, V.: Parallel methods for matrix multiplication. In: 2012 Summer School on Concurrency, pp. 1–50, St. Petersburg, Russia, 22–29 Aug 2012

# Author Index




X.-S. Yang et al. (eds.), *Proceedings of Fifth International Congress on Information and Communication Technology*, Advances in Intelligent Systems and Computing 1183, https://doi.org/10.1007/978-981-15-5856-6

The manufacturer's authorised representative in the EU is Springer Nature Customer Service Centre GmbH, Europaplatz 3, 69115 Heidelberg, Germany. If you have any concerns regarding our products, please contact ProductSafety@springernature.com

Printed and bound by CPI Group (UK) Ltd, Croydon, CR0 4YY
30/09/2026
02225292-0016